797,885 Books
are available to read at

Forgotten Books

www.ForgottenBooks.com

Forgotten Books' App
Available for mobile, tablet & eReader

ISBN 978-1-5280-0073-4
PIBN 10922361

This book is a reproduction of an important historical work. Forgotten Books uses state-of-the-art technology to digitally reconstruct the work, preserving the original format whilst repairing imperfections present in the aged copy. In rare cases, an imperfection in the original, such as a blemish or missing page, may be replicated in our edition. We do, however, repair the vast majority of imperfections successfully; any imperfections that remain are intentionally left to preserve the state of such historical works.

Forgotten Books is a registered trademark of FB &c Ltd.
Copyright © 2017 FB &c Ltd.
FB &c Ltd, Dalton House, 60 Windsor Avenue, London, SW19 2RR.
Company number 08720141. Registered in England and Wales.

For support please visit www.forgottenbooks.com

1 MONTH OF FREE READING

at

www.ForgottenBooks.com

By purchasing this book you are eligible for one month membership to ForgottenBooks.com, giving you unlimited access to our entire collection of over 700,000 titles via our web site and mobile apps.

To claim your free month visit: www.forgottenbooks.com/free922361

* Offer is valid for 45 days from date of purchase. Terms and conditions apply.

English
Français
Deutsche
Italiano
Español
Português

www.forgottenbooks.com

Mythology Photography **Fiction**
Fishing Christianity **Art** Cooking
Essays Buddhism Freemasonry
Medicine **Biology** Music **Ancient Egypt** Evolution Carpentry Physics
Dance Geology **Mathematics** Fitness
Shakespeare **Folklore** Yoga Marketing
Confidence Immortality Biographies
Poetry **Psychology** Witchcraft
Electronics Chemistry History **Law**
Accounting **Philosophy** Anthropology
Alchemy Drama Quantum Mechanics
Atheism Sexual Health **Ancient History**
Entrepreneurship Languages Sport
Paleontology Needlework Islam
Metaphysics Investment Archaeology
Parenting Statistics Criminology
Motivational

QUANTITY SURVEYING

FOR THE USE OF

SURVEYORS, ARCHITECTS, ENGINEERS
AND BUILDERS

By J LEANING

AUTHOR OF 'THE CONDUCT OF BUILDING WORKS AND DUTIES OF A CLERK OF WORKS;
A COMPLETE SET OF CONTRACT DOCUMENTS, BUILDING SPECIFICATIONS, ETC.'

FIFTH EDITION, REVISED AND ENLARGED

London
E. & F. N. SPON, Limited, 125 STRAND

New York
SPON & CHAMBERLAIN, 123 LIBERTY STREET

1904

PREFACE

TO

THE FIFTH EDITION.

THIS edition is the result of a careful revision, in course of which many new items have been inserted.

The section on prices has been entirely re-written and greatly extended. The greater part of this section, up to and including the trade of joiner, was published a few years ago in a weekly journal. Prices have altered since then, and are now revised. The importance of the subject to the quantity surveyor affords a good reason for the comparatively large bulk of this section.

Machinery Engines, Boilers, Electric Lighting, etc., have in the past been frequently left to the specialist, and the amount of their estimates included in the bill of quantities as provisional sums. There is a prevailing tendency to obtain competitive prices for these, and to supply quantities for them, although they may still be the subject of separate estimates. References to works of this character will be found in this edition.

A schedule for pulling down has also been included.

Short Sections on Arbitration from the surveyor's point of view, Charges and Deficient Quantities, and some new examples of Taking Off, may probably prove useful to the student.

PREFACE TO THE FIRST EDITION.

THE nucleus of the following pages appeared in the 'British Architect' in the years 1878 and 1879. These papers have since that publication been revised and much increased in volume.

The change of practice which has been in course of development for some years, and the gradual settlement of a methodical and reasonable system which is not completely explained in any book hitherto published, would seem to warrant the assumption that this manual will not be unwelcome.

The value of orderly procedure cannot easily be overrated, and the writer has endeavoured to indicate various ways by which this can be ensured. The sections on the Order of Taking Off on various principles, Office Practice, the Settlement of Variations, the Treatment of Works of Alteration and Restoration, the Law, the Forms of Bills and Schedules of Prices and the Examples of Collections and Taking Off will, the writer trusts, prove useful to the student, and tend to increase the number of those men (daily becoming more numerous) who, in the face of adverse opinion and opposing ignorance, hold to the belief that a thing worth doing at all is worth doing well.

PREFACE TO THE THIRD EDITION.

ALTHOUGH an effort has been made to exclude information extraneous to the subject, this edition will be found more voluminous than the last. This increase is to some extent referable to the introduction after each direction for measurement, of the item as it would appear in a bill of quantities. These items make the directions clearer, and show more completely the distinctions which should be observed.

Specifications having been dealt with in a separate book, the chapter relating to them has been altered but little. It should, however, be noticed that many of the bill items before referred to may, with very little alteration of phraseology, be used in specifications.

The lists of items of Taking Off in the several orders furnished will assist the student to broad views of the subject; if he should desire to see examples of the *office work* of the quantity surveyor, "A complete set of contract documents," referred to on the title-page, presents such examples in facsimile; for this reason it has been considered inexpedient to repeat such a work in this book.

Some recent important decisions have been added to the chapter on law. Their finality is questionable, and there are still many doubtful points as to the legal relations of the quantity surveyor.

The chapter on cubing and approximate estimates has been extended. For this the importance of the subject to the quantity surveyor is, perhaps, a reasonable excuse.

Reference to the index will show that this edition comprises a great number of small items of advice and information which have not appeared in the former ones.

CONTENTS.

CHAPTER		PAGE
I.	General Directions	1
II.	Order of Taking Off	25
III.	Modes of Measurement	44
IV.	Squaring the Dimensions, Abstracting and Billing	296
V.	Restorations	378
VI.	Errors, etc. in Estimates—Schedule of Prices—Adjustment of Accounts	387
VII.	Specifications	436
VIII.	Prices	441
IX.	The Law as it Affects Quantity Surveyors	760
X.	Order of Taking Off if the Old Method be Adopted	797
XI.	Northern Practice	800
XII.	Examples of Collections	820
XIII.	Examples of Taking off	829
XIV.	The Present System of Estimating	896
Index		915

QUANTITY SURVEYING.

CHAPTER I.

GENERAL DIRECTIONS.

THE diversity of usage in the profession of quantity surveying (see page 13), and the existence of a number of manuals on the subject, none of which can claim to be complete or entirely reliable,* have led the writer to the conclusion that a treatise which should present the most generally recognised modes of measurement and office practice, and furnish good reasons for the rejection of vicious methods, would prove acceptable to the profession generally, tend to a settlement of disputed modes, and help the establishment of a system which should be generally adopted.

Tables and formulæ have been purposely excluded from this book. The student will find all he may require in Hurst's 'Architectural Surveyor's Handbook.'

The feeling which has been entertained by some architects, that the supplying of quantities for works adds to their cost, is, it is believed, giving place to the impression that, whatever may be the result in that respect, their supply is just and correct in principle.

A well-prepared set of quantities is a valuable adjunct to a contract, especially in the final settlement of the accounts, and in nearly every instance the advantage is worth very much more than the surveyor's fees.

If work is to be well done the quantities are absolutely essen-

---

* These for the most part are wanting in system (a very important element in the consideration of the subject), and the information they afford is mixed with facts relating to construction with which every person who professes to take out quantities should be familiar, and which are more properly taught in books which specially treat of practical architecture.

tial; there is no more fruitful source of bad work than is found in the attempts of builders to recoup themselves for losses from error in their original estimates.

It has, moreover, become the practice with the majority of builders of good position to refuse to tender unless quantities are supplied.

The decision in the case of Priestley *v.* Stone (reported in Hudson's 'Building Contracts'), which establishes the principle that the quantity surveyor is not legally liable to the builder for errors in quantities, shows the importance to contractors of a thorough training of the quantity surveyor for his work.

The honesty of the parties who dispense with quantities in order to keep down the cost of a building may be fairly questioned. The liability to error in pricing an estimate is considerable, and in the cases where quantities are not supplied there is added the further risk of the production of a bill of quantities by persons who, in the majority of cases, have had hardly any special training for the work.

It is also most unfair to give a number of builders (sometimes twelve or more) the trouble of preparing estimates when only one tender can be accepted.

It is a disadvantage incident to the attempt to dispense with quantities, that they are under such circumstances liable to be prepared by a surveyor as a speculation, and builders are ready enough to save themselves trouble by accepting them. Some of the evils of such a condition of things are as follows: the surveyor is not usually the kind of person whom the architect would select; the commission charged is generally larger; and the surveyor not having been nominated by the architect, neither he nor his charges are under the architect's control.

A builder ostensibly tendering without quantities, not infrequently includes a charge in the amount of his tender, as he commonly employs a surveyor to make his estimate.

The employer should obtain neither more nor less work than is included in the estimate, consequently the quantities should form a part of the contract; and if a stipulation is made in the bill that a priced copy of it shall be deposited with the architect, the prices of which shall form the bases of the rates of any variations, the advantage to the proprietor is usually great. Sometimes it is pro-

GENERAL DIRECTIONS.

vided that the deposited bill shall remain sealed, as in some cases in which this has not been done the architect has omitted from the executed work many of the things on which the contractor would have obtained the largest profit. It will be necessary to examine the bill before sealing, to see that all the prices have been extended, the casts correct, etc.

Some public bodies insist upon the delivery, at the same time as the tender, of the priced bill of the quantities upon which such tender is based.

The object to be attained is the production of a bill of quantities which shall afford a uniform basis for competition, give the builder all the information that he may require to make his estimate, and be sufficiently clear for any surveyor (not necessarily the one who produced it) to readily adjust variations therefrom. As a contractor prices many estimates before he obtains a contract, the bill should be as short as is consistent with clearness and facility of reference during the progress of the building.

A prejudice exists in some minds against the elaboration which the more accomplished quantity surveyors affect. This can only reasonably be objected to when the items are of no value. Every item, however small, which has a value should be presented in the bills. The builder may then price them or not, as he may prefer. Often, as in the case of labours on stonework and extra labours on joinery, he does not price them, and suffers in consequence.

Another consideration is the surveyor's prime obligation to protect the building owner from claims for extra work—the war of wits between the quantity surveyor and the astute builder's clerk yearly increases in severity. In the offices of many contractors every item of the bill of quantities supplied is checked, as the work proceeds, by the actual quantity of material sent to the building, and any imperfection in a description is made the foundation of an extra charge.

The experienced surveyor will, however, to some extent regulate the amount of detail in his bills by the quality of the work.

It may be admitted that with inexperienced builders the increase in the number of items will produce a corresponding increase in the amount of the tenders, but the unworkmanlike results which many sets of quantities present are quite as likely to do so. The adoption of rough methods, such as low class

QUANTITY SURVEYING.

ESTIMATE, WOODCOTE.

Provisions. *Excavator.*

Burke's mosaic
paving as dead

Supl.

Excavation 9 in. deep
to general surface,
sepg. vegetable soil, and
wheeling and depg.
where directed at an
average distance of
2 runs.

1	3	Wilkinson's		
7	4	ct. paving.		
11	5	80	9	
4	2	169	0	
6	2	435	11	
95	4			
		685	8	
·125	8			
		76 yds. 2 ft.		
14 yds.				

4376 8

486 yds. 3 ft.

Run trench
for pipe.

20 0
25 0

45 0

15 yds.

Provide for
carving two
stone tablets,
£15.

Provide for
carving corbels
at 10/ each.

2
14
5

21

Cube.

Excavation to
basement trenches,
and wheel a. b.

17 6
78 0
272 6
54 2

422 2

15 yds. 17 ft.

Cube.

Concrete as
described in
trenches.

587 6
565 11
15 0

1168 5

43 yds. 7 ft.

Cube.

Excavation to
surface trenches,
and wheel a. b.

Provide for
two W. I. finials
and fixing £8.

587 6
565 11
15 0
22 0
5 8
16 6

1212 7

44 yds. 25 ft.

Do. P. F.
and R.

587 6
565 11
15 0
22 0
5 8
16 6

1212 7

44 yds 25 ft.

Supl.

Concrete 9 in.
thick, levelled to
falls to receive
paving.

455 6
190 4

645 10

71 yds. 7 ft.

GENERAL DIRECTIONS.

January 1886.

Bricklayer.

Reduced Brickwork in mortar.

1 B.	1½ B.	Deduct		⅔ B. sleeper walls built honey-comb.	Brick vaulting in two ½ B. rings in ct.	
		1 B.	1½ B.			
78 4	39 2					
792 2	87 1	43 0	12 0			
87 1	493 4	114 10	7 6			
1259 6	1210 9	93 3	13 6			
2 0	204 7	70 9	6 0	260 4	73 0	
50 0	1 0	39 11	36 0	108 6	42 9	Damp-proof course ½ in. thick of Seyssell asphalte.
2269 1	2035 11	59 10		368 10	115 9	
421 7	75 0	75 0				
1847 6	1960 11	421 7				
Ddt. ⅓rd 615 10	1231 8				Ro. cutting on B.	761 9
1231 8	3192 7					6 0
	11 rods 201 ft.					48 11
					29 6	5 11
		⅔ B. partn. in ct.		Deduct	43 9	822 7
		54 9			33 6	
		96 3		23 7		
		56 11		19 11	106 9	
		133 3		43 10		
		341 2		87 4		
		87 4				
		253 10				

⅔ B. trimmer arch in ct. and levelling up w. concrete.	Fair struck joint and 2ce L.W.
	804 0
	61 3
	865 3
12 3	96 yds. 2 ft.
7 0	
13 2	
32 5	

builders favour, is akin to the complaisance of those persons who when in vulgar company imitate its vulgarities, and is surely unworthy of building experts, who should rather be arbiters than mere followers of a fashion.

The systems of measurement in various parts of the country are very different. The adoption of a code of rules by the Manchester Society of Architects, which differs in many respects from the best London practice, is to be deplored, for though some differences are at present unavoidable (local customs being difficult to change), yet united action on the part of that and other societies would have resulted in the removal of some of the divergencies if they had thought it advisable, which it is presumed they did not. One recognised system of measurement throughout the kingdom should certainly be the object which the profession should strive to realise.

To take off quantities with facility, the surveyor should have a thorough knowledge of construction, of mensuration of superficies and solids and of their development, a knowledge of the architectural styles, their characteristic differences, and the technical names of their parts. He should be well acquainted with the provisions of the Building Act and by-laws of the Metropolis, or of the locality in which the work is to be done. He will be practically the technical adviser of the architect upon all matters of construction and detail. A knowledge of book-keeeping is absolutely necessary to the methodical adjustment of the accounts of a building.

The general course of procedure in the preparation of a set of quantities is the following:—

Taking off, i.e. measuring drawings to scale, and making notes of the dimensions on paper ruled as below.

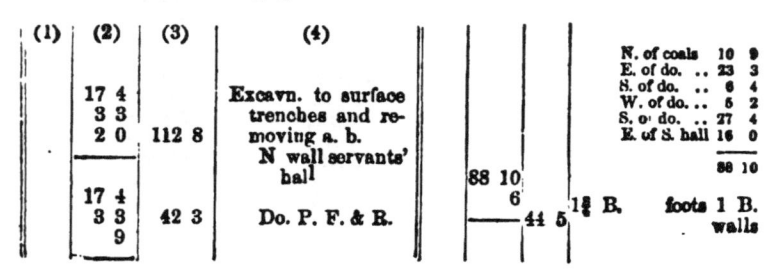

(1)	(2)	(3)	(4)				
						N. of coals	10 9
						E. of do. ..	23 3
	17 4		Excavn. to surface			S. of do. ..	6 4
	3 3		trenches and re-			W. of do. ...	5 2
	2 0	112 8	moving a. b.			S. of do. ..	27 4
			N wall servants'	88 10		E. of S. hall	16 0
			hall	6			88 10
	17 4				1½ B.		
	3 3	42 3	Do. P. F. & B.	44 5		foots	1 B. walls
	9						

Column 1 is usually described as the timesing column; 2, the dimension column; 3, the squaring column; 4, the description column.

Some surveyors write dimensions from drawings in measuring books, but it is not advisable to use measuring books, except for work at a building.

Squaring the dimensions, i.e. calculating the products of the dimensions, and writing the results with ink in the squaring column. Squarer also checks the "wastes." These calculations should be afterwards checked by another person, ticked in red if correct, or altered in red if wrong.

Abstracting, i.e. transferring the results and descriptions from the dimension sheets to the abstract, as on pp. 4 and 5, in the order of the proposed bill.

The vertical lines of abstract paper are frequently disregarded in practice so far as the writing of headings is concerned, but their adoption is essential to the columns of figures.

Reducing the abstracts, i.e. casting the columns of figures and reducing them to the denominations in common use, as yards, squares, etc. Casting may be done by juniors, but the checking of casts by an experienced man. Avoid the casting of columns by two persons at one time, a fruitful source of mistakes.

Billing, i.e. transferring from the dimension sheets in the customary order to paper ruled as below—which, if abstracting is properly done, will be the order of the abstract—the items therein.

yds.	ft.	in.		Continued.	£	
335	-	-	supl.	Excavate to general surface about 12 in. deep, (averaged) wheel and deposit within an average distance of 5 runs.		
153	-	-	cube	Ditto to surface trenches and wheel and deposit as before.		

Taking off is seldom checked, but should be.

Abstracting is checked by a clerk, who ticks in red every item from the dimensions. Billing is checked in the same manner.

There are a few general rules which it will be advantageous to remember in the preparation of quantities.

The adoption of one unvarying system facilitates the identification of dimensions, when the whole of the collateral circumstances are forgotten. Take off the dimensions always in the same order. This is the surest protection from error.

QUANTITY SURVEYING.

When dimensions in different drawings do not agree, those of the plan should always be preferred.

Take always, first, the length; second, the breadth; third, the depth. In collecting dimensions begin at one particular angle of the building, and proceed always in the same direction from that point in succeeding collections.

Prefer always to take the largest dimensions you can find, and make deductions afterwards for voids, "wants," etc., to taking a quantity piece by piece; the former plan better preserves you from the risk of the quantities being short.

Always clearly distinguish in your dimensions between voids and wants. The latter are portions of work measured in excess; as when one measures an area square for convenience, and deducts a triangle from it, as sketch; spaces enclosed by dotted lines indicate wants.

Keep all the dimensions as clear and distinct as possible; do not squeeze dimensions together in order to spare paper; closeness of writing is a frequent cause of error.

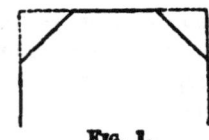

Fig. 1.

In all cases clearly describe the position of each item of the work, keeping such description well to the right-hand edge of your paper, and this is especially necessary when the surveyor writes the specification.

Description of position may be distinguished from description intended for the bill by underlining it or parenthesis. Thus—

```
        10  0
         4  0
        ─────  40  0  1½ B.                    West of kitchen.
or,
        10  0
         4  0
        ─────  40  0  1½ B.                    (West of kitchen).
```

Do not use a rule which has several scales on it; if an eighth scale, let it have nothing but an eighth scale on both edges; if a quarter, nothing but a quarter. Errors have been often made by applying to the drawing the wrong edge of a rule with various scales on it.

When a dimension is made up by the addition of several smaller dimensions, show the addition on the right-hand side of

GENERAL DIRECTIONS. 9

the description column (on "waste," as it is called), or (though this is not so well) write after the description of the work, "collected." Thus—

		61 0						25 0	East Wall.
		10 0		1¼ B.				5 0	South ditto.
		———			Still Room.		25 0	West ditto.
or,								6 0	North ditto.
		61 0						———	
		10 0		1¼ B.				61 0	
		———			Still Room.				
					Collected.				

In adopting the total of a waste, if less than half an inch, reject it; if half an inch or over call it another inch, thus: 10′ 3¼″ would be transferred to the dimension column as 10′ 3″; 10′ 3½″ as 10′ 4″.

In collecting on waste, use the exact dimensions, thus:—

$$\begin{array}{rr} 20 & 3\tfrac{1}{4} \\ 10 & 2\tfrac{1}{2} \\ 8 & 1\tfrac{3}{4} \\ \hline 38 & 7\tfrac{1}{2} \end{array}$$

In taking off dimensions, collect lengths as much as possible; time is thus saved in the squaring and abstracting. The more expert the measurer the fewer dimensions does he take off to produce the result.

Collections are disliked by some surveyors, on the ground that if a mistake is made in calculation it is generally much larger than when the dimensions are more separated. The advantages of collections, however, more than counterbalance the risk.

Observe that, as a rule, collections on waste need only be made when their total is required for one of the factors of a subsequent multiplication; in all other cases it is better to write them directly in the dimension column.

Before commencing your measurement inquire as to what details are to be supplied to you, as, if you do not, you will possibly find when you have well advanced with your work on the small-scale drawings, that details furnished will render parts of it useless.

Carefully examine the drawings and compare them to see whether the plans and sections agree as to thickness of walls, heights, etc.

Read the specification through, making notes of any descriptions which appear doubtful, or of any omissions which occur to you.

Find out, if possible, the points of the compass, as walls are more conveniently described as north, south, etc., than in any other way. The writer prefers, in default of the ability to discover the real north point, to treat the plan in question like a map, calling the top north, etc., and though this conclusion may be wrong, it is to be looked upon as a mere expedient which facilitates his work. A note on his dimensions stating that he has assumed this will insure him from misconstruction in the future.

If the drawings have not been figured, the surveyor will save himself much trouble if, before he commences his dimensions, he figures in pencil the lengths and breadths of rooms, the lengths of external walls, and the heights of the stories; and during the process of taking off, when calculating a dimension, write it on the drawing in pencil, as it may be useful later.

If the bedrooms have not been numbered, or if a system of numbering has not been carried by the architect through the whole building, number them in pencil consecutively, 1, 2, 3, 4, etc.; make a rough tracing of the arrangement and the numbers to preserve with the dimensions. This will save labour in the description of the situation of the rooms. These figures are simply for the surveyor's own use. Their adoption may, however, be suggested to the architect, as they facilitate the writing of the specification. Avoid the use of more than one series of numbers in the same building.

Make notes of any alterations of arrangement that may be adopted in the course of taking off, and if not easily described, make tracings of them, or draw them in pencil on the original drawing.

When a building appears naturally to divide itself into two (or more) separate blocks, keep the dimensions of each block distinct, following the same order of taking off, and the same entire process, as if each were a separate building. The parts which repeat may then be simply doubled, but the figures used for this purpose (the "twicing") should be red, to show that the doubling is not a part of the same block.

Some surveyors use red ink for the timesing of any series of dimensions.

When taking off oriels, bays, turrets, etc., in stone or terra-cotta, the work will be easier if the surveyor makes a plan of each course on tracing paper. When the girth of mouldings is doubtful draw them full size, and a drawing to a larger scale of any part of the work obscurely shown by the general drawing will help the surveyor to realise the possibilities; these details should be carefully preserved with the other notes.

Measure all detached buildings, outbuildings, or boundary walls last, and keep the dimensions relating to them separate, as the whole or part of them are not unfrequently omitted in course of erection, and such alterations can be much more easily adjusted if the dimensions be kept distinct.

Do not describe unnecessary things, such as the width of the rails or stiles of a door, unless there is something unusual about them; framed and braced doors and gates do vary in the width of their stiles and rails.

In measuring finishings, refer to the dimensions of the deductions of openings; by this means another check takes place, and the same order being maintained in both, reference is easier.

In altering dimensions after the squaring has been done, take the precaution of crossing out the figures in the squaring column, and, besides this, put a mark in pencil in the margin to attract attention. If an alteration is made after the dimensions are abstracted, the whole process of altering should be done at once by two persons, one making the alteration and the other checking it.

In "taking off" openings in a large building it is absolutely necessary, and it is in the case of all buildings (large or small) advisable, to put a mark on the drawings (in pencil) showing which openings have been dealt with. A good plan is as follows (Fig. 2); a line through the opening signifies opening deducted; a shorter line across it on outside of wall, facings deducted; line on inside of wall across first line, plastering deducted; a line between two last, finishings measured. By these means any doubt of the kind can be readily solved. Some surveyors use a small pencil tick instead.

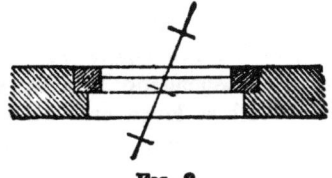

Fig. 2.

Some marks are especially necessary when several men are engaged on the same set of drawings, and the work is divided into carcase and finishings.

Write the title of the work at the top left-hand corner of every sheet of dimensions; the danger of mixture with dimensions of other work is thus avoided.

Number every column of dimensions. This is handier for reference than numbering each page.

Write an index when all is taken off, which may be fastened to the front of the set of dimensions, thus—

Cube of Building	..	..	..	1	
D. S. Fees	..	..	..	2	
Surface Digging	..	..	..	2	&c.

If the papers are bound when the work is finished, the following order is a reasonable one: Specification, Dimensions, Bills, Variations, Memoranda, Correspondence. Abstracts and Draft Bills are not bound, but should be preserved. Sometimes the papers are bound before the variations on the contract are adjusted, and it is best, when the work is large, to separate the dimensions into comparatively small volumes of not more than 300 pages, as it is frequently the case that several assistants should be working on them at one time, and this is impossible if the dimensions are in one volume: the reasonable course is *then* to cut the volume into sections.

For similar reasons, in the measurement of a building or of variations on a contract, the measuring books should not much exceed fifty pages.

Before the papers are bound, it is convenient to keep them and the correspondence in a box allotted to the building.

With a large practice the risk of mislaying papers is great. Observe the rule of putting one set of papers away before dealing with another.

The practice of "taking off" by a surveyor without any check is frequent, and objectionable. The dimensions should always be examined item by item with the drawings by another surveyor. When circumstances prevent this, the surveyor should himself look through the dimensions after several days have elapsed; he will then probably discover errors or inconsistencies which will more than repay him for the trouble thus taken.

There are various methods of "taking off" quantities; all, except for unimportant differences, referable to the three following categories. (*a*) Taking off each trade separately, as the whole of the excavation, the whole of the brickwork, etc., keeping each trade separate in the dimensions. (*b*) Taking off the work in two divisions—1st, the construction or carcase, comprising excavator, bricklayer, mason, slater, carpenter, etc.; 2nd, the finishings, comprising joiner, ironmonger, plasterer, glazier, painter, paperhanger, etc. (*c*) A modification of (*b*), which discards the division between carcase and finishings to a great extent and groups the dimensions as hereafter explained.

The former method (*a*) is now but little used. Its advantages are, the comparatively few dimensions and the consequent saving of labour in squaring and abstracting them. It is said also that it favours the writing of the specification. Its disadvantage is, the separation in the dimensions of the items of a particular piece of work so completely that in the adjustment of a contract involving omissions the work of picking out each item is often considerable and liable to be incomplete.

The second method (*b*) is the one which has been adopted in this work. The labour is certainly greater, but the superior clearness obtained in the event of variations much more than compensates for the additional labour at first.

Method (*a*) is sometimes used when it is desired to write the bills immediately after the taking off, and before the whole of the dimensions are completed.

It frequently happens that another dimension occurs exactly like one which has been previously taken. It is then convenient either to "dot it on," as it is called, or to "times" the first dimension. The first dimension being 2/5 feet, read 5 feet twice, the "timesing" would be thus, 2/2/5 feet, read 5 feet twice-twice and the dotting on thus, 2·2/5 feet, read 5 feet four times, in this case equivalent. The process requires, in either case, great care, as it is a frequent cause of error. In all cases where a dimension is dotted on, make a reference in the dimensions, where it would properly be, to the column to which it is added, and add a description of its position at that part of the dimensions where it is dotted on. It will sometimes be convenient to dot on $\frac{1}{2}$, $\frac{1}{3}$, or $\frac{1}{4}$ times, etc. When items are likely to be frequently repeated, "dotting on" will often prove more convenient than timesing.

A precaution sometimes taken, if the surveyor has a series of dimensions which will require "timesing," is to put the number or some mark in pencil opposite the first of the series, putting the number in ink to the whole of the items after they are taken off. This will be seen by the abstractor, if the surveyor, taking off, should forget to attach the number, and an error will thus be avoided. Always finish a series before you "times" it.

It is generally admitted, and well to remember, that quantities prepared by inexperienced surveyors are deficient in quantity and in completeness of description.

Keep in mind the principle of making one collection of dimensions serve for several kinds of work. In many instances one collection will be sufficient for excavation, concrete, footings and walls; and, with slight modification, one for skirting, plastering, cornices; one for eaves fillet, cutting to eaves or extra slating, fascia, and in many more cases which the student may, with little thought, discover for himself.

Much labour may be saved by averaging. Care must be taken that the things averaged are of the same relative value, and that the result arrived at is truly an average. In the case of items measured superficially one dimension should be constant, items measured cubically two. Always show the process "on waste." (See remarks on averaging in Chapter IV., section ABSTRACTING.)

It will not infrequently occur that a result obtained by elaborate labour will be further from a correct one than that produced by averaging; as a general rule, it may be assumed that the less detail involved in the process the less liability to error.

As the surveyor proceeds, if he is supplied with a specification, he should run a vertical line in pencil through each clause as it is dealt with, and he will save time if he corrects each clause requiring correction as soon as he has finished the dimensions relating to it; but this is not always practicable. Always take off in your own order, not that of specification. If the specification is long, and has not been indexed, the surveyor may avoid future waste of time by making an index; and this will be specially useful if the specification (as often happens) is ill-arranged. But the most convenient practice, when architect and quantity surveyor know each

GENERAL DIRECTIONS. 15

other's ways, is for the latter to write the specification after the quantities have been prepared; the architect, before the quantities are begun, furnishing notes of any special arrangements he may require.

In measuring triangles, the dimensions are more easily identified if, instead of halving the base or perpendicular dimension of the triangle, both be entered in full, and ½ put in front of them, thus—

| ½ | 20 0
10 0 | 100 0 | 1 B. | | Gable. |

instead of

| | 20 0
5 0 | 100 0 | 1 B. | | Gable. |

For the method of writing circles and semicircles see p. 65.

The following are the methods of writing dimensions of various surfaces and solidities. The circle is sufficient to identify such as are cubic dimensions in abstracting.

The solidity of a pyramid 5 feet base, 15 feet vertical height :—

| (*Base*) .. .. | ⅓ | 5 0
5 0
15 0 | 125 0 | spire |
| (*Perpendicular height*) .. .. | | | | |

The surface of a pyramid is treated as a number of triangles.

The solidity of a cone 5 feet diameter at base, 15 feet vertical height :—

| (*Base*) .. .. | ⅓ | ⟨5 0⟩
15 0 | 98 2 | spire |
| (*Perpendicular height*) .. .. | | | | |

The surface of a cone 5 feet diameter at base, 15 feet vertical height, 15 feet 10 inches slant height :—

| (*Circumference*) .. .. | ⅓ | 15 8
15 10 | 124 0 | spire |
| (*Slant height*).. .. .. | | | | |

The solidity of a sphere 5 feet diameter :—

| (*Base of circumscribing cylinder*) .. .. | ⅓ | ⟨5 0⟩
5 0 | 65 6 | |
| (*Height of ditto*) .. .. | | | | |

The surface of a sphere 5 feet diameter :—

(*Circumference*)	.	15 8		
(*Height of circumscribing cylinder*) ..		5 0	78 4	

The solidity of a cylinder 5 feet diameter, 10 feet vertical height :—

⟨5 0⟩			
10 0	196 5	Dig and cart.	

Well.

The usual method of measuring the length of a hip rafter is as follows: A B C is a plan of the hipped end of a roof; B D is the vertical height drawn at right angles to A B; A D is the length of hip. Observe that if the roof plane is at an angle of 45° with the horizon, E B and B D will be equal, and you may take your dimension from the plan without reference to the section.

To avoid repetition, and to show that an item is to be abstracted in two operations, the dimensions are written as follows :—

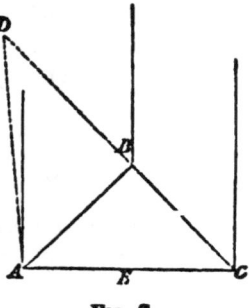

Fig. 3.

4 0		
3 0	12 0	B. P. P. ground O. S.
		in 1 square.
		and
		B. P. P. ground O. S. and cut to shapes.
		in 1 square.

When dimensions are given of a cesspool or similar receptacle, say if in clear, as " 9 inches by 9 inches by 12 inches all in clear."

A general description of the materials, etc. (see Chapter IV., section BILLING), to form the preamble to each trade, may be written in the dimensions, or may be written as a heading to the bill direct from the specification; and this latter course is best, as much writing in the dimensions is thus saved. For items involving a long description refer to specification thus: "No. 1 Dresser as described. See p. 30 specification." The preliminary bill may always be written direct from the specification, or from a former bill. Head the dimensions with a description of the work, the place, the name of the proprietor, the name of the architect, and the date thus :—

Estimate for the erection of a house and offices at Godalming, in the county of Surrey, for John Smith, Esq. (with Mr ...., Surveyor).
WILLIAM BROWN, Esq., Architect.
March 1878. 2 Fenchurch Chambers, E.C.

If the quantities are prepared jointly with another surveyor, insert the name as shown in parenthesis.

Some surveyors commence their dimensions with the cubic content of the building. A number of such calculations for different buildings are very useful data for the comparison of prices. The dimensions should always be taken on the same principles.

The letters a. b. (as before) or the words "as last" will frequently save repetitions of descriptions, but use these with care.

Writing may often be saved by marking a previous item with a letter, and referring to it thus :—

| 8 0 |
| 7 0 | 21 0 | 2¼ in. door as A col. 48 dims.

Before beginning to "take off," head a sheet of paper with the name of the work, and the word "Queries"; rule a vertical line down the middle of the sheet, write your queries on the one side of the line as they occur in the course of your work, leaving the other half of your paper blank for the answers, and write these down as you obtain them either by further inspection of drawings and specification, or inquiry of the architect.

Do not trust to your memory. If you discover that you have forgotten to take an item in its proper order, do it on the first opportunity; nothing contributes so much to a clear head as the avoidance of an accumulation of small items of arrears of work.

All the items which will be found in the different following

c

sections could not apply to one building, they are introduced for the purpose of clearly showing the system.

After the quantities have been taken off, read through the specification, to see if all has been taken.

When the quantities are required for a large building, and it is found necessary for a number of men to take off at one time, it is the usual practice to make several sets of tracings (one for each man) so that the inevitable interruption of each other's work when several are working on the same set of drawings, may be avoided. The expense thus incurred is amply repaid.

There are other well-known methods of reproducing drawings of the same size as the originals, which are generally cheaper than tracings if several copies are required, such as the ferro-prussiate process, white lines on a blue paper; the ferro-gallic process, dark lines on white paper; the aniline process, which reproduces the lines of the drawing on a special slightly tinted paper; and photo-lithography, for which ordinary drawing paper may be used. These two latter kinds of copy can be coloured in the ordinary way.

When more than one person is engaged in taking off a set of quantities, the liability to error is much increased, as each man is apt to assume that one or the other has taken off a particular section of the work, and the way to obviate the risk is for one man to take the responsibility, and supervise this. It is a wise precaution to make a list of all the sections and to put against each item the initials of the man who will deal with it. Cross out each item as it is done. Supervisor should take off the painting from the dimensions of the whole work, when they are complete, and he will discover mistakes and discrepancies, which he can remedy. If the necessity of checking taking off should arise, and he cannot follow the dimensions, let him take it off in his own way and compare the results.

Other considerations arise when the work of taking off for a large building is divided among a number of men.

The main objects are such a classification of the work in the dimensions and allotment of it to the workers as shall accelerate the taking off, facilitate the settlement of variations, and assist the valuation of the work for certificates.

Surveyors of experience know that no rules of practice can be uniformly applied, but there are several principles of action which may generally be adopted with advantage.

First as to acceleration of taking off. If we take as an example an asylum or a hospital, we shall have in the one establishment several distinct types of building—the wards, the offices, the corridors, the houses of the staff, the sanitary work, the drains. It will generally be most advantageous, in the case of the wards, to let one man do the roofs *of all* the wards, another the whole of the brickwork, another the whole of the joinery and finishings, rather than allot one ward in all trades to each man, for (and more especially when there is no specification) it is obvious that, according to the first-named arrangement, the minutiæ of construction and finish in one branch of the work being repeated throughout the wards, only one man would spend the time necessary for its investigation, instead of several, as would be the case if each man dealt with a complete ward.

The work of such a building when divided among a number of assistants is facilitated by having one block taken off complete in an approved manner by one assistant and instructing the others to adopt the same order and system with the succeeding blocks.

Probably each of the sections other than wards may be most conveniently taken off complete in all trades by one man.

Second. As to facilitating settlement of variations. The drains are so frequently varied that they should be taken off in all trades as a separate set of dimensions, and should form a separate bill. The wards might form a separate bill.

Each of the officers' houses should form a separate bill; the boundary walls and gates should form a separate bill.

In the allotment of brickwork to its particular sections, let the dominant building include the whole of its necessary walls.

If the machinery and apparatus of laundry, the sanitary work, the heating work, or the ventilation is the subject of a provision or a sub-contract, the builders' work in connection with either should form a separate bill.

The dimensions of the work in all trades below the damp-proof course should be kept distinct and together, as foundations are frequently varied in execution.

Third. As to valuation for certificates, when a building is large, the valuation for certificates is helped by keeping separate the dimensions of the work above and below certain levels, as between one floor and another, and by abstracting such section separately, as, for instance, all work below first floor on one set of abstracts, all work above first floor on another, and although the work on one abstract would probably be transferred to the other, so as to make one bill of the whole, the material for rapid separate calculations of approximate value would be thus ready for the surveyor's use as soon as occasion arose, or abstracts for separate blocks may be adopted.

The recommendation as to separate bills before mentioned, involving, as they do, separate abstracts, will afford the surveyor similar assistance.

Each man engaged should make any requisite notes on his dimensions for the guidance of the others.

It will be obvious that doubts will arise at those parts of the building where one man's work is divided from another's. The following is a specimen of such a note.

> "*Work in Corridors.*—In the general store, coal store, visiting room and medical assistant officer's quarters, the wall of building is measured including digging and deductions, but the plastering in corridors is not measured; the lead gutters between these roofs and corridor roofs are measured, but lead gutters to corridor roofs against walls are not measured. Facing is measured on the walls above roofs of corridors."

Mere notes should be written across both dimension and description columns, not confined to the description column.

Future reference to the dimensions will be much easier if headings are written in the body of the dimensions wherever the work appears to naturally divide itself, thus: "Main roof," "Roof of laundry," "Fittings," etc.; the indexing will also be facilitated by this practice.

Much time may be saved by employing a *careful* clerk to write the dimensions to your dictation. Keep a watchful eye on the page as it is written. When you desire to have a series of dimensions added together, begin with the words "on waste," and when the clerk has finished the series use the word "dimension"; the clerk will then add them together, state the result of the cast, and write it in the dimension column.

The received method of dictating the following $\overline{2/3/7\cdot0}$ 1 inch by 1 inch deal fillet rough, is seven feet three times twice inch by inch deal fillet rough.

When in taking off you make a sudden change in the thickness of work of the same description, underline it thus: "1½ four panel square door."

Items should not leave the reader in doubt as to whether part of the work is included in the prices or not. To avoid this, use the words, "elsewhere taken," thus: "1½ inch Honduras mahogany w.c. seat fixed with brass screws and cups *elsewhere taken*." It will sometimes happen that there are certain things which the surveyor is compelled by circumstances to take, but which he knows must of necessity be altered in execution. If the surveyor keeps a list of these items he will find it an assistance when he is required to settle variations.

Work in narrow widths, small quantities, or short lengths, should be so described and kept separate; but where the quantity in the bill is so small as to speak for itself, the specific words "in small quantities" need not be used. Such work should always be billed in feet.

"Deduct and add," is a convenient form of words when a small part of a large total quantity differs from the remainder, moreover, the process saves dimensions, and is less liable to error than a measurement of small pieces. Digging is often treated thus. A large dimension is taken of "Dig fill and ram," and a small quantity of the same is afterwards measured as "Ddt. and add Dig and Cart." Moulding on stone would be first measured as "Moulding" and followed by an item "Ddt. and add" "moulding stopped."

The words "Extra for" and "Extra only" are useful, but require judgment in their application. What the extra is upon should clearly appear. It is obvious that some item has been previously measured and appears elsewhere in the bill, or the "Extra for" may appear "written short."

| | | No. | 10 | Extra for bends in 4" pipes .. .. | | £ | s. | d. |

Shows that a lineal dimension of 4" drain has been measured, including the length of these bends.

| rods. | ft. | in. | | | | £ | s. | d. |
| 5 | 200 | – | supl. | Reduced brickwork, extra only in cement | | | | |

Shows that a larger quantity of brickwork appears elsewhere in mortar in the bill, and that this item represents the difference between cement and mortar.

ft.	in.			£	s.	d.
1850	–	supl.	Extra on common brickwork for facing of best red Fareham facing bricks, and raking out and pointing with a neatly struck bevelled joint in cement			

Shows that the ordinary brickwork appears elsewhere, and that this item represents the extra value for the better bricks and the pointing.

When describing an article from a special manufacturer, his name and address should appear in the item of the bill, but the address need not be given in more than one item—adopt as nearly as possible the description in the trade list.

The loss and inconvenience which surveyors have sometimes suffered by injury to drawings by ink, fire, etc., have led many to insure them against such damage. The practice is a good one and may preserve the surveyor from heavy loss. Some surveyors refuse to prepare quantities from drawings in pencil; if he does he should retain tracings of them, or have them finished by the architect's clerk in his own office; an inadvertent alteration may be important.

The surveyor will endeavour to measure the *exact* quantity of work in the building he deals with. Perhaps excess in a bill of quantities is not so frequent as it was a few years ago, but it is not rare even now to see 5 per cent. taken off at the end of a bill by the estimator for full quantities. No doubt the closeness of the quantities will vary with the temperament of the surveyor (the personal equation as it has been called) who prepares them, but the general principal now is to give the exact quantity to the best of the surveyor's ability. Mr. Rickman said in his paper on 'Building Risks,' read before the Surveyors' Institution: "Probably it will be near the mark to state that in very careful quantities taken from general drawings only, there is an excess of $\frac{1}{2}$ to 1 per cent. (and there ought not to be more), and that there are various labours taken which a builder tendering will consider either not imperative or included in the general description of the work to the extent of possibly 1 per cent.; these points are commonly discounted by the successful tenderer."

The following extract from 'Quantities and Quantity Practice,' Transactions of Institution of Surveyors, 1880, as to the extent of detail in taking off, may be recommended to the student:—

"Reverting to the subject of greater or less minuteness of detail, it will be obvious that in some cases it would be advisable to go into very much more detail than in others. It occasionally happens that a long description, illustrated by a sketch in the margin of the bill, will best help the builder in arriving at the cost of some feature or element in the work. Sometimes, on the other hand, the better way is to measure the work in considerable detail, so that, although its individuality is lost in the bill of quantities, its components are added to those of others and priced with them. Sometimes, again, from the fact of only one object of a kind existing, it may, with advantage, be numbered, as a slight difference between actual and estimated cost is not increased by multiplication; whereas, if there were several features of the same kind, greater detail should be observed, as a trifling error in the price of one would obviously become magnified. Another illustration of this principle is met with in large works like hospitals, workhouses, etc., where there is a large amount of repetition as regards the features, such as many windows of the same size and kind, and it is easy to see that great attention to minutiæ is necessary where one window only is actually measured and multiplied many times, as that which would be inappreciable in the one, becomes serious in the many."

The student of quantity surveying may derive advantage from a familiarity with the literature of the subject. With the exception of a few references to the measurement of builders' work in Peter Nicholson's 'Architectural Dictionary,' Reid's 'Young Surveyor's Preceptor' is the first methodical statement of modern methods. It contains a complete set of drawings of a house, specifications, dimensions, abstracts and bills, and although much of the practice there shown is obsolete, the book may be read with advantage. This was succeeded by Dobson's 'Students' Guide to Measuring and Valuing.' In Mr. Hurst's 'Architectural Surveyor's Handbook,' the section 'The Measurement of Builders' Work' was a considerable advance upon the published information on the subject. Since then Fletcher's 'Quantities' and Leaning's 'Quantity Surveying' have further exemplified the practice. A list of modes of measurement is attached to the specification of the Houses of Parliament (*see* Donaldson's 'Handbook of Specifications') In the published official schedules of prices of H.M. Office of Works, the War Office, the London County Council, the School Board for London, etc., each trade commences with directions for measurement. Several of the provincial professional societies have also formulated systems of measurement.

CHAPTER II.

ORDER OF TAKING OFF.

THIS order is referred to as (*b*) at page 13 in Chapter I.

The items in this section are furnished as a general guide, and do not profess to comprise everything which arises in the course of practice.

The principle to be observed in "taking off" is the avoidance of the loss of time, liability to error, and uncertainty as to the stage of the work arrived at, which is a consequence of mixing things up too much. The arrangement of the following list is believed to meet some of these objections. The similarity of its order to that adopted in the measurement of a building, for which it is well suited, may possibly be a recommendation to some, and it certainly favours the division of the taking off among a number of assistants.

In looking through dimensions some time after they have been taken off, be very careful to make no alteration without mature consideration, for although a dimension may appear at first sight to be wrong, it will most frequently prove on further thought to be correct.

Many of the arrangements suggested in the following pages are applicable only to a large work, in a small one they would only produce confusion in the guise of order.

It will generally be necessary for the surveyor to visit the site of the proposed work. In cases of new buildings, he should discover if possible whether sand or gravel suitable for the works can be obtained on the ground; should observe the access, and if exceptionally difficult, should describe it in the preliminary bill; should note any flues of adjoining buildings that may require raising, any parts of abutting property exposed by the pulling down of party walls or boundary walls, necessitating inclosure by contractor or special shoring; should measure the lengths of frontage if paved, and describe as " . . . feet run. Take up street paving and deposit and relay, and make good at completion, or pay parish authorities for so doing." In cases of alterations to old buildings, see if temporary floors are required (sometimes necessary when business is carried on during construction of a new floor); temporary ceilings or roofs (when new stories are constructed over occupied premises); temporary inclosures (to separate occupied part of building from part to be altered). In all cases measure the quantity and describe. See what arrangements are necessary for temporary supply of gas or water, where these are interfered with. Note what work requires casing and protecting, as stone staircases, chimney pieces, etc.

A visit to the building is better deferred until the surveyor has taken off all the work which can be measured from the drawings, he can then settle all the points noted, and will better know what to look for.

Some surveyors, however, visit the building first, taking the drawings with them and making notes in a measuring book of everything which will be imitated or repeated, the actual taking off being afterwards done at the office.

In the preparation of quantities for the completion of buildings in carcase, particular care must be exercised in the observation of the condition of the work, much of which is left in various stages of progress. The quantities should include cleaning down all the work, both new and old; renewing work which has possibly already failed; removal of hoarding, making good paving up to building frontage, etc. If a hoarding is already fixed, state what alteration will be required to adapt it to contractor's purpose.

When there is a necessity for application to the freeholder, district surveyor, Ecclesiastical Commission, Education Department, or other authority, it should be made before the quantities are commenced.

When the surveyor receives the drawings he should, before preparing the quantities, examine them, and if he discovers any infraction of local acts or by-laws, unsafe construction, or interference with the rights of light and air of adjacent buildings (as when a higher building is proposed on the site of an old and lower one) he should inform the architect, and if the architect has omitted to have the façades of the whole building measured and drawn, and in some cases photographed, he may render good service by reminding him of the omission.

CARCASE—

Cube of building.
Area for district surveyor's fee, if in Metropolitan area.
Any work to party-walls complete.
Digging to surface.
Ditto to basement.
Ditto filled in and rammed.
Ditto to basement trenches. External walls. *Collect the lengths a.*
Concrete to basement trenches. External walls. *Collection a.*

    Cases will sometimes arise in which it will be the simplest course to measure one wall at a time, beginning with the footings and completing it to the top.

Footings. External walls. *Collection a.*
External walls up to a certain level (say ground floor). *Collection a.*
Digging concrete and brickwork of projections on external walls up to the same level, but not chimney breasts.

    Measure everything up to one level if possible, before taking any work above that level, as to level of damp-proof course or ground floor.

Iron joists cover stones. Stone for corbelling. Oversailing courses, incidental to the support of the brickwork, should be taken as they occur.

Incidental cuttings are best taken as they occur. Also preparation of old walls to receive new. Cutting and bonding new work to old. Extra only in cement, etc.

Digging to basement trenches. Internal walls. *Collection b.*
Concrete to trenches. Internals walls. *Collection b.*
Footings. Internal walls. *Collection b.*
Internal walls up to a certain level (say ground floor). *Collection b.*
Digging concrete and brickwork of projections on internal walls up to the same level, but not chimney breasts.
Digging to surface trenches. External walls, i.e. walls where there is no basement. *Collection c.*
Concrete to surface trenches. External walls. *Collection c.*
Footings. External walls. *Collection c.*
Walls, external, up to a certain level (say ground floor). *Collection c.*

Digging concrete and brickwork of projections on external walls up to the same level, but not chimney breasts.
Digging to surface trenches. Internal walls. *Collection d.*
Concrete to surface trenches. Internal walls. *Collection d.*
Footings. Internal walls. *Collection d.*
Walls, internal, up to a certain level (say ground floor). *Collection d.*
Digging concrete and brickwork of projections on internal walls up to the same level.
Strutting and planking to basement.
Strutting and planking to trenches. *See collections of walls.*
Damp-proof course. *See collections of walls. See also example of a collection of damp-proof course.*
External walls from last level up to first floor level. *See former collections.*
Internal walls from last level up to first floor level. *See former collections.*
External walls, from first floor level, floor by floor to top.
Brick fronts to dormers and gables.
Internal walls, from first floor level, floor by floor to top.
Deduct wants as they occur.
Projections of chimney breasts, *one at a time*, commencing with concrete and finishing with shafts.

    If surveyor follows these up from the ground, tracing the flues, he is sure to get the breasts of the right size.

    When the building is large, writing will be saved by making a sketch on the dimensions of the disposition of the chimney stacks, and referring to them by letters thus :—

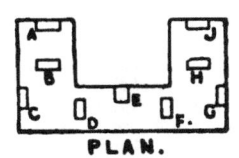

PLAN.

Hoop iron bond. *Collect. See collections of walls.*
Brickwork extra only in cement for last.

FIRES—

Beginning with topmost story. Deduct brickwork of opening.
All the openings on one story of similar width and depth may be collected.
Add chimney bars. *Collect.*
Segmental arches. "
Flues, parget and core. *Collect.*
Flue-pipes. *Collect.*
Soot doors and sweeping flues.
Fender walls. *Collect.*
Centering to trimmers. *Collect.*
Trimmer arch and levelling up with concrete. *Collect.*
Skewback cutting. *Collect.*
Feather-edged springer. *Collect.*
Filleting soffits of trimmers.
Hearths and back-hearths. *Collect.*
Notchings for chimneypieces. *Collect.*
Curbs, tile hearths, cement-floated face concrete.
Chimney-pots or chimney moulds or bafflers. *Collect.*
Stoves and setting.
Coppers and setting.
Stone cover to copper setting.
Copper-lids.
Chimneypieces.
If a provision for stoves and chimneypieces, make a collection as shown in examples of collections.
Fixing only chimneypieces.
If setting stoves is not in provision take particulars of materials for setting only.

INTERNAL OPENINGS—

To be referred to when taking off the joinery.
Beginning with topmost story.
Deduct brickwork. *Collect or average the openings where possible.*
Add lintel. *Collect.*
Relieving arch (or segmental arch and centering where no lintel).
Wood bricks, pads, or fixing blocks.
Steps or thresholds (if any). *Collect the various widths.*
Ends cut and pinned or jointed.
Frame bed and point (if any).
Iron dowels and mortises.

ARCHWAYS AND RECESSES—

Deduct brickwork.
Centering.
Arch.

FACINGS—*Collect.*

Facings to chimney-stacks, taking stone dressings, if any, at the same time, also extra projecting brickwork, and extra only in cement.
Gable copings (stone or brick), finials, kneelers, bonders.
General facings. *Collect the lengths.*
Strings and brick cornices, and extra brickwork for their projection, quoins, plinths, tablets, niches.
Excavation, brickwork, and facing for extra thickness of walls to form plinths, etc. The walls having been previously measured minus the plinth and attached sleeper wall.
Ashlar facing and general stonework, or external cement work, except to openings.
Terra-cotta, if any, except to openings of doors and windows.

EXTERNAL OPENINGS—

Beginning with topmost story and floor by floor.
Deduct brickwork.
Deduct facings.
Add jambs, stone or brick, mullions, transoms, lintels, centering, strutting.
Arches, cutting to facings, rough cutting. *The openings may often be averaged.*
Sill, fair ends, window sills made up and pointed.
Facing made good to ends of window sills.
Lintel. *Collect.*
Relieving arch.
Frame bedded and pointed or screeded.
Wood bricks or pads.
Iron dowels and mortises.
Stone steps, including brickwork supporting them, and the landings and railings to external flights taken with the doorways to which they lead.
Ends cut and pinned.
Areas complete in all trades, with the basement windows. *Collect the walls of areas where possible.*

Pavings—

Beginning with basement.
Hard dry rubbish.
Concrete.
Cement-floated face.
Pavings of all kinds.
Iron joists connected with vaulting.
Ends cut and pinned.

Templates.
Centering and horsing to vaulting.
Vaulting.
Cutting.
Raking out and pointing to soffits.
Stone staircases complete in all trades.

Ground Floors—
Brickwork of oversailing to receive plates. *Collect.*
Excavation for sleeper walls. *Collection e.*
Concrete for sleeper walls. *Collection e.*
Footings for sleeper walls. *Collection e.*
Remaining brickwork for sleeper walls. *Collection e.*
Extra thickness of walls to receive plates.
Excavation, footings, remaining brickwork. *Collection f.*
Damp-proof course.
Plates. *Collection f.*
Joists. *Collect.*
Concrete under wooden floors.
Air bricks and openings through walls.

Upper Floors—
Templates, felt or lead pads, girders, wood or iron.
Girders of wood or iron.
Brick piers, bases and iron columns.
Ends cut and pinned.
Iron flitches, with the wooden girders.
Iron tension rods, etc., ditto.
Iron bolts, ditto.
Joists. *Collect.*
Trimmers. *Collect.*
Corbels, or templates for trimmers.
Deductions of joists for voids.
Strutting. *Collect.*
Sound-boarding and pugging.
Deductions of ditto.

Fire Proof Floors—
Concrete and digging.
Brick piers.
Stone bases.
Iron or steel columns or stancheons.
Wedging up bases and running with cement.
Templates and ends cut and pinned.
Iron or steel girders.
Bolts to caps and bases.
Iron or steel intermediate bearers.
Centering and horsing, or hanging.
Concrete.

Partitions—
Floor by floor, beginning with topmost story.
Heads and sills. *Collect.*
Interties. *Collect.*
Posts. „
Quarters. „
Braces. „
Bridging-pieces. *Collect.*
Ironwork in straps and bolts.
Partitions trussed, same order as last.
Brick-nogged partitions. *Collect.*
Deduct openings.
Heads and sills. *Collect.*
Quarters.
Nogging-pieces.

ROOFS—
Templates.
Felt or lead pads.
Ends of timbers cut and pinned.
Trusses. Tie beams.
Principal rafters.
King or Queen posts.
Struts.
Collars.
Circular ribs.
Wrought face.
Ironwork and fixing.
Straps.
Bolts.
Gibs and keys.
King or Queen heads.
Hoisting and fixing roof trusses.
Templates.
Ends cut and pinned.
Purlins. *Collect.*
Cleats.
Scarfings and bolts to purlins.
Plates. *Collect.*
Ridges. „
Scarfings and bolts to ridges.
Tile or slate ridges.
Ends fitted.
Ends splayed and fitted.
Rough chamfer on fir.
Ridge rolls. *Collect.*
Lead to ridges.
Hips. *Collect.*
Hip rolls. „
Rough chamfer on fir. *Collect.*
Lead to hips.
Bossed ends to rolls.
Cutting to slating or tiling.
Cutting to boarding or battens.
Bossing intersections of ridges and hips.

Hip tiles or soakers.
Close cut and mitred hip.
Valleys. *Collect.*
Valley boards. *Collection j.*
Tilting fillets.
Valley fillets.
Cutting to boarding.
Cutting to slating or tiling.
Lead to ditto.
Valley tiles.
Dragon pieces and bolts.
Rafters. *Collect.*
Sprockets. "
Cut and shaped and wrought ends to rafters.
Eaves boards. *Collect.*
Eaves fillet. "
Slating or tiling for eaves.
Tilting fillet. *Collect.*
Battens or boarding and felt.
Slating or tiling.
Gutter boards and bearers.
Rebated drips.
Short lengths of roll.
Gussets.
Lead to gutters.
Cesspools. *Collect.*
Lead to ditto.
Extra labour and solder to cesspools.
Socket pipes from ditto.
Perforations in brickwork for ditto.
Copper wire or galvanised iron covers to cesspools.
Flashings to gutters. *Collection g.*
Raking out, wedging and pointing. *Collection g.*

GENERAL FLASHINGS to main roofs—

Flashings.
Stepped flashings or soakers.
Tilting fillets.
Raking out and pointing.
Fillets to secret gutters. *Collection h.*
Lead to ditto. *Collection h.*
Extra labour to secret gutters.
Flashings to ditto. *Collection h.*
Raking out, wedging and pointing. *Collection h.*

CHEEKS—

Boarding and quarters and lead.
Soldered dots.
Flashings.
Copper nailing.
Vertical tiling.
Tilting fillets.
Soakers.

CHIMNEYS—

Deduct tiling or slating, and boarding of battens, and felt.
Add for cuttings to ditto.
Chimney gutters. *Collection l.*
Short lengths of roll.
Gusset pieces.
Flashing boards. *Collection l.*
Tilting fillets. "
Lead to gutters, cover flashings, aprons, soakers, stepped flashings. *Collect.*
Bossed ends to rolls.
Rake out, wedge and point, flashings. *Collect.*
Soakers.

DORMERS WITH BRICK FRONTS—

All trades except finishings and the brickwork, which last is measured with general brickwork.
Deduct slating or tiling, and boarding or battens, and felt.
Deduct rafters.
Plates. *Collect.*
Valleys or footing-pieces, valley boards, valley fillets. *Collect.*
Lead to valleys or valley tiles, cutting to boarding or battens and to tiling.
Trimmers.
Add ridge. *Collection m.*
Ridge roll. "
Chamfer on flr. *Collection m.*
Lead to roll. "
Bossed ends to ditto. "
Tile ridges.
Ends splayed and fitted, and lead soaker.
Lead soakers.
Rafters. *Collect.*
Eaves board. "
Tilting fillet. "
Tiling or slating for eaves.
Sprockets. *Collect.*
Cut and wrought ends to rafters.
Tiling or slating, and boarding and felt.
Boarding and backings to cheeks.
Cutting to boarding. *Collect.*
Lead or other covering to cheeks.
Copper nailing. *Collect.*
Brass screws and soldered dots and sinkings in boarding.
Flashings. *Collection n.*
Raking out, wedging, and pointing ditto. *Collection n.*
Ceiling joists.

ORDER OF TAKING OFF.

CEILING JOISTS—
Plates. *Collect.*
Ceiling joists. „
Stretchers. „
Hangers. „
Ways in roofs. „
Traps in ceilings complete.
Ladders.

EXTERNAL TRAPS—
Deduct rafters, slating or tiling, battens or boarding, felt.
Add slating or tiling for cuttings.
Trimmers.
Boarding or battens for ditto.
Trap and rima.
Lead covering.
Copper nailing. *Collect.*
Bossed angles.
Fastenings.
Curb, and labour on same.
Internal lining.
Lead covering to curb.
Copper nailing. *Collect.*
Flashing boards. „
Tilting fillets. „
Gutter board and bearers.
Lead to gutter.
Deduct plastering.
Add plaster quirk. *Collect.*

LANTERN LIGHTS, &c.
These vary greatly in detail, but generally as to traps and dormers.
Snow boards. *Collect.*
"Cat" or roof ladders.

EAVES GUTTERS AND RAIN-WATER PIPES—
Fascias.
Eaves gutters. *Collect.*
Stopped ends. „
Angles. „
Outlets. „
Covers to outlets. „
Rain-water pipes. „
Swan necks. „
Heads. „
Covers to heads. „
Shoes.
Tee pieces.
Bends.
Plinth bends. „
Loose bands.
Holder bats.

Cuttings to brick or stone strings. *Collect.*
Connections with drain. *Collect.*

WATER SUPPLY—
Well.
Well cover.
Bearers to cisterns, iron or wood.
Templates.
Ends cut and pin.
Cisterns of iron.
Cisterns of wood.
Lead lining.
Soldered angle. *Collect.*
Copper nailing. „
Standing waste. „
Brass washer and waste, and fly nut and hole in cistern.
Lead under waste.
Boiler screw and overflow.
Perforations in brick or tile.
Cistern covers.
Joiner attend plumber for each cistern.
Perforations in brickwork for pipe.
Chase for pipe, pipe casing, plaster quirk. *Collect.*

SINKS—
Stone, stoneware or porcelain.
Bearers.
Perforations for traps.
Cutting and pinning sink to brickwork.
Brass grate.
Trap.
Waste.
Perforations in brickwork for pipes.
Joiner attend plumber to sink.
Wall tiling or cement around sink.

WOODEN SINKS—
Bearers.
Ends cut and pin.
Woodwork of sink.
Lead lining.
Soldered angle. *Collect.*
Copper nailing. „
Wooden capping.
Brass grate and plug, and rim and soldering.
Lead waste.
Lead cones.
Lead trap and cap and screw.
Chase. *Collect.*
Pipe casing. *Collect.*
Plaster quirk. „
Perforations in brickwork and making good.
Deal top, if any.

Perforation for sink.
Joiner attend plumber to sink.
Enclosure of space beneath, complete.
Draining boards.
Covering with lead.
Copper nailing. *Collect.*
Flashing boards, and covering with lead.
Tiling or cement around.

LAVATORIES—
Apparatus and valves complete, rough deal tops and bearers.
Perforation for basin.
Slate or marble top, and perforation.
Rounded or moulded edge. *Collect.*
Skirting and angles. *Collect.*
Basin and washer, plug and chain.
Waste.
Trap.
Connection with drain.
Overflow pipe and joints.
Chase. *Collect.*
Pipe casing. *Collect.*
Perforation in brickwork for pipe.
Enclosure of space beneath, complete.
Shelves and bearers.
Joiner attend plumber to lavatory.

BATHS—
Deal cradle for bath.
Bath and valves.
Waste and joints.
Connection with another pipe, or as the case may be.
Perforations for pipes.
Pipe for overflow and joints.
Lead safe.
Joiner attend plumber to safe.
Lengths, bossed angle.
Pipe for waste from safe and joints.
Dribble pipe or weighted copper flap and soldering.
Wooden top.
Rounded or moulded edge. *Collect.*
Deduct perforation for bath.
Perforation in bath top.
Add rounded or moulded edge.
Enclosure to front and ends. *Collect.*
Extra for door.
Fastenings.
Skirting and mitres to top. *Collect.*
Joiner attend plumber to bath.

WATER-CLOSETS—
Apparatus.
Anti-D trap.
Soil pipe, including joints.

Bends.
Extra joints.
Cap or hood and straps to vertical pipe.
Connection with drain, and brass thimble or tail piece.
Perforations for pipes.
Ventilating pipe and joints (anti-syphonage).
Perforations in slating or tiling.
Lead slate.
Set of flushings.
Lead safe.
Bossed angles.
Dribble pipe, if any.
Waste from safe and joints and copper flap.
Joiner attend plumber to safe.
Seat and riser and bearers and flap and frame.
Butts.
Hole, cut and dished for pan.
Ditto cut and beaded for handle.
Extra for seat and riser to remove easily.
Grounds.
Skirting and mitres and cut or mitred ends, or back and elbows. *Collect.*
Waste preventer and its appurtenances.
Flushing pipe.
Overflow pipe.
Hole in wall.
Joiner attend plumber to w.c. and its w.p. cistern.
Paper boxes.
Candle brackets.
Pot cupboards.

SUPPLY PIPES—
Cistern connector with union and fly nut and joint to supply.
Lead pipe to each draw-off w.c., apparatus, sink, &c. *Collect.*
     In measuring supply pipes from cisterns, take the longest length first and then its branches.
Branch joints as they occur.
Perforations in brickwork as they occur.
Cocks as they occur, at the end of each supply.
Chase. *Collect.*
Pipe casing. *Collect.*
Plaster quirk.   „
Connection with main, paying fees, and making good roads and footways.
Lead pipe from main to supply the cisterns. *Collect.*
Trench for pipe. *Collect.*
Ball valve.

Perforations in brickwork for pipe.
Chase. *Collect.*
Pipe casing. *Collect.*
Plaster quirk. „
Stop-cock in each case, with the supply in which it occurs.
Chamber for stop-cock, if outside of building.
Water meter.
Brick chamber and cover.
Pump.
Lead suction pipe and joints, &c.
Rising main. *Collect.*
Warning pipe.
Joiner attend plumber to pump.

HOT WATER SUPPLY—

Hot water cistern or cylinder and bearers, and expansion pipe.
Pipes. *Collect.* Beginning with the flow and return.
Cocks with each supply as they occur.
Perforations in brickwork for pipes.
Chases. *Collect.*
Pipe casing. *Collect.*
Plaster quirk. „
Safety valve.
Attendance on hot water fitter.

DRAINS—

Cesspools, rain-water tanks, gullies, grease-traps, dumb wells.
Inspection pits complete.
When there are several inspection pits or turning chambers they may be distinguished by letters A, B, C, etc., or by numbers.
Drain pipes, with their various depths of digging, bends, junctions, diminishing-pieces, syphon-traps. *Collect.*
Connection with sewer and pay fees.

FINISHINGS—

Floor by floor, beginning with topmost floor.

FLOORS—

Flooring.
„ for cuttings. *Collect.*
Flooring in small quantities and bearers. *Collect.*
Steps complete from one level to another.
Skirtings. *Collect.*
Mitres. „

Irregular ditto. *Collect.*
Fitted ends. „
Housings.
Cement skirtings.
Mitres. „
Stopped ends. „
Returned mitred ends. *Collect.*
Cement or wooden dados, and cappings.

DORMERS—

Dormers complete when there is no brickwork to them, with the windows of topmost floor.

WINDOWS—

May often be averaged.
Refer to the dimensions for the deductions of the openings. The surveyor will thus ensure the correct dimensions, and will be less likely to forget any of them.
Deduct plastering and paper.
Add sashes and frames.
Moulded horns.
Iron tongue and white lead and groove in oak, and groove in stone. *Collect.*
Sash fastening.
Lifts.
Top fastenings, or eye and plate.
Linings. *Collect.*
Window boards and bearers. *Collect.*
Notched, returned, and mitred ends. *Collect.*
Window-nosing. *Collect.*
Returned and mitred ends. *Collect.*
Window backs.
Canvassing and painting backs of ditto.
Rendering in cement behind ditto.
Long notchings of linings over ditto.
Beaded capping.
Grounds.
Architraves and their bases. *Collect.*
Plinth to window back.
Housings to ditto.
Glass.
Painting to frame and squares.
Shutters and boxings.

DOORS—

May often be averaged.
Refer to dimensions of deduction of brickwork.
Deduct plastering and paper.
Add door.
Butts or hinges.
Lock.

D

Fastenings.
Finger plates.
Linings. *Collect.*
Grounds. „
Dovetailed backings, if any.
Architraves and bases. *Collect.*

CASEMENTS—

May often be averaged.
Deduct plastering and paper.
Add frame.
Transome.
Casements.
Fastenings, water-bar, butts, centres.
Linings. *Collect.*
Grounds. „
Architraves and bases. *Collect.*
Glass.
Painting to frame and squares.

FITTINGS—

Not included with water supply, as cupboards, shelves, dressers, etc.

WOODEN STAIRCASES—

Treads and risers.
Winders.
Housings.
Moulded, returned, and mitred ends to steps.
Cut brackets.
Curtail ends and veneered fronts to risers.
Bull-nosed or quadrant ends.
Ends notched and fitted to newels.
Wall string, ramps, wreaths, mitres. *Collect.*
Outer string, ditto. *Collect.*
Landings and bearers.
Nosings.
Apron linings. *Collect.*
Newels, wood and iron.
Turnings to newels.
Balusters, wood and iron. *Collect.*
Iron stays.
Hand rail and core rail. *Collect.*
Scrolls to both. *Collect.*
Ramp to both. „
Wreaths „ „
Handrail brackets.
Plastering to soffits and whitening. *Collect.*
Ditto ditto flueing. *Collect.*

Quirks.
Boarding to soffits.
Spandril framings.
Doors and ironmongery in ditto.

PLASTERING AND PAPER—Taken together.

First, all the apartments and next the passages, floor by floor.
Ceilings and whitening.
Cornices and bracketing. *Collect.*
Mitres. *Collect.*
Irregular ditto. *Collect.*
Enrichments.
Returned and mitred ends. *Collect.*
Centre flowers. *Collect.*
Partitions. *Collect.*
Walls. *Collect.*
Cement angles.
Archways and recesses.
Deduct plastering.
Add plastering and mouldings to reveals.
There is seldom much advantage in collecting the plastering of walls of attics; the variety of heights is often so great that it is better to measure these one room at a time.

GASFITTER—

Pipes. *Collect.*
Syphon.
Meter.
Shelf or enclosure for meter.
Short length of pipe and connections.
Main cock.
Pay fees and make good footway.
Perforations in brickwork, as they occur in measuring pipes.
Attendance.
Provision for fittings.
Fixing of fittings.

BELLHANGER—

Bells.
Pulls.
Bell-boards.
Attendance.
Electric bells usually a provision.

PAINTING—

VENTILATION—

In all trades.

The following order of taking off, referred to as (*c*) in the introduction, is not quite so favourable to the division of the work between a number of assistants as (*b*), but it has advantages, inasmuch as it deals with various parts of the work completely, so that no return to them is required. Its advantage in the case of variations is considerable, as in the event of the omission of a window or a door a *continuous* series of dimensions can be entirely omitted. Moreover a particular group of dimensions is completely considered and finished at one time. These are its chief points of contrast to (*b*).

The suggested preliminary works cannot be more advantageously dealt with than as suggested in the previous section of this Chapter. The differences will be readily seen by comparing the two lists.

Area for district surveyor's fees, if in Metropolitan area.
Work to party-walls complete.
Digging to surface.
Ditto to basement.
Ditto filled in and rammed.
Digging to trenches of external walls, basement.
Concrete to ditto.
Footings of ditto.
Walls of ditto up to a certain level, as ground floor, or top of plinth or damp-proof course.
Digging, concrete, footings and brickwork of projections on these walls up to the same level as last.
All the cuttings on this section of the work.
Digging and concrete to trenches of internal walls of basement, footings of ditto.
Brickwork up to the same level as before.
Digging concrete footings and brickwork of projections on these walls up to same level as before.
All the cuttings on last.
Digging and concrete to surface trenches (i.e. walls where there is no basement), external walls.
Footings and brickwork up to the same level as before.
Digging, concrete, footings and brickwork up to same level, of all projections on these walls.
All the cuttings on these walls.
Digging and concrete to surface trenches, internal walls.

Footings and brickwork up to same level as before.
Digging concrete footings and brickwork up to same level, of all projections on these walls.
All the cuttings on these walls.
Strutting and planking to basement.
Strutting and planking to the foregoing trenches.
Damp-proof course.
External walls other than basement, from last level up as far as they continue of one thickness, projections on these walls.
Internal walls from first level up to same level as the external walls.
Projections on these walls.
All the cuttings on these walls.
External walls from last level to base of gables, or up to roof plate, or as far as they continue of one thickness.
Gables of external walls.
Projections on these walls.
Internal walls from last level to base of gables, or up to roof plate, or as far as they continue of one thickness.
Gables of internal walls.
Projections on these walls.
All the cuttings on these walls.
Projections of chimney breasts one at a time, commencing with concrete and finishing with the shafts.
Hoop-iron bond.
Brickwork extra only in cement, take as it occurs in the foregoing stages.

FIRES—

Deduct brickwork of opening.
Chimney-bars.
Segmental arches.
Flues, parget and core.
Flue-pipes.
Soot doors, and sweeping flues.
Chimney pots.
Centering to trimmers.
Trimmer arches.
Skewback cutting.
Feather-edged springers.
Filleting soffits of trimmers.
Fender walls.
Hearths and back-hearths.
Curbs.
Notchings of hearths.
Stoves and setting.
Coppers and setting.
Stone cover to copper setting.
Copper lids.
Chimneypieces and fixing.
Painting to chimneypieces.
Blacking stoves.

INTERNAL DOORS—

Floor by floor.
Deduct brickwork.
Lintel.
Relieving arch, or segmental arch, and centering.
Wood bricks or fixing blocks.
Frame bed and point.
Flooring in the opening or threshold, or step.
Ends of steps, cut and pin or joint.
Door.
Ironmongery.
Linings or frame.
Iron dowels.
Architraves and grounds.
Cement or plaster reveals.
Deduct plastering.
Painting.

FACINGS—

Facings.
Facings to chimney stacks taking stone dressings, if any, at the same time, also extra projecting brickwork, and extra only in cement.
Gable copings (stone or brick), finials, kneelers, bonders.
General facing. (*Collect the lengths*.)
Strings and brick cornices and extra brickwork for their projection, quoins, plinths, tablets, niches.
Excavation, brickwork and facing for extra thickness of walls to form plinths.
Ashlar facing and general stonework or cement work except to openings.
Terra-cotta except to openings of doors and windows.

EXTERNAL DOORS—

In the same order as internal doors.
Deduct facings.
Stone steps, landings, railings, etc., with the doorways.
Add jambs in stone or brick, stone mullions, transomes, lintels, centering and strutting.

WINDOWS—

Floor by floor.
Deduct brickwork.
Deduct facing.
Centering to external arches.
Arches or stone lintels, and deduct brickwork.
Cutting to facing.
Facing to reveals or stone dressings, and deduct brickwork.
Sills.
Fair ends.
Make good facings to window sills.
Lintel.
Relieving arch.
Wood bricks.
Frames, bed and point.
Window frames and sashes.
Iron tongue and grooves in oak and stone.
Moulded horns.
Extra labours on sash or frame.
Ironmongery.
Glass.
Window boards or window nosing.
Notched, returned, and mitred ends.
Window backs.
Canvassing and painting backs of do.
Rendering in cement behind do.
Long notchings of linings over do.
Beaded capping.
Plinth to window back.
Housings to do.
Linings.
Architraves and grounds.
Deduct plastering and paper.
Plaster or cement reveals.
Rough chamfer on brickwork.
Painting.

ARCHWAY AND RECESSES—

Floor by floor.
Deduct brickwork.
Centering.
Arch.
Plastering and mouldings to reveals and angles.
Deduct plastering.

PAVINGS—

Hard dry rubbish.
Concrete.
Cement-floated face.
Pavings.
Iron joists for vaulting or concrete floors, painting on iron.
Ends cut and pinned.
Templates.
Iron columns, commencing with the digging, concrete, brickwork and stone base, and painting.
Centering.
Vaulting or concrete.
Cutting on brickwork.
Raking out and pointing soffits.

STONE STAIRCASES—

Complete in all trades, including painting.

GROUND FLOORS—

Oversailing to receive plates.
Sleeper walls.
Concrete under floors.
Plates.
Joists.
Air bricks and painting.
Floor boarding.
Deductions of ditto.

UPPER FLOORS—

Floor by floor.
Digging, concrete.
Brick piers, stone bases and iron columns, supporting floors, and painting.
Templates, felt or lead pads.
Girders, wood or iron, and painting.
Flitches and bolts and painting.
Tension rods and painting.
Joists.
Trimmers.
Corbels or templates for trimmers.
Strutting.
Sound boarding and pugging.
Deductions of ditto.
Floor boarding.
Flooring in same quantities, and bearers.
Deductions of ditto.
Steps at change of level of floors.

FIREPROOF FLOORS—

Concrete and digging.
Brick piers.
Stone bases.
Iron or steel columns or stanchcons.
Wedging up bases and running with cement.
Templates and ends cut and pinned.
Iron or steel girders.
Bolts to caps and bases.
Iron or steel intermediate bearers.
Centering and horsing or hanging.
Concrete.

PARTITIONS—

Floor by floor.
Heads and sills.
Interties.
Posts.
Quarters.
Braces.
Bridging-pieces.
Ironwork and painting.

TRUSSED PARTITIONS—

Similar order to last.

BRICK-NOGGED PARTITIONS—

The brickwork.
Deduct openings.
Heads and sills.
Posts.
Quarters.
Nogging pieces.

ROOFS—

Templates.
Felt or lead pads.
Ends timber cut and pinned.

TRUSSES—

Tie beams.
Principal rafters.
King or Queen posts.
Struts.
Collars.
Circular ribs.

Wrought face and painting, staining or varnishing.
Ironwork and painting.
Hoisting and fixing roof trusses.
Templates.
Ends cut and pinned.
Purlins.
Wrought face and painting, staining or varnishing.
Cleats.
Scarfings and bolts and painting.
Plates.
Ridges.
Wrought face and painting, staining or varnishing.
Scarfings and bolts and painting.
Tile or slate ridges.
Ends fitted.
Ends splayed and fitted.
Rough chamfer on fir.
Ridge roll.
Lead.
Hips.
Rough chamfer.
Roll.
Lead.
Bossed ends to rolls.
Bossed intersections of ridge and hip.
Hip tiles or soakers.
Cutting to tiling or slating.
Cutting to boarding or battens.
Dragon pieces and bolts.

VALLEYS—
Timber.
Cutting to slating or tiling.
Cutting to boarding or battens.
Valley boarding.
Valley fillets.
Tilting fillets.
Valley tiles.
Lead.

RAFTERS, BOARDING AND SLATING—
Rafters.
Sprockets.
Cut ends to rafters.
Eaves fillet or eaves board.
Extra slating or tiling for eaves.
Slating or tiling.
Battens, boarding, felt.

GUTTERS—
Gutter boards and bearers.
Drips.
Rolls.
Gussets.
Lead.

Cesspools.
Lead.
Extra labour and solder.
Wire or other covers.
Socket pipes.
Holes for ditto.
Flashings to gutters.
Rake out, wedge and point.

GENERAL FLASHINGS TO MAIN ROOF—
Tilting fillets.
Flashings.
Stepped flashings, or soakers.
Raking out and pointing.
Soakers.
Fillets to secret gutters.
Lead to ditto.
Extra labour to secret gutters.
Flashings to ditto.
Raking out, wedging and pointing.

CHEEKS—
Boarding and quarters.
Lead.
Copper nailing.
Soldered dots.
Flashings.
Vertical tiling.
Tilting fillets.
Soakers.

CHIMNEYS—
Deduct tiling, slating, boarding, battens, felt.
Cuttings to ditto.
Gutters.
Rolls.
Gussets.
Lead.
Bossed ends to rolls.
Tilting fillets.
Flashings.
Rake out, wedge and point.

LEAD APRONS—
Lead.
Rake out, wedge and point.
Tilting fillets.
Stepped flashings.
Soakers.
Rake out, wedge and point.

DORMERS WITH BRICK FRONTS—
Deduct slating, tiling, rafters, boarding, battens, felt.
Trimmers.

Plates.
Valleys or footing pieces.
Cutting to boarding, battens, and tiling or slating.
Valley boards, valley fillets, tilting fillets.
Lead or valley tiles.
Cutting to boarding, battens or tiling.
Wooden ridge.
Roll.
Rough chamfers on fir.
Lead to roll.
Or tile ridge.
Ends splayed and fitted, and lead soaker.
Rafters.
Cut ends or sprockets.
Boarding or slating and tiling or battens.
Extra tiling or slating for eaves.
Eaves board.
Eaves fillet.
Boarding and backing to cheeks.
Cutting to ditto.
Tiling or slating.
Cutting to ditto.
Lead to cheeks.
Copper nailing.
Soldered dots.
Flashings or soakers.
Stepped ditto.
Rake out, wedge, and point.
Tilting fillets.
Ceiling joists.
Sill.
Frame.
Casements.
Ironmongery.
Glass.
Lead apron.
Linings.
Architrave and grounds.
Window board.
Plastering of ceiling and cheeks.
Deduction of plastering.
Painting.

CEILING JOISTS—
Plates.
Stretchers.
Hangers.
Joists.

MISCELLANEOUS ADJUNCTS OF ROOF—
Ways in roof.
Traps in ceilings, and ironmongery and painting.
Ladders and painting.
Snow-boards and painting.
Cat ladders and painting.

EXTERNAL TRAPS—
Deduct: rafters, slating or tiling, battens, boarding, felt.
Trimmers.
Slating or tiling for cutting.
Boarding or battens for cutting.
Trap and rims.
Lead covering.
Copper nailing.
Bossed angles.
Ironmongery.
Curb and labours on it.
Lead covering to curb.
Copper nailing.
Flashing boards or tilting fillets.
Gutter board and bearers.
Lead.
Internal lining.
Deduct plastering.
Plaster quirk.
Painting.

LANTERN LIGHTS, FLÈCHES, DORMERS—
Take complete in all trades, including painting.

EAVES GUTTERS AND RAIN-WATER PIPES—
Fascias.
Painting of ditto.
Eaves gutters.
Stopped ends.
Angles.
Outlets and wire covers.
Painting ditto.
Rain-water pipes.
Swan necks.
Heads and wire covers.
Shoes.
Tee pieces.
Bends.
Plinth bends.
Loose bands.
Holderbats.
Painting.
Cuttings to brick or stone strings.
Connections with drain.

WATER SUPPLY—
Well.
Well cover and painting.
Bearers of iron or wood for cistern and painting.

Templates.
Ends cut and pin.
Cisterns of iron.
Cisterns of wood.
Lead lining.
Soldered angle.
Copper nailing.
Standing waste or apparatus for emptying.
Brass washer and waste with fly nut and union.
Lead under waste.
Boiler screw and overflow.
Hole in brickwork.
Ditto in tile or slate and lead tile or slate.
Cistern covers and painting.
Holes in brickwork for pipe.
Chase for pipe, pipe casing and painting, plaster quirk.
Joiner attend plumber to cistern.

SINKS—
Stone, stoneware, or porcelain.
Bearers, wood or brick.
Cut and pin sink to brickwork.
Holes for gratings.
Brass grate.
Trap.
Lead waste.
Holes in brickwork for pipes.
Joiner attend plumber to sink.
Tiling or cement around.

WOODEN SINKS—
Bearers.
Ends cut and pin.
Woodwork of sink.
Lead lining.
Soldered angle.
Copper nailing.
Wooden capping.
Brass grate and plug and rim.
Lead cones.
Lead trap with cap and screw.
Lead waste.
Chase.
Holes in brickwork.
Pipe casing.
Plaster quirk.
Deal top.
Perforation.
Joiner attend plumber.
Enclosure of space beneath.
Draining boards.
Lead covering.
Copper nailing.

Painting.
Tiling or cement around.
Flashing board and covering with lead.

LAVATORIES—
Apparatus and valves complete.
Rough deal top and bearers.
Holes for basin and taps.
Slate or marble top and perforations.
Rounded or moulded edge.
Skirting and angles.
Basin and washer, plug and chain.
Waste.
Lead trap with cap and screw.
Overflow pipe and joints.
Chase.
Pipe casing and painting.
Hole in brickwork for pipe.
Enclosure complete and painting.
Shelf and bearers.
Joiner attend plumber.

BATHS—
Deal cradle.
Bath and valves.
Lead trap.
Lead waste.
Connection with another pipe, or, as the case may be.
Overflow and joints.
Lead safe.
Joiner attendance to safe.
Lengths, bossed angles.
Pipe for waste from safe and joints.
Copper flap and joint.
Wooden top and enclosure.
Extra for door.
Ironmongery.
Rounded or moulded edge.
Deduct perforation for bath.
Perforation in bath top.
Rounded or moulded edge.
Skirting and mitres to top.
Joiner, attendance to bath.
Painting or French polishing.

WATER-CLOSETS—
Apparatus.
*Anti-*D trap.
Soil pipe and joints.
Bends.
Extra joints.
Connection with drain, and brass thimble or tail piece.
Holes in brickwork.
Ventilating pipe and joints.
Holes in slate or tiling.

Lead, slate or tile, and solder joint.
Cap or hood and straps to vertical pipe.
Lead safes.
Bossed angles.
Waste from safe and copper flap.
Holes in brickwork.
Joiner attend plumber to safe.
Seat and riser.
Flap and frame.
Butts.
Hole for pan.
Ditto for handle.
Or seat for pedestal closet, p.c.
Extra for seat and riser made to remove easily.
Grounds.
Skirting and mitres, or back and elbows.
Waste-preventing cistern and its appurtenances.
Overflow pipe.
Hole in wall.
Flushing pipe.
Joiner attend plumber to w.c. and waste preventing cistern.
Paper boxes.
Candle brackets.
Pot cupboards.
Painting or French polishing.

SUPPLY PIPES—

Cistern connector with union and fly nut and joint, and hole in cistern.
Lead pipe to each draw-off, w.c., sink, bath, lavatory, with valve at end of each branch.
Branch joints as they occur.
Holes in brickwork.
Chases.
Pipe casing.
Plaster quirk.
Connection with main.
Rising main.
Ball valve.
Chase.
Pipe casing.
Plaster quirks.
Stop-cocks.
Chambers for ditto.
Water meter and chamber and cover.
Painting.

PUMPS—

Pump.
Lead suction and joints.
Foot valve and rose.
Rising main.

Warning pipe.
Holes in brickwork.
Joiner attend plumber to pump.

HOT WATER SUPPLY—

Hot water cistern or cylinder, and bearers and expansion pipe.
Flow and return.
Safety valve.
Branch pipes and valves.
Holes in brickwork.
Chases.
Pipe casing.
Plaster quirk.
Painting.
Attendance on hot water fitter.

DRAINS, CESSPOOLS—

Rain-water tanks.
Cesspools, grease traps, dumb wells.
Gullies.
Inspection pits, complete.
Drain pipes of various sizes and depths.
Connection with sewer and pay fees.

FLOORS—

Flooring for cuttings.
 „   in small quantities and bearers.
Steps complete from one level to another.

SKIRTINGS AND DADOES—

Floor by floor.
Skirtings.
Mitres.
Ditto irregular.
Fitted ends.
Housings.
Wooden dadoes.
Cement skirtings.
Mitres.
Ditto irregular.
Stopped ends.
Cement dadoes.
Painting.

FITTINGS—

Cupboards.
Shelves.
Dressers.
Painting.

STAIRCASES—

Wooden staircases in all trades.

PLASTERING AND PAPER—

Floor by floor.
First all the rooms, then the passages.
Ceilings and whitening.

Cornices.
Bracketing.
Mitres.
Ditto, irregular.
Enrichments.
Returned and mitred ends.
Centre flowers.
Walls and paper.
Partitions and paper.
Cement angles.

GASFITTER—
Pipes.
Syphon.
Meter.
Shelf or enclosure for meter.
Short length of pipe and connections.
Main cock.
Pay fees and make good footway.
Holes in brickwork.

Attendance.
Fittings and fixing.

BELLHANGER—
Bells.
Pulls.
Bell boards.
Attendance.
Painting.

ELECTRIC BELLS—

PAINTING—
Such painting as may not have been dealt with in preceding sections.

VENTILATION—
Work in all trades.

Although the order of taking off, shown in the foregoing pages, is desirable, its adoption is not always practicable, pressure by building owner or architect will often necessitate unusual speed. In such a case the time necessarily spent in examining the drawings may be saved by beginning with sundry finishings as doors, windows, etc. instead of excavation and brickwork. The measurement of these will give the surveyor a general idea of the building as he proceeds and the excavation and brickwork may be taken off later.

Those things for which the same dimensions may be used should be taken off together, as floor covering and ceilings.

With the plastering of the walls and partitions, skirtings, cornices, dado rails, picture mouldings, etc.

If the sizes of rooms are marked in pencil on the drawings, they will prove useful later in the measurement of the brickwork.

A less experienced assistant may be employed in taking off all the windows and doors. If necessary, the principal may take off the dimensions of one window and one door of each type as a pattern to be followed. By confining the junior to the completion of the whole of a simple section, such as doors or windows, he will work with more speed than if employed on parts of different sections.

The principal or an experienced assistant may measure the difficult parts as roofs, excavation, brickwork, facings, stonework, water supply, drains, gas, bells and heating.

ORDER OF TAKING OFF.

The man who takes off a complicated roof with rooms in it, should also take off the plastering of such rooms.

The excavation and brickwork being left to take off last. Several trades can be billed while these dimensions are abstracted.

If several assistants are employed on the abstracts of the same building, one of them may abstract water supply, drains, staircases, gas, bells, as they do not interfere with the general abstract.

Measuring in the irregular manner above described exposes the surveyor of large practice to much danger of mislaying dimensions as they cannot be paged consecutively until the whole of the dimensions are taken off.

Each sheet of each section can, however, be paged at the top as windows 1, 2, 3, 4, water supply 1, 2, 3, 4, 5, etc.

The assistant in charge of the abstract will keep a list in the form below.

Mr. Jones.—Windows 1, 2, 3, 4, 5, 6, 7, etc.
„ Floors 1, 2, 3, 4.
Mr. Smith.—Water supply 1, 2, 3, 4, 5.
„ Drains 1, 2.
„ Stairs 1, 2, 3.

The sheets as received by the assistant in charge are entered in the list, ticked when abstracted and cross-ticked when checked. By this method sheets cannot be left out altogether and the state of the abstract can be seen at a glance.

CHAPTER III.

MODES OF MEASUREMENT.

AFTER each clause of direction as to mode of measurement will be found a specimen item, as it would appear in the bill.

ATTENDANCES.—Attendances are generally met by providing a sum, and the work involved is either charged day account or measured at completion; but inquiry will often give surveyor sufficient information to enable him to measure the work beforehand, and include it in quantities.

DISTRICT SURVEYOR'S FEES.—State the number of squares in area and the number of stories in height, counting the basement as one. (See Chapter IV., section BILLING, Preliminary Bill.)

Observe where party-walls are raised or otherwise altered, that half the full fee will be chargeable for adjoining house (see the Schedule of Fees, London Building Act). When a building is of intricate plan, or the application of the Schedules of the Act is doubtful, the district surveyor may be consulted with advantage, or a sum provided.

Where any particular part of the work is of uncertain quantity, and it is nevertheless necessary that it be measured, it is usual and convenient to include such a quantity as the surveyor deems sufficient, which is kept separate in the bill and followed by the words "as provision," the real quantity used being adjusted at the settlement of accounts.

Bill.

sqrs.	ft.	in.			£	s.	d.
2	50	–	supl.	1" rough boarding as ways in roof " *as provision* "			

In cases where the drawings are to a very small scale, and no details are supplied, the surveyor should make a detail of any part which cannot be exactly measured from the small scale

WORKS AT THE BUILDING. 45

drawing, especially of mouldings, the girth of which cannot otherwise be accurately obtained. Use good-sized pieces of paper, small pieces are likely to be lost; and preserve them with the dimensions.

WORKS AT THE BUILDING.—When the contemplated building is partly new and involves alterations to old work, notes of the particulars of the alterations must be taken at the building in the usual dimension book; it is most convenient to take out all that can be taken from the drawings before going to the building, making notes as work proceeds of what to observe when visiting the building. When parts of an item are in book and part on dimensions, write a reference in each to the other part.

Bill.

yds.	ft.	in.	supl.		£	s.	d.
				"All the cutting of openings to include removing and carting away rubbish and any needling or shoring"			
				"The making good after cutting of openings to be in hard stocks and cement, and the arches to include centering"			
			A	"Allow for cutting opening 3' 0" × 7' 0" in 1½-brick wall, for inserting lintel and pads, elsewhere taken, for making good brickwork around, and for making good plastering on both sides after fixing joinery" ..			
				"Allow for cutting two openings 4' 0" × 9' 9" in 1½-brick wall for inserting semicircular arches in two rings, and making good brickwork"			
				"Allow for cutting opening 2' 9" × 2' 0" in 2-brick wall, for inserting stone lintel (elsewhere taken), and making good brickwork"			
				"Allow for cutting opening 15' 3" × 11' 0" in 1½-brick eastern wall of study, for shoring up and needling wall over, for supporting floor, inserting two rolled joists and cover stone (elsewhere taken), and for making good all works disturbed"			
				Often earlier items may be referred to by a letter, thus:—			
				Allow for cutting opening 3' 6" × 8' 0" in 2-brick wall and all as A.			

PROVISIONS.—Observe to take templates, painting, or any collateral work, as unloading, hoisting, or fixing.

When you decide to provide a sum for a particular portion of the work, as tiling, patent cement, paving, etc., the quantity should be measured and sent to the manufacturer for his estimate. The sum thus obtained can then be adopted, plus a certain sum for attendance. The protection of the work of sub-contractors and responsibility for its damage should be imposed upon the general contractor. (See also Chapter IV., section BILLING, Provisions.)

When a provision is to be P.C. deduct the trade discount and quote the difference, or take the price without alteration from the list and call it list price.

It is advisable to treat all provisional sums uniformly throughout the bill, either all as P.C's. or all as list prices.

A P.C. need not necessarily have a manufacturer's name attached, a list price should always have it.

EXCAVATOR.

DREDGING AND REMOVING.—The quantity being usually uncertain, the contractor tenders at a price per ton; state that it includes barging and finding a deposit, or barging and depositing within a certain number of yards, and state the method of depositing.

Bill.

Tons.				£	s.	d.
800			Dredging and removing mud from bed of river, barging away, and finding a deposit			

PILING.—Measure the piles per foot cube; describe the kind of timber and as including fixing. Keep separate those below and above 144 square inches in section, and describe them in bill accordingly.

Bill.

ft.	in.			£	s.	d.
1250	—	cube	Fir in piles not exceeding 144 square inches in section and 30 feet in length, and to include heading, pointing and fixing, and depositing ready for driving			
1300	—	,,	Ditto, exceeding 30 feet			
500	—	,,	Ditto, exceeding 144 square inches in section, not exceeding 30 feet in length			
640	—	,,	Ditto, exceeding 30 feet			

DRIVING PILES.—Each, state scantling, the length measured from point of shoe to finished top of head before driving, and the depth of driving measured from surface of ground to point of shoe.

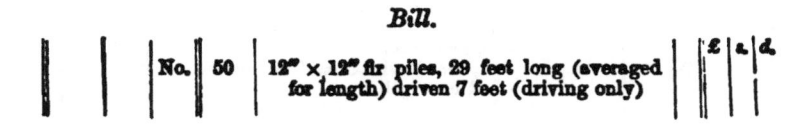

Bill.

| | | No. | 50 | 12″ × 12″ fir piles, 29 feet long (averaged for length) driven 7 feet (driving only) | £ | s. | d. |

Number the cutting off of the heads, stating size of pile; if below water state it, and describe the depth.

Bill.

| | | No. | 50 | Cutting off heads of piles 12″ × 12″ .. .. | £ | s. | d. |
| | | ″ | 20 | Ditto, 12″ × 12″, 2 feet below water surface at low water .. .. | | | |

WALING PIECES at per foot cube, and describe as including fixing. Number the shoes to piles, stating size of pile and weight of each shoe. State that they include fixing, and all nails, spikes, staples, etc., or a clause about nails, spikes, etc., may appear in preamble of bill.

Bill.

	ft.	in.				£	s.	d.
	540	—	cube	Fir in waling pieces and fixing and bolting to piling (bolts elsewhere taken)				
	No.	100		Wrought-iron shoes, weight 30 lbs. each, in two varieties, including pointing pile and fitting and fixing with 1″ diameter screws..				

BOLTS.—Number and describe, as directed in Smith.

STRAPS.—Measure per foot run, afterwards reduce to weight, as described in Smith.

COFFER DAMS.—Measure the piles and timber per foot cube as "use and waste in coffer dams, including all material," "carriage, fixing and removal." State if piles are 3 inches or 6 inches thick, as the case may be, and whether edges are shot. Also whether bevelled or rebated, or measure the bevelling separately.

Measure the ironwork as above, but describe as for "use and waste."

Bill.

ft.	in.			£	s.	d.
800	-	cube	Use and waste of *material* in coffer dam, including all material, labour, carriage, fixing and *removal* .. Fir in piles 6" thick, and not exceeding 20 feet in length .. 3200 feet run bevelled edge to 6" piles ..			

Measure the clay-puddle at per yard cube if over 12 inches thick, and per yard superficial if not exceeding 12 inches, and describe; state if mixed with chopped straw, and allow for ramming and removal.

Bill.

yds.	ft.	in.			£	s.	d.
1000	-	-	supl.	Clay puddle of well-tempered clay, 6" thick and filling in and ramming in layers to coffer dam .. or,			
166	-	-	cube	Clay puddle of well-tempered clay, and filling in and ramming in layers to coffer dam ..			

DIGGING AND CONCRETE.—Digging is measured as before excavation, and although earth always increases in bulk by removal, no allowance is made in quantity.

The nature of the soil should be stated if involving unusual trouble, as very stiff clay or rock, or if including grubbing up old foundations. And if the soil is likely to be liquid a special clause should be inserted in the quantities to ensure the liability of the contractor. (See Kirk and Randall *v.* East and West India Dock Company.)

If the building is on the site of an old one with a basement, the excavation should be measured from street level, as the rubbish from pulling down will have filled the cellars, or nearly so. The way the excavation has been measured should be stated, and a clause inserted, "Contractor to make his own allowance for any voids that may exist."

yds.	ft.	in.			£	s.	d.
				Bill.			
400	–	–	cube	Digging to basement from surface to a depth of 11 feet, wheeling or throwing out and carting away. (Note the basement is filled with rubbish from the pulling down. Contractor to make his own allowance for any voids that may exist)			

State if basketed (as is necessary in confined situations); if wheeled; if wheeled more than one run (20 yards); and how many runs, also if carted away, or deposited on the site.

Keep excavation beyond 6 feet in depth, beyond 12 feet, beyond 18 feet, etc., separate. State the depth and keep separate the various multiples of 6 feet, as 12 feet, 18 feet, etc., and to depths over 6 feet, say "including staging."

In measuring trenches where there is concrete, measure the trench the width of the latter, except in cases where the projection of the concrete is less than 6 inches on each side beyond the lowest course of footings; in that case, as in that of trenches without concrete, 6 inches on each side must be allowed.

If no strutting and planking is taken, further allowance must be made for slopes where the excavation is deep; but this is very rarely done, it is better to take strutting and planking. If done it should be stated in bill.

It is usual to commence measurement of excavation by removing the surface all over the site, commonly about 12 inches deep.

Observe whether the ground is truly level; if not, and it is desired to bring it to a level, take off the soil to an *average* depth.

SURFACE DIGGING.—Where earth is removed over a surface it should be measured superficially, if not more than 12 inches deep, stating the depth; where more than 12 inches deep, measure by the cubic yard. State if it includes separation of vegetable soil.

The digging to basements should be taken as digging to basement . . . feet deep, throwing out, or basketing, or wheeling, and carting away, or spreading. If the basement is paved, take this excavation down to the bottom of the paving; or if there is concrete beneath it, to the bottom of the concrete.

The space between outer face of wall and outer edge of trench by the depth is to be next taken and deducted from the last, and added with the description, "excavation and returning, filling in and ramming."

Bill.

yds.	ft.	in.			£	s.	d.
1240	–	–	supl.	Digging to general surface 9" deep (averaged), wheeling 2 runs and depositing, including separation of vegetable soil			

DIGGING TO TRENCHES, *at per yard cube.*—State if to basement trenches, i.e. trenches below level of general basement excavation, or to surface trenches, i.e. trenches below level of general surface.

Describe as "excavation and carting away," or "wheeling and spreading," or "digging to trenches, part filled in and rammed, and the remainder carted away." The former being the proper description for the part to receive concrete, the latter for the part of trench to receive brickwork.

Sometimes, when the depth of filling in is considerable, earth from the foundations is used for filling with a layer of brick or stone rubbish over it. The item would appear in the bill as follows:—

Bill.

yds.	ft.	in.			£	s.	d.
200	–	–	cube	Fill in with approved material from excavation in layers to make up levels, well ram and roll			

Under this arrangement the carting away is best excluded from the description in the bill of the items of excavator, and separately stated as follows:—

Bill.

yds.	ft.	in.			£	s.	d.
350	–	–	cube	Fill earth into carts and cart away			

EXCAVATOR. 51

The foregoing would be the amount of difference between the quantity dug and quantity filled in and rammed.

In the measurement of trenches, collect the lengths of external *walls* and internal walls respectively, and be careful to put opposite each dimension a description of its position. These dimensions will be used for lengths of excavation, concrete, footings and walls. This is the more usual practice, the difference between the lengths of excavation, concrete, and brickwork being, as a rule, but small. If it be desired to take the quantity with more exactness, a foundation plan must be made showing widths of trenches. For ordinary buildings, like dwelling-houses, the former course is recommended, but for a large building with thick walls a foundation plan should be used.

EXAMPLE OF COLLECTION OF WALLS.

External Walls.

	1½ B.	2 B.	2½ B.
South of dining-room ..	18 0	12 0	
West of ditto	14 6		
Bay (collected)	23 6		
Remainder of west wall ..	4 3		
North of west wall		15 9	4 6

If it should be necessary to separate the part carted away from the part returned, filled in and rammed, it is quite easily arrived at. In the sketch (Fig. 4) the wall occupies three-eighths of the trench; about three-eighths of the whole width measured above the concrete multiplied by the depth from the surface to half way down the footings is a close approximation to the quantity that would be carted away, and the remainder would be filled in. It is obvious that the part of the trench filled with concrete admits of no return of the earth.

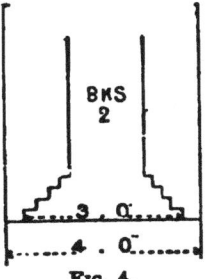

FIG. 4.

The dimensions of 100 feet run of digging and concrete, measured in the foregoing manner, assuming the depth from surface to top of concrete to be 4 feet and the concrete 12 inches deep, would be as follows.

E 2

QUANTITY SURVEYING.

100 0 / 4 0 / 5 0	2000 0	Dig surface trenches and cart away.
100 0 / 2 6 / 3 6	250 0	Ddt. and Add Dig surface trenches, fill and ram.
100 0 / 4 0 / 1 0	400 0	Concrete as described in trenches.

A more exact method is as follows:—

100 0 / 4 0 / 1 0	400 0	Dig surface trenches and cart away, and Concrete as described.
100 0 / 4 0 / 4 0	1600 0	Dig S. T. F. and R.
100 0 / 2 5 / 1 0	241 8	Ddt. Dig S. T. F. and R. and Add Dig S. T. and cart.
100 0 / 1 6 / 3 0	450 0	

Side calculation:
1 10¼
3 0
———
4 10¼
———
2 5¼

When foundations are stepped, it will generally be most convenient to collect the whole round of the walls for the minimum depth, and afterwards take the extra depths one at a time.

At the points where the steps occur, observe that concrete must be taken between the bottom of that of wall at upper level and the top of that at lower level, as sketch (Fig. 5).

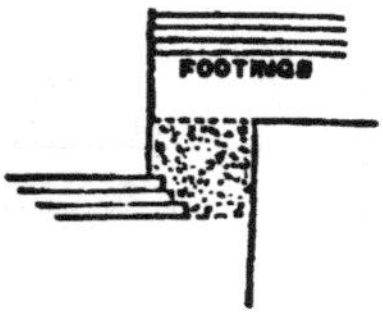

Excavation in ground where buildings have previously existed should be described as "including any necessary grubbing up of old foundations"; or in

FIG. 5.

the preamble of bill, "The digging to trenches to include any necessary grubbing up of old foundations."

If excavation is in small quantities to underpinning, state it and keep it separate.

Observe that where trenches are very close together, the earth between them cannot be left in; the surveyor must allow for removing it and filling in the space.

When it is required to dig a trench close to a heavy building adjoining, even though it may be the wall of an intended basement, it will be necessary to measure the trench as dug down from the ground level before the remainder of the basement excavation, and strutted and planked the whole depth.

Bill.

yds.	ft.	in.			£	s.	d.
100	–	–	cube	Dig surface trenches not exceeding 6 feet in depth and cart away			
100	–	–	"	Ditto, from 6 to 12 feet deep			
100	–	–	"	Dig basement trenches and cart away			
100	–	–	"	Ditto, in small quantities in underpinning			

DIGGING FOR DRAINS is better included with the drains, but whether separated (rarely done) or not, state the *average* depth of excavation and the size of pipe.

Observe that these depths must be taken at *regular* intervals, or it will be no average. Great errors have arisen from taking the depths without observing this rule.

Measure trench for water or gas pipes at per foot run, stating average depth, and describe as trench for pipe.

Bill.

yds.	ft.	in.			£	s.	d.
100	–	–	run	Dig trench for pipe, return, fill and ram			

For the computation of excavation in large quantities, as for railways, the prismoidal formula must be used, and the published earthwork tables will greatly facilitate the surveyor's work.

STRUTTING AND PLANKING TO SIDES OF EXCAVATION, *per foot run*, stating depth of excavation.—This should always be measured to basements, and if doubtful of its necessity, add the words "if necessary" to the description, but in any case measure it all; or it may be measured per foot super.

STRUTTING AND PLANKING TO TRENCHES, *at per foot run*, described as "one side only measured," stating width and depth of trench, or an item may appear in preamble to the bill of excavation, "all excavation to trenches to include strutting and planking, if necessary," in which case none need be measured, but it should always be measured when the trenches exceed 4 feet in depth.

When concrete has no earth to support it "rough boarding to edge of concrete" should be measured superficially.

Bill.

ft.	in.			£ s. d.
500	–	supl.	Strutting and planking to basement 12 feet deep	
500	–	run	Strutting and planking to trench 4 feet wide and 5 feet deep (one side only measured)	
500	–	„	Ditto, 4' 9" wide and 6 feet deep	
			Some surveyors say—	
500	–	supl.	Strutting and planking (if required) to basement 12 feet deep, &c.	

Strutting and planking to holes 3 or 4 feet square is best numbered, stating length, width and depth of holes.

Bill.

				£ s. d.
No.	3		Strutting and planking to hole 4 feet × 4 feet and 5 feet deep	

CLAY PUDDLING, *at per cubic yard*, if more than 12 inches thick; if not exceeding 12 inches thick, per yard superficial; if filled in over arches or hoisted, state the height of hoisting.

Bill.

yds.	ft.	in.			£ s. d.
100	–	–	cube	Puddling of well-tempered clay rammed in layers	
100	–	–	supl.	Puddling of well-tempered clay average 9" thick, and hoisting and filling in over arches 12 feet above street level	

EXCAVATOR. 55

CONCRETE, *at per yard cube*, where exceeding 12 inches thick; where not exceeding 12 inches, per yard superficial; if thrown from stages, state that it includes stages. State if in trenches, if levelled or to falls, and if measured per yard cube, measure the levelling separately at per yard superficial, except in trenches which includes levelling, if rammed state it in description. State if filled in over arches, in which case average the depth. If hoisted, state height of hoisting. Over trimmer arches, concrete is included (see p. 81). State whether of lime or cement.

The description of its composition is preferably written in the preamble of the bill of excavator.

When a separate contractor does the concrete work, a superficial dimension of "levelling and making up" to receive tile or other paving will be required to the surface, and to the soffit "make good and dub out as may be necessary to receive plastering."

Bill.

yds.	ft.	in.			£	s.	d
100	–	–	cube	Concrete as described in trenches			
100	–	–	supl.	Ditto 6" thick levelled to receive paving..			
100	–		„	Ditto 6" „ „ falls			

Concrete to Fireproof Floors.—Measure at per yard superficial, stating its thickness, composition and height from ground of each floor, or the limit of height may be stated.

Bill.

yds.	ft.	in.			£	s.	d
100	–	–	supl.	Cement concrete 6" thick, and filling in between joists 12 feet from street level			
				or,			
300	–	–	„	Breeze concrete 9" thick as described, and filling in between joists from 12 to 40 feet above street level			

The casing of iron girders and similar members is best measured per foot run, stating the finished sizes, how finished, and what else is included.

Bill.

ft.	in.		Casing to iron stringers and joists to carry floors and landings, to average 11" × 8" in section, including the necessary galvanised iron wire foundations and all wooden casings and moulds, and finishing with fine Portland cement faces and chamfered edges, as sketch, including all necessary stops and mitres	£	s.	d.
90	–	run				

Fig. 6.

EXPANSION BOARDS, *at per foot run.*—When the sizes of concrete arches render it necessary (over 9 feet span), an item as follows should appear: "allow for supplying, fixing, and removing expansion boards to edges of concrete arching." Also feet run, "allow for grouting in after removal of expansion boards and pointing with mortar (or cement)."

HARD DRY RUBBISH.—*Measure as described for concrete.*

Bill.

yds.	ft.	in.			£	s.	d.
100	–	–	cube	Hard dry brick or stone rubbish, and filling in and ramming..			
100	–	–	supl.	Ditto, 9" thick and ditto			
100	–	–	„	Ditto, 9" thick to falls..			

LEVELLING AND CONSOLIDATING GROUND, *at per yard superficial.*

Bill.

yds.	ft.	in.			£	s.	d.
100	–	–	supl.	Level and ram surface			

POST HOLES.—*State size of post and depth of hole.*

Bill.

		No.	10	Dig hole 5 feet deep for 12" × 12" post, and fill in and ram	£	s.	d.

DIGGING TO CESSPOOLS is usually taken with the cesspool and included in the description; if not, measure as for basements, stating depth of excavation (see also p. 89).

DRAINS, *at per foot run.*—State the kind, if tested, if "Town made," how jointed, if the joints are puddled around with clay, if opercular or half-socketed, or half-socketed at intervals, stating the distance (or measure as plain pipes and count the half-socketed pipes as extra on the plain), if Stanford's patent; also state the average depth of digging and include with the description.

Any length of pipe under 2 feet must be called 2 feet.

Number and describe as "extra for" bends, junctions, double junctions, diminishing pieces, siphon traps, the length of these bends, junctions, &c., having been previously measured in the length of the drains.

And observe that junctions and diminishing pieces are designated by the size of the main pipe of which they form a part. Thus, the junction shown in sketch (Fig. 7) is a 9-inch junction.

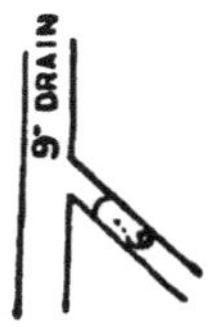

Number gullies, grease traps, &c., stating size and number in trade list of manufacturer. Include in description digging, bedding in concrete and connecting with drain.

Number the connection with sewer, and fees.

Fig. 7.

If the architect has not shown the drains on his plan, the surveyor should, on a tracing of the ground plan, draw the drains, marking upon it all the rain-water pipes and waste pipes, as he measures them, and should supply a copy to the architect.

Drains embedded in concrete or laid on concrete may be measured, pipe, concrete and digging together; but in that case the length should be measured net, and the fact stated in the bill of quantity. The detail of the embedding or envelope of concrete may appear in the preamble of the drains bill.

Bill.

ft.	in.			£	s.	d.
100	–	run	4" drain and concrete bed, and digging 3 feet deep			
100	–	„	4" drain and enveloping with concrete, and digging 3 feet deep			

The best way to deal with a system of drains is to begin with the gullies, cesspools, inspection or turning chambers.

After taking all the gullies and inspection pits, etc., it will be found most convenient to take off the longest length of drain on the plan, the digging being calculated as an average depth, i.e. a mean between the depths at each end of the length. The part in a public way should be kept separate, and if the depth of sewer is unknown, described as including deep-digging in roadway; but the depth may generally be discovered by enquiry at the office of the Local Authority. Next take all the bends on this line of drain. Then, beginning at the shallower end of the same line of drain, take each branch, including junction and bends.

Proceed with each main line of pipes in the order of its length, then its branches as before, with the extras on them.

Inspection pits should be taken out in detail, the work kept separate, and described as in small quantities. The making up of bottom is best numbered, stating size and average thickness. The iron covers should be described by a number in a trade list. (See also Chapter XIII.)

The following in small quantities in No. 5 inspection pits :—

Bill.

yds.	ft.	in.			£	s.	d.
17	–	–	cube	Dig and cart away			
		No.	5	Strutting and planking to manhole, average 4' 9" × 4' 9" × 4' deep			
13	–	–	supl.	Portland cement concrete (1 to 6) 6" thick			
		No.	5	Make up bottom of inspection pit 2' 3" × 2' 3" with cement concrete, average 9" thick, finished with Portland cement trowelled, including making good to edges of channels			
	3	–	run	4" half-round white glazed main channel and bedding in concrete (elsewhere taken) and pointing in cement			
	7	–	„	6" ditto and ditto			
		No.	2	6" white glazed short main channel bend and ditto			
		„	1	6" ditto long ditto and ditto			
		„	1	4" ditto short branch, three-quarter round channel bends and ditto			
		„	1	4" white glazed long branch three-quarter round channel bends, as last			
		„	1	6" ditto short ditto and ditto			
	160	–	supl.	Reduced brickwork in cement			
13	–	–	„	Joints of brickwork struck fair			
	45	–	run	Cement trowelled square skirting 9" wide No. 20 mitres to ditto			
		No.	9	Broad's (South Wharf, Paddington) channel shoe and gully-trap, with galvanised iron grids with one inlet and bedding in, and including cement and connecting to drain			

EXCAVATOR.

yds.	ft.	in.	No.		£	s.	d.
			1	Ditto with two inlets and ditto			
			5	Tylor and Sons $\frac{182}{41}$ galvanised cast iron air-tight manhole covers and frames, with opening size 24″ × 24″, and setting in cement			

In a large system of drainage, it will be found convenient to number the inspection pits consecutively, or to designate them by letters A, B, C. etc.

Bill.

ft.	in.			£	s.	d.
100	–	run	4″ drain, and digging 2 feet deep			
88	–	″	4″ ″ ″ 3 ″			
120	–	″	6″ ″ ″ 2 ″			
66	–	″	6″ ″ ″ 4 ″			
60	–	″	6″ ″ ″ ″ in roadway, and taking up and making good road and footways to the satisfaction of the Vestry Surveyor, including any hoarding, watching and lighting			
	No.	3	Extra for 4″ bends			
	″	6	″ ″ 6″ ″			
	″	6	″ ″ 4″ junctions..			
	″	3	″ ″ 6″ ″			
	″	2	″ ″ 4″ double junctions			
	″	2	″ ″ 4″ diminishing pipes			
	″	10	Doulton's (Lambeth) figure 15 yard gully with dished cover and galvanised iron grating, and digging, bedding on and including cement concrete and connecting with drain			
			Allow for connecting drain with sewer, and supply flap-trap or other requirement, or pay Vestry for so doing and pay fees ..			

CHANNEL PIPES.—Measure by the foot run, calling anything less than 12 inches another foot. State the kinds as "white enamelled," "brown glazed," etc.; state diameter and how jointed, and if any particular manufacturer is desired, give his name and address. Number the bends, junctions, etc., but not "extra for."

The bends, although designated in the trade list by letters which correspond with certain varieties of curves, may be referred to two categories, short and long, both of which vary a little in price, but when the surveyor takes off drains from an eighth scale plan, or makes a plan, as he frequently does, nicer distinctions are impossible. When a large-scale plan of the inspection pits is furnished, the curves may be compared with a trade list and identified; but such a detail rarely appears at the quantity-taking stage, and would probably be varied in the doing of the work.

Bill.

ft.	in.			£	s.	d.
22	–	run	Broad's (South Wharf, Paddington) 4" white enamelled channel pipe, and bedding and jointing in cement			
28	–	,,	Ditto 6"			
	No.	3	4" white enamelled channel junctions, and bedding and jointing with cement ..			
	,,	4	6" ditto			
	,,	2	4" ditto, double junctions			
	,,	2	6" ditto			
	,,	10	4" short channel bends			
	,,	8	6" ditto			
	,,	10	4" long channel bends			
	,,	6	6" ditto			
	,,	1	Winser's (Buckingham Palace Road) 6" white glazed drain chute, and bedding and jointing with cement, and cutting and fitting brickwork around..			

IRON DRAINS AND DIGGING per foot run. Measure net.

State the bore, the jointing, the depth of digging, and if laid on concrete describe its thickness.

Number the bends, junctions, taper pieces, access pipes, junction boxes, channel pipes and bends.

Bill.

ft.	in.			£	s.	d.
–	–	run	5" drain cement concrete bed 6" thick and digging 3 feet deep averaged			
–	–	,,	5" drain and ditto 5 feet deep averaged ..			
–	–	,,	6" ,, ,, ,, 4 ,, ,, ..			
	No.	6	Extra for 5" bends			
	,,	3	,, 6" ,, ...			
	,,	4	,, 5" junctions			
	,,	6	,, 6" ,, ...			
	,,	2	,, 5" diminishing pieces			
10	–	run	5" channel pipe and bedding in concrete (elsewhere taken) and making good cement rendering to edges			
9	–	,,	6" ditto			
	No.	10	5" branch channel bends 30" girth and ditto			
		3	6" main channel bends 15" girth and ditto			

BRICKLAYER.

In a large building it will be found of advantage to the surveyor in the collection of internal walls to divide his work into sections. He will usually find several walls which pretty clearly perform this function. He will first take these walls, and then those in each section thus formed. (See also Chapter XIII.)

In making collections of walls, measure each wall to its extremity through the wall transverse to it; if this is invariably done, the surveyor will never be in doubt. The external walls should be collected, commencing at an angle of the building and going regularly round to the starting point.

Compare the plan of each floor with that of the floor above, to see if any brickwork is shown on the upper floor for which there is no support below, as in such case provision must be made for it, either by thickening walls or introducing girders, etc.

Examine the drawings in order to omit from your collection at ground level the parts which belong to projections such as bays on one story only.

Count the fireplaces, see that the proper number of flues are shown on each floor, and that the chimney breasts or backs are of sufficient projection to contain them.

Chimney breasts wider on one floor than on the floor below must be supported by stone or brick corbelling. Turning an arch over the recess on the floor below is a convenient expedient often adopted.

Observe whether shafts are drawn large enough to accommodate the requisite flues.

A figured diagram (for permanent use) of chimney stacks to contain various numbers of flues is useful; from it the stacks can be readily figured on the roof plan.

Where there are extra thicknesses of walls, to receive plates of ground joists or to form plinths, it is better, as these usually only occur at parts of the walls, to collect them afterwards, measuring the wall at first (in the collection) of the thickness required minus the thickness of that part required for plinths or sleepers.

Observe that plinths are usually carried up from the footings, and that the width of footings or concrete is rarely increased because of them.

Where arches occur over passages, measure the work as though the wall was continuous across the passage, deducting the voids in the usual order.

State in all cases whether brickwork is in mortar or cement.

It is the usual practice where only a small part of the work is built in cement to measure the whole of the brickwork as though it were in mortar, and to again measure the parts in cement, calling it "Brickwork extra only in cement;" but very often it is more convenient to measure it in one operation as brickwork in cement—it can then be dealt with in two places on the abstract.

Brickwork is measured superficially, stating opposite the dimensions the number of bricks in thickness, or measured cubically where in very thick or irregularly shaped walls. The whole is reduced to the superficial rod = $5\frac{1}{2}$ yards by $5\frac{1}{2}$ yards, or 272 feet 3 inches, but always taken as 272 feet $1\frac{1}{2}$ brick thick, except $4\frac{1}{2}$ inches and 9 inches (as page 62).

Brickwork in large or irregular shaped masses is often most conveniently measured in cubic feet, and is so abstracted, although billed, as usual, as "reduced brickwork."

If brickwork is over 60 feet from the ground, it should be kept separate, and all above that level divided into heights of 20 feet, described as "60 feet to 80 feet," "80 feet to 100 feet," etc.

Brickwork in small quantities in filling in of openings of old work or in similar positions should be kept separate and described as "in small quantities inserted."

This may be measured the net size of the openings, in which case the item should include "*Extra labour and materials, cutting and bonding,*" or the brickwork should be measured larger than the opening, and should include only "*Extra labour, cutting and bonding.*" The same consideration arises when new walls are built in an old building, and the treatment and its alternative are the same.

Keep brickwork in backing to stonework separate, and state that it includes the necessary cutting and fitting to stonework, or this cutting and fitting may be mentioned in the preamble of the bill.

In cases where a wall is faced with stone, it is generally most convenient to measure the whole thickness of the wall including the stonework (but not including the projections of the latter), making the deductions from the brickwork when the stone is measured.

Keep work in raising an old wall separate; if any work is built " overhand " it should be stated.

Keep work to old walls distinct from that to new work.

If brickwork is in small quantities to underpinning, state it, and whether in cement, and keep it separate. Also state if it includes iron wedges or pinning up with slates, or if any timber will be left in.

It should be observed that the bricks used in some of the northern and midland counties will frequently rise as much as

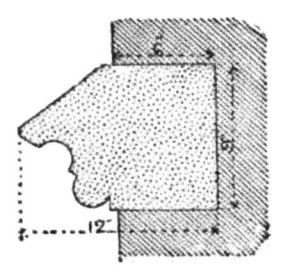

Fig. 8.

13 inches and upwards to the four courses instead of the London average of 12 inches; this will affect both the brickwork and the stonework bonded with it.

If gault bricks are used, the walls will finish of a greater thickness than they would if built with ordinary bricks; a three-brick wall would be about 2 feet 4 inches thick. The work may be measured as of the ordinary thickness and a percentage added to the abstract before billing, or measured by the cubic foot to the exact finished dimensions.

Measure walls to the top of the wallplate where the latter is not more than 3 inches thick; where over 3 inches allow that height for labour of bedding, or the brickwork may be measured net and a lineal dimension taken of " bedding plate."

Allow 3 inches by the thickness of the wall for raking cuttings to gables.

Stonework is not deducted unless 6 inches in height or over. In cases where it is deducted it is better to do so after taking the dimensions of the stone, usually the same dimensions serve for both. Often the stone exceeds the deduction of brickwork, and in some instances it will save labour to fix some proportion for the deduction, as ½, ⅓, etc., otherwise a new dimension must be taken for the deduction. The following instance shows the above-mentioned process (see Fig. 8):—

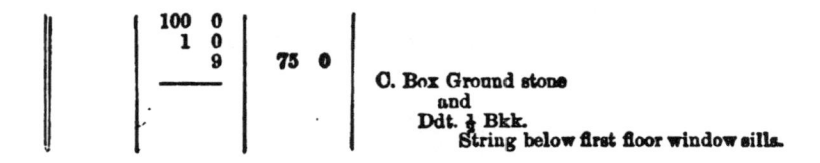

	100 0 1 0 9	75 0	C. Box Ground stone and Ddt. ⅓ Bkk. String below first floor window sills.

Sometimes in the case of a brick wall faced with stone, when it is desired that the brickwork in backing to masonry shall be kept separate, it will be convenient to begin by deducting the whole area occupied by masonry by the total thickness of the wall from the ordinary brickwork, then to measure the stonework, and afterwards the brickwork, as brickwork in backing to stonework.

Where the stonework of a building consists principally of quoins it is sometimes not deducted, but the course adopted should be clearly stated in the bill.

No deductions are made for flues under 2 feet square in diameter, but a flue of such size would certainly be for a furnace chimney. In such cases measure the shaft and deduct the void; keep the brickwork separate and state that it has been measured net, and what it is. Flues at an angle with the horizon of less than 45° must have soot doors provided.

A separate bill of all trades is generally most convenient for a large furnace chimney, even when it is connected with a building; the work measured in the usual way, but the total height stated.

Where semicircular or circular superficial or cubical dimensions occur, they are treated as in the following examples:—

BRICKLAYER.

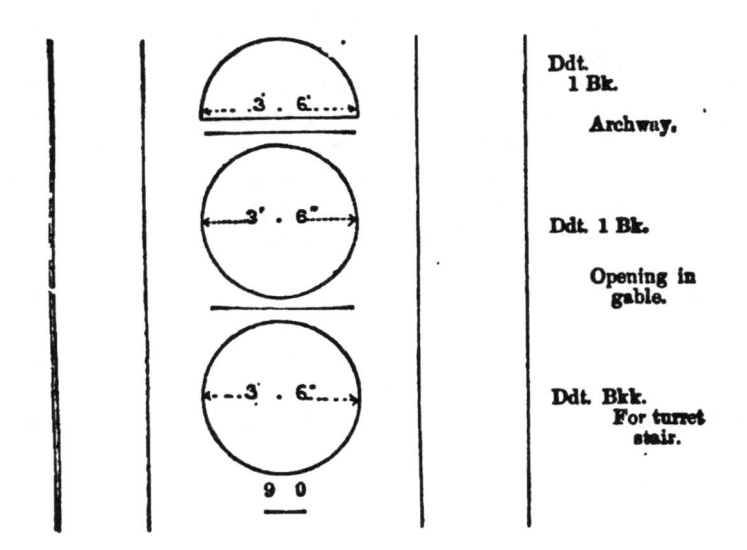

3' . 6" (semicircle)	Ddt. 1 Bk. Archway.
3' . 6" (circle)	Ddt. 1 Bk. Opening in gable.
3' . 6" (circle) 9 0	Ddt. Bkk. For turret stair.

Circular segments may be dealt with as follows :—

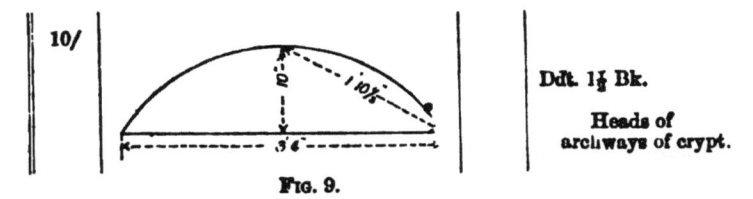

10/ [Fig. 9: segment, height, 1'10½", 3'4"] Ddt. 1½ Bk. Heads of archways of crypt.

Fig. 9.

but commonly the width by the mean height is sufficiently near.

In dealing with large Gothic arches the dimensions are best shown as below:—

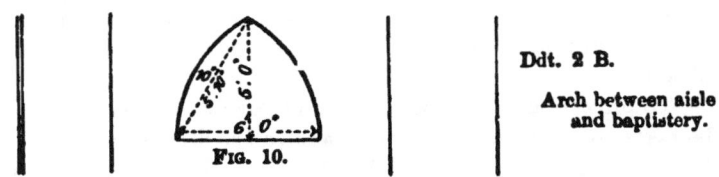

[Fig. 10: Gothic arch, 10", 6'0", 6'0"] Ddt. 2 B. Arch between aisle and baptistery.

Fig. 10.

To fireplaces deduct the chimney opening only. The theory is

F

66   QUANTITY SURVEYING.

to deduct ash-holes of coppers, but it is rarely done, and in the writer's opinion should not be. See "setting coppers" in works numbered: it is better to give setting coppers as an item, including brickwork. The majority of deductions will depend upon the size of the joinery to be fixed in the openings. The *general* rule to be observed as to deductions of brickwork for internal doorways is to allow 3 inches in width and 3 inches in height beyond the finished size of door; for external door-openings and window-openings to receive solid frames, the clear dimensions between reveals and between sill and head for the external part, the same width plus 4 inches, and the same height plus 3 inches for the internal part. The deduction for window-openings intended to receive sashes and frames will be for the external part as last described, the same width plus 9 inches and the same height plus 3 inches for the in-

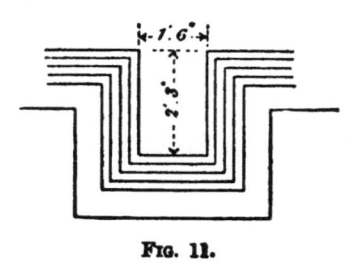

FIG. 11.

ternal part. Deductions for openings with segmental heads should have their height measured up to the springing of the arch only, if the rise does not exceed 1 inch to a foot of span; if it does exceed that the mean height must be taken.

When frames are put in from the outside, deduct the width between the external reveals by the height from sill to springing, for the inside 2 inches less in width and 3 inches less in height. See also deduction of a window in a hollow wall, EXAMPLES OF TAKING OFF.

To projections the same number of courses of footings should be taken as to the adjacent walls, and in the dimensions it will be found most convenient to take the dimension of projection as the item of *length* as follows, assuming the depth from surface to bottom of concrete to be 4 feet.

BRICKLAYER.

2 3					
4 0					
1 0	9	0	Dig surface trenches and cart and Concrete.		
2 3					
4 0					
3 0	27	0	Dig surface trenches, fill and ram.		
				2¼ B. top course of footings.	
2 3					
2 5				4 B. bottom do.	
1 0					
	5	5	Ditt. and	2)6¼	
2 3			Add	3¼ = 2' 5".	
1 6					
2 0	6	9	Dig surface trenches and cart.		
2 3					
1 0	2	3	3¼ B.	foots.	
2 3					
2 0	4	6	2 B.	up to ground line.	

In the measurement of brickwork turn to the collection of the lengths of the walls of basement, if there is one, and measure first the external walls up to a convenient level, such as ground-floor level, then the internal walls up to the same level.

After measuring the walls up to a certain level measure the projections up to the same level, then measure the walls where there is no basement, bringing them up to the same level as last.

Proceed with the measurement of each floor in the same manner until you have arrived at the top.

Finish measuring the whole of the brickwork up to a certain level before you begin to measure anything above that level.

Chimney breasts and shafts are better left until the whole of the general brickwork has been taken off, then take off one stack at a time, beginning with the digging and concrete, and finishing with the shaft.

Deductions of openings and voids should also be left until the whole of the general brickwork has been taken off. " Wants ' may be deducted as they occur.

F 2

Footings are averaged thus: By adding the top and bottom courses together and dividing by 2, the dimensions as below :—

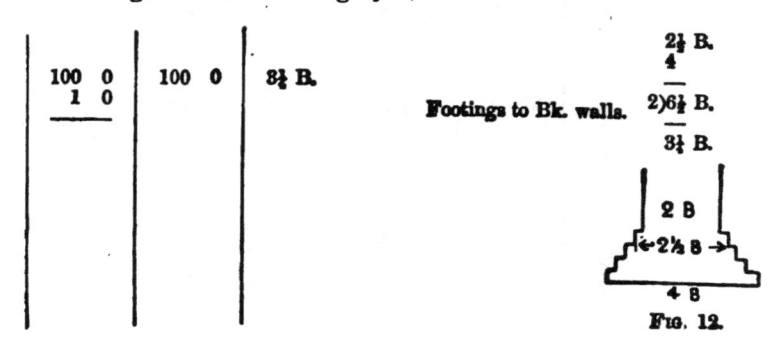

100 0 1 0	100 0	3¼ B.	Footings to Bk. walls.

$$\begin{array}{r} 2\frac{1}{4} \text{ B.}\\ 4 \\ \hline 2\overline{)6\frac{1}{4}} \text{ B.} \\ 3\frac{1}{4} \text{ B.} \end{array}$$

2 B
←2½ B→
4 B

Fig. 12.

Observe within the jurisdiction of the London Building Act to take as many courses of footings as are prescribed by the schedule of the Act; the drawings will sometimes show fewer.

The number of courses prescribed equals in number the half bricks in thickness of base of wall, and the average width of footings is as follows :—

½ brick wall average	..	..	..	..	..	1 brick
1 ″ ″	..	..	..	..	..	1¾ ″
1½ ″ ″	..	..	..	..	..	2½ ″
2 ″ ″	..	..	..	..	..	3¼ ″
2½ ″ ″	..	..	..	..	..	4 ″
3 ″ ″	..	..	..	..	..	4¾ ″

Brick walls 9 inches in thickness and not plastered are kept separate and described as "reduced brickwork in one-brick walls, fair both sides." Do not insert the word "struck" as is sometimes done. If fair one side only reduce as ordinary brickwork.

A one-brick wall faced both sides is best measured superficial, describing the thickness, and the facing and pointing. This course is specially necessary with glazed facings, as the headers are glazed on both ends.

Half-brick partitions and half-brick sleeper walls are kept separate and not reduced; if built "honeycomb" or "pigeon hole" state it.

Brickwork circular on plan, where over 25 feet radius, should be measured as common brickwork first, and then one face measured and described as "extra labour to circular face on brickwork, one face only measured." Where under 25 feet radius keep the brickwork separate, state the radius, that it is in wall circular on plan, and that it includes any necessary cutting, and take a template, stating length, or say that the work includes templates.

When walls are built battering, keep them separate and state how many inches they batter for each foot in height. When one face only batters, measure a superficial item of extra for battering face, giving the same particulars; the brickwork may go into bill with the general brickwork. Observe to take the rough cutting where it changes from battering to perpendicular or the opposite.

Bill.

rods.	ft.	in.			£	s	d
40	185	—	supl.	Reduced brickwork in mortar			
10	14	—	,,	Ditto, extra only in cement..			
20	15	—	,,	Ditto in cement as backing to masonry ..			
1	15	—	,,	Ditto in one-brick wall in mortar, fair both sides			
1	90	—	,,	Reduced brickwork in cement, in small quantities in filling in of openings, including extra labour and materials, cutting and bonding to old			
2	25	—	,,	Ditto in cement in raising on old walls ..			
3	55	—	,,	Ditto in cement in small quantities in underpinning			
2	120	—	,,	Do. do. in vaulting in two half-brick rings in cement			
				100-ft. run, extra labour and waste to fair-cut groin point			
	6257	—	,,	Hollow wall of two thicknesses of brickwork 9" and 4½" respectively, with 2½" cavity, bonded with galvanised wrought-iron wall ties four to each superficial yard, and weighing 60 lbs. per hundred, and allow for keeping the hollow clear of droppings of mortar and rubbish, and for leaving openings at bottom of hollow, and for cleaning out hollow and filling up openings at completion			
	576	—	,,	Half-brick sleeper walls built honeycomb in mortar			
	573	—	,,	Half-brick partition in cement			
	68	—	,,	Half-brick trimmer arch in cement, and levelling up with fine concrete			

Measure hollow walls solid as reduced brickwork, state the thickness, make no deduction for cavity, but state its width, the kind of ties used, the number to each superficial yard, and whether haybands or movable boards are placed along hollow to keep out falling rubbish. The hollow usually commences above the damp-proof course.

Measure vaulting at per rod reduced. Take the mean girth by the length. State the thickness and whether in mortar or cement. Observe whether it runs beyond the end walls enclosing the area to be covered.

Measure groin point at per foot run; describe as labour and waste to groin point of vaulting, stating thickness of vault. (See Bill on preceding page).

Brickwork filled in between stone ribs, as in groined roofs, is usually measured by the foot superficial, stating the thickness and the kind of bricks and including the cutting; the item in the bill would be similar to the following:—

Bill.

ft.	in.			£	s.	d.
254	–	supl.	Half-brick groin cells in small spandrils of red bricks, as described for the facing set in mortar, the soffits slightly arched or cambered and neatly cut and fitted to the stone ribs, including cutting, cleaning off and pointing (or joints struck as the work proceeds), and all requisite centering or laths			

BRICK SEWERS.—In the taking off of brick sewers great care must be exercised. They are usually public works, are often of considerable extent, and a small error, either of measurement or judgment, becomes a serious matter when it affects a mile or two of sewer. It will generally be best to measure the manholes and lampholes first, and afterwards the sewers.

The vertical section generally to a small scale will show the contour of the ground, and usually a red line below it indicating the bottom of the invert. The digging will be separately measured; the dimensions of depth being taken at *regular* distances of not more than 50 feet, added together and averaged.

The digging of that part of the trench which receives the brick sewer will be either carted away or deposited. The remainder will be dug, got out, wheeled, deposited, wheeled back or thrown back, filled in and rammed.

The brickwork will be measured in feet and reduced to standard rods, and described as in egg-shaped sewer, giving a sketch in the bill with the dimensions figured. Describe the bricks whether in mortar or cement, and how finished inside, whether purpose-made arch bricks are used.

Where passing through fields it will generally be necessary to take a superficial dimension of taking up turf, rolling and relaying, and in roads an item of cutting through road formation.

The method of *building* brick sewers is to work from each end of the line towards the middle, using a few comparatively short lengths of centre, which are shifted as the work proceeds, but it is better to give the whole length in the bill as—

Bill.

yds.	ft.	in.			£	s.	d.
850	–	–	run	Centering and moulds as may be required to brick sewer 2' 2" wide and 3' 3" high, the arch 3' 5" girth			

The contractor can then make his own computation of the quantity required. (See also the preamble to a bill of sewers Chapter IV., section BILLING.)

Although the practice of measuring and billing excavation as before removal has long been settled, both by custom and legal decision, it is a general usage with surveyors to state in a bill of quantities for sewers how the digging has been taken.

Ironwork is usually so described as to correspond with specimens to be seen at the offices of the Local Authority.

The books of Bailey Denton, Baldwin Latham, Molesworth, and Moore, afford much valuable information about sewers, as well as tables of the quantity of brickwork in a lineal yard of various sizes; these are useful as a check on results, but should not supersede a careful measurement and calculation.

Bill.

yds.	ft.	in.		
2693	-	-	cube	Dig trench from surface to the average total depth of 13 feet (varying from 9' 8" to 14' 9") including trimming sides and bottom of trench, staging, casting, or otherwise getting out the earth .. ..
15	-	-	,,	Ditto, but circular on plan .. ..
6065	-	-	,,	Ditto, as first, but to an average of 11 feet (varying from 7 to 16 feet) .. .. ..
6200	-	-	,,	Partly fill in from banks, and partly return, wheel and fill into trenches and ram as described .. ..
2573	-	-	,,	Fill into carts and cart away surplus earth
4	-	-	,,	Dig trench to a total depth of 5 feet from surface, get out earth, fill and cart away
7	-	-	,,	Ditto to a total depth of 8 feet, and ditto..
3	-	-	,,	Ditto to a total depth of 5 feet, get out earth, return and ram .. ..
7	-	-	,,	Ditto to a total depth of 9 feet, get out earth, fill and cart away .. ..
3	-	-	,,	Ditto to a total depth of 9 feet, get out earth, return and ram .. ..
215	-	-	supl.	Cutting through road formation, and making good as described .. .. ..
1640	-	-	,,	Cutting and taking up turf, rolling up and depositing and relaying, the relaying to be delayed to such time as the engineer shall direct.. .. ..
3	-	-	run	Cutting through road formation for trench for 12" pipe, and making good as described .. ..
11	-	-	,,	Cutting through road formation for trench for 18" pipe, and making good as described .. ..
8	-	-	,,	Take up old 6" drain and cart away, no digging .. ..
5	-	-	,,	Dig trench average 4 feet deep and take up and cart away 15" pipe, including getting out earth, returning, filling in and ramming, planking and strutting trench, and carting away surplus earth
4	-	-	,,	Ditto, 8 feet deep and 12" pipe, and ditto
-	12	-	,,	Thoroughly clean and relay, and joint, as before described, old 18" pipes, supplying new in place of any that may be broken or not satisfactory to the engineer .. ..
.	20	-	,,	6" drain, and laying and jointing as described .. ..
-	26	-	,,	15" ditto.. .. ..
		No.	1	Fair splayed end to 12" pipe .. .. ..
		,,	1	Ditto 15" ditto .. ..

Continued £

BRICKLAYER.

					Continued	£
			No.	1	Extra for 6" bend	
			"	1	Ditto 12" ditto	
			"	5	Ditto 15" ditto	
			"	1	Make good junction of new and old 6" drain	
			"	1	Ditto 15" ditto	
			"	5	Upright shafts as lamp-holes, average length 5' 6", formed of 9" glazed stoneware socketed pipes, except the bottom length which shall be a 12" taper-piece, all jointed in cement and embedded in and including cement concrete 2' 5" diameter in all, and allow for the necessary boarding and apparatus	
			"	40	Take up house drains and connections each for a length of 10 feet, and relay and reconnect with sewer with valve trap and galvanised iron flap, and any necessary new bends, junctions, or straight pipes in place of those defective or disapproved by engineer (*as provision*)..	
yds.	ft.	in.				
6	–	–	cube		Cement concrete in trenches as described..	
	5	–	supl.		Level surface of cement concrete to falls in manholes in small quantities	
1219	–	–	run		Planking and strutting to trench 4' 2" wide, and average 11' 0" deep	
450	–	–	"		Ditto 4' 4" wide, and average 12' 6" deep	
3	–	–	"		Ditto all as last, but circular	
			No.	1	Planking and strutting to excavation for manhole 5' 0" × 3' 0" and 5' 0" deep ..	
			"	5	Ditto 6' 9" × 4' 9" and 11' 0" deep (averaged)	

Gault Brickwork, as described.

rods						
132	173	–	supl.		Reduced brickwork in one brick in cement laid and worked as described, in egg-shaped sewer (as sketch), the joints of inside faces to be neatly struck as the work proceeds	
	60	–	"		Ditto, circular on plan, about 6' 6" radius	
	5	–	"		Ditto, 8 feet radius	
	10	–	"		Ditto, in sewer as first, but slightly diminishing	
	12	–	"		Egg-shaped sewer, all as first, but half-brick thick..	
	16	–	"		Half-brick in cement, semicircular channel ramped	
			No.	1	Connection of 18" pipe with one brickwork, including extra for two half-brick rings as eyelet, including all cuttings	
			"	1	Ditto 12" pipe with egg-shaped sewer, and ditto	

Continued £

Staffordshire Brickwork, as described.

rods	ft.	in.		
1	124	–	supl.	Reduced brickwork in one-brick walls in cement in small quantities in manholes, the inside face neatly struck as the work proceeds .. ..
	22	–	”	Ditto, circular, to 12″ internal radius ..
	70	–	”	Arch in cement in two half-brick rings in small quantities in manholes .. .. ..
	40	–	”	Clean off soffit of arch and point with cement .. ..
	41	–	”	Rough cutting .. ..
	209	–	run	Labour, rough, oversail or set back, one course circular .. ..
	8	–	”	Rough cut bird'smouth circular .. ..
	81	–	”	Circular cutting .. ..
		No.	1	Make good junction of 18″ pipe with old one-brick manhole, and making good with brickwork in cement as eyelet in two half-brick rings .. ..
		”	1	Make good junction of 6″ pipe with new one-brick manhole, and all as last ..
		”	1	Ditto 12″ pipe, ditto .. ..

Yorkshire Stone, finely tooled where exposed.

	ft.	in.		
	37	–	supl.	3″ tooled one side cover and bedding in cement .. ..
	2	–	”	3″ ditto triangular measured net .. ..
	30	–	run	Jointed edge to 3″ .. ..
	31	–	”	Finely-tooled edge to 3″ .. ..
	3	–	”	3″ ditto circular sunk .. ..
	3	–	”	Sunk and finely-tooled edge to 3″ .. ..
	3	–	”	Ditto 4″ .. ..
		No.	5	4″ covers tooled both sides 2′ 9″ × 3′ 0″, with fair tooled perforation 12″ diameter, and bedding in cement .. ..

BRICKLAYER.

Continued.

Granite.

yds.	ft.	in.		
			No. 8	Rings of four courses of 4" cubes around circular manhole cover 24" diameter, grouted with cement on and including cement concrete 6" thick, and the necessary digging .. ..
–	40	–	supl.	Centering to semi-arch.. .. ..
1212	–	–	run	Centering and moulds to brick sewer 2' 2" wide and 3' 3" high, the arch 3' 5" girth
450	–	–	„	Ditto 2' 4" wide and 3' 6" high, the arch 3' 8" girth .. ..
3	–	–	„	Ditto all as last, but circular on plan ..
	5	–	supl.	Portland cement plain face, ¾" thick (two of sand to one of cement), trowelled to falls in small quantities in manholes, including making good to edges of channels

Cast Iron.

				The ironwork to match in pattern, quality and weight, the similar articles at present used by the Vestry. Samples may be seen at the engineer's office .. ..
			No. 5	Lamphole covers and dirt boxes, and bedding in cement .. ..
			„ 5	Circular manhole covers and frames 2' 4" diameter, with galvanised iron dirt box with perforated sides, and bedding in cement .. ..
			„ 5	Sluice frame for 2' 2" sewer, including bolts and fixing to blue Staffordshire brickwork, and neatly cutting and fitting gault brickwork of sewer against frame 4' 2" × 2' 11" extreme .. ..

Wrought Iron.

			„ 5	½" eye-bolts 9" long, one end caulked and built into brickwork .. ..
			„ 32	Foot irons of ½" × 1¼" metal, and building into brickwork .. ..

Clean and Paint three Oils on Iron.

			„ 5	Eye-bolts .. ..
			„ 32	Foot irons .. ..
			„ 5	Lamphole covers and frames .. .. ..
			„ 5	Manhole covers and do. .. .. ..
			„ 5	Sluice frames .. ..

Carried to Tender .. .. .. £

QUANTITY SURVEYING.

If invert blocks are used for the bottom of the sewer, measure per foot run. State width of block and radius of invert, and, if possible, a number from a trade-list, and state how filled and jointed.

Bill.

ft.	in.			£	s.	d.
100	–	run	Doulton's No. 13 stoneware invert blocks 1' 10" wide, 9" radius of invert, and filling hollows with fine lias lime concrete as described, and bedding and jointing with cement			

WORK MEASURED BY THE SUPERFICIAL YARD.

LIMEWHITING is usually taken with joints of brickwork, struck fair. Raking out and pointing to soffits is taken with limewhiting, when limewhited. Observe that raking out, &c., is impossible if the vaulting is in cement. Describe as cleaning off, and making good. It is commonly most convenient to separate the joints struck fair and the limewhiting.

Bill.

yds.	ft.	in.			£	s.	d.
540	–	–	supl.	Joints of brickwork struck fair			
1840	–	–	"	Twice limewhite			

BRICK-NOGGING.—State that the timbers (taken with partitions) are not deducted, and are "elsewhere taken," whether in mortar or cement, whether bricks flat or on edge, and if including nogging pieces. (See p. 154.)

Bill.

yds.	ft.	in.			£	s.	d.
187	–	–	supl.	Brick-nogged partitions 4½" thick in mortar, and allow for all cutting (quarters and nogging-pieces elsewhere taken)			

CEMENT FLOATED FACE to receive the pavings. State thickness and keep separate that on walls and floor.

BRICKLAYER.

Bill.

yds.	ft.	in.			£	s.	d.
18	–	–	supl.	Portland cement floated face, 1" thick, to receive tile pavings			
25	–	–	„	Ditto ¾" thick on walls to receive tiling ..			

The foregoing items are sometimes billed in Plasterer.

CEMENT PAVING.—State thickness, if finished in pure cement, if to falls. The patent cement pavings, like Wilkinson's, are usually included as a provisional sum.

Bill.

yds.	ft.	in.			£	s.	d.
94	–	–	supl.	Portland cement paving 1½" thick, floated and finished in pure cement			

TAR PAVING.—State thickness and if covered with siftings of Derbyshire spar, or obtain an estimate and include the amount in the estimate as a provision.

Bill.

yds.	ft.	in.			£	s.	d.
110	–	–	supl.	Hobman's tar paving, or other approved tar paving equal thereto, 2¼" thick, rolled and levelled to falls, and finished with fine Derbyshire spar, and allow for dressing six months after laying with fine tar, and for keeping in good repair and condition for twelve months after laying ..			

BRICK PAVING.—Describe quality of bricks, if flat or on edge, if herring-bone, if bedded in mortar or sand, if jointed or grouted with cement.

Measure cuttings at per foot run where the sides of the area covered are not at right angles, or when the paving is herring-bone.

If paving is to falls, state it; and measure a running length of intersection of falls, if the paving falls towards the centre, as in loose boxes.

78 QUANTITY SURVEYING.

Bill.

yds.	ft.	in.			£	s.	d.
30	–	–	supl.	Paving of hard stocks on edge, bedded in mortar and grouted with cement .. .. 24' 0" run, raking, cutting and waste			
76	–	–	„	Paving of 3" best blue Staffordshire grooved paving bricks in cement .. .. 41' 0" run, raking, cutting and waste .. .. 50' 0" run, raking, cutting and waste to intersection of falls			

WALL TILING.—Measure the exact area covered, and describe as "measured net." State the size, colour and quality of the tiles, and how fixed, and if from any particular manufacturer give the address. Measure at per foot run, cutting around openings occurring in the surface measured, up the angles and to the upper edges, where the height is not a multiple of the size of the tile. Measure angle tiles per foot run, and take cutting and waste to each edge.

Measure moulded capping per foot run, number the mitres, fitted ends, quadrant angles, etc.

Bill.

yds.	ft.	in.			£	s.	d.
24	–	–	supl.	Minton's best 6" white glazed tiles, set and pointed in Parian cement (measured net) 100' 0" run, cutting and waste to edges .. 20' 0" ditto, raking, ditto .. .. 100' 0" ditto, cutting and waste to internal angles			
	75	–	run	Minton's best white glazed angle tile set and pointed in Parian cement, and allow for cutting and waste on both edges			
	100	–	„	Tile capping P.C. 6d. per foot run at manufactory, and setting and pointing in Parian cement			
		No. 30		Fitted ends			
		„ 21		Mitres			
		„ 6		Quadrant angles 4" girth			

FLOOR TILING.—Measure as for wall tiling, state price per yard superficial at *manufactory*, and that it shall include packing, carriage, profit and fixing. Measure at per foot run, the cutting around voids and to edges. If the number of colours in a tesselated pavement can be stated it is better, as it regulates the cost of laying.

BRICKLAYER.

Bill.

yds.	ft.	in.			£	s.	d.
26	–	–	supl.	Maw & Co.'s (Benthall Works, Jackfield, Shropshire) floor tiling, P.C. 10s. per yard at manufactory, and allow for packing, carriage, profit and laying in cement, measured net .. .. 35' 0" run, cutting and waste to edges 20' 0" ditto, raking			
18	–	–	"	Tiling, p. c. 20s. per yard at manufactory, and allow as last, in small quantities in hearths, measured net .. .. 18' 0" run, cutting and waste to edges			

Some surveyors bill wall tiling and all tile paving, except common quarries, at the end of the plasterer's bill; plasterers usually fix them unless the manufacturer does it. Some of the tile merchants keep men who do nothing else but lay tiles, and they generally do it very well.

ASPHALTE PAVING.—State thickness, kind, if to falls, if vertical, if on arches, if laid by any particular company's own men, if in two thicknesses.

Measure forming channel at per foot run extra upon the area measured, number making good to bases, gullies, &c.

Measure skirting at per foot run, state thickness, height, whether square or chamfered, and number the mitres, stopped ends, &c.

Measure angle fillet at per foot run.

Bill.

yds.	ft.	in.			£	s.	d.
350	–	–	supl.	Seyssel Asphalte Co.'s (38 Poultry, E.C.) asphalte paving 1" thick, laid in two thicknesses to falls			
				94' 0" run. Extra labour forming gutter			
202	–	–	run	Ditto ¾" skirting 6" high, and allow for angle fillet and letting into joint of brickwork ¼" including raking out mortar joint			
				No 84 mitres..			
70	–		"	Extra for forming shallow channel 9" wide			
			No. 4	Making good to column bases 9" diameter			
			" 4	Labour and materials to outlets through 14" wall			

VERTICAL DAMP-PROOF COURSE OF ASPHALTE.—State maker, thickness, whether put on in two thicknesses, and include raking out joints of brickwork.

QUANTITY SURVEYING.

Bill.

yds.	ft.	in.			£	s.	d.
20	–	–	supl.	Vertical damp-proof course 1" thick, of mineral asphalte put on in two thicknesses, and including raking out joints of brickwork			
	100	–	run	Ditto on three courses of footings, including arrises and angles and ditto			

WORK MEASURED AT PER FOOT SUPERFICIAL.

ROUGH CUTTING.—Measure it to all rakes, skewbacks, and other positions where the brickwork requires cutting. If cuttings are fair they are taken with the facings. If under 6 inches wide measure per foot run. (See facings, p. 98.)

Rough cutting to brickwork bonded with masonry is usually covered by a clause in the preamble of the bricklayer's bill. A lineal dimension of cutting to facings is assumed to be 4½ inches wide. Consequently to the upper edge of a 9-inch brick parapet to a gable surmounted by a coping there would be a superficial dimension of the length by 5 inches of *rough* cutting besides.

Bill.

ft.	in.			£	s.	d.
95	–	supl.	Rough cutting to rakes, skewbacks, &c. ..			
110	–	,,	Extra labour and materials, cutting and binding new brickwork to old in cement..			

THICKENING OLD WALLS, per foot superficial, stating how bonded and including the cutting and bonding, and give a sketch.

Bill.

ft.	in.			£	s.	d.
149	0	supl.	Thickening old wall with hard stocks in cement of three courses, headers bonded to and five courses of stretchers built against old wall alternately, including cutting and bonding, making good and grouting with cement as sketch			
			or,			
149	0	,,	Thickening old wall 4¼" with hard stocks and cement, and allow extra labour and materials, cutting and bonding to old			

BRICKLAYER.

When more than a mere facing of brickwork is built against an old wall, as a chimney breast for instance, measure the brickwork clear of the old work in the ordinary way, and take a similar superficial dimension of "extra labour and materials, cutting and bonding new brickwork in cement to old."

ROUGH RELIEVING ARCHES.—Some surveyors simply take the superficial dimension of the rough cutting, which is billed with the other rough cuttings, others measure the quantity in them and describe them as "reduced brickwork, extra only in rough relieving arches." Observe that if the cutting be taken it will be extrados, intrados and skewbacks, and if in cement, a dimension brickwork extra only in cement. The best way is to number them describing their length, width of soffit, and depth (about 6 inches longer than the lintel is about the average length), and describe them as "extra labour, cutting and waste to relieving arches 5 feet by 1 brick by 1 brick," or, as the case may be, average abstract. (See p. 308.)

ROUGH SEGMENTAL OR SEMICIRCULAR ARCHES are measured in same manner as last.

Bill.

No.	45	Extra labour, cutting and waste to relieving arches 5' 3" × 1 brick × 1 brick (averaged)	
„	24	Ditto, segmental arches 3' 0" × 1½ brick × 1 brick (averaged)	
		or,	
ft.	in.		
145	—	supl.	Reduced brickwork extra only in rough relieving arches
72	—	„	Ditto, segmental

Quantities roughly prepared sometimes describe the brickwork as "reduced brickwork in mortar, including all cuttings and relieving and segmental arches."

TRIMMER ARCHES.—Measure the length between the trimmers by the width of the hearth plus 3 inches whether in mortar or cement, and if levelled up with and include the concrete with the item.

Bill.

ft.	in.		
60	—	supl.	Half-brick trimmer arches in cement and levelling up with fine concrete

EXTRA LABOUR AND MATERIALS, CUTTING AND BONDING NEW WALL (STATING THICKNESS) TO OLD.—In this case measure new work up to face of old wall only, or sufficient brickwork may be measured to allow for bonding, and the words "and materials" omitted, but the former way is preferable.

Bill.

	ft.	in.			£	s.	d.
	25	–	run	Extra labour and materials, cutting and bonding new 1-brick wall to old			
	30	–	,,	Ditto 1½-brick			
	25	–	,,	Extra labour, cutting and bonding new 1-brick wall to old			

In the last case measure sufficient brickwork to allow for bonding.

CHASE CUT AND PARGET.

Bill.

	ft.	in.			£	s.	d.
	15	–	run	Cut and parget, chase in old wall for 1-brick new wall			
	20	–	,,	Ditto 1½-brick ditto			

FLUE PIPES.—State size, material, how jointed, and that they include cutting and fitting brickwork around. Count the bends as extras on the pipe.

Bill.

	ft.	in.			£	s.	d.
	100	–	run	9" terra-cotta flue pipes, and bedding and jointing in mortar, and cutting and fitting brickwork around. .. .. No. 10 extra for bends			

FLUE PLATES.—Measure per foot run, state whose make, and their size, and if tarred.

Bill.

	ft.	in.			£	s.	d.
	100	–	run	Boyd's (Hendry and Patterson, Marlborough Mews, Oxford Street, London), 14" by 12½" iron flue plates, and twice tarring and building in			

BRICKLAYER. 85

"LABOUR ROUGH OVERSAIL OR SET BACK ONE COURSE."—A course of brickwork 20 feet in length projecting from the face of a wall would appear in the bill as 40-feet run, the brickwork measured in the usual way and included in the item of reduced brickwork.

Bill.

ft.	in.				£	s.	d.
40	-	run	Labour oversail or set back one course ..				

CORES TO COLUMNS.—Describe as "Brick in cement core to shafts to columns, 14 inches mean diameter, including cutting." If diminished or swelled state it.

Bill.

ft.	in.				£	s.	d.
40	-	run	Brick in cement core to diminished shafts of columns, 14" mean diameter, including all cutting				

LEVEL AND PREPARE OLD WALL TO RECEIVE NEW WORK.—State thickness of wall.

Bill.

ft.	in.				£	s.	d.
70	-	run	Level and prepare old 1½-brick wall to receive new work				
22	-	„	Ditto 2-brick				

EXTRA LABOUR TO BEAM FILLING.—State height and thickness of such part of the wall as it affects.

Bill.

ft.	in.				£	s.	d.
100	-		Extra labour to beam filling 6" high to 9" wall				

WORK MEASURED AT PER YARD RUN.

IRON HOOPING.—Give size, length may be obtained from collections of walls, and add 5 per cent. to length to allow for laps. State if tarred and sanded, if Tyerman's patent, and that it includes laying in walls. Make no deduction for the parts cut out, and measure right through openings. Measure one line of hooping to each half brick in thickness of wall at the level specified. This length would be multiplied by the number of tiers; it is usual to state the weight as well as the length.

Bill.

yds.	ft.	in.			£	s.	d.
333	–	–	run	1½" × ¼" No. 15 B. W. gauge hoop-iron, lapped together at joints and angles, and laying in walls by bricklayer (weight, 3 cwt. 1 qr.)			

WORKS NUMBERED.

FRAMES, BED AND POINT.—State if screeded in. Keep window frames separate if the sills are bedded in white-lead. Call frames over 24 feet superficial large; over 36 feet, extra large. In the case of loophole frames measure at per foot run. Some surveyors measure all frames per foot run.

Bill.

		No.	20	Frame bed in lime and hair, and point with cement	£	s.	d.
		"	24	Ditto, large			
		"	6	Ditto, extra large			
ft.	in.			or,			
100	–		run	Frame bed and point in lime and hair, and point with cement			

PARGET AND CORE SMOKE FLUES.—When of a common length the dimensions need not be mentioned, but if very long—over 50 feet—their length should be averaged and stated, or instead of *each* *per foot run.*

Bill.

		No.	10	Parget and core flues	£	s.	d.
		"	10	Parget and core flues 90 feet long (averaged)			
ft.	in.			or,			
900	–		run	Parget and core flue			

PARGET AND CORE VENTILATING FLUES.—Average their length and state size, or instead of *each per foot run.*

Bill.

| | No. | 8 | Parget and core ventilating flues 14" × 14" and 40 feet long (averaged) .. .. .. | £ | s. | d. |

SETTING STOVES OR RANGES.—State kind of stove. Give size of openings, and state that it includes the necessary firebricks. If slow combustion stoves, such as Barnard and Bishop's, state that they include fire-lumps also, and if any extra firebricks. Average the widths of similar stoves.

SETTING COPPERS.—State capacity of copper, whether rendered, or faced with brickwork externally, if circular front, and that it includes connection with flue and setting of furnace work. If a stone cover, state what stone, the thickness, its finish, and the perforation.

Bill.

	No.	6	Set register stoves in hard stocks and mortar to 3 feet opening (averaged)	£	s.	d.
	"	1	Ditto range to 5 feet opening, and form flues and supply any necessary firebricks ..			
	"	4	Set slow combustion stoves in hard stocks and mortar, and set any extra firebricks supplied with them 18" fire-bars			
	"	1	Set 20-gallon copper and furnace work in hard stocks in mortar and connect with flue, and render and float the brickwork with Portland cement, including all labour and materials			

If loose tiles are supplied with the stoves, a superficial quantity of setting tiles and cement floated face must be measured.

WINDOW SILLS.—Where very large sills, as in Gothic work, take "sills bedded hollow, made up and pointed at completion," counting each light of the window as one sill.

Bill.

| | No. | 45 | Bed sills hollow and make up and point at completion .. .. | £ | s. | d. |

Where this item is not taken, take "making good to window sills," and bill with the facings.

CUT AND PIN ENDS OF STEPS, THRESHOLDS, &c.—State whether to new or old walls or to facings, or to stonework.

NOTE.—"Cut and pin" should be taken to any ends of timber or stone which go into a wall, as in the majority of cases, whether built in and made good to, or *cut away for*, the labour is about equal.

Ends of stone treads and risers may be described together as ends of steps cut and pinned.

Sometimes a claim is made for cutting and pinning to *old* walls during the progress of a building, the wall in question being one of the new walls built under the contract. This should not be allowed; and moreover, if the contract prices are to be applied, is opposed to the spirit of that condition.

Bill.

			£	s.	d.
No.	74	Ends of steps, thresholds, curbs, &c., cut and pinned in cement			
,,	20	Ditto to old walls			
,,	10	Ditto, iron joists			
,,	12	Ditto, girders			

SETTING AIR-BRICKS OR GRATINGS AND FORMING AND RENDERING OPENINGS.—State thickness of wall and size of grating, and if grating is more than 9 inches long, that it is to include the necessary cover stones, and whether the opening is cranked.

Small metal items which are fixed by bricklayer are most conveniently billed with the labour to them.

Bill.

			£	s.	d.
No.	20	9" × 3" cast-iron ornamental air-brick and fixing, and forming and rendering with cement opening through 2½-brick wall (averaged)			
,,	14	18" × 12" cast-iron ornamental grating ⅜" thick, Macfarlane's (Glasgow) No. —, and fixing, and forming and rendering with cement cranked opening through brick walls (average) 3 bricks thick, and including any necessary self-faced York cover stones			

CHIMNEY POTS AND FIXING. — State height and material, or provide a sum for each, or state number from a trade list. State that they are bedded and flaunched with cement.

Bill.

	No.	10	Chimney pots, Brown's (Braintree), No. 230, and bedding and flaunching with cement	£	s.	d.
			or,			
	No.	10	Chimney pots, p.c. 4s. 6d. each at manufactory, and allow for packing, carriage, profit, and bedding and flaunching with cement			

SMALL CESSPOOLS, GREASE TRAPS, &c.—Describe length, breadth, and depth in clear, thickness of brickwork, if in cement, if rendered in cement, cover, grating, if any, how many ends of pipes made good thereto, and connection or connections, stating how many with drain, and state that they include digging.

Rain-water tanks are best taken out in detail, and billed in all trades as a separate section after the drains.

Inspection pits and turning chambers are best taken out in detail, and the work to them included in the bill "Drains in all Trades."

Bill.

	No.	1	Brick cesspool 14" × 14" and 12" deep all in clear of 9" brickwork in cement, with brick flat bottom on 6" of cement concrete and 2½" self-faced York cover, including digging and connecting with drain	£	s.	d.

PERFORATIONS IN WALLS FOR PIPES AND MAKING GOOD.—State thickness of wall. Thickness to be averaged in abstract.

Distinguish small pipes (up to 2 inches) from large (beyond 2 inches).

EYELETS, stating thickness of wall and what they are for, as "for 6-inch drain pipe through 1½-brick wall (averaged)."

Bill.

	No.	10	Perforations for pipes in walls 1¼-bricks thick (averaged), and making good ..	£	s.	d.
	„	10	Eyelets for 4" drain in walls 2-bricks thick (averaged), and making good..			

WELLS.—Describe in one item, state diameter and depth in clear, if steined dry or in mortar or cement, or part in either, and that it includes digging. The curb may be included in the description. Sometimes in addition to this the quantities are given in detail " written short." (See Chapter IV., section BILLING.)

Treat large cesspools in the same manner.

Bill.

	No.	1	Brick cesspool 6' 0" diameter and 12 feet deep, both in clear, of 9" brickwork, steined dry, the bottom 6" thick in cement, on cement concrete 9" thick, domed over in 9" brickwork in cement, with eye 24" diameter in same covered with a 3" tooled York cover 3' 0" × 3' 0" with strong iron ring and fanged eye let in and leaded, including curb, digging and connecting with drain, all labour and materials complete	£	s.	d.

UNDERPINNING.—The measurement of underpinning requires considerable judgment and experience, and the work is best kept as a separate section of the bill. Both the digging and brickwork should be described as in small quantities and short lengths in underpinning, and the description of the digging should include strutting and planking and the removal of old concrete.

Measure the cutting away of projecting footings on one side of the wall at per foot run, stating the number of courses.

Measure the pinning up between the new and old work at per foot run, stating the thickness of the wall.

In almost all cases of underpinning the trench must be dug from the level of the paving or ground adjoining, to the bottom of the new concrete, even although a new basement is contemplated, the remainder of the basement digging being done afterwards; and

BRICKLAYER. 91

the trench must be of sufficient width on the building owner's side of the wall for the men to work in it—not exceeding 10 feet in depth 2 feet will be enough, beyond that depth 2 feet 6 inches will be required.

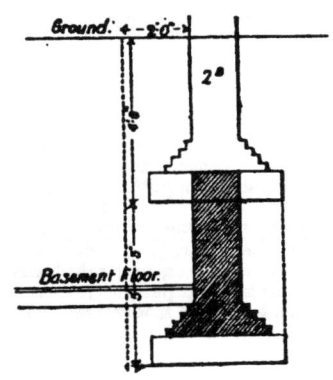

For such a case as shown by the above sketch, assuming the length of the wall at 20 feet, the dimensions would be as follows. The necessary shoring would be covered by a general clause in the preliminary bill.

```
20  0
 2  0
 4  6
 ———    180  0  Dig basement trench and cart.

20  0
 4  9
 5  3
 ———    498  9  Dig B.T. part fill and ram in S.L. and S.Q. in underpinning,
                   including strutting and planking and any part of it which
20  0              may necessarily be left in.
 4  0
 1  0
 ———     80  0  Ddt.
                                or,
                     ⎫  Add, Dig B.T. and cart in S.L. and S.Q. in underpinning.
20  0                ⎪                                              1 10½
 2  5                ⎪                                              3  0
 1  0                ⎬                                             ——————
 ———     48  4       ⎪                                              4 10½
                     ⎪                                             ——————
20  0                ⎪                                              2  5¼
 1  6                ⎪
 3  3                ⎪
 ———     97  6      ⎭
```

length midway between the soffit and top of arch. Describe the bricks and the pointing, and whether straight, cambered or Gothic, if set in cement or putty, and how pointed. Measure the cuttings to facings with the arches. According to this treatment the ordinary facings should not be deducted.

Bill.

ft.	in.			£	s.	d.
100	–	supl.	Extra *on picked stock facings* for gauged segmental arch in Brown's best red rubbers set in cement, and raked out and tuck-pointed with putty			
			or,			
100	–	,,	Extra *on common brickwork* for gauged segmental arches in Brown's best red rubbers set in cement, and raked out and tuck-pointed with putty .. .. This treatment involves the deduction of the facing			

When the bricks for dressings are of better bricks than the general facings, measure the former and deduct the latter. For quoins, average the width of the two faces, and multiply by the height. A running length of "extra for bonding quoin with common facings" should also be taken, or a general clause may be put in the bill stating that the better facings are to include bonding with the commoner ones.

If a better kind of facing is confined to bands and quoins in small proportion to the general facing, it should be measured in the usual way, and the common facing which has been previously measured over it deducted; when facing is in glazed bricks it should be similarly treated.

Bill.

ft.	in.			£	s.	d.
150	–		Facings of Lawrence (Bracknell) best red facing bricks laid Flemish bond in small quantities in quoins and bands, and raking out and pointing with blue ash mortar..			

Observe that facing in coloured mortar is invariably *raked out and pointed.*

BRICKWORK IN RUBBERS CLOSELY SET IN SHELLAC OR PUTTY FOR CARVING, measure per foot superficial as extra on common brick-

BRICKLAYER.

work, usually projected from the general surface of the wall. Measure the projection extra as common brickwork.

Bill.

ft.	in.			£	s.	d.
20	–	supl.	Extra on common brickwork for rubbed and gauged facings of Lawrence's best red rubbers, set Flemish bond, in shellac (or putty) for carving			

BATTERING FACE *at per foot superficial*, state amount of batter.

Bill.

ft.	in.			£	s.	d.
40	–	supl.	Extra on red facings for battering face 3" to each foot of rise, including all cutting and waste			

DIAPERS.—Measure these at per foot run if not exceeding 12 inches in width, per foot superficial if over 12 inches; describe as "Extra on common brickwork," deducting the general facing. Say how bricks are arranged, if any of them project, and how much, and what setting and pointing. They are best selected by number from a trade list.

Bill.

ft.	in.			£	s.	d.
40	–	supl.	Extra on common brickwork for diaper of Brown's (Braintree) No. 428 diaper bricks pointed to match facings			

CARVING ON BRICKWORK varies so much in quality that it is best treated as a provision.

Bill.

			£	s.	d.
		Provide for carving 20 feet supl. of gauged brickwork	20	0	0

CORNICES AND STRING COURSES are measured in two ways, extra on facings and extra on common brickwork. If treated as *extra on facings*, the projection is measured as ordinary brickwork; if in cement, the extra only in cement is also measured. The string course or cornice is girt and billed with the similar facing. The general facing behind it is deducted, and the lineal length of the feature measured, and the mitres, &c., counted.

QUANTITY SURVEYING.

Bill.

ft.	in.			£	s.	d.
50	–	run	Extra on facings for one course of moulded bricks, all headers as string 3" high, including setting out and raking out, and pointing to match facings .. .. .. No. 20 mitres .. .. „ 4 ditto irregular .. .. „ 4 returned mitred ends			

IF TREATED AS EXTRA ON COMMON BRICKWORK, the projection is measured as ordinary brickwork, and if in cement the extra only in cement is also measured, the general facing behind it is deducted, and the lineal length of the feature measured, and the mitres, etc., counted.

Bill.

ft.	in.			£	s.	d.
35	–	run	Extra on common brickwork for labour, moulded bricks and facing for one course of moulded red bricks, all headers as string 3" high, including setting out, raking out, and pointing to match facings No. 20 mitres .. .. „ 4 ditto irregular .. .. „ 4 returned mitred ends .. .. .. This latter method is less liable to misapprehension by the builder.			

If the maker of the moulded bricks is named in a preamble to the section facings, and the height of the member stated in the item, the numbers in the trade list need not be quoted, as the value of the various sections varies but little. If the bricks are enriched, the price is higher than for ordinary facing bricks, and the numbers in the trade list should be stated.

Bill.

ft.	in.			£	s.	d.
35	–	run	Extra on common brickwork for labour, moulded bricks, and facing for cornice, in red bricks 12" high, 8" projection of one course plain, one course (No. 44) enriched, two courses moulded, including setting out, raking out, and pointing to match facings .. .. No. 4 mitres .. .. „ 2 returned mitre ends .. .. „ 2 stopped ends			

BRICKLAYER.

OVERSAILING COURSES OF FACING are described as oversail or set back one course of facings, if in short lengths say so, and make the item include square angles. Measure the projection as common brickwork.

Bill.

ft.	in.			£	s.	d.
130	–	run	Labour oversail or set back one course of facings, including square angles			

MOULDED ANGLES AND SPLAYS *per foot run.*—These are commonly to reveals of openings. Distinguish between those cut and rubbed, and those supplied ready made, and between circular and straight. Count the mitres and stops.

Bill.

ft.	in.			£	s.	d.
100	–	run	Extra on facing for splayed angle 3" wide			
100	–	,,	Ditto ditto, circular			
			No. 20 mitres			
			,, 15 ditto, irregular			
			,, 20 cut and rubbed moulded stops			
100	–	,,	Extra on facing for moulded angle 4" girth			
100	–	,,	Ditto ditto, circular			
			No. 10 mitres			
			,, 10 ditto, irregular			

MOULDED COURSES AS STRINGS OR CORNICES, when cut and rubbed would be treated in a similar manner to the foregoing, but describing the work as in rubbers.

Bill.

ft.	in.			£	s.	.
35	–	run	Extra on *common brickwork* for best red rubbers, as cornice 12" high, 8" projection, moulded 18" girth, set in putty ..			
			No. 4 mitres			
			,, 2 returned mitred ends			
			,, 2 stopped ends			

CUT AND RUBBED MOULDINGS on brick would be measured as *extra on facing* per foot superficial when over 12 inches girth, per foot run when not exceeding 12 inches girth.

H

Bill.

ft.	in.			£	s.	d.
20	–	run	*Extra on facings* for cut and rubbed moulding 9″ girth, raked out and pointed to match facings			
			No. 2 mitres			
			„ 2 moulded stops			

BRICK ON EDGE COPING *at per foot run.*—Measure as "extra on brickwork for brick on edge, and double plain tile creasing in cement," and state the thickness of wall, and if cement fillet on each side (this will have been measured before as common brickwork).

Bill.

ft.	in.			£	s.	d.
100	–	run	*Extra on brickwork* for brick on edge, and double plain tile creasing all in cement to 9″ wall, and cement fillet on each side			
24	–	„	Ditto, circular to ramps			
			No. 3 angles			

Or the ordinary brickwork may be measured up to under side of creasing, and the foregoing description altered by omitting the words "extra on brickwork for."

COPINGS OF ORNAMENTAL BRICKS, *at per foot run.*—State the number from a trade list, state how set, count the angles and irregular angles, and whether purpose-made, and if so give their length each way.

Bill.

ft.	in.			£	s.	d.
100	–	run	Brown's (Braintree) No. 316, coping 13″ by 4½″, and bedding and jointing in cement			
			No. 6 purpose-made angles, 15″ by 15″, and ditto			
			No. 4 ditto irregular, 14″ by 14″..			
			No. 3 stopped ends..			

PILASTERS, *per foot run.*—Measure the brickwork in the projection and the extra facing, and bill it with the general brickwork and facing; describe the labour as extra on facing, stating width of face and the projection.

BRICKLAYER.

Bill.

ft.	in.			£	s.	d.
100	–	run	Extra labour on facing for forming pilaster 9" wide and 2¼" projection, including all cutting and waste			

Brick sills per foot run, state how many courses, if on edge, if in cement, if all headers, and the depth of reveal, and describe as including fitted ends.

Bill.

ft.	in.			£	s.	d.
55	–	run	Window sill of red-splayed bricks on edge, all headers in cement to 4½" reveal, including fitted ends			

It is the practice of some surveyors to measure all the better facings as "extra on common facings," but this leads to confusion.

Measure at per foot run.—Fair cut and rubbed splay not exceeding 12 inches, stating width over 12 inches per foot superficial.

Bill.

ft.	in.			£	s.	d.
25	–	run	*Extra on facings* for labour to cut and rubbed splay 9" wide, including two arrises .. .. No. 4 mitres			

Measure at per foot run, "circular and skewback cutting," "fair cut and rubbed squint quoin," "fair cut and rubbed birdsmouth," "mouldings under 12 inches girth." Extra labour plumbing angles of battering facing, stating the amount of batter in each foot of rise.

Bill.

ft.	in.			£	s.	d.
110	–	run	Circular and skewback cutting to facing ..			
15	–	,,	Labour fair cut and rubbed squint quoin or birdsmouth			
44	–	,,	Ditto to battering face 1" batter to 12" rise			

H 2

KEY BLOCKS, *number extra on arch.*—Measure the extra brickwork, state width, height, projection and the return into reveal.

Bill.

				£	s.	d.
	No.	6	Extra on gauged arch for rubbed and gauged key block, 6" wide, 16" high, 1¼" projection, 5¼" soffit			

CORBELS, *number as extra on facing.*—Measure the extra brickwork, state width, height, projection, girth of moulding, whether rubbed and gauged.

Bill.

				£	s.	d.
	No.	6	Extra on facing for rubbed and gauged corbels to pilasters, 13" wide, 4¼" projection, 9" high, moulded, 15" girth, with returned and mitred ends			

ORNAMENTAL PANELS, *number as extra on common brickwork.*—Deduct the facings, state size, and number from a trade list, and how set.

Bill.

				£	s.	d.
	No.	3	Extra on common brickwork for Brown's (Braintree) No. 181 ornamental panel 18" by 18", and setting in cement, including cutting and fitting facing around, and raking out and pointing to match facing			

ORNAMENTAL APRONS BELOW WINDOW SILLS, *number as extra on common brickwork.*—Measure the brickwork in projection, state length, height and projection, and give sketch.

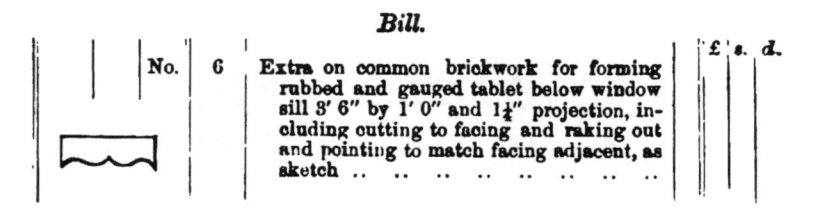

Bill.

				£	s.	d.
	No.	6	Extra on common brickwork for forming rubbed and gauged tablet below window sill 3' 6" by 1' 0" and 1¼" projection, including cutting to facing and raking out and pointing to match facing adjacent, as sketch			

BRICKLAYER.

GLAZED BRICK FACING.—State quality, as best or second quality salt glazed, etc., maker's name, the colour, whether "measured net as seen," how set and pointed.

Measure cutting per foot run, as to raking edges to junctions of one colour with another, edges next treads and risers of steps, etc.

Observe that squint quoins and birds-mouths cannot be cut and that arch bricks must be purpose-made.

Bill.

ft.	in.			£	s.	d.
500	–	supl.	Extra on common brickwork for Cliff and Sons (Wortley) best ivory-white glazed facing, set and pointed in parian cement			
25	–	”	Extra on ditto for purpose-made arch, segmental			
20	–	run	Extra on ditto for bull-nosed angle			
20	–	”	Extra on ditto for squint quoin			
90	–	”	Extra on ditto for purpose-made splayed angle bricks to octagonal piers, including fair vertical cutting and waste			
50	–	”	Raking, cutting and waste			
50	–	”	Circular cutting and waste			
50	–	”	Cutting and waste to edge next plain treads and risers of stairs			

HOLES THROUGH WALLS.—State thickness of wall (average), whether faced one or both sides, and the kind of facing; if holes are for pipes distinguish small from large. They should be billed with the particular kind of facings in which they occur.

Bill.

				£	s.	d.
No.	10		Holes for small pipe through 1½ Bk. (average) wall faced one side, and make good			
”	10		Ditto for large pipe, and ditto			
”	10		Holes for small pipe through 2 Bk. (average) wall faced both sides, and make good			
”	10		Ditto for large pipe, and ditto			

TERRA-COTTA, *per foot cube.*—Each piece is measured as a cube of the size which will contain it. Only two kinds of terra-cotta are recognised in the trade, one kind *plain* or *moulded*, the other *enriched*; keep the two kinds separate. State the colour, the general thickness of the terra-cotta, if chambered, the composition of filling in, if of Portland or Roman cement, or cement concrete, and whether filled in before or after fixing, the kind and colour of the mortar in which it is to be set. Make no deduction of the brickwork, but state this fact in the bill, and although the greater proportion of the terra-cotta will be chambered, it is measured as if solid. State if pieces of hoop iron are used for the bonding of blocks together, and if galvanised, or whether the blocks are joggled together, and if it includes modelling.

Take care to arrange for plenty of vertical joints. If the plan of a moulded jamb can be divided into several pieces the adjusting of the lines of moulding will be so much the easier.

Where in considerable quantity, an estimate is sometimes obtained from a modeller for the modelling, and the amount for modelling included in the estimate as a provision; or the models for enriched work only may be thus treated (the estimate being obtained from a stone carver or a decorative plasterer), the general contractor preparing and supplying the remainder.

A model must be taken for every piece of terra-cotta, except where exact repetitions of shape; but observe that in some cases the same model will be usable for two pieces apparently different, as for right and left jambs of an opening when moulded on one face only. Plain pieces can generally be reversed, in which case one model serves for two pieces.

If a separate contract is entered into with the terra-cotta manufacturer, it will be stipulated that certain proportions of the material shall be delivered by given dates, as " all the terra-cotta below first floor," that " between first and second floor level," &c., and it will be necessary to keep these sections distinct in dimensions, abstracts and bills.

Per foot run, " clean off reveal of terra-cotta door and window jambs to receive wooden frames."

" Clean up and straighten groove for lead lights."

Number notchings, ends cut and pinned, copper cramps or dowels and mortises, &c.

Although the contract may be a separate one, the manufacturer should price every item of the bill.

Bill.

ft.	in.			
1350	–	cube		Terra-cotta in mortar, including hoisting and setting at various levels .. .. ..
400	–	,,		Ditto, enriched .. . .
285	–	,,		Ditto, extra only for setting in cement ..
20	–	run		Clean off reveal or door and window jambs to receive frames .. ..
110	–	,,		Clean up and straighten grooves for lead lights .. ..
		No.	4	Cut or form mortises for dowels .. .. ..
		,,	3	Ditto, perforations 6" diameter through 9" terra-cotta .. ..
		,,	6	Mortises and lead plugs .. ..
		,,	210	Copper cramps 8" long, weight ¼ lb. each, and letting in and running with cement
				About 100 models will be required for the plain and moulded work, and about 80 for the enriched.

The foregoing arrangements are the most reasonable when the intended quantity is large; when small the pieces may be numbered, stating the size and giving sketch, or may be selected from a trade list, as Doulton of Lambeth or Edwards of Ruabon.

Bill.

ft.	in.			
100	–	run		Cornice 18" (average) on bed 12" high, 9" projection, moulded 18" girth to detail, and hoisting and setting in fine mortar ..
				No. 2 mitres .. ..
				,, 2 returned mitred ends .. ..
		No.	3	Pier caps, 14" × 14" × 12", as sketch ..
		,,	2	Edwards' (Ruabon) No. 391 window heads, and allow for packing, carriage, profit, and setting in fine mortar .. .. ..
		,,	4	Doulton's (Lambeth) No. 49A terminals, and ditto and setting in cement .. ..

FAIENCE.—When in the form of tiles, cappings or skirtings, and used as a wall lining, it may be selected at such manufacturers as Burmantoft's or Doulton's at a price per yard at manufactory, and treated in the same manner as a tile wall lining.

When in large quantity and specially designed it must be measured and carefully described, brought into a bill of quantities, and sent to the manufacturer to price; a sum may then be provided in the bill of quantities; each of the items should be priced for the convenience of adjusting variations. The same amount of care must be exercised as to arrangement of times of delivery, and the definition of the respective duties of general contractor and separate contractor as is recommended for terra-cotta.

CONCRETE BUILDINGS.—Measure the concrete per yard cube, state how it is mixed, and the proportions of the materials and that it is in walling; keep the 9-inch and the 4½-inch walls separate, and measure them at per superficial foot. Number the openings as "extra labour to forming openings average ... feet superficial each, with square (or segmental, as the case may be) heads in walls ... inches thick, including moulds and centering."

Lintels of angle iron are often specified for openings in the walls of concrete buildings; these would be taken with the openings and reduced to weight as described for iron joists. The outer faces of external walls are generally finished " plain face in Portland cement," which should be described as " on concrete walls elsewhere taken." The internal plastering, instead of " render float and set," will be " float and set on concrete walls."

Bill.

yds.	ft.	in.			£	s.	d.
400	–	–	cube	Concrete as described in walls, including all necessary boarding and apparatus			
25	–	–	supl.	Ditto in 4½" partitions..			
40	–	–	„	Ditto 6" ditto..			
50	–	–	„	Ditto 9" walls			
		No.	10	Extra labour forming openings with segmental heads externally in 12" wall, average 3' 0" × 6' 9", including all centering and boarding			
		„	20	Ditto, with square heads in 4½" partitions and ditto, 2' 9" × 6' 9" (averaged)..			
		„	10	Ditto, 6" ditto and ditto 3' 0" × 7' 0" ditto			
		„	12	Ditto, 9" walls and ditto 3' 3" × 7' 3" ditto			

MASON.

The usual London practice of measuring stonework, and by far the best method, is to take out all the labours upon it; this is the only way to arrive at the exact value of the work in question. The labour of the surveyor is, however, frequently not appreciated as it should be, the builder simply reading through the labours or looking at the drawings, and afterwards pricing the items of cube stone at a price which, in his opinion, will pay for stone and labour, and leaving the labour items unpriced. In such a case the estimator would read the list of labours and regulate his price mainly by the proportion of the moulded work to the other labours.

Another way, which has arisen as a consequence of the uncertainty as to the extent to which preliminary faces have been measured by the surveyor (and this varies very much), is to omit beds and joints and sawing entirely, and to describe the stone as including "all plain beds and joints, and preliminary faces"; when this course is adopted every other labour is measured as it finishes, including *sunk* joints and *sunk* beds, and as to these labours there is rarely any difference of opinion.

A reasonable average of beds and joints to each cubic foot of stone is in Classic and Classic Renaissance work 1½ superficial feet, Gothic work 2 superficial feet.

Another method is to take out the stone, including labour, and to divide it into a few main items, each composed of stone upon which the labour is similar, and giving sketches to the more ornate parts, as "stone and labour in chamfered jambs"; "ditto in moulded ditto"; "ditto in chamfered plinths and strings, etc."; "ditto in sedilia, as sketch, etc."

To measure stone properly, the surveyor should know all the processes through which the stone must pass.

State in all cases how finished, whether rubbed, dragged or tooled, if any stones are in scantling lengths (6 feet long and over for most stones except Mansfield), if hoisted over 40 feet, after which divide into heights of 20 feet, describing as 40 feet to 60 feet 60 feet to 80, etc., if bedded in mortar or cement, or the stone may all be put together and the extra hoisting charged, as "100 feet 0 inches cube, extra hoisting to stone between 40 and 60 feet of height," or as the case may be.

This division of heights is sometimes modified when a well-defined feature of the building occurs within a few feet of the limit, but the real state of the case should be clearly stated.

State if any of the labour or materials is in small pieces, although the aggregate quantity may be large.

The adherence to the order of length, breadth, depth, in taking off is more important in this trade than in any other, for the sake of after identification of dimensions.

Measure all stone not exceeding 3 inches thick per foot superficial.

In cases where plastering stops against quoins, jambs, etc., as in the interior of some churches, the surveyor must observe to measure the stone sufficiently large to leave it flush with surface of plaster. Sometimes, in case of the restoration of an old building, the stones of jambs are kept flush with the inside surface of the walling, and the plastering stands its whole thickness in front of the stone; in such a case a lineal length of arris and ¾ inch or 1 inch return must be measured.

CUBE STONE, *measured per foot cube.*—The stone is measured, the net size of a cube which will contain the proposed finished stone; claims for waste are frequently made, and builders will argue for an allowance beyond this, but the custom is firmly established, and although two stones of irregular shape may often be cut out of one cube (see preamble of a masonry schedule of the War Department), that treatment of the measurement can only be adopted by an express stipulation in the quantities or schedule. When any dimension includes a fraction of an inch call it another inch, thus 10½ × 11½ × 11½ should be called 11" × 12" × 12". In all cases describe the stone as cube stone, including hoisting and setting. Whether it is set in mortar or cement is prescribed by the preamble of the bill.

Bill.

ft.	in.			£	s.	d.
540	–	cube	Stone as described, including hoisting and setting..			
55	–	,,	Ditto in scantling lengths ..			
396	–	,,	Ditto and hoisting between 40 and 60 feet from street level			
40	–	,,	Ditto in scantling lengths and ditto ..			
75	–	,,	Extra only for setting in cement			
150	–	,,	Extra only for hoisting between 40 and 60 feet			

In some parts of the country the facing bricks are so large that four courses will rise 13 inches and upwards instead of 12 inches. Quoins and jamb stones will be affected accordingly. Observe that the height of each stone agrees with a certain number of courses.

In taking off intricate stonework, the surveyor will sometimes find the labour of description of position saved by numbering the stones in pencil, but if he adopts these numbers he should make a tracing showing the stones with the numbers, to deposit with his dimensions.

In taking off a long length of string or plinth, &c., where there are exceptional pieces of stone, as splayed angles or stones of extra size, take all these before taking the main length, the bed of which latter may be averaged.

All labours (except preparatory faces) are measured over the finished surface shown, except beds and joints, in which one face measured represents two faces.

When a number of stones are measured in one dimension state on your dimensions how many there are.

In all cases of "stone and labour" state what faces of the stone are rubbed, tooled, &c.

As the stone of different kinds and its labour appears in categories under different *headings*, any reference in the bill item to the kind of stone is unnecessary.

WASTE IN CONVERSION.—The waste on stone varies with its quality and the character of the architecture; some of this waste is allowed for in all systems of measurement by the general rule of measuring all stones as square. The rules of the Manchester Society of Architects (elsewhere quoted) prescribe an allowance of an "inch each way beyond the net dimensions of each block when worked." The London surveyor measures the stone net, and sometimes states it thus in the preamble of the mason's bill:—

"The stone is measured net as set, and no allowance has been made for waste."

HALF-SAWING *measured per foot superficial, described as "all measured."*—It is the surface produced on each of the two pieces of stone which a saw divides in converting the block of stone as raised from the quarry into rough cubes; the foreman examines the quarry block when it comes to the works, and decides for what purposes it will best cut up. Supposing that he decides to cut the stone for jambs of windows, he would settle the varieties of depth required as 12 inches, 15 inches, etc., and direct it to be sawn into slabs of those thicknesses. The zinc moulds for the various sizes of beds would then be marked on the faces of the slabs, and the stone again sawn into pieces, each a trifle bigger than the plan of the mould.

The sawing of many of the oolites is done with a toothed saw. The harder stones, as York, Portland, etc., are sawn with a saw without teeth, by the aid of sand and water. Others, still harder, cannot be sawn, but are split. Even some soft stones have so much grit in them that they soon render a toothed saw useless, and must be sawn with a wet one. After the block is divided, many of the surfaces of the cubes will be simply the quarry faces untouched, other pieces will be sawn all round.

Bill.

ft.	in.				£	s.	d.
100	–	supl.	Half-sawing			
150	–	„	Labour to back			

The examples of the measurement of stonework in Reid's 'Young Surveyor's Preceptor,' published in 1848 (a book to which all the succeeding writers on systematic quantity surveying owe considerable obligations), commence by taking half-sawing to all the six faces of the original cube, and over some of these same surfaces measures plain work in addition. Very few surveyors measure half-sawing now. They call such faces half-bed or half-joint.

Other surveyors take half-sawing to all sawn surfaces to which no other labour has been taken. This will usually confine it to the back of the stone, and as this face is very frequently the quarry face, many surveyors describe it as labour to back, and the estimator prices it at a lower rate than half-sawing.

STOPPED WORK.—When a labour on a piece of stone is not continued for the whole length of such stone, it is said to be stopped. The splay on a Gothic window sill of one or two stones, the lower part of the moulding on a jamb which runs down to such a splay, the mouldings or splays on the stool of a mullion, are instances of stopped work. Sometimes such work is stopped at both ends; the difference is of such small value that the distinction is not worth notice. Stopping will qualify both superficial and lineal dimensions.

Bill.

ft.	in.			£	s.	d.
45	–	supl.	Sunk face stopped			
10	–	„	Ditto in short lengths			
18	–	run	Moulding 3″ girth stopped..			
5	–	„	Ditto in short lengths			
			No. 4 stopped ends			

PREPARATORY FACES.—Preparatory faces comprise any surfaces required to set out upon, such as plain face on tracery, plain work all around small mullions and transomes, rough sunk work to the face of undercut mouldings to sunk angles of cornices and to sunk work above a certain depth.

In arguments about labour on stonework these are the chief points of difference, and because of this, various methods have been adopted in the attempt to avoid the difficulty.

One expedient is to exclude them by a clause in the preamble of the mason's bill, "all finished surfaces to include preparatory labours," and to measure only the beds and joints and the labours on the finished surfaces. Another way is to exclude the beds and joints also from the specific measurement, including them with the description of the stone, thus :—

Bill.

ft.	in.			£	s.	d.
1800	–	cube	Stone, including hoisting and setting, preliminary faces and plain beds and joints Or by a clause in the preamble to the bill, "All stone to include plain beds and joints and preliminary faces," and in the bill..			
1800	–	,,	Stone, including hoisting and setting ..			

SAWING.—Sawing is always considered to be included for any face on which any finished labour has been taken. If the attempt were made to measure it exactly it could not be done, as the sizes of blocks of stone (and consequently the sawing) vary considerably.

ROUGH SUNK FACE.—Rough sunk face occurs when mouldings are undercut thus, measured on the face A to B, Fig. 15. In cases where a stone has a re-entering angle, as Fig. 16, the girth from A to B by the height and to the splays of sills which are 3 inches or more below the general face of the stone.

FIG. 15.

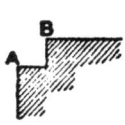

FIG. 16.

PLAIN FACE.—Plain face to all four faces of mullions, afterwards the finished faces of moulded or sunk work would be measured, the plain face on the finished mullion would be covered by the preparatory face. Both faces of the stone in tracery heads are measured as plain face, and are also preparatory, but these should not be excluded from the bill even when the above stipulation as to preparatory faces is used.

Although sawing done with judgment reduces the labour on the beds and joints, it is not generally sufficient to enable the mason to square the stonework without further labour; in soft stones with the chisel and the drag, in hard stones with the chisel.

BEDS AND JOINTS *measured per foot superficial, described as " one face measured for two."*—Half-bed is the upper and lower surface of a stone, half-joint is the surface of its sides. Bed is made up of the upper and lower surface of a stone in contact with a stone above or below it, the two touching surfaces being two half-beds. Joint is made up of the surface of the side of a stone in contact with a stone adjoining, the touching surfaces being two half-joints.

It is the custom to take a bed and a joint to each stone, which will be equal to half-bed or joint on four out of six of the surfaces of the cube, that is, to top, to bottom and to two sides. When plain work occurs on a surface coincident with one already measured as half-bed or joint, measure the plain work and deduct half-joint or half-bed. In the measurement of the stonework of a building it will be convenient sometimes to take a dimension of half-bed, at others of bed, sometimes a dimension of half-joint, at others of joint. The beds and joints are abstracted in one item and described as "plain beds and joints, one face measured for two." The *half*-beds and *half*-joints abstracted together, divided by two, and the result added to the total of beds and joints, thus:—

Bed and Joint.	Half-beds and Joints.
20 0	10|0
25 0	13|0
3 0	2|0
2 0	14|0
50 0	2)39|0
19 6	
	19|6
69 6	

In long lengths of stonework, as in plinths, strings, cornices, &c., where the joints are not indicated on the drawings, allow one to each 3 feet in length.

For mullions and small columns take *two* beds to each stone. To save dimensions the bed and joint are not infrequently taken together (see Chapter XIII.).

Bill.

ft.	in.				£	s.	d.
1540	–	supl.	Labour in plain beds and joints, *one face measured for two*			

When there is much stone in a building the quantity surveyor should be supplied with tracings of the details of the stonework, upon which he should mark (preferably in red) the bed of each stone of the size when it comes to the banker.

Some surveyors call the surface above described as *half*-beds and joints, *bed and joint*. The above quantity would appear in their bills thus, the amount being exactly double.

Bill.

ft.	in.				£	s.	d.
3080	–	supl.	Labour in plain beds and joints, *each surface measured*			

ROUGH SUNK WORK *measured per foot superficial, described as* "*all measured.*"—This should be measured in addition to the sunk face to the general face of all mouldings or sunk faces which are 3 inches or more below the general surface of the stone.

Bill.

ft.	in.				£	s.	d.
35	–	supl.	Rough sunk work, *all measured*			

SUNK BEDS AND JOINTS *measured per foot superficial, described as* "*all measured.*"—These are the beds and joints which are sunk below the general surface of a true cube. Some schedules define the treatment of arch stones in this respect thus: "ordinary arch stones to be considered as having one plain bed and one sunk bed," and in the treatment of a stone of irregular shape it is necessary to decide which bed shall be treated as a plain one, for on this decision the description of the other surfaces depends.

Bill.

ft.	in.				£	s.	d.
75	–	supl.	Sunk beds and joints, *all measured*			

CIRCULAR BEDS AND JOINTS *measured per foot superficial, described as* "*all measured.*"—These are beds and joints which are sunk below the general surface of a true cube, and may be either concave or convex. The extrados and intrados of a relieving arch are a familiar illustration.

Bill.

	ft.	in.			£	s.	d.
	22	–	supl.	Circular bed and joint, all measured ..			

CIRCULAR SUNK JOINT *measured per foot superficial, described as "all measured.*—These are joints which are sunk below the surface of a circular face; for example, a rebate for a frame in the soffit of an arch; if stopped say so, and keep it separate.

Bill.

	ft.	in.			£	s.	d.
	37	–	supl.	Circular sunk joint, all measured			
	10	–	„	Ditto stopped			

PLAIN WORK *measured per foot superficial, and described as "all measured."*—This is taken to all *exposed* plain surfaces, like the exposed plain surfaces of a true cube. When the face involved is identical with either of the four faces dealt with in the regulation beds and joints, deduct a corresponding quantity of half-bed or half-joint, i.e. wherever plain face is taken no bed or joint should be allowed for that face, it includes it. (See description of beds and joints.) In all cases state the kind of finish, as plain work dragged, plain work rubbed, plain work tooled, &c.

It is customary to measure plain work on both faces of the original stones out of which tracery is worked, this is necessary for setting-out upon.

Also on all four faces of the original stone out of which a mullion is worked.

Plain work in widths not exceeding 3 inches is measured per foot run, stating the width, and if stopped describe it so.

Bill.

	ft.	in.			£	s.	d.
	40	–	supl.	Plain work rubbed, all measured			
	25	–	run	Plain rubbed margin 3″ wide			
	12	–	„	Ditto 2″ wide, stopped..			
	15	–	„	Ditto 2″ wide, sunk and stopped (as in a sunk panel)			
				No. 4 mitres			

SUNK WORK *measured per foot superficial, described as "all measured."*—This is work to fair surfaces sunk below the general surface of a true cube. The *girth* by the length is measured; when the sinking cannot be worked straight through the stone it is called sunk face stopped.

Weatherings to string courses and cornices, rebates, channels, are familiar instances of sunk work. Sunk work not exceeding 3 inches wide is measured lineally, and the width stated. If stopped state it, and keep it separate.

Bill.

ft.	in.			£	s.	d.
95	–	supl.	Sunk work, all measured			
45	–	,,	Ditto stopped ..			
50	–	run	3" sunk margin			
20	–	,,	3" ditto stopped			

CIRCULAR WORK *measured per foot superficial, and described as "all measured."*—This is taken to convex surfaces, such as to shafts of columns. If stopped state it, and keep it separate.

Bill.

ft.	in.			£	s.	d.
44	–	supl.	Circular work, all measured			
22	–	,,	Ditto stopped ..			

CIRCULAR WORK SUNK *measured per foot superficial, and described as "all measured."*—This labour is taken to concave surfaces, such as soffits of arches, and the concave surfaces of copings or curbs, circular on plan or elevation. If stopped state it, and keep it separate.

Bill.

ft.	in.			£	s.	d.
37	–	supl.	Circular work sunk, all measured			
36	–	,,	Ditto stopped			

CIRCULAR CIRCULAR WORK *measured per foot superficial, and described as "all measured."*—This labour is taken to the convex surfaces of domes or spheres. If stopped state it, and keep it separate.

Bill.

ft.	in.			£	s.	d.
21	–	supl.	Circular circular work, all measured ..			
11	–	,,	Ditto stopped ..			

CIRCULAR CIRCULAR WORK SUNK *measured per foot superficial, and described as "all measured."*—This labour is taken to the concave surfaces of domes or niche heads. If stopped state it, and keep it separate.

Bill.

ft.	in.			£	s.	d.
24	–	supl.	Circular circular work sunk, all measured			
13	–	,,	Ditto stopped			

MOULDED WORK *measured per foot superficial, and described as "all measured."*—This labour is taken to the profiles of cornices, to strings, caps and bases, panel mouldings, &c. Undercut mouldings or those on stones which contain mitres should always be described as stopped.

Mouldings not exceeding 6 inches in girth are measured per foot run and the girth stated.

In measuring the superficial quantity of moulding on cornices or strings with external mitres, measure the extreme length along the "nib" of the moulding, as that length of moulding is worked and partially cut away afterwards to form the mitre.

Mouldings not exceeding 9 inches in length would be described as in short lengths, and kept separate.

Bill.

ft.	in.			£	s.	d.
100	–	supl.	Moulded work, all measured			
20	–	,,	Ditto stopped			
10	–	,,	Ditto in short lengths			
20	–	,,	Ditto ditto stopped			
25	–	,,	Ditto circular			
12	–	,,	Ditto ditto stopped			
10	–	,,	Ditto ditto in short lengths			
5	–	,,	Ditto ditto ditto stopped			
18	–	,,	Ditto ditto sunk			
10	–	,,	Ditto ditto stopped			
5	–	,,	Ditto ditto in short lengths			
5	–	,,	Ditto ditto stopped			
6	–	,,	Moulded work, circular, continuous in small quantities in caps and bases			
18	–	run	Moulding 4" girth			
10	–	,,	Ditto 4" ditto stopped			
9	–	,,	Ditto 4" ditto circular			
			No. 5 mitres			
			,, 2 returned mitred ends			
20	–	,,	Ditto 6" girth			

WORK MEASURED AT PER FOOT RUN.

Throat.—State if it includes stops, if not, number the stops.

Bill.

ft.	in.			£	s.	d.
95	–	run	Labour throat, including stops			
			or,			
95	–	,,	Labour throat			
			No. 20 stops			

Rebate or moulding not exceeding 6 inches girth. State if it includes stops.

Bill.

ft.	in.			£	s.	d.
26	–	run	Labour to rebate 4" girth			
10	–	,,	Ditto stopped			
			No. 10 stops			
84	–	,,	Moulding 5" girth			
20	–	,,	Ditto stopped			
			No. 3 mitres			
			,, 2 stopped ends on splay			
			,, 2 external mitres, with one 6" return and one stopped end to each			

Back Joint.—State thickness of stone.

Bill.

ft.	in.			£	s.	d.
10	–	run	Labour back joint to 4" landing			

Mitre to Splay either per foot run or by number averaged on abstract for width.

Bill.

ft.	in.			£	s.	d.
20	–	run	Labour mitre to splay			
			or,			
No.	40		Mitres to splay, 6" wide (averaged)			

Chamfer not Exceeding 3 Inches Wide.—State width, and if stopped state it, and keep it separate.

Bill.

ft.	in.			£	s.	d.
74	–	run	Labour to chamfer 2" wide..			
25	–	,,	Ditto stopped			
10	–	,,	Ditto circular			
			No. 6 mitres			
			,, 10 splayed stops			

JOGGLES.—State kind.

Bill.

	ft.	in.		
	74	–	run	Double arris joggle and cement.. .. ..
				or,
	74	–	,,	Double arris joggle, pebbles and cement ..
	20	–	,,	Joggle joint in cement to 6″ landing.. ..
	25	–	,,	Ditto on solid in cement to 6″ landing ..

£ s. d.

GROOVE FOR LEAD LIGHTS AND POINTING.— The rebate on sills is included with it.

Bill.

	ft.	in.		
	140	–	run	Groove or rebate for lead lights, and pointing on both sides with cement.. .. ..

£ s. d.

GROOVE FOR FLASHING AND BURNING IN.

Bill.

	ft.	in.		
	5	–	run	Cut groove for flashing and burn in .. ..

£ s. d.

RUSTIC GROOVE.—State width and depth.

Bill.

	ft.	in.		
	420	–	run	Labour to rustic groove 1½″ wide and 1″ deep .. ..
	15	–	,,	Ditto circular .. ..
	20	–	,,	Ditto in short lengths .. ..
	10	–	,,	Ditto ditto circular .. ..

£ s. d.

MARGINS.—State width, if stopped, if circular, if sunk. Number the mitres. Plain work not exceeding 3 inches wide, is called plain margin.

Bill.

	ft.	in.		
	15	–	run	Plain margin 2″ wide .. ..
	10	–	,,	Ditto sunk and stopped .. ..
				No. 4 mitres .. ..
	15	–	,,	Plain margin 1¼″ wide, circular.. .. ' ..
	3	–	,,	Ditto stopped .. ..

£ s. d.

ITEMS NUMBERED.

PERFORATIONS.— State diameter and thickness of stone. If square, if rebated, whether fair or rough.

Bill.

	No.				£	s.	d.
	,,	2	Rough perforations through 3" stone for 4" pipe				
	,,	1	Fair rebated ditto in 4" stone for 14" coal plate				
	,,	2	Fair square perforations through 6" stone for 4" × 3" pipe				

MITRES TO MOULDING.—State girth. STOPPED ENDS TO MOULDINGS.—State girth. STOPPED ENDS ON SPLAY TO MOULDINGS.—State girth. SPLAYED OR MOULDED STOPS TO CHAMFER.—State width of chamfer. MITRES TO CHAMFER.—State width of chamfer.

These are generally "written short" after the lineal item to which they apply. When the moulding or chamfer exceeds 6 inches girth, this cannot be done, and such items would be billed with Numbers.

Bill.

	No.	100	Mitres to moulding 13" girth, averaged ..	£	s.	d.
	,,	70	Stopped ends to mouldings 11" girth, averaged			
	,,	75	Mitres to splay 9" wide			
	,,	20	Moulded stops to splay 7" wide			

MORTISES AND SULPHUR OR LEAD AND RUNNING, stating what for. LARGE DITTO.

Bill.

	No.	110	Mortises for balusters and lead and running	£	s.	d.
	,,	6	Ditto large for newels			

Note.—This order of the words, "and lead and running," is more exact than "and running with lead." In the districts where it is customary to let the work to separate trades, it settles which trade shall supply the lead for such purposes.

Iron Dowels and Mortises.

Bill.

| | No. | 20 | Iron dowels and mortises .. ..
Described in the taking off as iron dowels and mortises in York and fir. | £ | s. | d. |
| | „ | 20 | Mortises in fir for dowels .. .. | | | |

Slate Dowels, Mortises and Cement.—State size of dowel.

Bill.

| | No. | 200 | 1" × 1" × 4" slate dowels, mortises and cement .. .. | £ | s. | d. |

Copper Cramps.—State weight and size, and mortises and lead and running.

Bill.

| | No. | 10 | Copper cramps 2" × ¾" and 12" long (weight 5 lbs. each), and letting in flush and lead and running .. .. | £ | s. | d. |

Copper Dowels.—State size and weight.

Bill.

| | No. | 2 | Copper dowels 1" × 1" and 12" long (weight 4 lbs. each), and mortises and cement .. | £ | s. | d. |

Galvanised Iron Cramps.—State size and weight.

Bill.

| | No. | 45 | Galvanised wrought-iron cramps 2" × ¾" and 15" long (weight 6¼ lbs. each), and letting in flush and running with sulphur | £ | s. | d. |

Mortises for Door Frames.

Bill.

| | No. | 74 | Mortises 6" × 3" and 8" deep for frames .. | £ | s. | d. |

LEAD PLUGS AND MORTISES.

Bill.

| | No. | 12 | Mortises and lead plugs .. .. | £ | s. | d. |

SMALL METAL ARTICLES CONNECTED WITH MASONRY.—Describe the labour with them.

Bill.

| | No. | 6 | Polished brass gratings 9″ × 9″ and ½″ thick, P.C. 7s. 6d. each at manufactory, and forming fair rebated opening through 9″ stone, and bedding in cement .. .. | £ | s. | d. |
| | „ | 4 | 6″ × 2″ × ½″ wrought-iron socket plates for bolts, and letting in flush, and lead and running .. .. | | | |

TEMPLATES.—State size, if rough, self-faced, tooled, or tooled or rubbed where exposed.

Bill.

| | No. | 22 | 9″ × 9″ × 3″ templates tooled where exposed .. .. | £ | s. | d. |
| | „ | 14 | 14″ × 9″ × 3″ ditto rubbed on longer edge and tooled elsewhere .. .. | | | |

BASES.—State size, and the labour on the various faces and elsewhere.

Bill.

| | No. | 4 | Column bases 18″ × 18″ × 6″, tooled on top and edges, and jointed on bed sunk 1″ deep for 12″ × 12″ base plate, and with two mortises for Lewis bolts and lead and running .. .. | £ | s. | d. |

CORBELS.—State size, and description of labour.

Bill.

| | No. | 7 | Corbels 18″ × 9″ × 9″, rounded on front edge, projecting 9″ from wall, tooled on all faces except back .. .. | £ | s. | d. |

SINK STONES.—State size and thickness, and how many holes,

if dished, if tooled or rubbed, if with iron grating include it with the description, and state if it is run with lead or cement.

Bill.

	No.	4	4" × 18" × 18" 5-hole sink stones finely tooled on top and edges, dished and set in cement	£	s.	d.
	„	4	4" × 15" × 15" cover stones finely tooled on top and edges and set in cement, dished, perforated and rebated for and including 6" × 6" × ½" wrought-iron grating			

SINKS.—State size and thickness, how many rounded corners, if perforated for bell trap, on how many sides cut and pinned to wall, whether on and including brick or stone piers, or measure at per foot superficial, and state rounded corner, perforation, and cutting and pinning separate, state if tooled or rubbed.

Bill.

	No.	1	5" rubbed York sink 4' 0" × 2' 0", sunk and with two quadrant corners and rebated hole for bell trap, pinned on back edge into brickwork on and including two half-brick piers in best white glazed bricks in cement, with semicircular front edges	£	s.	d.

CHIMNEYPIECES.—For these provide a sum, to include fixing, or take the fixing separately and state that it is to include the necessary cramps, or they may be measured in detail.

Bill.

		No.	2	Enamelled slate boxed chimneypieces to 36" opening, P.C. 3*l*. 10*s*. 0*d*. at manufactory, and allow for packing, carriage, profit and fixing	£	s.	d.
				or, Provide for two slate chimneypieces, 7*l*. P.C. at manufactory, and allow for packing, carriage and profit			
ft.	in.	„	2	Fixing only slate boxed chimneypieces to 36" opening, and allow for any necessary cramps			
26	—	run		1½" × 7" rubbed and chamfered mantel and jambs, and allow for any necessary cramps			
12	—	„		1½" × 9" ditto shelf rubbed both sides .. No. 2 small quadrant corners			

WORK USUALLY MEASURED AND DESCRIBED AS STONE AND LABOUR.

ASHLAR *per foot superficial.*—State the *average* thickness; for instance, if the courses are alternately 9 inches and 4½ inches the average would be 6¾ inches and would be called 7 inches; if there are bond stones state their average length, width and height, and how many to a superficial yard, and how the stone is finished on face and back; describe as including all beds and joints, or the ashlar may be measured as stone and labour, as for other stonework, i.e. stone per foot cube, labour per foot superficial.

Bill.

ft.	in.			£	s.	d
110	-	supl.	Ashlar, average 7" on bed, rubbed on face, sawn at back, and including all beds and joints, stone and labour, and all cutting and fitting to dressings			

FILLING IN TO GROINED ROOFS *per foot superficial.*—State the kind of stone, the width and thickness of the courses, what it is set in, how finished, and state that the price includes cutting and centering. An alternative is to take out all the labours, which is often a very intricate process and generally unsatisfactory, as most estimaters would prefer to have it presented in the other way, as follows:—

Bill.

ft.	in.			£	s.	d.
946	-	supl.	Bath stone filling in to cells of groining in spandrils of various sizes, in mortar, in courses 2¼" to 5" wide and 4½" thick, slightly arched or cambered, the face finely dragged, cleaned off and pointed (or the joints struck as the work proceeds), including all cutting, waste and fitting to the stone ribs, and all necessary centering or laths			

The ribs and bosses would be measured in detail.

STONE TRACERY HEADS *per foot superficial.*—These are sometimes measured per foot superficial; state the kind of stone, its thickness, and what other elements it includes; it is by this treatment usually measured square, if measured net the fact must be mentioned.

Bill.

ft.	in.			£	s.	d.
143	–	supl.	Bath stone sunk tracery-heads 7″ thick, including all dowels, cement plugs, and strutting and ribbing, measured square			
70	–	,,	Ditto 9″ thick			
39	–	,,	Bath stone cusped and moulded tracery heads 8″ thick, and all as last			

CROSSES, FINIALS, and similar work may be numbered, stating size, and giving a sketch in the bill. They may either be billed as *stone and labour*, or the stone may be included in the cube stone and the item may appear as labour only.

Bill.

			£	s.	d.
No.	2	*Stone and labour* in crosses 3′ 0″ × 3′ 6″ × 6″ thick, diminished in thickness from bottom to top, with foliated ends to the arms and circular pierced ring to centre, stop-chamfered on all arrises with pyramidal stops, and fixing with and including 1″ × 1″ copper dowels 18″ long, and mortises and running with lead, as sketch			
		or,			
No.	2	*Labour only* to crosses, etc.			

BOSTING.—To enriched mouldings not exceeding 6 inches girth at per foot run; when exceeding 6 inches girth at per foot superficial.

Number bases, terminals, finials, caps, etc., giving the three dimensions.

Or, the bosting may be included with the carving, which is probably the better way if contractor does both.

CARVING.—The value of this depends so much upon the skill of the workmen employed that the usual method is to provide a sum for it, an estimate being first obtained from a carver selected by the architect.

If it should be necessary to measure it, measure it as described for bosting and refer to sketches in margin, but take care that it is carving and not mere mason's work.

The ordinary mason will do almost any work, however intricate, so long as it is not enriched.

Bill.

ft.	in.			£	s.	d.
40	–	supl.	Bosting and carving diaper in low relief ..			
49	–	run	Ditto enrichment 4" girth, including mitres			
50	–		Ditto 6" girth, ditto			
No.	7/4		Ditto foliated capitals to columns, 12"×12" and 15" high, extreme dimensions ..			
"	6		Ditto Gothic crocketed finials 15"×15" and 24" high, extreme dimensions			

If the carving is measured, a clause should appear in the bill of the carving, "Allow for supplying models of any parts of the carving that the architect may direct to be modelled, and for offering them up previous to the work being done." If a provisional sum, it should be large enough to include models.

YORKSHIRE STONE.

The use of Yorkshire stone in London for dressings is exceedingly rare; when so used the same principles of measurement may be adopted as directed for other kinds of stone. Its uses in London are for such things as pavings, landings, steps, thresholds, copings, hearths, window sills, templates, bases for columns and stancheons. As nearly all these are worked at the quarry, it is customary to bill the stone and labour in one item, except in the case of such labours as must be done later.

State how finished, self-faced, tooled, quarry worked, or rubbed.

ROUGH YORK AS CORBEL.—State thickness, finish (as self-faced or tooled) and that it includes building in.

Bill.

ft.	in.			£	s.	d.
100	–	supl.	3" self-faced corbel, and building in			

PAVINGS *per foot superficial.*—State thickness, if in parallel courses, if the stones are to be not less than a certain number of superficial feet each, if bedded in mortar, if bedded and jointed in cement; it is usual to measure irregular shaped stones as square stones of the smallest size out of which it is possible to obtain them. It is sometimes convenient where stones are of triangular shape, and many of them, to number them, and state the extreme dimensions, or measure net and describe as in "triangular shapes measured net."

Measure at per foot run tooled edges, coped edges, rubbed edges, sunk edges, splayed edges, notching, splayed cutting, and waste, circular cutting and waste ; in all cases state the thickness of the stone.

LANDINGS *at per foot superficial.*—When not over 30 feet superficial. Any stone of 15 feet superficial or over should be called landing, also any stone of whatever area if more than 3 inches thick. Measure portion tailed into wall, and allow in the measurement for each joggle joint.

When over 6 feet in length, state the exact size, as 7 feet by 4 feet, state thickness and how finished.

When landings are thinner than the rise of step, observe that a riser must be taken and a stoppel rebate in the edge of landing to receive it.

Measure tooled or rubbed edges (exposed) per foot run.

Bill.

ft.	in.			£	s.	d.
120	–	supl.	3" tooled paving in parallel courses (no stone less than 14 feet superficial) bedded in mortar and jointed in cement			
28	–	,,	2¼" rubbed paving ..			
50	–	,,	2¼" ditto in triangular shapes, measured net			
	No.	4	Pieces of 2¼" rubbed paving of triangular shape 4' 3" × 2' 6", and bedding and jointing with cement			
94	–	supl.	3" tooled both sides landings			
56	–	,,	4" rubbed both sides ditto ..			
70	–	,,	4" ditto in stones over 30 feet superficial each			
	No.	2	4" rubbed one side and tooled the other, landing each 7' 6" × 4' 3"			
	,,	4	6" rubbed both sides, landings each 8' 0" × 5' 0"			
40	–	run	Labour tooled edge to 3" paving			
12	–	,,	Ditto rubbed edge to 4" landing			
5	–	,,	Ditto circular sunk and rubbed edge to 6" landing			
12	–	,,	Ditto moulded edge to 6" landing			
4	–	,,	Ditto circular ..			
7	–	,,	Ditto stop-splayed edge to 4" landing, including stops			
10	–	,,	Raking, cutting and waste to 3" paving ..			
10	–	,,	Joggle joint on solid to 6" landing			
4	–	,,	7½" × 5¼" riser rubbed all round, splay-rebated, and twice splayed, as sketch ..			

MASON.

TREADS AND RISERS *per foot superficial.*—State thickness, describe whether rubbed or otherwise, take the tooled edges separately, or include them in the description, keep winders separate, measure them net, and describe them as such.

Bill.

ft.	in.			£	s.	d.
74	–	supl.	2" tooled treads and risers, including tooled or jointed edges			
20	–	,,	2" ditto in winders, measured net			

HEARTHS *per foot superficial.*—State thickness, describe whether tooled or rubbed, and as including jointed edges. Number the notchings to chimneypieces, state whether single or double. Hearths should never be measured less than 12 inches longer than width of opening and 15 inches wide. They are generally measured 18 inches longer than the width of opening and 18 inches wide. If hearth and back hearth in one, measure the notching per foot run.

Bill.

ft.	in.			£	s.	d.
44	–	supl.	2" rubbed hearths and jointed edges ..			
		No. 12	Notchings for chimneypieces			
		,, 2	Ditto double			

SHELVES *per foot superficial.*—State thickness, take tooled edges per foot run, or include them in description.

Bill.

ft.	in.			£	s.	d.
34	–	supl.	2" self-faced shelves			
17	–	run	Tooled edge			
			or,			
34	–	supl.	2" self-faced shelves, including tooled edges			

COVER STONES *per foot superficial.*—State thickness, state if bedded and jointed in mortar or cement, either measure tooled or coped edges per foot run or include in description.

Bill.

ft.	in.			£	s.	d.
45	–	supl.	3" self-faced cover stone, including coped (or tooled) edges, and bedding and jointing in cement			
60	–	,,	3" tooled cover stone, and bedding and jointing with cement			
80	–	run	Tooled edge			

WORK MEASURED AT PER FOOT RUN.

In dealing with a long length of either of the following items, if the description does not include joints, take two "jointed ends" where each joint occurs, stating sectional size of stone.

Assume that joints of curbs and copings, in long lengths, are about 8 feet apart. The joints of curbs, copings and steps in ordinary cases would be regulated by the rule as to scantling lengths, the stones generally being measured as under 6 feet in length where convenient.

CURBS.—State size, if joints joggled with cement, or plugged with lead, and whether joints are included, if rounded, rebated, chamfered, or throated. Take fair throated ends where they occur.

Bill.

ft.	in.			£	s.	d.
40	–	run	9" × 4" curb rubbed on top and two edges, including joints and bedding and jointing with cement			
			No. 2 fair ends			
			„ 4 sunk joints			

COPINGS.—State size, if weathered, twice weathered, throated, twice throated, circular on plan or elevation, if circular state that the size given is finished size, state the radius, how jointed, and if bedded in cement, and if including joints.

Number the angles or knees, and state if out of solid and the size of stone out of which they are obtained, sunk joints, fair ends.

Bill.

ft.	in.			£	s.	d.
95	–	run	12" × 2½" parallel coping, twice throated, bedded and jointed in cement, including joints			
			No. 6 sunk joints			
			„ 4 fair throated ends			
104	–	„	16" × 8" rubbed feather-edge coping, throated both edges, bedded and jointed in cement, including joints			
			No. 20 sunk joints			
			„ 6 fair throated ends			
			„ 6 apex stones out of solid, sunk 1½" on two faces, 18" × 16" × 12" extreme dimensions, and to match the coping, including joints..			

STEPS.—State extreme size, whether solid or spandril, on which faces worked, if sunk rebated, if back jointed, if rubbed all round. State if moulded, and if spandril, whether with square wall hold, in which case number them. Steps of spandril section should generally be illustrated by a sketch in bill.

Number the winders, giving extreme size, and state if soffits are flueing.

Number steps with curtail ends, stating size and describing the labour upon them.

Sometimes the plan of staircase and the sections of the steps will make it necessary to measure the steps as directed for general masonry, i.e. cube stone, beds and joints, moulding, etc., but this is very rarely done, although the text-books describe this method of measuring.

If they are measured per foot cube, the clause as to two stones of triangular section being cut out of one stone of rectangular section must be inserted in the preamble of the mason's bill. Generally, the measurement per foot run for the fliers and numbering the winders is to be preferred.

Bill.

ft.	in.			£	s.	d
105	–	run	13″ × 8″ spandril steps rubbed all round, splayed and splay rebated (as sketch) ..			
5	–	„	No. 26 fair ends .. .. 13″ × 7″ step, rubbed top and two edges .. No. 1 fair semicircular end			
	No.	20	15″ × 8″ spandril steps with moulded nosings rubbed all round, splayed and splay rebated, and with two square ends for building in (as sketch), and 4′ 0″ long ..			

				£	s.	d.
No.	6	Ditto, but spandril-shaped on plan as winders, 6' 0" × 18" × 8" extreme dimensions, and with flueing soffits				
„	1	15" × 8" step, 5' 6" long, with moulded nosing, rubbed top and two edges, splayed and underjointed, and with one moulded curtail end and one square end for building in				

WINDOW SILLS of the usual section. — State size, if sunk, weathered, throated, if grooved for iron tongue (to every joint take two jointed ends).

Number fair ends.

Bill.

ft.	in.			£	s.	d.
50	–	run	11" × 4" rubbed, sunk, weathered and throated sills			
24	–	„	11" × 4" ditto in scantling lengths			
			No. 20 fair ends			
			„ 10 sunk joints			
			„ 2 jointed ends			

THRESHOLDS.—State size, and that they include jointed edges.

Bill.

ft.	in.			£	s.	d.
35	–	run	3" × 9" rubbed thresholds and jointed edges			
20	–	„	3" × 14" ditto			

GRANITE.

In London and its vicinity, granite when in considerable quantity is usually separately tendered for, and would in such case form a separate bill.

Even when the amount is inserted as a provision it should be measured and billed and every item priced by the person tendering, because of the convenience of that course in the event of variations.

Granite is measured as described for other stones, except that the beds and joints are *all* measured instead of "one for two."

MASON.

Describe whether "axed," "finely axed," "nidged," or "sparrow picked" on face.

In cases such as corbels or bases, number the stone and describe all the labours on it in one item, and bill it as "stone and labour."

Take polishing by the foot superficial.

When granite is a provision, the bill must state clearly who is to fix, unload, get in, and protect; and if it is desired that the general contractor shall fix, the items must appear in the general bill headed "Fixing only, including unloading, depositing and protecting." Granite used in London is nearly always quarry worked.

Bill.

Grey Aberdeen Granite.

ft.	in.			£	s.	d.
128	—	cube	Stone as described			
7	—	,,	Ditto in 6' 6" lengths			
10	—	,,	Ditto in 9' 0" lengths			
175	—	supl.	Roughly axed back			
12	—	,,	Ditto sunk			
360	—	,,	Finely axed beds and joints (*all measured*)			
49	—	,,	Ditto sunk			
2	—	,,	Finely axed plain face..			
5	—	,,	Ditto sunk and stopped			
9	—	,,	Ditto circular sunk			
228	—	,,	High shine polished face			
21	—	,,	Ditto circular			
16	—	,,	Ditto sunk and stopped			
75	—	run	High shine polished margin ½" wide ..			
53	—	,,	Ditto 1" wide			
4	—	,,	Ditto 2" wide			
53	—	,,	High shine polished moulding 3" girth ..			
			No. 2 mitres			
			,, 2, 2" returns with one internal and one external mitre to each			
56	—	,,	High shine polished moulding 4" girth ..			
41	—	,,	Ditto stopped			
35	—	,,	Ditto circular			
			No. 4 mitres			
			,, 4 stopped ends on splay			
			,, 10 moulded stops			
			,, 2, 2" returns with one internal and one external mitre to each			
			,, 4, stopped ends to rebate 5" girth ..			
108	—	,,	Double V-groove in joints for cement and pebble joggles			
			Total	£		

Bill.

Fixing only.

ft.	in.			£	s.	d.
225	—	cube	Setting only granite, including all rough cutting on brickwork (the brickwork deducted), and all casing and cleaning down at completion			
84	—	„	Ditto in courses 9" high			
150	—	supl.	Building in granite, dressings average 10" thick, and casing and protecting same, the brickwork not deducted			
28	—	run	Clean up reveals against joinery			
55	—	„	Clean up joggles			

A few of the simpler processes of working stones after the first operation of sawing the quarry blocks into pieces as nearly as possible of the sizes and shapes of the finished pieces, may here be referred to. When the stone in question is a "free stone," like Bath, Teynton, Doulting, Painswick, &c., the further superfluous stone is *sawn* off with a mason's ordinary toothed saw, but when the stone is Portland and such harder stones, it must be removed by the tool, a much longer process.

The working of any stone must begin by the production of one plane surface, if it does not already exist. In the case of a piece of stone cut out of a large block this would generally be a sawn face, which would require but little more labour to produce a plain surface sufficiently true to square from. On this surface is laid the zinc mould, which is of the shape of the plan of the intended stone, and its outline is marked on the surface.

The working of a true cube fair all round with the chisel would begin with the top plain face first referred to. Assuming it to be true, on this would be marked two lines crossing at right angles, and by measurement a rectangle of the intended size would be marked out. A chisel draft would then be worked along one side on the end of the stone, and by the use of the mason's square other similar drafts would be worked along the other edges of this "end," and a second plain face produced.

If the surface selected as the starting point is not true, there are two ways of making it so. For a large stone a chisel draft is worked along two edges of the top, two chisel drafts diagonal to the sides, and crossing each other, would then be worked. Two other drafts are then worked, one at each end of the stone.

The stone between these drafts is then worked off and the whole surface tested by a straight-edge, Fig. 17. Another way (generally used for small stones) is to sink a chisel draft, tested by a straight-edge or rule, along one edge of the top surface, another chisel draft is worked along the edge parallel to the last, using a rule of the same size as the first. When the upper edges of the two rules are in the same plane the drafts are each in the same

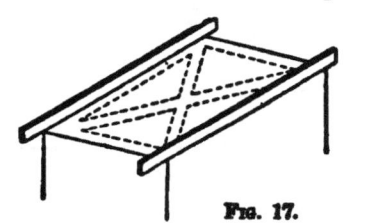

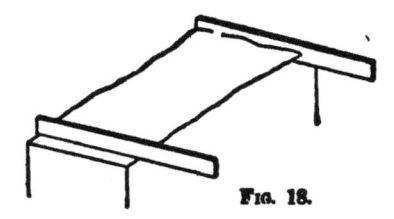

Fig. 17.

Fig. 18.

plane, and the rough stone between the drafts is worked off, Fig. 18, or a line may be worked in the same plane all round the stone. The stone above this line is then removed and a chisel draft worked along the edges of the top face, the remainder is levelled with two equal straight edges as before.

When a surface not at a right angle with the first plane is required, it is produced by using a bevel instead of a square.

When a winding or twisted surface is required, the process is similar to those before described, but one ordinary rule

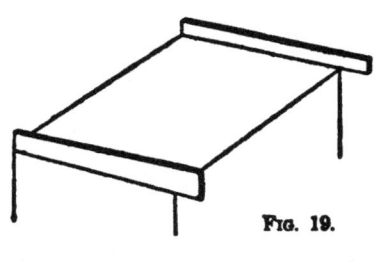

Fig. 19.

and one rule wider at one end than the other, called a twisting rule, is used, Fig. 19.

For a curb of which the upper face is of circular section the stone is first brought roughly by the chisel or saw to a parallelopiped, and the ends, bed and one side worked at right angles to each other, labour, technically half-bed, so that the mould may be applied to each end of the stone; the contour of the ends is then marked and formed on each end by a chisel draft, the sectional shape is then obtained by removing the stone between these drafts and testing by a straight-edge, Fig. 20.

K 2

132   *QUANTITY SURVEYING.*

For a moulded cornice or string course, Fig. 21, the stone would first be brought roughly by the chisel or saw to a parallelopiped, and the ends, bed and one side worked at right angles to each other, the labour on these faces would be half-bed or half-joint; the mould would then be applied to each end of the stone, the contour marked and formed on each end by a chisel draft, the triangular pieces of stone ABC DEF removed by the hand-saw if free stone, and by the chisel if hard stone, the stone then removed between the drafts to the contour of the moulding and tested by the straight-edge.

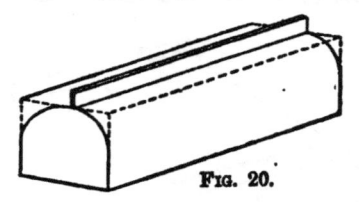

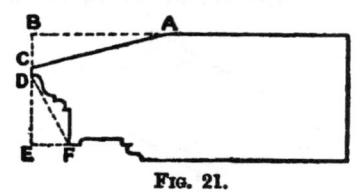

FIG. 20. FIG. 21.

The voussoirs of arches would, wherever possible, be sawn nearly of the shape required, Fig. 22, out of a slab previously sawn, of thickness equal to the dimension of the arch from back to front, and the circular faces worked afterwards, the front and back faces of stone being worked first, the mould applied and marked on each, a chisel draft sunk of the curve required, the circular faces worked at right angles to the front and back face by the use of a square and a straight-edge.

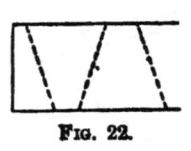

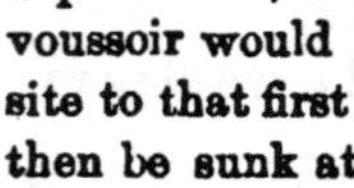

FIG. 22.

In the case of an unsawn stone, the workman would first work a plain bed, then the two ends. On these last the shape of the voussoir would be marked from the template, then the bed opposite to that first named would be worked, curved chisel drafts would then be sunk at the ends of the stone, the rough stone knocked off to produce extrados and intrados and tested by straight-edges.

The stones forming the courses of a dome would first be sawn into cubes or parallelopipeds, then worked roughly as if forming parts of an upright circular ring. The other surfaces would be set out on the plain surfaces and brought to shape or tested by circular rules or straight-edges. Each course would require a special set of winding rules.

The nature of the beds of each course which are circular circular (the top being concave and the lower convex) may be conceived as produced by the revolution of an inverted cone the apex of which in a hemispherical dome is identical with the centre of the plane at the base of the dome.

In work of this kind curved templates and curved rules (concave or convex) are frequently used with advantage. For the wing walls of bridges which are circular on plan, every stone involves special twisting rules, a bevel and a face mould.

There are very few stones which cannot be worked from two preparatory faces, consequently the orthodox half-bed and half-joint is sufficient.

WALLER.

RUBBLE WALLING.

In the stone districts the majority of the walls are entirely built of local rubble for the commoner buildings, and with a thin backing of brickwork for the better ones. The walls are rarely built less than 18 inches thick, and those of less thickness always cost a higher price per cubic yard, so that the saving by adopting thinner walls is but small. In exposed situations, there should be a cavity between the stone and the brick, but often the brickwork is bonded to the stone. In some districts, rubble walls are measured by the perch of 18 superficial feet, reduced to a thickness of 24 inches, in others by the rod of 36 superficial yards reduced to 24 inches in thickness, in others by the perch of 16½ superficial feet of 18 inches.

In Glasgow and its neighbourhood, by the superficial yard, 24 inches thick, walls exceeding that thickness being reduced to it, and those under 24 inches classed according to their respective thicknesses. In Ireland, by the perch of 21 feet run, stating height and thickness, or by the square perch of 21 feet superficial at a standard thickness of 18 inches. In some districts no deduction is made for openings. The London surveyor, however, invariably measures rubble by the cubic yard, keeping any walls under 18 inches thick separate, and stating their thickness.

The various kinds of coursing and their descriptions are illustrated in 'Notes on Building Construction' (Longmans), and Seddons, 'Builders' Work' (Batsford).

In the case of rubble walls where stone of the same kind as that used for the rubble is used for the facings and dressings, the best plan is to measure the whole of the walling at per yard cube. The facing per foot superficial as extra on rubble, giving a description of the coursing, pointing, &c. The dressings as extra on facings, stating what they are, as jamb, string, quoins, &c., and stating the labours upon the stone, and that it includes joints, beds and all labours.

Where rubble walls with comparatively thin brick backing are used, measure the *whole* wall as solid rubble and afterwards measure the brick backing as extra thereon per foot superficial. If hollow, measure the cavity in the cubic content and describe it with the extra for backing.

RUBBLE FOUNDATIONS *per yard cube.*—State whether in mortar or cement, whether random, random coursed, or coursed, and if the stones are of unusual size.

Walls under 18 inches thick should be measured per yard superficial. Describe fully, and state if fair both sides.

FACINGS *per foot superficial.*—State all as last, how finished, also the average thickness and the number of through stones per yard superficial, whether galleted joints, the kind of mortar, and how pointed.

ARCHES.—*Number extra only on facings,* stating the length and depth (including cutting in the description), and the kind of arch, as segmental, pointed, semicircular, etc.

Take levelling up, wherever a damp-proof course or level bed of masonry occurs, or mention it in the preamble of the bill.

Some architects insist upon the internal angles of rubble walls being solid, Fig. 23, in which case an extra labour should be measured lineal.

FIG. 23.

Bill.

yds.	ft.	in.			£	s.	d.
100	–	–	cube	Irregular coursed rubble brought up to level courses in mortar, with one through stone 14" square on face to each superficial yard			
50	–	–	supl.	Ditto in 16" walls			
20	–	–	,,	Ditto in 10" parapet walls fair both sides..			
	143	–	,,	Raking rough cutting			
	1505	–	,,	Extra on rubble walling for local stone facing roughly hammer-dressed, and raking out and pointing with blue ash mortar			

yds.	ft.	in.			£ s. d.
	190	–	run	Extra on facing for dressing quoin, with a tooled chisel draft 1" wide on each return	
	526	–	,,	Levelling wall for stonework 12" on bed (averaged)	
	710	–	,,	Ditto 20" wall for damp-proof course ..	
	No.	35		Extra on facing for segmental-pointed relieving arch (averaged) 5' 0" long, 12" on bed and 10" high, including all cuttings	
	1760	–	supl.	Extra on rubble walling for brick backing three courses, stretchers, and one of headers alternately	
	1425	–	,,	Extra on rubble walling for half brick internal lining, with 2" cavity between it and the stone, bonded together every sixth course with Jennings' patent No. 6 stoneware bonding bricks 2 feet apart, and allow for leaving openings at bottom of hollow for properly cleaning out hollow, and filling up openings at completion ..	

The rubble most commonly used in London is of Kentish rag, this is measured by the foot superficial; as it is more generally used as a facing with a thick brick backing, the walls should be measured, first their whole thickness as of brick, and afterwards the stone measured superficially of an average thickness and described as "extra on brickwork for facing," describing the coursing and pointing.

The extras would be described as for other kinds of rubble. The dressings are commonly of some kind of freestone.

Bill.

ft.	in.			£ s. d.
1354	–	supl.	Extra on brickwork for facing of Kentish rag stone in irregular hammer-dressed courses, average 7" on bed, in mortar, raked out and pointed with a bevelled joint in blue ash mortar	

FLINT FACING, *measured per foot superficial.*

Bill.

ft.	in.			£ s. d.
1210	–	supl.	Extra on brickwork for facings of approved knapped flints, set and pointed in cement	
54	–	,,	Extra on brickwork for facing of black flints 3" × 3", accurately squared on beds and joints, and face filled in to panels of stonework set in cement with a close joint and neatly pointed with cement, including all cutting and fitting in small quantities	

ARTIFICIAL STONE.

GRANOLITHIC STONE.—There are several firms who make this material, and it may either be treated as a provision, or measured and put into the bill as items; it is usual to prescribe the maker. Whether treated as a provision or not, it should be measured and every item priced, either by the specialist or the general contractor.

Copings, strings and cornices may be measured per foot run, stating thickness and width.

Pavings, landings, &c., may be measured per foot superficial, pier caps, finials, steps, &c., numbered and described.

Generally the treatment should be like that suggested for items of stone and labour.

As the manufacturers adopt different methods, some sending such things as landings ready cast to the building, others casting them *in situ*, it will be advisable to enquire of the particular maker what his process may be, as attendance, centering, chases, &c., will be affected thereby.

For the bill, the items given for stone and all labour, in Mason, will be a sufficient guide.

SLATER.

SLATING.—*Per square of* 100 *feet superficial.*

Measure the exposed surface, describe the kind of slates, if in promiscuous sizes, if from any particular quarry, the gauge or lap, the kind of nails, whether composition or otherwise, and number to each slate, and weight of nails per thousand if of copper, if screwed with copper screws, if torched or pointed underneath, if circular or vertical, or in small quantities. When in surfaces of irregular shapes, measure the exact superficial content and describe as measured "net." If diagonal or fancy keep separate.

Keep circular slating separate, and state the radius of the curve.

Keep circular slating to conical roofs separate, and state the radius of the curve at eaves.

It is better to describe the size of the slates as well as their designation.

Slating to steep roofs, as spires or roofs similar to spires, costs more than ordinary slating and should be billed separately.

It is an established custom to allow a certain superficial quantity to cover extra *labour* as well as waste. As to eaves, hips and valleys, and cuttings around voids, this is not in accordance with the best principles of quantity taking, i.e. that extra labour and waste on anything should be so described; the allowance of superficial quantity for troublesome work is deceiving. As a *general* rule labour and waste on any work should be described in a bill of quantities, and the contractor can then judge its value for himself.

The practice as to deductions varies. The merchant deducts all voids and measures the cutting around. Many London surveyors make no deduction unless it exceeds a yard superficial.

Some of the schedules of prices of public bodies define their intention thus:

" Deductions will be made for all openings above 5 feet in area, as chimney shafts, skylights, dormer-windows or the like, but no allowance will be given for what is called workmanship for any openings." In Manchester, nothing is deducted under 6 feet superficial.

It is best to deduct all voids.

Allow for cuttings to deductions the length of the edges of the void by 6 inches.

Allow for cutting to all irregular lines of edge the length by 6 inches.

No allowance is made for cutting when the slating abuts against a wall or gable if the plan is at right angles.

No allowance is made for cutting at ridge.

A roof however hipped has the same area as the simplest roof with gables at ends, provided that the pitches are similar.

Allow for double course to eaves and edges of curbs, the length by the gauge of the particular denomination of slates used.

This, in the case of Countesses, to a $2\frac{1}{2}$-inch lap would be $8\frac{3}{4}$ inches, but the majority of surveyors allow 12 inches.

If the lap is prescribed, the gauge, or exposed part of the slate, is readily found. Thus, for Countess slates, 20 × 10 inches to a $2\frac{1}{2}$-inch lap, the gauge would be half the difference between $2\frac{1}{2}$ inches and 20 inches, i.e. $8\frac{3}{4}$ inches.

Observe that where vertical slating occurs in dormer cheeks, the allowance for both eaves and cutting should be made along the lower edge adjoining the slope of main roof A to A, Fig. 24.

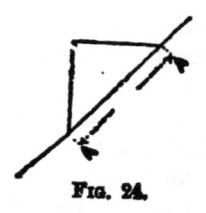

Fig. 24.

The slate merchants always stipulate for an allowance for eaves equal to one-half the length of the particular size of slate used, by the length of the eaves, i.e. for Countesses (20 × 10 inches), 10 inches; Duchesses (24 × 12 inches), 12 inches, etc.

Allow for cutting to hips the length by 6 inches on each side.
Allow for cutting to valleys the length by 6 inches on each side.

When there is a large number of dormers, skylights, etc., to a building, the space they occupy should be deducted, and some such item as the following should appear in the bill: "No. 24, allow for slater attending on plumber and making good after him around skylights and dormers." When there is nothing uncommon this item need not be taken.

For Westmoreland slating (ton slating), the following allowances are common:—Eaves, the length by 12 inches; hips and valleys, the length by 18 inches.

Bill.

sqrs.	ft.	in.			£	s.	d.
40	25	–	supl.	Best Bangor Countess slating (20" × 10") laid to a 2½" lap with two 1¼" copper nails, weight 10 lbs. per thousand, to each slate			
6	55	–	„	Ditto to steep roofs			
10	15	–	„	Best Bangor Doubles slating (12" × 8") laid to a 4½" gauge, with two 1¼" copper nails, as last, circular to an average radius of 10 feet			
55	5	–	„	Best green Westmoreland slating in promiscuous sizes laid to a 3" lap in courses diminishing from eaves to ridge, with two 1¼" copper nails to each slate			

Slating in Bands.—*Per square of* 100 *feet superficial.*

If part of the slating is in bands of slates laid diagonally or with rounded ends, the proportion of diagonal to straight may be stated, or the whole surface may be first measured as ordinary slating, then the bands measured and deducted from it and the two kinds billed separately.

SLATER.

Bill.

sqrs.	ft.	in.			£	s.	d.
40	25	–	supl.	Best Bangor Countess slating laid to a 2¼" lap, with two 1½" copper nails, weight 10 lb. per thousand, to each slate, about one-quarter of the whole laid diagonally in bands, and allow for all cutting and waste			

WORK MEASURED AT PER FOOT RUN.

"DOUBLE COURSE OF SLATES (where no gable parapet) AND BEDDING IN CEMENT AND CEMENT HOLLOW FILLET TO VERGE."

Bill.

ft.	in.			£	s.	d.
40	–	run	Double course of slates and bedding and slate soffit and hollow fillet, all in cement, to verge .. .. No. 5 apex mitres			

CLOSE CUT AND MITRED HIP.

Bill.

ft.	in.			£	s.	d.
74	–	„	Extra labour for close cut and mitred hip and bedding in red-lead cement and screwing with copper screws			

SLATE RIDGE AND HIPS.—Describe the diameter of roll and thickness and width of wings, if screwed with copper screws, if bedded in red-lead cement.

Number the fitted ends and intersections of ridges and hips.

Bill.

ft.	in.			£	s.	d.
120	–	run	Slate ridge of 2¼" roll screwed, with 2" copper screws and ½" sawn slate wings 6" wide, in lengths of not less than 4 feet, bedded and jointed with cement .. .. No. 6 ends fitted .. .. „ 4 ditto, splayed and fitted .. .. .. „ 2 mitres .. .. „ 1 fourway intersection			

Number holes for pipes and making good—distinguish between small and large pipes. Holes for iron stays and making good. Glass slates, stating size and thickness of glass.

Bill.

	No.	20	Holes through slating for small pipe and making good	£	s.	d.
	,,	20	Ditto large pipe			
	,,	40	20" by 10" pieces of Hartley's small pattern fluted rolled plate glass (as slates), each drilled and screwed with two copper screws			

NEW TRADES RULES FOR MEASURING SLATING.

The leading firms of slaters who make sub-contracts with builders (and it may be here remarked that the majority of contractors sublet slating) have agreed upon certain rules which are here given, not that the surveyor is recommended to adopt them, the rules elsewhere stated being at present those almost uniformly observed by London surveyors.

RULES OF MEASUREMENT

For every description of slating, including Westmoreland.

Eaves	The length by half the length of full-sized slate used at eaves over the undereaves.
Cuttings	The length by 6 inches on each side.
Valleys	Ditto by 6 inches on ditto.
Hips	Ditto by 6 inches on ditto.
Dormer sides	Ditto by 6 inches on ditto.
Skylights and other openings	Ditto by 6 inches on ditto.
Chimneys	Ditto by 6 inches on ditto.
All irregular or run out cuttings, except eaves	The length by 6 inches wide.
Circular cuttings	Ditto by 12 inches wide.
Mitre cuttings to hips or valleys	The running length for both sides extra to the usual measurement of 6 inches each side.
Deductions	All openings containing more than 4 super. feet. Hinged skylights not to be deducted.
Circular work	Measure on face as other slating, and allow one-third or one-half extra in price according to radius—if circular on face or elevation add one-fourth extra.

SLATE MASON.

Schedule of Charges.

Bedding eaves, labour only	per foot run,	1d.
Filleting ditto	ditto	1d.
Undercloak of slate to verge bedded in cement, but not including cement	ditto	4d.
Mitred hip, including requisite wide slates, and bedding in oil cement	ditto	9d.
Ditto to Westmoreland and ditto	ditto	1s.
Fixing only zinc soakers, including zinc nail	per dozen	6d.
Ditto, if lead and copper nail	ditto	8d.
Slater	per hour	11d.
Labourer	ditto	7d.
Vertical slating extra	per square	1s.

Including shop time. Travelling expenses extra.

SLATE MASON.

SLATE SLABS *per foot superficial*, allow in the measurement any portion let into wall. State whether rubbed or sawn, and if rubbed on both sides, or enamelled one or both sides. Give the size of each slab if more than 5 feet long and if more than 2 feet 6 inches in width. State if bedded against walls, and if in cement, if screwed with copper screws to and including oak plugs.

Or slate slabs may be billed in accordance with the merchant's classification, or described as not exceeding....feet superficial; the classification is as follows:

From 3 feet to 5 feet long and 1 foot to 2 feet 6 inches wide.
From 5 feet to 7 feet long and 1 foot to 3 feet 6 inches wide.
From 7 feet to 9 feet long and 1 foot to 4 feet wide.
From 9 feet to 10 feet long and 1 foot to 5 feet wide.

WORK MEASURED AT PER FOOT RUN.

ROUNDED EDGE, state thickness of slate. BEVELLED EDGE, state thickness of slate. SAWN EDGE, state thickness of slate. GROOVE. REBATE, state girth. REBATED JOINT and state if in cement or oil cement, and to what thickness of slate. DOUBLE GROOVED JOINT AND COPPER TONGUE, and state if in cement or oil cement, and to what thickness of slate.

SLATE SKIRTINGS.—State height, thickness and finish, whether bedded in cement, whether fixed with brass screws and including plugging. *Number* mitres or rebated angles.

SLATE LOUVRES per foot run, state width, thickness and finish. Number ends housed on rake, or any labours to ends.

Bill.

ft.	in.				£	s.	d.
54	–	supl.		1¼" sawn and rubbed both sides slate shelves not exceeding 5 ft. long and 2' 6" wide (or not exceeding 12' 6" supl.) including rubbed edges			
ft.	in.	No.	1	1¼" do. triangular shaped 4' 0" × 4' 0" extreme dimensions			
20	–	run		Rebated joint and red-lead cement to 1¼" shelf			
10	–	„		Rubbed rounded edge to 1¼"			
15	–	„		⅞" × 7" rubbed and chamfered skirting, fixed with and including copper screws, with countersunk heads and oak plugs in brickwork			
		No.	6	rebated angles			
		„	6	rubbed and shaped ends			
20	–	„		1¼" × ¼" copper tongue, and two grooves in slate, and bedding in red-lead cement			

CHANNELS *per foot run.*—State finish, size, and the labours.

Bill.

ft.	in.				£	s.	d.
10	–	run		9" × 6" rubbed slate channel, hollowed to falls, and bedding in cement			
		No.	2	Rounded stopped ends to channel			
		„	1	Rebated perforation for 6" grating			
		„	1	6" brass heavy grating with rim, and bedding in cement			

WORKS NUMBERED.

ROUNDED CORNERS TO SLABS, state girth and the thickness of slate. Holes for pipes, and state whether for large or small pipes and the thickness of slate. PERFORATIONS FOR BASINS, state diameter and thickness of slate, and if rebated, rounded or dished, if for screws state if counter-sunk and size of screw. NOTCHINGS (stating thickness of slate and size). CANTILEVERS, stating thickness, length and width, and if shaped give sketch. LEAD PLUGS AND SCREWS, state if brass or copper screws and size of screw.

BRASS OR GALVANISED IRON ANGLE PLATES AND SCREWS, state size of plates and number and size of screws.

LAVATORY TOPS.—State size and thickness, and whether perforated and with rounded edge or of irregular shape.

SLATE CISTERNS.—State length, width and depth, all in clear, or to hold water so many inches deep, or the number of gallons it shall hold, stating thickness and description of slate, if grooved together, if put together in red-lead cement, and describing the ironwork and including in the description all holes cut for supplies and wastes, or take the holes separately. State the height of hoisting and include fixing.

CHIMNEYPIECES.—For enamelled slate chimneypieces a sum is generally provided, and the fixing included in the amount, or the fixing may be stated separately, and that it is to include all necessary cramps. They may also be conveniently designated by a number in a trade list.

Bill.

No.			£	s.	d.
No.	4	Rubbed and rounded quadrant corners 6" girth to 1¼" slate			
„	20	Countersunk perforations through 1¼" slate for 2" screw..			
„	2	Rebated rubbed perforations for basins 14" diameter in 1¼" slate			
„	7	Rubbed notchings 9" girth (averaged) in 1¼" slate			
„	20	Lead plugs and mortises, and 1½" copper screw, and driving			
„	18	Sets of two ¼" brass angle plates, 2" × 2" and 4" long, each set including 3 holes through 1¼" slate and 3" brass bolts, with heads and nuts; six mortises and lead plugs in slate, and six 1¼" brass screws, the heads countersunk..			
„	2	Rubbed cantilevers, out of 15" × 6" × 2", shaped, diminished, and twice chamfered			
„	1	1½" rubbed slate lavatory top, with moulded front edge, and perforation for 12" basin with rounded edge, 4' 0" × 21", and fixing on deal top (elsewhere taken)			
„	1	1¼" rubbed both sides slate cistern to hold 150 gallons, grooved together, jointed in red-lead cement, and bolted together with ¼" galvanised iron bolts, with all necessary holes, and hoisting and fixing 40 feet above street level			
„	1	Black enamelled slate chimneypiece, P.C. 5l. at manufactory, and allow for packing, profit, carriage and fixing with the necessary iron cramps .. .5			

STONE SLATES *measured at per square of* 100 *feet superficial.*—These must always have laths whether there is boarding or not, and they are almost always bedded in lime and hair.

The superficial allowances in the districts where they are used are, for eaves, 12 inches by the length ; hips and valleys, 24 inches by the length ; cuttings, 6 inches by the length. If, however, the same system of allowances be adopted as recommended for Welsh slating, such a clause should appear in the bill as may be found in Chapter IV., section BILLING. "In all trades the London mode of measuring has been adopted, etc."

It is generally reasonable to say where stone slates are to be quarried.

Bill.

sqrs.	ft.	in.			£	s.	d.
95	10	–	supl.	Stone slating of the best quality from Finch's quarries (Barton, near Winchcomb, Gloucestershire) in promiscuous sizes, laid diminishing from eaves to ridge, bedded in lime and hair to a 3" lap, and each slate nailed with two 2" galvanised steel nails on and including 1¼" × 1½" sawn fir laths			

STONE RIDGES.—Describe the stone, how finished, the size, thickness and length of the stones, and how jointed, number the intersections, hip-ends, etc.

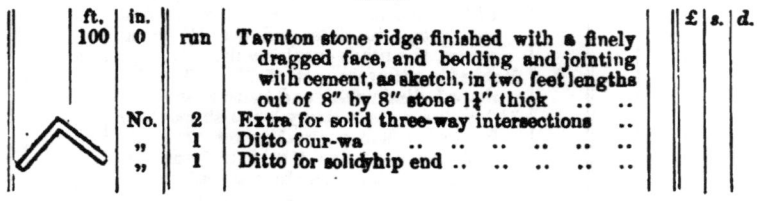

Bill.

	ft.	in.			£	s.	d.
	100	0	run	Taynton stone ridge finished with a finely dragged face, and bedding and jointing with cement, as sketch, in two feet lengths out of 8" by 8" stone 1¼" thick			
	No.	2		Extra for solid three-way intersections ..			
	,,	1		Ditto four-wa			
	,,	1		Ditto for solid hip end			

TILER.

PLAIN TILING, *per square of* 100 *feet superficial.*
Describe the kind and gauge of the tiling, if laid with gauge diminishing from eaves to ridge, whether laid dry or pointed, or torched, the nails or pins, if of oak, iron, galvanised iron, the laths (described and included with the tiling), if vertical or circular, and in the latter case what radius, if screwed with copper screws, if bedded on straw, if laid in lime and hair mortar.

Deduct chimneys, skylights, etc., and allow 3 inches by the length around same for cuttings.

Allow for cutting to irregular angles 3 inches by the length.

Allow for eaves the length by 6 inches.

Allow for cutting to hips the length by 6 inches, viz. 3 inches on each side.

Allow for cutting to valleys the length by 6 inches on each side; or where purpose-made valley or hip tiles are used, let the cutting be included in the description, thus: "Extra for purpose-made valley tiles to course and bond with general tiling, and allow for all cutting and waste." Observe that where hip or valley tiles are used for roofs of different pitch the tiling must be laid to different gauges, so that the tiles on one slope may course with those on the other. The quantity of each gauge will be billed separately.

VERTICAL TILING *per square of* 100 *feet superficial.*—Describe kind of tiles, the gauge, the battens, and if plugged, and whether the tiles are secured with screws, copper screws, or tenter-hooks, or if nailed to joints of brickwork. If with bands of ornamental tiles, state what proportion in quantity they bear to the plain ones —either a fraction or a percentage. Make the same allowance for cuttings as to roof tiling.

MEASURE AT PER FOOT RUN.

EXTRA ON TILING FOR PURPOSE-MADE HIP OR VALLEY TILES, and state if bedded in cement (or mortar, state if to various slopes.

Observe that this will not obviate the necessity of making allowance for cutting; the use of valley tiles where the pitches of the intersecting roofs vary is exceedingly inconvenient. EXTRA FOR BEDDING EAVES IN CEMENT. EXTRA VALUE OF "TILE AND HALF" AT VERGE TO BREAK JOINT. State if bedded in cement or mortar, if with cement fillet underneath. FILLETING. State if mortar or cement. In good work, extra for tile and half is taken to all edges of slopes which adjoin brickwork or other vertical faces and on each side of hips and valleys. RIDGE TILING, describe by reference to a trade list, state if to varied angles, number the splayed and fitted ends (where lower ridges run into plane of main roof). INTERSECTIONS, and state of how many ridges.

Number HIP HOOKS.—State if ornamental, if of unusual shape, give sketch, and if possible the weight. Holes for pipes, distinguishing small pipes from large.

TILE HIP KNOBS, giving sketch or providing a sum for each. APEX TILES to top of hips, allow for the fixing.

Where ridges or hip knobs are to be obtained from a particular manufacturer, or if to special design, it is best to state it.

Bill.

sqrs.	ft.	in.			£	s.	d.
55	55	–	supl.	Best Broseley tiling of true shape and even colour, free from fire cracks and other defects, laid to a 3¼ gauge, with two stout 1½″ zinc nails to each tile, on and including 2¼″ × ⅞″ sawn fir laths			
10	10	–	,,	Ditto to steep tower roofs			
	422	–	run	Extra on tiling for purpose-made hip and valley tiles accurately fitted to slopes of roof, and to course and bond with general tiling (and allow for cutting and waste to tiling)			
				If the words in parenthesis are adopted, the superficial allowance for cutting would not be measured.			
	100	–	,,	Extra for bedding eaves in cement			
	420	–	,,	Extra for tile and half to break joint.. ..			
	100	–	,,	Cement filleting			
	104	–	,,	Tile ridge, Edwards' (Ruabon) No. 26, and bedding and jointing in cement			
				No 6 ends splayed and fitted			
				,, 4 ends fitted			
				,, 4 three-way intersections, *purpose-made*			
				,, 2 hipped ends, *ditto*			

sqrs.	ft.	in.			£	s	d
			2	Edwards' (Ruabon) No. 4 hip knobs, and setting in cement			
25	15	–	supl.	Best Broseley tiling and lathing, as before, but 4" gauge fixed vertically, and the laths plugged to brickwork			
	74	–	,,	Ditto, but without laths, fixed to boarding with copper screws in small quantities to dormer cheeks			
	90	–	run	Extra on ditto for angle tiles, to course and bond with the general tiling			

NEW TRADE RULES FOR MEASURING TILING.

The leading firms of tilers who make sub-contracts with builders (and it may be here remarked that the majority of contractors sublet tiling) have agreed upon certain rules which are here given, not that the surveyor is recommended to adopt them, the rules elsewhere stated being at present those almost uniformly observed by London surveyors.

RULES OF MEASUREMENT
For Broseley, Reading and Yorkshire Tiling.

Dripping eaves	6 inches per foot run.
Valleys	6 inches each side per foot run.
Hips	6 inches ditto ditto.
Dormer sides	6 inches per foot run on each side.
Skylights, &c.	Ditto ditto ditto.
Run out walls	6 inches per foot run.
Circular cuttings..	12 ditto ditto.
Circular work	Measure on face as other tiling, and allow one half or double extra in price according to radius—if circular on face in elevation add one-fourth extra.

Schedule of Charges.

Mitred hips, extra on labour	10d. per foot run.
Mitred valleys, extra on labour	10d. per foot run.
Under verges, bedded in cement, but not including cement..	6d. per foot run.
Verges, bedded and pointed in cement, but not including cement..	2d. per foot run.
Bedding eaves, labour only	2d. per foot run.
Tiler..	per hour 11d.
Labourer	per hour 7d.

Travelling expenses extra. Including shop time. Vertical tiling extra.

PANTILING *per square of* 100 *feet superficial.* Describe the gauge, the laths, and whether pointed inside or out, or both.

Allow the length by 12 inches for cutting to hips and valleys; for all other cuttings the length by 6 inches.

Note whether plain tile eaves are used, if so, add one and deduct the other, and describe it as " plain tiling in narrow widths to eaves of pantiling."

MEASURE AT PER FOOT RUN.

PLAIN TILE HEADING, HIP AND RIDGE TILES, and state if bedded in mortar or cement. FILLETING.

Number HIP HOOKS, and describe, stating weight.

Bill.

sqrs.	ft.	in.			£	s.	d.
28	25	–	supl.	Local pantiles laid to a 10″ gauge in lime and hair, on and including 2¼″ × 1¼″ sawn fir laths			
	40	–	run	Plain tile heading to pantile roof, bedded in cement			
	80	–	″	Plain tile eaves to a 4″ gauge, nailed with 1½″ galvanised iron nails, on and including laths as last, four courses wide, the lowest laid double			
	100	–	″	Pantile ridge and hip, bedded and jointed with cement			
		No.	4	Wrought-iron hip hooks, weight 2½ lbs. each, and screwing to fir			

CARPENTER.

The broad distinction between carpentry and joinery is that the latter is for the most part prepared at the bench and brought to the building ready to fix; the former is usually prepared and fixed at the building.

The surveyor should have ready for use paper scales of convenient lengths, each with divisions marked upon it corresponding with the centre lines of joists, rafters, or quarters at the usual distance apart (12 inches) for 2-inch, 2¼-inch, 2½-inch, and 3-inch joists respectively, for scales of ½ inch and ¼ inch to a foot, thus:—

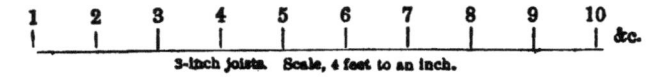

1 2 3 4 5 6 7 8 9 10 &c.

3-inch joists. Scale, 4 feet to an inch.

CARPENTER.

Such a scale applied to a drawing will show at once the number of timbers necessary. Or, he may construct a table on the principle of the following one, which may be extended to any extent.

No.	1¼″	2″	2¼″	2½″	2⅝″	3″
	Length of Room or Building.						
	ft. in.	ft. in.	ft. in.	ft. in.	ft. in.	ft. in.
2	1 1	1 2	1 2	1 2	1 3	1 3
3	2 3	2 4	2 4	2 5	2 5	2 6
4	3 4	3 6	3 7	3 7	3 8	3 9
5	4 6	4 8	4 9	4 10	4 11	5 0
6	5 7	5 10	5 11	6 0	6 2	6 3
7	6 9	7 0	7 1	7 3	7 4	7 6
8	7 10	8 2	8 3	8 5	8 7	8 9
9	9 0	9 4	9 5	9 8	9 10	10 0
10	10 1	10 6	10 7	10 10	11 1	11 3
11	11 3	11 8	11 9	12 1	12 3	12 6
12	12 4	12 10	13 0	13 3	13 6	13 9
13	13 6	14 0	14 2	14 6	14 9	15 0
14	14 7	15 2	15 4	15 8	16 0	16 3
15	15 9	16 4	16 6	16 11	17 2	17 6
16	16 10	17 6	17 9	18 1	18 5	18 9
17	18 0	18 8	18 10	19 4	19 8	20 0
18	19 1	19 10	20 0	20 6	20 11	21 3
19	20 3	21 0	21 2	21 9	22 1	22 6
20	21 4	22 2	22 5	22 11	23 4	23 9
21	22 6	23 4	23 7	24 2	24 7	25 0
22	23 7	24 6	24 9	25 4	25 10	26 3
23	24 9	25 8	25 11	26 7	27 0	27 6

These scales or tables will save him some time and liability to error in the calculation of the number of rafters or joists required in a given space. He must not take for granted that the number shown on the drawings is the correct one.

Where the work is framed, make allowance in the length for tenons, and observe that in most cases the tenon passes right through the timber which is mortised. In this respect carpentry differs from joinery, tenons not being measured in the latter.

Wrought timbers may be either measured at per foot cube, as fir framed, in floors or roofs as the case may be, and the "planing on fir" measured superficially, or may be measured at once, where there is much planing as in exposed roof trusses, as fir wrought and framed; the latter method usually produces a lower price.

Measure all timbers not exceeding 3 inches square at *per foot run*, stating what they are; also up to 4½ by 3½ when in small quantities.

When there is anything special about the fixing of timbers they are better billed in lineal dimensions.

Bill.

ft.	in.			£	s.	d.
520	–	run	3″ × 2¼″ fir framed ceiling joists			
44	–	,,	4½″ × 3½″ fir framed quarters in short lengths and small quantities			
120	–	,,	3¼″ × 2¼″ fir joists in short lengths, framed at each end in stepped galleries			
254	–	,,	3¼″ × 2¼″ ditto, fixed radiating			

Specified sizes of timber (in the absence of any stipulation to the contrary) are held to mean those sizes minus the waste caused by the saw cuts.

Where timbers are specified to be " finished sizes when fixed," no allowance is made in the measurement beyond the specified scantlings, the waste in sawing will be met by a clause in the preamble to the bill of the trade, stating that the sizes are to be " finished sizes," and that no allowance has been made for this in the measurement.

If the timbers are wrought and "finished sizes" allow ¼ of an inch for each wrought face for loss in planing, thus—4 inches by 3 inches wrought all round will appear in the dimensions as 4¼ inches by 3¼ inches, or the timbers may be measured without allowance for waste, and a superficial dimension taken of " wrought face on fir including waste."

In measuring wrought face on a plate, observe to take it on three sides, although only one exposed. Working one face only leaves rough edges.

Keep different kinds of wood, as fir, oak, teak, &c., separate.

Where chamfers, mouldings, rebates, &c., are stopped, state it.

Keep separate any part of the work which is in small quantities or inserted, or both.

Some surveyors measure the timber in floors and partitions by the superficial foot of surface, using the tables in the published price-books, which state the number of cubic feet in a square—a practice not to be commended.

MEASURE AT PER FOOT CUBE.

FIR IN PLATES.—Add to the length 6 inches for lap when 20 feet long or over. Add 12" to the length at angles.

Dragon ties would be billed with the fir framed in roofs. Measure with the plates, and take bolts at the same time.

Measure circular plates per foot run, and state the radius of the curve.

FIR IN LINTELS.—If framed, keep separate; in default of instructions allow 9 inches longer than width of internal opening.

WOOD BRICKS.—Take these as a running dimension; as they are usually all of the same scantling, they can be cubed after they are collected in the abstract. In the absence of instructions, assume that they are 2 feet apart.

Bill.

	ft.	in.				£	s.	d.
	110	—	cube	Fir in plates, lintels and wood bricks	..			

FIR IN GROUND JOISTS AND SLEEPERS, kept separate from other floor joists.

Bill.

	ft.	in.				£	s.	d.
	421	—	cube	Fir in ground joists				

FIR FRAMED IN FLOORS.—All floors are taken as framed except ground floors. Take trimmers and deduct joists to voids. Collect the joists of the various scantlings.

The different lengths of joists in a room, in consequence of projection of chimney breasts, are better measured as they present themselves instead of deducting for chimney breasts. (See COLLECTION OF JOISTS, Chapter XII.)

Bill.

	ft.	in.				£	s.	d.
	522	—	cube	Fir framed in floors				

FIR IN GIRDERS.—State if sawn down and reversed, if bolted, if fitted to iron. Take flitches, bolts, and all labours upon the girder at the same time.

Ordinary girders and binders (i.e. not sawn down and bolted) would be billed with the item "fir framed in floors."

Bill.

ft.	in.			£	s.	d.
73	–	cube	Fir framed in girders, sawn down, reversed and bolted			
25	–	,,	Ditto fitted to iron and bolted			

Measure all timbers 2 inches thick and under at *per foot superficial.*

Bill.

ft.	in.			£	s.	d.
70	–	supl.	1¼" fir framed ridge			
44	–	,,	2" ditto hip			

For scarfings to purlins, &c., where the scarfing is not over a principal or other bearer, allow a length equal to four times the depth of the timber every 20 feet; take the necessary bolts or straps to the scarfings and let an item appear in bill.

Bill.

				£	s.	d.
	No.	20	Extra labour on 8" × 5" purlin for scarf 2' 8" long, including four ¾" wrought-iron bolts 10" long, with heads, nuts, washers and fixing			

All timbers over 25 feet in length, or more than 15 inches in depth, should be kept separate.

Bill.

ft.	in.			£	s.	d.
124	–	cube	Fir in joists 9" deep, and between 27 and 28 feet in length			

Keep separate items of fir fitted to iron, as to iron joists or stanchions.

Bill.

ft.	in.			£	s.	d.
115	–	run	4½" × 3" fir plate, fitted and bolted to iron			

Ironwork in straps and bolts should be measured with the timbers to which it is attached; it is either described as "ironwork in straps and bolts, including perforations and fixing by carpenter," or

CARPENTER.

as "straps and bolts," and the fixing made a separate item in the Carpenter's bill. The former way is the better one. (See preamble Smith and Founder, Chapter IV., section BILLING.)

Bill.

Fixing only, Ironwork.

No.	40	⅞" bolts, with heads, nuts and washers, 11" long (averaged)	£	s.	d.
"	25	¾" ditto 15" long			
"	80	Straps average 3 feet long			
"	20	King heads average weight 64 lbs. each			
"	4	Iron flitches 24' 3" long (averaged), and 12" deep			

Collect joists, rafters, plates, &c., as much as possible, as it saves much labour in the squaring and abstracting.

An alternative to collecting on the dimensions when there are timbers of similar size, is to write them lineally and leave them to be squared on the abstract, as in the case of wall plates, which may be written in the dimensions in either of the following ways:—

	100 0			Fir plates.
	4½		or,	
	3			4½" × 3" fir plates.
	100 0			

FIR FRAMED IN ROOFS, as Purlins, Ridges, Valleys, Hips, Dragon pieces, Rafters, Pole-plates, Angle-ties, all collected into one cubic item of bill.

Observe that in the case of a hipped roof, the rafters may be collected from the extreme length A to B, the ends of roof being disregarded, except that one rafter shall be added for centre of each end, as a roof,

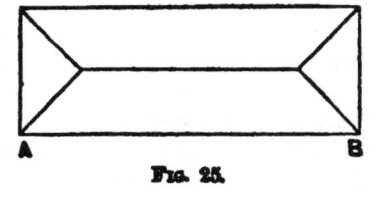

FIG. 25.

however hipped, has the same surface as a gabled roof, provided that the slopes are similar.

Bill.

| ft. | in. | | | £ | s. | d. |
| 644 | — | cube | Fir framed in roofs | | | |

QUANTITY SURVEYING.

FIR-FRAMED IN ROOF TRUSSES.—Keep separate. Take the king or queen posts at their largest scantling. It is a practice with some surveyors to deduct one shoulder from the king posts and half a shoulder from the queen posts; but this should not be done unless the piece cut out is as much as 3 inches wide and as much as 3 feet in length, and should be measured 3 inches short of the length between the shoulders. The majority of surveyors, and rightly, deduct nothing.

Bill.

	ft.	in.			£	s.	d.
	212	–	cube	Fir framed in roof trusses (including hoisting), or the words in parentheses may be omitted and a separate item given for hoisting (see items numbered)			

CEILING JOISTS.—STRETCHERS AND HANGERS.—If small, the rule as to running them would apply. (See introduction to this trade.)

Bill.

	ft.	in.			£	s.	d.
	110	–	cube	Fir framed ceiling joists			

FIR FRAMED IN QUARTER PARTITIONS.—Quarters, heads and sills; posts, interties, braces, &c. (See example for collection, Chapter XII.) If trussed, state it, and keep separate. Measure the nogging pieces at *per foot run*, stating size. *Number* the bearers or bridging pieces to support sill, and give their scantling and length, and describe them as framed between joists. (These are taken in cases where the sill runs parallel with joists, and would be not more than 2 feet 6 inches apart.)

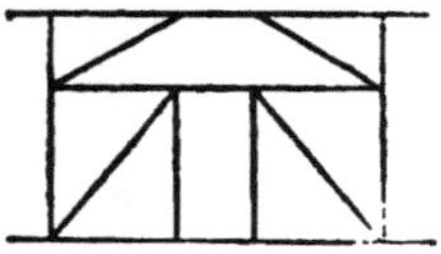

In default of special directions for brick-nogged partitions, take the quarters 3 feet apart, the nogging pieces 2 feet apart.

An outline sketch, on the margin of the dimensions, of the arrangement of timbers in trussed partitions is useful; single lines with the scantling of each timber marked, are sufficient, Fig. 26.

FIG. 26.

Bill.

	ft.	in.			£	s.	d.
	722	–	cube	Fir framed in quarter partitions..			
	75	–	,,	Ditto trussed			
	110	–	run	4" × 2" nogging pieces			
		No.	52	4" × 3" fir bridging pieces about 15" long, framed at each end between joists ..			

Some surveyors measure brick-nogged partitions, brickwork and timber together.

Bill.

ft.	in.			£	s.	d.
120	–	supl.	4½" brick-nogged partition, including timber and nogging pieces, the timbers measured in			

The alternative is to bill the timber with the fir framed in quarter partitions, and to measure the brickwork separately.

HALF TIMBERING.—Measure the length, including the tenons; when diminished, measure the extreme width; the circular struts (if any) may be treated as described for circular ribs (see p. 163).

Measure separately labour to mouldings, sunk edges, etc.

Bill.

ft.	in.			£	s.	d.
533	–	cube	Selected clean fir timber wrought on all exposed faces and grooved where required to receive plastering, framed to detail drawings and put together in white-lead, and with oak pins to half timbered portions of upper part of building, the scantlings measured net, and allow all waste for planing			

Sometimes it may be advisable to bill such timbers lineally in some such form as follows, *as a separate section*.

Bill.

ft.	in.			£	s.	d.
			Best well-seasoned English oak, sawn, framed together in white lead, and pinned together with stout projecting oak pins			
100	–	supl.	Planing, including waste			
129	–	run	6" × 3¼" grooved one edge			
27	–	,,	6" × 3¼" ditto built into brickwork, including all cutting			
219	–	,,	6" × 3¼" ditto grooved both edges			
53	–	,,	6" × 3¼" ditto ditto, built into brickwork, including all cutting..			
93	–	,,	9" × 4" grooved one edge			
20	–	,,	9" × 4" ditto built into brickwork, including all cutting			

ft.	in.				£	s.	d.
113	–	run	Sawn sunk edge to 4" oak				
40	–	„	Labour moulding 9" girth				
	No.	100	Fair notchings, 2" wide, to form battlements, in moulding 6" girth				
	„	10	Moulded stops				

MEASURE AT PER SQUARE OF 100 FEET SUPERFICIAL.

BOARDING TO ROOFS.—State the thickness, if edges shot, if traversed, if wrought one or both sides, if beaded or V-jointed one side, if grooved, if matched, if tongued with hoop-iron. If in small quantities state it, and keep separate. If boarding is laid on felt keep it separate. If to circular roofs or cupolas, keep it separate, and describe it as in narrow widths. To conical roofs, state the radius of the curve at the eaves. To very small curves it will require saw-kerfing, and sometimes to be as thin as half an inch.

Allow for cuttings 3 inches by the length on each side of all hips and valleys, to irregular angles and around all deductions the length by 3 inches.

Measure splayed edge at the junction of the boarding with hips, valleys and ridges.

Bill.

sqr.	ft.	in.			£	s.	d.
40	10	–	supl.	1" rough boarding, edges shot, for roofs ..			
10	5	–	„	1" ditto wrought one side			
4	5	–	„	1¼" wrought one side, grooved, tongued and V-jointed boarding to roofs			
4	30	–	„	1¼" wrought one side, grooved and tongued with iron tongues, boarding to roofs ..			
5	20	–	„	¾" rough boarding in 4½" widths to conical roof 7 feet radius at eaves			
	150	–	run	Splayed edge to 1" boarding			

BOARDING TO ROOF CHEEKS AND DORMER CHEEKS.—Measure net and so describe it, and say what it is for. In the case of roof cheeks it is often nailed to a partition, part of which is inside the building and which would be measured with the general quarter partitions. To dormer cheeks the quarters may be included with the description.

CARPENTER.

Bill.

	ft.	in.			£	s.	d.
	100	–	supl.	1" rough boarding, traversed to receive lead, in triangular shapes and small quantities, measured net as roof cheeks			
	100	–	,,	1" ditto and small fir framed quarters, as dormer cheeks			

BATTENING FOR SLATES.—State the size of battens and for what slates they are spaced. Allow for waste as directed for roof boarding. If for slating on walls, describe them as plugged.

Bill.

sqrs.	ft.	in.			£	s.	d.
42	35	–	supl.	2¼" × ⅞" deal battens spaced for countess slating			
40	–	–	,,	2¼" × ⅞" ditto for vertical slating, and plugging to brickwork			

BOARDING TO FLATS, as for roofs, but state that it is to flats and that it includes firrings.

If the firrings are more than 3 inches deep, call them deep firrings, or state the depth of firrings or take them separately as a superficial dimension and state their average depth.

If in small quantities to very small flats or to tops of dormers, keep it separate and say what it is for.

Bill.

sqrs.	ft.	in.			£	s.	d.
20	15	–	supl.	1" rough boarding to flats traversed for lead and firrings to falls			
10	–	–	,,	1" ditto, and firrings, average 4" deep ..			
10	–	–	,,	Firrings to flat, average 5" deep of 2¼" fir..			
	100	–	,,	1" rough boarding to small flat, traversed for lead..			

CENTERING.

To VAULTS *at per square of* 100 *feet.*—State that it includes horsing, where it is possible to strut it up from floor, and state height of story. If the vaulting is groined keep the centering separate, and state that it is to groined vaulting.

Measure at per foot run. EXTRA FOR GROIN POINT to centering. EXTRA FOR RAKING CUTTING and waste to centering.

Bill.

sqrs.	ft.	in.			£	s.	d.
75	5	–	supl.	Flat boarded centering for concrete floors and horsing 12 feet			
10	10	–	,,	Ditto and horsing 10 feet			
12	55	–	,,	Centering to vault and horsing 7 feet ..			
5	40	–	,,	Ditto groined			
				30 feet run, extra for groin point			
				20 ditto ditto, raking, cutting and waste ..			

To FIRE-PROOF FLOORS *at per square of* 100 *feet.*—State that it includes horsing where it is possible to strut it up from floor, and state height of story or offer the alternative of hanging-irons.

Measure at per foot run raking, cutting and waste.

Measure at per foot run boarded edge to concrete, stating thickness of concrete, necessary at sides of openings for traps, lifts, staircases, etc., unless they are formed by the structural ironwork.

Bill.

sqrs.	ft.	in.			£	s.	d.
100	5	–	supl.	Flat boarded centering for concrete floor, and horsing 11′ or hanging irons			
				100′ run, raking, cutting and waste			
	140	–	run	Boarded edge to concrete floor 9″ thick ..			

To APERTURES *at per foot superficial.*—Where the soffits exceed 9 inches in width, state if the soffits are straight, segmental, semicircular, pointed, or segmental-pointed.

At per foot run where the soffits do not exceed 9 inches in width, stating the kind of curve as before, and these are sometimes described as turning pieces.

In the case of relieving arches to facings, centering will be required if the facing below them is filled in after they are constructed.

Centres to arches of small span (18 inches and under) or to circular openings are usually numbered.

CARPENTER.

Bill.

ft.	in.				£	s.	d.
69	–	supl		Centering segmental soffit			
24	–	,,		Ditto segmental-pointed ditto			
70	–	run		Ditto 4½" flat soffit			
20	–	,,		Ditto 4½" segmental ditto			
25	–	,,		Ditto 9" semicircular soffit			
106	–	,,		Ditto 9" Gothic-pointed ditto			
	No	2		Centres to semicircular openings 18" span, 9" soffit			
	,,	1		Ditto to circular opening 24" diameter, 4½" soffit with four notchings for key-blocks			

Some surveyors keep centering to *gauged* arches separate, as it is not unfrequently close boarded and more expensive than centering constructed of fillets nailed across the ribs, lagged centering as it is called.

Observe that chimney bars will serve for centres to openings of fire-places.

Sometimes centering is unavoidably left in position, in such case it must be so described.

' To TRIMMERS *at per foot superficial.*—State if the centres are left in for lathing to, or number "filleting soffits of trimmers."

Bill.

ft.	in.				£	s.	d.
42	–	supl.		Centering for trimmers			
	No.	5		Filleting soffits of trimmers for lathing to			

NUMBER. CENTRES to openings over 10 feet span, stating width of opening, girth, width of soffit and kind of curve.

Bill.

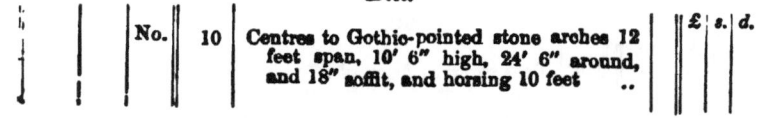

	No.	10	Centres to Gothic-pointed stone arches 12 feet span, 10' 6" high, 24' 6" around, and 18" soffit, and horsing 10 feet ..	£	s.	d.

STRUTTING TO STONE LINTELS. (In windows of several lights take one to each light.) STRUTTING AND RIBBING TO TRACERIED WINDOWS, giving clear width and height, and stating the number of lights or averaging the superficial contents, and describing as "strutting and ribbing to traceried windows . . . . feet superficial each (averaged)," or state the width and height.

Bill.

| | No. | 45 | Strutting to stone lintels .. .. | | £ | s. | d. |
| | „ | 6 | Strutting and ribbing to three-light window, with Gothic-pointed traceried head 5 feet wide, 15' 6" high .. .. | | | | |

EXTRA ON CENTRES FOR NOTCHING FOR KEYSTONES. EXTRA ON ENDS OF CENTRES FOR FITTING SAME TO JAMBS.—Where centres can be re-used, count each instance of taking down and refixing, but it is seldom possible to do this, or take all as new and by a clause in the bill call the contractor's attention to the possibility of re-using some of them.

Bill.

| | No. | 20 | Extra on centres, 9" soffit for notchings for keystones .. .. | | £ | s. | d. |
| | „ | 10 | Ends of centres, 12" soffit fitted to splayed jambs .. .. | | | | |

FENCES, *at per lineal rod of 16½ feet, or per foot run.*—State if deal or oak. Give size of posts (length and scantling). State if butts are charred. State distances apart of posts. Give size of rails; state if square, arris or "cant," and their number. Describe gravel boards. Describe the pales, state whether cleft or sawn, and the total height of the fence. State what parts are wrought or if merely sawn. As pales are generally cleft very thin, if a substantial fence is required their thickness should be stated. Number the digging of the post-holes.

Bill.

ft.	in.				£	s.	d.
100	—	run	Deal wrought fence of 5" × 5" posts, with large butts, and 8 feet apart, two 4" × 3" arris rails, and ¾" × 3" pointed pales 3" apart, nailed with galvanised steel nails, and 5 feet high				
100	—	„	Oak fence of 6" × 6" sawn posts 9 feet apart, with substantial charred butts 30" in ground, three 4" × 3" sawn arris rails, stout cleft pales nailed with galvanised iron nails, two rows of No. 16 B.W.G. galvanised hoop iron, and 1¼" × 9" sawn gravel plank, and 6 feet high in all ..				

CARPENTER.

SOUND BOARDING, state thickness and that the space occupied by the joists is included in the measurement; give size of fillets, and state if single or double fillets. In practice the joists are never deducted.

Bill.

sqrs.	ft.				£	s.	d.
29	50		supl.	¾" sound boarding and stout deal fillets, the joists not deducted			
10	25	—	„	¾" ditto and double fillets			

WALL BATTENING.—State size of battens and distance apart, and if plugged.

Bill.

sqrs.	ft.	in.			£	s.	d.
20	15	—	supl.	2½" × 1" deal battens, spaced 2 feet apart to partitions			
10	5	—	„	2½" × 1" ditto plugged to walls			

When intended to receive boarding, they are frequently included in the description, thus:—

Bill.

sqrs.	ft.	in.			£	s.	d.
10	5	—	supl.	¾" matched and beaded boarding in 7" widths, on and including 2½" × 1" deal battens 2 feet apart, plugged to walls ..			

WEATHER BOARDING.—State thickness, if wrought both sides, if beaded, the width of boards and the lap; measure the net finished surface; and let part of the description be "allow for laps."

Bill.

sqrs.	ft.	in.			£	s.	d.
18	45	—	supl.	1" weather boarding, wrought one side, measured net, and allow for laps			
5	15	—	„	1" ditto beaded one edge			

MEASURE AT PER FOOT SUPERFICIAL.

WROUGHT FACE ON FIR, INCLUDING WASTE, where no allowance has been made for size of timber lost in the planing.—Take planing on fir when an allowance has been made; measure at same time mouldings, rebates, &c., nor exceeding 6 inches girth per foot run over 6 inches per foot superficial.

Bill.

ft.	in.			£	s.	d.
100	–	supl.	Wrought face on fir, including waste ..			
100	–	„	Labour planing on fir			

EAVES BOARDS, LEAR BOARDS, FLASHING BOARDS AND VALLEY BOARDS.—State thickness and if feather edged, state that the valley boards include the necessary splaying and fitting. Deduct an equal superficial quantity of slating battens.

Bill.

ft.	in.			£	s.	d.
76	–	supl.	1″ deal valley boards, splayed and fitted ..			
			or per foot run.			
108	–	run	1″ × 9″ valley boards, splayed and fitted			
74	–	„	1″ × 7″ feather-edged eaves board and fitting			

FELT.—State the kind, describe the nails, and that it is measured net, and that allowance is to be made for laps. Make the same allowances for cutting as on roof boarding.

Bill.

ft.	in.			£	s.	d.
940	–	supl.	Stout inodorous felt to roofs, measured net, and allow for laps and nailing with clout nails			
500	–	„	Boiler felt, 16 oz. per sheet (Oroggon's), to roofs, measured net, and allow for laps and nailing with clout nails			

If boiler felt is required, as this varies very much in quality, state the weight per foot superficial in ounces, or the weight per sheet, stating size of sheet, and give name of patentee.

FIRRINGS TO CUPOLAS, DOMES, ETC.—These are generally nailed to the backs of raking rafters, and the outer edge cut to the contour of the roof. Where each piece is not more than 5 or 6 feet in length, take the running length measured along back of rafter by the *extreme* width; when more than this, divide it into similar lengths. Measure the circular cutting, stating the thickness of the timber.

Bill.

ft.	in.			£	s.	d.
74	–	supl.	2" deal in firrings to ogee roof of cupola, and nailing to backs of rafters			
			108' 0" run, labour circular sawing			

CIRCULAR RIBS TO ROOF TRUSSES.—Measure the full quantity of timber out of which the ribs are cut. Keep them separate. State thickness, if wrought and framed, and if including joints, say so; if not, take the joints separately, and in either case say if the joints are dowelled together. If more than 14 inches deep, keep separate and so describe it.

Ribs of short length in one piece may be numbered.

Bill.

ft.	in.			£	s.	d
154	–	supl.	2½" wrought selected fir, framed in circular ribs			
74	–	„	2½" ditto in fir 18" deep			
108	–	run	Grooved and rebated joint to 2½" rib, and dowelling with oak dowels			
70	–	„	Labour, circular sunk, wrought and twice moulded, 2" girth edge to 2½" fir			
		No. 6	Fir struts, 5 feet long out of 8" × 12", twice chamfered on exposed edges, framed at each end and pinned with oak pins, as sketch			

Or the timber may be billed to include a certain proportion of the labours, thus:—

QUANTITY SURVEYING.

Bill.

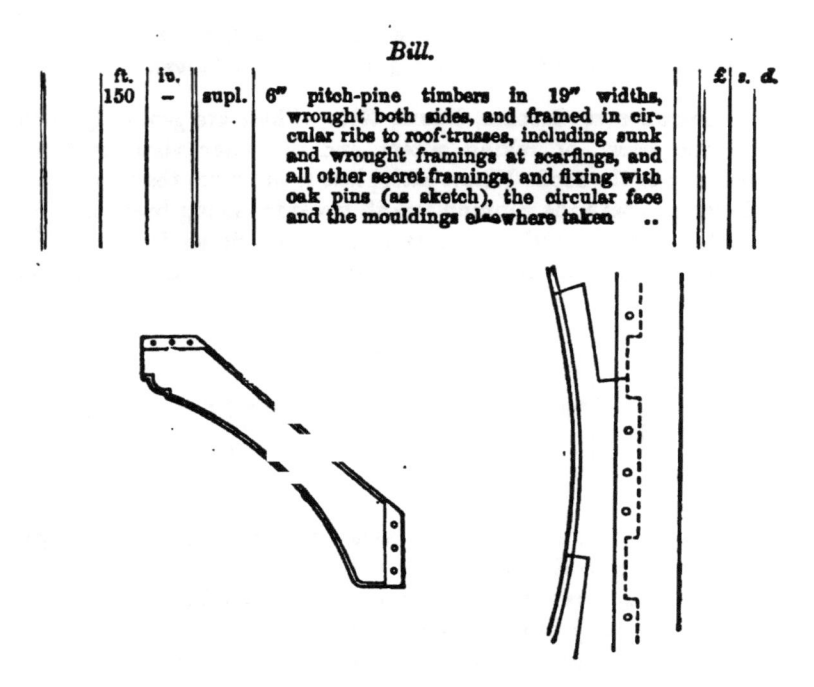

ft.	in.			£	s.	d.
150	–	supl.	6″ pitch-pine timbers in 19″ widths, wrought both sides, and framed in circular ribs to roof-trusses, including sunk and wrought framings at scarfings, and all other secret framings, and fixing with oak pins (as sketch), the circular face and the mouldings elsewhere taken ..			

Ribs in Thicknesses.—Where circular ribs are formed of thicknesses of deal, measure the net superficial area of them, stating total thickness, and describe as ribs of certain thickness bolted together and measured net. Take the cutting to edges separately. Take the bolts (if any) at the same time.

Bill.

ft.	in.			£	s.	d.
90	–	supl.	2″ deal ribs to roof in two thicknesses of 1″ deal, wrought and screwed together, measured net			
			40′ 0″ run, circular sunk and wrought edge to ditto			

Fascias.—State the thickness, and if rough or wrought one or both sides, if beaded, but generally better measured lineal. Number the mitres, fitted ends, etc.

CARPENTER.

Bill.

ft.	in.			£	s.	d.
74	–	supl.	1" deal rough fascia			
178	–	run	1" × 5" rough fascia			
120	–	,,	1" × 6" wrought one side and beaded fascia			
			No. 2 mitres			
			,, 6 ends splayed and fitted			

SOFFITS.—State thickness, and measure the labours separately. If less than 12 inches wide, measure per foot run, and include the labours in the description.

Bill.

ft.	in.			£	s.	d.
100	–	supl.	1" wrought one side cross-tongued soffit ..			
			No. 2 mitres to 18" × 1" soffit			
			,, 5 ditto irregular			
100	–	run	1" × 9" wrought one side soffit, one edge tongued the other scribed to brickwork..			
			No. 5 mitres			
			,, 6 ditto irregular			
			,, 2 ends splayed and fitted			

BRACKETING TO CORNICES.—Measure the girth of the face of the bracketing by the length. Take the run of fillet to attach brackets to, stating size. Number the angle brackets, and describe as "extra for." Keep that which is circular on plan or in small quantities separate. State thickness of the wood used.

Bracketing is sometimes measured at per foot run, stating the girth of cornice for which it is required.

In the absence of details of cornices about two-thirds of the girth of the cornice may be taken for the bracketing, but as a rule none will be required for cornices under 12 inches girth.

Bill.

ft.	in.			£	s.	d.
149	–	run	1¼" deal bracketing for cornices 12" girth			
15	–	,,	1¼" ditto, circular..			
			No. 6 extra for angle brackets			
75	–	supl.	1" deal bracketing for cornices			
			No. 4 extra for angle brackets 15" girth ..			

CRADLING, and state what to, as entablatures, and if fitted to iron.

Bill.

	ft.	in.			£	s.	d.
	93	–	supl.	Deal cradling fitted to iron girders			

FILLETING AND COUNTERLATHING TO PARTITIONS, including the quarters in the measurement, and stating that such has been done.

Bill.

	ft.	in.			£	s.	d.
	226	–	supl.	Filleting and counterlathing to quarter partitions			
	54	–	,,	Ditto fixed between quarters, the quarters measured in			

GUTTER BOARDS AND BEARERS.—State thickness of boarding and whether bearers are framed.

Where gutters do not exceed 6 inches wide take them at per foot run, stating the average width.

Number. "EXTRA TO REBATED DRIPS IN GUTTER," or "SHORT LENGTHS OF REBATED DRIP," SHORT LENGTHS OF ROLL. The two latter may be so described up to a length of 30 inches, over that length, measure at per foot run. CESSPOOLS, stating thickness of deal and length, width and depth, all in clear. GUSSET PIECES, stating thickness and size.

Where parapet gutters occur take rough fillet (giving size) and plugging to wall to receive ends of bearers.

Bill.

	ft.	in.			£	s.	d.
	110	–	supl.	1¼" gutter boards and framed bearers ..			
	84	–	run	1¼" ditto, average 6" wide			
	42	–	,,	1¼" cross rebated drip			
	140	–	,,	2" deal roll for lead			
		No.	10	Short lengths 1¼" cross-rebated drip			
		,,	22	Ditto 2" deal roll for lead			
		,,	5	1¼" deal dovetailed cesspools 10"×10"×6" all in clear, holed and fitted			
		,,	8	1" gusset pieces 9" × 9"			

SNOW BOARDS.—State size of battens and bearers, and the distance apart of both, and that they are to be made movable, and in . . . feet lengths. Where the gutters are under a foot wide *at per foot run*, stating average width, and describing as before.

Bill.

ft.	in.			£	s.	d.
175	–	supl.	Deal wrought snow boards, of 1¼″ × 2¼″ laths, about 1″ apart, on and including 3¼″ × 2″ out bearers about 4 feet apart, hollowed for passage of water, all screwed together and fitted in short lengths to remove..			

MEASURE AT PER FOOT RUN.

The CIRCULAR SUNK EDGE TO RIBS, CHAMFER, REBATE, GROOVE, stating girth or width.

VALLEY FILLET, TILTING FILLET, EAVES FILLET, FEATHER-EDGED SPRINGER, DRIP TO FLATS, giving size in each case.

LABOUR TO MOULDINGS, *per foot run.*—State girth, number the stops. Distinguish between circular or straight. Measure mouldings over 6 inches girth per foot superficial.

MOULDINGS.—State girth of moulding, and *what out of*, as 4 inches by 5 inches; or a clause in preamble of bill may state that all mouldings shall be finished sizes.

HIP AND RIDGE ROLLS.—State size, if birdsmouthed, if spiked on. Take two rough chamfers to ridge and hips if roll is birdsmouthed. Number mitres and intersections.

STRUTTING.—State if herring-bone and give size, or if solid give thickness; state depth of joists, and that the joists are measured in the length. In practice they always are measured in.

FEATHER-EDGED SPRINGER to trimmers, state size.

Bill.

ft.	in.			£	s.	d.
124	–	run	Labour circular sunk and wrought edge to 3" fir			
72	–	,,	Ditto and twice chamfered 1" wide			
104	–	,,	Ditto groove			
720	–	,,	Ditto chamfer 1¼" wide			
79	–	,,	Ditto circular..			
20	–	,,	Ditto chamfer 1½" wide, stopped 4 moulded stops			
10	–	,,	Ditto rebate, 2¼" girth..			
24	–	,,	Ditto stopped, including stops			
104	–	,,	Tilting fillet			
95	–	,,	8" × 2" twice splayed valley fillet			
116	–	,,	4" × 2½" splayed eaves fillet			
30	–	,,	4½" × 3" feather-edged springer			
25	–	,,	4" × 3" cornice, moulded 6" girth finished			
			or,			
25	–	,,	4½" × 3¼" cornice, moulded 6" girth.. .. No. 4 mitres .. .. ,, 1 end splayed and fitted			
43	–	,,	2" deal roll for lead spiked			
73	–	,,	2" ditto birdsmouthed .. .. No. 4 mitres .. .. ,, 2 three-way intersections .. .. ,, 2 four-way ditto			
108	–	,,	2" herring-bone strutting, accurately fitted and nailed to 9" joists (the joists measured in)			
140	–	,,	2" ditto to 11" joists			

PADS OR PALLETTES, *as fixing for joinery.*—State length, width and thickness, and if creosoted.

Bill.

				£	s.	d.
No.	150	¾" × 4½" × 9" deal creosoted pads and building in as wood bricks				

NUMBER.

SPROCKETS.—State what size two are cut out of, or the length, breadth and thickness of each.

Bill.

				£	s.	d.
No.	64	Deal wrought sprockets, two out of 2" × 6" × 24", and nailing to rafters				
,,	20	Ditto, but the upper edge curved				

CUT ENDS TO RAFTERS.—State if moulded, and if wrought for what length, and the size of rafters.

Bill.

| | No. | 101 | Ends of 4″ × 2½″ rafters, wrought for a length of 15″, and shaped to detail .. | £ | s. | d. |

EXTRA LABOUR TO SCARFINGS, stating size of timber, and include the bolts, stating their number and size, billing all together.

Bill.

| | No. | 10 | Extra labour on 8″ × 5″ purlins for scarfings about 32″ long, including four ¾″ bolts 9″ long, and fixing .. .. | £ | s. | d. |

HIP KNOB.—State size, and how many rafters are framed into each, if firred out or turned, describe it.

Bill.

| | No. | 2 | 5″ × 5″ hip knobs 5 feet long, framed to ridge, and four hips turned for a length of 2 feet, firred out at base for a height of about 2 feet, and prepared to receive lead .. .. | £ | s. | d. |

CLEATS.—State size and if wrought.

Bill.

| | No. | 20 | Fir wrought cleats for purlins 12″ × 6″ × 6″, and spiking .. .. | £ | s. | d. |

HOISTING AND FIXING ROOF TRUSSES.—State width and height of truss, and distance from ground to ridge when fixed. (Some surveyors only take this item to very large trusses.) Bell cots and flèches are often hoisted after they are framed, and should be numbered in a similar manner.

Bill.

	No.				£ s. d.
		10	Hoisting and fixing roof trusses 22' × 12', 30 feet from ground to ridge		
	"	1	Hoisting and fixing flèche 10' × 10' × 18', 59 feet from ground to finial		
	"	1	Allow for extra framing, hoisting and fixing to spire roof 8' 6" × 8' 6" and 20 feet high, the apex 65 feet from ground ..		

JOINER.

Keep the different kinds of wood separate.

Describe all work as cross-tongued where over 9 inches in width, if in deal or pitch pine; if in American walnut, 12 inches.

Where the work is in small quantities, or short lengths, or circular, keep it separate, and observe generally that very small things where possible should rather be numbered than measured by the foot, superficial or lineal.

Circular work to be described as "flat sweep" when the rise is ½ an inch to a foot of chord line; quick sweep when more than that. Where in circular cupboard fronts, wreathed strings, etc., state that the work is to include cylinder.

The custom of describing the radius instead of using the terms quick or flat sweep is becoming very general, and is perhaps to be preferred.

Observe the distinction between circular work of various kinds, as curved work bent in fixing, curved work in cylinders, work curved on plan, as to ribs, curved work glued up in thicknesses, etc.

Circular fillets may be either bent circular, cut circular one edge, or cut circular both edges.

If work is screwed instead of nailed, if secretly nailed, if fixed with brass screws and cups, state it.

If work is to be of selected deal and to be stained, keep it separate from the parts painted.

Where work is unusually constructed and cannot be quite intelligibly described, give a sketch.

In collecting dimensions on waste, half inches will be used in widths only, but in lengths and in transferring the resulting length to the dimension column call any part of an inch 1 inch.

Backings are sometimes included with the article measured, state in all cases if framed or dovetailed.

Take plugging at per foot run to the edge of all joinery next brickwork, where there are no grounds or other attachment, and add the word plugged to the description of any other joinery adjoining brickwork which is fixed by plugging, but observe where wood bricks have been taken for fixing of frames, &c., the plugging will not be necessary.

Notching should be measured to any work notched to fit a projection, the thickness of the wood and the girth of the notching to be stated; notchings to materials of same thickness may be averaged.

Where work is of irregular shape measure it net (that is, the exact superficial quantity), and take scribing at per foot run to the irregular sides, stating the thickness of the wood. Scribing is sometimes claimed unjustly, as for edges of shelves against plastering. Where the plan is not irregular this should not be allowed, as the face of plastering, if floated, should be sufficiently near to a true plane to obviate the necessity of scribing; but if the plaster is not floated scribing may be allowed.

Wrought both sides includes wrought edges, wrought one side includes one edge wrought.

Observe to use the words tongued or rebated in their proper places; a familiar instance of their misapplication is in the case of window boards and window nosings which are rebated thus, Fig. 27, and are frequently described as tongued. The meaning of the word tongued is, however, generally received as a rebated edge and a groove to receive it.

Fig. 27.

Let the words "splayed edge" always be used when the whole thickness is dealt with, "chamfered edge" when part only. Thus :—

Fig. 28.

Where joinery is described to be of "finished sizes," no allowance is usually made in the measurement, but it should be stated in a clause of the preamble of the bill; it is, however, better, in addition to the statement in preamble, to give in the bill the actual thickness of the work as $1\frac{1}{4}$ inch floor when reputed $1\frac{1}{2}$ inch floor is intended, &c.

Joiners' work is generally described as of the thickness of the stuff it is produced from. A 1½-inch door will measure about 1‑7/16 inch.

Deal framing wrought one side will measure ⅛th less than the reputed thickness, wrought both sides 3/16ths. Wainscot, wrought both sides, ¼ to ½ an inch less.

In some cases, where the sizes generally are not intended to be finished sizes, an architect produces details of work which must of necessity be finished of the size shown, or the design would be altered; in such cases the surveyor must state it, as, for instance, "20 feet run, 5 inches by 4 inches (finished) frame, rebated and moulded 3 inches girth," and where much work is moulded to detail finished sizes should be specified. An alternative is to describe the size of the original section of the wood used.

Bill.

ft.	in.			£	s.	d.
20	–	run	5" × 4" (finished) frame rebated and moulded, 3" girth			
			or,			
20	–	„	5¼" × 4¼" frame, rebated and moulded, 3" girth			

The same principle should be applied to mouldings shown by detail, or a clause in the bill may state that all mouldings shall be finished sizes. The sooner this anomaly is got rid of the better; it is to be wished that all architects would describe their work as of finished sizes. Scottish architects nearly always do so.

In measuring running lengths of labours, as to moulding, begin with the words "Labour to," as otherwise the item may be mistaken by the abstractor for an item of labour and material.

In describing joinery, square framed is always held to mean square framed both sides, but in all other cases, as moulded, bead, flush, etc., it is better to describe as "moulded both sides," bead-flush both sides, moulded and square, bead-flush and square, etc.

Some country builders erroneously argue that square framed only indicates the rectangular shape of the piece of framing, not its section.

Where the panels of framings are very small, or of irregular shape, it should be stated, and the average superficial contents given: thus, "1¼-inch moulded and square bath enclosure one panel high, the panels averaging 12 inches superficial each."

All doors and framings under 5 feet in height must be kept separate and described as dwarf, unless from their nature they cannot be otherwise, as window backs.

When framings are rough on one side and square framed, no notice need be taken in the description; if flush framed, state the thickness of the wood in the panels.

In measuring finished framings nothing extra is allowed for the tenons, but in cases where a frame or similar work is measured in separate parts the tenons must be measured in the length.

When a piece of work has various labours on it, mouldings, rebates, &c., it should either be run, including all the labours in the description, or measured superficial, omitting from the description the labours in question, and measuring them per foot run separately.

Take the ironmongery with the joinery to which it belongs.

French polishing is not unfrequently included in the joiner's bill, with the description of the work intended to be polished.

In describing mouldings to doors or framings, be careful to state if they are on solid, and if they are stopped.

When an edge is chamfered, state whether once or twice chamfered, and in such things as architraves whether once or twice moulded, as Fig. 29.

FIG. 29.

If any mouldings are to detail state it, otherwise machine-made mouldings are intended. If mouldings have unusually small members state it.

Small mouldings up to 3-inch girth may have the mitres included with the description, and if the purpose of the moulding is stated, the item can easily be priced, but if there are many more mitres than are usual for such a purpose, separate them.

Short returns of moulding not exceeding 6 inches long should be described with their mitres, thus :—

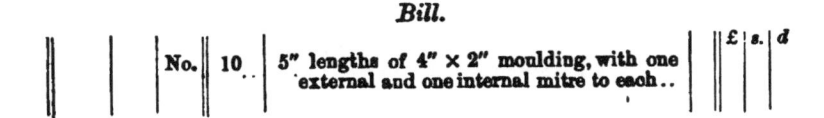

Bill.

| | | No. | 10 | 5" lengths of 4" × 2" moulding, with one external and one internal mitre to each.. | | £ | s. | d |

Similar work, like neckings to pilasters, etc., may be treated thus :—

Bill.

| | No. | 5 | 9″ lengths of 2″ × 1½″ moulding, with two 2¼″ returns, two external mitres and two internal mitres, and rebated on back-edge and groove cross-grain as neckings to pilasters .. .. | £ | s. | d. |

Sometimes much labour may be saved by averaging the sizes of doors or windows.

When a piece of joinery is of small size and with much labour on it, the surveyor will usually find it best to measure it at per foot run, or to number it; as if he measure it at per foot superficial, the extra labour will make a large number of items.

In describing labour to edges of joinery, state if any part is "cross grain," and keep it separate; it is sometimes expedient, when the labour is of small value, to keep it all together and describe as "part cross grain."

FITTED ENDS.—When a piece of wood is sawn off to a required length, the edge will usually be ragged, and must be planed. This is described as a fitted end. The planing is necessary when it stops against a wrought wooden surface, as nicety of junction is necessary, and it often happens, besides, that the plane of such surface is not truly horizontal.

NOTCHED AND FITTED ENDS would involve a notching in addition.

ENDS FITTED TO MOULDING are similar ends which intersect with a moulding.

Fitted ends are always written short, following the lineal item upon which they are extra.

At other times they are incorporated with the lineal item, thus :—

Bill.

| ft. | in. | | | £ | s. | d. |
| 1840 | 0 | run | 1¼″ × 2¼″ moulded architrave, including mitres and fitted ends .. .. | | | |

Judgment must be exercised as to whether the measurement of work shall be lineal or superficial; very often a lineal quantity is easier to price than a superficial one; for instance, an item of

JOINER.

"100 feet 0 inches supl. of 1 inch lining, tongued at angles, rebated one edge and staff beaded the other," would not afford much information as to the quantity of rebate or staff bead; the absence of the words *cross-tongued* would show that it did not exceed 9 inches in width, but if those words were inserted the item might be of any width; on the other hand, "100 feet run, 1 inch × 8 inches lining rebated one edge and staff beaded the other, and tongued at angles," is definite enough. However, when labours are of the ordinary character, like rebates or beads, the question is not important, but when the work is more expensive it is worth consideration, and the item should either be run and all its labours described with it, or the material measured superficial and the labours separately measured. There is never any objection to measuring lineally anything 9 inches wide and under, after that it comes into the category cross-tongued, and as a general rule should be measured superficially.

Deal intended for staining should be kept separate and described as "selected deal."

FLOORS, *per square of* 100 *feet superficial.*—State the quality and thickness, and whether deal or batten. If of unusual widths state it, whether rough, edges shot, folding, straight joint, splayed headings, tongued headings; if dowelled, if grooved, if side nailed, if tongued, and the kind of tongues, if traversed and cleaned off at completion, if covered with sawdust after laying.

Take separately all flooring in recesses and doorways, and describe as "in small quantities and bearers."

Allow 3 inches by the length for all raking cutting, or measure a running length of cutting and waste.

Deduct chimney breasts, projections and voids, but not hearths.

Number fittings to circular corners, columns, etc.

Bill.

sqrs.	ft.	in.			£	s.	d.
10	15	–	supl.	¾" rough sub-floor			
60	25	–	,,	1¼" yellow batten floor, laid straight joint, with splayed headings and mitred borders to hearths (hearths not deducted), punched, puttied, traversed and cleaned off at completion			
10	5	–	,,	1¼" ditto, grooved and tongued with 1¼" galvanised hoop iron			
	70	–	,,	1¼" ditto, in small quantities, and bearers			

SINKINGS FOR MATS.—Measure a running length of "extra on 1¼-inch floor for glued and mitred border; and 1 inch by 1½-inch wrought fillet rebated both edges and two grooves in floor," or as the case may be.

When the sinking is in a tile floor, the border will be of iron or slate.

Bill.

ft.	in.			£	s.	d.
16	–	run	Extra on 1¼" floor for glued and mitred border ..			
16	–	,,	1" × 2" wrought fillet and mitres			
16	–	,,	¼" × 2" wrought-iron border to mat sinking ..			
			No. 4 forged angles			

Measure at per foot run—

SCRIBING EDGE OF FLOOR TO STONE PAVING where it adjoins the wooden floor.

SCRIBING FLOOR to brickwork where actually scribed.

MITRED BORDER TO HEARTHS if the slabs are deducted, but this is rarely done. A common plan is not to deduct the slab, the excess of floor being considered sufficient set-off to the mitred border; however, the hearth may be deducted, and the mitred border measured.

Bill.

ft.	in.			£	s.	d.
24	–	run	Labour scribing 1¼" floor to edge of stone paving			
20	–	,,	Extra on 1¼" floor for glued and mitred border to hearths			

WOOD BLOCK FLOORS, *measured per yard superficial.*—The patent kinds are generally treated as a provision, and they should be laid by the patenteé; observe that cutting all around walls is charged extra.

For wood block floors of the ordinary kind state what wood, the size of the blocks, how laid, as herring-bone, etc., whether bordered, whether traversed and cleaned off, whether blocks are gauged.

A cement floated face, to receive the paving, should always be taken.

Bill.

yds.	ft.	in.			£	s.	d.
84	–	–	supl.	Wood block floor of 18″ × 3″ × 3″ pitch-pine strips wrought all round and at ends, and carefully gauged, the lower half dipped in hot Stockholm tar, laid herring-bone, traversed and cleaned off at completion, grouted with fine dry Portland cement brushed into the joints, left clean and free from grit, measured net, including all cuttings, against walls			

PARQUET FLOORS.—For these, a sum of money is usually provided, or they may be measured and the price per foot superficial stated. A sub-floor must be measured to receive the parquetry.

In the case of very good parquetry borders, the parquetry is sometimes laid on several thicknesses of well-seasoned pine, the layers crossing each other, the whole glued together, and the joists reduced in depth to receive it.

Bill.

ft.	in.			£	s.	d.
		provide	For 270 feet superficial of parquet floor, and laying and wax polishing 40*l*. P.C. ..			
250	–	supl.	Parquet floor, P.O. 2s. per foot, including laying and allow for all cutting and waste and wax polishing in the best manner			
250	–	,,	Parquet floor P.O. 2s. per foot, exclusive of laying on, and including a backing of three thicknesses of ¼″ thoroughly seasoned pine in narrow widths crossing each other and glued together, and allow for all cutting and waste			

SKIRTINGS, *at per foot run.*—State the thickness, the height, and if it includes backings and narrow grounds. (These latter are sometimes included in the description.) If it is tongued to floor; and measure similar length of groove in floor in position, part cross grain. If the moulding is of greater size than usual, state its girth. If in two or more pieces, state the size of each and how put together, and state size of moulding and total height. When plinths are planted on (as to window backs) describe them as wrought both sides.

For very ornate skirtings it is better to give a sketch in addition to the description.

Number TONGUED AND MITRED ANGLES, IRREGULAR ditto (only necessary to distinguish these last when a large skirting of several pieces), HOUSINGS (as to architrave bases), FITTED ENDS (as to chimneypieces), RETURNED ENDS, TONGUED AND MITRED ENDS, HEADING JOINTS.

In collecting skirting some surveyors measure the net length as finished; other surveyors measure them across the doorways, and deduct that length with the other deductions related to such doorway. There is some show of reason for this, a doorway is often omitted, and the result of the omission of the whole series of dimensions relating to such doorway will be the restoration of the skirting if it has been measured across the opening.

The running length around the room is sufficient for the quantity, where in very long lengths heading joints occur, but where the length is under 20 feet these are usually disregarded.

Bill.

ft.	in.			£	s.	d.
			All skirtings to include backings			
6	–	run	$\frac{3}{4}$" × 4" square skirting, and mitres			
110	–	”	1" × 7" torus skirting			
			No. 12 mitres			
			„ 10 fitted ends			
			„ 15 housings			
343	–	”	Skirting in three pieces (as sketch), moulding out of 2$\frac{1}{4}$" × 5", surbase 1" × 6", base 1$\frac{1}{4}$" × 5", all rebated and grooved together, the lower edge rebated and let into floor, and 15$\frac{1}{4}$" high in all			
			No. 8 fitted ends			
			„ 43 mitres			
			„ 70 irregular ditto			
			„ 38 housings			

DADOS, *per foot superficial.*—Take the round of the room, minus openings, and at the same time deduct the plastering behind it, the same dimensions serving both purposes. Measure at per foot run, capping, housings tongued and mitred, or rebated and grooved angles (internal), rebated, grooved and moulded angles (external).

JOINER.

Bill.

ft.	in.			£	s.	d.
210	–	supl.	1" deal wrot. one side, V-jointed, grooved and tongued boarding in 4" widths, as dado, fixed to grounds, elsewhere taken			
48	–	run	Labour, housing of 1"			
72	–	„	Labour, rebated and grooved angle to 1" ..			
64	–	„	Labour, rebated, grooved and chamfered angle to 1"..			
52	–	„	3½" × 2" capping, rebated and moulded, 3¼" girth			
			No. 10 mitres..			
			„ 6 housing			
			„ 5 fitted ends..			

GROUNDS, *per foot run.*—State thickness. If under 3 inches in width call them simply narrow grounds, in all other cases state width and thickness, if framed (as behind architraves), splayed, grooved, beaded, or chamfered, plugged, fixed with wall hooks. State if circular on plan or circular both edges, or if skeleton. Where over 3 inches in width they are sometimes measured at *per foot superficial*. In good work grounds should be taken to edges of all work where it comes against the plastering, and framed wherever it is practicable. To openings they are always framed.

Pilasters are often fixed to skeleton grounds; and although much of the surface is void, measure the whole surface superficial including the voids.

Bill.

ft.	in.			£	s.	d.
225	–	run	1" × 3" splayed grounds			
1125	–	„	1" × 3" ditto, plugged			
52	–	„	1" × 3" ditto, fixed with wall hooks.. ..			
137	–	„	1" × 3" twice splayed grounds			
1677	–	„	1" × 3" framed and splayed ditto ..			
212	–	„	1" × 3" ditto, plugged			
10	–	„	1" × 3" ditto, circular both edges			
70	–	supl.	1¼" wrought and framed skeleton grounds to receive pilasters			

PIPE CASING, *per foot run.*—State thickness, and say whether fixed with screws or brass cups and screws. If boxed measure per foot superficial. Pipe casing is often provisional.

Bill.

ft.	in.			£	s.	d.
54	–	run	1¼" wrought, rebated and beaded grounds, and 1" wrought pipe casing, fixed with brass cups and screws to remove			
40	–	supl.	1¼" ditto, boxed			
			No. 4 extra on ditto for small doors hung with and including 2" brass butts and brass knob turnbuckles			

SKYLIGHTS, *per foot superficial.*—State thickness, if chamfered or moulded bar, approximate distance apart of bars, if put together in white-lead, if screwed with brass screws.

Per foot run.—Labour to throat. To splayed edges if splayed, stating thickness of the skylight.

Number the sets of fillets for condensation, state if teak, oak, or deal, if screwed with brass screws, and the average length in each set, or sinkings for the same purpose in bottom rail, stating size and depth.

Bill.

ft.	in.			£	s.	d.
74	–	supl.	2" deal moulded skylight, with moulded bars about 13" apart, throated all round and fixed with brass screws to curbs ..			
	No,	50	Sinkings ¼" deep in bottom rail of skylight for escape of condensed water (as sketch)			
	"	20	Strips of boiler felt 13" long, 5" wide, nailed with copper nails to bottom rails of skylights			
	"	45	Sets of two ¾" × 1" × 6" teak wrought fillets, and nailing to bottom rails of skylights			

SKYLIGHT CURBS, *per foot superficial.*—State thickness, whether dovetailed, put together in white-lead, beaded or grooved.

Bill.

ft.	in.			£	s.	d.
10	0	supl.	2" wrought skylight curb, cross-tongued, staff beaded, grooved and tongued at angles			

JOINER.

BORROWED LIGHTS AND FANLIGHTS, *per foot superficial.*—State thickness, whether moulded or chamfered, whether in single squares, in squares with moulded bars, in small squares, if extra large bars, if with loose beads or mouldings, if with segmental heads measured square. Keep the semicircular ones separate, and state that they are measured square.

Bill.

ft.	in.			£	s.	d.
20	–	supl.	2" moulded fixed casement in single square			
10	–	„	2" ditto, but with segmental head measured square			
15	–	„	2" moulded fixed casement with 1½" moulded bars in small squares			

SASHES AND FRAMES, *per foot superficial.*—To the height of the external opening from top of stone sill to soffit of arch add 3 inches for the height and to the width between the external reveals 9 inches for the width.

Where with semicircular, or pointed heads, keep the part above the springing separate, and state that it is "in semicircular (or pointed) heads to sashes and frames measured square," and state the number of frames; if segmental headed, measure all together and describe as "sashes and frames with segmental heads measured square."

If circular on plan keep separate and state it.

Number any sash and frame not exceeding 12 feet superficial, stating the extreme dimensions of breadth and height.

State the thickness of outside, inside and back linings, and of pulley stiles, and whether the pulley stiles are of a different wood, the size of sill, and whether of oak or teak; whether sunk, double sunk, weathered, throated, check-throated; the thickness of sashes whether moulded or chamfered, if hung with iron or lead weights, if weighted for plate glass, if with parting slips of stout zinc; the description of lines and pulleys, if single or double hung or fixed, if the frames are grooved all round for linings, if with margins at sides or all round, if in single squares, if in small squares (under 1 foot superficial), if the bars are unusually thick give their size, if in Venetian frames state which of the lights are hung and which fixed, and keep these frames separate.

If the sash weights are hinged state it; if specially good sash lines are required, state the maker's name and the number in his trade list.

Measure at per foot run any unusual labours to bottom or meeting rails as rebates or grooves, groove in sill, iron tongue, and state if galvanised, painted, or bedded in white-lead.

Deep beads to sash frames, state lengths (averaged), size and labours; and if the bottom rail of sash is made deeper in consequence, measure, if per foot run, as extra on sash, stating thickness.

If sashes and frames are fitted to stone mullions and transoms, they frequently require a greater width of lining. This may be either measured at per foot run as "extra" on sashes and frames for extra wide linings to mullion or transom, or included in the general description.

Number moulded horns to sashes, if the mouldings and rebates are stopped to form these say so, stating thickness of sash, sash fastenings, sash lifts or hooks. State if lifts are sunk. Brass eyes and plates. Poles and ends ("long arms"), stating length of pole, its diameter, and of what wood, and the kind of end. Sets of lines, pulleys and hooks or cleats, describing their use and the height of top of sash from floor, and whether iron or brass work.

The large windows of public buildings sometimes combine double-hung sashes with casements in one opening; in such a case the various types should be collected and the quantities in detail of one of each kind attached to the joiner's bill, designating each type by letters, as A, B, C, &c. If we adopt the sketch, Fig. 80, as A and assume a quantity the process would be as follows.

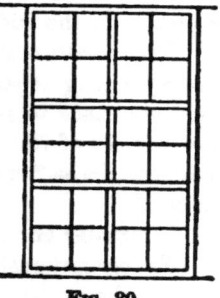

Fig. 30.

In the section sashes and frames of the joiner's bill would appear an item.

ft. in.   £ s. d.
890 0 supl. Sashes and frames as detail A at end of bill ..

At the end of bill would appear—

Detail A.   £ s.

Detail of one window to opening 7' 0" × 11' 6" extreme, in two lights with cased frame, the lower part with double-hung sashes and cased mullion, the upper part with casements hung on centres, with solid transom and mullion, 80' 6" supl.

Here will follow the quantities for one opening in the usual order.

Total divided by 80' 6" supl. .. .. £ ———

Price per foot supl. .. .. ———

JOINER.

The estimator will thus be furnished with a rate per foot superficial to apply to the item 890 feet as above.

Bill.

ft.	in.			£	s.	d.
100	–	supl.	Deal cased frames of 1" inside and outside linings, 1¾" pulley stiles, ½" back linings, with proper beads and parting slips, all rebated and grooved together, and 3¼" teak double-sunk, weathered and check-throated sills, and 2" moulded sashes with large moulded bars in small squares, double hung with Austen's No. 8 patent best superfine quality finely plaited thread lines equal to sample, and Gibbon's (Wolverhampton) No. 20 best quality patent pulleys with solid brass fronts and wheels, steel axles and gun-metal bushes and sides, net price 36s. per dozen, gross price 42s., and iron weights, the frame grooved all around for finishings in 5 complete frames			
100	–	,,	Ditto in two lights, with deal-cased mullion in three frames			
100	–	run	Labour, splay-rebated and grooved bottom rail to 2" sashes			
100	–	,,	Labour, splay-rebated meeting rail to 2" sashes			
100	–	,,	Extra on 2" sash for making bottom rail 6" wide			
18	–	,,	Extra on sash frame for mullion being 7" wide			
		No. 20	Moulded horns to 2" sashes, including stopping, moulding and rebate, so that horns may be the full thickness..			
		,, 10	Extra on bottom inside bead (average 4 feet long) of sash frame for being 8" deep, including tonguing to sill and framing the ends to inside lining			
		,, 1	Sash and frame as before, but 2' 6" × 4' 0" extreme			
		,, 1	1½" ovolo moulded casement hung on butts (elsewhere taken) in and including 4½" × 3" wrought frame rebated, chamfered and beaded, and 3" oak sunk and weathered sill, 2' 9" × 2' 3" extreme dimensions			

An alternative method, when part of the window has sashes hung with lines and weights and part as casement, is to measure the whole sash and frame per foot superficial and the casements as extra thus:—

Bill.

				£	s.	d.
No.	20		Extra on sashes and frames as described for casements 2′ 9″ × 3′ 6″ for being in small squares for being hung on butts to open inwards, including cutting the inside beads of the sash frame and screwing and mitreing them to the casement			

SHOP SASHES, *per foot run.*—Measure the length of bar, rail head, bead, guard-bead, etc. State if they are moulded, how many times rebated, etc., and state the size of each part.

Bill.

ft.	in.			£	s.	d.
114	–	run	$\frac{3}{4}$″ × $\frac{1}{2}$″ bead for glass and mitres, fixed with and including brass cups and screws			
76	–	,,	$1\frac{1}{4}$″ × $\frac{3}{4}$″ guard-bead .. .. No. 6 mitres			
24	–	,,	$2\frac{1}{4}$″ × $1\frac{1}{2}$″ framed sash bar, twice rebated and twice ovolo moulded			
19	–	,,	$2\frac{1}{4}$″ × $2\frac{1}{4}$″ ditto			
22	–	,,	$2\frac{1}{4}$″ × 2″ sash stile and head, framed, rebated and ovolo moulded			
7	–	,,	$2\frac{1}{4}$″ × 3″ bottom rail to sash, framed, grooved, rebated and ovolo moulded ..			

WINDOW LININGS, *per foot superficial.*—State thickness, whether rebated to frames, beaded, tongued at angles, rounded, on splay; if moulded, if framed, and in how many panels the set; if under 6 inches wide measure at per foot run; and state the labour upon them, soffits with splayed ends should be measured to their extreme points, and described as with splayed ends. (*See* also Jamb-Linings.)

Bill.

ft.	in.			£	d.
45	–	supl.	All linings to include backings .. .. .. 1″ window linings, cross-tongued and tongued at angles		
105	–	run	1″ × 9″ ditto, rebated one edge and tongued at angles		
66	–	,,	1″ × 5″ ditto, ditto and ovolo moulded ..		

WINDOW BOARDS AND BEARERS, *per foot superficial.*—State the

JOINER.

thickness, whether rebated, rounded, with moulding tongued under, giving size of moulding, if under 9 inches wide measure at per foot run.

Number the fitted ends, the notched, returned and mitred ends.

Bill.

ft.	in.			£	s.	d.
54	–	supl.	1¼" window board, cross-tongued, rebated and rounded and bearers			
76	–	run	1¼"×7" window board, rebated and rounded, and bearers			
			No. 40 notched, returned and mitred ends			
		„	10 fitted ends			
20	–	„	1¼" × 9" window board, rebated and rounded, and with small moulding rebated and let in beneath			
			No. 6 notched, returned and mitred ends ..			
		„	2 ditto on splay			

WINDOW NOSINGS, *per foot run.*—State width and thickness, and if with moulding tongued under, if rebated to frame, if rounded.

Number the fitted ends, the returned and mitred ends.

In good work separate the window board or window nosing and the moulding beneath.

Bill.

ft.	in.			£	s.	d.
64	–	run	Labour to stopped groove			
64	–	„	1¼" × 1¼" bed-moulding, rebated one edge			
			No. 18 returned mitred ends			
65	–	„	2¼" × 1¼" window nosing, rebated and rounded			
			No. 18 returned mitred ends			

ARCHITRAVES, *per foot run.*—Measure round the outer edge of the architraves for the length, including the plinth (if any) in the measurement of the length. State the size. If in two pieces, state it, and give the size of each piece. If of unusual section give a sketch. Include the mitres in the description, when in deal, but in hard woods take " tongued, mitred and screwed angles."

Number the plinths, and describe as " extra on architrave," stating width, height and thickness, having first measured their height in the length of the architrave. State if moulded or splayed. If the architrave is dowelled or dovetailed and screwed to the base, include it with the description of base.

When a part of an architrave only is to be used, because of limited space, measure as a whole architrave and take scribing in addition, as the whole architrave will have been produced by the machinery.

Bill.

ft.	in.			£	s.	d.
422	–	run	2¾″ × 1¼″ moulded architrave and mitres No. 64 extra on ditto for 2½″ × 1½″ × 9″ moulded bases, dovetailed and screwed..			

SOLID FRAMES AND CASEMENTS.—Measure the frame and transom and oak sill at per foot run, giving size. State if rebated, beaded, staff-beaded, moulded (give girth of moulding). Make the same allowance as in door frames (which see, p. 194). Describe transom and sill as framed. Measure the casements at per foot superficial. State thickness, if moulded or chamfered, if fixed, hung on hinges or centres, hung folding (in which case add an inch to the width for rebate), if in small squares.

Per foot run.—Labour to water hollow around frame or casement. Labour to hook rebate. Water bar, stating size, if galvanised, if bedded in white-lead, if patent. Weather fillet, state size, if rebated, weathered, or moulded.

Number the casement fastenings, the stays, patent water bar (stating whose patent), stating length, and if for folding casements; the pairs of centres, stating if bushed; flush bolts, giving length and width; espagnolette fastenings, stating height of casement.

Casements and frames 12 feet superficial and under should be numbered.

It will sometimes be necessary to measure beads to casements separately; they are best billed in sets, and their lengths may be averaged as follows. State if fixed with and including brass cups and screws.

Bill.

	No.	10	Sets of beads to casements, each containing 9 feet run (averaged) and including mitres	£	s.	d.
	„	5	Sets of cut beads to swing casements, each containing 12 feet run (averaged) and including mitres			

Bill.

ft.	in.			£	s.	d.
220	–	supl.	2" moulded casements in single squares, hung with butts			
64	–	run	Labour to hook rebate to 2" casements ..			
42	–	„	3" × 2" moulded and twice rebated weather fillet (as sketch) and fixing with screws			
712	–	„	5" × 3¾" frame, rebated 3" girth, moulded 2¾" girth, and three times grooved ..			
252	–	„	5" × 4½" mullion framed, twice rebated 3" girth, twice moulded 2¾" girth, and four times grooved			
173	–	„	5" × 4½" transom framed, twice rebated 3" girth, moulded 2¾" girth, splayed 3" wide and four times grooved			
231	–	„	7" × 3" oak sill framed, grooved, check-throated, twice sunk-weathered			
			No. 8 irregular framed angles, with 6" joint screw			
			No. 16 fair ends			
	No.	1	1¾" ovolo moulded casement in single square, hung on butts in and including 4½" × 3" rebated, beaded and chamfered frame, 4½" × 3" oak sunk, and weathered sill 2' 9" × 3' 0" outside dimensions			

Casements and frames are measured over all by same measurers, but it is better to dissect them.

WINDOW BACKS AND ELBOWS, *per foot superficial.*—State thickness, in how many panels, if square framed, moulded, flush-framed, keyed, if on splay. If canvassed and painted at back include in description or measure it at per foot superficial.

Per foot run.—The capping, state width, thickness, if tongued, if rounded, if moulded, or with moulding tongued under, or simply call it beaded capping, if the common beaded capping and housings. Number the fitted ends and housings.

Bill.

ft.	in.			£	s.	d.
66	–	sup.	Canvas and glueing to window backs and painting twice in red-lead			
66	–	„	1¼" framed and moulded window back one panel high			
30	–	run	2" × ¾" beaded capping and mitres			
			No. 10 fitted ends			
			„ 4 housings			

BOXING SHUTTERS, *per foot superficial.*—State the thickness, the number of panels, if moulded, if moulded and square, moulded and bead flush, bead butt, in how many heights hung. Add 1 inch to the width for each of the rebates. Keep the back flaps separate.

State that they include all rebates and splay rebates.

Number the elbow caps, and the clearing pieces or blind rails, stating size of each, the shutter bars, giving length, shutter knobs, shutter latches, pairs of butts, pairs of back flaps, stating size in each case.

SHUTTER BOXINGS, *per foot superficial.*—State thickness, if framed, rebated, beaded, splayed, how many times grooved.

BACK LININGS, *per foot superficial.*—State thickness, if panelled, how many panels in height, if moulded, square framed, bead-butt, bead flush, &c. State if splayed.

RETURN LININGS, *per foot run.*—State thickness, and if rebated or grooved.

Bill.

ft.	in.			£	s.	d.
84	–	supl.	1" bead-butt back linings, three panels high			
135	–	„	1" bead-flush and square back flaps, four panels high in two heights, hung folding, including rebates and splay-rebates			
77	–	„	1¼" moulded and bead-flush front shutters, four panels high, hung folding in two heights, and ditto			
68	–	run	6" × 1" return lining, rebated both edges and tongued at angles			
69	–	„	8" × 1½" boxing grounds framed, splay-rebated and staff-beaded			

	No.	12	10" lengths of 1¼" × 3" blind rail framed and staff-beaded	£	s.	d.
	„	12	1" elbow caps 20" × 12" extreme, both ends splayed, housed on three sides and the front edge rounded			
	„	12	¾" soffits to boxings 20" × 11" extreme, both ends splayed, housed in all round..			

SLIDING SHUTTERS.—The only proper way is to measure these in detail, the shutters at per foot superficial, stating the kind of pulleys, lines and weights, the thickness, and if moulded, bead flush, &c., the pulley stiles, heads, &c., at per foot run, describing the labour upon them; or per foot run deal-cased frame for single (or double) hung shutters, describing the various parts, as linings and pulley stiles, as directed for sash frames.

Some surveyors measure shutter and frame together as they do deal-cased frames and sashes, but this is an inexact method.

Number the flush rings or drop rings, the lifts, pairs of butts, thumb-screws, &c.

Measure the flap at per foot run, stating thickness and width, and if rounded or moulded.

Sometimes the deal-cased frame for the shutter is placed inside a sash frame, and in such case its inside lining would be omitted from the description as unnecessary.

Bill.

ft.	in.			£	s.	d.
10	—	supl.	1¼" three-panel, moulded both sides, shutter hung with and including best flax lines, brass axle pulleys and iron weights ..			
4	—	run	3" × 2" framed head to deal-cased frame..			
14	—	„	Deal-cased frame for single hung shutter of 1" inside, and outside linings 1¼", pulley stiles and ½" back linings, all rebated and grooved together			
4	—	„	3" × 1¼" flap, with rounded nosing and hung with butts..			

LIFTING SHUTTERS, *per foot superficial.*—State thickness, if square framed, moulded, &c. State if rebated (as for shop shutters), and allow for the rebate in the measurement.

Per foot run.—The moulding or fillet, if any, on face of door to receive shutter, giving size and description.

Number the thumb-screws. The stubs and plates. The shutter shoes, state if screw heads are countersunk or if the shoes are patent.

Bill.

ft.	in.			£	s.	d.
8	-	supl.	1½" two-panel, bead flush and square lifting shutter..			
55	-	,,	1¼" bead butt, both sides rebated and beaded, shop shutters in narrow widths			
3	-	run	1⅜" × 1½" wrought fillet, with 2" × ⅜" bead on face, and screwing to door .. No. 2 fair ends			

REVOLVING SHUTTERS.—These are best treated as a provisional sum. The trade lists of the well-known manufacturers show the allowances required beyond the visible measurement for grooves and coil, usually 2 inches more than the "sight" width and 1 foot beyond the "sight" height and the minimum superficial quantity for which they charge (as anything less than 20 feet as 20 feet). Measure the letting in and forming grooves at per foot run.

Measure coil casing at per foot superficial.

Provide a sum for attendance in all trades.

Bill.

ft.	in.			£	s.	d.
100	-	supl.	Self-coiling revolving shutter, P.C. 3s. 6d. per foot supl. at manufactory in one shutter, and allow for carriage, profit, fixing and attendance .. .. or, Provide for 100 feet supl. of self-coiling revolving shutter 17l. 10s. 0d., and allow for carriage, profit, fixing and attendance			

DOORS.—Refer to the dimensions of deductions of brickwork, and for external doorways assume the doors to be of the same size as the deduction of external openings. For internal doorways assume the doors to be 3 inches less in width and height respectively than the deduction.

Keep folding doors separate. Allow 1 inch in the width for the rebate. If the rebates are specially moulded or hook rebated take the extra labour at per foot run. Measure doors with segmental or segmental pointed heads as if square, taking the extreme dimensions, but stating that they are segmental or segmental pointed-headed measured square.

Separate the semicircular heads of doors from the part square, and describe the quantity as in semicircular heads to doors of description as the case may be.

Where both leaves of folding doors are intended to open simultaneously, take "sympathetic hinges." State whether glazed doors have diminished stiles. Where doors are to be covered with cloth state it, and include the cloth in the description. The doors must be described as flush framed.

To doors hung with rising butts, take run of splayed edge to top rail, and in very good work to head lining, a dimension of labour to splay . . . inches wide.

When doors or framings are polished, take to edge of each lock and hinge stile "$\frac{1}{2}$-inch mahogany (or other wood) dovetailed slip to edge of 2$\frac{1}{4}$ inches door and letting in and glueing," to hide the tenons.

Note the difficulty of making doors with curved heads to swing, because liable to catch against soffits of reveals.

Measure separately any mouldings planted in, and take the same length of labour to stopped groove.

DOORS LEDGED, *per foot superficial.*—State if tongued, grooved and beaded, if braced. Almost invariably hung with cross garnet hinges or strap hinges.

An "inch proper ledged door," means a door of 1-inch ledges covered with 1-inch boarding; a "$\frac{3}{4}$-inch proper ledged door," $\frac{3}{4}$-inch boarding and $\frac{3}{4}$-inch ledges, etc.; if anything different to this usage is intended, it must be described.

Bill.

ft.	in.				£	s.	d.
65	–	supl.	1$\frac{1}{2}$" proper ledged door				
42	–	„	1$\frac{1}{4}$" ditto braced				

DOORS FRAMED AND BRACED, *per foot superficial.*—State the *total* thickness, if cross-braced, if stop-chamfered including stops, if *covered* with boarding, if *filled in* with boarding, state thickness of boarding, if in narrow widths, if grooved, tongued, and V jointed or beaded one or both sides, if filled in diagonally. If preferred, the chamfering may be omitted from the description of the door and measured separately.

If the door is filled in with boarding and converted on the inside into a number of panels state it, and give their number and average size. If the boarding and framing are of different wood state it, and in such case describe the doors as of skeleton framing .... inches thick, covered with boarding, giving description and thickness.

Number the fastenings, pairs of hinges, etc.

It is advisable to state the width of stiles and rails of framed and braced doors, either in the preamble of the Joiner's bill or in the item.

Bill.

ft.	in.			£	s.	d.
64	–	supl.	2" framed and braced door, the stiles, top rail and braces of batten width, bottom rail 11" wide, lock rail 9" wide. stop chamfered, filled in with 1" matched and beaded both sides boarding in 4¼" widths..			
84	–	„	2½" framed door in two panels outside, converted on the inside into 8 panels. The bottom rail 11" wide, the lock rail 9" wide, the stiles and other rails of batten width, filled in with 1" matched and beaded both sides boarding in 4¼" widths			

PANELLED DOORS, *per foot superficial.*—State thickness, the number of panels, if square framed, bead flush, bead butt, moulded, bolection moulded, if the latter state if mouldings are rebated to stiles and rails, if mouldings are tongued in or tongued and mitred at angles, state it.

If bolection mouldings are moulded on their outer edge, it may either form a part of the general description or that labour may be separately measured.

If folding state the number of panels "*per set*" doors, not the number in each leaf, and allow in the width for the rebate.

State if " prepared with and including " shifting beads for glass, and if the beads are secured with brass screws or brass screws and cups, or if the doors are prepared to receive sashes. In this latter case measure the sashes in addition and describe them as " fitted to panels of doors." If in small squares, or with extra large bars, state it and give the size of the bars. State if any of the panels are filled-in with and including wire gauze or perforated zinc, and include the beads in the description.

In the absence of a drawing of glazed doors, assume that the top of the middle rail is 3 feet 2 inches from floor.

Sometimes brass cups and screws are taken separately and described as brass cups and screws and driving, stating the length of screw.

Bill.

ft.	in.			£	s.	d.
40	—	supl.	1¼" four-panel square-framed doors			
12	—	"	1¼" square framed dwarf doors in two panels, the set hung folding			
22	—	"	2" three-panel, moulded both sides, door, the upper panels with diminished stiles, rebated and prepared for glass, with and including shifting mouldings screwed with screws and cups			
32	—	"	2¼" 10-panel door, the lower six panels bolection moulded both sides, and with sunk and mitred margins on one side to form raised panels, the upper four panels moulded both sides, rebated and prepared for glass, with and including mitred mouldings screwed with brass cups and screws (elsewhere taken)			

GATES, *per foot superficial.*—State thickness, and describe generally as for framed and braced doors (which see, above).

Per foot run.—The capping. State if beaded or moulded, and if of teak or oak.

Number wicket and describe as "extra for wicket," giving size and description; take the fastenings.

Bill.

ft.	in.			£	s.	d.
60	—	supl.	2¼" framed and braced gates, the bottom and lock rails 11" wide, the other rails, stiles and braces in batten widths, filled in with 1¼" matched and beaded both sides boarding in 4½" widths, and hung folding			
			No. 1 extra for forming and hanging wicket about 2 feet × 4' 6", including rebating the edges of both gate and wicket all round			
10	—	run	4" × 2" capping to gates grooved 3½" girth, moulded both edges, bedded in white-lead and fixed with screws			
			No. 2 fair ends, rebated and splayed . ..			

GATE POSTS.—Stating size and if the butts are charred.
Number the out ends, and if ornamental give sketch.
Measure rebates per foot run, state girth and if stopped.

Bill.

		No.	4	Oak wrought posts 12″ × 12″ and 10 feet total length, with large charred butts 3 feet in ground, and with cut pyramidal tops		£	s	d.

In the case of an ornamental gate it is advisable to give a sketch and state the kind of gate and the number of feet of stile, rail, braces, with sizes and labours that it contains, written short in bill.

DOOR FRAMES, *per foot run.*—State size; if wrought, rebated, chamfered, or beaded, and how many times rebated, chamfered, etc. To the collected length of the sides and width of the door add four times the width of the frame, and 6 inches for the two horns. Where tenoned into sill, allow further 2 inches for each tenon. Where not tenoned, take "iron dowels and mortises in fir and York (or other) stone"; the latter usual for all external doors.

Often a detail drawing of a frame shows the finished size, although finished sizes are not specified; in such case, the size out of which it is obtained should be used as a dimension.

Where frames are wrought, rebated and beaded, they are sometimes (very rarely) measured at per foot cube and described as fir *proper* door frames, but the former method is the most convenient and best.

Where frames are semicircular or pointed headed, measure the frame to 3 inches above the springing as straight, and the girth of the outer edge of the circular part from the springing as circular. If the head is to show segmental outside and square inside, the size out of which the head is obtained should be stated, and it should be kept separate.

Number the heading joints and sets of oak keys and wedges, or handrail screws, and fixing.

Describe circular parts of frame as of 5-inch by 4-inch *finished*, or "out of 9 inch by 5 inch," or as the case may be.

If frames are in two or more pieces, cross-tongued and glued together, state it and give sketch. These are generally of hard wood on a deal core.

Number "extra for irregular framings" to angles which are not square. The term frame covers square angles only, state the scantling of frame.

If oak sills to door frames, measure at per foot run and state size and labours.

Bill.

ft.	in.			£	s.	d
38	–	run	4½" × 3" frame, rebated and twice beaded			
5	–	,,	4½" × 3" ditto, circular to segmental head out of 4½" × 9"			
			No. 2 extra for irregular framings			
4	–	,,	7" × 4½" head framed, rebated and beaded with segmental soffit			
28	–	,,	5" × 4" frame, rebated and ovolo moulded, 1½" girth			
10	–	,,	5" × 4" ditto, circular to semi-head out of 10" × 5"			
		,,	No. 2 joints and 6" joint screws			
40	–		5½" × 4" frame in four pieces, tongued and glued together with oak cross tongues, of yellow deal core 3¾" × 3", two pieces of 3½" × 1¼" wainscot, rebated one edge, and one piece of wainscot 5¾" × 2", rebated and moulded, 1½" girth (as sketch)			

Oak Foot-boards to Loophole Frames *per foot superficial.*—State thickness, if wrought both sides, if cross-tongued.

Number the sets of strap hinges and chains, stating size and weight, and how many bolts.

Number the bow handles, stating size and weight.

Bill.

ft.	in.			£	s.	d
16	–	supl.	3" oak wrot. both sides, and mortise clamped foot-boards, with rounded edges, and hung			

Shelves, *per foot superficial.*—State thickness, whether wrought one or both sides; the description usually includes bearers. Brackets, if of iron, state size or p.c., and describe them as plugged to wall. Wooden gibbet or gallows brackets, state size of the material and the length in each bracket. Brackets of Tee iron and pinning to wall, and the number of screws to each. When over 20 feet in length, take ship-lap joint and allow 2 inches in the length. Similar joint but part cross-grain to intersections of shelves.

Bill.

ft.	in.			£	s.	d.
100	–	supl.	1" shelf wrought both sides and chamfered bearers			
100	–	,,	1¼" ditto cross-tongued and ditto			
5	–	run	Ship-lap joint to 1¼" shelf			
10	–	,,	,, ,, part cross-grain			
	No.	5	Gibbet brackets, wrought, and each containing 6 feet of 2" × 2½" deal, and plugging to wall			
	,,	6	Iron shelf brackets, p.c. 1s. each, fixed with screws and plugging to wall			
	,,	10	Pieces of 2" × 2" × ¼" and ⅜" tee-steel 21" long, each with two screws with countersunk heads, and cutting and pinning to brickwork			

LATTICE SHELVES, *per foot superficial.*—State size of laths, and their distance apart. Measure the bearers separately.

Bill.

ft.	in.			£	s.	d.
100	–	supl.	Deal lattice shelves of 1¼" × 2" laths spaced 1" apart			
50	–	run	3" × 2" framed bearers for shelves			
50	–	,,	3" × 3" ,, ,, ,, ,, ...			

JAMB LININGS, *per foot superficial or per foot run.*—Add to the collected length of two sides and top of the door four times the thickness of the linings and 1 inch for horns for the length. State thickness, if on splay, if single or double rebated, if tongued at angles, if once or twice beaded and moulded, if panelled (and state the number of panels in the set), if tongued on one or both edges. Where in two pieces, it is better to measure the lining at per foot run, describing the pieces. State in all cases that they are to include backings, and if they are dovetailed backings, or the dovetailed backings may be numbered.

JOINER.

Bill.

ft.	in.			£	s.	d
70	–	supl.	1¼" jamb linings, double rebated, twice beaded, tongued at angles			
110	–	„	1¼" ditto, rebated, twice beaded, framed and moulded in eight panels the set, and tongued at angles			
44	–	run	1¼" × 7" ditto, rounded both edges, tongued at angles with and including ½" × 2¼" stop, nailed on to form rebate			
50	–	„	1¼" × 11" ditto, cross-tongued, twice narrow chamfered, double rebated, tongued at angles			

BOARDING TO WALLS AND CEILINGS, *per square superficial.*—State thickness, widths of the battens, how jointed, size of the grounds and their distance apart when on walls. Measure raking, cutting and waste, or circular cutting and waste per foot run.

Take tongued angles at internal angles, and tongued and staff-beaded or chamfered angles at salient angles.

Describe boarding to dado as "In short lengths to dado."

Bill.

sqrs.	ft.	in.			£	s.	d.
10	75	–	supl.	¾" matched and beaded boarding in 4" widths to ceiling			
20	55	–	„	¾" grooved, tongued and V-jointed boarding in 3¼" widths to walls, on and including 3" × ¾" deal grounds about 2 feet apart plugged to walls			
	40	–	run	Tongued angle to ¾" ..			
	40	–	„	Tongued and staff-beaded angle to ¾" ..			
	50	–	„	Raking, cutting and waste to ¾" matched and beaded boarding			
	20	–	„	Circular cutting and waste to ditto			

FRAMINGS, *per foot superficial.*—State thickness, if panelled, if in small or irregular shaped panels, the number of panels in height, if spandril-shaped, in that case measured net and so described, if prepared for glass, or with bars as sash. Deduct the doorways, or if the doors are uniform with the framing, let the measurement include the door, and number them, and describe as "extra for forming four-panel door in same," the item following the framing in the bill and written short; in either case, take the stops, rebates, and ironmongery in addition.

Number extra on framings, describing them, as for instance, to undersides of girders, stating size of girder. Notchings, giving their girth and stating thickness of framing.

Per foot run.—Rebate on framing or on edge of doors. Door stops, stating size and if rounded, twice rounded, chamfered. The mitres to be included in the description.

Fig. 31.

Where quadrant corners occur, measure the framings short of them and measure the corner per foot run: thus, "8 feet run, solid quadrant corner 6-inch girth out of 2¼-inch deal, beaded and grooved, both edges as sketch," Fig. 31, or as the case may be.

Bill.

ft.	in.			£	s.	d.
20	–	supl.	1¼" moulded and square spandril framing, measured net			
108	–	,,	2" moulded both sides partition, three panels high			
75	–	,,	2" moulded and square framing, two panels high, the upper panel with moulded bars in small squares for glass			
			No. 2 extra for forming three-panel door in same			
			,, 2 extra for framing top rail and bars, for passage of 14" × 12" girder			

CUPBOARD FRONTS, *per foot superficial.*—Measure the whole surface for the front; measure the doors and deduct them from the front, the two being usually different; state thickness and description. Call any front less than 5 feet high, dwarf. The ends, state thickness and description.

Bill.

ft.	in.			£	s.	d.
40	–	supl.	1¼" framed and beaded cupboard front ..			
20	–	,,	1¼" ditto in two heights			
20	–	,,	1¼" ditto dwarf			
15	–	,,	1¼" square framed cupboard ends, three panels high			
			Doors as ordinary doors			

W.C. FITTINGS, *per foot superficial.*—Observe that 20 inches is the least width that should be taken for a good w.c. seat. Where seat and riser are plain they may be kept together, where there is a difference between the two keep them separate. State the thickness, and include in the description deal-framed bearers, or the bearers may be separately measured; if the bearers are dovetailed state it.

Seats for pedestal closets are best described from a trade list.

The flap and frame *per foot superficial.* State if mortise clamped, or mortise and mitre clamped, if frame is beaded, moulded, or chamfered. Number "button blocks," "thumb cuttings," etc.

BACK AND ELBOWS, *per foot superficial.*—State thickness and if square-framed or moulded.

Per foot run.—Skirting, giving description. Moulded nosing tongued on (state girth). Rounded edge to flap or seat. Capping to back and elbows, state size and if beaded or moulded. Tongued and mitred angle to elbows. Grounds. State size in all cases.

Number mitres to skirting; holes cut and dished for pan; holes cut and beaded or chamfered for handle. "Extra for seat and riser made to remove easily, with oak button blocks and brass cups and screws" (or otherwise). Joiner attend plumber to w.c. Joiner attend plumber to safe. Paper boxes, giving description.

The ironmongery.

As these items will be under a heading, Mahogany, the word does not appear in the following bill:—

Bill.

ft.	in.			£	s.	d.
21	–	supl.	1¼" w.c. seat and riser, and deal-framed bearers			
13	–	,,	1¼" mortise and mitre-clamped flap, and beaded frame			
7	–	run	Labour, rounded edge to 1¼"			
7	–	,,	2¼" × 2", moulding and tonguing to edge of flap			
28	–	,,	7" × ¾" moulded skirting			
			No. 6 mitres			
			,, 6 shaped ends			
	No.	2	Seat-holes, cut, shaped and dished			
	,,	2	Holes, cut and beaded for handles			
	,,	2	Sets, w.c. fittings made to remove easily, with oak button blocks and brass cups and screws			
	,,	2	Cutting frame of flap and hanging with and including 2" brass butts, and forming small ¾" deal paper-box under seat			
	,,	2	Bolding's (Davies Street, London, W.) No. 131 polished double-wood seat, for pedestal closet and fixing to glazed brickwork			

BATH FITTINGS, *per foot superficial.*—The top: state thickness, and that it is framed. Deduct the opening, and state that it is measured net, or number the top, giving description and stating size. The enclosure, stating thickness, if panelled, if moulded, if in very small panels. The step and riser, describing thickness. State if any part is screwed with brass cups and screws to remove, or count the cups and screws.

Per foot run.—Skirting. Labours to rounded or moulded edges.

Number, extra on framing (of enclosure) for small door for access to cocks. Perforation in top. Deal framed cradle for bath. Joiner attend plumber to bath. Joiner attend plumber to safe. Quadrant angles to rounded edge.

Bill.

ft.	in.			£	s.	d.
13	–	supl.	1¼" bath top, fixed with screws, measured net			
			or,			
	No.	1	1¼" ditto, 7' 0" × 3' 0", with rounded edge and shaped perforation, also with rounded edge, fixed with brass cups and screws to remove			
14	–	supl.	1¼" framed bath enclosure, moulded and square, one panel high			
			No. 1, extra for small door, hung with and including 2" brass butts and brass knob, turnbuckle both door and frame, rebated all round			
7	–	run	Labour rounded edge to 1¼"			
13	–	„	9" × 1" moulded skirting			
			No. 2 tongued and mitred angles			
			„ 2 shaped ends			
	No.	1	Stout deal cradle for bath			
	„	1	Extra for set of bath fittings, fitted to remove easily, with oak button blocks, and brass cups and screws			
	„	1	Labour to perforation for bath in 1¼" top, part semicircular, with small quadrant corners and rounded edge			
	„	1	Joiner attend plumber to bath			
	„	1	Ditto to safe			

CISTERNS AND SINKS, *per foot superficial.*—The bottom, state thickness, and describe as screwed, state if wrought one or both sides. The sides, state thickness, and describe as dovetailed, state if wrought both sides.

JOINER.

Small draw-off sinks may be numbered.

ft.	in.		Bill.	£	s.	d.
25	–	supl.	1¼" sink sides, wrought one side, cross-tongued and secret dovetailed, prepared for lead			
15	–	,,	1½" sink bottom, wrought one side, cross-tongued, screwed and prepared for lead			
		No. 1	Joiner attend plumber to sink			

			Bill.	£	s.	d.
		No. 3	Pail sinks 14" × 14" and 2" deep (all in clear) of quadrant plan, of 1¼" dovetailed sides and 2" bottom, screwed on, perforated for grating, and all prepared to receive lead			

PERFORATIONS FOR WASTES, or other purposes. State thickness of wood and size of pipe or perforation. They are generally rough, but if fair say so.

			Bill.	£	s.	d.
		No. 20	Holes for pipes ¾" to 2" through 1" deal ..			
		,, 2	Dished holes for 3" pipe through 1¼" deal			

SHELF FOR GAS-METER, give description and how fixed. It is generally sufficient to state the capacity of gas-meter.

			Bill.	£	s.	d.
		No. 1	Stout deal shelf and brackets for 40-light gas-meter, and fixing to brickwork ..			

LADDERS TO TRAPS, state width and length, and size of sides and rounds; state if folding, and describe the ironwork.

			Bill.	£	s.	d.
		No. 1	Wrought ladder 10 feet long and 15" wide, of 3" × 2" sides, and 2" × 2" rounds, 9" apart, dovetailed and screwed, and including wrought-iron hooks and eyes for fastening			

PLATE RACKS: State size and description of parts.

			Bill.	£	s.	d.
	No.	1	Deal-framed plate rack of 1½" × 2" frame, and ¾" diameter round bars, 2½" apart, 5 feet long, and to take three rows of full-sized plates, and fixing to brickwork or, when no particulars are given,			
	„	1	Strongly-framed plate rack 5 feet long, 3' 6" high, to detail and fixing to brickwork			

DRESSERS: State length and height, and give description, or provide a sum. The latter is the more frequent practice.

Provide for dresser, including fixing and painting, 5*l*., or an item of complete description as it would appear in a specification, thus:—

			Bill.	£	s.	d.
	No.	1	Deal dresser 8' 0" long and 8' 0" high, of 2" top 1' 9" wide, 2½" × 2½" framed legs and rails, four dovetailed drawers with ½" bottoms, ⅞" sides and 1" beaded fronts, all glued and blocked, and with hardwood runners and two japanned wooden knobs to each drawer, 1¼" cut and shaped standards, four tiers of 1" shelves arris grooved for plates, ⅞" fascia, ⅞" top, 2" × 1¼" moulding as cornice, ⅞" matched and beaded back, 1" pot-board and bearers, and 4 dozen cup-hooks			

Or the complete detail may be taken out and billed as "One dresser containing as follows," followed by the items written short.

Draining Boards: state size and thickness, and that it is blocked up to fall, and describe the grooving. If to be covered with lead state it.

Shelves, butler's pantry fittings, cupboards, should be taken out in detail.

ft.	in.		Bill.	£	s.	d.
5	0	supl.	1½" draining board, with grooves ¼" deep, 1" wide and ¼" apart, prepared to receive lead, and blocked up to falls			

JOINER.

COPPER LIDS.—State what wood, thickness and diameter, if in two thicknesses, if pinned with oak pins, if dowelled together, and describe the handle.

Bill.

| | | No. | 1 | Deal copper lid 24" diameter, of two thicknesses of ⅜" pinned and dowelled together with oak pins, and deal shouldered handle pinned on with oak pins . . . . . . | £ | s. | d. |

PILASTERS, *per foot superficial.*—If in more than one piece measure the girth by the height, state thickness, if glued and blocked, how fixed if not nailed, as screwed. If slotted screws, they should be numbered, and the work described as "fixed with slotted screws elsewhere taken." Measure the salient angles per foot run, stating the labour. The caps, bases and necking are best numbered thus: "No. 1. Base to pilaster 1½ inch by 15 inches and 12 inches high, moulded 6-inch girth, with two external tongued and screwed mitres and two 6-inch returns, the upper edge rebated, and the whole glued and blocked."

If the pilaster is in one piece, i.e. the returns only formed by the thickness of the stuff, they may be run, stating width, thickness and labour, in one item. Measure flutes or reeds per foot run, and count the stops.

Bill.

ft.	in.				£	s.	d.
30	–	supl.		1¼" pilasters, glued and blocked			
150	–	run		Labour to flute 1" × ¼" stopped			
		No.	40	Rounded stops			
		No.	8	19" lengths of 1¼" × 1¼" moulding, rebated on back edge with two 8½" returns, two external mitres, and two fitted ends as necking to pilaster			
		"	3	21" lengths of 1¾" × 9" skirting, moulded 4" girth, and rebated with two 4½" returns, two external tongued and mitred angles, and two fitted ends as base to pilaster			

COLUMNS OF WOOD are best numbered with complete description; state how constructed, height, girth, cap and base; if any part is carved describe those parts as prepared for carving, and provide a sum for carving. If with entasis state it

Those 6 inches diameter and under are usually solid, those of larger diameter put together in sections glued and blocked; the

former may be described in one, the latter in pieces. An alternative to the provision for carving may be found in the following items.

Bill.

ft.	in.				£	s.	d
		No.	6	Ionic-turned diminished columns of selected deal 6" extreme diameter, with moulded base 7" × 7" and 4" high, and carved and enriched cap 9½" × 9½" and 3¼" high, 5' 3" high in all			
18	–		run	In two turned diminished fluted column shafts of 2" best selected yellow pine, glued together (as sketch above) 12" extreme diameter and 9 feet high, rebated at each end to receive caps and bases ..			
		No.	2	Moulded and enriched Roman Ionic column caps 1' 7" × 1' 7" and 7" high, rebated at bottom, and fitting and glueing to shafts			
				Moulded attic bases to column 1' 2" × 1' 2" and 7" high, rebated at top and fitting to shafts			

STAIRCASES, see that there is sufficient headway, not less than 6 feet 6 inches. If the treads and risers are of different woods keep them separate.

Where the steps are curved on plan the risers must be veneered, and so described and kept separate.

TREADS, RISERS AND FLYERS, *per foot superficial.*—Measure the width of steps, including the housings (¾-inch each), by the whole length of the area occupied by them (the flight) on plan, plus the height from floor to floor. To this length add 1 inch for each tread where nosings are rounded and 1½ inch where moulded.

State the thickness of treads and risers, if glued, blocked and bracketed, if grooved and rebated together, if screwed together, the number and size of carriages, if rounded nosings, if moulded nosings, if with moulding tongued under, whether prepared for close or cut strings.

WINDERS, *per foot superficial.*—For the treads measure the size on plan, the width by the length, collect the length of the risers, multiply that length by the height from top of tread to top of tread, plus 1 inch each for rounded nosings and 1½ inch each for moulding nosings. Describe them in the same way as for flyers, and state in description that they are cross-tongued and measured net, and keep them separate from the flyers.

Number housings to treads and risers, stating whether rounded or moulded nosings. Housing to winders. Ends of treads notched and fitted to newels. Returned rounded or moulded and mitred nosings to steps, state if circular on plan. Cut brackets, state thickness, size, and if circular on plan. Curtail end to bottom step, and state length of step for veneered front to riser.

LANDINGS, *per foot superficial.*—State thickness, describe as cross-tongued and including bearers.

WALL STRINGS, *per foot run.*—State thickness, if moulded, if in two pieces give size of each and state how put together, if parts of the string are ramped for a length of 2 feet or thereabouts keep them separate and take a heading joint at each end, but if there are only short ramps, shorter than those last mentioned, describe them as "extra for short ramps," having previously measured them in the length of string.

Number the tongued and mitred angles, heading joints, fitted ends, returned moulded ends.

OUTER STRINGS, *per foot run.*—State thickness, if framed, if cut and mitred, beaded, sunk, grooved, moulded, chamfered, if circular or wreathed, if in either of the two latter cases "glued up in thicknesses," and state size of well-hole or the radius of the curve. Measure applied mouldings per foot run.

Number ends framed on splay, if the string has not been described as framed.

NEWELS, *per foot run.*—State size, describing as wrought and framed, state if chamfered, moulded, stop moulded, and how many times, or the labours last mentioned may be separately measured and the newel billed as wrought and framed.

Number stops splayed or moulded to moulding or chamfer; the turnings to newels, stating length; turnings to pendants, stating length; finials describe, and if put on, state how fixed and if of different wood.

HANDRAILS, *per foot run.*—State the size, whether rounded or moulded, if framed, if continuous, if ramped, wreathed, level circular, if fitted to iron core; if the moulding is of uncommon section give a sketch.

Number ends framed, if not described as framed, ditto on rake; joints and handrail screws, and state if joints are dowelled; lengths of wreathed or circular handrail where under 12-inch well-hole, stating length, and including the two joints and handrail screws in the description. Scrolls, stating size and including the joint and handrail screw, turned and mitred newel cap and joint, and handrail screw, and short ramp. Balusters, stating size and average length, if turned, and labour generally, if dove-tailed to step, if housed to handrail. Iron balusters, with flapped ends, state size and length, what parts are let in flush, and if the heads of screws are countersunk.

APRON LINING, *per foot superficial.*—State thickness, if moulded, beaded, sunk, rebated, grooved, circular on plan.

Number short lengths of circular or wreathed, state length, width and all labours, and include two heading joints in the description, and state radius; tongued and mitred angles.

NOSINGS TO EDGES OF LANDINGS, *per foot run.*—The superficies measured with the landing and described as "extra for nosing," etc. State thickness, width and labours, if tongued to edge of floor. *Observe* that in counting the housings of steps the housing of this nosing should be counted as a housing to step and included with them.

JOINER.

Bill.

Staircases in Deal.

ft.	in.			£	s.	d
111	–	supl.	1½″ treads, with rounded nosings and 1″ risers, all rebated and grooved together, glued, blocked and bracketed on and including two strong fir carriages, and prepared for close strings			
108	–	”	1¼″ ditto ditto, &c., all as last, but with 1¼″ × 1″ moulding, tongued under and prepared for cut strings on and including three strong fir carriages			
25	–		1¼″ ditto in winders, cross-tongued and measured net			
14	–		1¼″ cross-tongued, landing and bearers ..			
9	–	run	Extra on 1¼″ floor, for 1¼″ nosing with 1¼″ × 1″ moulding, rebated and let in beneath, and glueing and tonguing to edge of floor			
			No. 2 small quadrant corners			
			,, 1, 20″ length of ditto circular to 12″ radius, including two fitted ends			
6	–	”	12″ × 1′ apron lining, cross-tongued, rebated one edge, grooved and beaded ..			
			No. 2 mitres			
			,, 2 housings			
23	–	”	2¼″ × 1½″ capping, grooved 2″ girth, twice moulded 2″ girth			
			No. 2 housings			
			,, 2 fitted ends on splay			
23	–	”	1¼″ framed and beaded outer string			
52	–	”	2″ outer string, wrought, framed and twice rebated 1″ girth			
52	–	”	2″ ditto, and cut and mitred			
			No. 1, 36″ length of ditto, wreathed to 12″ well-hole, and with two cross-tongued heading joints			
70	–	”	1¼″ moulded wall, string and backings ..			
10	–	”	1¼″ ditto, but the backings plugged ..			
5	–	”	1¼″ ditto, ramped			
			No. 10 mitres			
			,, 2 heading joints			
			,, 4 extra for short ramps, with one heading joint to each			
44	–	”	2¼″ "mopstick" handrail framed			
32	–	”	8″ × 2¼″ rounded handrail, including dowelled heading joints and handrail screws			
16	–	”	4″ × 4″ newel, wrought, framed and four times stop-chamfered			
			No. 16 moulded stops to 1″ chamfer			
			,, 2 ends turned for a length of 6″ as pendants			
			No. 3 ends turned, cut and shaped (as sketch), including moulded stops to chamfer as finials			

ft.	in.			
		No.	10	No. 4 turnings 24" long (averaged) .. .. " 5 turned, cut and moulded top 4"× 4" × 7", with wainscot ball with strong pin glued in, and the ball French polished 2" turned balusters, 26" long, with moulded stops on caps and bases, housed to deal string and wainscot handrail .. ..
		"	29	2" ditto on rake .. ..
		"	37	Ends of steps with rounded nosings, housed to strings .. ..
		"	28	Ditto, with moulded nosings, ditto .. ..
		"	28	Ditto, with moulded nosings, cut, mitred and returned .. ..
		"	6	Ends of winders, ditto .. ..
		"	4	Ditto circular to 12" radius .. ..
		"	6	Ends of steps with moulded nosings, notched and fitted to newels .. ..
		"	1	Extra for semicircular end, and veneered front to riser of step, with moulded nosing, the step 1' 10" wide and 4' 6" long

Staircases in Mahogany, French polished.

ft.	in.			
23	—	run		4¾" × 3¼" moulded and framed handrail, including dowelled heading joints and handrail screws .. ..
48	—	"		4¾" × 3¼" moulded handrail, continuous, including dowelled heading joints and handrail screws .. ..
				No. 2 extra for short ramps .. .. .. " 1, 48" length of ditto, wreathed to 18" well-hole .. .. No. 1 turned and moulded newel cap 7" diameter, with short ramp and heading joint .. ..

ATTENDANCES.—To every sink, safe, w.c., bath, lavatory, or cistern, an item should be taken "joiner attend plumber to sink," or as the case may be.

Bill.

Attendances.

No.		Joiner attend plumber to Baths	
"	2	Sinks	
"	2	W.c.'s	
"	2	Cisterns	
		Joiner attend bellhanger, cutting away and making good after him to No. 13 bells with No. 13 pulls, and include the necessary boards for bell runners and for fixing the floor boards and joinery over same to remove with brass cups and screws ..	

JOINER.

DADO AND PICTURE RAILS, *per foot run.*—State size and labour. Number mitres, housings, fitted ends. Take at the same time a similar length of twice splayed ground.

When there is much labour on a rail, for either of these purposes, a sketch is advisable.

Bill.

ft.	in.			£	s.	d.
100	–	run	4″ × 2½″ chair rail moulded both edges ..			
			No. 20 mitres			
			„ 10 ditto, irregular			
			„ 10 fitted ends			
			„ 10 splayed and fitted ends			

DEAL CORNICES, *per foot run.*—State the girth of the moulding, how it is put together, and the thickness of the wood, and include backings with the description. If the moulding is out of a comparatively thin piece of wood, set diagonally, call it "sprung."

Number the mitres, housings, etc.

Bill.

ft.	in.			£	s.	d.
20	–	run	Cornice moulded 18″ girth, 12″ high, 9″ projection of 2″ deal, all rebated together, glued and blocked, and including ⅞″ cover board 9″ wide, and deal backings			
			No. 2 mitres			
			„ 2 housings			
			„ 4–9″ lengths with one external and one internal mitre to each			
20	–	run	6″ × 2″ deal sprung moulding moulded 10″ girth as cornice, splayed both edges ..			
			No. 6 mitres			

CHURCH FITTINGS.—The joinery for church fittings will present little difficulty; the same principles will apply to its measurement as to the work dealt with in the foregoing pages.

The seats, book-boards, etc., will be measured at *per superficial foot*, and the extra labours upon them measured at *per foot run.*

The boarding for filling in between top and bottom rails or between seats and floor should be described as in 12-inch or 14-inch lengths, as the case may be, and what it is to be used for.

P

Bench ends and standards are most conveniently numbered, giving a sketch and stating the thickness, size and labour on each.

Ornamental framings, as stall fronts, may be measured at *per foot run*, stating thickness and height, and giving sketch and description, or the labour in a given length of it may be stated.

In all cases provide in the estimate for a pattern bench, including its carriage to and from the building.

Bill.

		£	s.	d.
Allow for the cost of preparing, delivering and fixing at the building a pattern bench complete, for the architect's approval before the others are made, and for carriage back to yard and for alterations if any				

Church fittings are best billed as a separate section.

For pulpits, lecterns, prayer desks, etc., a sum is usually provided, but where not thus treated, the quantity of material in each should be kept separate, so that the cost of the article in question may be seen without trouble.

FITTINGS.—If in large quantity, as the shelves, dressers, cupboards, etc., of a large house, should be made a separate section in the bill; it is not fair to bill them with the general joinery, as they are worth a larger price. Take them out in as much detail as possible.

IRONMONGERY.

Measure the ironmongery with the joinery to which it belongs.

Where it is essential that the ironmongery should be of very good quality it is a frequent practice to specify the prime cost of each of the articles, or to select a manufacturer to whom a list of the ironmongery is sent to affix his prices, the total sum being then included in the estimate as a provision, in such a case the "fixing only" appearing in the joiner's bill.

State in all cases whether ironmongery is of brass, iron, or gun-metal, and keep ironmongery and fixing to deal separate from that fixed to hard woods, with which latter include that fixed to pitch pine.

When an article of ironmongery does not appear in any trade list it must be described as "purpose made."

Articles of ornamental ironmongery may often be described with advantage by the number and price in a trade list, thus—

Bill.

				£ s. d.
	No.	10	Hill's (100A Queen Victoria Street), No. 1075 brass grip handles, list price 3s. 9d.	

IRON BUTTS OR BACK FLAPS.—Give the size, measuring along the knuckle, if the kind known as broad, state width and height, state if rising or projecting butts, state if with steel washers and pins, if with face plates, and if the face plates are engraved.

Bill.

				£ s. d.
	No.	10	Pairs 2½" wrought-iron butts	
	„	6	Ditto 3½" × 4" best wrought-iron welded butts	
	„	6	Ditto 5" wrought-iron projecting butts	

CROSS-GARNET HINGES. — State the length measured from knuckle to point.

Bill.

				£ s. d.
	No.	5	Pairs 15" cross-garnet hinges	
	„	6	Ditto 18" extra strong cross-garnet hinges	
	„	2	Ditto 18" light welded water-joint hinges	

STRAP HINGES.—State length, size of iron, if with back straps, if with fanged or double-fanged hooks, if with screwed plates, if bolted state the size of bolts and the number to each hinge. Observe to take the labour to letting into stone and lead and running where fixed to stonework. Sometimes it may be convenient to state the weight per pair.

Bill.

	No.				£ s. d.
		2	Pairs wrought-iron strap hinges 18" long, of 2" × ½" iron, with back straps and plates, fixed with screws with countersunk heads		
	„	2	Ditto ditto, but with double-fanged hooks and eyes		
	„	1	Ditto 3" × ½" strap hinges 6' 0" long, with back straps 3 feet long, with cups instead of eyes, and double-fanged hooks with steel points, each hinge fixed with six ⅜" screw bolts, with octagon heads, nuts and washers		

PATENT HINGES.—Describe the kind, the thickness of the door they are for, if spring hinges, including filling boxes with neat's-foot oil, and in all cases adjusting at completion. For spring hinges take a wood cradle or oak block, and letting in where in a wooden floor, and letting into stone and running with lead where a stone floor.

Bill.

	No.				£ s. d.
		2	Pairs Cottam's (2 Winsley Street, Oxford Street, London), 48" improved spherical hinges, fixed with and including bolts ..		
	„	2	Sets Archibald Smith's (48 Leicester Square, London), double-action swing centres for 2" doors, including letting into oak (or stone), filling the boxes with glycerine and adjusting at completion ..		
			No. 2 oak blocks, for swing hinges and framing between joists		
	„	1	Pair japanned iron (spring hinge and blank hinge), 7" double-action helical spring butts for 2½" door		

ORNAMENTAL HINGES.— If cast-iron hinge-fronts they may generally be most conveniently selected from an illustrated trade catalogue, the number of the pattern and name of manufacturer being stated in the quantities. If wrought-iron ornamental

IRONMONGERY.

hinges to design it will be better to provide a sum per pair, the fixing being either taken separately or included in the amount provided.

The sum to be provided is the more likely to be reasonable if the detail drawing be sent to an ironworker for a price, and if sent to several of equal ability the advantage of a competitive price may be obtained.

Bill.

				£ s. d.
No.	4	Provide for 8 pairs of wrought-iron hinges and bolts for fixing 15*l.*, and allow for packing, carriage, fixing and profit .. Pairs Hill's (100A Queen Victoria Street, London), ornamental iron hinge fronts, No. 877, list price 11s. per pair		

BOLTS.—State length (measured by the length of the rod of the bolt), if tower, barrel, bright rod, necked, brass mounted, square, monkey-tailed. These last are measured by their total length, and the sectional size should be stated. Number the brass or iron thimbles or floor-plates for bolts, and state whether let into wood or stone, and in the latter case whether run with lead or cement.

Bill.

			£ s. d.
No.	8	4" iron necked bolts	
„	2	4" iron flat spring bolts	
„	2	6" iron barrel bolts	
„	5	9" ditto	
„	1	¾" monkey-tail bolt, 16" long	
„	1	1" ditto, 48" long	
„	2	Iron thimbles for bolts, and letting into York and lead and running	

FLUSH BOLTS.—State length and width.

Bill.

			£ s. d.
No.	4	¾" brass flush bolts 4" long	
„	3	¾" ditto, 9" long	

ESPAGNOLETTE BOLTS.—State the height of the doors they are intended to secure, or their length, and whether of brass or iron.

Bill.

	No.	2	¼" malleable iron Berlin black espagnolette bolts, 6 feet long	£	s.	d.
	„	2	Burnished brass espagnolette bolts, 6' 6" long with 1" rods			

SMITH'S SILL BAR.—State length and whether to single or folding casements, and observe that they cannot be applied to a casement under 2 inches in thickness.

Bill.

	No.	2	Archibald Smith and Stevens' (48 Leicester Square, London), patent weather-tight Janus sill bar, for 2" casement opening outwards, 3 feet long, and fixing to deal casement and oak sill	£	s.	d.
	„	2	Ditto, 4' 6" long to folding casements ..			

SMITH'S WEATHER-TIGHT CASEMENT FASTENINGS.—State length, and whether to single or folding casements.

Bill.

	No.	1	Archibald Smith and Stevens' (48 Leicester Square, London) patent weather-tight Janus fastening, 6' 6" long for folding casements	£	s.	d.

LOCKS.—State the size, the kind, as iron rim, mortise, drawback, if three bolt, two bolt, or one bolt (two bolt is the usual kind), dead or stock, if upright, if rebated or half rebated (the two latter for folding doors), the kind of furniture. In cases where a number of doors occur and each side has a different kind of furniture, it will be most convenient to take the locks and furniture separate, numbering each side of the door as a half set of the furniture required, and describing the locks as "without furniture." State if locks are "en suite," and if so number the master keys.

In cases of oak stock locks, with wrought-iron mountings, to design, it will be better to provide a sum, taking the fixing separately or including the fixing in the amount provided.

Cupboard locks are described by their height in inches; all other locks by their length in inches.

If locks are copper or brass warded, state it.

In measuring mortise locks after they are fixed the length from the centre of the spindle to the edge of the door, plus 1 inch, will be the length of the lock.

As iron rim locks are sold at a price per dozen including furniture, the furniture may be included with the description of the lock. Mortise locks are sold at a price per dozen exclusive of furniture, and the locks and furniture appear separately in the bill.

In good work take with each mortise lock, "door prepared with double tenons for mortise lock."

KNOBS.—State diameter, and whether metal, porcelain, ebony, etc., and describe as sets or half sets of furniture.

If patent spindles, state whose patent.

Bill.

No.	6	8" iron cupboard locks	£	s.	d.
„	14	4" brass spring latches, P.O. 5s. each ..			
„	6	5" brass mortise latches, with brass furniture			
„	4	6" iron rim locks, and Mace's patent brass furniture			
„	18	Hobbs, Hart & Co. (76 Cheapside, London), 7" machine-made, fine finish, two-bolt mortise locks			
„	12	Ditto, but half-rebated			
„	18	Sets of Hobbs, Hart & Co.'s solid steel slotted spindle furniture for mortise locks, with 2¼" strong brass knobs			
„	4½	Sets of Hill's (100A Queen Victoria Street) No. 489 fluted brass furniture, with Hill's patent spindles for mortise locks			

FINGER PLATES.—Describe them and state whether long or short. Number each *one*, not each set—the latter term is misleading.

Bill.

No.	20	Sandeman's (15 Boro', London), No. 4492 short finger-plates	£	s.	d.
„	20	Ditto No. 4492 long ditto			

SASH FASTENINGS.—State length and quality.

Bill.

| | | No. | 15 | 2¼" strong brass bronzed sash fastenings .. | | £ | s. | d. |
| | | „ | 10 | 3" Hopkinson's patent brass sash fastenings | | | | |

SASH HANDLES OR LIFTS.—State length and kind, or give P.C. each. Observe that they may be dispensed with when there are bars in the sashes.

Bill.

| | | No. | 10 | 3" brass sash lifts .. .. | | £ | s. | d. |
| | | „ | 10 | 4" brass sash handles .. .. | | | | |

SASH CENTRES.—Number the pairs, state length and quality; observe that a pair of sash centres is two centres and two slotted plates.

Bill.

		No.	10	Pairs 2½" japanned iron sash centres ..		£	s.	d.
		„	10	Ditto 3" brass ditto				
		„	10	Ditto 3" brass ditto, with steel bushes ..				

SLOTTED SCREWS, each. Brass screws (state length) and brass slotted plates, including all labour for slotting on architraves or as the case may be; average 18 inches apart.

Bill.

| | | No. | 305 | 1½" brass screws and brass slotted plates, including all labour for slotting on architraves, mouldings and other joinery .. | | £ | s. | d. |

PLUMBER.

A sheet of lead is ostensibly 7 feet by 30 feet, usually about 6 feet 10 inches wide, and of length varying from 25 feet to 30 feet. The lead for gutters and flats is cut lengthwise of the sheet, that for flashings, hips, ridges, &c., across the sheet.

If a large work, keep external and internal plumbing separate. Measure all the sheet lead by the foot superficial, and bill it by weight.

Gutters, flats and flashings are billed together. Stepped flashings and soakers should be kept separate, as there is more trouble in fixing.

Lead to curved roofs, ornamental turrets and flèches should be kept separate and billed in separate items.

Measure per foot run, dressing lead to mouldings, stating girth, and number the mitres.

EXTERNALLY.

FLATS AND GUTTERS.—Measure the lead by the foot superficial, stating weight, allow in the width for turning up under slating (usually 9 inches from bottom of gutter, but very often more), and against walls (usually 6 inches from bottom of gutter). Allow for 1½-inch drips 6 inches, and for 2-inch drips 8 inches; for 1½-inch rolls 6 inches, and for 2-inch rolls 8 inches, beyond the original measurement, all over the surface. Assume always that drips are not more than 10 feet apart, as when in longer lengths there is much risk from expansion and contraction. Collect rolls and drips as much as possible.

It is advisable always to set out the gutters and calculate their widths, so as to test the widths shown on the drawings. (See " Examples of Taking Off.")

FLASHINGS, *per foot superficial, stating weight.*—5 inches to 6 inches wide usually. In these an allowance must be made for tacks and passings, add 4 inches for every 7 feet in length, and 6 inches for every angle. This will be found to be practically correct, and will cover the weight of tacks; or, allow 3 inches or 4 inches as may be directed every 7 feet in length for passings, and take tacks 6-inch or 7-inch by 2-inch at distances of 3 feet 6 inches apart.

APRONS, *per foot superficial, stating weight.*—Usually 12 inches wide. Make the same allowance in length as for flashings.

RIDGES, HIPS AND VALLEYS, *per foot superficial.*—The two former usually 18 inches wide, the latter 20 inches wide. Allow for clips or tacks and passings as for flashings.

Observe that where ridges stop against a roof plane as in the case of dormers—a " soaker " must be taken about 18 inches square.

STEPPED FLASHING.- Usually 12 inches wide; over soakers

usually 7 inches wide. Make the same allowance as for horizontal flashings. When soakers are specified instead of stepped flashings, a piece of lead should be taken to each course of slates or tiles, allowing the same lap as the slates or tiles, and to course with them, running 4 inches under slating or tiling, and turning up 4 inches against wall.

The length of each soaker to slates or tiles, measured up the roof slope, will be equal to the gauge plus the lap; thus, to tiles of a $3\frac{1}{2}$-inch gauge, the lap will be $3\frac{1}{2}$ inches, and the length of soaker 7 inches. To countess slates (20 inches × 10 inches) to a $2\frac{1}{4}$-inch lap; the gauge will be $8\frac{3}{4}$ and the length of the soaker $11\frac{1}{4}$ inches; the surveyor would call this 12 inches. Soakers to the edge of a roof covering adjoining a gable parapet, would be covered with a stepped flashing usually measured as 7 inches wide.

In the case of slating in promiscuous sizes obtain the average gauge and add the *lap* as follows:—

Average for Soakers.

	in.	
E slope B.R. 2	3	*Slates*, average gauge
	5	
	7	
E. ditto of dormitory 2	4	
	4	
	$4\frac{1}{2}$	
	8	
So. of 4.	$3\frac{1}{2}$	
	4	
	$4\frac{1}{2}$	
	5	
	9	
	12)$61\frac{1}{2}$	
Av. gauge	$5\frac{1}{8}$	
Lap	$2\frac{1}{4}$	
	$7\frac{3}{8}$ say $7\frac{3}{4}$	

Where roofs are covered with lead forming "lean-to's" against walls, as to the aisle roofs of some churches, lead tacks, about 9 inches by 3 inches, should be taken, wedged at one end into the wall, and soldered to the lead covering, one to each bay formed by the rolls, otherwise the lead is apt to slip downwards.

PLUMBER.

Bill.

cwts.	qrs.	lbs.		£	s.	d.
35	1	7	Milled lead and labour in gutters, flats and flashings			
17	2	14	Ditto in covering to ogee cupola			
12	3	21	Milled lead (labour elsewhere taken) in covering to mouldings, crockets and finials			
27	3	0	Milled lead and labour in stepped flashings			
10	2	7	Ditto in soakers, and fixing by tiler			

Measure per foot run.—Lead or oak wedging to flashings. Keep that in stepped flashings separate; in either case take the same length as that taken for the flashings. Copper nailing, state whether " open," " close " (1 inch apart), or " very close." Bedding edge of flashings in white-lead. Labour to vandyked or serrated edges. Soldered angle. Extra labour, dressing lead over fillet to form secret gutter. Extra labour, dressing lead over glass. Welted edge. Welted lap. Labour dressing lead to mouldings, stating the girth of moulding. Rain-water pipe, state size of bore, weight per foot superficial of the lead used, if tacked, if collared, if fixed in chase. Where a flashing is grooved into a horizontal instead of a vertical face, take " burning in " to flashings in stonework only.

Bill.

ft.	in.			£	s.	d.
40	–	run	Bedding edges of flashings in white-lead ..			
150	–	„	Lead wedging to flashing			
72	–	„	Ditto, stepped			
22	–	„	Copper nailing			
76	–	„	Ditto, close			
24	–	„	Burning in flashing			
10	–	„	Soldered angle			
64	–	„	Extra labour dressing lead over fillet to form secret gutter			
15	–	„	Labour to welted edge			
25	–	„	Labour dressing lead to moulding 3" girth No. 10 bossed mitres			
34	–	„	Ditto to moulding 6" girth .. .. .. No. 2 bossed mitres .. .. „ 3 four-way intersections .. .. .. „ 2 three-way ditto			
10	–	„	Labour escalloped edge to 7 lbs. lead			
15	–	„	Ditto corrugated edge to ditto			
70	–	„	4" rain-water pipe, weight 9 lbs. per foot run, including tacks and joints and nailing with wrought-iron rose-headed nails to and including teak plugs in brickwork No. 6 extra for bends			

Number. Extra labour and solder to cesspools. Lead, iron, or copper wire covers to cesspools, or to outlets of eaves gutters. Lead heads of rain-water pipes (often a sum provided for these), copper wire covers to ditto. Lengths of socket pipe, stating size of bore, length in inches, whether "once," "twice," or "all bent," and the number of soldered joints to each, and the weight per foot superficial of lead it is made from, or the weight per foot run. Bossed ends to rolls. Bossed intersections to rolls. Bossed cross-intersections of roll with ridge. Soldered dots and brass screws. Bossing and dressing lead to crockets, stating extreme dimensions. Sets of flashings and soldering to pipes passing through roof, giving weight per foot of the lead used, and the size of pipe, or the total weight of the set. Bossing lead out of solid over bases of finials, vanes, etc., describing each, and giving sketch. Some surveyors say " plumber and labourer, . . . hours each, as *provision,*" to such an item as the foregoing, the charge for the actual time spent being adjusted at completion. Lead hood or bonnet and straps out of . . . lbs. lead to (ventilating) pipe, state size of pipe.

Bill.

No.			£	s.	d.
	50	Labour bossing and dressing 7 lbs. lead to carved crockets, 10" long, 5" wide, 5" projection			
„	102	Ditto bossed ends to rolls			
„	10	Ditto bossed three-way intersections to rolls			
„	14	Ditto bossed four-way ditto			
„	2	Ditto, bossing 7 lbs. lead to turned and moulded finial 6" diameter with disc terminal 12" diameter— 4 feet high in all			
„	10	Bossing lead to 2" returns of 9" moulding with two mitres to each			
„	2	Bossed irregular angles and welted lap, and copper nailing to mitres of moulding, 30" girth			
„	6	Extra labour and solder to cesspools			
„	50	Soldered dots and brass screws			
„	10	Strong copper wire covers to outlets of eaves gutters			
„	6	Ditto to cesspools			
„	4	Ditto to rain-water heads			
„	10	24" lengths (averaged) 4" drawn lead socket pipe 7 lbs. per foot run, all bent, one end tafted and soldered to ces-pool			
„	4	7 lbs. lead, slate and soldering to 4" ventilating pipe			

No.	4	Bolding's (Davies Street, W.) "Simplex" lead ventilating cap, and fitting to 4" pipe	£	s.	d.
"	1	Hood and straps out of 7 lbs. lead and soldering to 4" ventilating pipe			
"	2	Cast lead heads to rain-water pipes P.C. £7 each, and solder joint to 4" pipe ..			

INTERNALLY.

The larger proportion of the internal work will be connected with the water supply.

Before measuring the water supply it will be necessary to obtain a copy of the regulations of the water company which supplies the district, you will then be sure of the prescribed weights of pipes, regulations as to wastes, waste preventers, brasswork, &c., and in cases where the building is to be supplied by meter the surveyor should ascertain whether the proprietor will supply his own meter or hire one of the water company at a rental.

Observe generally that where a water meter is used no waste preventers will be required, though even then some of the water companies insist upon their use.

In all items relating to pipes state the bore of pipe, except to copper and brass pipes, which are described by their external diameter, and where a cock or short length of pipe requires tafted ends, soldered joints, soldered ends, or bends, include them with the article. Observe that this clause applies to many of the items under the head of "*numbers*."

Do not measure short lengths of pipe over 2-inch bore by the lineal foot, but number them (up to 4 feet in length), stating the length.

The weight of pipes, up to 2-inch bore, will be settled by a clause in the bill of quantities. Where of greater bore state the weight per foot superficial of the lead they are made from, or the weight per foot run if drawn pipes.

Where very good cocks or valves or apparatus are required, it is well to state in the bill the name of the manufacturer.

The technical descriptions and illustrations of most of the articles of plumber's brasswork are given in the trade catalogues of

Bolding, Tyler and Sons, Farmiloe, and others. Often a number from a trade list is useful to define the quality.

In a large work it will be advisable to make a complete list of the cisterns and their supplies, and the apparatus connected with each, and cross off each as it is measured. This saves time, although apparently cumbrous, thus:—

ATTIC FLOOR:
Cisterns.

1 Drinking water	waste	supply	overflow.
1 Rain water	waste		overflow.
2 W.C. cisterns	waste	supply	overflow.
1 W.C. south of bedroom 2		soil-pipe	supply.
1 „ north of ditto 3		ditto	ditto.

	Hot Water.	Cold.	Rain Water.	Waste.
Housemaid's Sink, next back stairs	1	1	1	1
Ditto, next Bathroom				

&c.

It will also be convenient and safe to have a tracing of the block plan of the building on which the wastes and rain-water pipes which deliver at ground level may be marked in readiness for the measurement of drains, or they may be marked on the drain plan, if there is one.

It will be seen by reference to the order of taking off that every lavatory, sink, or w.c. apparatus is followed by all its adjuncts except the supplies. In measuring the supplies from cistern, first take the one which runs to the apparatus furthest from the cistern, and finish with the valve. Then the branches from that pipe, beginning with the one nearest to the cistern, proceed in the same way with each main line of pipes and its branches, then the rising main.

Some surveyors adopt the following order in taking off a water supply, and it is a very clear and reasonable system.

Tapping main.
Rising main.
Cistern and all its adjuncts.
Longest service from cistern.
Apparatus supplied by last, with all its adjuncts.

Branches from last pipe (beginning with the one nearest cistern).
Each apparatus as it is reached, with all its adjuncts.
Follow with other main services, with the branches from each, and apparatus, all as before described.

Whether measuring from drawings or the actual work, adopt a system and adhere to it rigidly, even at the cost of many journeys up and down stairs, which will no doubt be urged as an argument against it.

LEAD IN SINKS, CISTERNS, AND SAFES, *per foot superficial.*—Keep the lead in safes separate from that in sinks and cisterns, to these two latter allow 1 inch for the lap where soldered angles occur, and a width equal to the thickness of the wooden sides for turning over at the top. The soldered joints to sinks should be taken to four vertical angles and two edges of bottom, the lead of ends and bottom being usually in one piece. To cisterns, take soldered angle to all angles if large, if small, treat as directed for sinks. Take copper nailing around the outer edge of the sides and ends, both to sinks and cisterns.

Measure at per foot run. Soldered angle, copper nailing. Lead pipes, state size of bore, and, if not defined by preamble of bill, weight per foot or yard; include the joints in the description except the branch joints, i.e. junctions of one pipe with a main line of pipe which are counted, and charged as "Extra soldered joints." Soil pipes; state bore, what lead out of (as 7 lbs.), whether tacked, collared, fixed in chase. Copper pipes; state external diameter, and whether perforated as for urinals. Haine's tin-encased pipe. Pipes laid in ground, fixed to walls, fixed to glazed facings of brick or tile, should form separate items of bill.

For measuring the diameter of pipes, if the necessity arise, pocket dividers may be adjusted in the form of callipers, and used in a similar manner; or callipers may be used.

Number branch joints to pipes.

Bends in lead pipes are of small importance of $1\frac{1}{4}$ inches diameter and under, if over that diameter, they are sometimes numbered. The counting of bends, when measuring an executed system of pipes, is easy. When measuring from drawings it is difficult, and in either case it occupies time with little ultimate advantage. Any estimator can allow for them in his price, but they should be a part of the description of the pipes.

Observe that branch joints are described by the size of the main pipe, thus a joint at the junction of a ¾-inch pipe with 1-inch pipe is called a "1-inch extra soldered joint."

Fire bends (if any), soldered ends, ends tafted and soldered. Lengths of pipe; state size, length, whether bent, and how many times bent, if jointed to iron, if soldered and pierced (as for dribble pipe). Trumpet-mouth waste; state length, average bore, what lead out of, if with beaded top, whether lead or galvanised iron. Extra for bends to soil pipes, stating size of pipe and how many soldered joints to each. Bonnets or cowls to air pipes; state weight per foot of lead used and size of pipe. Lead collars and soldering to soil or waste pipes; state size of pipe and weight per foot of the lead used. Lead D traps; state if full size, what lead out of, if with brass screw cap and lining, state size of latter, state if trap is soldered to safe. Cast-lead traps; state size, whether with brass screw cap, state size of latter. Observe that where waste pipes deliver over gratings in the open air you may safely dispense with traps. Brass screw unions; state size, and if with fly-nut, and whether to iron or slate, and whether jointed to iron or lead pipe. Brass washers and wastes, all as last. Brass cistern connectors and all as last. Tee pieces to copper pipe, state size of pipe, include the brazed joints. Stopped ends to ditto. Cocks; state size, if plated, brass, iron, or gun-metal, stop cocks or bib cocks, round or square way, screw bottom, screw boss, if with loose key or spanner, if high-pressure screw-down loose valve, diaphragm, if screwed for iron, if engraved, and what, as "HOT," etc. Observe that air vessels are not necessary when screw-down valves are used. Ball cocks or valves; state size, and if copper or zinc ball and stem. Sets of bath valves; give full description. Copper clips, and brass screws.

In all cases, state whether brasswork is jointed to iron, lead, or slate, and wherever possible include the joint in the description of the article.

PUMPS.—Give description, and P.C. price if possible, or refer by number to a trade catalogue, and include joints with description; take suction pipe and rose and warning pipe.

WROUGHT-IRON CISTERNS. — State if galvanised, the length, width, depth in clear, or to hold water ... feet deep, or if the space to receive cistern will allow of variation of size, and shape is unimportant, give the alternative, "or to hold ... gallons," as stock sizes are cheaper than special. Where the usual trade iron cistern is all that is required, no particulars of thickness, stays, etc., need be given, but where a better kind or larger size is required state the thickness of iron, the size of angle iron around top and at angles, the size of stays if any, include the " necessary perforations " and small angle stay for ball valve or cock in the description, and state the height from the ground of the hoisting.

A system has arisen of supplying iron cisterns at $\frac{1}{8}$ inch, or $\frac{1}{4}$ inch, or $\frac{1}{2}$ inch (as the case may be), "full" or "bare" as to thickness, and as in the case where there are a number of cisterns this may make a large difference in cost, the surveyor should see that his employer obtains what is specified.

Observe that in calculating the content of cisterns 3 inches less than the absolute depth should be taken as their capacity.

The holes for pipes may be either counted and separately billed, or they may be included with the description of the cistern.

W.C. VALVE APPARATUS.—Provide a sum for each, and give address of manufacturer. State colour and kind of basin, size of valve; if with regulator, and what kind; if with looking-glass bottom valve; if copper pan; the metal of sunk dish, and description of handle; include the joints and fixing in description.

PEDESTAL CLOSETS.—Define by a number in a trade list; these often include waste-preventer and seat.

Where there are a number of w.c.'s in one establishment it is a good plan to add to the description " to be set in order by manufacturer (or patentee) at completion."

LAVATORY APPARATUS.—It is better to provide a sum for these, or give a full description and address of maker. Include all joints and unions and fixing with the description. If the marble or slate top is not to be supplied by the manufacturer of apparatus, state in description that it is " elsewhere taken."

URINALS.—Give description as before, and include the joints and fixing.

BATHS. — State if full sized or what size, if iron, copper, porcelain. Include joints. If porcelain baths have extra inlets, state it; as these baths are very heavy state the height of hoisting and observe that their great weight will often necessitate extra strength in the floor timbers. Take a wooden cradle for each metal bath if enclosed.

Baths and their valves are often described by numbers from a trade list.

WATER METERS.—Provide a sum to include fixing and joints. If outside the building take a brick chamber and cover to enclose it.

WASTE PREVENTERS.—Give descriptions, and include joints and fixing.

ATTENDANCE.—It is customary to take an attendance by joiner upon plumber to each apparatus. If the intended course of the pipes is known take holes through walls, stating in each case thickness of wall (afterwards averaged); if the course is not known insert an item "allow for attendance, cutting away and making good in all trades after plumber (attendance to apparatus elsewhere taken)."

Bill.

cwts.	qrs.	lbs.			£	s.	d.
7	3	7		Milled lead and labour in safes and sinks			
	2	7		Ditto and labour in covering grooved draining board			
	ft.	in.					
	18	–	run	Close copper nailing			
	11	–	"	Soldered angle			
				No. 16, 4" lengths of ditto			
				20 bossed angles, 4" long, to safes			
	20	–	"	¾" middling lead pipe, including bends, joints and fixing..			
	14	–	"	1" ditto			
	99	–	"	1¼" ditto			
	75	–	"	1½" ditto			
	25	–	"	2" ditto			

ft.	in.			£	s.	d.
3	–	run	½" strong lead pipe, including bends, joints and fixing			
125	–	„	¾" ditto			
207	–	„	1" ditto			
25	–	„	1" ditto, and digging in roadway, *as provision*			
13	–	„	1¼" strong lead pipe, as before			
27	–	„	1½" ditto..			
12	–	„	3" ditto			
75	–	„	4" soil pipe, weight 9 lbs. per foot run, including joints, tacked and collared, and fixed with strong wrought-iron nails to and including oak plugs in brickwork ..			
			No. 6 extra for bends			
	No.	2	Extra to ¾" soldered ends			
	„	4	Ditto 1" ditto..			
	„	1	½" extra soldered joints			
	„	10	¾" ditto			
	„	6	1" ditto			
	„	3	2" ditto			
	„	2	Ends of 1" pipe tafted and soldered to safe			
	„	1	Ditto 1¼" ditto			
	„	3	24" lengths of 1" middling lead pipe, all bent, one end tafted and soldered, the other end with soldered joint to soil pipe			
	„	2	36" lengths of 3" lead soil pipe, 7 lbs. per foot run, all bent, one end tafted and nailed to floor, the other end with 4" soldered joint			
	„	1	1¼" to 2½" trumpet-mouthed standing waste 2' 6" long out of 7 lbs. lead, and soldering to brass waste			
	„	2	Connections of soil pipe with drain, including lead flange out of 7 lbs. lead soldered to 4" pipe fitted to socket of drain pipe and sealed down with cement			
	„	3	¾" brass cistern connector for iron cistern, with fly-nut, and union and soldered joint			
	„	2	1" ditto			
	„	2	¾" boiler screws, and joint to iron cistern, and soldered joint			
	„	2	½" long ditto			
	„	1	1¼" brass washer and waste, with fly-nut and union, and soldered joint and joint to iron cistern			
	„	2	Tyler and sons (2 Newgate Street, London) ¾" gun-metal loose valve screw-down high-pressure bib cocks, strong waterworks pattern, with screw boss and soldered joint			
	„	2	¾" ditto as last, but screwed for iron, and with joint to iron pipe			
	„	1	1" bib cock, as first described			
	„	2	¾" brass high-pressure screw-down stop cocks, with unions both ends, and soldered joints			

No.			£ s. d.
No.	1	1" gun-metal ditto strong water-works pattern, and ditto	
,,	1	Shanks' (46 Cannon Street, London) bath plate for 1" supplies and waste, with discharge pipe and grating tested and stamped by the New River Company, and joints and fixing	
,,	1	2¼" brass Butler's pantry washer plug, and extra strong brass chain, and soldering to sink and soldered joint	
,,	2	4" heavy brass thimble or tail piece to soil pipe and soldered joint, and connection with gully	
,,	2	Bolding's No. 92 3-gallon "Tranquil" syphon water-waste preventing cistern, with strong brass chain and porcelain pull, including soldered joints to supply, overflow and flushing pipe, and fixing on and including galvanised iron ornamental brackets	
,,	2	Jennings' (Stangate, Lambeth) figure 4 housemaid's sink 3' 6" long, with improved square top, flushing rim, earthenware slop sink with loose grating, 3" drawn lead trap and ⅜" hot and cold taps, and all joints and fixing	
,,	1	Bolding's strong copper bath 5' 6" long, japanned white inside, with unions for overflow and waste, and joints and fixing	
,,	1	Doulton's (Lambeth) flush-out closet with white basin and trap, P.C. 12s. 6d., and joints and fixing	
,,	3	Bolding's No. 22 new pattern white Kenon pedestal closet, with fixed slop top in one piece of earthenware, and joints and fixing	
,,	1	Galvanised wrought-iron riveted cistern, size 2' 6" × 2' 6" × 3' 0", and fixing ..	
,,	1	Ditto, 6' 0" × 4' 0" × 4' 6", of ¼" boiler plate with angle iron around top, bottom and angles, and one cross stay, and all necessary holes for pipes, and hoisting and fixing 50 feet above pavement level	
		Allow for paying fees to water company for connection with main, for supplying screw ferrule, and for making good roads and footways to the satisfaction of the local authorities	

LEAD, SOIL AND VENTILATING PIPES, *per foot run*, including tacks, joints and fixing.—State bore, weight per foot or yard run, how fixed, if anything unusual about it, if tacks are ornamental to design or of cast lead describe them (with the pipe), state if fixed in chase. Number the terminal or labour to top of pipe. Take connection with drain, often with brass thimble and soldered joint.

Count the bends. Number the short branches from w.c. apparatus into vertical pipe, describing the number of bends and brass thimble (often required at junction with arm of pan) with each.

IRON, SOIL AND VENTILATING PIPES, *per foot run.*—State bore, how fixed, how jointed. State if fixed in chase. Number the terminals. Take connection with drain, count the bends, the branch pieces, the connection with drain. The branches from pans into the vertical pipe would be in lead with brass thimbles at junction with the iron.

Bill.

ft.	in.			£	s.	d.
100	–	run	4" heavy galvanised cast-iron soil pipe, with ears cast on, jointed in molten lead, and fixed with galvanised wrought-iron rose-headed nails to hard wood plugs in brickwork			
			No. 6, extra for bends			
			„ 3, extra for branch pieces			
			„ 3, extra for swan necks, 6" projection..			
			„ 3, 3½" heavy brass thimbles and molten lead joint to 4" iron pipe, and cement joint to drain			
No.	6		36" lengths averaged of 4" lead soil pipe with three bends, one brass thimble and soldered joint and molten lead joint to iron pipe at one end, and brass thimble and soldered joint and red lead joint to pan at the other			

WRAPPING PIPES, *per foot run.*—State the nature of the wrapping and what it is secured with, the diameter of the pipe need not be mentioned.

Bill.

ft.	in.			£	s.	d.
200	–	run	Wrapping pipes with Croggon's matted felt, well lapped and bound on with copper wire			
200	–	„	Wrapping pipes with a thickness of 2" of boiler felt, cut in strips, wound spirally round the pipe and tied with string ..			

PACKING CHASES WITH SLAG WOOL, *per foot run.*—State size of chase, and the consistency of the slag wool after packing.

Bill.

ft.	in.			£	s.	d.
100	–	run	Pack chase 9" × 4½" with slag wool to a consistency of 12 lbs. per cubic foot ..			

HOT-WATER SYSTEM.

This work is rarely done as the surveyor measures it, in which case it would be remeasured at completion. As the price of the pipe, including fittings, such as tees, bends, etc., is quite well known, they are generally measured together; their separation involves much trouble and small advantage.

It is generally most convenient to have the whole of the items of the hot water system in the same part of the bill of quantities, they can then be easily omitted if a remeasurement should be necessary.

Measure the pipes, per foot run. State the size, quality, if galvanised, whether including fittings, how jointed.

Observe that the connections with baths, lavatories, etc., must be of strong lead pipe.

Bill.

Hot Water Supply.

ft.	in.				£	s.	d.
250	-	run		$\frac{3}{4}$" iron pipe as described			
100	-	,,		1" ditto			
	No.		4	Tyler and Son's $\frac{3}{4}$" gun-metal loose valve screw-down high-pressure bib cocks, strong water-works pattern, with screw boss and union, engraved Hot, and joint to iron pipe			
	,,		2	Emanuel and Son's (57 Marylebone Lane, London) $\frac{1}{2}$" polished brass valve, with spoke knob and china tablet lettered, and soldered joint			
	,,		1	1" brass strong waterworks pattern high-pressure screw-down stop cock, with unions both ends and soldered joints			
	,,		1	Bolding's $\frac{3}{4}$" No. 829 dead weight safety valve, and joint to iron			
	,,		1	Drilling boiler of range for 1" flow and return, and for connecting, including short lengths of pipe, back-nuts and joints			
				No. 1 galvanised wrought-iron hot-water cylinder, of $\frac{1}{2}$" plate, to hold 50 gallons, with bolted top, brass unions and joints, and fixing			
				Allow for testing the hot-water system, and for leaving perfect and in working order at completion of the contract			
				Allow for attendance, cutting away and making good, in all trades after hot-water engineer			

Number the hot-water cylinder, or cistern, stating its capacity, specially mention the manhole, brass unions and soldered joints.

The Taps. State size and say whether screwed for iron stop cock, as last, and state whether with brass unions at each end.

Safety valve describe, and a P.C. may be stated. Item for drilling boiler of range and connecting flow and return to it. Item for attendance and making good.

ZINCWORKER.

ZINC.—Measure at per foot superficial in the same manner and with the same allowances as for lead, except that the sheets being only 7 feet and 8 feet in length and 2 feet 8 inches and 3 feet wide, more drips will be required. State the gauge, and whether English or "Vieille Montagne," or as may be.

If the zinc is corrugated, state it, and measure as if plain, making no allowance for the corrugating. State how it has been measured, and the way the sheets are connected.

The zincworker usually charges his work at a price per foot superficial, including everything except soldered shields to ends of rolls; these should be numbered.

Observe that the "rolls" should be described (in the carpenter's bill) as twice splayed fillets, stating the size.

Bill.

ft.	in.			£	s.	d.
754	–	supl.	No. 15 zinc, as described in flats			
352	–	,,	,, 12 zinc, ditto in flashings			
174	–	,,	,, 12 zinc, ditto in stepped flashings ..			
	No.	155	Soldered shields to rolls			
194	–	run	3" rain-water pipe out of No. 15 zinc, with ears soldered on and fixed with strong galvanised iron nails to brickwork, measured net			
			No. 10 shoes			
			,, 8 swan necks 15" projection			
			,, 10 plinth bends 2¼" projection			
102	–	,,	4" Half round eaves gutter out of No. 15 zinc, including stout zinc clips nailed to the rafters, measured net			
			No. 18 stopped ends			
			,, 10 outlets			
			,, 4 angles			

Measure gutters and rain-water pipes by the foot run, stating the size and out of what gauge zinc, and that they are measured net.

Number the angles, outlets, etc., to gutter, and the heads, shoes, bends, etc., to the rain-water pipes.

Articles of ornamental stamped zinc may be described by numbers from a trade list.

COPPERSMITH.

Measure copper per foot superficial. Allow for turning up against vertical faces 6 inches up slopes of roof, measured from the sole of the gutter 9 or 10 inches, or as may be specified. Allow rolls 2 feet 9 inches from centre to centre (1¾ inch by 1½ inch) 7 inches extra for each.

Number the ends of rolls (corresponding with bossed ends in leadwork) as extra labour and copper to ends of rolls welted all around.

Number the ends of rolls where stopped against vertical faces as extra labour and copper to saddle clips welted to flashing.

Number the ends of rolls stopping against a drip with another roll, continuing the line on the upper level, as extra labour and copper to end of roll welted all around, and saddle clip to roll at lower level.

For drips 1½ inch high, allow 2½ inches extra; 2 inches high, 3 inches extra.

Flashings usually 6 inches wide, the lower edge welted, laps 3 inches or 4 inches, as to lead flashings. Tacks are not required.

Stepped flashings 12 inches wide, the lower edge welted, laps as last described.

Copper flats are often constructed without wooden rolls, for which double welts are substituted about the same distance apart. Allow for each welt 4 inches.

As the sheets are usually 4 feet long, either drips or double welts must be allowed at a distance consistent with this. For double welts allow 12 inches extra in the measurement, and measure the extra labour per foot run.

Measure per foot run. Labour to double welt, close copper nailing, brazed angle (in cases where soldered, angle would occur in leadwork), escalloped edge, serrated edge, corrugated edge, etc.

COPPERSMITH.

Bill.

ft.	in.			£	s.	d.
500	–	supl.	Copper and labour in gutters, flats and flashings			
200	–	,,	Ditto in stepped flashings			
20	–	run	Close copper nailing			
50	–	,,	Labour to double welted lap			
20	–	,,	Ditto escalloped edge			
	No.	20	Extra labour and copper to ends of rolls welted all around			
	,,	40	Extra labour and copper to saddle-clips welted to flashing			
	,,	40	Extra labour and copper to ends of rolls welted all around, and saddle-clip to roll at lower level			

PLASTERER.

Keep external work separate from internal work.

Measure plastering generally by the yard superficial. In all work, except cement and stucco work, narrow widths may be included with the general work. Measure walls and partitions from top of skirting grounds to ceiling where not plastered down to floor; where they are, measure from floor to ceiling. If only rendered behind skirtings, measure as if ordinary plastering down to floor, the difference in value is so small as not to be worth the distinction.

Where plastering is on old walls, take "hacking off old plastering, and raking out joints to form key" for new, and where the surface is likely to be very uneven, say in the description, "allow for any necessary dubbing with tiles and cement," and keep it separate from that on new walls.

If part of the plastering is on old, and part on new walls, the dubbing must be separately measured.

If it is at any time necessary to produce a face more than 1 inch beyond the face of brickwork, add to the description the words "and dubbing out with tiles and cement."

If any plastering is to be more than 1 inch thick it must be stated.

Keep circular plastering separate from straight.

Keep plastering in small quantities separate from the rest, and measure and bill by the foot superficial instead of by the yard; separate pieces not exceeding a superficial yard are thus described.

If any plastering is finished in gauged stuff it must be stated in the description, and say what it is gauged with and proportions of the material.

Deduct openings clear of the grounds, or clear of the openings where there are no grounds.

Make no deductions for chimneypieces, as the quantity not deducted is considered equivalent to the making good; or deduct plastering, and add ... making good to chimneypieces. The former is the usual practice.

The case of plastering to specially high stories may be met by a clause in the preliminary bill, which see.

When cement work, as in reveals, etc., is stopped against stone, terra-cotta, or brick facings, state it.

PUGGING, *per yard superficial.*—State the thickness and of what composed. Measure the whole surface, including the space occupied by joists, and state that the joists are not deducted. Deduct chimney breasts, hearths and voids.

COUNTERLATHING AND FILLETS AND TWICE ROUGH RENDERING TO PARTITIONS, *per yard superficial.*

COUNTERLATHING should be measured to all timbers, such as lintels, finished with a plastered face which would be lath, plaster, float and set in narrow widths."

PLASTERING TO WALLS, *at per yard superficial.*—State if render, render and set, or render, float and set, if circular keep separate and bill in feet instead of yards.

PLASTERING TO PARTITIONS, *at per yard superficial.*—State if *lath, plaster and set*, or *lath, plaster, float and set*, if "lath and half" or "double laths." The foregoing descriptions in italics will be sufficient when the lathing is of the common thickness. Deductions as to walls.

PLASTERING IN NARROW WIDTHS means surfaces not exceeding 12 inches in width; in cement work, these are always stated separately in the bill; in ordinary plastering, the addition to the general item of the words "including narrow widths" is sufficient.

STUCCO, *per yard superficial.*—State if "render, float and finish, trowelled or bastard stucco for paint," or "lath, plaster, float and finish, trowelled or bastard stucco for paint." State if floated with a felt float. If circular, if in narrow widths or smallquantities, measure at per *foot* superficial.

Bill.

yds.	ft.	in.			£	s.	d.
76	-	-	supl.	Twice whiten ceilings and soffits			
254	-	-	,,	Pugging 2" thick of lime, sand and chopped hay (the joists not deducted)			
72	-	-	,,	Lathing and stout deal fillets nailed to the quarters, and rough render one coat on both sides between quarters of partitions, the quarters not deducted			
110	-	-	,,	Rake out joints of old brickwork to form key for plastering			
15	-	-	,,	Dubbing out 1" thick in tiles and cement on old walls			
10	-	-	,,	Render and set walls			
75	-	-	,,	Render float and set ditto			
64	-	-	,,	Ditto, circular			
110	-	-	,,	Render float and finish trowelled stucco for paint			
142	-	-	,,	Lath, plaster and set partitions			
425	-	-	,,	Lath, plaster, float and set ditto			
	9	-	,,	Ditto, circular			
400	-	-	,,	Render float and set walls gauged with equal quantities of lime and cement ..			

PLASTERING TO CEILINGS AND SOFFITS, *at per yard superficial.*—As described for partitions, but state that it is to ceilings and soffits. Keep sloping ceilings separate. Keep circular ceilings separate. If between rafters state it; measure the rafters in, and state that they are not deducted. If in panels, describe as " in panels not exceeding ... yards each." Keep flueing soffits separate. The whitening will be measured with the plastering, but separate in the bill. The theory is to deduct from the ceiling the projection of cornice at one side and end, but this is rarely done unless the cornice is unusually large (24 inches girth). If groined, measure as circular and measure the groin-point per foot run.

Bill.

yds.	ft.	in.			£	s.	d.
100	-	-	supl.	Lath, plaster, float and set ceilings and soffits			
64	-	-	,,	Ditto, sloping..			
75	-	-	,,	Ditto, circular			
29	-	-	,,	Ditto, between rafters, the timbers measured in			
	75	-	,,	Lath, plaster, float and set sloping ceiling in panels, not exceeding 4 feet superficial each			
	20	-	,,	Lath, plaster, float and set flueing soffits ..			

QUANTITY SURVEYING.

Measure at per foot run.—Reveals not exceeding 9 inches wide. Where with an arris, describe both together as arris and ... inch return, arris, beads, chamfers, rounded angles, of the two latter, stating girth or width. Arrises to salient angles of walls or partitions.

Bill.

ft.	in.			£	s.	d.
20	–	run	Arris			
73	–	,,	Arris and 7" return			
42	–	,,	Bead and two quirks, 3" girth in all			
17	–	,,	Chamfer 2" wide, and arrises			
22	–	,,	Rounded angle, 6" girth			

INCISED PLASTERING, *per yard superficial*, or a price per yard should be provided. Give a sketch to show the kind of work. If in panels number them, giving the size and a sketch. Bill in feet. The material will be defined by the sub-heading under which it appears, as ordinary plastering, Portland cement, &c.

Bill.

ft.	in.			£	s.	d.
15	–	supl.	Labour and materials to incised plastering, to design			
No.	4		Ditto in panels 18" × 12" ditto			

ROUGH CAST, *per yard superficial.*—State composition whether on brick or lath, describe the laths, if in small quantities measure per foot superficial.

Bill.

yds.	ft.	in.			£	s.	d.
750	–	–	supl.	Rough cast, made with clean sharp washed shingle and sand, and Portland cement in approved proportions			
	643	–	,,	Lath with oak laths as described, plaster and float in lime and hair, and finish with rough cast, made with clean sharp washed shingle and sand, and Portland cement in approved proportions, between timbers the panels averaging 12" wide in clear, and many to irregular shapes, measured net, and clean timbers ready for painting			

PLASTERING ON METAL LATHING, *per yard superficial.*—State whether lime or cement, and its finish. Describe the metal lathing and how fixed.

Bill.

yds.	ft.				£	s.	d.
50	–	supl.	Expanded metal, as lathing, fixed with galvanised iron nails and plaster, float and set, including narrow widths to partitions				
50	–	„	Jhilmil patent metal lathing (Hayward Bros. and Eckstein, Union Street, Borough) fixed with galvanised iron nails, including narrow widths to ceilings..				

CORNICES.—Where not exceeding 12 inches in girth by *the foot run*, where exceeding 12 inches in girth by *the foot superficial* If the cornices do not exceed 12 inches in girth the length of the walls will be sufficiently near for the length of the cornice; where they exceed 12 inches in girth the mean length must be taken.

Where cornices are bracketed keep them separate, and state that they are "on lath." Where cornices are circular, state radius and keep them separate.

Where cornices are enriched measure the whole girth of the cornice, including the space which will be occupied by the enrichment, as moulded work.

Coves, in cases where they are not run in the same process which produces the cornice, should be measured per foot superficial, and described as coves; otherwise measure and describe them as cornices.

It is customary where the girth of a cornice is measured from the mould used to run it, to allow 1 inch beyond the actual girth on the mould. Make the same allowance when measuring from a detail drawing. This allowance is to cover the expense of running a screed on the ceiling. See also preamble to Plasterer's bill, Chapter IV., section BILLING. In the absence of details, it will be reasonable to assume that the height of the wall and width of ceiling covered by cornice added together equal three-quarters of the girth, that one-third of this is on wall and two-thirds on ceiling.

It is not usual to make any distinction for short lengths of cornice, where of the usual character, as to returns of chimney breasts, but where from the nature of the work there are many; anything under 18 inches in length should be described as in short lengths, and kept separate.

Number all the mitres, irregular mitres, stopped ends, stopped ends on splay, returned and mitred ends, stating girth of cornice, circular corners, stating girth of corner and cornice.

ENRICHMENTS, *per foot run.*—State the girth, and whether to design. Keep circular separate, and state the radius.

Number the mitres, and state whether specially modelled.

Stopped ends, etc., as to cornices.

In all cases of enrichments state that modelling is to be included.

Bill.

ft.	in.			£	s.	d.
70	–	supl.	Plain face on lath, including dubbing out 1¼" to form panel margin			
402	–	„	Moulded cornice			
86	–	run	Ditto 6" girth..			
			No. 4 mitres			
86	–	„	Ditto 9" girth..			
			No. 16 mitres			
251	–	„	Ditto 10" girth			
			No. 29 mitres			
			„ 1 stopped end on splay			
			„ 2 mitred returned ends..			
1065	–	„	Ditto 12" girth			
47	–	„	Ditto 12" girth on lath..			
			No. 145 mitres			
			„ 2 returned mitred ends			
80	–	„	Ditto 12" girth, on lath with two enrichments 3" girth			
			No. 8 mitres			
			„ 2 irregular ditto			
			„ 3 quadrant corners 9" girth			
70	–	„	Enrichment 7" girth to detail			
			No. 6 mitres			
74	–	„	Ditto 9" girth			
			No. 4 mitres			
			„ 2 stopped ends..			
	No.	46	Mitres to cornice, average 15" girth			
	„	12	Ditto, irregular			

Moulding ceiling ribs, *per foot run.*—The ceiling will be measured over all, and billed with the ordinary ceilings. State girth and that they are to include cutting to form key and making good after. If net girth state it. If not, add to girth 2 inches for screeds. Keep circular ribs separate, stating the radius.

PLASTERER. 239

Number the mitres, stopped ends, etc. It will be found most convenient where ribs intersect with others, or with the top-member of a cornice, to count and describe the mitres as "half-mitres."

In the case of ceilings with ribs, an item should appear in the bill something like the following: "Allow for setting out ceiling 29 feet by 21 feet ('to irregular plan,' if irregular) in geometrical panels to detail."

Bill.

ft.	in.			£	s.	d.
112	–	run	Ceiling ribs 1¼" × 2¼", moulded 5" girth, including cutting to form key and making good after			
105	–	,,	Ditto, circular			
			No. 412 half mitres			
			Allow for setting out ceiling 18' 0" × 11' 0", to geometrical plan to detail			

COFFERED CEILINGS IN FINE PLASTER.—The plain ceiling would be measured at per foot superficial over the stiles and rails and described as in panels, keeping those not exceeding one yard superficial separate and billing them separately. Stiles and rails if plain with plain returns per foot run, stating width and depth of returns and if dubbed.

MOULDINGS ON ANGLES, *per foot run.*—Stating girth and counting the mitres.

MOULDINGS IN PANELS.—State girth and that it is in panels, and count the mitres.

Bill.

yds.	ft.	in.		*Coffered Ceiling in fine Plaster.*	£	s.	d.
30	–	–	supl.	Lath, plaster, float and set in panels			
	130	–	,,	Ditto in panels not exceeding one yard superficial			
	45	–	run	Stile 6" wide with two 1½" returns, including dubbing			
	35	–	,,	Ditto, elliptical on plan, ditto			
	130	–	,,	Moulding 3" girth			
	60	–	,,	Ditto 3" elliptical on plan			
				No. 8 mitres			
				,, 24 ditto, irregular			

CIRCULAR WORK.—Keep circular work separate. If to columns

keep separate, describe its purpose, and state if columns are diminished or with entasis.

Generally at per foot run.—Plaster quirk. Making good plastering to slate shelves, etc. Mouldings not exceeding 12 inches girth, stating the girth and counting the mitres.

Bill.

ft.	in.			£	s.	d.
70	–	run	Plaster quirk			
104	–	,,	Moulding 3″ girth			
			No. 24 mitres..			
			,, 2 stopped ends			
			,, 4 ditto on splay			
26	–	,,	Make good plastering to edges of slate shelves			

Generally number, making good plastering to ends of bearers, cantilevers, handrails, &c., and to those things which have been fixed after plastering was done.

Lengths of mouldings, modillions, paterae, caps to columns or pilasters, stating size and giving full description and sketch, and state that they include modelling, or the repetition of the allusion to modelling may be avoided by the insertion of a clause in the preamble of the bill.

Bill.

			£	s.	d.
No.	4	18″ lengths of moulding 9″ girth, with two external mitres, two internal mitres, and two internal irregular mitres to each as caps to angle pilasters, including modelling			
,,	1	Cast centre flower 4′ 6″ diameter to design, including cutting away to form key and making good			
,,	2	Enriched Ionic caps to columns, with moulded abacus and necking 10″ × 10″ and 10″ high			
,,	2	Foliated caps to piers 14″ wide, with one 4½″ and one 14″ return, and 5″ high ..			

PORTLAND CEMENT, *per yard superficial*.—Measure as for other plastering. State whether "rendered," plain-face, render and float, or trowelled, and whether on brick or lath, and if jointed to imitate stone. Describe anything under 2 feet wide as in narrow widths.

Measure all arrises, rustic grooves, etc., and mouldings not exceeding 12-inch girth, at per foot run.

PLASTERER.

Measure mouldings over 12 inches girth at per foot superficial.

Mouldings, arrises, etc., not exceeding 18 inches long should be described as in short lengths and kept separate.

If mouldings are flush, state it.

Observe that all cement work forming plain faces between rustic grooves must be described as in narrow widths.

Generally per foot run. Skirtings, state height and whether square, flush-beaded, chamfered, or moulded. If dubbed out, state how much projection. Number mitres returned and mitred ends, stopped ends, quadrant corners, stating girth.

Bill.

yds.	ft.	in.			£	s.	d.
70	–	–	supl.	Portland cement, plain face			
	40	–	„	Ditto, in narrow widths			
	990	–	„	Ditto, 1¼" thick, in narrow widths to form rustics			
62	–	–	„	Ditto, trowelled			
	63	–	„	Ditto, in narrow widths			
	105	–	„	Portland cement moulding			
	20	–	„	Ditto in short lengths			
	65	–	run	Arris			
	54	–	„	Ditto, slightly chamfered			
	6	–	„	Mitre to splay			
	10	–	„	Arris, and two 2" returns			
	25	–	„	Ditto, and one 2" and one 9" return			
	380	–	„	Flush bead, and two quirks 1½" wide in all No. 40 mitres			
	520	–	„	Rustic groove, 1½" wide and ¾" deep			
	247	–	„	Splay 4" wide, including arrises No. 40 mitres „ 20 ditto irregular			
	25	–	„	Reveal 6" wide and arris No. 6 mitres „ 6 stopped ends			
	75	–	„	Moulding 6" girth			
	12	–	„	Ditto in short lengths No. 4 mitres „ 2 stopped ends			
	194	–	„	Square skirting 7" high, including mitres, stopped ends, etc.			
	67	–	„	Ditto raking to stairs			
	13	–	„	Ditto ramped to ditto No. 14, extra for short ramps			
	95	–	„	Skirting moulded 3" girth, 12" high in all No. 10 mitres „ 2 ditto, irregular „ 4, 9" lengths, with one mitre and one stopped end to each			

R

Angle, state if with one or two returns, and the width of the returns; state if chamfered or rounded.

Angle beads, and state if with one or two quirks. Chamfered angles, state width. Splays, state width and that they include two arrises, and describe returns if any. Reveals, describe width and include arris.

The repetition of the words Portland cement in each item may be avoided by arranging a subsection of the bill headed Portland Cement.

Weatherings not exceeding 12 inches wide per foot run. State width, and that they include dubbing and arrises.

Those over 12 inches wide are measured superficially and the arrises taken separately.

Bill.

yds.	ft.	in.			£	s.	d.
	100	–	supl.	Weathering, including dubbing with tiles and cement			
	80	–	run	Arris			
	100	–	,,	Weathering 9" wide, including dubbing with tiles, and cement and arris			

Generally numbers. Mitres to skirtings, splays, mouldings, etc. Stopped ends and mitres to splays, junctions of circular and straight splays, etc. In all cases state girth or width.

Many of these are "written short" after the lineal dimension to which they belong.

Bill.

			£	s.	d.
No.	10	Make good to ends of iron bars			
,,	20	Mitres to mouldings 19" girth (averaged)			
,,	6	Splayed ends to ditto 21" ditto			
,,	18	Stopped ends to ditto 12" ditto			
,,	16	Ditto on splay 21" ditto			
,,	8	Moulded stops to moulding 15" ditto			
,,	2	Returned and mitred ends to moulding 16" girth			
,,	20	17" lengths of flush moulding 4" girth, with four irregular mitres			
,,	45	Quoins (as sketch) 22" long, with 12" return and 15" high, including dubbing and projecting 1" from general plastering			
,,	6	Dubbing out for and forming plain key stones 1' 2" × 1' 8", and 2" projection			

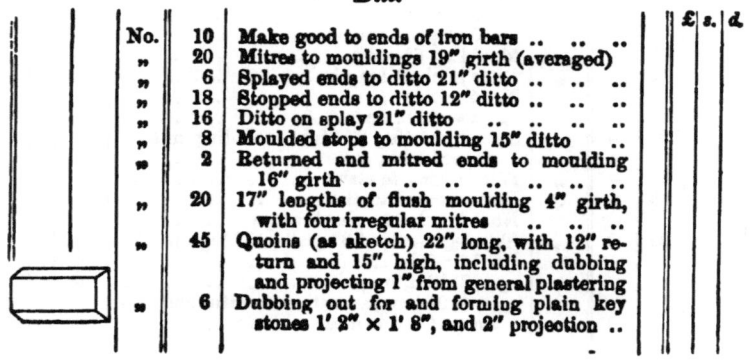

PLASTERER.

Short lengths of mouldings with mitres, returns, etc., are sometimes numbered, stating girth lengths, lengths of returns, mitres, etc., and their purpose.

Circular lengths may be similarly treated.

Bill.

No.	5	10" lengths of moulding 2" girth with two 2" returns, two external mitres, and two internal mitres as necking to pilaster	£	s.	d.
"	5	10" lengths of plain plinth with moulding 3" girth, 9" high in all, with two 2½" returns, two external mitres, and two stopped ends as base to pilaster			
"	5	14" lengths of moulding 6" girth with two 4" returns, two external mitres, and two internal mitres as cap to pilaster			

MARTIN'S, KEEN'S, OR PARIAN CEMENT.—Measure as other cement work. State whether on backings of coarse cement of the same kind or on backings of Portland cement, and whether finished for paint.

WHITENING AND COLOURING, *per yard superficial.*—State if claircolled, if stippled, whether once or twice whitened or coloured, and whether on plastering or brick. Measure the foregoing with the plastering on which it is done.

DISTEMPERING TO CORNICES, *per foot run*, measured with the cornices. State whether once or twice, whether in one or more tints, and the girth of the cornice.

Bill.

yds.	ft.	in.			£	s.	a
423	–	–	supl.	Twice whiten ceilings			
140	–	–	"	Claircolle, and twice distemper ceilings a tint			
640	–	–	"	Ditto walls pink			
140	–	–	run	Ditto cornice 12" girth, and pick out two members in another tint			

Whitening and distempering are more often done by painters than by plasterers, and many surveyors bill such items in the Painter's bill.

FIBROUS PLASTER, *per yard superficial.*—State whether on walls or ceilings, how fixed and how finished, and what with, and whether it includes battens. Measure it net and to include all cuttings, and describe it so.

Bill.

yds.	ft.	in.			£	s.	d.
80	–	–	supl.	Fibrous plaster slabs of the best quality, fixed with 1¼" galvanised iron screws to and including 2" × 1" grounds about 2 feet apart, plugged to walls, stopped and finished with one coat of setting stuff, measured net, including narrow widths, and allow for all cutting and waste ..			
44	–	–	„	Fibrous plaster slabs of the best quality as ceiling, fixed with 1¼" galvanised iron screws to the joists, stopped and finished with a thin coat of Parian cement, measured net, including narrow widths, and allow for all cutting and waste			

FOUNDER AND SMITH.

Keep wrought and cast work separate.

In all cases of heavy work measure the superficial quantity of iron and state the thickness in the dimensions, so as to obtain the weight.

For girders, columns, stanchions and heavy pieces of ironwork generally, state the height of hoisting and what trades they are fixed by.

Keep the different kinds of articles separate, and in addition to the weight state what the weight comprises, as, so many columns or stanchions of such a height.

Take a pattern to every variety of casting and an alteration to pattern where the pattern is altered, as in the instance of the shortening of a column.

If artistic wrought ironwork of high finish is required, a sum should be *provided* in the quantities founded on an estimate from an ironworker to be selected.

In measuring labours on iron, be careful to state if any of the work is to iron in position, as for instance " No. . . . rivets ⅝-inch diameter and including holes through one thickness of ¼-inch and of ⅜-inch iron *in position.*"

Observe to measure the iron in the internal rounded angles of castings (featherings), or make a note to add a percentage to the weight on the abstract. 2½ per cent. addition is the ordinary custom.

The dimensions of iron columns should always appear in the dimensions immediately preceding those of the work they support, and at the same time should be measured all their adjuncts, as brick piers, stone bases, etc.

IRON COLUMNS OR STANCHIONS.—Measure the iron superficially, stating the thickness so that it may afterwards be reduced to weight.

Measure planing per foot superficial and describe it as in small quantities.

Bill.

Cast Iron.

cwts.	qrs.	lbs.			£	s.	d.
45	2	7		In 7 hollow columns, and hoisting and fixing at various levels, not exceeding 40 feet above street paving			
				No. 1 pattern for hollow column 8 feet high, with moulded base, cap and necking, and four brackets			
				No. 3 alterations to ditto			
70	3	14		In 5 stanchions of H section, and hoisting and fixing at ground floor level			
				No. 1 pattern to stanchion of H section, with square cap and base, and four stiffeners			
		No.	10	Wedging up bases of stanchions 15″ × 15″ with iron wedges, 1″ from stone bases, and running 1″ thick with pure Portland cement			

EAVES GUTTERS, *per foot run.*—The most frequent practice is to measure by the yard, any dimension less than 3 feet being called a yard. Some surveyors measure the exact length, but if this is done it should be described as "measured net," and including all short lengths. State size, if jointed in red-lead, if bolted at joints, if screwed to fascia or rafters' feet, if bedded on brickwork and blocked up to falls. If circular, state the radius. Describe the brackets or clips if of unusual pattern.

If it be desired to adopt a gutter of any particular manufacture.

state the name of manufacturer, and the number in their trade catalogue. State if "heavy."

Number. The angles, outlets, stopped ends, returned and mitred ends. The brackets, if of wrought iron, stating the weight, and if ornamental give a sketch, and state how they are fixed, or a trade catalogue may be referred to.

If angles are irregular and not to be found in a trade catalogue, call them irregular purpose-made.

Bill.

ft.	in.			£	s.	d.
51	–	run	4" half round eaves gutter, with wrought-iron brackets screwed to feet of rafters.. No. 4 angles .. .. „ 4 outlets .. .. „ 2 stopped ends			
114	–		Macfarlane's (Glasgow) 4¼" × 3½" No. 12 eaves gutter, bolted and jointed with red-lead cement, and fixed with stout screws to fascia .. .. No. 44 stopped ends .. .. „ 4 angles.. .. .. „ 2 ditto, irregular .. .. „ 23 outlets cast on to back of eaves gutter, with small quadrant bend into rain-water pipe			

RAIN-WATER PIPES, *per foot run.*—Measure in the same way as described for eaves gutters. Where a manufacturer is selected, give the number in trade catalogue as described for eaves gutters. State size, if "heavy," whether round or square, if set up in red-lead or iron cement, if joints are plugged with tow, if with ears cast on, if nailed with rose head or ornamental nails, if with oak or teak plugs in the brick or stone work. If with ornamental bands give price each, or state number in trade catalogue, and keep them separate from the pipe.

If measured net, describe as measured net including all short lengths.

Number shoes, plinth bends (state projection), swan necks (state projection), heads, if ornamental, giving price or number in trade catalogue, Y pieces. State in all cases the size of pipe. If either of these are described in the bill as "extra for," their length must first be measured in with the ordinary pipe.

FOUNDER AND SMITH.

If either of the foregoing require to be purpose-made state it, and that it is to include pattern. Sometimes troublesome bends are made in heavy lead pipe, generally at a less cost than in iron.

Observe that angle bends to rectangular pipes are only made for equal sided pipes.

Bill.

ft.	in.			£	s.	d.
69	–	run	3" round rain-water pipe, with ears cast on, and fixing with wrought-iron rose-headed nails to brickwork			
			No. 5 shoes			
			„ 2 elbows			
			„ 5 extra for plinth beads, 2¼" projection			
			„ 4 extra for swan necks, 15" projection			
342	–	„	Macfarlane's (Glasgow) 4"×3" heavy rain-water pipe, fixed with loose bands (elsewhere taken), with wrought-iron nails to and including teak plugs in brickwork			
			No. 14 extra for bends			
			„ 11 plinth bands, 3" projection			
			„ 26 shoes			
			„ 117, No. 6 Macfarlane's loose bands ..			
			„ 5, No. 13, ditto heads			

KING OR QUEEN HEADS, *number*.—Give weight of each—State that it is to include pattern.

Bill.

			£	s.	d.
No.	6	King heads, weight 112 lbs. each, including pattern and fixing by carpenter			
„	40	Corbels, weight 6 lbs. each, and building in by bricklayer			

COAL PLATES, *number*.—State diameter, whether plain or illuminating, and include wrought-iron hook, chain, staple and padlock, and protecting ring if any.

Bill.

		£	s.	d.
1	Hayward Bros.' (Union Street, London, S.E.) patent 14" illuminating coal plate with protecting ring, stout chain, hook and staple, and fixing to brickwork, and running ring with cement			

PAVEMENT LIGHTS, *number*.—State size, if illuminating, state how glazed, if prismatic, if with round or square lenses, if with irregular angles or irregular shaped.

If patent, give address of manufacturer and number in trade catalogue.

The adoption of stock sizes will save time and expense.

Bill.

No.	2	Hayward Bros.' (Union Street, London, S.E.) pavement lights glazed with 4" × 3" convex lenses, section 2 C.C., the lenses bedded in red-lead, and the frames bedded in cement, 3' 9" × 1' 7"	£	s.	d.
,,	2	Ditto, to irregular shape, 3' 0" × 1' 4" extreme			
,,	,,	Ditto, semicircular, 3' 0" diameter			

GRATINGS OR PANELS.—State size and thickness, or refer to a number in a trade book.

Bill.

No.	5	⅜" gratings, with 1" bars about 2¼" apart, and 1" border, and coating while hot with a mixture of boiling tar and pitch, including pattern and fixing, and running with cement, 3' 0" × 1' 3"	£	s.	d.
,,	2	Macfarlane's (Glasgow) No. 389 ornamental panel 1' 9" × 1' 9", and fixing by brick-layer, and forming and rendering opening with cement through 1½-brick wall			

STOVES.—These are usually dealt with as a provision. The setting and blacking taken separately, but billed with the general work. If separately stated in the smith's bill, describe them at a P.C. or list price.

Bill.

	3	Stoves P.C. 3*l*. each, and add for carriage and profit	£	s.	d.
	1	Range P.C. 15*l*., etc., ditto			

ORNAMENTAL BALUSTERS OR NEWELS.—Generally selected from a trade catalogue and described by a number.

Bill.

No.				£	s.	d.
No.	8		Macfarlane's (Glasgow) No. 629 baluster and fixing, and riveting to core rail ..			
„	2		Ditto newel No. 671, and fixing and riveting to core rail			

STABLE FITTINGS.—The trade catalogues of the St. Pancras Ironwork Co., Musgrave, or Hayward Bros. and Eckstein, which are all well illustrated, show examples of all the requirements of a stable and harness room.

Any article may be described by the number in the catalogue. The bill should clearly show the intention as to carriage, profit and fixing; some of the more fragile articles like bit-cases will also require packing. Either list price, or P.C., which is the list price minus the trade discount, may be adopted.

Take the ends cut and pinned, ends cut and fitted, and mitres to such things as rails; and the boarding to wall linings, and filling in to stall divisions, or loose box enclosures.

Take stone bases for stall posts, or blocks of concrete for posts, with self-fixing bases.

State whether fittings are plain, in which case the painting must be measured, or enamelled.

Bill.

ft.	in.			£	s.	d.
			Stable Fittings to be supplied by the St. Pancras Ironwork Co., St. Pancras Road, London.			
			Allow on all stable fittings for carriage, profit and fixing			
127	–	run	Wrought-iron bottom sill, list price 1s. 3d. per foot run			
			No. 5 extra for shifting pieces			
			„ 10 ends cut and fitted			
			„ 4 mitres			
88	–	„	Wrought-iron moulded top capping for 1" wall lining			
			No. 10 ends cut and fitted			
			„ 4 mitres			

ft.	in.			£	s.	d.
32	–	run	Wrought-iron ventilating grating for loose box, with cast-iron top capping and intermediate sill 2' 0" deep for 2" boarding, list price 6s. 6d. per foot run			
	No.	1	No. 507 iron wall recess for water-cocks and drain pot, and fixing, including cutting to brickwork and paving			
	„	4	Cast-iron grating, list price 8s. 6d., and fixing			
	„	2	Set of iron and brasswork for loose box door, comprising ventilating grating 2' 0" deep, improved hangings and No. 148 patent safety latch, list price 42s. per set			
	„	4	No. 380A improved quadrant movable ventilator, with slide-bar fastening, list price 2l. 5s.			
	„	2	4" diameter loose box post, with self-fixing base, list price 3l. 8s., and preparing to hang door to it			
	„	2	4" ditto, list price 3l. 3s., and preparing with slot for latch			
	„	2	Patent manger fittings with wrought-iron hay-rack, sloping grid, brass plug and washer, etc., list price 3l. 5s.			
	„	1	Set of ironwork for stall divisions, including cast-iron extra strong moulded ramp, wrought-iron sill for 2" boarding, and wrought-iron post 4" diameter, with self-fixing base, list price 3l. 10s. 6d.			
	„	2	Patent manger fitting 6' 0" long, with extra large barrel front, wrought-iron hay-rack, improved shoes and bearers for fixing, brass plug and washer, brass shackles for halter tyings, etc., list price 3l. 17s. 6d. ..			

ROLLED JOISTS IN SMALL OR MEDIUM SIZES, UP TO 9 INCHES DEEP, *at per foot run*, billed *at per cwt.*—State if to exact lengths, if they are hoisted to various heights. State the weight in lbs. per foot run. If over 30 feet long, state it and keep them separate.

LARGE ROLLED JOISTS, *per foot run*, billed *at per cwt.*—State the number of joists comprised in the weight, and the height of hoisting.

Many large girders are constructed of rolled joists of stock sizes with a bottom or top plate or both bolted on; they should be separately billed.

Iron or steel joists are usually selected by their weight per foot run from an iron merchant's list. Iron joists or girders should always in the dimensions immediately precede the work they support, as a wall or floor.

Preserve the calculations made to settle the size of joists, with the dimensions.

Bill.

cwts.	qrs.	lbs.		£	s.	d.
44	3	21	Rolled joists not exceeding 9" deep, cut to exact lengths, hoisting to various levels, and fixing by bricklayer			
26	2	14	Ditto not exceeding 12" deep, and ditto ..			
27	1	7	Ditto 18" deep in two joists, and hoisting and fixing 60 feet above street level by bricklayer			
20	–	–	In one combination girder of two 10" × 5" joists, and 1" top and bottom plate, bolted together, and hoisting and fixing 25 feet from ground level..			

RIVETED GIRDERS, measure the iron *per foot superficial*, stating thickness of iron, bill *at per cwt.*—Add 5 per cent. of the total weight in abstract for the weight of rivets of the usual distance apart ("4-inch pitch").

Five per cent. is the usual allowance, but in deep lattice girders 2½ per cent. is sometimes sufficient.

Riveted stanchions are measured in a similar manner.

Bill.

cwts.	qrs.	lbs.		£	s.	d.
26	1	7	In one riveted girder 23 feet long, of boiler plate, with angle irons and stiffeners, and hoisting and fixing 24 feet above ground level			

HEADS OF RIVETS COUNTERSUNK AT BEARINGS.—State number of rivets.

Bill.

				£	s.	d.
No.	10	Countersink set of six rivets to bearing of steel girder				
"	20	Ditto set of eight ditto				

FLITCH PLATES, *at per foot superficial*, billed *at per cwt.*—State that they include perforations, or the perforations may be taken separately, stating their diameter and the thickness of the iron.

Bill.

cwts.	qrs.	lbs.		£	s.	d.
22	2	14	In rolled iron flitches, including perforations and fixing by carpenter			

CHIMNEY AND BEARING BARS, *at per foot run*, billed *at per cwt.*—Generally measured about 2 feet longer than opening, and ½ inch by 2½ inches. State size, and if cambered and caulked, and add to description "and fixing by bricklayer."

Bill.

cwts.	qrs.	lbs.		£	s.	d.
2	0	21	In chimney and bearing bars, cambered and caulked, and fixing by bricklayer ..			

STRAPS AND BOLTS, *at per foot run*, billed *at per cwt.*—Describe as "bolts and straps, including perforations." Only bolts over 24 inches long should be included with the foregoing. Some surveyors allow in the measurement six times the diameter of the bolt for the head and nut, but it is better to make an item "extra for large head nut and washer to 1¼-inch bolt," or as the case may be.

Bill.

cwts.	qrs.	lbs.			£	s.	d.
2	1	7		In bolts and straps, including perforations and fixing by carpenter			
		No.	8	Extra for sets of large heads, nuts and washers to 1¼" bolts			

SADDLE-BARS, *per foot run.*—State also the total weight in bill. State if round or square, the size of section, and if galvanised. Rings of saddle-bar are best numbered, stating the number of points to each.

When a provision is made for stained glass those windows in which it occurs may be omitted from the measurement of saddle bars; the persons who supply the glass prefer to supply them also.

Bill.

ft.	in.			£	s.	d.
242	-	run	¾" diameter galvanized iron saddle-bars (weight 1 cwt. 1 qr. 21 lbs.)			

HOOP IRON.—See Bricklayer.

HANDRAILS, *at per foot run*, billed *at per. cwt.*—In bill, state length, size and section, as well as weight; describe as framed. Keep the ramped or wreathed parts separate, but treat in the same manner. Number the scrolls as "extra labour" the iron being measured in with the straight rail.

Bill.

qrs.	lbs.		£	s.	d.
3	21	In 2" × ½" framed handrail (25 feet run)			
1	-	2" × ½" ditto, wreathed (6 feet run)			
		No. 1 extra labour forging scroll			
		„ 1 end forged, perforated, countersunk and screwed			
		„ 2 extra for forging small ramps			

AREA GRATINGS, the bars and rails *at per foot run*, stating size in the dimensions afterwards reduced to weight billed *at per cwt.*—Describe them as *framed*. "Framed" implies all the necessary perforations, riveting, etc.

Bill.

cwts.	qrs.	lbs.		£	s.	d.
1	1	21	In one framed area grating of ¾" × 1½" frame and 1½" × ⅜" bars			

GUARD BARS, *at per foot run*, billed *at per cwt.*—As for area gratings. Number the forged ends (generally pointed), stating the size of bar.

Bill.

cwts.	qrs.	lbs.		£	s.	d.
3	-	14	In framed guard bars, and fixing by joiner No. 22 ends of 1½" × ½" bar, forged, and with countersunk perforation and large screw .. .. No. 2 ends of ⅞" bar, pointed			

NEWELS OR BALUSTERS, *at per foot run*, stating size and whether round or square, billed *at per cwt*. Describe as "framed balusters." Increased size to parts of bars or rails are numbered and described as "extra labour and iron" and a sketch assists the description.

Bill.

cwts.	qrs.	lbs.		£	s.	d.
2	2	7	In ⅞" diameter framed balusters			
2	1	14	In 1½" × 1½" framed newels			

As many estimators price handrail, balusters and newels at a uniform price per cwt., some surveyors put all together in one item as follows.

Bill.

cwts.	qrs.	lbs.			£	s.	d.
5	–	14		In framed balustrade			
		No.	20	Extra labour and iron on 2" × ¼" rail for increasing to 3¼" × 3¼", where 1¼" standard passes through, as sketch ..			

CORE RAILS, *at per foot run*.—In the bill state length and size, as well as total weight. Keep the wreathed or ramped parts separate. Number the extra labours.

Bill.

qrs.	lbs.			£	s.	d.
1	21	In 1" × ¼" core rail, with countersunk holes and screws (60 feet run)				
1	–	In 1" × ¼" ditto, wreathed (30 feet run) ..				
		No. 10 forging short lengths of level circular				
		„ 8 forging short ramps				
		„ 1 forging end as newel cap 3" diameter, extra metal and labour..				

Iron Railings.—If of cast iron, they may be selected from a trade list at a price per foot run ; take the mortises and ends cut and pinned, and painting.

Ornamental iron railings of wrought iron are most conveniently treated as a provision. Measure in addition the cutting of mortises, the ends cut and pinned, and the painting.

Ordinary wrought-iron railings may be reduced to weight. Measure the lengths and sizes of all the various rails and bars, and the extra labours on them.

Bill.

Wrought Iron Railings.

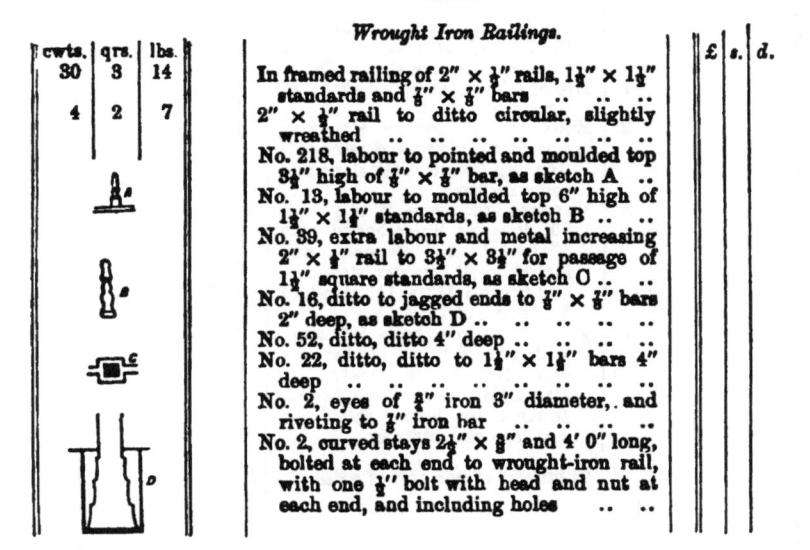

cwts.	qrs.	lbs.		£	s.	d.
30	3	14	In framed railing of 2" × $\frac{1}{2}$" rails, 1$\frac{1}{2}$" × 1$\frac{1}{2}$" standards and $\frac{7}{8}$" × $\frac{7}{8}$" bars			
4	2	7	2" × $\frac{1}{2}$" rail to ditto circular, slightly wreathed			
 A			No. 218, labour to pointed and moulded top 3$\frac{1}{2}$" high of $\frac{7}{8}$" × $\frac{7}{8}$" bar, as sketch A ..			
 B			No. 13, labour to moulded top 6" high of 1$\frac{1}{2}$" × 1$\frac{1}{2}$" standards, as sketch B			
			No. 39, extra labour and metal increasing 2" × $\frac{1}{2}$" rail to 3$\frac{1}{2}$" × 3$\frac{1}{2}$" for passage of 1$\frac{1}{2}$" square standards, as sketch C			
			No. 16, ditto to jagged ends to $\frac{7}{8}$" × $\frac{7}{8}$" bars 2" deep, as sketch D			
 C			No. 52, ditto, ditto 4" deep			
			No. 22, ditto, ditto to 1$\frac{1}{2}$" × 1$\frac{1}{2}$" bars 4" deep			
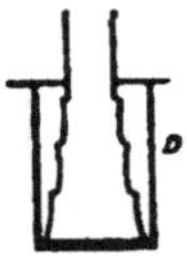 D			No. 2, eyes of $\frac{3}{4}$" iron 3" diameter, . and riveting to $\frac{7}{8}$" iron bar			
			No. 2, curved stays 2$\frac{1}{2}$" × $\frac{3}{8}$" and 4' 0" long, bolted at each end to wrought-iron rail, with one $\frac{1}{2}$" bolt with head and nut at each end, and including holes			

Wrought-Iron Gates.—If ornamental, these are best treated as a provision. They would be fixed by the maker, but the work of the other trades connected with them should be measured and billed with the general work.

Ordinary iron gates may be treated as indicated by the following bill.

Bill.

cwts.	qrs.	lbs.			£	s.	d.	£	s.	d.
				Continued						
		No.	2	Pairs of wrought-iron gates, *each pair* containing as follows:—						
				Detail of one pair.						
1	3	–		In wrought-iron framed gate, of 2" × ½" rails, ⅞" square bars and 1½" square standards ..						
		No.	6	Forging and notching 2" × ½" rail, as sketch						
		"	8	Labour to pointed and moulded top, 3½" high, of ⅞" × ⅞" bar, as former sketch						
		"	2	Labour to moulded top, 6" high, of 1½" × 1½" standards, as former sketch						
		"	2	Extra labour and metal, increasing 2" × ½" rail to 3½" × 3½" for passage of square standards ..						
		"	1	Eye of ⅝" iron, and riveting to ⅞" square bar						
		"	2	Ends of 1½" square standards turned as pivot, and wrought-iron cup let into York stone and run with lead						
		"	2	Wrought-iron straps, 2" × ½" and 20" long, as hinge-bolted with ½" bolt to 1½" standard, and forged to fit circular standard, and turning 1½" × 1½" standard for length of 2¼"						
		"	2	¾" wrought-iron bolts 27" long, with wrought iron slotted eye, and two bands riveted to standards, and wrought-iron thimble, and letting into York, and lead and running						
				Total cost of one pair of gates .. £						
				End of Detail.						
				Carried to Summary				£		

Iron Roof Trusses.—Measure the various sections of bar, rod, angle or tee iron, etc., per foot run and bill by weight, stating the number of trusses and the sizes of their parts. The shoes are usually of cast iron, these appear as numbers, give description and weight of each. It is best to put them all together in the bill as a sub-section headed Iron Roof Trusses.

FOUNDER AND SMITH.

Bill.

Iron Roof Trusses.

cwts.	qrs.	lbs.			£	s.	d.
135	1	21		In No. 8 roof trusses of 4" × 3", angle iron rafters 3½" × 3½", struts and rods varying from ½" to 1¼" diameter, and hoisting and fixing about 32 feet from ground to ridge			
	No.	9		Ends of ½" bolt, forged as eye, and perforated for 1" bolt			
	„	27		Ditto ¾" ditto and ditto			
	„	16		Ditto 1" bolt, ditto for ¾" bolt			
	„	24		Ditto 1¼" bolt, ditto for 1" bolt			
	„	8		Forging ends of ½" rod as fork, beating out ends to form eye for 1" bolt			
	„	15		Ditto 1" rod ditto			
	„	36		Ends of 2¼" × ½" bar forged, beaten out and perforated for ¾" bolt			
	„	48		Ditto 2½" × ¾" ditto for 1" bolt			
	„	32		Ditto 3½ × 3½" tee iron, ditto for ½" bolt ..			
	„	8		Large nuts to ½" bolt			
	„	8		Ditto ¾" ditto			
	„	34		Screw ends and nuts to ¾" bolt			
	„	8		Unions and screw ends to ¾" bolt ..			
	„	16		1" Lewis bolts 7" long, with nuts and screw ends			
	„	32		2¼" lengths of 3" × 3" angle iron as cleats, and riveting with two rivets to ₁₆" iron, and one perforation, and ½" bolt 7" long with head and nut			
	„	16		Cast-iron shoes to roof trusses, weight 28 lbs. each, including pattern			

ANGLE, TEE, OR H STEEL OR IRON BEARERS.—Measure at per foot run, reduce to weight in the bill and state what they are for, as purlins, etc., and it is better to keep the various sections and the various sizes of sections, as separate items in the bill.

Bill.

cwts.	qrs	lbs.			£	s.	d.
20	2	14	in.	5" × 4" × ½" tee as rafter			
10	1	7	„	3½" × 1¾" × ⅜" and ₁₆" angle as purlin ..			
25	1	21	„	6" × 2" I joist as purlin			

Numbers. Holes rimed out as bolt holes (this only occurs in cast work). Perforations, state diameter and the thickness of iron. Forged ends, state size of bar. Sets of ornamental heads, nuts, and washers to bolts, stating diameter of bolt. This item to bolts already measured at per foot run, and no allowance having been made for their weight. Bolts up to 24 inches long averaged,

s

keeping separate those not exceeding and exceeding 12 inches long, and keeping the various diameters separate; state if with ornamental heads, nuts and washers.

Forged angles to steel joists, purlins and the like, stating their nature and the size of the bearer.

BOLTS, *Number*.—State diameter and length, and the kind of nut and washer. Keep separate those not exceeding 12 inches long from the rest.

Bolts should be measured from the inside of head to the point of the screw.

Bill.

	No.	20	Holes rimed out	£	s.	d.
	,,	10	Holes through $\frac{1}{2}$" iron for $\frac{3}{4}$" bolts			
	,,	12	Ditto $\frac{7}{8}$" ditto for 1" ditto			
	,,	6	Forging ends of 2" × $\frac{1}{2}$" bar, and hole for $\frac{3}{4}$" bolt			
	,,	6	Sets of ornamental heads, nuts and washers to 1$\frac{1}{4}$" bolt			
	,,	37	$\frac{3}{4}$" bolts, average 8" long, not exceeding 12" long, with heads, nuts and washers and fixing by carpenter			
	,,	14	$\frac{7}{8}$" ditto, average 8" long, and ditto			
	,,	25	$\frac{1}{2}$" bolts, average 15" long, all over 12" long, with heads, nuts and washers, and fixing by carpenter			
	,,	12	$\frac{7}{8}$" ditto average 18" long, and ditto			
	,,	6	$\frac{1}{4}$" bolts 9" long, with head and handrail nut, and fixing by carpenter			
	,,	10	Sets of gibs and keys			
	,,	50	3" ornamental cup-washers to detail			

CONNECTIONS, *Number*.—Ends of joists cut, fitted and connected with cross-joists, such an item as follows is common. "No. .... ends of iron joists, 2$\frac{1}{4}$ inches by 7 inches, cut and fitted to iron and riveted with 6-inch length of 4-inch by 4-inch angle iron to web of joists of similar section." If notched or forged, it is important to describe such labour. Include the bolts or rivets and perforations with the item.

Bill.

	No.	2	Fitting end of 8" × 5" joist to 12" × 6" transverse joist, and riveting together with two 6" lengths of 2" × 2" angle iron, including perforations and rivets ..	£	s.	d.

The various connections and labours to ends of rolled joists are illustrated in Dorman and Long's trade list.

IRON CASEMENTS, *Number.*—Stating name of manufacturer, the size of each, and how fixed. For dimensions give sight measurement, i.e. the size exposed to view externally, and state number of squares.

Bill.

| | No. | 3 | Burt and Potts' (York Street, Westminster) wrought iron casements and frames, in single squares, section 1, quality 2, with hinges, gun-metal stays and fastenings, and fixing and bedding in white-lead to wooden frames 1' 0" × 2' 0", sight measurement .. .. | £ | s. | d. |
| | „ | 10 | Ditto, 1' 8" × 1' 11" .. .. | | | |

When iron or steel casements are not of a well recognised type such as are supplied by a well-known specialist, state size and shape, whether square, semicircular headed, segmental headed, the thickness, the sections of the various parts, the number of squares, the kind of fastenings.

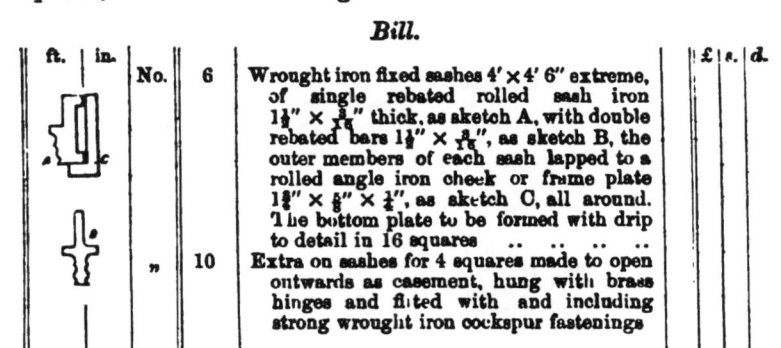

Bill.

| ft. | in. | No. | 6 | Wrought iron fixed sashes 4' × 4' 6" extreme, of single rebated rolled sash iron 1¼" × ¾" thick, as sketch A, with double rebated bars 1⅜" × ¾", as sketch B, the outer members of each sash lapped to a rolled angle iron cheek or frame plate 1⅞" × ⅝" × ¼", as sketch C, all around. The bottom plate to be formed with drip to detail in 16 squares .. .. | £ | s. | d. |
| | | „ | 10 | Extra on sashes for 4 squares made to open outwards as casement, hung with brass hinges and fitted with and including strong wrought iron cockspur fastenings | | | |

FELT PADS.—State size and kind of felt.

Bill.

| | No. | 20 | Pads of stout asphalted felt .. .. .. | £ | s | d. |
| | „ | 12 | 12" × 12" pads of boiler felt, 16 oz. per sheet .. .. | | | |

RIVETS.—State diameter, how many thicknesses of iron they pass through, and that they include holes, and as these (as a general rule) are separately measured only when the work is "in position," state it.

Bill.

		No.	20	Rivets, and two perforations in ¼″ iron to each *in position*		£	s.	d.

FIRE MAINS.—These are best billed as a separate section comprising all trades.

The various applications of mains, hydrants and cisterns are illustrated in the trade lists of Merryweather and Sons; Shand, Mason and Co., and others; Beck and Co. also give illustrations of the appliances.

Measure the pipes per foot run. Number the bends, crosspieces, etc., as " Extra for." Number the cocks, hydrants, sluicevalve, pressure-gauge, etc.

Bill.

yds.	ft.					£	s.	d.
26	—	run		Dig trench for main and return, fill and ram				
		No.	5	Holes through 5″ concrete floor for 4″ pipe, and making good				
		„	1	Ditto through 2½-brick wall for ditto, and ditto				
		„	1	Chamber 1′ 8″ × 1′ 0″ and 2′ 0″ deep, of half-brick sides in cement, on, and including cement concrete 3′ 0″ × 3′ 8″ and 4″ thick, including digging 3′ 0″ deep ..				
		„	1	Cast-iron surface-box, about 12″ × 8″, with chained cover and bedding on brickwork, and embedding in cement concrete about 2′ 4″ × 1′ 8″ and 9″ thick, the surface finished fine				
4		run		2½″ × 2½″ × ¼″ angle iron cut to exact lengths, and fixing by bricklayer as bearer				
		„	8	Strong wrought iron brackets for 4″ main, and fixing to brickwork				
59		run		4″ cast-iron main and joints as described, and coated inside and outside with Dr. Angus Smith's composition				
82		„		4″ ditto, coated on inside only with Dr. Angus Smith's composition, and fixing with and including wrought-iron bolted saddle clips (as sketch) piuned into brickwork				
				No. 2 caps to ditto				
				„ 2 extra to eighth bends				
		„	1	Special cast-iron shoe to support vertical main, and fitting and fixing same				
		„	6	Extra on 4″ main for hydrant tee-piece, with screwed boss for hydrant, and another beneath it at a convenient level above the floor for bucket-cock				

No.				£	s.	d.
No.	1	4" socket cross pipe, with three sockets and joints and one plug				
,,	6	¾" gun-metal bib-cocks screwed for iron with screwed nozzle for hose-tap and chain				
,,	1	Merryweather's or Shand's best quality 4" sluice-valve, having four gun-metal facings, fitted into turned grooves with gun-metal screw and nut tested to a pressure of 200 lbs., the joints flanged and bolted with short spigot and faucet ends, and including joints and fixing, the spindles of valve finished square for loose key, including a large loose wrought-iron tee-key with steel socket, and with steel hook on handle for lifting the iron cover				
,,	6	Merryweather's (figure 403) hydrants of polished gun-metal, the outlets of 2½" internal diameter, with London Brigade thread, wheel, horizontal spindle, loose cap and chain				
,,	1	Best quality hydraulic pressure gauge, and highest pressure indicator and lock dial-face, and connection with 4" iron main by a short length of copper tube, fly-nut and union, and a brass stop-valve (all labour and material)				

GASFITTER.

The intentions of the architect and building owner are so often uncertain at the time of the preparation of quantities, that the gasfitting is almost always remeasured at the completion of the building.

Frequently, therefore, the whole of the gasfitting is put into the bill "as provision" of ... feet of the various sizes of pipe.

State the quality of the tubing ("iron barrel") and if galvanised, and whether the price is to include tees, elbows, connections, bends, etc., and whether the joints are set up in red-lead cement. Also whether it includes attendances and fixing floor boards and joinery with brass cups and screws to remove.

State that the gasfitting is to be executed to the satisfaction of the company's inspector.

PIPING, *per foot run*, giving size.

Number generally. Brackets or pendants, usually P.C. so much each. They are best described by numbers from a trade list. Gas meter, stating whether wet or dry, and the number of lights, and include stamping by company's inspector, and for carriage. When the gas fittings are treated as a provision (nearly always the case), *fixing only* so many brackets, pendants, etc., should be taken.

Count the number of proposed lights, and provide a meter equal to about ⅗ of the number. Thus 100 lights should have a 60-light meter. Reference to a price book will show the capacity of the stock gas-meter.

The meter is generally hired from the company at an annual rental, in such case it will not appear in the quantities.

Number. Tees, elbows, diminishing sockets, screw caps, etc., if not included in general description of piping, but they are usually.

Length of lead pipe and connection with main and meter.

Gas main cock, stating size and that it is jointed to iron pipe.

Item for attendance.

Bill.

ft.	in.	run	GASFITTER—*All as provision.*	£	s.	d.
400	–	„	¾" pipe, as described			
250	–	„	½" ditto			
100	–	„	1" ditto .. .. „			
	No.	1	1" brass gas main cock, with iron spanner and connections with iron pipe			
	„	1	Short length of 1" strong lead pipe, and joints and connections with main and meter			
	„	1	Glover's best dry gas meter for 40 lights, and allow for testing and stamping by gas company, and carriage, profit and fixing Allow for giving notice to gas company, for paying any charges for bringing gas-pipes from main to meter, for testing pipes and leaving perfect at completion, attending upon, cutting away for, and making good after gasfitter in all trades, and making good roads and footways to the satisfaction of the local authorities ..			
			Fixing only, including nipples.			
	No.		16 gas brackets, and allow for 3" polished mahogany rose			
	„		3 gas pendants, and allow for ceiling plates or,			
	„		16 Everad's (29 Drury Lane, London) No. 2795 ¼" gas brackets, and fixing with nipple and 3" polished mahogany rose ..			
	„		3 Ditto No. 2632 three-light pendant, and fixing			

As gas fitting is sometimes valued by the number of points and the contractor finds no difficulty in pricing a bill produced on that principle, the surveyor may measure the pipe from main to the front boundary of the site, i.e. the pipe beneath road and footway, the pipe thence to house, and the rising main, and count the points. The bill would then appear as follows:—

Bill.

ft.	in.				£	s.	d.
			The pipes to be Russell's patent welded gas barrel, including all bends, elbows, tees, and connections, all jointed in red-lead cement, and painting 2 oils after fixing. The diameter of the branch pipes to be as follows: Pipes supplying 1 to 3 lights, ⅜″ diam. Ditto 4 to 9 lights ½″ ditto Ditto 10 to 20 lights, ¾″ ditto				
12	—	run	1″ main and fixing with wall hooks				
50	—	,,	1″ ditto and laying in ground				
20	—	,,	1″ ditto and digging and laying in roadway, and protecting and filling in, including watching and lighting and making good as required. *Provisional*				
			Laying on gas to various points from 1″ main with pipes as described				
	No.	22	Short length of 1″ lead pipe, and brass connections with main and meter, and joints				
	,,	1	1″ brass gas-main cock, with loose key and joints to iron pipe				
	,,	1	Give notice to gas company and pay their fees (if any), and fix a properly stamped meter for the requisite number of lights				
			Allow for attendance upon gasfitter in all trades, and for fixing floor-boards and joinery over pipes with brass screws and cups to remove				

BELLHANGER.

Count the number of bells and the number of pulls. Describe the carriages. State weight of bells (average each), if with pendulums and indicators, the gauge of the copper wire, if in concealed zinc or copper tubes, with brass cranks, steel springs and brass tee plates, and if the lever pulls are fixed with iron boxes and mouth-pieces. As the estimator prices at so much a pull, the various parts are not measured in detail.

Provide a sum for the pulls.

Number the bell boards. State thickness, if moulded or beaded and for how many bells. If numbers are painted below bells (instead of indicators) measure the writing in inches, and state if plain or ornamental writing, or number the figures, stating their height.

In cases of electric or pneumatic bells it will be most convenient to obtain an estimate from a manufacturer, and to provide a sum in the estimate. If taken out, see section ELECTRIC BELLS.

State in an item that the carpenter and joiner are to attend upon, cut away for, and make good after bellhanger to so many bells with so many pulls (and to include the boards for bell runners, &c., and fixing the floor boards with brass cups and screws to remove).

A similar item for bricklayer's attendance, omitting the words of the foregoing paragraph which are in parenthesis.

A similar item for mason's attendance, if any is required.

Or the whole of the cutting away and attendance may be put into one item.

Bill.

| No. | 7 | Bells as described, with six lever pulls and two sunk plate pulls (a sum provided for the pulls) .. .. Provide for six lever pulls and two sunk plate pulls 8l., and allow for carriage, profit and fixing.. .. .. Allow for attendance on bellhanger and for cutting away and making good in all trades .. .. | £ | s. | d. |

If the fixing of the pulls is not included in the provision, state the fixing as follows.

Bill.

			Fixing only new pulls.	£	s.	d.
No.	6	Lever pulls			
,,	2	Sunk plate pulls			

BELLHANGER.

ELECTRIC BELLS.—Count the number of pulls. Describe the wires, how insulated, how covered, the kind of roses, the backs, the springs, the contacts, the bells, the indicators, the battery. The greater part of the above particulars will appear in the preamble. See section BILLING.

The pulls and presses are best described at "P.C. . . . each at manufactory."

Bill.

Ground floor.

		s. d.	
Principal entrance, one 4½" bronzed press	P.C.	10 6	
Trade entrance, one 3½" ditto	"	8 6	
Study, one 3" china button	"	2 6	To ring and indicate in kitchen with 2½" bell.
Dining room, one 3" ditto	"	2 6	
Drawing room, one 3" ditto	"	2 6	
Morning room, one 3" ditto	"	2 6	
Drawing room, one 3" ditto	"	2 6	To ring and indicate on second floor landing with 2½" bell.
Hall, one 3" ditto	"	2 6	

First floor.

Bed room 4, one pear press	"	6 6	
" 6, one 3" china button	"	3 0	To ring and indicate in kitchen with 2½" bell.
" 8, one 3" ditto	"	3 0	
" 10, one 3" ditto	"	3 0	
" 11, one 3" ditto	"	3 0	
" 4, one pear press	"	6 6	To ring and indicate in room 16, second floor, with 2½" bell.
" 5, one ditto	"	6 6	
Landing, one 3" china button	"	3 0	To ring and indicate on second floor landing with 2½" bell.

| No. | 1 | Switch connection inside front entrance to transfer current from kitchen bell to bell on second floor landing .. .. .. | £ s. d. |

Allow for attendance on electrician, cutting away and making good in all trades, also for fixing floors and joinery over wires with brass cups and screws to remove, and for supplying and fixing all necessary wooden casings fixed with brass cups and screws, for wires which would otherwise be exposed.

SPEAKING TUBES, *at per foot run*.—Assume that the mouth-pieces are placed about 4 feet 9 inches from floor. The pipes and fittings vary so much that it is better to obtain an estimate from a selected manufacturer, and provide a sum founded upon his estimate.

Attendances may be included in the sum provided, or may be described in the bill, stating the number of feet run.

If a measurement is preferred, measure the length of the pipe, describing its size. The brass connections, the mouth-pieces and whistles (for these latter a P.C. may be stated). Describe the mouth-piece cases and flexible tubes.

Bill.

ft.	in.			£	s.	d.
100	0	run	¾" stout composition pipe, including bends, soldered joints, wall hooks and fixing ..			
8	0	,,	¾" flexible tube, braided with mohair and silk			
			No. 4 ¾" brass elbows, with two soldered joints			
			No. 4 ¾" brass unions with two soldered joints			
			No. 4 ¾" walnut screwed mouth-pieces, each with whistle and indicator			
			No. 4 brass holders for flexible tube			
			No. 2 1" mahogany French polished mouth-piece cases, with ivory tablet for two pipes			
			Allow for attendance on speaking-tube maker, cutting away and making good in all trades, and for fixing floors and joinery over pipes, with brass cups and screws to remove			

ELECTRIC LIGHTING.—In the majority of cases this is a separate contract. If it is included in the general contract the most convenient treatment is similar to that suggested for electric bells.

The simpler conditions (*a*) are those in which an electric lighting company exists in the locality, in which case little more than wiring is involved. The alternative (*b*) is a complete installation as in the case of a country house, in which case power of some kind is involved and an apparatus of accumulators.

(*a*). The regulations of the public company supplying the district must be obtained and followed.

The lights must be marked on the plans, each point indicated thus **X**.

It is not necessary to measure all the lengths of wire nor the casings unless the latter are intended to imitate the details of the joinery which is sometimes required, such as casings, serving as cappings of dados, picture-rails, skirtings, wooden cornices and the like.

A general preamble describing wires, casings, &c.

The number of the points and the number of lights at each point.

The price of the fittings P.C. and their fixing.

Give notice and pay fees to company.

Bill.

No.			£	s.	d.
„	70	Wire for fittings of one-light each in the positions directed and in the manner described (Note for these points 85 switches will be required)			
„	50	Ditto, two lights each			
„	2	Switch boards each for 50 lights of 1¼" Valencia slate rubbed, sanded and enamelled both sides, fixed with stout screws with ebonite rings and collars to and including 3" by 1¾" teak fillets plugged to wall, enclosed in teak case, French polished inside and out of 1¼" rims dovetailed at angles with 1¼", one panel door rebated for glass with moulded beads screwed with brass cups and screws, and glazed with British polished plate of the best glazing quality bedded in indiarubber. The door hung with 3" brass butts and fitted with a brass lock, P.C. 4s. and connecting..			
„	1	Ditto for 70 lights and ditto..			
„	1	Distributing eight-way switch board, with switches and fuses on all poles, and all as last			
		Fittings.			
		Such as brackets and pendants			
		Allow for giving notice to Electrical Supply Company and paying their fees and charges..			
		Allow for attending upon, cutting away for and making good after electrical engineer in all trades for wiring to 120 points ..			

(*b*). The engine, whether gas, steam, or oil, may be described from a trade list by a code word or by adopting the phraseology of the list. Describe as P.C. at manufactory and allow for packing, carriage and profit.

The dynamo. Describe from a trade list as P.C. at manufactory and allow for carriage, packing and profit.

The accumulators. Describe from a trade list as P.C. at manufactory, and allow for carriage, packing and profit.

The switch-boards and cut-out boards. Describe from a trade list as P.C. at manufactory, and allow for packing, carriage and profit.

The fittings. Describe as P.C. at manufactory and allow for carriage, packing and profit.

Mark the lights on the plans.

Number the points and describe the number of lights to each point.

The setting of the boilers; if a steam engine the steam pipes; the bed for engine and dynamo; and the water supply to the boiler would be measured in the ordinary way.

Attendance on the engineer and making good.

The arrangements here suggested are consistent with the inclusion of the work in a general building contract.

If a separate contractor, conditions as to time, payment, &c. should be inserted. If part of the building contract, the general conditions will apply.

If, as is frequently the case, the electric wiring is a provision, clauses similar to the following would be used.

Bill.

	£	s.	d.
Electric Wirings and Fittings.			
Provide the sum of.... P.C. for electric wiring to be executed by an electrician appointed by the architect			
Allow for attending upon, cutting away for and making good after electrician in all trades in wiring to No. 85 points			
Provide for 40 one-light brackets, 20 one-light pendants and 25 two-light pendants, 60*l*.			
Fixing only.			
40 one-light brackets			
20 one-light pendants			
25 two-light ditto			

GLAZIER.

The surveyor in measuring should observe where glass should be ground, as in borrowed lights, w.c. windows, etc., and he should call the attention of the architect to dark corridors or rooms where there are no windows, when the inadvertence can be obviated by glazing door panels or the introduction of windows, borrowed lights, or fanlights.

All glass should be measured per foot superficial, the extreme dimensions being measured, whether of square, circular, or irregular shape; any part of an inch is called an inch in measuring.

Measure the whole of the glass in each opening in a single dimension where possible.

If openings are square, and the joinery is common sashes and frames, the quantity of glass would be very nearly as follows:— The size of the external openings, minus 2 inches of the width and 6 inches of the height.

In all cases state if glazed in indiarubber or washleather.

The practice is now almost uniform of describing sheet glass as cut to shapes, instead of measuring the circular cutting and risk by the linear foot.

Glass in skylights, if in long squares, should be kept separate, stating the lengths, but dividing into categories of . . . . . . feet superficial.

The term "not exceeding . . . . feet" is better than "under. . . . feet."

SHEET GLASS.—No distinction need be made between sizes not exceeding 2 feet superficial. Keep separate the various sizes above 2 feet, 4 feet, 6 feet, etc. State the number of ounces per foot superficial, as 15 oz., 21 oz., 26 oz. Keep panes of irregular shape separate, and describe them as " cut to shapes"; keep separate, glass over 42 inches long and describe length; state if ground on one or both sides; if "matted"; if coloured, stating what colour; if fluted.

FLUTED ROLLED PLATE.—State thickness and if any particular maker and kind, as fluted, rolled, and whether large or small pattern; state quantity in each square as for sheet.

ROUGH PLATE, sometimes described as hammered. State thickness and quantity in a square as for sheet.

POLISHED PLATE GLASS.—State thickness and quality, and the superficial quantity in a square, commencing with not exceeding 2 feet superficial, and keeping each size separate, as follows, 3, 4, 5, 6, 7, 8, 10, 12, 14, 16, 18, 20, 25, 30 feet, etc. State if cut to shapes, or bent, or ground, and keep separate, or the bent glass may be billed with the ordinary glass and the bending separately stated. Observe that British polished plate is polished down to *about* ¼-inch out of ⅜-inch, but it is of a mere average thickness. Exact thickness must be paid an extra price for, as it must be gauged, and must be billed separately and so described.

Measure at per foot superficial. Embossing or enamelling, and state the prime cost per foot superficial. Observe that embossers charge the whole square, although part only may be embossed. Bending, state superficial content, kind of glass, and radius of curve. Silvering, and observe to take flannel and wood to back of it, and keep the glass separate as plate glass of best silvering quality.

LEAD LIGHTS, *per foot superficial.*—It will usually be sufficient to allow half an inch around the light to be glazed beyond the clear width between the mullions or tracery, but reckon all fractional parts of inches as inches; describe the lead, if in quarries, if bordered, if geometrical, the description of glass, if the lights are secured with copper wire or copper bands. Keep the parts in cusped or pointed heads, or in tracery, separate, measuring to the extreme points, and stating that it is measured square.

If it should happen that a number of triangular lights occur in a building, keep them separate and state what they are and that they are measured square.

Some of the lead light manufacturers charge lights *under* 12 inches wide as 12 inches wide; but this fact may be disregarded, if it is described as measured net, or the surveyor may measure them at per foot run, keeping each width separate and stating the width.

State if lead lights have to be put in from the outside of building. Sometimes required in old buildings where the stanchions are inside the windows, and their removal is inexpedient.

BEVELLING EDGE OF PLATE GLASS, *per foot run.*—State width, and describe it as polished; keep circular separate from the straight.

Where the glazing is ornamental, and the quarries are painted or enamelled, it is better to obtain an estimate from a glass painter to be selected, and either include the total sum as a provision, or describe as P.C. 3s. per foot, or as the case may be. State whether the provision includes saddle bars.

Number. Copper clips and brass screws to skylights. Lead clips and screws. Engraved corners, stating size and colour and the kind of glass.

Iron casements; giving description, or provide so much each and refer to trade list, and include packing, carriage and fixing, and state whether fixed to stone or wood.

NOTE.—In case of dispute as to the description of glass, look for bubbles in its substance; if in sheet glass they are oval, if in plate glass, spherical. Where glass is supposed to be crown, look to the edges of the squares, when by careful examination faint concentric waves may be seen.

Bill.

ft.	in.			£	s.	d.
54	–	supl.	15 oz. sheet, not exceeding 2 feet in a square, and glazing			
84	–	„	21 oz. ditto, 2' 0" ditto..			
181	–	„	21 oz. ditto, 4' 0" ditto..			
95	–	„	21 oz. ditto, 6' 0" ditto..			
18	–	„	21 oz. ditto, ground one side, 4' 0" ditto			
33	–	„	21 oz. ditto, 6' 0" ditto..			
5	–	„	Hartley's ¼" small pattern fluted rolled plate, not exceeding 3 feet in a square, and glazing			
10	–	„	Ditto 4' 0", and glazing in indiarubber			
180	–	„	British polished plate glass, not exceeding 2' 0" in a square, and glazing..			
70	–		Ditto, 3' 0" ditto..			
50	–		Ditto, 4' 0" ditto, and glazing in wash leather..			
70	–	„	Bending British polished plate glass..			
25	–	„	Enamelling ditto, P.C. 3s. 6d. per foot supl.			
20	–	run	Circular cutting and risk to British polished plate glass			
74	–	supl.	British polished plate, not exceeding 4' 0" in a square, and glazing with beads, elsewhere taken			
10	–	„	British polished plate glass of the best silvering quality, not exceeding 4' 0" in a square, and silvering and glazing			
90	–	„	Stout lead quarry lights, P.C. 5s. per foot at manufactory, and allow for profit, packing, carriage, and fixing with strong copper bands to saddle-bars			

ft.	in.			£	s.	d.
174	–	supl.	Stout lead quarry lights in ornamental geometrical patterns of four varieties, glazed with stout rolled cathedral plate in varied tints with two narrow borders of white and ruby respectively, and fixing as last			
	No.	20	Stout copper clips and brass screws			
	,,	10	Cast lead clips and ditto			
	,,	4	Burt and Potts' (York Street, Westminster) section 10, quality 1, wrought-iron casements under 8 feet high, and fixing and bedding in white-lead to wooden frames			
	,,	8	Ditto under 4 feet, with two cross-bars Allow for leaving all glass clean and perfect at completion..			

PATENT ROOF GLAZING, *per foot superficial.*—In some systems the wooden bars are supplied by the patentee, in others the general contractor supplies them. Some systems comprise metal bars supplied by the patentee. Nearly all the makers supply illustrations and surveyors should consult them.

Such glazing is often the subject of a provision.

MUFFLED, VENETIAN-RIPPLED, MURANESE, may be classified in sizes, as described for polished plate.

ROLLED CATHEDRAL GLASS.—State if white or coloured, and classify sizes as for sheet glass.

SHEET CATHEDRAL GLASS.—State whether ordinary or pot metal and number of ounces per foot, and classify sizes as for sheet glass.

FIGURED ROLLED GLASS.—State whether white, tinted or pot metal, classify sizes as for sheet glass.

BRILLIANT CUT PLATE GLASS is best described, glass and cutting together. Classify the sizes as for British polished plate, and the brilliant cutting at per foot.

Bill.

ft.	in.			£	s.	d.
100	–	supl.	Best British polished plate, not exceeding 6 ft. in a square, and add brilliant cutting, P.C. 7s. 6d. per foot supl., and glazing..		.	

PAPERHANGER.

This trade may generally be measured with the plastering. The dimensions would be written thus:—

| 108 0 | |
| 10 0 | R. F. and S. walls, and Paper at 1s. and hanging. |

Measure the superficial area of the parts of the walls papered, and deduct as for plastering. The quantity may be obtained from the dimensions of that trade. The *usual practice* is to allow one piece in seven for waste, and as a piece of paper should equal 63 feet superficial, i.e. 12 yards long, 21 inches wide, the total superficial quantity in feet divided by 54 should give the quantity required. But a piece of paper rarely exceeds 11 yards in length = 58 feet superficial, it is therefore fairer to use 50 feet as a divisor. Reckon any part of a piece as a whole piece. If any part is on ceilings keep it separate.

French papers are 18 inches wide, and rarely exceed 9 yards in length = 40 feet 6 inches superficial; a deduction of $\frac{1}{7}$th gives a divisor of about 35 feet.

The sizing and the general preparation of the walls may be charged as "extra for" at per piece, or may be included in the preamble of the bill.

State the P.C. per piece, if hung in blocks, if lined out.

In the case of ornamental papering, Lincrusta-Walton, Anaglypta, or Japanese paper, measure the net surface covered with each kind, as before described, and the borders at per dozen yards run. As these vary in width and length, surveyor must obtain particulars before he reduces them. In the case of these papers a P.C. per superficial yard is the simplest method of dealing with them. The waste is mentioned in the item of the bill.

Number. The pieces of paper sized and varnished, describe the varnish, and state how many times varnished.

In all cases include the hanging with the description.

CANVAS AND BATTENS, *at per yard superficial.* State size of battens, their distance apart, and whether they are plugged.

Bill.

vds.	ft.	in.				
			No.	165	Pieces of paper, P.C. 1s. 6d. per piece, and hanging	
			,,	35	Ditto 2s. 6d. ditto	
			,,	34	Ditto 4s. 6d. ditto	
			,,	20	Pieces of marble paper, P.C. 4s. per piece, and hanging in blocks, and including marking lines as joints	
			,,	20	Ditto, extra for sizing and twice varnishing with the best paper varnish	
75	–	–	supl.		Lincrusta-Walton, P.C. 5s. per yard supl., measured net, and allow for cutting, waste and hanging	
	6	doz.	run		Border 3" wide, P.C. 3s. per dozen, and hanging	
80	–	–	supl.		Stout canvas nailed to and including 2¼" × ¾" battens 18" apart plugged to wall, and stoutest lining paper, and hanging with butt joints	

PAINTER.

In measuring painting on the building the object is to obtain in the readiest way the whole surface which has been painted.

Keep painting on old work as a separate section.

It is the custom to add edges to the height only, returns to the width only. This may be illustrated by the example of a 2-inch moulded both sides door 3 feet by 7 feet, with architraves 6 inches wide projecting 1 inch beyond the face of plastering, and jamb linings 11 inches wide double rebated—the door painted both sides.

			Width across architraves	4' 0"	Height from floor to top of architrave.	7 6
			Two returns of architraves	2"	Two edges of door.	
			Two panels	2"		4
				4' 4"		7 10
2/ 4 4						
7 10		Knot, prime, stop and paint 3 oils.				
			Door and architraves	11"		7 0
17 0			Two rebates	1"		7 6
						8 6
1 0		Ditto.	Jamb linings	12"		
						17 0

A piece of moulded both sides framing 10 feet by 10 feet, where no edges are exposed may be measured as follows. Treating the result (although the product of three factors) as a superficial dimension, we have included an allowance for edges.

```
2/ 10  0
   10  0
    1  3   Knot, prime, stop and paint 4 oils,
              or
2/ 10  0
   10  0   Knot, prime, stop and paint 4 oils, and add edges.
```

If the painting on ironwork provided is not included in the amount of provision, observe that it is often painted two oils before it is supplied.

In taking off the painting from the dimensions, the surveyor has another opportunity of checking the work, and if he is watchful will probably find errors or discrepancies.

Painting is measured in various ways, either of which will produce similar results. In deciding which plan to adopt, it is well to bear in mind the manner in which it will affect the convenience of possible variations.

Some of the methods are as follows:—

1st. Abstracting the quantity directly from the bills in two columns, one in which no allowance is required for edges, the other in which it is required, and adding a proportion to one column (usually one-seventh) for edges.

2nd. Abstracting directly from the dimensions into two columns of the abstract to one of which one-seventh is added as before, the other column being for work upon which no allowance is required, i.e. where no edges are exposed. With this method it is well to put against the original dimension, in the description column, the kind of painting which has been taken, thus: as—

⊙ = four oils, ⊙ and varnish, = four oils and varnish, &c.

3rd. (and best). Look through the dimensions from the beginning (after they are squared), and repeat the dimensions of the squaring column, multiplying by a figure, according to the kind of work, as 1 foot for plain where no edges, 1 foot 2 inches where there are edges, 1 foot 3 inches where square framed or moulded. Put opposite each dimension the number of the column of the original dimensions from which it has been taken, thus—

```
  2/ |  149 0 |
     |    1 2 |
              |  347 8  |    (2)        Col. 122
```

If this course be adopted the painting to the sashes, frames and squares had better be taken at the same time as the joinery is measured, as one sometimes finds a difficulty (the drawings having been sent home perhaps) in determining the number of squares in a sash.

State if knotted, stopped, and how many oils (the priming coat counting for one oil), if in extra colours, if in party colours, whether on iron, wood, brick, or cement, and keep each kind separate.

Measure at per yard superficial all work except that hereafter named. If painting on iron (say four oils) is specified to be "two oils before, and two oils after fixing," it should be kept separate.

Keep work to roofs and ceilings separate.

Per foot run. Shelf edge, skylight bar and rail, measuring each side. Rail and bar. Include in this description such things as balusters, and things of an approximate size. Skirting. Include in this description anything of similar size not exceeding 12 inches wide which is cut in both edges. Eaves gutter. Describe as "in and out." Rain-water pipe, and state if heated and coated inside with tar. Iron railing, stating the height, whether ornamental or plain; it is more convenient, when a price per foot run is provided for railing, to make the provision include the painting. Bar heated and dipped in oil. Roof timbers, giving average size.

Number. Coal plates and chains. Air-bricks. Gratings and frames. State if one or both sides. Rain-water pipe heads, in and out. Bolt heads. Hinges. Latches, etc., blacked. Touching up vanes or finials. Squares per dozen (counting two for one to allow for the two sides), where over 2 feet superficial call them large squares, where sashes are in one square describe them as "sheets" (counting two for one), where over 8 feet superficial call them "large," where over 13 feet, "extra large." Window frames (counting two for one to allow for the two sides) over 24 feet superficial describe as "large frames," over 36 feet superficial as "extra large frames."

OILING AND RUBBING, *at per foot superficial.*—State how many times oiled and on what wood.

STAINING, SIZING AND VARNISHING, *at per yard superficial.*—Describe the stain. State if oil stain. Describe the varnish and how many times varnished.

FRENCH POLISHING, *per foot superficial.*—Handrails per foot run and if to be covered with holland or paper state it.

WRITING, *per inch lineal.*—Thus twelve letters 1½ inch high equal 18 inches of writing, or they may be numbered, stating the height. State if plain and if ornamental, what kind and the colour, and if shaded.

VARNISHING, *per yard superficial.*—State if " on natural wood " or on painted work.

GRAINING AND VARNISHING, *per yard superficial.*—Measure as for plain painting, and describe as " extra for." State if combed, if grained and over grained, and describe the wood intended to be imitated, if varnished once or twice, and describe the varnish.

MARBLING, *per foot superficial.*—Describe as " extra for," describe what marble is to be imitated, if sized, if once or twice varnished.

FLATTING.—Measure as for plain painting and describe as " extra for."

If decoration is a separate contract, the taking off and refixing of ironmongery is best done by general contractor.

PICKING OUT MOULDINGS, *per foot run.*—State whether one or more tints, and the girth of the moulding.

Bill.

On iron.

yds.	ft.	in.			£ s. d
101	-	-	supl.	*Two oils* before fixing	
	894	-	run	Bar and rail	
		No.	2	14" × 9" panels both sides	
15	-	-	supl.	*Two oils before fixing, and two oils after fixing*	
	536	-	run	Rail ..	
	129	-	„	Eaves gutter, in and out	
	429	-	„	Rain-water pipe, and heat and coat on the inside with purified gas tar	
				No. 4 heads to ditto, inside and out	
		No.	16	Air bricks both sides	
		„	20	Handrail brackets	
		„	3	Pairs, strap hinges out in black	
		„	10	Stoves blacked	

QUANTITY SURVEYING.

yds.	ft.	in.		£	s.	d.
			Clean, rub down, and paint two oils on old work			
468	–	–	supl. On woodwork			
	324	–	run Bar and rail			
	91	–	,, Rail cut in both edges			
	614	–	,, Skirting			
		No. 42	Frames			
		,, 30	Ditto, large			
		,, 4	Ditto, extra large			
		,, 9	Dozen squares			
		,, 6	Ditto, large			
		,, 2	Ditto, extra large			
634	–	–	supl. Knot, prime, stop and paint three oils on woodwork			
	318	–	run Rail			
	774	–	,, Skylight bar and rail			
	2150	–	,, Skirting			
		No. 54	Frames			
		,, 5	Ditto, large			
		,, 1	Ditto, extra large			
		,, 6	Casements			
		,, 10	Dozen squares			
		,, 1½	Ditto, large			
		,, ½	Ditto, extra large			
80	–	–	supl. Knot, prime, stop and paint four oils on woodwork			
	489	–	run Skylight bar and rail			
	123	–	,, Skirting			
		No. 48	Frames			
		,, 1	Ditto, large			
		,, 1	Ditto, extra large			
		,, 6	Casements			
		,, 6	Casement edges			
		,, 10	Dozen squares			
		,, 1½	Ditto, large			
		,, ½	Ditto, extra large			
75	–	–	supl. Knot, prime, stop and paint four oils, the last parti-colours on woodwork			
	114	–	run Skirting			
			French polishing, including protection from damage			
	85	–	supl. On mahogany			
	27	–	run Skirting			
	1079	–	supl. On wainscot			
	101	–	run Ditto, handrail			
75	–	–	supl. Stain, size and twice varnish on woodwork.			
	101	–	run Skirting			
		110	Plain writing			
			Allow for touching up at completion, and leaving perfect			

WHITENING CEILINGS, *per yard superficial.*—This is frequently done by painters.

MACHINERY AND PIPES. 279

DISTEMPER WALLS, *per yard superficial.*—State if clearcolle and if once or twice distemper. If other than cream, drab or fawn, it must be stated, as anything beyond this is an extra colour. If Duresco, the colour need not be mentioned. Cornices picked out, state the girth and how many tints.

Bill.

yds.	ft.	in.			£	s.	d.
100	–	–	supl.	Clearcolle and whiten ceiling			
100	–	–	,,	Clearcolle and distemper walls, a tint ..			
50	–	–	,,	Ditto, extra colour (French grey)			
50	–	–	,,	Twice distemper with Duresco			
	100	–	run	Twice distemper cornice, 12" girth, picked out in two tints			

MACHINERY AND PIPES.

Although the mechanical engineering of a building is very frequently arranged with a particular manufacturing engineer, and the amount of his estimate included in the general estimate as a provisional sum, yet this is not always the case. The architect decides to obtain competitive estimates, and this, if the tenders are to be made on a uniform basis, involves the supply of quantities.

If there are no quantities, each contractor should describe clearly what he proposes to do; and the surveyor should see the accepted man, and find out what work will be required from the general contractor.

Mechanical engineers appear to prefer the presentation of the work in a good many sections to the condensation characteristic of ordinary builder's quantities, for instance, after the boiler they would make a separate section in the bill for each of the following.

BLOW-OFF PIPE IN CHANNEL, i.e. the pipe followed by all its bends and connections.

SURFACE BLOW-OFF, i.e. the pipe, etc., as last.

VAPOUR PIPE, i.e. the pipe, etc., as last.

STEAM PUMP.—Steam supply to pump, i.e. the pipe followed by all its bends and connections and valves.

EXHAUST FROM PUMP, i.e. the pipe followed by all its bends and connections, steam trap, etc.

The motive is probably to facilitate the settlement of variations, but there is no better reason in the case of engineering quantities than in building quantities. The notes of the quantity surveyor are just as easily referred to in one case as the other.

There are, however, a number of main sections into which a bill of mechanical engineering work may very reasonably be divided.

The engines.
The apparatus.
The steam-pipes and coils, and valves.
The shafting and gear, and driving belts.

The cold water supplies, mains and wastes.
The hot-water supply, pipes and valves.

ATTENDANCES.—Some architects specify that the engineer shall do his own cutting away and making good. It is rarely so well done by him as by the general contractor, and often the so-called making good has to be done again.

The builder's work involved by the mechanical engineer's work is often conveniently treated as a separate section of the general contractor's bill.

A reasonable order of taking off is as follows:—

WORK TO BOILERS.

Blow-off pipes to do.
Vapour pipe from safety valves.
Bends, connections, etc.

STEAM PUMP VALVE.

Brackets to do.
Steam supply to do.
Bends, connections, etc.
Exhaust pipe.
Bends, connections, etc.
Steam trap.
Engine and fixing.
Engine foundation.
Steam supply to engine.
Bends, connections, etc.
Condense pipe from cylinder.
Bends, connections, etc.
Exhaust pipe from engine.
Bends, connections, etc.

FEED-WATER HEATER.

Exhaust from do.
Brackets, bends, connections, etc.

LAUNDRY MACHINERY AND ADJUNCTS AND WASTES.

Shafting, pulleys and driving belts.
Steam services to apparatus, cast and wrought, and valves.
Steam heating pipes and coils and valves
Hot-water supply pipes and valves.
Cold-water supply pipes to apparatus and valves.
Cold water supply from main.

STEAM BOILERS.—The quantities for these are not taken out, and they are nearly always a separate estimate. Their connection with engines and pipes is usually included in the general engineering estimate. One way of presenting the work is as follows:—

MACHINERY AND PIPES.

Bill.

		£	s.	d.
The boilers will be supplied with all their mountings and fittings, and will be set by the special contractors engaged for that purpose, but the contractor for these works to couple up these boilers and to take all responsibility after doing so until the period prescribed for his maintenance is expired				

Boiler mountings are illustrated in the trade list of Messrs. Hopkinson, Huddersfield; and others.

Boilers are most conveniently treated as a provision.

This may be founded on the result of a competitive tender to obtain which a *specification* may be sent to several engineers.

A type of specification may be found in Leaning's Building Specifications.

Note.—The clause as to "Leaving temporary openings in roofs, walls, &c.," see section BILLING.

State clearly to what extent the general contractor shall cart, deliver and get into position the material supplied by engineer and attend upon him during the fixing of his work, and make good afterwards.

Bill.

		£	s.	d.
Provide for two Lancashire boilers delivered at station of the Railway £				
The boilers will be delivered at the station of the Railway, contractor to unload, load on to carriage and convey to the building, unload, get into building and deposit in position ready for the engineer				
Allow for attendance, cutting away for and making good after engineer in all trades to two boilers and their ironwork and appurtenances				
Allow for supplying water and attendance in the testing of the boilers				

ENGINES.—Describe from a trade list, state whether vertical or horizontal, the nominal horse-power, diameter of cylinder and length of stroke, and whether it includes any extras as feed-pump, feed-water heater, or any parts in duplicate, and state that the engine includes foundation bolts and plates. Many trade lists designate each variety of engine by a code word which may be mentioned.

Bill.

| | No. | 1 | Tangyes' horizontal steam engine, 100 lbs. series, with cylinder 11" diameter, 12 h.p., 24" stroke, without feed pump, title "paeseto," and including foundation bolts and plates .. .. | £ | s. | d. |

Engines of all kinds are illustrated in the lists of Tangyes, 35 Queen Victoria Street, London; Robey, Lincoln; Crossley Bros., Manchester; and others.

CAST-IRON TANKS. *Number.*—Measure the cast iron in feet and inches and reduce to weight. Measure the wrought iron in stays and bolts, and reduce to weight. State size of tank in clear, and the height of hoisting.

Measure the waste and overflow.

Bill.

| | No. | 1 | Cast-iron flanged and bolted tank 13' 6" × 6' 6" × 4' 0" deep in clear, of 1" plates, with stiffening webs and rounded vertical and bottom angles, and the joints truly planed and caulked in iron cement, including wrought-iron stays and bolts, four bosses cast on for supplies and wastes, and pattern and all holes cast and rimed for bolts, supplies and wastes, weight 126 cwt. of cast iron and 9 cwt. of wrought iron, and allow for hoisting and fixing 41 feet above street level .. .. | £ | s. | d. |

CAST-IRON STEAM-PIPES, *per foot run.*—State diameter and whether ordinary or special lengths. Weight should be dealt with in preamble of bill. State if flanges are faced, how jointed and fixed. Count the bends, crosses and elbows.

Number millboard and red-lead joints, asbestos joints, flanges and flanged connections.

Measure the trench for pipe per foot run.

The various pipes and connections and the stock sizes are

MACHINERY AND PIPES.

illustrated in such catalogues as Bailey, Pegg and Co., Bankside, London.

The pipes are reduced to weight before billing, but state the length also.

The courses of all pipes should be indicated on a special plan, so that the bends, joints and connections may be counted, and the distinctions between ordinary and special lengths of pipe noted.

Bill.

cwts.	qrs.	lbs.				£	s.	d.
			In		3" cast flanged steam-pipes with faced flanges and in ordinary lengths and fixing .... feet run			
			,,		3" ditto special lengths and ditto......feet run			
			,,		4" ditto ordinary lengths and ditto....feet run			
			,,		4" ditto special lengths and ditto......feet run			
			No.	2	3" cast flanged steam bends			
			,,	3	3" × 3" × 2¼" cast flanged steam tees ..			
			,,	3	3" × 3" × 2" ditto..			
			,,	1	3" blank flange			
			,,	8	2¼" flanged connections to blow-off cocks on boilers...			
			,,	3	2" wrought flanges drilled and bolted to tees for surface blow-off pipes			
			,,	3	Millboard and red-lead joints to 2" pipes ..			
			,,	6	Ditto 2¼" ditto			
			,,	16	Ditto 3" ditto			
			,,	10	3" asbestos joints			
			,,	1	4" bent pipe about 6' 6" long			
			,,	1	4" S-pipe to special sweep and about 4' 0" long			

VALVES.—State bore, whether brass or gun-metal, whether screwed for iron, whether with unions. State whether for cold water, hot water or steam.

Bill.

			No.	65	¾" Sir William Thomson's gun-metal hot water valves screwed for iron	£	s.	d.
				2	¾" gun-metal steam wheel valves with unions screwed for iron			
				8	1¼" Tylor's fig. $\frac{178}{2}$ gun-metal hot water stop valves screwed for iron			
				2	3" screw down gun-metal wheel valves with flanged ends and joints			

QUANTITY SURVEYING.

		No.	4	3" Hopkinson's best quality diminishing valve, marked for pressure of 5 to 15 lb. at 2¼ lb. intervals	£	s.	d.
			4	Hopkinson's Bourdon pressure gauge with syphon and stop-cock			
			4	1" dead weight safety valves, loaded to blow off at such pressure as may be directed			
			6	3" copper flanged expansion joints of approved make			

SHAFTING AND PULLEYS, *per foot run*.—State diameter. Number the pairs of fast collars, the loose collars, the couplings, the wall boxes, the wall brackets, the driving pulleys, the guide pulleys.

Bill.

ft.	in.				£	s.	d.
				The price for shafting to include hangers, which are to be bolted up to the rolled steel joists of floor above, and hard wood packing pieces properly fitted. The shafting to be fixed perfectly level and true, and the beds for all the levels to be planed ..			
40	-	run		2¼" bright turned steel shafting			
15	-	„		3" ditto			
		No.	3	Pairs, forged steel fast collars to 2¼" shafting			
		„	1	Pair ditto 3" ditto			
		„	2	Loose collars to 2¼" shafting			
		„	1	Pair of faced and turned flanged couplings to 2¼" shafting			
		„	2	Cast-iron wall boxes built into walls, and fitted with two 2¼" plummer blocks with gun-metal bearings and lubricators ..			
		„	2	Ditto, with two 3" plummer blocks, etc. ..			
		„	1	Wall-box pedestal, built into wall and fitted with gun-metal bearings and lubricators for 2¼" shaft			
		„	1	Ditto, 3" shaft			
		„	1	Best turned cast-iron driving pulley for engine 48" × 7"..			
		„	1	Ditto for line shaft 48" × 7"			
		„	2	Ditto for counter-shaft 18" × 6"..			
		„	2	Pairs best turned guide pulleys, with bracketed facings bolted to wall including bolts			
		„	1	Pair of mortise mitre wheels, 20" diameter, with hornbeam teeth properly geared, one wheel fitted on to short length of shafting			

STEAM COILS.—Pipes per foot run, reduced to weight before billing, but state the length also.

Number the bends, flaps, asbestos joints, etc.

Bill.

cwt.	qrs.	lbs.				£	s.	d.
			In		4" cast flanged steam pipes in ordinary lengths, in coils to drying closets....feet run			
			No.	52	4" cast flanged syphon bends			
				4	4" blank flaps..			
				160	4" asbestos joints			
				4	Blank flanges drilled and tapped for 1" wrought pipes			

DRIVING BELTS, *per foot run.*—State width, and whether single or double leather. Number the lacings, i.e. the connecting of the ends to make the band continuous.

Bill.

ft.	in.				£	s.	d.
		run		Best single leather belt 2" wide..			
		"		Ditto 2½" wide			
		"		Ditto 3" wide			
		"		Best sewn double leather belting 6" wide..			
		No.	"	Lacings to 2" single leather belt			
			"	Ditto 2½" ditto			
			"	Ditto 3" ditto			

LAUNDRY MACHINERY.—The illustrated catalogues of Bradford, London, and Manlove, Alliott and Co., London, will afford the necessary information as to requirements. The apparatus supplied by them is frequently the subject of a provisional sum.

As hydro-extractors, washing machines, washing-box enclosures, washing troughs, boiling troughs, waste troughs, soap-boilers, drying horses, mangles, ironing tables, etc.

Bill.

		No.	1	Bradford's (140 High Holborn) patent No. 5 wrought-iron Woodenway injector washing machine, with wood washing-compartment, parallel wringing-rollers and new automatic reversing gear	£	s.	d.
		"	1	Galvanised wrought-iron boiling-tub 2' 9" diameter and 2' 6" deep, of No. 9 B.W.G. iron, riveted with ⅝" rivets, 1¼" pitch, to 1¾" angle-iron at bottom and 2" angle-iron around back. Fitted with and including			

			£ s. d.

    1½" gun-metal waste and nut, fixed in bottom, with short lengths of 1½" galvanised iron waste, with Tylor and Son's Fig. $\frac{178}{2}$ 1¼" stop-valve fixed therein. A coil of ¾" stout copper finely perforated pipe, fastened to bottom with copper clips. A false bottom of $\frac{1}{16}$ galvanised iron, closely perforated with ¼" holes, supported by angle-iron ring riveted to boiler, the whole on, and including galvanised iron riveted stand of 2" × ⅜" ring, on five legs, 1¼" diameter .. .. ..

No. 1   Galvanised wrought-iron riveted soap-boiler, 18" diameter and 18" deep, of No. 10 B.W.G. iron, with 1¼" angle-iron around top and bottom, and strong cover hung with brass hinges, and fitted with brass handle, and ¼" stout copper perforated pipe for steam, connected through sides with brass unions and boiler screw to, and including a ½" Tylor's Fig. $\frac{178}{2}$ stop-valve .. ..

„ 1   1½" best quality rubbed and sanded slate rinsing tank, 6' 0" × 2' 6" × 2' 8", in two compartments, grooved together and jointed in red-lead cement, bolted with nine ½" galvanised iron bolts, with heads, nuts and washers, two 3" × 1½" 7 lb. lead trumpet-mouthed wastes, each with gun-metal washer and waste, short length of 1½" galvanised iron pipe, one 2¼" gun-metal waste, washer and plug, and strong brass chain, on and including three supports, 2' 6" × 4½" × 9" of white glazed bricks in cement, with rounded ends ..

„ 1   Manlove, Alliott & Co.'s (57 Gracechurch Street) 32" hydro-extractor, with overhead friction cone gearing, fast and loose pulleys, screw striking gear, galvanised iron wire cylinder, steel vertical shaft, gun-metal bushes and brake complete ..

„ 2   Bradford's No. 44 mangling machines, with 36" sycamore rollers, and levers and weights .. ..

„ 2   Sets of eight draw-out drying horses (sixteen horses in all), with framework as follows:—a cast-iron bed-plate to fit edge of coil-pit; a cast-iron head plate, bolted to and including 8" × 4" × 20 lb. rolled steel joist; cast-iron upright division bars, with rebates to stop front and back plates of horses, bolted to bed and head plates; overhead runners of 2½" × 2½" × $\frac{7}{16}$" tee-steel passing through head plate, and supported at front by an 8" × 4" × 20 lb.

rolled steel joist. Each horse to be 6' 6" long and 6 feet high, made of wrought-iron front and back plates, 15" × ¼", properly flattened and planed on edges with 1" white deal fronts, with rounded edges, screwed to the front plates and desiccated; seven hanging rails of 1" iron steam barrel screwed to front and back plates, and with back nuts on each side of front and back plates; each horse to be over-running, with two cast-iron bored and turned wheels, 9" diameter, with double wheel-brackets. A 1½" stout brass bar handle, 16" long, with heavy cast-iron ends. All the bearings to have oil holes drilled and countersunk, and the whole of the metal to be galvanised .. ..

One horse of each set to have 12" metal-cased thermometer, with bulb, passing into drying chamber and properly protected .. ..

Engine beds or beds for machines are mostly of cement concrete.

Measure the digging and concrete at per yard cube and describe the latter as in bed for engine or machine, as the case may be.

Measure boarding to edge of that part of concrete which is above the floor.

Take the rendering of face at per foot superficial.

Number lengths of iron barrel fixed vertically in the concrete for the holding-down bolts to pass through.

If a stone slab, immediately under the engine, measure at per foot superficial, state the kind of stone, thickness and finish.

Number the bolt-holes—the holes left in concrete for screwing up the bolts, and the making good.

The engineer should furnish a working drawing for an engine bed, such as he may require

Bill.

yds	ft.					£	s.	d.
1¾	–	–	cube	Dig surface, truck and cart (or basket and cart)				
3½	–	–	,,	Cement concrete in engine bed				
		No.	6	Leave holes inside of concrete bed for screwing up bolts and fill in with concrete and make good				
	38	–	supl.	Boarded edge to concrete				

ft.	No.			£	s.	d.
	No.	6	18" lengths of 1¼" iron barrel and bedding in concrete..			
	”	6	9" by 9" by ½" wrought-iron plates with hole in each as washer and bedding in concrete			
	”	1	6" York slab rubbed top and edges, and the arrises chamfered 1¼" wide, 5' 6"× 3' 0", and six holes for 1" bolts, and bedding in cement..			
38	–	supl.	Render and float in Portland cement in narrow widths on concrete			
8	–	run	Arris			
19	–	”	Arris and 3" return			
			The engine will be delivered at station of the railway. Allow for unloading from truck, delivering into carriage, carting to building, unloading, getting into building and into position, and supplying attendance to engineer in fixing and making good			

SETTING OF STEAM BOILERS.—The engineer will supply working drawings of the setting for the surveyor's use.

The setting may be either measured in detail or treated as a day account.

If measured in detail, the work is most conveniently billed as a separate section.

Measure the brickwork net and state that it has been so treated.

The facing and the seating and other blocks to be measured as extra on the ordinary brickwork.

Poulton's is incomparably the best system of setting. They will on application send a statement of the requisite number of each kind of block required for the setting of a particular sized boiler.

Bill.

Work in Setting Lancashire Steam Boiler.   £ s. d.

All descriptions of labour and materials to be as described in General Bill.

Allow for supplying all necessary centres and templates for setting boiler 30' 0" long, and 8' 0" diameter.

Allow for attending upon boiler-fixer, and for building in all fire doors, bearing bars, soot doors, &c., in fixing boiler as last described.

yds.	ft.				£	s	d.
				The fireclay to be the best Stourbridge fireclay from an approved firm.			
				The seating blocks, flue covers, cross wall and partition blocks shall be of fireclay supplied by Poulton and Son, Reading, and of their patent. The numbers and letters attached to the items are from their trade list.			
				The white glazed bricks to be Cliff's second quality.			
			cube	Excavate to trenches, not exceeding 6′ 0″ in depth, and cart away			
			supl.	Labour forming surface of ground to slope			
			,,	Strutting and planking to side of trench ..			
			,,	Ditto circular on plan			
			cube	Cement concrete in trenches			
			,,	Ditto and filling in over arches to channels beneath boilers			
			supl.	Labour forming surface of concrete to segmental contour			
			,,	Ditto circular on plan			
			,,	Cement concrete 9″ thick beneath channels			
rods.	ft.		,,	Ditto 9″ thick to falls			
			supl.	Reduced brickwork in pink Aylesford wire-cut gault bricks in boiler setting, including all cuttings and measured net			
			,,	9″ brick bottom to flue, of bricks as last ..			
			,,	9″ ditto circular on plan, including all cutting and waste			
			,,	Rough cutting			
			run	Skewback cutting 5″ wide			
			,,	Ditto 9″ ditto			
			supl.	Extra on ordinary brickwork for facing of fire-bricks set in fireclay and neatly pointed..			
			,,	Ditto circular on plan			
			,,	Ditto for facing of white glazed bricks set in fireclay and raked out and pointed with cement			
				Extra on common brickwork and allow for all cutting.			
		No.	26	No. 1 long curvilinear boiler seating blocks			
		,,	26	,, 1 short ditto			
		,,	28	,, 1 long flue cover, 9″ span			
		,,	28	,, 1 short ditto			
		,,	19	,, 6 partition wall blocks			
		,,	13	,, 6A ditto			
		,,	22	,, 6 recess wall blocks			
		,,	52	,, 5 front cross wall blocks			
		,,	22	,, 5 division wall blocks			
		,,	21	,, 5A ditto			
		,,	36	,, 5 fire bridges			
		,,	8	,, 3 anti-expansion down-takes			
		,,	2	,, 6B pyrometer blocks and lids			

USE AND WASTE.

Shoring where not provided for by a clause in the bill, should be measured and charged as "use and waste."

Measure shoring at per foot cube. Number the sets of wedges, describe size and whether oak or fir. Measure the ironwork as described in section FOUNDER AND SMITH.

Needling is best included with the particular item of cutting away which requires it.

Bill.

ft.	in.			£	s.	d.
100	–	cube	Use and waste of timber in shoring, including all wedges, hoop-iron and labour			
	No.	6	Pairs, 18″ × 12″ oak wedges			

Shoring rarely appears in quantities in the above form, it is more generally provided for by a clause, as follows.

"All cutting of openings to include any necessary needling or shoring." See also section BILLING, preliminary bill.

In the absence of an express stipulation that the timber used in shoring shall be new, old, if it serves its purpose, should be accepted.

VENTILATION AND WARMING.

This work, if for a large building, is generally a provision, but the surveyor should see the engineer to whom it is to be entrusted, and make inquiry as to the requisite constructive arrangements, so that he may provide for them in the quantities by measuring all but apparatus in the usual way.

The attendance upon the engineer is most fairly met by the provision of such a sum as the surveyor may think will cover the work, the exact amount to be adjusted at completion as a day account.

Bill.

				£	s.	d.
			Provide for ventilation to be done by £500.			
			Allow for attendance upon, cutting away for, and making good after ventilating engineer £50.			

CREDITS.

Where an old building is to be pulled down to make room for a new one there are several courses open to the architect. Where none of the old materials are to be used in the new building it is better either to sell them as they stand to a dealer in old building materials (this is often arranged by tender), or, if too large for one man to deal with, and worth the expenditure for an auctioneer's charges, they are sold at auction in lots; in either case the purchase involving pulling down and clearing away rubbish by the purchaser, as also any necessary shoring by him to adjoining buildings.

Observe that the purchaser will usually only pull down to ground level, unless an express stipulation is made to the contrary.

A wise precaution is to insist upon the deposit, by the purchaser of an old building, of a sum of money to be forfeited if he should fail to finish his contract.

Another way is to arrange that the old materials shall be the general contractor's property, and in such a case a bill of credits should be prepared. It is obvious that the measurement of old materials to produce a bill of credits should be conducted in a different manner to that of new material, as old buildings frequently hardly pay for pulling down. The most valuable material is usually lead. Where there is to be no clerk of works, the exact weight should be as nearly as possible arrived at by careful measurement. It is always best to assume for the bill that the lead weighs 1 lb. per foot superficial less than the original weight. Where it is intended to have a clerk of works it is better to measure the lead sufficiently carefully to obtain a near approximation for the purpose of the tender, and by a clause in the bill state that the lead will be weighed as it comes from the roofs, and that the contractor is to state in his tender how much per cwt. he will allow for it, after deducting the expense of weighing, the exact quantity being adjusted at the settlement of the accounts. Some surveyors weight the whole at 4 lbs. per foot superficial, and state this in a clause of the bill. The allowance for waste on old lead is 4 lbs. per cwt. In the case where only parts of a building are to be removed, the items must be measured and enumerated.

A few examples of the method of procedure are as follows :—
brickwork may be measured at per rod superficial; roofing may be described as so many squares of slating and boarding and roof timbers; flooring as so many squares of flooring and joists.

QUARTER PARTITIONS.—As so many squares.

Other items in some such condensed form as the following :—
"One set of sashes and frames to opening 4 feet by 6 feet 6 inches, with linings and finishings, York sill, &c."

"One four-panel square door 2 feet 9 inches by 6 feet 6 inches, with linings and finishings." In all cases state the position of the material in the old building.

"Remove and credit roof of present third floor, about 32 feet by 56 feet, comprising 5 roof trusses, 6 dormer windows, 7 skylights, 13 squares of slating and boarding and rafters, and about 45 cwts. of lead."

Where the whole of a small building is to be removed its position and general description will be sufficient without measuring.

Bill.

				£	s.	d.
			Pull down and credit the coal store in kitchen yard			

Where any of the old materials are to be re-used, and their use is dependent upon the decision of the architect, they should be stacked in such a manner that they may be conveniently examined and the rejected parts carted away, and such parts as are left measured and valued by the surveyor and charged to the builder.

Where old bricks are intended to be re-used and a quantity are stacked, a proportion of which are likely to be rejected, it is better to take down a cubic yard of them from the stack and settle upon a certain proportion per yard which may be used, and such proportion may be charged to the builder as before.

See also the section on BILLING, page 315.

The foregoing pages will give a general idea of the principal methods of measurement, and prove some guide as to the way to deal with analogous cases, but in the course of practice there will arise many points which only a thorough knowledge of construction and mature experience will enable the surveyor to treat. Besides these, he will not unfrequently meet with arches without abutment; girders, joists and beams inadequate to the support of the proposed weight; roof-trusses which would inevitably thrust out the walls they were intended to rest upon; walls apparently standing upon nothing; suggestions of impossible jointing of stone-work; impracticable staircases, &c.; all of which anomalies as a building expert it will be his duty to rectify.

The surveyor should also be able to judge of the possibility of working particular materials in the manner specified, and if necessary, suggest the substitution of others.

Although, in the erection of a building, a division of responsibility is as much as possible to be avoided, yet it should be remembered that where there is a *large* quantity of work for which a builder would necessarily obtain a sub-contract, as iron-work or glass, it will sometimes produce a considerable saving to obtain separate tenders for it, and the bills of quantities may be arranged with that view.

ABBREVIATIONS.

As a general rule it is best to avoid abbreviations, except those which are universally understood and used; frequently even those can only be interpreted by their context; they are as follows. They may be used in dimensions and abstracts, but not in bills:—

GENERAL.

A. B.	As before.
B. S.	Both sides.
B. W. G.	Birmingham wire gauge.
Ct.	Cement.
Cir.	Circular.
Co.	Course.
Ddt.	Deduct.
Dia.	Diameter.
E. T.	Elsewhere taken.
E. O.	Extra only.
Incg.	Including.
Irreg.	Irregular.
Lab.	Labour.
Mo.	Moulded.
M. G.	Make good.
N. W.	Narrow widths.
O. S.	One side.
P. M.	Purpose made.
P. C.	Prime cost.
Ro.	Rough.
Semi.	Semicircular.
S. E.	Stopped ends.
Segl.	Segmental.
S. L.	Short lengths.
S. Q.	Small quantities.
1ce.	Once.
2ce.	Twice.
3ce.	Thrice.
4ce.	Four times, &c.
L	Angle.
⌀	$\tfrac{3}{4}''$ diameter.
Wrot.	Wrought.

EXCAVATOR.

P. F. I. & R.	Part filled in and rammed.

BRICKLAYER.

B.	Brick, as 1 B, 2 B, &c.
B. & P.	Bed and point.
B. M.	Birdsmouth (R.C.B.M. or F. C. B. M.).
Co.	Course.
C. & P.	Cut and pin.
Chy.	Chimney.
D. P. C.	Damp-proof course.
E. O. C.	Extra only in cement.
F. C.	Fair cutting.
Foots	Footings.
Gd. arch	Gauged arch.
H. I. B.	Hoop-iron bond.
Ptd. arch	Pointed arch.
Relg. A.	Relieving arch.
R. O. W. & P. F.	Rake out, wedge and point flashings.
R. C.	Rough cutting.
R. C. B. M.	Rough cut birdsmouth.
P. & C.	Parget and core.
Segl. A.	Segmental arch.
S. Q.	Squint coins (R.C.S.Q. or F. C. S. Q.).
S. B. C.	Skewback cutting.

MASON.

B. & J.	Bed and joint.
Cir. F.	Circular face.
Mo. F.	Moulded face.
P. F.	Plain face.
R. S.	Rough sunk face.
S. F.	Sunk face.
Yk.	York.

CARPENTER AND JOINER.

Archve.	Architrave.
B. S.	Both sides.
B. F.	Bead flush.
B. B.	Bead butt.
B. A. P.	Brass axle pulleys.
Centg.	Centering.
Chfd.	Chamfered.
D. C. F.	Deal cased frames.
D. H.	Double hung.
F. E. S.	Feather-edged springer.
Fd.	Framed.
F. S.	Flat sweep.
H. B. S.	Herring-bone strutting.
I. W.	Iron weights.
N. R. M. E.	Notched returned mitred ends.
O. S. & W. Sills	Oak sunk and weathered sills.
O. S.	One side.
O. G.	Ogee.
Q. S.	Quick sweep.
Q. P.	Quarter partition.
R. M. E.	Returned mitred ends.
Rebd.	Rebated.
Sqr.	Square.
S. H.	Single hung.
Wrot.	Wrought.
W. B.	Wood brick.
× Tongd.	Cross-tongued.

PLASTERER.

Ct.	Cement.
Dist.	Distemper.
K. Ct.	Keene's cement.
L. & P.	Lath and plaster.
L. P. & S.	Lath, plaster and set.
L. P. F. & S.	Lath, plaster, float and set.
L. W.	Lime white.
P. Mo. C.	Plaster moulded cornice.
P. P. C.	Plain plaster cornice.
Par. ct.	Parian cement.
Portd. ct.	Portland cement.
R.	Render.
R. & S.	Render and set.
R. F. & S.	Render, float and set.
Wh.	Whiten.
Weathg.	Weathering.

FOUNDER AND SMITH.

C. I.	Cast iron.
Galvd.	Galvanised.
H. N. W.	Head nut, and washer.
R. I. J.	Rolled iron joists.
R. W. P.	Rain-water pipe.
W. I.	Wrought iron.

PLUMBER.

C. C. N.	Close copper nailing.
H. P.	High pressure.
S. J.	Soldered joint.
T. M.	Trumpet mouthed.

PAINTER.

F.	Flat.
G.	Grain.
K.	Knot.
2 O. or ②	Two oils.
P.	Prime.
S.	Stop.
V.	Varnish.

GLAZIER.

B. P. P.	British polished plate

CHAPTER IV.

SQUARING THE DIMENSIONS, ABSTRACTING AND BILLING.

AFTER the dimensions are taken off, they should be squared ready for abstracting. They should be squared by one person, who should put his figures in the squaring column in black ink. They should be checked by another person, who should tick every dimension correctly squared with red ink, and where the calculation is wrong, make the alteration in red ink. This is a better plan than for the first man to square in pencil, and the second to ink in the dimensions. The result of squaring in ink is greater carefulness. Do not neglect to check the corrections.

The squaring of dimensions may be done by boys, but the checking should be done by a careful and capable assistant.

Areas of circles and semicircles will be obtained from the published tables without calculation. 'Laxton's Price Book' contains these.

There are various short ways of squaring dimensions; at the risk of appearing prolix, a few well-known instances may be mentioned :—

15′ × 4″ × 3″ = 1′ 3″ cube.
15′ × 6″ × 2″ = 1′ 3″ „
Dividing the 15 feet by 12 gives the same result in each case.
Analogous cases will frequently be discovered by an observant person.

The working may often be cleared of fractions by multiplying the "times" into one of the dimensions instead of into the total.

The student as he becomes familiar with dimensions will discover many labour-saving expedients; one is varying the order of his factors, thus :—

$$\frac{\begin{array}{c}8/16\quad 0\\3\\\overline{2\tfrac{1}{4}}\end{array}}{} = 6'\ 8''$$

3″ × 8 = 2 ft., 2¼″ × 2 ft. = 5′, 16′ 0″ × 5″ = 6′ 8″, a cumbrous expression of a rapid mental process.

The person who is squaring should look for instances of dimensions omitted; he will sometimes find work which is ostensibly measured by the foot cube with only two dimensions, thus, $\frac{5\ \ 0}{4\ \ 0}$; superficial dimensions with only one, thus, $\underline{5\ \ 0}$; he should call the attention of the "taker off" to such as these.

Carry the result of *every* item into the squaring columns. The practice of neglecting to do so in the case of "numbers" is frequent, and is a fruitful source of error.

ABSTRACTING.

It is hardly necessary to say that the items of an abstract will be in the order of the bill. Consequently the form of bill, elsewhere given, may be consulted for that order.

Small quantities of work and separate sections may *sometimes* be *billed* from the dimensions, and the labour of abstracting saved. In such a case write on the margin of the dimension sheet "Bill Direct." The practice is, however, dangerous, unless a reference to the column appear on the abstract.

In abstracting, an unchanging order of procedure is as important as in "taking off."

The usual method is to abstract one trade at a time, and this, taking into consideration the pressure which is commonly put upon the surveyor, is the most convenient course; but there is an additional element of safety if the abstractor can commence abstracting at the beginning of the dimensions, and proceed with the dimensions *seriatim*, irrespective of distinction of trade. If there is any error in the "taking off" and he is observant, he is by following this course, much more likely to discover it.

If this cannot be done, not more than one person should employed to abstract one trade, nor more than one person

QUANTITY SURVEYING.

The abstractor should adopt one unvarying practice. First write the dimension in the column of the abstract, then cross it out on the dimensions. Let each line show clearly its beginning and end, see Fig. 32.

This will be a guide to each distinct process. When there are two or more processes to one item of dimensions, requiring two or more lines, put the lines in the squaring column last, see Fig. 33.

N.B.—The dotted lines indicate red ink.

At the bottom of every column, as soon as the abstracting of that column is complete, the abstractor should put a tick in black ink. When the trades are each abstracted by a different clerk, much time will be saved in this way, for instead of looking down every column of dimensions to see if all is abstracted, a glance at the bottom of the column will be all that will be necessary.

Fig. 32.

When there are several items to one dimension, it will be a good practice to abstract them in the order in which the builder does them.

The clerk who checks should first read the item on the

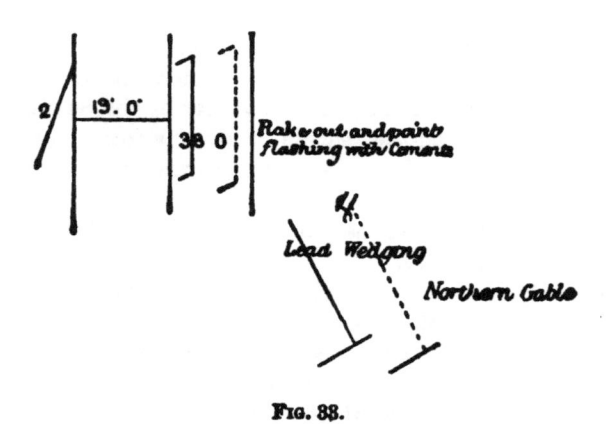

Fig. 33.

dimensions; secondly, find it on the abstract and tick it in red; thirdly, cross out the dimension in red. If he does not always proceed in one order he will find himself not unfrequently uncertain whether he has crossed out the item or not, as dimensions frequently recur. When the checker cannot find a dimension in its proper place in the abstract he should write it in red at the bottom of its proper column, and he should write opposite to it the page or column of dimensions, from which it is derived, thus, 125 feet 6 inches (14), so that it may be the easier checked. When he has checked the whole sheet of abstract, he will probably find the item in another column without a tick, he can then cross it out. When every dimension of a column is checked he should put a tick in red at the bottom of the column.

The tendency of the inexperienced abstractor is to use too little paper, and he will frequently find, when perhaps not more than half-way through the set of dimensions, that he has no room for any fresh items without using supplementary sheets.

In the early part of his practice he should use plenty of paper.

The trouble of abstracting may be much reduced by taking advantage of the folding of the paper which is in the middle of each sheet, so that two frequently recurring items, although separated by a number of other items, may, by a mere turning over of a half of the paper, be consecutively abstracted.

It is the practice of some surveyors to put opposite to each dimension the number of the column from which it is taken, thus:—

Rough Cutting.	
44·9	(54)
87·3	(58)

This greatly facilitates reference, and in the case of large works saves much time, for it sometimes happens that after an abstract has been ostensibly checked, some item is discovered without a tick against it, and time is lost in attempting to find the dimensions: by stating the column this is avoided. In abstracting from dimensions where one regular and known order of taking off is adopted, the abstracting of the column number is not so essential, but in measuring from the actual work the order followed depends so much upon accidental circumstances that the exact place of a dimension in the measuring book is (without this practice) very difficult to determine. Write references to books thus: $\frac{50}{110}\frac{\text{(page)}}{\text{(book)}}$.

When a dimension is corrected in the abstract, the number of the column should always be placed against it, so that the correction may be easily verified. The corrections should always be checked.

In abstracting a work of alteration, deal with the dimension books of work on site as early as possible, so that the billing may not be delayed.

When an item in the abstract is in the wrong place refer to it by some mark (as a star) in the right place.

To shorten the bill, averaging on the abstract should as much as possible be adopted. It should be observed that, where it is desired to produce an average size, one of the dimensions of each superficial item entering into the average should be similar to that of the others, for cubic dimensions two, as otherwise the result would be incorrect. When the conditions are as follows it will be right :—

SASHES AND FRAMES AS DESCRIBED.

```
             ft.  in.    ft.  in.
      2 = 4   0  ×  6    0
      2 = 4   0  ×  8    0
      4 = 8   0  × 14    0
Average  2   0  ×  3    6
```

The correct method, where the dimensions vary, is to find the superficial contents of each, and average them, thus :—

```
        ft.  in.  ft.  in.    ft.  in.
  3/    1    8  ×  2    6  = 12    6  superficial.
  2/    2    0  ×  3    0  = 12    0       ,,
                           5)24    6       ,,
                              4   10  average.
```

See also section, "Examples of Abstracting," p. 308.

Some examples of suitable items for averaging in various trades are as follows :—

BRICKLAYER.—The lengths of ventilating or smoke flues when measured by the foot run.

The thickness of the walls for eyelets and perforations.

The length, depth and thickness of arches when numbered.

MASON.—The size of perforations through stone of the same kind and thickness.

Girth of mitres, stopped ends, etc., to mouldings and splays.

Size of stone templates of the same thickness.

SLATE MASON.—Girth of notchings, size of sinkings, and perforations.

CARPENTER.—Size of cesspools.

JOINER.—Size of small casements and frames or sashes and frames.

Girth of notchings, lengths of balusters.

PLUMBER.—Short lengths of pipes where similar labours.

PLASTERER.—Girth of mitres, stopped ends, etc., to mouldings and splays.

FOUNDER AND SMITH.—Length of bolts of each diameter.

Height of hoisting of iron columns, girders, etc.

The order of the abstracts, and consequently of the bills, is as follows:—

No. 1. Preliminary and provisions.
" 2. Excavator and drains.
" 3. Bricklayer and waller.
" 4. Mason.
" 5. Slater, tiler and slate (or marble) mason.
" 6. Carpenter.
" 7. Joiner and ironmonger.
" 8. Plumber and zincworker.
No. 9. Plasterer.
" 10. Founder and Smith.
" 11. Gasfitter.
" 12. Bellhanger.
" 13. Glazier.
" 14. Paperhanger.
" 15. Painter.
" 16. Separate estimates.

A few *general* rules for abstracting should be remembered.

Write the name of the work and the trade at top of each sheet, and on the first sheet write " In . . . . sheets "; a sheet of abstract has occasionally been mislaid with disastrous consequences.

If there should be more than one abstract to a trade, number them consecutively, No. 1, No. 2, etc.

Commence the abstract of each trade with the leading item of that trade.

Then follow with *cubic* dimensions, *superficial* dimensions, *lineal* dimensions, *numbers*.

Items of labour only should always precede those which involve labour and material.

In each section commence with the item of the smallest value. In lineal dimensions—carpenter and joiner—this will be disregarded, size will regulate the process, as 1 inch × 2 inches, 1 inch × 2¼ inches, 1 inch × 2½ inches, etc.; but the order of size may sometimes be departed from, as 1½ × 2 rough fillet, 1½ × 3 ditto, 1½ × 3½ ditto, etc., recommencing with a smaller size, 1¼ × 2¼, chamfered fillet, etc. (See also pp. 304, 305.)

When an item has been measured lineally and there are extra labours upon it, it is better that they should immediately follow the lineal dimensions (see "writing short," p. 316.)

In casting abstracts, deducting or averaging, use blue ink.

For checking use red ink.

The examples of abstracts will sufficiently explain themselves. (See examples, p. 307.)

Where the description of an item is long it need not be completely copied in the abstract, but the description may be partly copied into the abstract, and the dimensions referred to by the number of the column of the dimensions for the remainder. Thus, "1 Dresser, see col. 54."

The abstractor should never alter the phraseology of any items in the dimensions without consulting the "taker off"; if he does so he is very liable to error.

Preliminary.—The first part of the abstract will be a collection of preliminary items, and it is a good plan to begin each of these with the word "allow," so as to clearly distinguish them from provisional items, the former being at the risk of the contractor, the latter being adjusted at completion according to proportion of amount or quantity expended or used. (See also p. 316.)

Items of allow, when long, may be referred to by column. Often they are items of notes made at the building, and it is convenient to write them into the bill *direct* from the book; this is best done by the man who makes the original notes.

Where the work consists partly of alterations the items of alteration should follow the general preliminary items.

It is often the case that preliminary items do not appear in the dimensions at all, nor in the abstract, such items being written direct from a former bill of similar work, and merely referred to in the dimensions.

Provisions.—The amounts provided are better placed together at the beginning of the series of bills. It is the practice of some surveyors to put provisional items at the end of the trade to which they belong. The former method saves trouble and is better, unless the work is tendered for by separate tradesmen, in which case the latter course must be adopted.

Where the precise quantity of the work which is intended to be executed, and for which the sum is provided, is known, it is better to state the quantity as well as the amount. For instance, "Provide for 20 feet of iron railing, 30*l.*"

Where the provision consists of material, it is better to place it in the same position in the bill as it would hold if it were not a provision, but write it in the manner following:—

	ft.	in.			
	100	-		Cube.	Fir framed in roofs. *As provision.*

Such an item would be dealt with at the final settlement, the total quantity in the bill being deducted and the quantity used measured and added.

Excavator.—Where there are but few items and few kinds of material the general rotation as to "cubes," "superficials," and "runs," may be maintained; where there are many it will save writing and be more convenient for pricing to adopt the order—cubic and superficial quantities of excavation, then cubic and superficial quantities of concrete, etc., then drains and cesspools.

Bricklayer.—In abstracting brickwork always prefer to abstract into the one and a half brick column, where equally convenient, as this saves reducing. The methods are as described below:—

50 feet superficial of half brick thick may be abstracted as 25 feet of one brick, or 16 feet 8 inches of one and a half brick.

50 feet superficial of two bricks thick as 100 feet of one brick.

50 feet superficial of two and a quarter bricks thick as 25 feet of one and a half brick and 50 feet of one and a half brick.

There is far less liability to error in the abstracting of brickwork if the deductions are placed in the position shown on the abstract (see pp. 4 and 5) than if arranged as follows:—

One brick.	One brick deduct.	One and a half brick.	One and a half brick deduct.

In abstracting, observe the principle of reducing on the abstract, not on the dimensions, the latter is a frequent cause of error. Some abstractors, to avoid making a heading for a cube brickwork, deduct ¼th from each item of it in the dimensions, thus reducing as they go on; this is not a good practice. Abstract it as cube brickwork—deduct ¼th from the total, and carry it to the reduced brickwork when preparing the abstract for billing.

Be careful that unwary assistants do not divide cubic dimensions by 9 instead of 27, and *vice versâ*.

Where there are various kinds of facings and a good number of items of extra labours upon the facing it is advisable to let an abstract of them follow the general abstract, but otherwise they may come in with the general items of the bill.

Small metal articles which involve bricklayers' work in the setting are usually billed with the bricklayers' work at the end of the numbers.

Mason.—Keep each of the various kinds of stone, with its labours, separate. Commence with the stone of least value.

Keep the items of "stone and labour" in a separate section.

Slater, Tiler, Slate and Marble Mason.—Keep the work to each trade separate, following the usual rotation—superficials, runs, numbers—beginning with the main item of the trade, as squares of slating or tiling.

Carpenter.—Abstract joists to flats with fir framed in roofs.

Keep the different kinds of wood separate, commencing with the ordinary items of cube fir.

In a small work it will be sufficient to adopt the common practice of arrangement—cubes, superficials, runs, numbers, according to value. In the case of a large work it is sometimes more convenient, and a saving of writing, to divide the work into sections, as where there is a large number of lineal items of "fir wrought and framed in white-lead" of various sizes and with various labours upon them; spires, flèches and dormers are often kept separate in the same manner with advantage.

Any separate section should follow the *numbers* of the general carpentry.

Where "fixing only" to ironwork has been taken it should come last in the abstract.

In abstracting things measured by the foot run and of various sizes, 2 inches by 3 inches, 4 inches by 6 inches, etc., take them in the order of their scantlings, the smallest first, disregarding the fact that some of the smaller ones are of more value than the larger, because of the greater amount of labour upon them.

Joiner and Ironmonger.—The usual order is as follows:— "Floors," "skirtings," "skylights and sashes and frames," "doors," "architraves and mouldings" (if a considerable number of items, if not abstract with sundries), "thicknesses and framings," "sundries," "staircases," "work in mahogany," "wainscot," etc., keeping each kind of wood separate; "ironmongery and fixing," or, where a sum has been provided, "fixing only ironmongery."

When two or three kinds of wood are used in the same building for similar purposes, it is sometimes convenient to let each section of soft wood in the bill be followed by the hard wood, as floors in deal, floors in wainscot, floors in teak, skirtings in deal, skirtings in wainscot, skirtings in teak, etc.

Plumber and Zincworker.—Abstract the lead first, under the heading of its own weight, as 4 lbs., 5 lbs., etc., and afterwards reduce it to cwts.; abstract together lead in gutters, flats, flashings, hips, ridges, valleys; abstract together stepped flashing and secret gutters; then labours to the lead, copper pipes, lead pipes, soldered joints, and labours to pipes, short lengths of pipe, traps of various kinds, brasswork, lavatories, w.c. apparatus, baths, urinals, cisterns.

When internal and external plumbery is kept separate the same order is adopted for each.

The zincworker may follow the plumber. Superficials, runs, numbers.

Plasterer.—Commence with the common plastering, superficials, runs, numbers, and keep the various kinds separate, then work in fine plaster.

Work in cement, keeping each kind, as Portland, Martin's, Keen's, separate.

Tile paving (except quarries) and wall tiling.

Plastering in narrow widths and small quantities should always be billed in feet.

Sometimes the external and internal plastering are separately billed, but only in a large work.

Founder and Smith.—Keep the cast and wrought iron separate. Reduce all to weight. The same constant of weight should be used for wrought iron and rolled iron.

The weights of iron, which have been measured in feet and inches, will be found in the current price books and Hurst's or Molesworth's handbooks.

Gasfitter.—Commence with the pipes, the smallest first; follow with fittings and their fixing according to their value, lengths of pipe, main cock, meter, attendance. *If the fittings are a provision*, fixing only gas fittings according to value.

Bellhanger.—Commence with the bells; follow with the pulls and fixing according to their value. If the pulls are a provision, a section of *fixing only*, according to their value, will follow here.

Glazier.—Commence with the glass of the smallest value and arrange it according to the superficial content of the squares, the smallest first; follow with each kind arranged in a similar manner, abstracting any extra labour on any particular kind of glass, following that glass, lead lights, copper or lead clips.

Painter.—Keep each kind of painting in a separate category arranged in the order superficial runs, numbers. The painting of smallest value first as: one oil on iron, two oils on iron, three oils on woodwork, four oils on woodwork, etc.

When the abstracting is completed, cast up all the columns, deduct the deductions, and reduce each item that requires reduction; let every process be checked, and the work will then be ready for billing. When columns are very long it will always be a saving of time to divide them into several casts. Let the calculations on the abstracts be exact, any adjustment as to measure or weight should be left until the time of billing.

The checking of casts should be done by a careful and capable assistant.

ABSTRACTING.

EXAMPLES TO ILLUSTRATE ABSTRACTING.
(See also form of Bill.)

BRICKWORK AND ITS REDUCTION.

" Bricklayer No. 1."

" For description of material see Col. 40 dims."

Reduced Stock Brickwork in Mortar.

Cube.	Deduct.	1 B.	Deduct.	1½ B.	Deduct.
100 0	Cube.	100 0	1 B.	100 0	1½ B.
100 0	—	100 0	—	100 0	—
100 0	10 0	100 0	10 0	100 0	10 0
—	10 0	—	10 0	—	10 0
300 0	10 0	300 0	10 0	300 0	10 0
30 0	—	30 0	—	80 0	—
—	30 0	—	30 0	—	30 0
270 0		270 0		270 0	
Ddt. ⅛th = 30 0		⅓rd = 90 0		180 0	
—		—		240 0	
240 0		180 0		—	
				690 0	

2 rods, 146 feet.

Hoop Iron.

Add 5 per cent. for laps.

1¼″ × ¹⁄₁₆″ hoop-iron bond and laying in walls.

100 0	
100 0	
100 0	
300 0	
5 per cent. = 15 0	
315 0	= 105 yards
·27	
85 lbs.	
0 3 1	

Billed as—

yds. ft. in.
105 - - run 1¼″ × ¹⁄₁₆″ hoop-iron bond and laying in walls (weight, 3 qrs. 0 lbs.)

x 2

QUANTITY SURVEYING

Perforations in brickwork.

Perforations in wall for pipe and making good.

10 = 20 bks.
10 = 20 „
10 = 15 „
―――――――
30 = 55 bks.
Average, 1¾ bks.

Billed as—

No. 30 Perforations in wall average 1¾ bricks thick for pipe and making good.

Arches.

Extra labour and materials cutting and waste to relieving arches.

	ft.	in.	B.	B.
10 =	100	0 ×	10 ×	15
10 =	100	0 ×	10 ×	15
10 =	100	0 ×	10 ×	15
30 =	300	0 ×	30 ×	45

10 0 × 1 B × 1½ B. averaged.

Ditto to segmental arches.

	ft.	in.	B.	B.
10 =	100	0 ×	10 ×	15
10 =	100	0 ×	10 ×	15
10 =	100	0 ×	10 ×	15
30 =	300	0 ×	30 ×	45

10 0 × 1 B. × 1½ B. averaged.

Billed as—

No. 30 Extra labour and materials cutting and waste to rough relieving arches.
  10 ft. 0 in. × 1 brick × 1½ brick averaged.
„ 30 Ditto, segmental, 10 ft. 0 in. × 1 brick × 1½-brick, ditto.

Labour on Facings—

Moulded cornice two courses high moulded 7" girth as sketch, col. 82.

		Mitres.	Stops.
100	0		
100	0	―	―
100	0	10	10
―		10	10
300	0	10	10
		―	―
		30	30

ABSTRACTING.

Billed as—

ft. in.
300 – run   Extra on facings for moulded cornice two courses high and 7 in. girth, as sketch.
No. 30 mitres.
„ 30 stops.

Method of arrangement by which the repetition to every item of the words "framed and wrought" is avoided.

Heading of section in bill—

Fir framed and wrought all round where required.

$2\frac{1}{2} \times 2\frac{1}{2}$ bearer.

100	0	Do. plugged.	
100	0		
100	0	100	0
—		100	0
300	0	100	0
		—	
		300	0

Lead.

Milled lead and labour in flats, gutters, flashings, &c.

4 lb.		5 lb.		6 lb.		7 lb.	
100	0	100	0	100	0	100	0
100	0	100	0	100	0	100	0
100	0	100	0	100	0	100	0
300	0	300	0	300	0	300	0
	4		5		6		7
1200 lb.		1500 lb.		1800 lb.		2100 lb.	
1500							
1800							
2100							
6600							

cwts. qrs. lbs.
58   3   20

QUANTITY SURVEYING.

Billed as—

cwts. qrs. lbs.
58   3   21   Milled lead and labour in flats, gutters and flashings.

Plastering.

Twice whiten ceilings and soffits.	Render float and set walls.	Ddt.	And 2ce whiten. Lath plaster float and set ceilings.	Ddt.	And 2ce whiten. Lath plaster float and set soffits.	Ddt.
270 0	100 0		100 0		100 0	
270 0	100 0	10 0	100 0		100 0	10 0
———	100 0	10 0	———		———	10 0
540 0	300 0	10 0	100 0	10 0	300 0	10 0
———	30 0	———	100 0	10 0	30 0	———
60 yds.	———	30 0	100 0	10 0	———	30 0
	270 0		———	———	270 0	
	———		300 0	30 0	———	
	30 yds.		30 0		30 yds.	
			———			
			270 0			
			———			
			30 yds.			

The words "twice whiten" being written in the foregoing manner over all the items which are whitened, the totals, as soon as all are abstracted, should be re-abstracted in the proper place ready for billing.

Mitres to
moulding.

	ft.	in.
100 =	50	0
100 =	100	0
100 =	66	8
300 =	216	8

8¾ in. average.

Billed as—

No. 300   Mitres to moulding average 9 in. girth.

Founder and Smith.

                                                    Exact Weight.
C. I. weighted at 38 lbs. per foot supl. of 1 in.   (37·50)
W. I.   „   41 lbs.   „   „   (40·32)

ABSTRACTING.

Cast Iron.

And add 2¼ per cent. for featherings.

In No. 10 girders and fixing 20 feet from ground level.	In No. 20 hollow columns and fixing at ground level.

×————————————×

Supl. ⅞ in.	Supl. ⅞ in.	Supl. ⅞ in.	Supl. 1 in.
100 0	100 0	100 0	100 0
100 0	100 0	100 0	100 0
100 0	100 0	100 0	100 0
(A) 300 0	(B) 300 0	300 0	300 0

Columns.
1″ collected.
150 0
300 0
375 0
600 0

1½ in.	2 in.	1425 0
100 0	100 0	38
100 0	100 0	54150 lbs.
100 0	100 0	2¼ % 1354
300 0	300 0	55504 lbs.

cwts. qrs. lbs.
495　2　8

Girders collected.

1″ iron supl.

(A) 150 0
(B) 225 0
———
375 0
38
——
14250 lbs.
2¼ % 356
——
14606 lbs.

cwts. qrs. lbs.
130　1　18

Pattern for girder 11 ft. 6 in. long and 9 in. wide as sketch col. 18

Alteration to do.

1

1

Pattern for hollow column with moulded cap, base and necking, and four brackets, 8 ft. high in all.

1

Alterations to do.

4

Billed as—

cwts. qrs. lbs.
130　1　21　In No. 10 panelled girders and hoisting and fixing 20 ft. from ground level.
No. 1 pattern for panelled girder 11 ft. 6 in. long and 9 in. wide, as sketch.
No. 1 alteration to ditto.

cwts.	qrs.	lbs.	
495	2	7	In No. 20 hollow columns and fixing at ground-floor level.
			No. 1 pattern for hollow column with moulded cap, base and necking, and four brackets, 8 ft. high in all.
			No. 4 alterations to ditto.

Wrought Iron.

Add 5 per cent. for rivets.

In No. 6 riveted girders.

Supl. ⅜ in.		Supl. ½ in.		Supl. ⅝ in.	
100	0	100	0	100	0
100	0	100	0	100	0
100	0	100	0	100	0
300	0	300	0	300	0
112	6	150	0	187	6

Supl. ¾ in.		Girders collected.	
100	0	Supl. 1 in.	
100	0	112	6
100	0	150	0
300	0	187	6
262	6	262	6
		712	6
		41	
		29213 lb.	
		1461 = 5 %	
		30674 lb.	
		cwts. qrs. lbs.	
		273 3 14	

Fixed up—

cwts.	qrs.	lbs.	
273	3	14	In No. 6 riveted girders and hoisting and fixing 20 feet above the ground level.

ABSTRACTING.

In rolled iron joists.

4″ × 3″ = 12 lbs. per ft.	4½″ × 3″ = 16 lbs. per ft.	9″ × 5″ = 20 lbs. per ft.
12 6	15 0	17 6
13 0	12 6	18 0
14 9	9 0	15 6
40 3	36 6	51 0
12 lb.	16 lb.	29 lb.
483 lb.	584 lb.	1479 lb.
584		
1479		
2546 lb.		

cwts. qrs. lbs.
 22   2   26

Billed as—

cwts. qrs. lbs.
 22   3   —   In rolled iron joists cut to exact lengths and hoisting to various levels and fixing by bricklayer.

Bolts.

Under 12 in. long		Over 12 in. long.
½ in. diameter screw bolts — long, with heads, nuts and washers, and fixing by carpenter.		¾ in. diameter screw bolts — long, with heads, nuts and washers, and fixing by carpenter.
in.	¼ in. ditto.	in.
10 = 100		10 = 200
9 = 90	in.	10 = 250
10 = 80	10 = 100	10 = 150
29 = 270	10 = 100	30 = 600
9 in. average.	10 = 100	20 in. average.
	30 = 300	
	10 in. average.	

Billed as—

No. 29 ½ in. diameter screw bolts, under 12 in. long, average 9 in. long, with heads, nuts and washers, and fixing by carpenter.
 „ 30 ¼ in. ditto, 10 in. long, and ditto.
 „ 30 ¾ in. ditto over, 12 in. long, average 20 in. long, and ditto.

Glass.

21 oz. sheet glass, and glazing in squares not exceeding ... feet supl.

2 ft. 0 in.	4 ft. 0 in.	6 ft. 0 in.
100 0	100 0	100 0
100 0	100 0	100 0
100 0	100 0	100 0
300 0	300 0	300 0

Billed as—

```
ft.   in.
300   -   supl.   21 oz. sheet glass in squares not exceeding
                  2 ft. 0 in., and glazing.
300   -    „      21 oz. ditto, 4 ft. 0 in., and ditto.
300   -    „      21 oz. ditto, 6 ft. 0 in., and ditto.
```

Paper.

```
          Paper
          P.C. 2s.
          per piece,
          and
          hanging.
          ─────────
          100   0
          100   0
          100   0
          ─────────
     54) 300   0
          ─────────
          5 pieces 30 ft.
```

Billed as—

No. 6 Pieces of paper P.C. 2s. per piece, and hanging.

When an item is transferred from one part of an abstract to another, it is customary with some to put a distinguishing sign, thus—

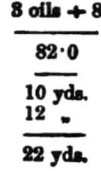

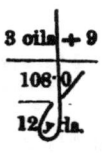

```
3 oils + 8              3 oils + 9
─────────               ─────────
  32·0                   108·0
─────────               ─────────
 10 yds.                  12 yds.
 12   „
─────────
 22 yds.
```

BILLING.

The general rule as to order of billing is to begin each trade with the principal item of that trade, as—Bricklayer, rods of brickwork; mason, cubic feet of stone; slater, squares of slating; carpenter, cubic feet of timber; joiner, squares of flooring; founder and smith, cwt. of iron; plumber, cwt. of lead, etc., and generally speaking, adhere as closely as possible to the order, cubes, suprs., runs, numbers.

Any total in the abstract having odd inches is billed as a foot when six inches or over, and where under six inches the inches are discarded, thus—4 feet 6 inches is called 5 feet, 4 feet 5 inches is called 4 feet. In work priced per yard cube $13\frac{1}{2}$ feet is called a yard, under $13\frac{1}{2}$ feet is rejected. Similarly, when work is priced at per square, parts of 5 feet are treated as 5 feet, or discarded, thus—108 feet would be called 1 square 10 feet, 107 feet 1 square 5 feet, 72 feet, 70 feet, 73 feet, 75 feet. This practice saves contractors much trouble in pricing, and in works of fair size the differences neutralise each other; but in *small* measured accounts it is better to adopt the exact quantities.

In plastering or painting, where half a yard superficial, call the dimension a yard; where under half a yard, disregard it.

Materials charged per cwt. proceed on the same principle, calling the lbs. 0, 7, 14, or 21, when less than a quarter.

Where there is any departure from the usual mode of measurement the fact should be stated in the preamble of the bill.

In items where scantlings are stated, put the smaller one first, as 6 inches by 8 inches, and write the name of the article immediately after the size, and before describing the labour, as, for instance, 6 inches by 8 inches *fascia*, moulded 6 inches girth.

Repeat the figures of scantlings or thicknesses when billing as

	3″ × 4″ wrought and framed rail.			
	3″ × 4″	,,	,,	twice splayed.
not				
	,, × ,,	,,	,,	twice splayed.

Numbers also should invariably be repeated even when similar.

Work described as "in small quantities" should always be billed in *feet*.

Where a number of items are in deal, and others in pitch pine, mahogany, etc., a heading will save the repetition of the words, deal, pitch pine, etc., thus: "Doors in *deal*," "Skirtings in deal," "Best Honduras mahogany," etc., and generally, if there are many items of a sort, judicious headings will save writing.

Gratings, air-bricks, etc., connected with the general construction, and which involve the labour of a bricklayer, mason, or carpenter, should be billed at the end of the trade with which they are connected.

In some cases it is found convenient to give the detailed quantities of a particular item, so that a separate amount may be arrived at; this is called "writing short" (see **A** next page). Another case in which the same plan is adopted is that of the extra labours on work measured lineally (see **B** next page). In cases like **A** inches may be stated.

In writing the bills keep the whole of one item on the same page, and do not write parts of the same word on two lines.

When it is desired that an article shall be supplied by a particular person, insert his name and address, with the description of the said article.

Carefully observe that the words *cube, superficial, run, number*, are written in where these divisions commence, as, to take an instance, the word *run* is sometimes omitted where the superficials terminate, and the lineal dimensions consequently appear to be superficial.

Observe that the words "and including" be used where required, as in describing tiling and lathing, write "*on and including* $1\frac{1}{2} \times \frac{3}{4}$ sawn fir laths," or in description of doors when glass is beaded in, "*with and including* mitred and screwed beads for glass."

Observe also the distinction between "as pattern" (out of stock, if not more definitely described), and "to detail" or "to design."

Take care that, so far as possible, the same word has but one meaning throughout the set of bills, especially the words "provide" and "allow." It is convenient to apply the word "provide" to those amounts or quantities which are provided, and which will be the subjects of future adjustment, and the word "allow" to all items the cost of which is at the contractor's risk, and not subject to interference. Parts of "allows" should consequently not be

BILLING. 317

A.—No. 1 STEPLADDER, AS DETAIL FOLLOWING, viz.—

ft.	in.			£	s.	d.	£	s.	d.
10	6		Cube, fir framed						
43	–		Superficial, planing						
17	6		„ 1¼-inch rough batten landing, with open joints ..						
3	6		„ 1¼-inch deal rough lining ..						
40	6		„ 1¼ „ „ treads ..						
13	6		„ 1½-inch deal rough treads, framed						
47	6		„ 2-inch deal rough framed strings						
37	–		Run, labour to rounded edge to 3 inches fir						
No.	84		Housings of treads						
„	8		Ogee cut ends to bearers, 6 inches by 3 inches						
„	9		Ends of bearers, etc., cut and pinned into wall						
			Amount of stepladder .. .. £						
			Carried to adjoining column.						

B.

ft.	in.		
50	–	run	3 inches by 9 inches window sill, tooled, sunk, weathered and throated.
			No. 20 fair ends.

deducted. (See p. 302.) It is not expedient to contract any of the words in a bill; write all at full length.

Sometimes labour is saved by billing items without abstracting; in such case write against them in dimensions "to bill direct."

Either *every* trade may have a separate bill, or two or three trades may be put into one bill, or the bill may be continuous; in either case each of the *trades* will be carried separately to summary, and when the bills are separate, each *bill* will have a heading similar to the following:—

Estimate,
 For proposed house and offices at Sutton, Surrey, for John Smith, Esq.
————, Esq., Architect,
24 Montague Square,
July 1878. London, W.C.

It is often convenient when part of the work is new and part alteration to old, to make a separate bill of the alteration, heading it "works on the site"; the builder in the latter case will have to take this bill only to the site.

A separate bill may often be made with advantage of the general contractor's work in connection with a sub-contract for ventilation, machinery, etc.

Head each trade with number and trade, thus—"No. 2. Excavator and drains."

The sketches should be drawn in the bill before any checking is done, so that they also may be checked.

When the work is in a district where other than London methods of measurement are used, it will be advisable if they are not adopted to say "all the modes of measurement and allowances in this bill of quantities are in accordance with the London practice."

Another preliminary clause sometimes used is as follows: "In all trades the London mode of measuring has been adopted, viz. all openings for doors, windows, etc., have been deducted from the gross totals, and all quantities of stone and other materials stated in the several bills are the net measurements. When fixed in the work, all allowances for waste, etc., must be considered in the prices."

BILL HEADINGS.—Much time may be saved by the preparation of a set of bill headings for office use, comprising all the varieties necessary for each trade, both simple and elaborate. Such clauses as may be applicable to a particular building may be marked, and a clerk will then write them as a preamble to each trade. This practice will often obviate the necessity of examining a number of old bills.

Lithographed sets of preambles to trades, suitable for an ordinary building, should also be kept. These vary but little, and they can be readily altered if necessary.

Preliminary Bill.—This should commence with clauses from the conditions which may affect the amount of tender, such as the following :—

"The site is on the north side of the road leading from Kew to Richmond, and opposite to the Cumberland Gate of Kew Gardens" (description of position will often be unnecessary beyond the heading of the bill).

"The building to be completed fit for occupation by the . . . day of . . . under a penalty of £ . . . per week as liquidated damages, delay consequent upon strikes only excepted." Sometimes, "in case of extra works, the time shall be extended one week for every £100 worth of additional work."

"Payments will be made to the contractor in sums of not less than £ . . . at the rate of 75 per cent. upon the value of the work executed until completion, when an additional 20 per cent. will be paid, and the remaining 5 per cent. at the expiration of six months."

The old "Conditions of Contract" published by the R.I.B.A. involve a different arrangement: "When the value of the works executed, and not included in any former certificates, shall from time to time amount to the sum of £ . . . , or otherwise at the architect's reasonable discretion, the contractors are to be entitled to receive payment at the rate of 80 per cent. upon such value until the difference between the percentage and the value of the works executed shall amount to 10 per cent. upon the amount of the contract, after which time the contractors are to be entitled to receive payment of the full value of all works executed and not included in any former payment, one moiety of the balance being paid on completion, and remainder in . . . months."

The payment and maintenance clauses are various, and have to be drawn to meet special cases. Architects' instructions usually regulate these.

"No part of the work to be let as task-work."

"A fully-priced copy of the estimate (*sometimes sealed*) is to be deposited with the architect within a week of signing contract, extras and omissions to be valued at the prices of the contract, *and any item of extra work which does not exactly agree with the descriptions of the original estimate to be valued at a price analogous thereto.*" Sometimes "the quantities will form part of the contract."

The clause commonly appears without the part in italics; when it does the builder will argue that the rates of the original estimate shall be only applied to such items of extra work as are exactly described in the bill of quantities.

This condition, if uniformly adopted, will produce in course of time an improvement in the system of tendering. When a large quantity of extra work has been anticipated, builders have sometimes priced their original estimate at an absurdly low rate, with the idea of making their profit by charging an exorbitant price for the extra work.

The School Board for London inserts the following clause. "A copy of the bill of quantities, fully-priced and moneyed out, which are to form part of this contract, is to be deposited with the works department by the contractor on the signing of the contract, to be used as a schedule of prices and for measuring and valuing the extras and omissions (if any) upon the contract. In adjusting extras or variations on the contract, the Board's measuring surveyor will measure on the same basis as that on which the quantities have been taken, and all labours not specially mentioned in the quantities will be taken as included in the prices of the various items."

The War Department stipulates. "The priced bill of quantities fully marked on the cover, so as to secure identification, shall be transmitted at the same time as the tender as a separate packet." The condition most comprehensive and useful would comprise the following stipulations. The deposit of the priced bill of quantities before the signing of the contract (preferably with the tender), its examination and comparison with the copy intended for sealing before signing the contract. A statement of who shall examine it, and that the sealing shall be after the examination.

"The contractor will be required to keep an approved foreman constantly upon the works."

"For the remaining conditions see the copy appended to the specification."

"Allow for insurance from fire to the amount of (two-thirds, or as the case may be) tender, in such manner as may be directed, and make good after fire or any other accident."

"Allow for supplying water for all the works, including fees, temporary plumbing, and storage of water."

"Allow for giving notices to all authorities requiring notice, and for supplying any drawings required, and paying all fees. (The fee for District Surveyor is for a building . . . stories in height, and not exceeding . . . squares in area.)" The foregoing part in parenthesis applies only within the jurisdiction of the London County Council.

Observe that where a new building is connected with old ones, District Surveyor's fees must be included for them as well as for the new ones; reference to the schedules of the Building Act will show how. Include also the fee for inspection under the new by-laws of the London County Council.

"Allow for each trade to attend upon all other trades, and for all jobbing connected therewith." This clause will not supersede the necessity of special items for attendance.

"Allow for all scaffolding, rods, etc., and stakes and labour in setting out works."

When the building is large, leave out the word scaffolding in previous clause and insert the following :—

"Allow for the necessary scaffolding for a building about . . . feet long, by . . . feet wide, and . . . feet from ground to ridge, with . . . chimney stacks rising to a height . . . feet from ground, and a turret rising to a height of . . . feet, and including an apartment . . . feet by . . . feet, and . . . feet high." (The latter clause when there is a very large room or hall involving internal scaffolding.)

"Allow for an office for clerk of works and the requisite firing, light and attendance, and for all sheds, etc., required for keeping materials under cover, and for carrying out the works." If any materials not usually protected are to be kept under cover they should be particularised, such as lime, facing bricks, etc.

"Allow for proper latrines for the workmen, for keeping same in a clean and decent condition, and for emptying as may be required, and removing at completion."

"Allow for enclosing the site with an approved post and rail fence, and for preventing the men from trespassing upon any other part of the ground beyond that enclosed." The foregoing in case of a park or garden.

"Allow for covering the walling during inclement weather, and for supplying all requisite temporary lights, doors, watershoots, covering to stonework and terra-cotta, tile pieces to steps, and any other requisite protection to the whole of the works."

"Allow for making good any injury to the building from any cause, and for making good pointing after injury from frost."

"Allow for making good any injury to adjoining buildings consequent upon these alterations, and for any necessary shoring."

"Allow for necessary watching and lighting."

"Allow for keeping foundations free from water, and for any temporary drainage, baling, or pumping that may be required."

"Allow for clearing away all dirt or rubbish and superfluous materials as they accumulate, for washing (or twice scrubbing) floors at completion, and for leaving the whole of the premises clean."

"Allow for affording facilities to any other parties employed upon the buildings, so that their works may proceed during the progress of the contract, and allow them use of ordinary scaffolding and ladders."

"Allow for casing and otherwise protecting any of the work done by other tradesmen, and be responsible for, and make good or pay for the making good of any work which may suffer for want of such casing or protection."

"Allow for keeping the works in proper repair for . . . months (usually six) after completion, and for making good any defects or imperfections which may arise during that period, and for making good pointing to roofs or walls after injury from frost."

"Allow for erecting, maintaining and altering as may be required, and afterwards removing, a hoarding, 6 feet in height, with the necessary gates and fastenings, fans, planked footway, post and rail fence, etc., to the satisfaction of the local authorities, for a length of . . . feet, with two returns . . . feet in length." The foregoing would apply in a public thoroughfare.

Sometimes supply and fix to front wall for its whole height from ceiling of ground floor to top of parapet a close boarded fan 2 feet 6 inches wide.

Sometimes the right of letting the hoarding to an advertisement contractor is reserved by the proprietor, and sometimes advertising is prohibited.

"Allow for carrying on the works while the present buildings are in use." This applies in case of additions.

In case of an erection on the site of an old building—

"Allow for emptying any old drains or cesspools that may be met with in course of excavation, and for filling in same with hard dry brick or stone rubbish, or lime core, well rammed."

"Allow for leaving temporary openings in roofs, walls or floors as may be required or directed for the getting in of cisterns, boilers, machinery, etc., and make good afterwards in all trades."

"The contractor to send in with his tender the names of two responsible sureties who are willing to be bound jointly and severally with him in the sum of £ . . . for the due completion of the contract and the bond to be executed on the signing of the contract."

"The contractor shall give notice to the local authorities for extra heavy traffic, and shall maintain and make good roads during and on completion of this contract to the satisfaction of the local authorities, or pay any legal claim they make." When the work is very large this is an important clause.

The foregoing is a general outline of the most frequent requirements, but in almost every building some special conditions will be required to meet the exigencies of the case.

The general rule is to embody in the preliminary bill any condition which is likely to affect the value of the work.

In the case of an estimate comprising works of alteration the items of alteration, i.e. such as are not measured for quantity, should follow here, and they should be arranged in such order that the contractor would, in reading the items, go through the building, beginning with the topmost floor, for instance, and dispose of it without any necessity for his return to that part of the building. In all cases state the exact position of the work.

Where there is a very large number of items of alteration some surveyors make a separate bill of them.

"Allow for pulling down," etc.; describe the pulling down. ("See Bill No. 20.") See also remarks on *credits* bill.

If any part of the work is in darkness, as in the case of sub-basements, it should be stated.

Sometimes the following, "allow for laying on temporary gas pipes to new basement, and for paying for the necessary gas, candles, etc."

Alterations.—When the estimate is mainly for the alteration of an old building, it will greatly assist the estimator to place at the beginning of the preliminary bill an item similar to the following :—

"The work consists of raising the building by an additional

story, of building a new kitchen and dining-room adjoining the north wall of the building, and a billiard-room, with conservatory over, adjoining the south wall, and of various small alterations and additions throughout the old building."

A small block plan, lettered A, B, C, etc. (and which may be referred to by these letters) is a useful appendage to a bill of quantities of alteration, and makes the items easier to understand.

"Insurance from fire, and the water company's fees will be paid by the proprietor, but contractor to allow for any temporary plumbing and storage of water."

"Allow for protecting all the old work from injury, and for supplying all requisite temporary roofs or tarpaulins during the time that the interior of the building is exposed."

"The whole of the inserted work, as mouldings, panellings, etc., to match old."

The following clauses will save writing if they are used before the items of alteration.

"The external size of the opening is stated in every case."

"All the items of cutting of openings to include removal and cartage of rubbish and the necessary needling and shoring."

"The making good after cutting of openings to be in hard stocks and cement, including centreing to arches and making good plastering after fixing of joinery."

"The whole of the making good of the plastering is to be in parian cement."

Special reference to scaffolding is also expedient, thus :—

"Allow for erecting a strong scaffold about 36 feet long and to a height of 50 feet from ground level as may be necessary, for the raising of additional story, and include cutting out and making good all putlog holes, and for removal of scaffold at completion."

"Allow for any necessary shoring and strutting while underpinning eastern wall and chimney breast about 22 feet run."

Work in difficult positions and small quantities.—In such cases it is sometimes advisable to insert a clause something like the following :—

"The contractor is referred to the drawings and specification, and to the site, as to the general character of the work; and he must allow in his prices for any extra cost he may consider involved by reason of the work being in detached positions, in small quantities, for difficulty of access and of working, and for any other cause."

Provisions.—The points connected with provisions which require attention are the *packing, carriage, profit and fixing*. Various methods of treatment of the same item appear below. Always state carefully what the sum provided is to cover.

To meet the question of profit it is sometimes stated that the amounts provided are to be paid net to any tradesman selected by the architect on his certificate.

Some surveyors insert the following clause :—

" Wherever the word ' allow ' occurs in this estimate the cost of the item shall be at the risk of the contractor, but wherever the words ' provide' or ' prime cost' are used, the amounts thus mentioned shall be paid by the contractor net to any tradesman selected by the architect on his certificate. If, therefore, the contractor desires a profit he must add it to the amount named in each case."

The following clauses are such as will define the position taken :—" Provide the following sums to be used as directed, or deducted if not required. If contractor desires a profit he must add it to the amount named in each case."

" The letters P.C. or the words prime cost, shall mean the price at the manufactory after deducting trade discount, but not discount for cash."

" Allow on all provisions for profit, packing, carriage and fixing."

Sometimes, " All the following provisions to include fixing, unless specially described otherwise."

Sometimes, " If the architect think fit he may direct that the provisional sums or any of them be retained out of the contract amount, and may certify to the building owners separately for the works for which such provisional sums are provided, in order that the firms executing such works may be paid by the building owners direct."

" Provide the sum of £100 for ten chimneypieces and allow also for packing, carriage, profit and fixing, including the necessary cramps."

" Provide the sum of £100 for ten chimneypieces and fixing."

" Provide the sum of £100 for ten chimneypieces (fixing elsewhere taken)."

"Provide for modelling of terra-cotta £ . . . "
"Provide for extra works £ . . . "

Some surveyors insert at the end of the list of provisions the following: "Allow on the total of the foregoing amounts for *packing, carriage, loading* and *unloading* in the form of a percentage." When thus treated the words in italics may be omitted from the foregoing items.

Submit the provided amounts to the architect for his approval before the bills are lithographed.

When the quantity of work for which a provision of money is made is not settled, it is not fair to add "allow for fixing."

It may be necessary to measure a considerable amount of work in connection with provisional sums, and in such cases as it may be anticipated it is convenient to make a separate bill of the provisions, commencing with the sums provided, and following with the collateral works in the usual order of a bill.

Some surveyors use the following: "In all cases where letters P.C. are made use of in this specification, they are intended to imply the published catalogue price, and the architect shall be empowered, if he thinks proper, to order the articles of any special manufacturer to the full value of the sum named"; others, "All provisional items shall include 10 per cent. profit for the contractor calculated on the net amount after deducting trade discount."

The difficulty sometimes experienced in inducing the builder to pay sums which are the subject of provision, although they may have been certified by the architect, has led to the adoption in some contracts of the following condition:—

"In all cases of provisional amounts for specific items, the architect shall be at liberty to pay such sums directly to the tradesman he may employ to do such work, and to deduct the sum provided for it from the amount of the contract."

The division of labour in the case of provisions requires careful stipulations. When a sum of money has been provided for constructional ironwork like the following—

"Provide for constructional ironwork and fixing £2000,"

and if it is intended that the general contractor shall unload, etc., some such clause as follows will appear in the bill of quantities of the general contract:

"Unload, get into building, deposit, hoist, and assist in fixing . . . tons of iron or steel joists, girders and columns."

"The iron and steel will be painted one coat before delivery."

When a sum of money is provided for terra-cotta delivered at the building, "provide for terra-cotta £ . . . ."

"Allow for assisting the manufacturer of the terra-cotta in setting out, and for furnishing any necessary particulars to manufacturer, and for unpacking, storing and protecting it, and making good any damage." The fixing only will be elsewhere described.

When a sum of money is provided for granite work, "Provide for granite work £. . ."

"Allow for assisting the granite merchant in setting out and for furnishing any necessary particulars to merchant, and for unpacking, storing and protecting it, and making good any damage."

"Allow for supplying water to the fibrous plaster manufacturer; for assisting to unload, store and protect the same to the extent of about 25 squares; for arranging as to time for the execution of the work, after the grounds are fixed; for the erection and use of scaffolding required by the manufacturer; for clearing away all rubbish when ordered by the clerk of works; and for leaving the whole in a clean and orderly condition. (The ceiling joists and other preparations are included in the carpenter's bill.)"

"Provide £100 P.C. for stone carving and moulds for ten panels and fifteen capitals of columns. The work to be done by a carver appointed by the architect."

"Allow for attendance on stone carver, allowing him the use of scaffolding, tarpaulins, etc., cleaning up and making good work up to carving at completion and for clearing away rubbish."

"Allow for carting models from carver's workshop, offering up and carting back to workshop."

The examples of preambles to trades which follow will, of course, be considerably modified in special cases, and by the specification of the particular work.

Piling, Coffer Dams, etc.—" The prices of piling and dams to include spikes, nails, oak treenails, and all workmanship and labour in preparing and connecting timbers together by lapping, notching, bevel or birdsmouth cuttings, including boring for bolts."

"The driving of piles to include use of pile engine, staging and hoops or rings."

"Allow for barges or floating stages, which will be necessary for a part of the work."

"The fir shall be sound Dantzic fir, free from sap and all defects, not less than 144 square inches in section properly squared, and from straight trees; all to be creosoted under pressure with not less than 8 lbs. of creosote to each cubic foot of timber; all cut ends, scarfs and surfaces of tops of piles shall be tarred with Stockholm tar. All piles to be driven with a ram weighing not less than 20 cwt., and the driving continued until each pile will not go down more than half an inch with ten blows of the ram falling 9 feet."

"Any pile split or otherwise damaged in driving shall not be drawn, but cut off at the level of the bed of stream, and another pile driven instead, as near as possible to that first mentioned, at contractor's expense."

"The net length only of the pile has been measured, contractor must allow for waste."

"Allow for any necessary barging, any extra cost for working in the water, and for any extra cost of labour involved in tide-work."

"Allow for all spikes, staples and dogs."

"Allow for all necessary pumping and baling."

"Contractor must make his own arrangements and allowances as to tide-work."

BRICK SEWERS.—"The work consists of the construction of a brick sewer running parallel with the existing northern main sewer, from a point about 600 feet north of the northern end of Victoria Road to the northern end of Lyndhurst Road."

"All materials supplied in carrying out the works mentioned in this specification shall be the very best of their several kinds, and to the approval of the engineer. Any material that may be disapproved shall be kept upon the works until their completion."

"The contractor shall pay to all persons engaged by him in carrying out the works, such wages as are generally accepted as current in the locality for each trade for competent workmen, and shall from time to time, whenever required so to do, produce to the Council sufficient evidence that such wages are paid by him, and also that none of the work is sublet."

"The contractor shall not sublet or let as task work any part of the work without the consent of the engineer in writing previously acquired."

"The work to be commenced within seven days from the receipt of a written notice from the clerk to the Council or the engineer, and to be completed and handed over to the Urban District Council within six months after the date of such notice, under penalty of 10*l*. per day as liquidated damages."

"Payments will be made to the contractor on the certificate of the engineer, at the rate of 80 per cent. on the value of work executed, and materials on the works. The first payment to be made when the engineer shall consider that 10 per cent. of the work contracted for has been done, and subsequent payments in similar proportions. Five per cent. three months after certificate of completion, and the remainder six months after completion."

"Each section of the works shall be commenced at such times and only such lengths shall be opened as the engineer may direct, and the execution of work shall proceed at such speed as he may direct."

"The contractor will be required to keep a competent and approved foreman constantly on the works, and he shall not be changed except with the engineer's approval."

"The contractor shall find two good and sufficient approved sureties in the penal sum of one-third the amount of the accepted tender for the due and complete fulfilment of the contract entered into by the said contractor."

"The contractor shall deposit with the engineer, within a week of signing the contract, a copy of the bill of quantities, upon which basis the tender has been prepared, fully priced and moneyed out, and the value of any deviation or deduction shall be calculated at the rates of the original estimate."

"The contractor shall send in with his tender samples of the Staffordshire blue bricks and seconds gault bricks, proposed to be used on the works."

"No earth, rubbish, or materials shall be deposited upon any pavements or crossings, under penalty of £5."

"For the remaining conditions, see the full copy appended to the specification."

"Allow for supplying water for all the works, including fees, any necessary temporary plumbing, and storage of water."

"Allow for giving notices to any persons requiring notice, and pay all fees."

"Allow for setting out the works and for supplying all rods, tapes, stakes, poles, labour and other matters in setting out the works, and also such as may be required by the engineer or his assistant to check the setting out."

"The works will be carried out under the supervision of a clerk of works. Allow for supplying and fitting up an approved movable office (10 feet × 8 feet) for his accommodation, with all requisite fittings, stove, firing, lighting, attendance, &c., maintaining same and removing during the progress of the works to such positions as may be directed."

"This office to be contractor's property at completion of the works."

"Allow for delivering, when called upon, at the engineer's office, samples of the Staffordshire blue bricks, seconds gault bricks, stock bricks, cement, gravel, sand, pipes and granite cubes, intended to be used on the works, for the approval of the engineer; all articles and materials used shall agree in every respect with the approved samples."

"Contractor to use great care to damage as little as possible the watercourses, hedges and fences which he may have to break through for the construction of the works. Allow for making good at completion, and for leaving in as good order as before disturbance."

"Allow for making ample provision in place of the watercourses where broken through during the progress of the works, so that the flow of water is in no way obstructed."

"Allow for preserving intact, properly slinging or holding in position, or raising or lowering where necessary all gas, water, or other pipes, plugs, boxes, &c., met with in course of the works, making good and leaving all perfect at completion."

"Allow for protecting the whole of the works included in this contract, and for making good or paying compensation for any damage or accident to persons or property, and all claims for anything that may be stolen, removed, or destroyed, also allow for making good all sewers, drains, gas-pipes, water-pipes, or other property broken or damaged by contractors, servants, agents, or workmen, by or in consequence of their operations, or in consequence of trespass committed by them, and whether such damage or defects may be or might have been discovered during the progress of the works, or whether payment may have been wholly or partially made on the works approved as having been properly done. And in case of any action or suit at law, or other proceedings being brought or taken against the Urban District Council, or any of their officers or servants in respect of any such damage or defects, or any loss, damage, or injury by reason thereof, or consequent therefrom, to fully indemnify the Urban District Council therefrom, and forthwith pay to them such sum as may be required."

"Allow for taking up or undoing any portion of the work executed (if the engineer should so order) for the purpose of ascertaining if such work has been done according to the terms of specification."

"Allow for affording the engineer or his representative every facility, and such tools and labour as he may require to examine and test the materials and work."

"The contractor will be required to enter into a proper contract and bond to carry out the work, to be prepared by the solicitors of the Urban District Council. Allow for the expense of its preparation."

"Allow for keeping the works in repair for six months after the date of certificate of completion, and for making good any defects or imperfections that may arise during that period."

"Allow for supplying, fixing and maintaining, during the progress of the works, such substantial and proper fences as may be necessary for guarding and protecting them from injury, as well as the public from accidents, as also a sufficient number of lights to properly light the works and fencing, and provide a watchman whenever a trench is open during the night."

"Allow for supplying, wherever necessary, proper planked footways with substantial handrails, to the engineer's satisfaction."

"Allow for keeping trenches free from water from whatever source, and for doing all baling or pumping required."

"Allow for any new sewers or house drains to be connected with the new sewers during the progress of the works, if the Council or their engineer shall require it, and all expenses consequent thereon."

"Allow, in commencing to dig trenches, for carefully taking up, laying aside and preserving for reinstatement all turf, soil, granite, gravel and all other surface material."

"Allow for removing from the works with all convenient speed all surplus ground, rubbish, materials, or other matters taken out of the trenches or elsewhere, and not required for use on the works."

"Allow for filling in all irregularities in the trenches resulting from bad workmanship or otherwise, with concrete or gravel firmly rammed in as the engineer may direct, and, where the bottom is soft, for excavating to a further depth, and filling in with cement concrete as described, well rammed."

"Allow in trenches for pipe sewers for taking out the ground under each socket so that no part of the socket shall touch the bottom of the trench, but each pipe shall have a firm and even bearing on the ground throughout its entire length. Should the soil be gravel, the hole for the socket shall be filled with soft puddled clay."

"The sand to be clean and sharp, free from all loam and clay, and well washed before using."

"The gravel to be clean and perfectly free from loam or clay."

"All hard core shall be composed of hard burnt clinkers, bricks, or other approved hard material; no rags, tins, or other perishable materials shall be used."

"The cement to be Portland of the best quality, from an approved manufacturer, to weigh not less than 114 lbs. per striked imperial bushel when poured lightly into the measure; to be slow setting, uniform in quality, grey in colour; when gauged of such fineness that at least 95 per cent. will pass through a sieve of 2500 meshes to the square inch, and when gauged pure in the proportion of 9 oz. of water to 40 oz. of cement, and on the following day placed in water and allowed to set for 7 days under water it shall withstand a tensile strain of at least 400 lbs. per square inch."

"The cement to be delivered perfectly fresh on the work, in such quantities only as from time to time directed by the engineer."

"The concrete to be composed of one part by measure of Portland cement, five parts by measure of clean gravel or broken stone, not larger than 1½ inch cubes, and one part of clean sand, the whole thoroughly mixed together on boards before any water is applied."

"The mortar to be gauged in the proportion of one part by measure of cement to two parts by measure of clean, sharp sand; no mortar or cement that has once set shall be used."

"Proper boxes shall be supplied and used for measuring the materials for mortar or concrete, and both shall be mixed on a proper timber staging."

"The bricks used in the construction of the brick sewers shall be hard burnt, wire cut, gault arch bricks, square and even in thickness, of the quality called seconds, and the sides radiating."

"The Staffordshire blue bricks shall be of the best quality."

"The stock bricks shall be sound, hard, well burnt, and truly shaped ringing stocks, free from all defects."

"The brickwork where possible shall be laid old English bond, all bed and cross joints are to be full of mortar, and no joint shall exceed a quarter of an inch in thickness, and all neatly struck finsh with the work as it proceeds."

"The drain pipes shall be of the best glazed stoneware, socketed, perfectly cylindrical, straight and free from blisters, flaws, cracks and other defects. All pipes shall be of dimensions as follows :—

Diameter.	Thickness.	Length in Work.	Depth of Socket.
Inches.	Inches.	Feet.	Inches.
4	½	2	1¾
6	¾	2	1¾
9	⅞	2' or 2' 6"	2¼
12	1	2' or 2' 6"	2½
15	1¼	2' or 2' 6"	2¾
18	1½	2' or 2' 6'	2¾

"The materials excavated shall be laid as compactly as practicable and neatly trimmed up, a space being left on each side of the trench of a width of not less than 2 feet as passage."

"When passing through the fields and gardens the contractor and his men shall keep within twenty-five feet on either side of this trench."

"The trenches for all sewers and drains shall be excavated in open cutting, the full width of trench."

."All joints in laying pipe-sewers, or drains, shall be made with one strand of tarred yarn and neat cement, the cement to thoroughly fill the space between the spigot and socket, and to be finished smooth on the outside. No part of the cement joint shall be made in water."

"Great care shall be taken that no cement or other material is left inside the pipes, every pipe must be cleaned before the next one is laid. If on examining the work at completion any cement or other material is found in the pipe, or that irregularities of any kind exist, the pipes shall be taken up and relaid by the contractor in a proper manner at his own cost."

"All sewers and drains shall, on completion, be perfectly watertight."

"On completion of each length of the work and after the same has been approved, the trenches shall be carefully filled in for a thickness of six inches above the brickwork of the sewer or the pipes with the finest material excavated, and the remaining portion in layers not exceeding six inches in thickness, extending the whole length to be filled in, and rammed with iron rammers; for every man filling in, two men shall be employed to ram, and so on in proportion for any number filling in."

"The trench shall be watered if the engineer considers it necessary."

"The surface of the trench in the fields when finished shall be left at a level of six inches above the adjoining land."

"In filling in the trenches in the highway, no clay shall be placed nearer than fifteen inches from the surface, but that portion of the trench shall be filled in with material as follows:—9 inches of hard core as before described, 3 inches of double screened ballast, and 3 inches of granite, similar to that forming the surface of the adjoining roadway."

"The surface materials preserved for re-use shall be screened, and so much of the same as may be considered suitable by the engineer shall be used for repairing that part of the road which has been broken up; but should there not be a sufficient quantity of screened material to make good the surface and reinstate it in as good a condition as before being disturbed, the contractor shall provide, and lay on at his own cost, the required quantity of fresh material, so as to assimilate it with the adjoining surface, whether the same be macadamised or otherwise."

"Provide for additional work, to be used as directed, or deducted if not required, £100."

"For the new sewer, which is 3 feet 10 inches across the widest part, the trench has been measured 4 feet 6 inches wide, and for the 9-inch pipe drains, 3 feet wide; anything beyond this must be allowed for by the contractor."

"Sometimes the old line of the sewers is nearly followed by the new, and in measuring the digging no deduction has been made for the space occupied by the old sewers, nor has any deduction been made from the measured quantity of earth carted away."

"The quantities stated of earth carted away represent solid cubic yards *in position*."

Excavator and Drains.—Commence with a description of materials, etc., something like the following:—

"The lime concrete to be composed of one part fresh ground stone lime, and six parts of clean ballast and sand, thrown from stages not less than six feet from the bottom of the trench," or well rammed in 6-inch layers.

"The cement concrete to be composed of four parts Thames ballast, one part clean washed sand, and one part Portland cement."

"The breeze concrete to be composed of one part Portland cement, two parts clean fine Thames ballast, and three parts clean coarse well-screened coke breeze."

"All excavation to *trenches* to include strutting and planking, and grubbing up old foundations, if required." The foregoing when the strutting and planking has not been measured.

"Sand and gravel sufficient for the builders' purposes may be obtained upon the site, but none to be carted away, and the contractor is to properly fill up any excavation made by him for that purpose," or "it is believed that sand and gravel, etc.," or "any sand or gravel found upon the site shall be allowed for at the market price of sand or gravel delivered, after deducting the cost of carting away a yard of earth measured before digging."

This is an item which will often save considerable expense, but the surveyor must be quite certain that the case is as he describes it.

The drains will follow the items of excavation and concrete, with a heading something like the following:—

Drains in all trades, as provision.

("As provision," if their course is not determined.)

"The drains are to be glazed stoneware socketed pipes, of the best quality, half-socketed at junctions (or every ten feet in length, or both), laid to falls in trenches, upon a well-rammed bottom, jointed in Portland cement, and the joints puddled around with clay. Including excavation, filling in, ramming, carting away (or otherwise removing) surplus earth, and any necessary strutting and planking, and cleaning out at completion." "Allow for carefully testing the drainage at completion by a water or other test to the satisfaction of the architect."

Or, "The drain pipes to be the best glazed stoneware socketed pipes, of approved London make, perfectly straight, truly cylindrical and perfectly smooth as to the interior glaze, to be tested for straightness by the insertion of a cylindrical plug of the full length of the pipe, and a quarter of an inch less in diameter than the pipe to be tested. This plug must pass quite freely through every pipe used."

"The channels and bends in manholes to be executed with white glazed channel pipes of Winser's, Broad's, or other approved make."

"The drains to be laid truly straight in line and gradient, so that a lighted candle held at one end of the pipe-line may be seen to be truly concentrical from the other end, the full bore of the pipe showing."

"The contractor will be required, at his own expense, to test the drains in the presence of the architect, or his assistant, at such time as he may appoint, by filling them with water as often as necessary, and proper stoppers, screw plugs, hose, etc., must be provided."

"If any drain is found to be leaky, or the test as to straightness is not satisfactory, it shall be taken out and relaid at the contractor's expense."

"The stoneware pipe drains, traps, etc., are to be jointed and filleted all round with neat Portland cement. The pipes are to be carefully and thoroughly cleaned out on the inside as the work of laying proceeds, and are to be laid on beds of Portland cement concrete 6 inches thick (7 to 1), and after being tested and approved by the architect are to be flaunched up with concrete on each side to half the diameter of the pipes, and again must be passed by him before being finally covered up, when the concrete shall finish 16 inches by 16 inches for 4-inch pipes, and 18 by 18 inches for 6-inch pipes."

"The prices for drains are to include for all necessary planking and strutting to the trenches, for keeping the excavations clear of water, for removing old drain pipes where exposed by new trenches, and for the removal of any soil or contaminated earth met with."

Note:—" The body of the pipe to be laid upon the concrete, and the concrete to be cut out under each collar to allow of sufficient space to make the joint."

Waller.—The local circumstances will often materially affect the arrangements as to walling.

"The rubble for walling shall be obtained from . . . quarry, and a quantity sufficient for the whole work shall be quarried and exposed to the weather within a month of the signing of the contract."

"The stone shall be free from sand or clay holes and all defects, and shall be selected for hardness."

"The walling shall be laid in good mortar in irregular courses on its quarry-bed. No stone less than 9 inches from front to back and the face roughly dressed, with large through stones for the footings, and with bond stones of length equal to the thickness of the wall, about 4 feet apart horizontally and 1 foot 6 inches apart vertically.

"The facing to include all cutting and fitting to the stone dressings."

Sometimes: "The contractor may quarry the stone free of cost at . . . quarry, about a mile from the proposed building, and contractor shall deposit the debris where directed on the adjacent ground, and so as not to interfere with the future working of the quarry"; or, "The rubble stone shall be obtained from Mr. Thompson's quarry, about 2 miles from the proposed building, where it will be quarried by him and stacked, and a charge of 3*s.* per yard cube made to contractor who shall cart it to the building."

All the other materials to be as described in bricklayer's bill. Sometimes " allow for building a piece of walling as described, about

2 yards superficial and 18 inches thick, as a specimen to be approved before commencing the work, and pull it down and clear it away when directed."

Iron Drains.—" The iron drains to be best quality cast-iron socket pipes in 9-feet lengths, cast vertically and of perfectly true bore; they shall be examined inside before laying and any rough projections which may exist shall be chipped down, to the satisfaction of the clerk of works, if this cannot be done they shall be rejected. The sockets shall be centred, bored, and faced at bottom to receive the spigots which shall be centred accurately, turned on the outside and cut off flush to fit the bored socket; the ends to go quite home and to be perfectly concentric with the socket and an annular space to be left for lead caulking $\frac{7}{10}$ wide and 3 inches deep all round."

"The pipes to be free from any wrought-iron pins, cast in, and to be well coated inside and out with Dr. Angus Smith's solution in such a manner that it will not chip off."

"The pipes shall be laid on piers of brick or cement concrete one cubic foot each and two to a pipe ranged in true gradient, the pipes, when laid on them, to be lined up, caulked with molten lead, run in without yarn, and afterwards set up and tested."

"The pipes shall be supplied in suitable lengths, so as to avoid cutting, but should cutting be required it shall be done with a proper cutter and not broken off with hammer or file."

"Caulking beads to be supplied and shrunk on to all cut lengths."

"The iron drains shall weigh the following weights per 9 feet lengths and shall be of the thickness below specified.

9" bore, 4 cwt. 0 qrs. 0 lbs., $\frac{1}{2}$" thickness
6" „   2 „   2 „ 0 „ $\frac{7}{16}$" „
5" „   2 „   0 „ 0 „ $\frac{7}{16}$" „
4" „   1 „   2 „ 0 „ $\frac{3}{8}$" „

"The bends, junctions, etc., shall be of similar weight."

Bricklayer.—" The bricks to be sound, hard, square, well burnt, truly shaped, and free from all defects, and equal to samples deposited with and approved by the architect before signing of contract."

" No soft or place bricks, broken bricks, or bats will be allowed to be used, except where required for bond."

"The lime to be freshly burnt Dorking or Merstham stone lime."

If selenitic lime—" The lime to be selenitic, used and mixed exactly in accordance with the company's printed instructions."

" The sand to be clean and sharp, and washed if required."

" The mortar to be composed of one part of lime to three parts of sand, mixed in a mortar mill, and in quantities sufficient only for the day's consumption."

" The cement to be the best Portland, weighing not less than 112 lbs. per bushel, and to bear a tensile strain of 600 lbs. on $1\frac{1}{2}$ inch square after being set in water seven days (or give maker's name), mixed in the proportions of one of cement to three of washed sand, and no cement that has once set is to be used."

See another description of cement in preamble to bill of Sewers.

" The bricks to be well wetted before being laid."

" The brickwork to be thoroughly flushed up with mortar, every joint to be filled (state if grouted), to be laid old English (or Flemish) bond; no four courses to rise more than 1 inch higher than the bricks when laid dry."

" The price of brickwork is to include for all rough cutting, and fitting and bonding of brickwork with stonework." The foregoing where ashlar or stone dressings form a part of the work.

" The red brick facings are all in small quantities to strings, quoins, reveals, etc., and in parts may be considered as nearly equal to rubbed and gauged work; the moulded courses are all in red bricks, the girth of same has been measured and included in the superficial quantity of facings. A considerable part of the moulded work is in short lengths, and all to include the extra labour in over-sailing." " The brickwork of projections is measured and included in general brickwork."

" The price for facings to include the bonding of the different kinds of facings with each other, and any cutting and fitting to stone dressings."

The foregoing item " the red brick facings," etc., is only to be used when the moulded courses are measured as " extra on facings."

" The moulded bricks to be of the usual stock patterns, but of perfect and true shape."

If the moulded bricks for strings, etc., cannot be obtained from the manufacturer's stock, insert the following: "All the moulded courses to be cut and rubbed to detail."

Sometimes "for the moulded strings, cornices, etc., the bricks may be moulded or cut at contractor's option."

If the moulded bricks are selected from a particular trade list, the maker may be mentioned thus: "The moulded bricks to be supplied by James Brown, Braintree."

Terra-cotta.—Where the quantity is small it will be sufficient to put the items into the bricklayer's bill; where a large quantity, make a separate bill of it; a description may be easily produced from the following clauses :—

Where the quantity is large it is the custom of some surveyors to arrange this work in two sections, one for "fixing only," which would be a part of the contract for the building; the other for the "manufacture and delivery," which would be a separate contract; but it is a better course to throw the whole responsibility upon the builder.

It should be remarked that no work requires a more stringent and careful *specification* than terra-cotta, but only those parts of the subject are dealt with in the following clauses which immediately affect the value of the work.

The preamble to each section should contain such conditions as follow :—

If for fixing only.—" The manufacturer to receive any necessary assistance from the contractor in setting out the terra-cotta, and to allow for the necessary workshops for the use of the modeller to an extent to be approved by the architect."

"The manufacturer will deliver the terra-cotta on the site," but the contractor to unpack it.

" The terra-cotta to be delivered on the site free from damage; any piece of it which may be damaged for want of protection during the progress of the building shall be removed by the contractor, and replaced by a perfect piece at the contractor's expense."

" The contractor shall make any objection that he may wish to make to any portion of the terra-cotta within ten days of its delivery at the building."

" The terra-cotta shall be thoroughly bonded with and course with the brickwork; to include all cutting, fitting and bonding of brickwork therewith; and no joint shall exceed one-quarter of an inch in thickness. The jointing to be in accordance with the detail drawings to be supplied. None of the original surfaces to be filed, rubbed, or chipped, unless they are to be concealed, and these surfaces are to be as little interfered with as may be." The chambers of the terra-cotta to be filled in with fine concrete (state its composition, and whether lime or cement concrete, or cement and sand, and its proportions). "Each piece of terra-cotta to be thoroughly soaked with water before the filling in of the chambers, and the brickwork and terra-cotta in connection therewith to be thoroughly flushed in with mortar."

"The terra-cotta to be set in fine mortar, neatly pointed as the work proceeds, cleaned down at completion, and left perfect." (Describe any peculiarities as to pointing.)

"All the vertical and horizontal arrises to be left exactly true and regular."

If for Manufacture and Delivery of Terra-cotta only.—" The whole of the terra-cotta to be thoroughly burned, of uniform colour, free from cracks or other defects, the arrises sharp and true, the enriched work clean and sharp, and the whole equal in finish to the original models. All jointed in accordance with the detail drawings and chambered as required, but so as in no place to leave a less thickness than 2 inches. No piece of terra-cotta shall measure more than 15 inches either way. The mortises, joggles, grooves, perforations, etc., to be prepared on each piece before the burning, and each piece to be so prepared as to require no filing, rubbing, or chipping after the firing. The whole to be equal in colour, truth of line and finish to samples, which are to be deposited with the architect before signing of contract."

"The terra-cotta is to be delivered on the site at the manufacturer's expense, and to be unpacked by the contractor, but the manufacturer shall be responsible for any damage that may occur to it before the unpacking."

Although the terra-cotta may be a separate estimate, the general contractor should have a copy of the bills for reference.

"The manufacturer shall set out the work with the contractor's assistance, and they shall together arrange as shall be necessary for the identification of the pieces of terra-cotta; but the manufacturer shall be responsible for the allowance for shrinkage and for the accuracy of the size of each finished piece, and shall mark each piece as arranged with the contractor."

"The models will be prepared for the manufacturer's use, but he shall perform all packing and carriage of them from the building to his works."

"The whole of the terra-cotta shall be delivered between . . . day of . . . and . . . day of . . . and the rate of delivery shall be in a regular proportion to the quantity, and the time commencing on the first mentioned day."

"If the manufacturer shall fail to deliver the terra-cotta at the stipulated rate, he shall pay or allow to the employers as and by way of liquidated or agreed damages, the sum of £ . . . per week for every week during which the building shall be delayed in consequence thereof."

"Any part of the work which shall prove defective between the date of its delivery and the end of six months after the completion of the building, shall be removed and reinstated at the manufacturer's expense."

Payments.—(Repeat the clause as to payments to be found in the preliminary bill.)

"Any alteration in the works is not to vitiate the contract, but is to be valued in accordance with the original estimate, and for this purpose a priced copy of quantities is to be deposited with the architect."

Models for Terra-cotta.—When the contractor prepares his own models, a clause should follow the quantities of terra-cotta in the bill. Thus, after the plain and moulded terra-cotta, " the foregoing quantity of terra-cotta will involve the preparation of (number) different models or moulds which are to be supplied by the contractor." After the enriched terra-cotta, " the foregoing quantity of terra-cotta will involve the preparation of (number) different models or moulds *which are to be supplied by the contractor* "; or if a provision has been made for the models of enriched work, say, " to be supplied by Messrs. . . . , and for which a sum is provided."

Concrete Buildings.— Describe materials and composition of concrete.

The concrete shall be mixed in the proportions of one of cement, two parts of clean sharp sand, and three parts of clean ballast, the stones to pass a 1½-inch ring, all thoroughly mixed and used before it sets.

No concrete which has once set shall be used.

All concrete exposed to hot sunshine shall be covered with wet sacks until well set, and all concrete intended to be raised upon shall be kept clean.

The tops of all walls and partitions shall be left rough, and shall be watered to receive the next raising.

The whole shall be grouted with a mixture of one of cement to two of sand.

No apparatus shall be shifted for at least 12 hours after the deposit of the concrete it supports.

"The prices to include all the necessary apparatus, centres, moulds, etc., required, and allow for fixing, altering, removing and making good after, and for building in all plugs, timbers, ends of stone, etc., as required."

Mason.—"The stone to be the best quality of its kind, free from vents, beds, sandholes and all other imperfections, finished with a finely rubbed (or dragged) face, set on its natural bed in fine mortar (or cement), cleaned down at completion, and left perfect."

"Every stone to hold its full length and height square to the back."

"Any stone which may be injured during the erection of the building shall be removed and replaced by the contractor at his own expense."

(It is sometimes provided that any stone injured shall be replaced, or the full cost of replacement deducted from the amount due to the contractor.)

See also clause relating to stones of triangular section P.

"The Yorkshire stone described to be rubbed to have the rubbing completed on the premises, and the arrises left perfect."

"All the mouldings to be worked to zinc moulds." Observe that any special point affecting any particular kind of stone should appear in the preamble, as when any particular bed is desired, as the "Scott" bed in Tisbury stone, or when it is necessary that the stone should be from any particular quarry, as the Waycroft for Portland stone, Parkspring or Spinkwell for Yorkshire stone, etc.

"The stone is measured net as set, and is to include sawing and waste."

Sometimes "the price for superficial labour to include arrises and those for moulded and sunk work shall include all preliminary faces."

If the use of quarry worked stone is intended, some such clause as the following will be required, if not a direct contract with the quarry owner:—

"The stone may be quarry worked, and its quality shall be equal to a sample deposited with the architect, but it shall receive any extra labour at the works which workmanlike finish may require; all to be finished with a neatly dragged face, and to include all stone, labour, pebble dowels, arris-joggles and cement."

If a direct contract with the quarry owner:—

"The quality shall be equal to a sample of stone and labour deposited with the architect. The stones shall be fitted together ready for setting, and shall include all necessary joggles and holes for dowels. The stones shall be marked with numbers for identification, shall be accompanied by a key plan, and shall be packed and delivered at the works at convenient times. All stone required below level of ground floor before . . . all stone required between ground and first floor before . . . The stone merchant shall be responsible for and make good any damage sustained in transit."

"The general contractor will unload and unpack and fix the stone and do any extra labour at the works which workmanlike finish may require."

When a Separate Contract for Granite.—"The granite to be equal to samples to be deposited with the architect, to be prepared in all respects ready for setting, each piece to be numbered or otherwise marked so as to indicate its position, and the price to include packing and carriage and delivery on the works. The contractor for the general works will unload and fix the granite, and any loss incurred in adapting the stones to fit their intended position to be deducted from any amount due on this contract."

Or, "The granite to be of an approved colour, equal to sample to be deposited with the architect, and the price to include preparing the whole, properly fitted together, ready for setting, with all necessary grooves, sinkings and holes for dowels, packing and delivery at the works at such times as will be found most convenient, and each piece properly marked for identification in setting."

Note.—" The unloading, hoisting and setting will be done by the contractor for the general works, but should any portion of the work require reworking, refitting, or notching, the cost of such work to be treated as a deduction due under this contract."

" The sizes of the various blocks are the net sizes required when fixed, no allowance having been made for waste or for the local mode of measurement."

" No preliminary faces have been given, excepting the beds and joints, which have been taken on the extreme sizes of the blocks, all labours are the net finished faces as they appear when fixed in the building."

COLOURED ARTIFICIAL STONE IN DRESSINGS.—" The work to be done by a specialist approved by the architect and the work to be ordered immediately after the signing of the contract, so that the stone may harden for use."

" The price to include for all patterns, moulds and models, and for setting, cleaning down and leaving clean and perfect at completion."

" The work is to be fixed and closely jointed in washed sand and Portland cement in equal proportions, and all surfaces kept clean and free from smearing."

" All artificial stone to be without core and of the same material throughout."

" A sample of the stone to be submitted to and approved by the architect."

" The stone to be tinted throughout its whole bulk and the colour to be approved by the architect."

" All transomes and heads to be strengthened by and to include $\frac{1}{2}$-inch iron rods."

" The joints of all weatherings to be saddle jointed. All joints of copings, sills, heads, strings, architraves and other horizontal members to be grooved and plugged with neat Portland cement to form joggles."

UNCOLOURED ARTIFICIAL STONE.—" Allow in the price of all stone described as constructed in situ for any necessary centering."

" All steps to be made on the bench at least three weeks before fixing."

Tiler.—No preamble

At end of bill,

"Allow for cleaning out gutters and leaving all roofs perfect and weatherproof at completion."

Slater.—No preamble.

At end of bill,

"Allow for cleaning out gutters and leaving all roofs perfect and weatherproof at completion."

"*Carpenter.*—"The timber to be of the best description, from Memel or Riga (or Dantzic), sawn die square, free from sap, shakes, large, loose, or dead knots, and all other defects, and sawn into scantlings immediately after signature of contract."

The teak to be of the best Moulmein," free from all defects.

"The oak to be of English growth, of the best quality and well seasoned" or (Dantzic oak).

(The whole of the specified or figured dimensions of timbers to be the finished sizes when fixed in the building.) See Carpenter: Modes of Measurement. Sometimes: "No allowance for waste has been made in these quantities."

When the proposed work is in one of the northern counties, it is advisable to say, "The prices for cube fir to include all labour and nails in framing and fixing," in which case labour and nails need not be measured separately.

Joiner.—" All the materials to be sound and well seasoned, free from shakes, loose or dead knots, and all other defects."

"The deals to be the best yellow Christiania or Onega."

"The whole to be carefully faced up for paint, or to be selected deal kept clean for staining, finished with the plane without the use of sandpaper."

"The teak to be the best Moulmein."

"The wainscot oak to be the best Memel," cut on the quarter.

"The mahogany to be the best quality of its kind, of good figure, and carefully prepared for French polishing."

"The American walnut to be of the best quality and even grain."

"All glued joints are to be cross-tongued."

"The whole of the work, except floors and skirtings, is to be wrought both sides, unless otherwise described."

(" The dimensions and thicknesses figured and measured in the bill of quantities to be the finished sizes when fixed.") See Joiner: modes of Measurement.

If the details show sizes of frames and mouldings which are clearly intended as finished sizes, and all the rest of the work is to have the usual allowance for working, a clause may appear in the preamble, " All mouldings and frames to finish the sizes stated in this bill."

Sometimes this : " The oak joinery to have double tenons to all framings where the thickness exceeds 1¼ inch. All oak window-backs, dadoes and similar framings are to have the mouldings screwed from the back. All oak doors are to have the mouldings framed and screwed together at angles, let into the stiles and rails, well glued in and fixed without nails or screws." The remainder of the oak joinery to be secretly fixed and the whole kept clean for polishing."

" All circular work to include the necessary cylinders."

" All the mahogany or other hard woods to be secretly fixed and kept clean for polishing."

" The pitch pine intended for varnishing to be from Savannah of best quality, of good figure, carefully selected, free from galls and other defects, finished with the plane or scraper without sandpaper, and kept clean."

Sometimes the following :—" If the joints of the flooring open $\frac{1}{18}$th of an inch before the payment of final balance, the flooring shall be taken up and relaid at contractor's expense."

" The floors to be covered thickly with dry sawdust after laying until completion."

" The flooring and joinery over all pipes and bell-wires to be fitted to remove, and screwed with brass cups and screws."

" The floor boards and other mill-prepared work to be stacked upon the ground within two weeks of signing the contract."

" The joiners' work to be put together without wedging up, immediately after signing contract, in a dry place, ready for the inspection of the architect, who is to be at liberty to visit contractor's workshops at all reasonable hours for that purpose."

Ironmonger.—" The ironmongery to be of the best quality and strongest description, all to be fixed with screws, the brasswork with brass screws. All iron butts to be wrought "

Sometimes, " All the locks to be subject to one mastership, and to have two master keys to pass. All the lock furniture to be . . . patent."

"The locks and latches to be Hobbs & Co.'s machine made, fine finish; all the lock and latch furniture to be Hobbs & Co.'s patent double spindle furniture."

If the ironmongery is a separate contract, "The whole of the ironmongery, except butts and cups and screws, to be supplied by Messrs. . . . ., but contractor will fix it."

Plumber.—The lead to be the best milled lead. "To include all solder joints, wall hooks, etc., necessary to the completion of works."

"The water supply to be executed to the satisfaction of the water company's inspector."

"All pipes above 1¼ inch diameter to have tacks."

All pipes to weigh the following weights per yard lineal in lbs. :—

	½ in.	¾ in.	1 in.	1¼ in.	1½ in.	2 in.
Strong	—	—	—	—	—	—
Middling	—	—	—	—	—	—

(Fill in weights for the various sizes.)

"All pipes to be thoroughly tested at completion, the soi pipes to be plugged at bottom and filled with water to the top, the traps to be temporarily covered with stout sheet lead soldered down and weighted. All cocks to be of the best quality high-pressure screw-down taps, with screw ferrules, and engraved Hot or Cold."

At end of bill,

"Allow for paying fees for connection with main, including screw ferrule, and for leaving the whole of the plumber's work in perfect order and working condition, and for making good roads and footways to the satisfaction of the local authorities."

Hot-water.—"The pipes to be best quality galvanised wrought-iron steam tubing, with all tees, bends, angles and connections, and short pieces, jointed in red-lead cement and fixed with wall hooks."

"Allow for testing pipes at completion to the satisfaction of the architect, and leave perfect."

Plasterer.—"The sand to be clean and sharp (free from salt, clay and other impurities), and washed if required."

"The lime to be fresh, well-burnt chalk lime; that for the setting coat run into putty at least one month before it is required for use;" or, "The lime to be selenitic, used in accordance with the company's instructions."

"The laths to be lath and half, rent out of the best red Baltic wood, butted at joints, the joints frequently broken, and nailed with wrought-iron nails."

"The laths for external work to be double English oak, rent out of straight hearty stuff, free from sap, well nailed at each bearing with strong galvanised iron nails and butt-jointed."

"The hair to be the best long back hair, well beaten, and used in such proportions as the architect shall direct."

Sometimes the exact proportions of materials are prescribed thus: "The stuff to be mixed in the following proportions:— 2½ yards of sand, 40 bushels of lime, one cwt. of hair."

"The Portland cement" (describe as in preamble to Bricklayer).

"The (Martin's or Parian or Keen's) cement to be on backings of coarse Martin's cement" (or Portland).

When cornices have been measured net girth, "the cornices have been measured the net girth of the moulding; no allowance has been made for screeds."

At end of bill,

"Allow for cutting out all blisters, and for making good any defects with plaster of Paris."

Founder and Smith.—" All the cast and wrought iron to be of the best quality, *and to include smith's work in fixing*, and all other tradesmen's work herein described."

This clause is used when fixing only ironwork is not stated in the bill.

Or, "The wrought iron to be equal in quality to the best Staffordshire, and capable of bearing a tensile strain of 22 tons per square inch of sectional area before fracture, and a cross-strain of 11 tons without permanent set."

"The castings to be of strong grey No. 3 pig-iron cast from second melting."

There is a great variety of description in the current specifications of iron.

"All the bolts to have Whitworth threads, of full diameter, and the nuts to fit perfectly."

Some surveyors insert, "The cast iron has been calculated at 38 lbs. and the wrought iron at 40 lbs. per foot superficial of an inch thick"; other surveyors, "both cast and wrought iron has been calculated at 40 lbs. per foot superficial of inch"; this is not recommended, the better way is to calculate at the proper weight.

"The weight of the steel (or iron) tees, angles, joists, etc., has been calculated from a list of trade sections, and contractor must make his own allowance for "rolling margin." The customary rolling margin is 2½ per cent. over or under the weights in the trade lists.

"The ironwork shall, if so directed, be subjected to a test not exceeding half the breaking weight at the contractor's expense, and the contractor shall replace any piece of defective ironwork with new."

Or, "The quantities are all to include the cost of proving, hoisting and all fixing complete, excepting where otherwise described."

Testing being expensive, the word proving shall not be inserted unless it is intended to carry it out.

The foregoing comprises most of the requirements for ordinary work. If the work is of large extent, a more stringent preamble will probably be necessary—something like the following.

"If the contractor does not intend to manufacture the constructional ironwork on his own premises and by his own men, or if the architect is not satisfied that he has sufficient means for the execution of the work, he shall submit to the architect for his approval the name or names of such firm or firms as he proposes to employ for this purpose, and arrangements shall be made whereby the architect or his representative shall have power and facilities to visit the works of the makers of the ironwork and inspect the work in progress, and at any time, either before or after the delivery, to reject such work or parts of it as may seem to him not to comply with the terms of the specification."

"The architect shall have the power to have any defective castings broken up at once if he thinks fit."

"Any sub-contractor shall be one who himself executes the whole of the work entrusted to him, and no further sub-contracting will be permitted."

"The architect or his representative shall have the right to select samples of the ironwork and to have any portion tested, the cost of which will not be charged to the contractor if the test should prove satisfactory, but if unsatisfactory the contractor shall be held liable for the cost of further tests and all expenses incurred."

" Provide for any tests which may be at the expense of the building owner . . . *l.*"

" The dimensions on the drawings are believed to be correct, but the contractor shall be responsible for their accuracy and shall examine them before putting the work in hand."

" The contractor shall supply, set up on the premises, and maintain until the conclusion of the work a weighing-machine to weigh up to two tons, shift it from time to time as may be required, and remove at the completion of the work."

" The contractor shall submit all patterns to the architect for approval before casting, and when more than one casting is required from the same pattern, shall submit the first casting before the remainder are made."

" The contractor, by the acceptance of the contract, shall be held to approve of the methods of construction adopted and the scantlings provided, and shall be held solely responsible for the strength and efficiency of the various works."

" The stanchions, base-plates, and other castings shall be clean, sound, smooth, free from flaws, holes, cinders, air blows and all imperfections, the whole to be slowly and carefully cooled to avoid internal strains in the metal and to be perfectly straight and true in shape. No lead or other plugging will be allowed. Test bars 3 feet 6 inches long, 2 inches deep and 1 inch thick, shall be cast from the same meltings as those for the general work and from as many meltings as the architect may direct; such bars when placed on bearings 3 feet apart shall bear 27 cwts. in the centre without breaking."

" The top bearing surfaces of all base-plates and the junction-flanges of all stanchions and columns shall be planed, and extra metal shall be supplied to allow for the reduction of the thickness by such planing. Each hole for a bolt to have a slightly raised boss cast on."

" The metal of the columns shall be concentric and of uniform thickness."

" The iron to be of the best quality, tough, fibrous, even and of uniform grain, the angles and tees equal to a tensile strain of

21 tons per square inch of sectional area, with a minimum reduction of area of 12 per cent., rod-iron and bolts equal to a strain of 28 tons per square inch and a reduction of 18 per cent."

"All bolts to be of the sizes indicated on the drawings, made of one piece of metal, and those ¾ inches diameter and upwards shall have hexagonal heads and nuts of proper proportions, the screwed ends of the bolts to project not less than ⅛ inch beyond the face of the nut."

"All steelwork to be of the best English manufacture, to bear the name of an English maker, and have an ultimate tensile strength of from 28 to 30 tons per square inch of section with an elongation of 20 per cent. in a length of 8 inches."

"The rolled sections shall be truly and cleanly rolled to the full sections, sizes and weights per foot, as drawn or described, and shall be free from scale, blisters, laminations, cracked edges and defects of every kind."

"The steel for the rivets shall be of mild quality and shall not contain more than ·2 per cent of carbon, the tensile strength not to exceed the limit of 25 to 27 tons per square inch of section, with an elongation of 50 per cent. in a length of 2 inches."

"The holes for rivets and bolts in steel girders and rolled joists shall be drilled and not punched; the riveting shall be done in the best manner, with the rivet-holes fair and true and exactly opposite to each other; the rivets to fill the holes tightly and completely, to have sound and well-formed heads, and snaps of uniform size. Any rivets loose or defective, or with burnt or cracked heads, shall be cut out and renewed with new or perfect ones."

"Thoroughly remove all scale to the bare metal and oil while hot with boiled linseed oil."

"The work will be inspected and approved in its oiled state, and no further coating shall be applied until the architect or his representative has passed the work."

"After the approval the whole of the work shall receive one thin coat of oxide of iron paint before delivery, and a second coat upon all exposed parts after fixing."

Fire Mains. — "The fire mains to be supplied and fixed by Messrs. Merryweather and Sons, of Greenwich Road, S.E., or Messrs. Shand, Mason and Co., of 75 Upper Ground Street, S.E., or other approved firm, in the best manner and left sound and perfect. They are to be fixed before the concrete floors are laid (or holes to be left for pipes), and to be of cast iron, 4 inches internal diameter, with sockets of ample depth for caulking, and each 9-foot length to weigh 1 cwt. 1 qr. 21 lbs., proved to a pressure of 200 lbs. on the square inch, and to be coated within and without, or within only, with Dr. Angus Smith's composition. The joints to be made with yarn and blue lead, well caulked with a sufficient depth of lead to render the joint sound and watertight under pressure."

Gasfitter.—" The pipes to be Russell's patent wrought-iron welded tubing, with all necessary tees, bends, angles and connections, and jointed in red-lead cement."

" All the gasfitting to be done to the satisfaction of the gas company's inspector, and to be left perfect."

At end of bill,

" Pay fees to gas company for connecting with main, and allow for all attendances, and cutting away and making good, and deal blocks, and for making good road and footways to the satisfaction of the local authorities." In good work add, " All floor boards and other joinery over pipes to be fitted to remove, with brass screws and cups."

Bellhanger.—" The bells to be . . . oz. bells (averaged) of varied tone, hung with best well-stretched copper wire in concealed zinc (or copper) tubing, with steel springs, cranks, pendulums and indicators, and with iron boxes and mouthpieces for the pulls."

Electric Bells.—" The bells to be electric, the wires to be No 20 B.W.G. copper of best quality insulated with gutta-percha, double-covered with cotton, and afterwards varnished, and to be concealed in zinc tubes. The buttons to have china roses, ebonite backs, German-silver springs and platinum contacts."

" The pear presses to be of hard wood, polished, with German-silver springs and platinum contacts, and each to have 9 feet of flexible silk cord and a stout hardwood rosette, polished."

" The bells and indicators to be in ⅞-inch Spanish mahogany cases of the best quality, French polished ; every bell which is connected with more than one room to have an indicator."

" The battery to be Leclanché, or other approved, and sufficiently powerful and efficient."

"The whole to be fixed in the best manner, and left in working order."

"The following fittings are to be supplied and fixed. The prices are gross list prices from Messrs. . . . catalogue."

At end of bill,

"Allow for all attendances, and cutting away and making good and fixing all the floors and joinery over pipes and bell-wires, with brass screws and cups."

Electric Lighting.—"The whole of the work shall be executed in the best and most workmanlike manner, with materials of the best and most approved qualities of their respective kinds, and shall be done strictly in accordance with the lighting company's wiring rules, the rules of the Royal Exchange Insurance Company, and the regulations of the Board of Trade and the London County Council."

"Allow for making all necessary insulation and other tests, together or in sections, to the satisfaction of the architect."

"Allow for setting out on outline plans supplied by the architect, and at the completion of the works; the sizes and positions of all pipes, wiring and other work done under the contract, and for supplying a complete schedule of the lights and fittings."

"All work is to be carried on in such manner and at such times as shall not interfere with the use of any portion of the building."

"The whole of the materials and apparatus, except the lamps, shall be suitable for a 200-volt supply."

"The mains and leads shall be of soft tinned copper, tinned before stranding, of a conductivity of 98 per cent. of pure copper, insulated with pure vulcanized india-rubber, braided and taped, and covered with a preservative solution, and shall be of the best quality and made by a manufacturer to be approved by the architect. No wire less than $\frac{3}{17}$ shall be used, and where a separate cut-out is not used a $\frac{7}{214}$ wire shall be brought up to the lighting point. All joints to be easily accessible and to be soldered with resin as a flux."

"Carry from the cut-out boards on each floor branch circuits which shall not exceed five amperes each, and of such sections that the drop in E.M.F. between any cut-out board and the furthest lamp on the circuit shall not exceed one volt when all are in use."

" Whenever the section of a conductor is reduced, a double-pole fuse shall be fixed."

" All switches shall be fireproof with substantial rubbing contacts, and of quick make and break type mounted on incombustible bases fixed to polished wood blocks."

"Supply teak dove-tailed wall blocks, and let into walls where required to receive switches, cut-outs, etc."

" The ceiling plates to have incombustible bases fixed to polished wood blocks, with a cut-out on one pole, a screw cover and approved cord grip."

" All wall and floor sockets shall be for 200 volts. The casings generally to be hard, well seasoned American white wood of best quality, free from knots, etc., and varnished inside with best shellac varnish, and painted two coats of oil colour outside."

" All casings to be fixed with brass cups and screws. Each conductor shall be placed in a separate groove, 1 inch apart for sub-mains and ¾ inch apart for remainder."

" The B.C. Edison 200-volt lamps for all the flexible pendants and plain brackets shall be supplied by electrician, who shall make good any lamps, broken or found defective during the progress of the work and the testing, and they shall be left perfect and in working order at completion."

" All positive leads shall be red, negative leads black. The positive and negative portions of the switch and cut-out boards to be completely insulated, and all connections so arranged as to be made from the front of the boards."

Zincworker.—" To include all soldering, nailing, beaded edges throughout, shields to rolls, etc., required."

" The zinc to be Vieille Montagne, laid in the best known manner, and as much as possible without soldering."

Coppersmith.—" The copper to be best selected strong sheets, 16 oz. per foot superficial, laid on the method introduced by Messrs. Ewart and Son, Euston Road, N.W.; the rolls finished with a double welt; the stopped ends welted all around, and the saddle ends welted to the sheets above and to the flashings. The edges of flashings to be welted."

Glazier.—" All the glass to be the best of its respective kind, free from all bubbles, scratches and other defects, cut to fit the rebates, carefully bedded, puttied and back puttied, and sprigged where required."

The British polished plate to be of best glazing quality and about ¼ inch thick, all edges to be blacked.

"The edges of ground glass to be sized."

If there are lead lights, "Allow for supplying the necessary templates for the lead glazing."

At end of bill,

"Allow for leaving glass clean and perfect at completion."

Paperhanger.—"All the walls to be carefully prepared to receive paper."

"All the paper to be hung with butt joints."

Painter.—"The materials to be of the best quality."

"The oil colours to be made with the best old white-lead and pure linseed oil, and to be mixed on the premises."

Sometimes, "The paint to be a different tint for each coat; no coat of paint is to be covered by another until it has been seen and passed by the architect or clerk of works."

"The varnish to be Mander's best copal."

"The stain to be Stephens' No. —. Stain let down with water to the required tint, and laid on with two brushes,"'or,

"The stain to be oil stain of approved tint," and as last.

"The work to be well rubbed down and faced up between each two coats of colour. The old paintwork, where cracked or blistered, to be burnt off or otherwise removed bare to the wood; and in such cases allow for knotting, priming, stopping and painting one oil in addition to the work taken in this bill."

At end of bill,

"Allow for touching up work at completion and for leaving all perfect."

Machinery and Pipes.—"The works are to be done in the most finished, substantial and workman-like manner with materials of the very best description, and are to be subject to the approval of the architect or his engineer, and in accordance with the specification and drawings and such other explanatory drawings as may be furnished from time to time."

"All steel and ironwork shall be of British manufacture and

must bear the maker's name thereon, and vouchers from the makers must be lodged with the architect."

"The contractor is to make good, at his own costs and charges, all damages done to other property or works by his workmen or otherwise."

"Any part of the work erected under his contract which does not stand all tests to the satisfaction of the architect or his engineer will be rejected, and must be replaced by the contractor at his own expense."

"The whole of the tests shall be conducted entirely at the discretion of the architect, in such manner and by such process as he may consider desirable, the cost of such testing, whether in fuel, labour, pumps or other machinery, etc., being defrayed solely by the contractor."

"The architect is empowered to refuse any materials that he may deem inferior, or any labour that he may consider unsound or unworkmanlike, and the contractor is to remove and amend the same upon notice being given to that effect."

"The contractor is to take all his own dimensions from the actual building, and is to be held by the acceptance of the contract to approve of the method adopted for the execution of the works, and shall for the space of twelve months from the date of the architect's final certificate of completion be responsible for, and make good at his own costs and charges, all defects which may arise or be discovered in the work executed by him, and shall, for the same period, keep the whole in good repair and thorough working order."

"All castings are to be of the best quality of tough grey iron, of approved brands; they must be sound, clean, free from cold shuts, sand cracks, air holes and all other defects, and must stand a tensile strain of nine tons per square inch."

"All forgings to be well finished and clean, and of best quality iron."

"All wrought tubes to be James Russell and Sons' (Wednesbury) best lap welded "steam" quality, galvanised; to be properly screwed and fitted, and all bends, elbows, tees, nipples, etc., are to be of similar quality."

"All cast-iron pipes are to be clean vertical castings, perfectly true and to bear the following weights, viz.:

										lb.
10 inch diam. per yard				..	..	..	..	..	..	264
9	,,	,,	,,	..	..	..	..	..	..	180
6	,,	,,	,,	..	..	..	..	..	..	132
5	,,	,,	,,	..	..	..	..	..	..	87
4	,,	,,	,,	..	..	..	..	..	..	66
3	,,	,,	,,	..	..	..	..	.	..	42
2	,,	,,	,,	..	..	..	..	..	..	30

and all bends and other fittings are to be of like thickness. All the pipes are to be provided, where necessary, with proper large bosses for drilling and tapping wrought tubes into same."

"All the works herein mentioned are to be quoted for as including fixing in the most substantial manner, as hereafter described, and to be painted—those pipes which are underground with Dr. Angus Smith's solution, and all other pipes and fittings, whether galvanised or plain, with two coats of Wolston's Torbay paint—and finished with a distinctive colour indicating its use, as hot, cold, steam, waste, etc."

"All machines, etc., in whole or in part, are to be properly prepared, the woodwork twice knotted and stopped, and the whole painted four coats of genuine white-lead oil colour."

"All pipes, etc., are to be carried on purpose-made cast or wrought-iron hangers or brackets, the former passing through the concrete floors to within 2 inches of top, and fitted with large plates and screws for regulating levels, etc., the latter properly cut and pinned to walls, and the prices quoted are to be inclusive of these and all labours in connection therewith, including the expenses of all carriage, delivery, hoisting and fixing, and for such building work as may be necessary."

"All charges for new patterns or altering old ones are to be included in the prices quoted."

"All flanges of pipes are to be faced, and holes drilled to fit accurately."

"All bolts, nuts, etc., are to be included, and shall be of the best S.C. crown iron, or other approved brand of equal quality."

"All bright work when fixed is to be painted with white lead and tallow for protection during progress, and at completion, or at such time as may be required by the architect, the whole of the work is to be cleaned, polished, and left in every respect perfect and in working order, and all superfluous material or rubbish arising from the work removed from the premises."

"All royalties to be included in contractor's prices."

"No omissions or variation in the work is to invalidate the contract, but the amount of such is to be added to or deducted from the contract sum by the architect or his engineer at the completion of the work, at exactly the same rate of charges made by him in his bill of quantities, a fully priced copy of which is to be deposited with the architect for that purpose."

"The contractor is, when directed by the architect on at least two occasions previous to the final delivery of the work, to fire up the boilers and set the whole of the machinery in motion for (say) three hours on each of the occasions mentioned, and as long as may be necessary, and on each of the occasions is to provide a full and complete staff of assistants to ensure a perfect starting and running of the entire instalment."

"The contractor shall supply with his tender a sample of each size of valve and air cock such as he proposes to use."

Separate Estimates.—When it is desired that the prices of certain parts of the work should appear separately in the tender, all the

No.—. ALTERNATIVE ESTIMATE FOR ROOF OF NORTH AISLE IN FIR.

ft.	in.			Fir selected and left clean for staining.	Oak as in Bill 2. folio 8.
				£ s. d.	£ s. d.
150	–	cube	Framed in trussed rafters pinned with oak pins		
100	–	,,	Ditto in roof trusses, and ditto ..		
500	–	supl.	Labour to planing..		
600	–	run	Labour to chamfer 1½ in. wide, stopped		
			No. 90 splayed stops		
			No. 40 labour to heads of bolts let in and pelleted		
			Value in oak		£
			Deduct value if in fir	£	
			Amount of reduction carried to form of tender		£

items of that work in all trades will form a separate bill, the items arranged in the same order as in a general estimate and carried to the summary. As to whether the total shall be added to the general total or not will depend upon the form of tender.

Alternative Estimates.—When an alternative price is required, as, for instance, for the execution of a roof in fir instead of oak, the roof must be measured first as of oak, and billed with the general estimate, and an alternative bill produced in some such form as on p. 359, written on doubled-columned bill paper.

The prices for the fir will be filled into the first column, and the prices for the oak repeated from the general bill in the second column.

It is most convenient for the adjustment of accounts to make a separate bill of the whole work in all trades comprised by an alternative.

Schedule of Prices.—As the contract which does not involve day work is an exceptional one, it will be always advisable to have a schedule of prices annexed to the form of tender. The general form is as follows:—

SCHEDULE FOR LABOUR PRICES IN DAY ACCOUNTS.

Excavator	Per hour
Bricklayer	,,
Ditto labourer	,,
Mason	,,
Ditto labourer	,,
Slater or tiler	,,
Ditto labourer	,,
Slate mason	,,
Ditto labourer	,,
Carpenter	,,
Ditto labourer	,,
Joiner	,,
Plumber	,,
Ditto mate	,,
Ditto labourer	,,
Plasterer	,,
Ditto labourer	,,
Ditto boy	,,
Smith	,,
Ditto labourer	,,
Gasfitter and bellhanger	,,
Ditto ditto labourer	,,
Glazier	,,
Painter	,,
Horse, cart and man	Per day.
Two horses, cart and man	,,

"Note.—The foregoing prices, and prices for day work generally, to include use of scaffold and all tackle, tools, tool sharpening, water, and general foreman's time. Subordinate foremen to be charged as ordinary workmen. Time for fixing and removing scaffolds will be allowed."

Summary.—A convenient form for the summary is as follows:—

Summary.

No.			£	s.	d.
No.	1	Preliminary and Provisions			
„	2	Excavator and Drains			
„	3	Bricklayer			
„	4	Mason			
„	5	Slater and Slate Mason (or Tiler)			
„	6	Carpenter			
„	7	Joiner and Ironmonger			
„	8	Plumber and Zincworker			
„	9	Plasterer			
„	10	Founder and Smith			
„	11	Gasfitter			
„	12	Bellhanger			
„	13	Glazier			
„	14	Paperhanger			
„	15	Painter			
„	16	Separate Estimate (Boundary Walls, or as the case may be)			
			£		
		Add for Surveyor's charges per cent. on the foregoing amount, to be paid on receipt of first instalment			
		Add for Lithography of Quantities and Expenses			
		Charge for preparing Bill of Credits			
		Carried to Tender (or Amount of Tender)	£		

Write at the bottom of summary the following clauses. Some surveyors write them also outside of summary.

"Tenders to be delivered on the form furnished herewith, at the architect's office, at or before 10 a.m. on Monday,   day of

"The drawings and specification may be seen at the architect's office after the 26th inst.

"The proprietor does not bind himself to accept the lowest or any tender." Sometimes. " and reserves to himself the right to accept any one of them."

"The priced bill of quantities shall be delivered in a separate envelope at the same time as the tender, endorsed with the name of the work and the name and address of the contractor."

Form of Tender.

For any work, unless of small size, it is advisable to have a form of tender prepared in some such form as follows :—

"Tender for the erection of house and offices at
"Godalming, Surrey, for John Smith, Esq."

"To William Brown, Esq.,
   "Architect,
       "2, Fenchurch Chambers, E.C.

"Sir,

"I am willing to contract for and perform the whole of the works required to be done in the erection of a house and offices at Godalming, Surrey, for John Smith, Esq., according to the drawings and specification and general conditions prepared by you for that purpose, and to your entire satisfaction, for the sum of   pounds."

Where several items are required to be stated, it should run thus, "for the undermentioned sums."

            "For   £
            "For   £
"or the whole for the sum of   £

"If the joinery be executed in deal, as per alternative estimate, Bill No. 17, I am willing to reduce the last-mentioned amount by the sum of pounds."

"And I am also willing to allow a deduction of   pounds for the old materials as per Bill No. 20 credits."

"And I am also willing to execute any day works at the prices for labour in the schedule attached hereto."

Sometimes, "and I am willing to complete the buildings within three months of the date of signing contract, instead of the time prescribed, in consideration of an addition of 5 per cent. on the foregoing amounts."

"And I also agree to sign the contract when called upon to do so within a month (or as case may be) from the date hereof.

"As witness   hand this   day of   , 1880.
                   " Signature _____
                   " Address   _____"

Sometimes: "Any alteration in this form of tender will prevent its being taken into consideration."

Credits Bill.—There are several methods of preparing a credits bill. One way is to commence with a clause.

No. 20 Credits.

"Take down, remove, and give credit for the whole of the following materials, after allowing for the expense of taking down and removing and clearing away rubbish: the whole of the old materials (except lead and timber) as are sound and approved, may be re-used in the new work."

Then the items should follow, in the order of the trades, stating their position in the manner before suggested in the remarks upon preliminary bill. According to this arrangement the total will be carried to the summary, where it will appear as—" Deduct Credits, Bill No. 20," or the amount may be stated in tender.

Another way is when no separation has been made between those items of alteration which involve a credit and those which do not. The bill of alterations is prepared in the usual way, and the items which involve a credit are marked with the word Credit in red ink in the margin of the bill, thus:—

"Take down and remove cross on gable of north porch and prepare original apex stone to receive new one."
(In foregoing item the old stone is assumed to be useless.)

"Credit."   "Take down, remove, and credit the roof of north porch, comprising about two squares of timbers and boarding and 15 cwt. of old lead, and prepare old walls to receive new roof and make good all works disturbed."

According to this arrangement an item appears in the summary as follows:—

Credit by value of old materials marked Credit in red on margin of Bill No. 3.

Another method is to prepare a bill with double columns, as on next page.

Special cases will arise requiring modifications of either of the foregoing methods.

Probably the client obtains a larger sum for the credits if the amount appears in the tender.

In billing works where there is more than one trade involved, as in a bill headed zincworker, in which there is joinery, write a preliminary clause to this effect, "include also joiner's work, herein described," or carpenter's or other tradesman's work, as the case may be. Write, "continued" opposite every total carried forward, and "carried to summary" to the total at end of each trade.

No. 20. PULL DOWN AND CREDIT.

rds.	ft.	in.		Pull Down. £ s. d.	Credit. £ s. d.
			"The old bricks so far as they are sound and approved may be used, but no bats or broken or sooty bricks will be allowed to be used."		
			Or, "The old bricks may be broken up to pass a 2" mesh, and after being well screened may be used instead of ballast in the concrete."		
			"The old slates may be used for damp-proof course."		
			"The old York stone may be used for templates."		
			"The old lead will be weighed as it comes from the roofs, contractor to supply scales and weights, and to cause it to be removed and weighed. The exact quantity will be adjusted at the settlement of the accounts and valued at the rate affixed to the item in this bill."		
			(State any other materials that may be re-used.)		
			"No other portion of the old materials to be used, but to be removed immediately after pulling down."		
			"Allow on all items of pulling down, for the removal and carting away of rubbish, and making good of all works disturbed."		
			"The building owner reserves the right to retain any portion of the old material at the prices allowed for them in the following bill."		

						£	s.	d.	£	s.	d
yds. 13	–	–		supl.	Clean and stack old bricks for reuse, basket and cart away rubbish.						
42	–	–		"	Brick-nogged partition						
20	–	–		"	Brick paving in stables						
	50	–		run	York stone coping 13 in. wide ..						
	40	–		"	York stone coping 18 in. wide ..						
			No.	1	Flight of twelve York steps, and landing back area						
sqrs. 21				supl.	Slating to main roof						
	30	–		run	Slate ridge						
10				supl.	Roof timbers and slating battens						
10				"	Old floor and joists						
10				"	4-in. fir framed partition						
			No.	1	Door and frame and linings back entrance						
			"	3	Sets of sashes and frames in western wall of larder						
					The staircase from third to fourth floor, consisting of twenty steps, with handrail and balusters to one side.						
yds. 100	–	–		supl.	Knock off plastering of old walls, and basket and cart away rubbish						
100	–	–		"	Knock off lathing and plastering to partitions, and basket and cart away rubbish						
100	–	–		"	Knock down old ceiling, and basket and cart away rubbish						
					Hot plate and hot closet in kitchen.						
cwts.				1	Range in kitchen						
20					Old lead						
					The w.c. pan and trap, first floor w.c., and all soil and supply pipes connected therewith.						
					Total for pulling down carried to preliminary bill, page 3	£					
					Total of credits to be carried to summary, and amount to be stated in tender				£		

For this arrangement.

In the preliminary bill a clause will appear:—
"Allow for pulling down, see Bill No. 20."
And in the summary a clause:—
"Deduct Credits, see Bill No. 20."

Put a tick at the bottom of each column of abstract when it is billed, in the same manner as suggested for the bottom of columns of dimensions.

When the draft bill is finished read it carefully through, and take care to understand every item. Observe that the words "cube," "superficial," "run," "No.," "continued," "carried to summary," are in their proper places. It is a good plan to emphasise these changes by leaving a blank line between cubes and superficials, etc. See that every trade is numbered and *in the summary*. If pressed for time, and equal attention cannot be given to every item, take particular notice of the larger or more valuable items, as it is in these that errors will be of the most importance. Always do this, however confident you may be of the carefulness of the checking.

Practice will enable the surveyor to detect errors and discrepancies readily. By the preservation of notes of various buildings the surveyor will know whether the quantity of brickwork is nearly correct. A generalisation as to the proportion of beds and joints to a given quantity of stone may be applied according to the character of the building. He will, among other things, compare the quantity of slating with the boarding or battens which receive it and with the felt, the ceilings with the floors, etc.

The surveyor should also be suspicious of any item which is presented in feet when by ordinary custom it is billed in yards, etc., of any extraordinary quantity of material billed in cubic yards, as the use of a divisor of 9 instead of 27 (in error) is not uncommon.

It will save much time in altering and correcting the lithographed copies of the bills if the surveyor reads (or causes to be read) the transfers (i.e. the lithographer's copy on paper prepared for transfer to the stones) as soon as they are ready, making corrections where the lithographer has made mistakes! It will be necessary to keep a list of them, and when the lithographed copies are received to see if the mistakes have been rectified. As all the copies are alike, this will be necessary only with respect to one of them.

The examination may be done by a boy and an experienced man, certainly not by two boys.

Count the copies to see that they agree with the lithographer's invoice. Fill in the charges to the summary, and count the number of bills in each set before they are despatched.

If errors are discovered after the bills are sent out, an addenda bill must be prepared and sent to the builders, but avoid this necessity if possible.

It is well to remember the function of a bill of quantities and that it is not a specification, it need never indicate position of work in a building except when the value of the item is affected thereby. The position of material, and the manner in which it has been measured, can be seen by referring to the dimensions, a copy of which should be supplied to the builder, but not until he applies for it. He has, however, no legal right to these if the quantities are not a part of the contract.

In some parts of the kingdom it is customary to combine the quantities and specification, making the bill of quantities serve for both. It is maintained by some that this can be done with less labour than the preparation of the two separate documents involves. The principle is to append to each item a statement of its position in the building.

This method makes a much longer bill, gives the contractor more labour in pricing, and is not to be recommended. The London surveyor never adopts the practice.

If tenders are advertised for and no quantities supplied, the selection of the builder may be deferred until after tenders are received, and there is no obligation to accept the lowest tender.

But where quantities are supplied, a selection is usually made before their issue, from the persons offering to tender, and in such cases it is usual to accept the lowest tender, the inference being that the eligibility of all was considered before sending to them. The result of the tender will be best if care is taken to select men who do the same class of work.

Any obscure directions in the following bill may in most cases be solved by reference to the items of bill which follow modes of measurement.

FORM OF A BILL.

Heading.

"No. 1. PRELIMINARY AND PROVISIONS."

yds.	ft.	in.			£	s	d.
				Contract conditions which affect the cost			
				Preliminary works			
				Allows items of alteration sometimes separately billed as "works on spot."			
				Provisions.			
				Preamble.			
				Provide for stoves £50			
				Ditto for chimneypieces and fixing £50 ..			
				Ditto for 850 yards of tile paving and fixing £400			
				Carried to summary	£		

No. 2. EXCAVATOR AND DRAINS.

yds.	ft.	in.			£	s.	d.
				Or a separate bill may be written of drains in all trades.			
				Preamble.			
			cube	Excavation of various kinds			
				Lime concrete, cubes and supls.			
				Cement ditto, cubes and supls.			
				Dry rubbish, cubes and supls.			
				Strutting and planking, supls. and runs ..			
				Drains.			
				Preamble.			
				Drains, commencing with the smallest, and including the digging thus:—			
	100	—	run	4" drain and 2 feet excavation, averaged..			
	100	—	″	4" ditto, 4 feet ditto			
				Follow with 6" pipe, and its various depths of digging.			
				Follow with the various sizes and depths in a similar way.			
		No.		Extra for bends of the various sizes, commencing with the smallest			
				Extra for junctions, ditto			
		″		Extra for taper pieces, ditto			
		″		Follow with double junctions, &c., in the same manner.			
				Ends of rain-water pipes and other pipes made good to drain, cesspools, grease traps, rain-water tanks, &c.			
				Connection with sewer and pay fees			
				Carried to summary	£		

No. 3. WALLER AND BRICKLAYER.

yds.	ft.	in.			£	s.	
			cube	Preamble.			
			cube	Rubble walling			
			"	Ditto of various thicknesses, 14" and under			
			supl.	Rough cutting			
			"	Extra for facing			
		runs	Nos.	Extras on facing			
			supl.	Extra on rubble for brick lining, or lining and cavity			

Follow with other extras in brick on the rubble—supls., runs, numbers.
Reduced brickwork in mortar .. ..
Ditto of various kinds in mortar .. ..
Ditto, extra only in cement.. .. ..
Half-brick partitions, brick-nogging, tile arches, &c. .. ..
Rough cutting, limewhiting .. ..
Damp-proof course .. ..
Pavings of cement, bricks, tiles, asphalt ..
Wall-tiling .. ..
(Or in Plasterer.)
Labour rough oversail, one course .. ..
Ditto rough chamfer on brick .. ..
Follow with cuttings to quoins, groin point, chases, copings.
1¾" × ⅛" hoop iron and laying in walls by bricklayer (weight 0 cwts. 3 qrs. 0 lbs.)
Ends of timber cut and pinned .. ..
Follow with ends of various things cut and pinned.
Frames bedded and pointed, arches (all except gauged, see facings) and numbers generally; labour and materials, setting stoves, ranges, soot doors, air bricks, and such small metal articles as include labour by bricklayer.. .. .. ..
"Allow for bricklayer to attend upon bell-hanger to No.... bells with...pulls" ..

Facings.

Description of facing *bricks* and pointing ..
"Extra on common brickwork for facings of picked stocks finished with a neatly struck joint as the work proceeds" ..
Follow with the various kinds of facings according to value.
Rake out cement joints, pointing, gauged arches, diapers, &c. .. ..
Labours on facings according to value ..
Follow with items of extra on facings, labour or extra labour and materials.
Follow with Nos. .... extra on facings.

Continued £

2 B

WALLER AND BRICKLAYER—continued.

cwts.	qrs.	lbs.		£	s.	d.
			Hoisting and Fixing only Iron Joists and Girders (when the fixing is not included in smith's bill, or when the ironwork is in a provisional sum).			
			Carried to summary .. .. .. £			

No. 4. MASON.
Preamble.

£ s. d.

Yorkshire Stone.

ft.	in.		
		supl.	Rough stone of various thicknesses .. ..
			Self-faced stone of various thicknesses ..
			Tooled stone of various thicknesses, the thinnest first .. ..
			Rubbed stone of various thicknesses, the thinnest first .. ..
			Lineal items of labour .. ..
			Lineal items of labour and material .. ..
			Follow with numbers of labour only, labour and materials, small metal articles fixed by mason.

Portland Stone.

Stone, including hoisting and setting ..
Ditto, in scantling lengths .. ..
Stone, including hoisting and setting from 40 to 60 feet above street level .. ..
Or, extra hoisting to stone from 40 to 60 feet above street level .. ..
Follow with the various heights of hoisting
Follow with the labour on the cube stone supls., runs, numbers.

Portland Stone and all Labour.

Supls., runs, numbers .. ..
Follow with cramps, dowels, and small metal articles fixed by mason.
Follow with the various kinds of stone, adopting the same arrangement as the foregoing.
"Allow for attending upon bellhanger to No. .... bells with No. .... pulls" ..

Bosting and Preparing for Carver in Portland Stone.

Supls., runs, numbers .. ..
Often the carving is a provision.

Carried to summary .. .. .. £

No. 5. SLATER AND SLATE MASON (OR TILER).

	ft.	in.		£	s.	d.

Chief item of slating .. ..
Follow with the supls. various kinds of slating .. ..
Follow with runs and numbers.
Clause for cleaning out gutters, &c. .. ..

Slate Mason.
Supls., runs, numbers .. ..
Small metal articles fixed by slate mason

    Carried to summary .. .. .. £

No. 6. CARPENTER.

sqrs.	ft.	in.		£	s.	d.

Preamble.
Fir in plates, lintels and wood bricks ..
Ditto, fitted to iron .. ..
Fir in ground joists .. ..
Fir framed in floors .. ..
Follow with the various descriptions of unwrought fir according to value.
Follow with the wrought fir, arranged in the same manner as above.
Hoisting and fixing roof trusses .. .. ..
Follow with centering, supls., runs, numbers.
Cradling, bracketing .. ..
Strutting to stone lintels, strutting and ribbing to traceried windows .. .. ..
Follow with general supls., runs, numbers.
Fixing only ironwork, numbers according to value, but these items will only occur when fixing is not included in smith's bill, and in such case make a note in smith's bill that fixing is in carpenter's bill.

    Carried to summary .. .. .. £

No. 7. JOINER AND IRONMONGER.

				£	s.	d.

Preamble.
    "Floors in Deal."
In order of value, commonest first.
Follow with extras on floors, as grooves and nosings, &c.
Follow with floors in pitch pine, floors in oak, &c., arranged in a similar manner.
Each division as floors, skirtings, doors, &c., may be followed by the similar categories in superior wood.

    Skirtings in Deal, including backings where required.
In order of value, the least value first.

                           Continued £

JOINER AND IRONMONGER—continued.

sqrs.	ft.	in.				£	s.	d.

Fanlights, Skylights, Sashes and Frames in Deal.

Fanlights, skylights, casements, according to thickness. Sashes and frames.
Lineal quantities of extra labour on sashes and frames, and numbers.
Number small casements and frames, or small sashes and frames.

Doors in Deal.

According to thickness, and in each thickness the doors of least value first.
Extra labours on doors, runs and numbers.

Thicknesses and Framings in Deal.

All linings to include backings.
Commencing with the least thickness.
This division includes framings, as partitions, window backs, cupboard fronts and supl. items which do not come into the other sections.
"*Fir framed and wrought all round where required*" (sometimes "all framed in white lead").
Posts, frames, door frames, according to scantling, the smallest first, irrespective of the quantity of labour on them.

Architraves and Mouldings in Deal.

According to scantling, the smallest first.
(In a small work the above may be included with the sundries.)

Sundries in Deal.

Runs and numbers, labours, labour and material, according to value.
This section comprises items not referable to other categories.

Attendances.

Joiner attend plumber to

	No.	2	Baths					
	„	2	Sinks					
	„	2	W.c.'s					
	„	2	Cisterns					

Joiner attend bellhanger, cutting away and making good after him to No. 13 bells with No. 13 pulls, and include the necessary boards for bell runners and for fixing the floor boards and joinery over same to remove with brass cups and screws ..

Continued £

JOINER AND IRONMONGER—continued.

sqrs.	ft.	in.		£	s.	d.

Staircases in Deal.

Items in the usual order supls., runs, numbers.

Staircases in Pitch Pine (or other superior wood).
Supls., runs, numbers .. ..

Staircases in Honduras Mahogany (including French polishing and carefully protecting until completion of works), supls., runs, numbers.
(Or the French polishing in painter's bill.)
Mahogany, wainscot, oak, teak, American walnut, or other superior woods follow in order of their value, keeping each wood distinct.

Ironmongery of the best quality and strongest description, all to be fixed with screws, the brasswork with brass screws.

To Deal.

Of the various articles begin with the smallest; the following is the recognised order of arrangement.
Iron butts, brass butts, iron back flaps, brass back flaps, iron hinges, brass hinges, iron bolts, brass bolts, iron latches, brass latches, iron locks, brass locks, sets of door furniture (where separately taken), window furniture, finger plates, general ironmongery, sets of lines and pulleys, general brasswork.

Ironmongery and Fixing to Mahogany or other Hardwood.

Same arrangement as for deal.

Carried to summary .. .. .. £

When a bill of joinery is very long, some surveyors make a subordinate summary at the end of that bill, making totals of each of the various sections in the body of the bill, thus :—

Summary.

Work in deal, page 10.
Ditto in pitch pine, page 30.
Ditto in wainscot, page 66.
Ironmongery, page 74.

Carried to summary .. .. .. £

cwts.	qrs.	lbs.		£	s.	d.

No. 8. PLUMBER AND ZINCWORKER.

Externally.

Preamble.
Milled lead and labour in gutters, flats and flashings .. ..
Ditto in stepped flashings .. ..
Follow with leadwork of various kinds, such as are reduced to weight.
Follow with labours on the foregoing lead
Supls., runs, numbers .. ..
Numbers of labour and material.. .. ..

Internally.

Preamble.
Milled lead and labour in covering to draining boards .. ..
Ditto in linings to sinks and safes .. ..
Follow with labours, soldered angle, copper nailings, middling pipes, according to size.
Strong lead pipe in a similar manner.
Iron pipe, copper pipe, and any extra labours on it in usual order of runs and numbers.
Soldered joints in order of size, soldered ends .. ..
Short lengths of pipe, traps, brasswork, lavatories, w.c. apparatus, baths, urinals, cisterns .. ..
Clause as to Water Company and fees.

Zincworker.

Supls., runs, numbers .. ..

Carried to summary .. .. .. £

No. 9. PLASTERER.

yds.	ft.	in.	Preamble.		£	s.	d.

Internally.

Distempering, colouring, pugging .. ..
General plastering according to value, thus:—
supl. Render and set walls .. ..
„ Render, float, and set ditto .. .. ..
„ Lath, plaster, float, and set partitions ..
„ Ditto, ceilings .. ..

Work in fine Plaster.

Supls., runs, numbers .. ..

Continued £

BILLING.

PLASTERER—continued.

yds.	ft.	in.		£	s.	d.

Work in Portland Cement.

Supls., runs, numbers.. .. ..

Work in Keene's Cement.

Supls., runs, numbers .. ..
The majority of surveyors bill encaustic floor tiling and wall tiling here instead of in Bricklayer.

Externally.

In the same order as internal plastering ..
Allow for cutting out blisters, &c. .. ..

Carried to summary .. .. ..

cwts.	qrs.	lbs.

No. 10. FOUNDER AND SMITH.

Preamble.

Cast Iron.

The various kinds of articles, priced at per cwt., such as columns and stancheons, weight and description, the pattern following each item.
Follow with runs (as eaves gutters, rain-water pipes, &c.) and numbers, in the latter case the patterns may often be included with description.

Wrought Iron.

Girders, rolled joists, flitch plates, chimney bars, framed gratings .. ..
Follow with runs and numbers.

Carried to summary .. .. ..

ft.	in.

No. 11. GASFITTER.

Preamble.
Pipe according to size, smallest first
Fittings according to value, but often a provisional sum.. .. ..
Lengths of pipe, main cock, syphon, meter, clause as to fees, &c. .. ..

Carried to summary .. .. ..

No. 12. BELLHANGER.

| No. | 12 | Bells with fifteen pulls .. .. |

Continued

BELLHANGER—*continued.*

"*Fixing only.*"

				£	s.	d.
No.	6	Ceiling pulls				
,,	4	Lever pulls				
,,	3	Box pulls				
,,	2	Sunk plate pulls, including letting into brickwork and making good				

The foregoing arrangement will involve a provision for the pulls, always the more satisfactory way.

Carried to summary .. .. .. £

No. 13. GLAZIER.

ft. in. Preamble.
supl. Sheet glass, according to size and weight..
,, Ditto, cut to shapes and ditto .. ..
,, Follow with the various qualities of glass in a similar manner.
,, Embossing or enamelling plate glass P.C., per foot .. ..
,, Lead lights and glazing .. ..
Copper clips and brass screws .. .. ..
Clause for leaving glass clean and perfect at completion .. ..

Carried to summary .. .. .. £

No. 14. PAPERHANGER.

Preamble.
No. Pieces of paper P.C., per piece, and add for hanging .. ..
Follow with other paper according to value.
No. Pieces of paper, extra for sizing and twice varnishing with best paper varnish ..

Carried to summary .. .. .. £

No. 15. PAINTER.

yds. ft. Preamble.
Two oils on iron, supls., runs, numbers ..
Follow with the various kinds of painting according to value and in the usual order to each.
Extra for graining oak and twice varnishing with the best copal varnish, supls., runs, numbers.
Allow for touching up, &c. .. ..

Carried to summary .. .. .. £

BILLING. 377

The usual endorsement of a bill is similar to the following :—

"Estimate for House and Offices,
    "Godalming, Surrey, for John Smith, Esq.

            "William Thomson, Architect,
                "14, Bedford Row, W.C.

"*July*, 1878."

            "Mason."
        (or "All Trades.")

        (Autograph) "John Green,
                "Surveyor,
                    "8, Westminster Chambers,
                        "S.W.

When two Surveyors are employed both should sign

CHAPTER V.

RESTORATIONS.

These relate generally to churches, and a statement of the common practice with respect to them will apply fairly well to all works of repair.

The extent to which the general directions of the specification will be enforced depends so much upon the ideas of the architect that work executed from the same specification under one man's supervision may cost 500*l.* and under another's 1000*l.*; and although the surveyor should adopt his share of responsibility, he should avoid taking more than his share, and in measuring the work he should, while giving the quantity, throw the decision as to its extent in other respects upon the builder.

The materials removed will be treated in the manner described in the remarks on credits.

The new work will be measured in the usual way.

The stonework will be described as in windows or doorways, or as the case may be, and as bonded with old work if such is the case.

The cutting away to insert new windows or doorways, item by item, stating thickness of wall and size of new opening, and that it includes making good to new stonework, and in what trades, and state in what part of the building.

A clause will usually be found, "*Cut out all decayed parts of general facing, make good with new, and point with mortar (or cement).*" In such a case use the words of the specification, and measure the whole area occupied by the facing.

In the case of worked stone give the *surperficial* area, where possible, of stonework to be repaired, and state the nature of the repair.

Measure at per foot run, repairs to strings, quoins, copings.

Number repairs to windows, doorways, pinnacles, finials, &c.

RESTORATIONS. 379

The repairs to stonework are almost always described in a general way, and when this is the case the surveyor's description should also be general, so as to throw the interpretation on the contractor. Nevertheless when particular stones are described to be replaced with new, they may be measured as for new work, but state that they are inserted.

As it is usually uncertain how much will be spent in such a work, it will generally be required to keep the work in different sections, so that they may be separately stated in the tender, as "works to tower," "works to south chapel," &c., putting under each head the work in all trades relating to the particular section.

Alternative estimates for oak and pitch pine or deal, &c., are frequently required. See remarks on alternative estimates.

In cases where it is impossible to measure the work beforehand it is fairest to provide such a sum of money or quantity of material as may in the opinion of the surveyor be necessary, adjusting the quantity at the settlement of accounts.

A few instances of the method of treatment of various items relating to the old work are as follows. The new work will be treated in the usual way.

"Remove and credit. The roof of north aisle, comprising two trusses, about three squares of boarding and roof timbers, and about 18 cwt. of old lead, and make good work disturbed by its removal."

yds. ft. in.
" 1000 – – supl.   Cut out all decayed parts of general facing, make good with new to match old, and point with cement."

sqrs. ft. in.
" 6 – – „   Take up and credit paving of nave and aisles, and remove earth for a depth of 6 inches to receive concrete, and deposit within a distance of two runs " (or cart away).

yds. ft. in.
" 1000 – – supl.   Remove plaster, limewhite, &c., from the walls, and general stonework, repair the stonework with stone of the same kind inserted in cement, and point the joints with fine mortar, and clear away rubbish."

ft. in.
' 1000 – run.   Examine, clean, and repair splayed plinths of nave and aisles and chancel, cutting out all decayed stones and inserting new where necessary to match old."

	ft.	in.	
" 50	–	run.	Cut out remains of old stone string 6 inches high and insert new (elsewhere taken), and make good (chancel)."
" No. 8.			Examine and repair pinnacles of tower, cutting out decayed stones and inserting new as required."

"Examine and repair eastern window of chancel, cutting out decayed stones and inserting new, remove glazing and saddle bars, and make good."

"Remove glazing and saddle bars of western window, north aisle, cut out the tracery of head, and prepare to receive new (elsewhere taken), remove wooden mullion and restore with new stone to match old, repair the remainder of stonework, and make good all works disturbed."

The general conditions and preliminary items of the bill will be as for new work.

There are few points specially to be observed in a bill of restoration, and the following should appear in the preliminary bill:—

"Any objects of interest which may be found during the progress of the works to become the property of the committee."

"Carefully case and protect from injury the mural tablets and monuments (enumerating and describing them), where necessary to remove them for the purposes of this work, place them in a safe position and refix them at completion, and make good around same, and contractor is to be responsible for and make good any damage that may occur to them during the progress of the works."

Or the foregoing clause may be made to include only those monuments which remain in position, and those removed treated something like the following:—

"Take down and deposit and protect the monument on north wall of north aisle, refix when the wall is rebuilt, and make good all work around same."

"Carefully take up the footstones, brasses, slabs, &c., in floor of nave, deposit and protect and refix same in such positions as may be directed, as follows." Enumerate them.

"Remove the gravestones or tombs which may interfere with the rebuilding of north aisle, deposit and protect and refix same, as follows." Enumerate them, or give the number of large and small ones.

"Allow for the removal and reinterment where directed by the churchwardens of any bodies that may be met with in course of excavation."

Provide a certain quantity of concrete for the filling in of any old graves that may be met with.

In measuring the plastering on the old walls take care to state "that it is to include any necessary dubbing," and if on old walls observe that a superficial quantity of raking out joints of old brickwork as key for plastering should be taken.

In measuring new glazing in old stonework take a running dimension of "clean out or rework, rebate, or groove for lead lights, and fill in old mortises."

Let the general items of pulling down and cutting new openings, &c., follow the conditions and general preliminary items.

Where any of the work is to match old, state it.

General Alterations.—It may be instructive to present here a few types of items which are of frequent occurrence in a bill of alterations, and which may serve to show their general treatment.

£ s. d.

"Allow for closing up the bottoms of No. 20 flues while raising the shafts for clearing away rubbish, and sweeping and coring the flues at completion"

"Allow for cutting opening for doorway 3' 0" × 7' 0" in 1½-brick wall, for inserting new lintel (elsewhere taken) and relieving arch, for making good brickwork in cement, and for making good plastering after fixing of finishings, flooring, and all other works disturbed on south side of drawing-room" ..

In case of openings cut the item will appear as above, and the lintel and finishings will be measured and treated as is usual with new work.

£ s. d.

"Allow for taking down doors of library and dining-room (three in all), taking off the ironmongery and making good after same, for nailing and glueing on pieces to heads where out of square, and for re-hanging in new linings, elsewhere taken" .. ..

In the cases of adapting old joinery it will generally be necessary to take off all the old ironmongery and substitute new. The new ironmongery will be taken in the usual way.

rods ft. £ s.
5 — sup¹ Reduced brickwork in cement in small quantities inserted, and including extra labour and materials, cutting and bonding to old work .. ..

QUANTITY SURVEYING.

In the case of openings filled in, measure the brickwork (the net opening) described as in last item.

Observe that cases are frequent in which the actual amount of pulling down desired and indicated on the drawings cannot be adhered to, as it would often cost more to support the work than to pull it all down and rebuild it, notably in cutting openings in old walls.

When old chimney breasts are cut down (as in a party wall) the remains of the old flues will require filling up, and this work should either be measured or mentioned in a general clause. The following are a few ordinary items.

		£	s.	d.
ft.				
90 supl.	"Cut back face of old wall to form toothing, and insert new brickwork average ¾-brick thick in cement and bond same to old wall"			
20 „	"Cut down projection of 1-brick chimney breast to old wall and prepare same for new work"			
100 run	Cut chase in old walls about 5" × 5" rough, render same in cement, and insert and bond new walls to old, including additional brickwork			
100 supl.	Old selected paving (taken from present basement) re-squared, re-tooled on upper face, laid in mortar, and jointed in cement			
100 cube.	"Fir framed joists inserted in small quantities"			
100 „	"Fir framed in quarter partition ditto"			
yds. ft.				
100 - supl.	Hack face of old walls and prepare for new plastering ..			
100 run.	Make good junction of new and old plastering and paint edge of old			
No 20.	Make good formation of new and old cornices average 10" girth			
100 - run	Make good plastering after removal of 9" wall and paint edges			
100 - „	Ditto quarter partition and ditto			

The walls of an old building are very often uneven or out of the perpendicular, and in such cases the preamble of the bill of plasterer should include a clause, "The price for plastering on the old brickwork to include for any necessary dubbing in tiles and cement, sufficient to produce a true face."

If hacking off plastering is not measured insert a clause, "Allow in the price of plastering for hacking off old and raking out joints of brickwork to form key."

In cases of repairing old plastering, there are two possible methods of treatment, one is to measure the specific parts of the work which in the surveyor's opinion require repair and present it thus:—

yds.
    supl.   "Strip off old paper, knock down old plastering, and reinstate with Parian cement in patches and allow for junction of new and old." This casts the responsibility of decision as to what repair may be necessary upon the surveyor which is dangerous. .. .. ..   £ s. d.

A better way is as follows, i.e. measuring the whole of the old remaining surface:—

yds.
    supl.   Strip off old paper, examine and repair defective plastering to walls, and replace where missing in Keene's cement. Cut out and remove old plastering where badly executed, bulged or damaged, and make good with Keene's cement .. ..
    „   „   Ditto all as last, but to partitions including making good lathing .. ..
    „   „   Ditto to ceilings .. ..

"*Note.*—The whole area of old plastered walls has been measured, and the contractor must examine the existing plastering to see how much new will be required."

Painting on old work will sometimes involve the burning off of certain parts. The surveyor should not attempt to distinguish such parts. He should measure the whole as clean, and 2 oils or 3 oils as the case may be, leaving the extent of burning off to the contractor's judgment, alluding to it in a clause of the preamble to the painter's bill. See section BILLING.

The painting in an old house, if good work is required, will involve the removal of the door and window furniture and possible relacquering.

A list should appear in the bill in the order of an ironmongery bill of "take off, label, relacquer and refix, including cartage."

Variations before acceptance of Tender.

After a tender has been delivered it is often found necessary to reduce it. There are various ways of doing this. To reduce the cubic content, the quality of the materials, the extent of the finishings.

The surveyor will, as a general rule, be asked to furnish a list of proposed alterations with each item priced.

The price per foot cube being calculated from the amount of tender and the cubic content at the commencement of the dimensions, the amount of saving effected may be easily arrived at. It should, however, be observed that the reduction of the height of the stories, or of the width of a building, will not produce a reduction so great as the number of cubic feet would seem to indicate. Whereas the omission of a part of the length of a building, or a part of it where roof, walls, floors, and finishings are all comprised, will correspond pretty accurately with the price per foot cube of the building as first projected.

Parts of the work which are not susceptible of cubic calculation, as covered ways, conservatories, boundary walls, etc., will be omitted by reference to the dimensions or bills; and it is in the case of variations that the advantage of their suggested separation in the dimensions and bill is apparent, as often the amount for a complete section of the bill may be omitted in one item.

Such calculations are much simpler and more certain if the builder will produce his original estimate.

As the result of these calculations will be reported to the client, and are but an approximate estimate, the amount of omissions had better be reduced by 10 per cent. and of additions increased by 10 per cent., as in the more careful preparation afterwards of the bill of additions and omissions it will inevitably happen that some differences will be made. The trouble and expense of a detailed calculation of items which may not be ultimately adopted will thus be saved. A list of the proposed variations should be supplied to the architect to submit to his client; but neither the items of alteration nor their estimated value should be communicated to the builder, as cases have been known (though, happily, rare) of the builder reducing in his copy of the quantities the prices of the items proposed to be omitted.

When it has been decided what shall be omitted the original dimensions must be carefully looked through, and the items affected copied as a fresh set of dimensions; where a series of dimensions as originally taken is to be omitted they may be referred to in the new dimensions by letters A, B, C, etc., putting the same letter at the beginning and end of each series in the original dimensions, thus :—

RESTORATIONS.

```
‖ A │ 20·0  │       │                    ‖
  │   │ 10·0  │ 200 – │ L. P. F. and S. ceiling
  │   │       │       │ and twice whiten.
  │   │       │       │           B. R. 3.
  │   │ 14·0  │       │
  │   │ 14·0  │ 196 – │ add
  │   │       │       │
  │   │       │       │           B. R. 4. ‖
  A
```

The reference in the new dimensions being as follows:—

Omit A to A, col. 40, ceilings B. R. 3 and 4. Sometimes it may be necessary to omit a part only, as for instance one door out of five: in such case the note would be "Omit ⅕th A to A col. 46, one door." Sometimes it will be convenient to omit only a part of a series, when the items not omitted may be marked thus ×, and the clause would run, "Omit A to A col. 52," "Except items marked thus ×." Always describe in the new dimensions what the omission is, as otherwise it will be difficult to distinguish without again referring to the original dimensions.

When the revision of an estimate is a very thorough one, although there may be many instances of the retention of a series of items of the first estimate, such series should be completely rewritten in the new set of dimensions and not referred to by letters.

Single dimensions may sometimes be transferred as follows, the equivalent of the squaring being written in the dimension column:—

```
    2/  12  6                    102  1
         4  1                      1  0
        ─────  102  1            ─────  102  1
```

In copying dimensions for the purpose of omitting them begin the series with the word *omissions*, and then copy them verbatim, additions and deductions, just as originally taken Be careful when measuring the additions to head that series with the word *additions*.

In abstracting omissions and additions it will occasionally be found that there is an item of deduction with nothing of the kind from which it can be deducted.

In such cases transfer all deductions of addition sheet to omission sheet as omissions.

2 c

Transfer all deductions of omission sheet to addition sheet as additions.

When the bill of variations is complete, if not before, the builder should produce his original estimate, and he and the surveyor together should affix the prices to each item, those to the omissions to be at the rate of the original estimate, those to the additions at analogous rates.

It will sometimes make the dimensions of a set of variations much clearer if the dimensions of omission and addition are written on the same dimension page side by side, using one set of columns for omissions and the other for additions.

CHAPTER VI.

ERRORS, ETC., IN ESTIMATES—SCHEDULE OF PRICES—ADJUSTMENT OF ACCOUNTS.

As a builder has sometimes been known to deliver a very low tender (perhaps several hundreds of pounds lower than the next above him), and has afterwards professed that he has made a mistake in his estimate to his own damage, it is perhaps not superfluous to consider the question of errors generally. In such a case as the foregoing the surveyor should request the builder to allow an examination of his original estimate, when he can generally form an opinion as to the genuineness of the plea. Where he has reason to suspect deceit, the whole of the builders originally tendering should have an opportunity of revising their tenders, or the work given to the contractor next above him in the list of tenders. Where the plea is a true one the case may generally be met by giving the builder a part of the difference, but after the contract is signed no claim for errors should be allowed.

It must not be forgotten that the signing of a form of tender agreeing to sign a contract when called upon to do so, if the building owner accepts that offer, is a legal contract, and its completion can be insisted on. Lewis v. Brass, L. R., 3, Q. B. D., 667 (1877).

Where a priced estimate is deposited, the rates of which are to be applied to items of variations, the moneying out of the bill should be carefully checked before the rates are used, as where a money item does not agree with its rate, as follows:—

	ft.	in.			s.	£	s.	d.
	100	–	cube.	Fir framed in roof.	3	10	0	0

The rate must be assumed in accord with the *result*, viz. 2s. per cubic foot.

Mistakes in the casting of columns will also affect the prices in a similar manner; for instance, the amount of the tender being 10,000*l.* and the true result of the casts being 10,500*l.*, the rates would be subject to a deduction of 5 per cent., as all parties are bound by the amount of tender. Observe also whether a deduction has been made of a certain percentage from the amount in summary to produce the amount of tender; if so, the rates may also be subjected to a deduction of that percentage, but it is sometimes so small as to be not worth the trouble.

In the adjustment of rates by a percentage the provisional *amounts* must be excluded from the calculation.

The treatment of these rates will depend upon a variety of considerations. One of the most important is whether or not the quantities are a part of the contract. See also The Law as it Affects Quantity Surveyors.

When the quantities are a part of the contract the builder will be entitled to payment by the building owner for the difference on any items imperfect in description or wanting in quantity, but he will not be allowed to revise the rates nor to alter the total except for the foregoing reasons.

The building owner will be entitled to deductions for excess in description or quantity, but he will not be entitled to revise the rates nor to alter the total except for the before-mentioned reasons.

It is necessary in all cases to remember that the fundamental principle of the present system of competitive tendering is finality. The building owner wants to know before the building is commenced what his expenditure will be. The tenders are totals and they are to be dealt with as such, and except for foregoing considerations are the builder's own affair; moreover, the decision as to whose tender shall be accepted is mainly based upon the relative totals.

But it is none the less necessary that the whole of the extensions and castings should be checked before the contract is signed. The most suitable person to do this is the quantity surveyor. It may be objected that one of the advantages of sealing up the deposited set of quantities will thus be lost, because the surveyor will be in possession of information which he may give to the architect to the builder's disadvantage. The surveyor need hardly be reminded that such work is confidential and should not be communicated; but the checking is an imperative necessity which will be more clearly shown by the following instances.

The probabilities of error are so numerous that it will be impossible to enumerate all the varieties that may arise; a few instances of those most frequently met with may, however, be considered.

It should be remembered that the total of the tender must always be treated as final, and that the rates of the original estimate are (as is mostly stipulated and properly so) to be applied to all omissions or additions of a kind similar to the original item.

1. *Items of feet inadvertently priced at the rate of a yard, and vice versa.*—These should not be interfered with; any adjustment opens the question of the judgment and motives of the builder. If a yard is priced at a rate which would only be reasonable for a foot, and the rate is increased, an argument would be afforded for the reduction of other rates, apparently too high, thus affecting the total and reopening the competitive question which is the keynote of the whole transaction.

In cases of variation these absurd prices often prove much to the builder's advantage, sometimes greatly to his disadvantage, but it is one of the imperfections of what is on the whole a convenient system. It should be remembered that when there are no quantities the builder (even more frequently than when there are) makes considerable mistakes but is bound by his bargain.

2. *Items which are priced at rates discordant with other rates*, such as circular work at a less price than straight, brickwork in cement at the price of brickwork extra only in cement, &c. The same principles apply as in item 1.

3. *Items neither priced nor extended.*—When the item is of small amount it is obvious that as a percentage it will affect either the total or the rates to but a small extent. Such a case is usually met by pricing the item at a current rate.

When the amount in question is a large one, it will be best to price it at a current rate and to reduce all the other rates by a percentage.

The attachment of a rate is necessary because of the possible increase or reduction of the original quantity during the progress of the work.

4. *Single items extended without a rate attached.*—In such a case

the rate may be calculated from the extension, the latter being one of the elements which produced the total.

5. *A series of items bracketed together by the estimator and one sum attached.*—This often occurs in the joiner's bill when a series of items of labour are thus treated. The most reasonable course is then to price the whole series at consistent current rates, extend them, and add the extensions together; if on comparison of this total with the total attached by the estimator it is higher, the difference must be calculated as a percentage and each of the rates reduced so as to make the totals agree; if it is lower, then all the rates must be increased by a similar process.

When quantities are not a part of the contract the same necessity for examination exists.

The builder will *not* be entitled to payment by the building owner for the difference on items imperfect in description or wanting in quantity, nor will he be allowed to revise the rates or alter the total.

The building owner will *not* be entitled to deductions for excess in description or quantity, nor will he be allowed to revise the rates or alter the total.

The treatment of the items 1 to 5 will apply in a similar way (under these conditions) as when quantities are a part of the contract.

In both cases the object is to make the bill of quantities complete and workable as a schedule for the adjustment of variations.

For this purpose their value is often great. If there is no condition in the contract for their use as a schedule of prices, the surveyor should endeavour to induce the builder to adopt them for that purpose; as a rule they are the fairest data for use in a settlement of accounts.

The foregoing cases presuppose that the original intentions of the drawings and specification have been observed. Ordinary variations from them, i. e. items not essential to the production of a complete building, would be measured and priced in the ordinary way.

When quantities are not the basis of the contract, the ordinary principle of law that the contractor shall supply a complete building must be considered, and if *essential* work has been omitted by the architect or his surveyor, the contractor is bound to do it without charge; the surveyor's interpretation does not supersede this principle. It is manifestly unfair that the building owner should have to pay for it. On the other hand, if such an item is omitted, it is equally unfair that he should derive no benefit from its omission.

The argument for this treatment is still stronger, if there is a condition in the contract that the builder shall do all things necessary to complete.

SCHEDULES OF PRICES.

When it is desired to commence a building, and there is not time for the preparation of an estimate in the usual way, it is a frequent practice to prepare a schedule of prices. Often the work of basement only is done by schedule, the drawings of quantities for the superstructure being completed during its progress. In such a case the surveyor carefully examines the drawings, and where there is a specification that also, or if a specification has not been prepared, he arranges the quality of the work with the architect. A schedule is then drawn up, commencing with conditions, and comprising every item which the surveyor (from the information he possesses) considers will be required. It is obvious that the character of the various items will vary according to the type of building. The preamble to each trade should set forth any special conditions as to, or modes of measurement, if any, and should describe the work and materials as in a bill of quantities. A copy is then forwarded to each builder, who prices the schedule and returns it to the architect, who then, by careful comparison and analysis of the prices, decides upon the person he will except.

There are a number of published schedules of prices, the possession of which will materially assist the surveyor in the preparation of similar ones; among others may be mentioned those of the War Department, H.M. Office of Works, the London County Council, the School Board for London,

It may be perhaps considered superfluous to caution the surveyor against the adding of the whole of the prices of each schedule together and comparing the results, but as this has been done, and is no test of the respective tenders, it is necessary to direct attention to it.

A proper analysis should be made, in the following manner:—

The surveyor should cube the building, and form an opinion as to its probable cost. He will then be able to judge of the extent of the main items of material and labour which will be required, as, so many rods of brickwork, so many feet of stone, so many feet of sashes and frames, etc.; and if this cannot be arrived at otherwise they may be roughly measured from the drawings, the approximate quantity being sufficient for the purpose.

The builders having seen the drawings, it will sometimes be found from an inspection of the schedules that an attempt has been made to entrap the unwary surveyor, by putting specially low prices to some of the items of which there will be but a small quantity used, as for masonry in a building for the most part of brick, for slating where of the roof covering the greater part is tiling, etc., thus producing a semblance of a low schedule.

The analysis may be conducted thus, assuming quantities as follows:—Ignore the shillings and pence if under ten shillings, if as much as ten shillings call it one pound. The labour prices may be averaged by adding together all the trades, as, excavator 7d., bricklayer 11d., mason 11d. = 9⅔d. average.

	Smith. £ s. d.	Smith. £	Brown. £ s. d.	Brown. £	Jones. £ s. d.	Jones. £
2000 yards excavation and carting	8 6	350	3 6	350	4 0	400
1000 yards concrete	7 0	350	6 6	325	7 6	375
40 rods brickwork	14 0 0	560	15 0 0	600	16 0 0	640
1000 ft. fir	2 9	138	2 7	129	3 0	150
100 squares flooring	1 10 0	150	1 8 0	140	1 9 0	145
Day work, 1000 hours	9	38	10	42	9½	40
Total		£1586		£1586		£1750

The surveyor when called upon to report, should state whose schedule is lowest, and by what percentage the remainder exceed him, and also that, "assuming that the building should cost so much (state sum), Smith's price would be (state sum), Brown's (state sum), Jones's (state sum)."

The form the schedule should take is generally indicated below. Its form bears a general resemblance to that of a bill of quantities.

A consecutive number to each item assists reference.

"*Conditions of tender and schedule of prices on which it is proposed to erect (describe building and its position, and name of proprietor).*"

"JOHN SMITH, Esq., Architect,

"March, 1878." "2 Fenchurch Chambers, E.C."

When no drawings have been prepared, a general description of the intended building, and its dimensions, should be given.

"The carcase of the building is to be completed and roofed in by (date), and completed by (date), or in case of non-fulfilment of such works by either date the contractor to forfeit as liquidated damages the sum of £ . . . per week. Delay consequent upon strikes only excepted."

"If at any time the architect is dissatisfied with the progress made with the works, as regards time, the contractor shall employ, at his direction, more workmen of their respective trades, or in case of the introduction of improper or inferior materials or workmanship the contractor shall alter them as directed by the architect, and in case the same be not done or altered within forty-eight hours of written notice being given by him to the contractor, the architect is hereby empowered to take the work out of the hands of the contractor, and to employ such other builder as he may think proper to complete the premises; and in the event of the contractor becoming bankrupt, the architect, after similar notice, shall have the same power."

"Payments will be made at the rate of . . . per cent. on the value of the works executed, on the certificate of the architect, in sums of not less than £ . . ., from time to time till a balance of

£ . . . is accumulated as a reserve, and after that time the amounts of such values to be paid in full, and the balance within one month after the architect shall have certified that the work contemplated under the schedule is completed to his satisfaction, with the exception of £ . . . to be retained for a longer period as hereinafter described."

"The proprietors retain for themselves the power of contracting with any other parties for fittings or separate portions of the work connected with the premises or for the completion of them at any stage they may think fit, and the contractor will be required to give every reasonable facility to such other contractors for carrying out their work."

"Should any unforeseen delay occur in carrying out the work (other than as herein alluded to) from causes not under the control of the contractor, it shall rest entirely with the architect to decide whether any and what pecuniary allowance shall be made to the contractor for such delay as respects the use of plant and scaffolding, or the waste of foreman's or other time, or loss of interest on capital, or from any other cause, but no allowance will be made for any rise in the general rate of wages or other prices during the continuance and till the completion of the works according to the above conditions, or during such additional time as may by any default of the contractor be requisite for the completion of them."

"The contractor alone will be held responsible, for all injury to life or property caused by carrying out the works from whatever cause arising, and also for all damage to the proprietors of this or adjoining properties, of whatever kind, caused by the carelessness or inadvertence of his workmen."

"The decision of the architect shall be binding on the contractor and on (name of proprietor) in all matters relating to the contract as to the interpretation of the drawings and specification, as to the quality of the materials and workmanship used, and as to any question which may arise in reference to the progress and conduct of the work, and as to the payment for the same."

"The whole of the materials and labour are to be of the best quality, and to be executed and completed in exact accordance with the drawings, details, specification, and instructions provided from time to time by (name of architect) or other the architect for the time being of the said (name of proprietor), and to his entire satisfaction."

"The works will be carried on under the supervision of a clerk of works, and every facility to be given to both the architect and his representatives for the full examination of all work or labour either before, during, or after their execution."

"The contractor will be required to keep a thoroughly competent foreman in constant attendance at the building as his representative."

"The works to be executed by proper workmen at daily wages, and none to be sub-let without the written consent of the architect, the architect to have the power of dismissing all workmen whom he may deem to have behaved improperly or to be incompetent."

"It is intended that the whole of the work shall be paid for on measured prices, but should any works in alterations be required to be executed at day prices, weekly accounts shall be sent in to the architect containing full particulars of the time and materials expended thereon, and the delivery of such weekly accounts within ten days of the expiration of each week shall be a condition precedent to payment of such week's work in any other manner than at the measured prices herein mentioned."

"The prices named are to include all cartage, labour and materials and the use of all tools, tool-sharpening, establishment charges, implements, scaffolding, ladders and machinery required for the proper execution of the whole."

"The prices for ordinary workmen for day labour shall also include all time expended by the foreman engaged in the superintendence of the same, unless he be engaged exclusively upon the work so charged."

"The works to be carried out in strict accordance with the London Building Act, with the by-laws of the Commissioners of Sewers, and with all other regulations of a public nature authorised by law."

(The foregoing in the metropolis. In the country the local Act and the town surveyor will be mentioned.)

"The following works shall be executed by the contractor as (*inter alia*) works necessary for carrying out the contract without any special payment for the same, but they shall be considered as part of his own expenses, which the profits to be obtained on the measured work will cover, viz.:—

"The giving of all requisite notices to (district surveyor and) all (other) authorities, and the furnishing of all particulars to such authorities." (Leave out words in parenthesis when not in metropolitan area.)

"The use of lines and stakes of all kinds in setting out the work, and all labour and superintendence thereon."

"The erection of an office for clerk of works, with all requisites for the same, together with all necessary firing and lighting for him during the progress of the works."

"The covering and protection of all walls from wet and frost on the occasion of any requisite suspension of the works."

"The covering and protection of all stone steps or protecting brick or stone work from injury during the progress of the works, and the removal of the same."

"All requisite temporary barricades or fences to doors or windows (exclusive of paper lights, if ordered by the clerk of works) and the requisite sheds or coverings to lime and facing bricks, etc."

"All plumber's work etc., in temporary water storage or supply, and gas service and use of braziers (the water company's charge for water for the works and the cost of gas, as sanctioned by the clerk of works, and the cost of fuel for drying the rooms, but not the labour connected with any of these, will be allowed)."

"The general attendance of each trade upon all others and the execution of all jobbing work and all requisite attendance and messages in connection with the superintendence of the same."

"The removal of rubbish and surplus materials as they accumulate, washing floors, if required, at completion, and leaving the whole premises in a neat and orderly condition."

"The contractor shall also, without further charge, keep in repair the whole of the work executed either at measured or day prices for a period of six months after the completion of the contract, during which period the sum of £ . . . shall be retained by

SCHEDULE. 397

(name of proprietor), and any defects appearing either in materials or workmanship shall be made good by the contractor at his own expense."

"The work will be measured up from time to time by Mr. . . . , of . . . , surveyor, on behalf of both parties, and will be priced by him at the rates of the selected schedule, subject to a reference to the architect in case of any dispute arising as to the interpretation of the schedule or the mode of its application."

"The surveyor's charges of . . . . per cent. and expenses to be divided equally between (name of proprietor) and the contractor, and to be paid by the latter as the works proceed."

SCHEDULE.

(*Note.*—Various items have been introduced to show convenient modes of treatment.)

"On all sums paid out of pocket for insurance fees and payment to local authorities, expense of water, of gas, to the extent sanctioned by clerk of works; of such fuel as may be used under the special instructions of the clerk of works for the purpose of drying the premises, on cost of tile paving, asphalting, ornamental glazing, stoves, chimneypieces, parqueterie, furniture and fittings, &c. (except trade fittings); and of any other works executed by workmen other than those of the contractor, during the progress of the work, to include remuneration for the use of scaffolding, temporary protections and casings, extra lighting and watchings, and other necessary accommodation (except time actually expended in unloading and fixing) an allowance or addition on the net amount certified by the architect" of .. .. Per cent.

"*Note.*—Any of the above amounts which may be paid by the contractor will be paid in full to the contractor as part of the next certificate to him."

"In all cases P.C. values to be interpreted as the amounts paid by contractor after deducting trade allowances, but not ordinary discount for cash."

"For the construction, alteration, and maintenance of all requisite hoarding, fences, gangways, rails, kerbs, coverings to footways, temporary ways over vaults, gates, doors, fastenings, painted notice-boards, and all other matters required by local authorities or considered necessary by the architect, and for removal of same on net frontage of building" .. .. Per foot run.

"*Use and waste* of timber in shoring to ground, or to buildings, and include cost of all wedges, hoop iron, &c., and labour complete" .. .. Per foot cube.

"Oiled paper lights if specially ordered by the architect" Per foot supl.

Reference to a price book as the schedule is prepared will help to preserve the surveyor from omission of necessary items of labour and material.

"*Excavator.*"

Preamble.

Excavation at various depths and of various kinds .. .. Per yard cube.
" Add if in underpinning.. .. .. ,,
" Fill in and ram around foundations selected hard dry material from site, or dry rubbish " .. ..
" Fill in and wheel one run " .. ..
" Extra runs " .. ..
" Fill and cart away " .. ..
" Basketing " .. ..
" Lime concrete (describe), including staging " .. .. ..

Concrete of various kinds and various thicknesses to follow here.

" Add in underpinning " .. .. ,,
" Clay puddle " .. .. ,,
" Excavator " .. .. Per hour.
" Gravel or ballast " .. .. Per yard cube.
" Sand " (Thames) .. .. ,,
" Ditto " (pit) .. ..
" Lime " (describe sort) of various sorts.. .. .. .. ,,
" Hard dry brick or stone rubbish .. .. ,,
" Portland cement " .. .. Per bushel.
" Strutting and planking to sides of larger excavations " .. Per foot supl.
" Strutting and planking to trenches (length of trench only to be measured) each foot in depth " .. .. Per foot run.
" Horse, cart and man " .. .. Per day.
" Two horses, cart and man " .. .. ,,

"*Drains.*"

Preamble.

Describe fully. Give the various sizes and the depths of the digging .. .. Per foot run.
" Extra for deep tunnelling." State average depth .. .. ,,

	Bends.	Single Junctions.	Double Junctions.	Siphons.	
" Extra on 4-inch drains for " .. .. Follow with the various sizes, as 6-inch, 9-inch, &c.					Each.

SCHEDULE.

"*Bricklayer.*"

Preamble.

Describe the labour and materials generally, as preamble to a bill of quantities.

The various kinds of brickwork	Per rod supl.
" Extra for hoisting above . . . feet "	"
" Add if in one brick walls fair both sides "	"
" Add if circular on plan "	"
" Add if in vaulting "	"

Follow with the various items, arranged as in a bricklayer's bill.

" Hoisting and fixing iron columns, girders, and joists up to 20 feet above footway "	Per ton.
Follow with the various heights	"
" Bricklayer "	Per hour.
" Labourer "	"

Various materials, as bricks per thousand, mortar per yard and hod, putty or fireclay per hod, cement mortar per yard and hod, &c.

"*Mason.*"

Preamble.

State clearly how the stone is to be measured.

Place the best stone first; assume in this case Portland stone.

To commence with a description of the stone, state that " the prices are to include moulds, trammels, &c., and those for labour to include arrises. The stone to be measured net as set, including hoisting and setting not exceeding 40 feet from level of footway " .... Per foot cube.

" Add, if exceeding 6 feet in length "	"
" Add on first item, if exceeding 8 feet in length "	
" Add on ditto, &c., if hoisted from 40 feet to 60 feet from level of footway "	

Follow with various heights.

Follow with labours arranged as in a bill of quantities.

" On the net amount paid for carving, to reimburse contractor for all scaffolding and cleaning up after same, an addition of "	Per cent.
" Bath stone of description, as Portland stone "	Per foot cube.
" Allow off price of labour on Portland stone "	Per cent.

Follow with the various stones with less labour in a like manner.

" Sound hard Yorkshire stone, as above "	Per foot cube.
" Add to price of labour on Portland stone "	Per cent.
" Mason "	Per hour.
" Ditto, setter "	"
" Ditto, labourer "	"

Follow with any material, as in previous trades.

In this trade time and writing may be saved by tabulating, thus:—

The following are a few examples.

	Diameter.				
	3½ in.	3 in.	2½ in.	2 in.	1½ in.	1 in. and under.
Drilling or jumping bolt or other holes through York, or Purbeck per foot run						

Proportion of the above rates if in Bath   ..   ..   ..
Ditto in Portland   ..   ..
Ditto in granite   ..   ..
Add percentage on the above done in position.

	9″ × 9″	9″ × 14″	9″ × 18″	9″ × 24″
3″ tooled templates, each				
4″ ditto				
6″ ditto				

"*Slater or Tiler.*"

Preamble.
The various kinds, as in a bill of quantities.
"Slater or tiler"   ..   ..   Per hour.
"Ditto, labourer"   ..   ..   "
Various slates or tiles   ..   ..   Per thousand.
Nails of various kinds   ..   ..   Per lb.

"*Slate Mason.*"

Preamble.
The tabulation of the items is in this trade specially useful.
Slabs sawn to any size required.

	Thickness.				
	½ in.	¾ in.	1 in.	1¼ in.	1½ in.
Slabs, quarry planed or self-faced					
Ditto, 16¼ feet supl. cut to size per foot supl.					
Ditto, 16¼ feet supl. and up to 30 feet .. .. &c.					

Grooving, labour to edges, holes, &c., may be similarly arranged.

A table of copper screws of various lengths is useful in this table.

SCHEDULES.

"*Carpenter.*"

Preamble.
Arrange items as in bill of quantities.
Labours (on the various hard woods) "extra on deal" .. Per cent.
" Fixing to ordinary straps and bolts, to floors, roofs, &c." Per cwt.
"Carpenter" .. .. Per hour.
"Labourer" .. .. "
"Fir and other woods, no labour" .. .. Per foot cube.
"Nails" .. .. Per cwt.
"Spikes" .. .. "

A table of various woods in thicknesses will sometimes be necessary.

				Thickness in inches.				
	½ in.	¾ in.	1 in.	1¼ in.	1½ in.	2 in.	2½ in.	3 in.	4 in.
Archangel, Christiania, or St. Petersburg, sawn to thicknesses per foot supl.								
Elm, English								
Oak, English ..								
Ditto, American ..								
&c.								

"*Joiner.*"

Preamble as to the same trade in quantities, and the items arranged in similar order.

But a modification of form will be required to meet the various thicknesses, and save items, thus:—

	½ in.	¾ in.	1 in.	1¼ in.	1½ in.	2 in.	2½ in.	3 in.
"Deal per foot superficial, with no labour in day account"								
Follow with the variations, as—								
"Ditto fixed with nails or screws								
"Ditto wrought one side" &c.								

The same arrangement will be required for doors, sashes, frames, &c.

Work in various more expensive woods, thus:—
"Add on prices of labour and materials in deal, for work in Honduras mahogany" .. .. Per cent.

2 D

"Joiner" .. .. Per hour.
"Labourer" .. .. "
"On all ironmongery, to include profit, carriage and fixing to deal in all cases, an addition on the net price to contractor of" .. .. Per cent.
"Ditto ditto for fixing to wainscot or other hard wood" .. "
A table of fillets is useful in this trade.
Fillets including fixing and fair ends.

	½ in.	¾ in.	1 in.	1¼ in.	1½ in.	2 in.
2" wide and under, rough						
Ditto wrought						
Ditto beaded or chamfered on one angle						
Ditto ditto two angles						
Ditto ditto rounded one edge						
Ditto ditto rounded both edges						
2" to 3" wide rough						
Ditto wrought						
&c.						

A table of screws is generally used.

	½ in.	¾ in.	1 in.	1¼ in.	1½ in.	2 in.	2¼ in.	2½ in.	3 in.	3½ in.
Iron screws per dozen										
Ditto and driving ditto										
Brass screws per dozen										
Ditto and driving ditto										

"*Founder and Smith.*"

Preamble as to the same trade in quantities, also "The prices for columns, girders and joists, to include all hoisting and fixing in all trades except the addition mentioned in bricklayer's schedule."

"The fixing of ironwork of roofs and floors is included in carpenter's schedule."

"All prices throughout to include smith's attendance and fixing and with the above exceptions the prices are to include fixing in all trades."

"Cast iron in girders, stancheons or columns, from 5 cwt. to a ton in each for the first casting, including pattern" .. .. Per cwt.

SCHEDULES.

"Cast iron in girders, stancheons or columns, from 5 cwt. to a ton in each for subsequent castings from same patterns" .. Per cwt.
"Ditto ditto, above 1 ton first casting" .. .. ,,
"Ditto ditto, other castings" .. .. ,,
Follow with the various articles billed by weight, &c.
"*Wrought iron.*" The various kinds .. .. Per cwt. or lb.
"Smith" .. .. Per hour.
"Labourer" .. .. ,,
Various materials.

If there is much variety in eaves gutters or rain-water pipes, they may be tabulated with advantage. Cast-iron pipes and their adjuncts are also better arranged in a table.

"*Bellhanger.*"

Preamble as to the same trade in quantities.
"Pulls to bells, fixing only." .. .. Each.
Bells (describe) .. .. ,,
"Bellhanger" .. .. Per hour.
Materials.

"*Gasfitter.*"

Preamble.
"Allowance off Russell's list for patent welded tubing, with all requisite bends, tees, nozzles, joints, &c., after allowing for fixing same, including cutting away, making good, and general attendance, testing pipes, and leaving perfect at completion" Per cent.
"Ditto, in day work" .. .. ,,
"*Note.*—The above prices to be applied likewise to similar pipes for water or other purposes. The pipes to be of the very best quality, equal to Russell's best, but no defective or inferior work will be passed on the plea that the ironwork is from Russell's."
"Gasfitter" .. .. Per hour.
"Labourer" .. .. ,,
Materials.
Pipes and their fittings are often tabulated.

"*Plasterer.*"

Preamble as to the same trade in quantities.
Items in the usual order of a bill.
"Cast enrichment (including modelling if 60-feet run is used)" .. .. Per foot run.
"Plaster screeds 6" by ⅜" .. .. ,,
"Plasterer" .. .. Per hour.
"Labourer" .. .. ,,
"Boy" .. .. ,,
Various materials, as lime, sand, hair, laths, nails, &c.
"Fixing only floor tiling of various sizes" .. .. .. Per yard supl.
"Ditto wall tiling .. .. ,,

"*Plumber.*"

Preamble.
Items in the usual order of a bill.
"Allowance on prime cost of brasswork, w.c. apparatus, &c., to include fixing but not joints" .. .. Per cent.
"Plumber" .. .. Per hour.
"Labourer" .. .. "
"Solder" .. .. Per lb.
Materials.
Lead pipes and joints may be tabulated.

"*Glazier.*"

Preamble as to the same trade in quantities.
Items in the usual order of a bill.
"Extra for bedding in wash leather" .. .. Per foot supl.
"Ditto   ditto   indiarubber" .. .. "
"Ditto   ditto   red-lead putty" .. .. "
"Glazier" .. .. Per hour.
"Putty" .. .. Per cwt.
Materials.

"*Painter.*"

Preamble as to the same trade in quantities.
Items in the usual order of a bill.
"Painter" .. .. Per hour.
"Labourer" .. .. "
Materials, as linseed oil, boiled oil, turps, size, &c.

"The above ruling prices are to be applied, with such variations as may be necessary, as the basis of valuation to all measured work as far as possible, but in cases to which no price is found applicable in the above schedule the work shall be priced at (rate) per cent. below the prices contained in 'Laxton's Builder's Price Book' for the year . . . , or for work not therein mentioned at (rate) per cent. above cost price, to be valued and assessed by the surveyor subject to the decision of the architect.".

"TENDER."

'To   Architect,
        (Address)
"Sir

. . . . . . willing to contract for and hereby undertake to execute the various works required in erecting (describe building and its position) according to the plans and specification prepared and to be prepared by you for that purpose to your entire satisfaction, and subject to the various stipulations and conditions hereinbefore written, at the several prices and rates hereinbefore specified.

"As witness . . . hand this . . . day of . . . ."
(Name)
(Address)

On the outside of the schedule, besides the usual endorsement, clauses of the following general tenor:—

"The schedules of the builders will be treated as confidential documents, and those which are not accepted will be returned under cover immediately."

"The plans and drawings showing the character of the building may be inspected at the architect's office."

"Tenders to be delivered at the architect's office before twelve o'clock at noon on (date)."

Another method adopted occasionally for schedules of prices is to prepare the schedule as before explained, the prices being filled in by the surveyor at prices rather above the current rates, so that tenders may all be at a percentage below them, and a clause is inserted thus:—

"The contractor to fill up the form of tender at the end of schedule, and to state thereafter the names of sureties proposed."

"No alteration to be made in the printed prices, the percentage to be general on the whole schedule."

At the end of schedule annex a tender, as follows:—

"To   Architect,
       (Address)
. . . . the undersigned, do hereby tender and undertake to execute, perform, and supply with the best materials and workmanship of their respective kinds, all and every such works, services, matters, and things as are enumerated in the schedules hereto annexed, and in quantities which may be required at (rate) per cent. (above or below, or if the contractor agrees to all the rates as they stand erase the words per cent.) the prices affixed to each item in the said schedule according to the terms and conditions annexed, and . . . hereby agree that . . . will enter into and execute a proper contract for giving effect to this tender, and that (we or I), and also the proposed sureties hereinafter named, will execute such instruments as you may require, founded on this tender and the conditions and schedule hereunto annexed, which form part of this tender, and that the bond to be entered into by (us) the contractor and the said proposed sureties shall be in the sum of   pounds, jointly and severally.

"As witness . . . hand this . . . day of . . .
       (Signature of contractor)
       (Address)

Names and addresses of proposed sureties follow here.

When this latter method is adopted it is obvious that analysis will not be required.

The necessary particulars as to modes of measurement are chiefly connected with masonry. The following examples will afford some guidance as to the requirements in such cases:—

Modes of Measurement prescribed for the new Houses of Parliament, Westminster.

MASON'S WORK.

CUBE STONE.—If square, to be measured the net size when worked; but where the stone is not of a square form, to be measured to the size of a square stone of the least extent required. Where the stones are of scantling lengths of 6 feet or upwards, to be measured separately from the ordinary cube stone.

DRAFTED BACKS.—The backs of the stones where drafted to be measured according to the surface actually shown.

PLAIN AND SUNK BEDS.—One plain bed only to be taken for each stone, except to mullions of windows, for which two beds are to be taken to each stone. Ordinary arch stones to be considered as having one plain bed and one sunk bed.

PLAIN AND SUNK JOINTS.—Not more than one plain joint to be taken for each stone having one or more plain joints. All sunk joints to be taken as they occur.

CHISELLED OR RUBBED FACES.—To be measured to the size actually shown on the external surface.

ROUGH SUNK.—To be taken when a large quantity of stone has to be removed, as in stop mouldings to sills, window heads and other similar work.

SUNK, CHISELLED, OR RUBBED FACES.—To be measured on the surface actually worked, adding the depth of the sinking.

STOPPED SINKING.—To be measured in such situations as do not permit the work to be carried straight through the stone, as in sills of windows and other similar work.

PREPARATORY LABOUR OR PLAIN FACE AS BED.—To be taken wherever it is necessary to produce a face for the purpose of setting out underwork, as in tracery heads and other similar works. This is also intended to apply to mullions of windows, one side and one edge of which are to be taken plain as bed.

SUNK, CHISELLED, OR RUBBED FACE IN SHORT LENGTHS TO HEXAGONAL CANOPIES.—To be measured as they occur, including arrises.

MOULDINGS.—To be girthed, the surface actually shown, the top bed, if weathered, only to be measured as sunk face.

MOULDINGS TO PANELLINGS.—To be girthed, including the backs of the panels.

CIRCULAR FACE TO SOFFIT OF CUSPS.—To be measured the whole thickness of the stone from back to front.

CIRCULAR FACE TO SOFFIT OF CUSPS IN PANELLING.—To be measured from the external face of the stone to the face of the panelling.

SUNK FACES TO TRACERY HEADS OF PANELLING.—To be measured net on the face, adding the depth of the sinking from the external face.

SUNK FACE IN MARGINS FOR EYES.—To be measured the extreme length and width.

CIRCULAR SUNK TO REBATED SOFFIT OF CUSPS.—To be measured from the external surface, adding the depth of the rebate.

MOULDINGS IN TRACERY.—The extreme lengths of the straight mouldings in the tracery of the window-heads to be measured through the mitres and junctions with other mouldings.

THROAT.—To be measured *per foot run*.

GROOVE FOR CEMENT.—To be measured *per foot run*.

GROOVE FOR METAL SASHES.—To be measured *per foot run*.

REBATE NOT EXCEEDING THREE INCHES GIRT.—To be measured *per foot run*.

MITRES TO SINKINGS.—To be numbered according to width.

MITRES AND RETURNS TO SINKINGS.—To be numbered according to the width of the sinking and length of the return.

MITRES TO MOULDINGS.—To be numbered according to the girt of the moulding.

MITRES TO LONG INTERSECTIONS OF CUSPED AND OTHER MOULDINGS.—To be numbered according to the girt of the moulding and length of run.

STOPPED ENDS OF MOULDINGS.—To be numbered according to girt of moulding.

STOPPED ENDS OF MOULDINGS ON SPLAYED SILLS AND SILLS OF PANELS.—To be numbered according to the girt of moulding and extreme length from top of sill to point of intersection.

ROUGH SINKINGS FOR CUSPED WINDOW HEADS AND SIMILAR SINKINGS.—To be numbered, taking the average area of the sinking and the full thickness of the stone.

HOLES PUNCHED.—To be numbered according to their area and depth.

SINKINGS TO FORM SHINGLES.—To be numbered as they occur, according to length, width and depth of sinking.

NOTCHINGS TO FORM EMBRASURES.—To be numbered according to their height, width and depth of sinking.

WATER JOINTS.—To be numbered according to their projection.

MITRES TO SOFFITS OF CUSPS IN TRACERY HEADS OF WINDOWS. —To be numbered according to their length, and taken the full thickness of the stone.

MITRES TO SOFFITS OF CUSPS IN SMALL TRACERY HEADS OF PANELLING.—To be numbered according to their length, measured from external face of stone to back of panelling.

POINTS TO CUSPS IN TRACERY HEADS OF WINDOWS.—To be numbered according to their length, and measured the whole thickness between the sunk faces.

POINTS TO CUSPS IN SMALL TRACERY HEADS OF PANELLING.— To be numbered according to their depth from sunk face to back of panelling.

SUNK AND MOULDED ŒILETS, EACH WITH ONE MITRE AND TWO LONG INTERSECTIONS.—To be numbered according to extreme size.

SMALL SUNK EYES.—To be numbered.

CRAMPS OUT OF SAW PLATE.—To be numbered according to length.

CAST-IRON CRAMPS.—To be numbered according to length and thickness.

PLUGS.—To be numbered according to length and size.

SMALL COPPER JOGGLES AND MORTISES.—To be numbered.

STONE JOGGLES AND MORTISES.—To be numbered.

JOGGLES TO VERTICAL JOINTS WITH PEBBLES IN CEMENT.—To be numbered according to size.

PAVINGS AND LANDINGS.—To be measured *per foot superficial*.

PERFORATIONS TO LANDINGS.—To be numbered according to size and the thickness of the stone.

BRICKLAYER'S WORK.

The brickwork to be measured according to the number of bricks in the thickness of each wall, deducting all openings, except pargeted flues.

CUTTING.—To be allowed for skewbacks of arches and surface cutting, but no cutting to be allowed for the interior of arches, excepting circular groined arches.

POINTING TO SOFFITS OF ARCHES AND LIME WHITING.—To be measured *per foot superficial.*

CEMENT TO BACK OF PARAPETS.—To be measured on the surface *per yard superficial.*

ROUGH SPLAY.—To be measured *per foot run* where it occurs.

GROIN POINTS.—To be measured *per foot run.*

POINTING TO LEAD FLASHINGS.—To be measured *per foot run.*

IRON HOOPING.—To be measured *per yard run* the quantity actually used.

CARPENTER'S WORK.

All framed timbers to be measured cube; the net quantity used in the work, fir or oak, in plates, corbels and lintels, to be measured separately.

BATTENING FOR SLATING.—To be measured on the surface of the roof *per square.*

BOARDING FOR LEAD.—To be measured on the surface *per foot superficial.*

VALLEY AND EAVES, BOARDS, GUTTERS AND BEARERS, AND BOARDING TO SIDES OF GUTTERS.—To be measured *per foot superficial.*

LABOUR TO ROUNDING RIDGES AND LABOUR TO REBATES.—To be measured *per foot run.*

LABOUR TO SPLAYED OR BEVELLED EDGE OR JOIST.—To be measured *per foot run.*

TILTING FILLETS AND ROLLS FOR LEAD.—To be measured *per foot run.*

REBATED DRIPS, ROUNDED HEADS TO ROLLS, SHORT ROUNDED ROLLS AND DOVETAILED CESSPOOLS.—To be numbered.

TIMBER PREPARED IN KYAN'S TANK, INCLUDING CARRIAGE TO AND FROM.—To be measured at *per load.*

FIXING BOLTS, STRAPS, AND CAST-IRON HEADS.—To be numbered.

FIXING CAST-IRON WORK.—To be included in the price of the same by weight.

WROUGHT IRON PROVIDED FOR BOLTS, STRAPS, &c.—To be charged according to the weight actually used.

CENTERING TO BRICK ARCHES.—The quantity to be measured on the soffit of the arch at *per square*.

FLEWING CENTERING TO POINTED APERTURES.—To be measured *per foot superficial*.

CENTERING TO STONE ARCHES.—To be measured on the soffit of the arch at *per square*.

SMITH AND FOUNDER'S WORK.

CAST-IRON WORK.—To be provided, proved and fixed complete, *at per cwt.* including patterns.

WROUGHT-IRON BOLTS, STRAPS AND TIES.—To be provided ready for fixing *at per cwt.*

LINSEED OIL RUBBED INTO GIRDERS.—To be measured at *per yard superficial*.

SLATER'S WORK.

SLATING.—To be measured on the surface at *per square*, allowing 1 foot for eaves, 1 foot for each valley and hip, and 6 inches for cutting to sides of dormers.

PLUMBER'S WORK.

LEAD.—To be provided and laid by weight, which weight is to be ascertained by admeasurement when the work is completed, the weight *per foot superficial* being previously ascertained in the sheet.

SOLDERED ANGLES.—To be measured *per foot run*.

DOTS, LEAD PLUGS, LEAD WEDGES AND SOCKET PIPES TO CESSPOOLS.—To be numbered.

CAST-IRON RAIN PIPES.—To be measured *per foot run*.

CAST-IRON HEADS AND SHOES.—To be numbered.

BEARINGS AND COLLARS.—To be numbered.

EAVES GUTTER.—To be measured *per foot run*.

PAINTER'S WORK.

PAINTING TO CAST-IRON WORK.—To be measured *per yard superficial.*
PAINTING TO STRAPS, BOLTS, ETC.—To be measured *per foot run.*
PAINTING BOLTS AND HEADS TO TRUSSES.—To be numbered.

PREAMBLE OF A MASONRY SCHEDULE OF THE WAR DEPARTMENT.

General Regulations and Specification for the Work.

1. The stone in masonry, steps, landings, sills, coping, flagging, hearths, channels and sinks, is to be placed with the grain or natural quarry-beds horizontal: in arches the grain is to radiate to their centres; in all curbs the grain is to be vertical.

2. In the admeasurement of stonework it is to be understood that those portions only will be paid for as moulded work that have members on them or irregular curves; for instance, in a cornice, as per sketch, Fig. 34, that portion from *a* to *b* will be paid as circular work, and from *b* to *c* as moulded work.

3. Plain work, circular plain work, sunk work, circular sunk work, moulded work, and circular moulded work, are to be paid for according to the following system:—The material being allowed for, at per foot cube, according to Item 1 or 2, and plain work, for faces so wrought, the other workmanship is to be paid for according to the items for sunk work, moulded work, &c., as the case may require. Beds and joints to be paid for as half plain work. The setting to be paid under Items 4 to 7 respectively, and the same quantity will be allowed as given for the stone.

FIG. 34.

4. Mantels and jambs are to be paid for by the measurement taken on the face only.

5. In measuring window sills of other dimensions than those

named in the Schedule, the full cubic quantity of stone is to be allowed; then the work upon them as follows, viz.:—The top superficies is to be taken as plain work, from the back edge to the sinking; from thence to the front edge is to be taken as sunk work; the two ends, the front edge and projecting part of the under side are to be taken as plain work.

6. The prices for steps include plain smooth work to tread, bed and riser; and any further work, such as back joint or rubbed work on face, riser and soffit, or plain work to soffit or back edge, will be paid for in addition. The net quantity only in winders will be allowed, ascertained by measurement of the extreme length by the mean width and thickness.

7. STONEWORK SUNK.—In stonework sunk, except in sawn stone alluded to in Art. 9, the net quantity of material obtained from the

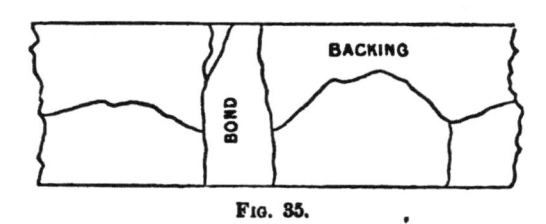

FIG. 35.

measurement from the extreme points, when wrought, as shown in the annexed sketch, Fig. 36, will be allowed according to Items 1 and 2 respectively, and the same quantity for setting under Items 4 to 7 respectively. Half plain work for beds and joints as in Art. 3, sunk work for the top, and plain work for the front edge. The dotted lines in diagram denote the quantity of stone allowed. The net quantity of beds on irregular ashlar stonework, where backed by brick or rubble work, and the net depth of the joints only as shown in Fig. 35, will be allowed.

8. CIRCULAR WORK.—In circular work, the material being measured, the extreme length, by extreme breadth and height (when wrought to the prescribed dimensions) being allowed for, as shown in the annexed sketches, Figs. 37, 38, 39 and 40, and the same quantity for setting; the top surface in each case to be allowed for plain work, when so wrought, the inner and outer faces, Figs. 39 and 40, and the edge of Figs. 37 and 38, as circular work; the beds as half plain work, and the radiated joints, Fig. 39,

SCHEDULES. 413

as rough sunk work, under Item 17. The dotted lines in the diagrams denote the quantity of stone allowed.

9. STONEWORK SAWN.—In stonework, where the stones are sawn one out of the other, the mean dimension only will be allowed. For instance, copings cut the one out of the other, the dimensions

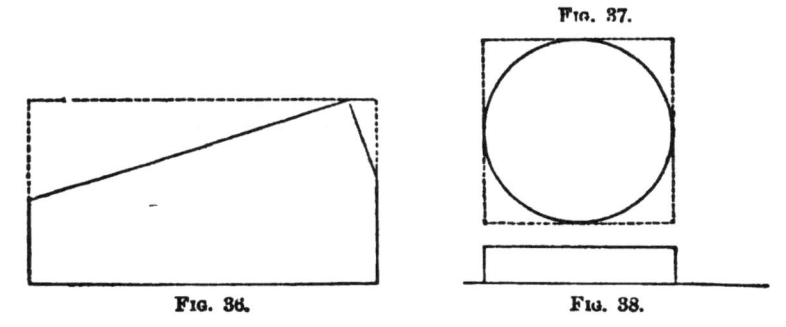

FIG. 37.

FIG. 36.

FIG. 38.

for the height to be taken in the centre, instead of at the extreme height when wrought to the mould, as shown by the annexed diagram, Fig. 41. The same quantity will be allowed for setting.

10. SUNK WORK.—By "sunk work" is to be understood the weathered part of a cornice or window sill, the weathered part of a

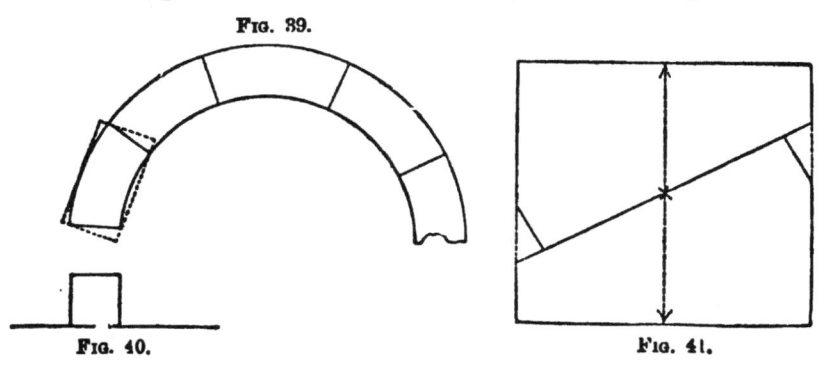

FIG. 39.

FIG. 40.

FIG. 41.

coping, as Figs. 36 and 41, the three faces of a diminished pilaster, and the sinking in a stone sink. In the latter case, the superficies of the bottom, added to that of the sides, will comprise the quantity of sunk work. In the case of perforations in sink stones, they will be paid for under Items 242 and 243.

11. CIRCULAR WORK.—By "circular" work is to be understood such as the concave and convex faces of curbs, as shown in Figs. 37 to 40, and the face of a parallel column.

12. SUNK CIRCULAR WORK.—By "sunk circular" work is to be understood the face of a diminished column, or the hollowed work of a channel which increases or diminishes as it extends, and all work of this character.

13. RUBBED WORK.—To the price of "rubbed work," in Items 19 to 24, will be allowed plain work under Items 9 and 10; and to the price of "rubbed" work on such as the three faces of a diminished pilaster, the weatherings of cornices and copings, and such work, denominated "sunk," the price of sunk work will be allowed; and likewise in rubbed cornices the price of moulded work will be allowed; and in rubbed circular plain, circular sunk, or moulded circular work, will be allowed respectively plain circular, sunk circular and moulded circular work.

14. By the terms "plain" work, "sunk" work, or "moulded" work, is to be understood as follows, viz. on Portland, York, or Scotgate Ash stone a tooled stroke over all the respective faces, with a draughted margin around each stone of ashlar work. On Purbeck, a broached or picked face, as may be ordered, with a draughted margin; on granite and Bramley Fall, a smooth tooled face; and on Bath, Caen, or Painswick, a combed or dragged face.

15. The portions of the General Regulations and Specification which are attached to the Bricklayer's Schedule, and are applicable, are to be considered as a part of this Schedule.

PULLING DOWN.

The pulling down of a building or buildings on the site of a proposed new one may be arranged in various ways. It may be sold in lots as it stands by an auctioneer; it may be tendered for by house-breakers; it may be pulled down by day-work, and various other works arranged by a schedule of prices; or it may be pulled down by day-work, the materials deposited in lots, and sold by auction.

It is often convenient to pull down an old building by schedule of prices, more especially when some parts of the old material are to be used in the new building.

This may be drawn (*a*) on the principle of charging time and material, or (*b*) items of labour and material measured.

(*a*) Requires a clerk of works to keep the time and check the materials. Use of scaffolding may be arranged for as a percentage upon the total spent; or hire of scaffolding arranged for; or the whole of the prices may include scaffolding.

(*b*) The schedule would be drawn like one for ordinary work, but composed of items such as are likely to be necessary in such special work. Some of the work would still of necessity be daywork.

In either case, prices may be agreed for such material as may not be required for the new building, and which will be the contractor's property.

(*a.*)

"*Conditions of Tender and Schedule of Prices on which it is proposed to pull down the building now standing between Nos.* 20 *and* 22 *John Street, Adelphi, for William Thompson, Esq.*"

"JAMES WILLIAMS, Architect,
20 Russell Square, W.C."

"January 1903."

"The whole of the work to be completed within two calendar months of the date of signing the contract, and in the case of default the contractor shall forfeit and pay as liquidated damages the sum of 10*l.* per week for every completed week during which such works shall remain incomplete."

"If at any time the architect is dissatisfied with the progress made with the works, the contractor shall employ more workmen at the architect's discretion, and in the event of his not doing so within forty-eight hours of written notice being given to him by the architect, the architect is hereby empowered to take the work out of the hands of the contractor, and to employ such other contractor as he may think proper, to complete the works."

"Payment will be made at the rate of 80 per cent. on the value of the works executed on the certificate of the architect in sums of not less than £ . . ., until a balance of £ . . . is accumulated, after which time the balance shall be paid in full, and the sum reserved within one month after completion."

"The decision of the architect shall be binding on the contractor and on the building owner in all matters relating to the contract, and as to any question which may arise with reference to the progress and conduct of the work, and as to the payment for the same."

"The contractor will be required to keep a competent foreman in constant attendance at the building as his representative."

"The architect to have the power of dismissing all workmen whom he may deem to have behaved improperly, or to be incompetent."

"Day accounts shall be delivered to the architect or clerk of works at latest during the week following that in which the work may have been done."

"The prices are to include the use of all tools, tool sharpening implements, scaffolding plant and ladders, required for the proper execution of the work."

"The prices for ordinary workmen for day labour shall also include the time expended by the foreman engaged in their superintendence, unless he be engaged exclusively upon the work so charged."

"The work will be valued and priced by Mr. . . . . ., surveyor of . . . . ., on behalf of both parties at the rates of this schedule, subject to a reference to the architect in case of any dispute as to the interpretation of this schedule, or the mode of its application."

"The surveyor's charges of 2½ per cent. and expenses to be divided equally between the building owner and the contractor, and the gross charges to be paid by the latter on each certificate as the work proceeds."

SCHEDULE.

	s. d.
"On all sums paid out of pocket for payments to local authorities and watchman	per cent.
"Use and waste of timber in shoring, exclusive of labour	per foot cube
"Hoarding, including fan, post and rail and footway to the satisfaction of the local authority, use and waste of material, exclusive of labour ..	per foot run
"Horse, cart and man	per day
"Two horses, cart and man	,,
"Carting rubbish or other material and finding a shoot, exclusive of loading	per yard cube

"Use of tarpaulins, including delivery and carting away:—

First Week.		Second Week.		Third Week.		After Third Week.	
Per day.	Per week.	Per day.	Per week.	Per day.	Per week.	Per day.	Per week.

"Excavator .. .. per hour
"Labourer .. .. ,,
"Scaffolder .. .. ,,
"Bricklayer .. .. ,,
"Carpenter .. ..
"Smith .. ..
"*Credits*, including loading and carting away—
 "Sound old slates, per thousand (of 1200)
 "York paving .. .. per 100 ft. supl.
 "Sound old bricks, uncleaned .. .. .. per thousand
 "Sound old fir timber .. .. per foot cube
 "Cast iron in stanchions and columns .. .. per cwt.
 "Wrought iron .. .. ,,
 "Old lead (gross weight), including use of scales
 and weighing .. .. ,,
 And any other materials of value .. .. .. ,,

"TENDER."

"To   Architect,
       (Address)
"Sir
      . . . willing to contract for and hereby undertake to execute the various works required in pulling down the building now standing between Nos. 20 and 22 John Street, Adelphi, to your entire satisfaction, and subject to the various stipulations and conditions hereinbefore written, at the several prices and rates hereinbefore specified.
      "As witness . . . . . hand . . . . . day of . . . . . 1903."

Witness	(Name)
(Name)	(Address)
(Address)	

(b.)

"*Conditions of Tender and Schedule of Prices on which it is proposed to pull down the building now standing between Nos. 20 and 22 John Street, Adelphi, for William Thompson, Esq.*"

"JAMES WILLIAMS, Architect,
"20 Russell Square, W.C."

"January 1903."
   Preliminary as (*a.*)

SCHEDULE.
                                                    s. d.

"On all sums paid out of pocket for payments
to local authorities and watchman .. .. per cent.

QUANTITY SURVEYING.

"Use and waste of timber in shoring to include cost of all wedges, hoop iron, etc., and labour complete, including removal .. .. ., ... per foot cube
"Hoarding, including fan, posts and rails and footway, to the satisfaction of the local authority, and removal .. .. per foot run
"Horse, cart and man .. .. per day
"Two horses, cart and man .. .. "
"Filling rubbish into carts, carting and finding a shoot .. .. per yard cube
"Use of tarpaulins, including delivery, fixing, removal and carting away :—

| First Week. || Second Week. || Third Week. || After Third Week. ||
Per day.	Per week.	Per day.	Per week.	Per day.	Per week.	Per day.	Per week.

"Excavator .. .. per hour
"Labourer .. .. "
"Scaffolder .. .. "
"Bricklayer .. .. "
"Carpenter .. ..
"Smith .. .. "
"Wheeling bricks to any part of site, cleaning and stacking .. .. per thousand
"Breaking old bricks into pieces to pass a 2-inch ring, and stacking in heaps where directed, measured after stacking .. .. per yard cube
"Filling rubbish into barrows, and wheeling ready for loading .. .. per yard cube
"Pulling down brickwork .. .. per rod reduced

"*Credits*, including loading and carting away—
"Sound old slates .. .. { per thousand of 1200
"York paving .. .. per 100 supl. ft.
"Sound old bricks, uncleaned .. .. .. per thousand
"Sound old fir timber .. .. per foot cube
"Cast iron in stanchions and columns .. .. per cwt.
"Wrought iron .. .. "
"Old lead (gross weight), including use of scales and weighing, ... "
"And any other material of value .. .. .. "

Tender as (*a*.)

ADJUSTMENT OF ACCOUNTS.

The adjustment of the accounts relating to a building is usually entrusted to the surveyor who prepared the original quantities, who then acts on behalf of both builder and architect. The alternative is two surveyors, one employed by the Building owner, the other by the Builder. The conditions which will facilitate the surveyor's work in such a case are the following :—

The bill of quantities deposited either with the architect or the quantity surveyor.

The schedule of prices (when there is one) also deposited.

When it has been made one of the conditions of the contract that no extra shall be allowed for unless a written order has been given by the architect, the builder should produce these orders, and where this condition is insisted upon by the architect only those works should be admitted for which orders are produced. When the builder, nevertheless, purposes to make a claim for such works the surveyor may measure them, but he should make a separate account of them, and should stipulate that the charges for their measurement shall be paid by the builder in the event of the claim being disallowed. The surveyor should at the same time inform both builder and architect that he does not by the measurement give any opinion as to the claim further than their quantity and price.

The surveyor, when requested to arrange the accounts, should obtain from the clerk of works, or the architect where there is no clerk of works, and the builder's foreman respectively, a list of works which they believe to be extra on the contract. Both foreman and clerk of works should be instructed to keep notes of extra works as the building proceeds. An appointment should then be made to meet on the works the architect, or his clerk of works, builder (or his surveyor), and the builder's foreman. The lists before mentioned should then be regularly gone through, and such items as are admitted to be extras, and require to be measured, entered in the measuring book as notes for measuring.

Before this meeting, or at some convenient opportunity during the measurement, look carefully through the bills and dimensions to see what omissions there are; see also what provisions require to be adjusted. *Notice* the way in which the original bill has been priced, if the builder has in his pricing disregarded all the general items of labour on stonework, your measurement will deal with the stone as including all labours. Facings to brickwork are also not infrequently priced in a similar manner. (See also section "Law of Quantity Surveying.") A list of these should also be written in the measuring book. As each item is measured draw a vertical line through the note of it, and put a reference against it to the page of the measuring book whereon it has been measured. Thus:—

"Provide for | gas fittings, 50*l*.; measured, p. $\frac{40}{150}$," &c.

When work requires to be kept in several sections it may be advisable to use a separate measuring book for each section.

Leave several pages at the commencement of each of the measuring books for an index, and when the book is filled make an index.

In a large measurement it will be found necessary to keep in a separate book a list of notes of work incomplete, and which it has been impossible to measure, so that the surveyor may return to them when they are ready to be measured.

If the building be a large one and the measurements extend over a considerable period, the surveyor should make tracings of the working drawings and colour with a wash of colour the parts as they are measured. The uncoloured portion will then show him at a glance what remains to be done. This expedient is specially useful when he has to make periodical measurements and reports, as for certificates.

After the space for index write a heading for the book, thus:—

"Measurement of Variations.

"House and offices at Sutton, Surrey, for John Smith, Esq.; Mr. Neal, clerk of works; Mr. Thompson, foreman, with Mr. J. Greene, surveyor.

"24th June, 1878. ——————— Esq., Architect."

Do not measure an omission if it comes to within 5 per cent. of the original estimate, as it is worth little more than the measuring charges; and it should be observed that the measurement of an extra is often pressed by either side, which the omission on the same part of the work much more than counterbalances.

The surveyor, when measuring. should have the dimensions called out at each measurement by the man using the rods or tape, and should look at the figures each time to see that it is called correctly; or he may, if he prefers it, measure the work himself, and have a clerk with him to book the dimensions. Follow the mode of measurement adopted in the original quantities as far as possible, so as to readily apply the prices.

The measurement having given you the additions, a reference to the dimensions will generally give you the omissions; nevertheless it may sometimes be more convenient to remeasure an omission; these may be referred to by letters as before recommended. See "Variations before acceptance of Tender."

For omissions use the word "omit," not "deduct." Where no quantities have been supplied, certain items of omission may be obtained from the builder's original estimate if he will produce it; if he declines, the work must be measured from the drawings and specification.

When a surveyor is employed by the builder to meet the architect's surveyor, both surveyors will book identical dimensions, will have the dimensions squared, and at convenient times after the measurement cause to be compared the respective books, and errors corrected; if not done at this stage errors will be abstracted, and they will be much more troublesome to rectify. Sometimes when surveyors measure in company, it is arranged that only one of them shall book; the trouble of comparing is thus avoided, and the other surveyor can, if he desires it, copy them afterwards. If both surveyors book, they should write identical dimensions and descriptions; and to ensure this one of the surveyors should state aloud the dimensions and descriptions. They should both write them in the same order; and it is convenient for facility of comparison that both dimension books should be precisely similar, page for page.

The measurement of an item such as an ornamental entrance

gate, or a piece of ornamental framing, will sometimes in the book take the following form: No. 1 gate 9′ 6″ by 5′ 0″ containing as follows, the *detail* being measured to arrive at a price. This detail need not be abstracted nor billed, it may be priced in the book and referred to thus: See detail $\frac{105}{44}$.

The surveyor for the architect should then abstract the contents of the books and forward them to the builder's surveyor to check, and this course would be taken whether there were one set of books or two; not infrequently the surveyors arrange that one only of them shall book, and this saves the comparing. The measuring books would of course be available at any time by either surveyor.

When there are omissions and additions on a particular section of the work, the abstractor should abstract the omissions and additions in the order as they occur in the book. He will thus much more readily detect any error that may exist than he will if he first abstracts the additions throughout the book and then recommences the book and abstracts the omissions.

Head each of the abstracts with the usual heading, and also write on each, at the upper left-hand corner, "omissions," or "additions," as the case may be, and the distinguishing numbers of the measuring books, both in red ink.

After the abstracts are checked, the measurement should be brought into bill, the work being either collected in the same manner as in a bill of quantities, or the particular work to each alteration kept separate, but the former way is preferable if there is no special reason for showing the total amount each item of extra work has involved. The latter way greatly increases the length of the account. If it is adopted the items may be usually billed direct without abstracting; any necessary abstracting can be done on the margin of the draft bill in small figures, and can be checked by the clerk who checks the bill, and at the same time. Its advantages are that comparative prices are more likely to be kept in view and discrepant ones less likely to occur.

If, however, the necessity should afterwards arise of showing the omission and addition involved by each alteration, the course first mentioned will subject the surveyor to the trouble of abstracting and billing the account afresh, and probably he will be unable to make any further charge.

Whichever way be adopted, let the omissions come first and the additions after, both being separately carried to the summary. Head each page with the word "omissions," or "additions" in red ink. The order and phraseology of the original bills should be adopted wherever it applies, as this will facilitate the pricing. In the case of omission, no more of the description in the original bill need be adopted than may be necessary to identify an item, and much writing may often be avoided by writing a part of a description and referring to the page of the original bill for the remainder.

Some public bodies will admit no charge for extra work unless it has been the subject of a written order. When this is the case, each order will probably have a distinguishing number. The work involved by each order, whether addition or omission, or both, should form a separate item of the bill, preceded by the number of the order.

In billing omissions the surveyor should keep the original bill before him, as the comparison of its items with those of the original bill will preserve him from the error of omitting too much.

The apposition of the items of omission or addition of a particular section of the work has the advantage of greater safety from error, which will be readily seen if the bill is thus arranged.

When there is a deposited bill of quantities the prices obtained from it are best written in red ink, as it is usually convenient to distinguish them from the prices which have been affixed independently; these latter should be written in pencil. After the rates are attached to the items, the bill should again be forwarded to the builder's surveyor that he may check both the billing and rates of items. Difference of opinion as to rates may be arranged at a meeting of the two surveyors, and the settled prices filled in with black ink as they are agreed on.

It will sometimes happen that in a bill of quantities, the prices of which are to be applied to a bill of variations, a series of items is bracketed together, and a total as 5*l.* placed against it. As the rate of each item of the series may be required in the variation bill, the surveyor will price each item at a current rate, then if the total of these tentative rates comes to more or less than 5*l*., they must be decreased or increased by a regular percentage or proportion. Thus if the pricing produces a total of 7*l*. 10*s*. the price of each item must be reduced by one-third.

It is argued by some surveyors that the builder should attach the prices to the bill [of variations and the architect's surveyor check them; their idea is that the builder may thus ask smaller rates, on some items, than the architect's surveyor might attach, and an advantage be thus gained by the client. If the architect's surveyor is incapable, or wishes to gain an advantage, this is the better course for him to adopt; but if he knows his business, and is desirous only to do justice, the course first suggested is preferable. Moreover the careful examination of a priced bill or schedule in order to attach the prices to the items of measured work is a good preparation for future argument upon them.

In cases of disputed prices the surveyor may be able to persuade the builder to produce his prime cost book, and if he is satisfied that there is no error in the statement he may, after adding a reasonable profit, adopt the price thus arrived at.

The items which produce the ostensible "prime cost" total should be carefully examined. Sometimes a ridiculously large percentage has been added for establishment charges.

If the method of pricing extra works has been agreed on, as for instance by the adoption of the rates of the original contract, nice distinctions of quantity and position may be excluded as it is reasonable to conclude that the contractor has informed himself of the character of the work before pricing his estimate.

When the rates have been agreed upon, the item should be carried into the money column, then checked and ticked in pencil, then checked again and the amounts written in ink. The second checker invariably finds errors, but this second checking is rarely done.

There are a few common precautions which should be observed in pricing and checking, write $\frac{1}{4}$ᵈ $\frac{1}{2}$ᵈ $\frac{3}{4}$ᵈ not $1/_4$—$1/_2$—$3/_4$, these are frequently mistaken for $1/_4$ˢᵈ—$1/_2$ˢᵈ—$3/_4$ˢᵈ.

See that every item is carried into the money column, both in items and casts. The clerk checking the casts should place a small piece of paper over the results, so that his calculations may be independent of the former man's work.

The tendency of increased experience is usually to incline the surveyor to more liberal treatment of the builder, it is well, however, to remember that his liberality is at his employer's expense.

The day accounts of extra work should be delivered by the builder within a week of the work being executed, and if the work is to be concealed as in foundations or other work not easily accessible, it should be examined and measured before it is covered.

Day accounts should only be allowed for works in which labour, and that not straightforward, is the principal item: as attendances upon other workmen, alterations of works already executed, pulling down, shoring, underpinning, etc. Work of the usual character, and which has entailed no special trouble, should not be charged day work but measured.

If a clerk of works is employed, he should examine and sign each day account before it is delivered to the architect or quantity surveyor. Where the work is extensive, it is the better practice for the quantity surveyor to take charge of the day accounts; he should carefully examine them as they are delivered, and when he discovers errors, return the day account at once to the builder for explanation or alteration.

When the materials and labour in a day account do not amount to more than the proper price for the piece of work measured as fixed, it may be accepted; it would be mere waste of time to measure it; but this is seldom the case; it will often be found that the charge for time alone amounts to more than the legitimate charge for both labour and materials.

The surveyor should be watchful of day accounts, and will find his trouble well repaid. The surveyor may also succeed in separating things which can be, from those which cannot be, measured.

In day accounts for pulling down, see that credit is given for the bricks and other materials.

Material comprised in day accounts is frequently charged by the builder as finished work instead of simple material, and the time is also charged for fixing. In the case of joinery in day account it should be kept separate, and described as "from Bench," and is worth a less price than similar work in the priced bills, as the latter includes fixing; "stone from Banker" is a similar case.

In measuring complex joinery, sections of the mouldings, frames, etc., may often be obtained, and will be found useful when valuing the items.

In measuring work from Bench or Banker, observe to omit such parts of the labour as are of necessity done in the course of fixing: as plugging, mitres, mortises, ends cut and pinned, etc.

When work has been altered it will be best to measure the work as originally made or fixed, allowing the alteration as a day account.

Foreman's time is not infrequently the subject of dispute. When there is a stipulation in the schedule of prices that the general foreman's time should be included in the price per hour for general workmen, or some other similar condition, the question is settled. Where no condition exists, and where work has been done requiring the foreman's exclusive superintendence, as in the case of small jobbing, executed when all but the men occupied upon that work have been discharged, or in special works requiring an undue proportion of his time, a fair amount of foreman's time should be allowed.

On examination of the preliminary bill the surveyor will discover some items which are priced, possibly at a sum altogether disproportionate to the work involved. If these items are "allows," i.e. at contractor's risk, they should not be deducted, for if they cost contractor many times as much as the sum he has in his estimate, he would be bound to do the work without extra charge, and by parity of reasoning, the items should not be interfered with if they should prove unexpectedly profitable; moreover, the contractor might insist upon the readjustment of other clauses of the same character if *any* were interfered with.

The adjustment of provisional sums is also a fruitful source of difference. The method of treatment will depend upon the way the provisions have been arranged in the original bill. In all cases the surveyor should insist upon the production of the original invoices, not copies, and if he is doubtful of these invoices the merchant's monthly statements and receipts.

Taking as an instance, a provision of 100*l.* for stoves under its various forms, its adjustment should be as follows :—The invoice is, say, 50*l*., the trade discount off stoves is 20 per cent., a fair profit to allow is 15 per cent.

When no special clause exists as to the deduction of provisional sums net, and the provision appears in the priced bill as—

| A. Provide for stoves 100*l*. .. .. .. | | 100 | 0 | 0 |

the setting will be charged as a day account unless the setting has been taken in the quantities. The 100*l*. will be deducted. 20 per cent. will be deducted from the amount of invoice, and 15 per cent. added; the amount will be £46.

When the item appears as below, contractor has fixed his profit at 5 per cent. :—

| B. Provide for stoves 100*l*. .. .. .. | | 105 | 0 | 0 |

deduct the 105*l*. and allow 50*l*., minus 20 per cent. plus 5 per cent. profit = £42.

Sometimes the item will appear as follows :—

| Provide for stoves 100*l*. and add profit .. | | 105 | 0 | 0 |

Treat this in the same manner as B.

When there is a condition for the deduction of provisional amounts net—

Deduct the amount stated in the specification without the profit, and add the amount of invoice less the trade discount—result £40.

The foregoing instances presuppose the expenditure of an amount for the purposes originally intended. There are, however, cases in which the whole amount is withdrawn from the contract and is spent directly by the architect or building owner, and in such case only the amount stated in the specification should be deducted from the contract sum, and the profit should not be interfered with.

Sometimes the builder fails to extend the amount of a provision in his estimate; the amount stated in the specification should nevertheless be deducted, and the amount actually expended allowed in the accounts.

Sometimes a contractor will take off a discount from a provided sum in a bill of quantities before extending it; the sum provided should nevertheless be deducted in the accounts.

P.C. means the amount paid by contractor after deducting trade allowance but not ordinary discount for cash, but disputes may be saved if a clause to this effect is put into the preliminary bill.

Within the metropolitan area many kinds of goods are delivered free of charge, but if carriage should be claimed, see the delivery note or some voucher for the payment, and allow it in addition to the foregoing allowances.

In some instances an exorbitant discount has been added to produce the amount of invoice as the result of collusion between builder and manufacturer. Nothing will meet such cases as these but great knowledge of prices, and the precaution of dealing only with firms of proved honesty.

A claim is sometimes made for profit on work omitted. As, if an original tender amounts to 10,000l., the omitted work 1000l., and the extra works 500l., the net profit which would have accrued upon 500l., the difference between extras and omissions, should properly be allowed; but observe to allow only net profit, 5 per cent. will generally be an ample allowance for such a claim, and also remember that in some cases it will be found that the omission of the particular part of the work has been an advantage to the builder rather than a loss. The surveyor may, however, if he prefer it, insert the following clause in the preliminary bill, "no allowance will be made for the loss of profit on omitted work."

If the contractor should assert that the building is larger, it must be carefully measured, as the allegation (made often enough in good faith) is sometimes a mistaken one, and it must be observed that the difference of value should not be settled by cubing.

Unless a special clause to meet such contingencies has been included in the schedule, claims may be made for water, tackle, hire of scaffolding and plant; tool sharpening, etc., and such claims must be allowed, but with careful consideration of the amounts.

Payment for lodgings of workmen is usually allowed if the charge is really incurred by the builder. Surveyor should see vouchers, if any exist. If the rate of wages has formed part of a schedule, lodging will not be allowed.

If works are stopped by the fault of the architect, or the building owner, for an unreasonable time, the contractor will have a just claim for hire of plant and wages of watchman.

In the course of practice exorbitant claims will be met with, usually founded on day accounts. If the surveyor is certain of the injustice of these, two arguments are usually available:—

Non-delivery of day accounts. The time prescribed for completion has been exceeded. The repudiation of the day accounts or the claim for liquidated damages will, as a rule, modify the demand.

The first of these pleas should, however, only be urged in cases of flagrant attempts at extortion.

In cases of claims for liquidated damages, the period claimed for may be reduced by wet days, and by delays caused by architect, as failure to furnish details or to give instructions when applied for, or by time occupied by extra works.

If the endeavours of the surveyor are after all insufficient to produce a settlement, the only alternative is a reference. As a preliminary, settle every point that can be settled and carefully define the points to be submitted by agreeing upon a list of them with the opposing surveyor; and, if possible, pledge the builder to a promise that he will make no claim beyond those agreed upon before the submission.

When the builder has a weak case, and is conscious that he has no useful witness, he may possibly propose to call no witnesses if the surveyor promise to call none. Such promise should, however, be carefully avoided.

The day work should be brought into bill either in one total or several as suggested for the measured work, which latter it should follow in either case.

If the surveyor who has measured the work can also abstract it, there will be far less liability to error.

He should keep the original bill before him when billing "Omissions," as he will sometimes find that the quantity of a particular item exceeds that in the original bill.

The checking of sub-contractors' accounts is generally involved in the settlement of a building account; when this involves a measurement it saves trouble to do it in the company of the sub-contractor or his agent, otherwise it may have to be done a second time.

In abstracting day accounts, the shortening of the bill as much as possible should be a leading motive; sacks, bushels, hods, etc., of cement being abstracted under their various headings, may all be afterwards reduced to bushels and so billed; the various measures of lime, sand and mortar, may be similarly treated; rough fir may be reduced to the foot cube before abstracting; deal of various thicknesses to the foot superficial.

Some builders' clerks strongly object to this concentration; doubtless it produces a smaller total.

Collect the time for each trade: as excavators, bricklayers, etc. with their labourers following them.

A.—Form of a Bill of Variations.

Variations on Contract.
House and offices at Sutton, Surrey, for John Smith, Esq.
———— Esq., Architect,
24 Montagu Square,
London, W.C

November 1878.

Omissions.

Begin with a preamble stating the leading items of the work.
The items trade by trade in the order of the original bill.
Carried to summary .. .. £ _____

Additions.

The items trade by trade in the order of the original bill.
Carried to summary .. .. £ _____

NOTE.—The trades will not be separately carried to summary, the omissions and additions respectively will form one total, and a heading to each trade is unnecessary, a small space between them is sufficient.

ACCOUNTS.

Day Account.

£

Begin with a synopsis something like the following :—
  "Cutting away and shoring to front of house, inserting iron girders, filling up openings of flat, building new bay window, removing old floor, &c."
" Bricklayer   150 hours " .. ..
" Ditto labourer 200   „   "   .. ..
" Mason   180   „   "   .. ..
" Ditto labourer 220   „   "   .. ..
  Follow in the same manner with the time of the several trades.
  Materials in the order of the trades in a manner similar to following :—
" 50 loads of rubbish carted away " .. ..
" 20 yards of sand " .. ..
" 20 bushels of Portland cement " .. ..
" 3 yards of stone lime " .. ..
" 3 loads of mortar " .. ..
" 3000 stocks " .. ..
" 50 ft. supl. 3-in. tooled York paving " .. ..
" 50 ft. cube fir," &c. .. ..
                              Carried to summary .. .. £ _____

If it should be necessary to show the amount of variation on any particular section of the work, the same order as the foregoing will be observed in each section, but the form will be similar to the following example :—

B.—WATER SUPPLY.

Omissions.

The measured work item by item in the usual order of the trades.

                              Carried to summary .. .. £ _____

Additions.

The measured work item by item in the usual order of the trades.

                              Carried to summary .. .. £ _____

Day Account.

Time and materials in the order before described.

                              Carried to summary .. .. £ _____

Summary (for A).

	£	s.	d.
Amount of contract	5300	0	0
Additions	1000	0	0
Day account	500	0	0
	£6800	0	0
Deduct omissions	500	0	0
	£6300	0	0
Add surveyor's charges, copies, and expenses	40	0	0
	£6340	0	0
Cash paid to builder on account	5100	0	0
Amount due to builder	£1240	0	0

Summary (for B).

	Omissions £ s. d.	Additions. £ s. d.
New bay window	—	100 0 0
Day account	—	21 0 0
Billiard room	—	540 0 0
Day account	—	70 0 0
Water supply	18 0 0	90 0 0
Day account	—	10 0 0
Laundry	94 0 0	120 0 0
Day account	—	20 0 0
	£97 0 0	
Deduct omissions		112 0 0
		£859 0 0
Surveyor's charges, copies and expenses		40 0 0
Total net additions		£899 0 0
Amount of tender		5000 0 0
		£5899 0 0
Cash paid to builder on account		5000 0 0
Amount due to builder		£899 0 0

In preparing the final statement for the client, the surveyor will be guided by the amount of detail that may be required, taking the summary A as a basis

ACCOUNTS.

House and offices at Sutton, Surrey, for John Smith, Esq.
November 1898.

Statement of Account.

	£	s.	d.	£	s.	d.
Amount of contract	5300	0	0			
Additions (stating of what they consist)	1540	0	0			
	£6840	0	0			
Omissions (stating of what they consist)	500	0	0			
	£6340	0	0—	6340	0	0
Cash paid to builder	5100	0	0			
Amount due to builder	£1240	0	0			

Amounts separately Certified for.

	£	s.	d.	£	s.	d.
Smith, for stoves	90	0	0			
Minton, for tile paving	50	0	0			
Brown, for gas fittings	100	0	0—	240	0	0
Total cost				£6580	0	0

Disputed Accounts.—The foregoing directions as to the adjustment of variations, presupposes the adoption of the ordinary course of practice, i.e. their measurement by the architect's surveyor in the company of the builder or his nominee, but when the architect adopts the undesirable way of asking the builder to deliver an account, it is often sent to the surveyor to check; in most cases this involves the measurement of many of the items to arrive at a conclusion. If the builder's attention can be called to palpable errors of measurement or overcharges, and he can be induced by the architect's surveyor to measure them in his company, one of the points of difference can be settled, and the only other consideration will be the prices, but the builder will often maintain that the account is correct and refuse to measure.

The next step is for the architect's surveyor to prepare an account on complete sheets of bill paper, the builder's account copied on one page and the surveyor's account on the opposite page. The difference on each item of claim is then easily seen and argued; sometimes the builder will not argue, but if he will not the same form is useful for presentation to a judge, or an arbitrator.

The surveyor should take special care that copies of all such

VICARAGE, CAVERSWALL, STAFFORDSHIRE
(BUILDER'S ACCOUNT).

Item 4.			Continued		£	140	10	6
yds.	ft.	in.	*Enclosure to Stairs.*					
48	6		2" moulded and square framing	supl.	1/6	3	12	9
10	—	—	2" moulded and square 2-panelled door	"	1/7	0	15	10
9	—	—	Labour, rebated grooved and staff beaded angle to 2"	run	—/4	0	3	0
24	—	—	Labour, rebated edge to 2"	"	—/2	0	4	0
	No.	1	Pairs 4" wrought-iron butts		1/6	0	1	6
	"	1	6" iron rim lock and brass furniture		7/6	0	7	6
12½	—	—	Paint 5 oils	supl.	1/8	1	0	10
						6	5	5

Item 5.			*General Brickwork.*					
rods.	ft.	in.						
18	68	—	Reduced brickwork in mortar	supl.	13/—/—	237	5	0
1½	—	—	Ditto extra only in cement	"	3/10/—	5	5	0
						242	10	0

VICARAGE, CAVERSWALL, STAFFORDSHIRE
(SURVEYOR'S STATEMENT).

Item 4.			Continued		£	120	9	8	Remarks.
yds.	ft.	in.	*Enclosure to Stairs.*						
36	—	—	1⅜" moulded and square framing	supl.	—/11	1	13	0	Only 1⅜" thick and very common deal.
10	6	—	1⅜" two-panel moulded and square door	"	1/—	0	10	6	
24	—	—	Labour, rebated edge to 1⅜"	run	—/1	0	2	0	
	No.	1	Pair 3" cast-iron butts		—/10	0	0	10	
	"	1	6" iron rim lock and brass furniture		5/—	0	5	0	
9⅓	—	—	Knot, prime, stop and 3 oils	supl.	1/—	0	9	4	Only 4 oils.
						3	0	8	

Item 5.			*General Brickwork.*						
rods.	ft.	in.							
18	68	—	Reduced brickwork in mortar	supl.	12/—/—	228	2	6	Bricks of inferior quality.
1½	—	—	Ditto extra only in cement	"	2/10/—	3	15	0	Cement inferior and deficient in quantity.
						231	17	6	

documents be carefully examined for clerical errors. Copies of such papers produced in solicitors' offices bristle with mistakes, and nothing can be more damaging to a case.

For facility of reference, number each item of the surveyor's statement and the builder's account with a corresponding number.

In the criticism of disputed items, the exact quantity had better be presented, as distinguished from the ordinary practice of billing, recommended in Section "Billing," i.e. the calling an item over 6 inches a foot, etc. An example of the kind of statement suggested is shown opposite.

A carefully prepared précis of the correspondence is very useful in a case of disputed account.

When a builder is ejected from an unfinished building, or when he abandons a work before it is finished, it will be necessary to measure the work undone and bring it into bill.

If the measurement can be made in the presence of the builder's nominee, much future trouble will be avoided.

Copies of this bill can be issued to various builders who will be invited to tender for the completion and the lowest tender accepted. This amount, plus any further claim for damages, should be deducted from the amount of the original contract, and the difference will be the amount payable to the first contractor. If the balance in hand of the contract sum is insufficient to pay the new contractor for the completion, plus the damages, the amount of such deficit would be recoverable from the first contractor.

The surveyor may be reminded of the important consideration that he is employed to save the architect trouble; consequently, appeals to him should, if possible, be avoided. The surveyor should endeavour to settle all issues with the builder before reporting the amount of a variation account to the architect. With this view he should, where possible, adopt reasonable compromises of the opposing claims, especially in matters of small amount. The adherence of the surveyor to his own conclusions will often produce litigation in which the expense to the building owner will, although the surveyor may establish his contention, possibly be as much as the amount in question—moreover giving the money to the builder is preferable to spending it in law. In most cases the architect will be much better pleased to escape disputes, and will probably discard a surveyor who frequently involves him in them.

CHAPTER VII.

SPECIFICATIONS.

As it is in many instances the practice to depute the writing of the specification to the surveyor, a few considerations relating to specifications may be of service.

It is always an advantage to have the specification written by the man who takes off the quantities, he checks his own work thereby, and it is better completed before the quantities are lithographed, so that any errors therein may be corrected. An addenda to quantities is generally the result of mismanagement or hurry, and unfavourably impresses both architect and builder.

Do not use the same word in more than one sense, and preserve a similar order in descriptions of similar things. The specification of a general principle will sometimes shorten specific directions, as for instance, "all timbers not exceeding 27 sectional inches shall have templates 9" × 9" × 3" except, &c."

The surveyor should remember, in writing his specifications, that it will save much future reference if he will embody *every* particular upon which the drawings and quantities give no information, otherwise he will be troubled with frequent questions as to the way he has measured the work.

Many of these points are suggested in the section "Taking off."

Be careful to state the scantling of everything of which it is necessary for the builder to know the size.

Conditions of contract are usually attached to the specification, but not always.

Various forms are given in 'Emden's Building Contracts,' and 'Hudson's Building Contracts,' with analyses and criticisms.

The revised conditions recently issued by the Royal Institute of British Architects are a result of much discussion, and the general principles upon which they are founded can hardly be improved upon.

The most convenient way of writing a specification from

dimensions is as follows:—Write the whole of the conditions, then the preliminary works and such preliminary clauses as cannot be properly referred to any particular trade, then a list of the provisions both of money, labour and material, and particularly observe to define the question of profit on provisions. The provisions are best kept together immediately after the general preliminary items, not distributed through the trades.

Then the preliminary of each trade. A part of this can be obtained from the bill of the trade, and will consist of a description of materials. Follow with any directions which apply generally, as "all iron butts to be wrought," "all window and door frames to be grooved where required to receive linings and finishings," "all floors to have glued and mitred borders to hearths," &c. It will be found that attention to this principle will save labour.

After preparing the first part of the description of each trade, the dimensions may be referred to for the rest, beginning at the first column of dimensions and going regularly through them, marking through the dimensions with a vertical line in pencil as you deal with each. The abstracts and bills will also be of considerable service in the writing of a specification, as they will present at a glance the various works requiring description. Care must be taken to make the premables to the trades in the specification agree with those in the bill.

Begin the description with either the top or bottom part of the building (the latter is preferable as being more convenient for the contractor), and maintain the principle throughout each trade, continuing the description upwards or downwards, as the case may be.

In the case of all buildings, except very small ones, it will be the best way to describe the joiner's work floor by floor, the attempt to describe several floors together usually results in confusion.

When a large quantity of work is distributed in different parts of a building, the surveyor will usually find it most convenient to commence the clause which refers to it with the description of the material, and follow with the places in which it occurs, he can then fill them in as he finds them in the dimensions, thus:— "Glaze with 21-ounce sheet glass," "the windows and fanlights of back front, the basement windows, the windows adjoining lighting court," &c.

The following list comprises a number of items which, although necessary to a specification, have not unfrequently been lost sight of:—

Excavator.—If there are various thicknesses of concrete, not shown on drawings, state thickness and position, and state the width to respective walls.

If a part of the earth is to be carted away, and part deposited, clearly define the proportion or quantity.

State the various depths of excavation, if not clearly shown.

Carefully describe the course of drains, or supply a plan of them as taken.

Bricklayer.—If the footings are not clearly shown, describe them. If in the metropolitan area they may be described as in accordance with the Building Act, if they are so; if not, they may be described with reference to the various thicknesses of the walls, as all $4\tfrac{1}{2}$-inch walls to have two courses, 9-inch walls three courses, &c.

Also in respect of plinths and sleepers, whether the footings of walls are widened because of them or not.

Describe in detail which parts of the work are to be built in cement. Some architects colour with a special colour, on the plan, all work intended to be built in cement. State the depths of relieving arches for various spans, &c.

Numbering each smoke and ventilating flue, or marking the latter V on the plan, sometimes makes the specification clearer.

Mason.—The average sizes of various stones on bed where their size is concealed. The average sizes on beds and average heights of quoins to angles and jambs, or a tracing may be used on which to mark the beds. See section " Taking off."

The principle upon which joggles, cramps, or dowels are to be used. Where the beds of stone are to be sunk, as sometimes in spires. The finish of the backs of stones where built into walls. The distance any stone is to be pinned into wall. The position and size of every template.

The principle upon which the length of steps and the length and width of hearths shall be regulated.

Carpenter.—Extent to which timbers lie on walls. The distance apart of wood bricks. If any partitions are trussed, describe the sizes of the various timbers and the bolts. A sketch of a trussed partition in single lines with the scantlings marked on it is sometimes useful. State the length of all straps and bolts. The distance apart of bolts to flitches. Describe the ironwork with the timber it secures.

Scantlings of all the timbers, and if of finished sizes.

Joiner.—Describe to what work grounds have been taken. State the finished sizes of all doors, i.e. width and height; and it is well, if the sizes of openings are figured in the drawings, to state whether they are the sizes of the doors or of the openings.

Describe all the thicknesses, and whether finished sizes.

A reasonable order for each story is floors, skirtings and dados, windows, doors, fittings, as cupboards, w.c. and bath fittings, &c.

Founder and Smith.—Describe how the rain-water pipes have been arranged, frequently a subject of question, or suggest that they be indicated on the drawings by the architect. Describe the positions, lengths and weights per foot run of all the iron joists. These are conveniently arranged as a table.

Plumber.—Describe fall and minimum width of gutter, height of turning up of lead, width of flashings and aprons and where they are intended to be, position of cisterns, and state what apparatus each cistern supplies, the lengths of the various socket pipes and the lead they are to be made of, the weight per yard lineal of lead pipes. Describing cisterns by letters A, B, C, &c., will be of assistance: thus, "from cistern A with $\frac{3}{4}$-inch brass connector, fly nut and union, $\frac{1}{2}$-inch lead pipe and $\frac{3}{4}$-inch bib cock, lay on the water to sink in scullery."

Gasfitter.—Describe position of gas meter. Describe positions of gas fittings, or cause them to be indicated on drawings. State size of the various pipes.

Glazier.—Thickness of rough plate glass. Weight of all sheet glass. Distance apart of saddle bars to lead lights. The position of iron casements or of casements intended to open may be shown thus × on the drawings.

And various other particulars which will present themselves to the mind of any observant person.

The foregoing list, though of course incomplete, indicates the general principle which it is desirable to follow.

It is well for the surveyor to keep in mind the fact that care in the description of the position of everything in the dimensions will very much assist the person who writes the specification.

When the surveyor has simply to correct the specification furnished by the architect, he should so amend and add to it that it shall require no extraneous explanation, except such as may be obtained from the drawings, and it should agree with the quantities.

The extent to which this correction should be carried requires judgment, as some architects resent interference with a document written by them, and in such cases the surveyor should alter as little as possible, and should consult the architect before doing anything.

Avoid such clauses as these in the same specification: "all walls not described to be plastered to be twice limewhited." See "Plasterer."

"Render, float and set all walls not described to be limewhited." See "Bricklayer."

These it will be seen do not assist each other.

Numbering each clause in conjunction with an index is convenient, especially when the specification is a long one.

CHAPTER VIII.

PRICES.

THE following pages deal rather with the *principles* of the pricing of building work than with absolute rates; these are constantly varying with the market, and with the position of the building to which they are to be applied, but as a particular item of prices must deal with material as one of its factors, the current cost of the material has been adopted for the item illustrated.

Notes of building construction are to be found in the well-known books; tables of weights and manufacturers' prices are to be found in the price books and the trade lists. Their repetition here has been as far as possible avoided, and only inserted to illustrate examples of valuation.

There is no part of the surveyor's work which requires so much ability and judgment as the assessment of prices; not only should the surveyor know the value of materials, but also the amount of time which should be spent in the production of the work; his experience should also enable him to modify his conclusion by the quality of its finish, and he should constantly bear in mind the fact that the difference between the quantity of work that a man can do and that which he does is usually great.

The foundation of all pricing must be the value of the labour and materials. The fluctuations of value of the latter from various causes are considerable, and the surveyor should watch and be familiar with all the changes of the market. Within the last ten years the cost of building has increased at least 25 per cent.

The surveyor may also keep note-books for each trade, arranged like the published builder's price-books; in them he should enter every price that is likely to be useful in future, notes of cost of working various materials, relative values, and time and material expended upon certain items of work.

In every case state the source from which the information is derived, and the date.

When a bill of quantities with the prices of the estimate attached has been deposited by the builder, the pricing of any extra work is much simplified, as where there is no price which will directly apply, proportionate prices can be pretty easily arrived at.

If the surveyor adopts the invariable principle of making a condition that the priced bill of quantities shall be deposited by the contractor, he will soon be in possession of much information as to prices which will be of great value in treating prices for other work under similar conditions.

The young surveyor will find it an advantage to know the cost and market price of the leading item of every trade, as a rod of brickwork, a cubic foot of each of the common stones, a cubic foot of fir, pitch-pine, oak, teak, etc., a foot superficial of 1-inch deal, mahogany and wainscot. He should be acquainted with the rough approximations to the truth which are current as to the relative values of straight and circular work, and of deal, wainscot and mahogany joinery. These will help him to analogous prices pretty readily, and readiness is essential.

Judicious pricing is the latest acquired faculty of the surveyor, and the severest test of his capability; perhaps it is his most difficult task to deduce analogies in the application to a measured account of the prices of a schedule prepared by an ignorant man. Whatever may be the system of measurement adopted, the principles of pricing are everywhere similar, and this may be a reasonable excuse for the adoption of many London illustrations. There is always an advantage in learning *one* system thoroughly, if coherent and complete, even though it may be in some of its details imperfect. In it the student possesses a definite starting-point—a base for the development of a system in which his individuality will find more ample scope. The constants of labour are derived from various sources, and if the student adopts the advice here given he will be able, as his knowledge increases, to make many for himself. Their value as a condensed register of experience is great, and the principle of their application deserving of more attention than it has hitherto commanded. The instances given hereafter of sums for water, attendances, etc., are derived from genuine estimates, and are the work of practised estimators.

That the study of the architect, the building surveyor, and the builder's clerk is in many respects similar, excuses the inclusion of some information which would seem the exclusive concern of the latter.

Builders' Price-Books.—The natural resort of the inexperienced in pricing is a builders' price-book, and while one must admit the value of the great store of information they contain, it is beyond question that they abound in difficulties for the neophyte: the rates of profit vary on different things, some of the prices are list prices, some with a large trade discount, some with a small one. In some of the books this trade discount has been modified; in some it has not. Moreover, no explanation is ever given in a price-book of the way a particular price is arrived at. Modifications to meet a special case are consequently impossible; the difficulty is increased by such artless statements as "8*d.* to 1*s.* 3*d.*" If the student buys all the current price-books, his last state is worse than the first, for he finds none of them alike, and even their diversities are not uniform in degree. A set of builders' prices published in the *Building News* (by George Stephenson) a few years ago, was commendable for the adoption of a uniform rate of profit of 10 per cent. A builders' price-book to be truly valuable should show the absolute cost of all materials and builder's work; but it should be remembered that no set of prices can be applied to any two buildings without some changes. Every work will possess some special features which will modify its value. There are various causes which contribute to increase the difficulty of framing a set of prices of universal application. Inefficient supervision by the contractor often enough makes the difference between a profit and a loss on a building; besides this, foolish and harassing interferences by the architect or his clerk of works, delay in furnishing details, and general indecision of superintendence, are all elements of value which no price-book can measure; but a builder will modify his profits in various degrees dependent upon the peculiarities of the circumstances, such as the locality, or the known idiosyncrasies of the architect.

Good and Bad Work.—The difference in value of good and bad workmanship is sometimes alleged, but often unreasonably, as reason for modification of price. Good work and high finish is mainly the result of a good system of business management, and a proper selection of men; it does not necessarily involve a greater outlay, and generally would be represented by a very small percentage of the cost. The traditions of particular businesses are so confirmed that some firms would find it as difficult to do good work as others to do bad.

Trade Discounts.—So much has been said or written about trade discounts that little need be said here on that subject, except that the student must learn all about them and take them, into consideration in the assessment of his prices. One may say in passing, that wholesale houses not connected with the building trades, with few exceptions, content themselves with giving a discount of 2½ per cent. for cash; the houses supplying building requisites vary in their practice, giving a discount of 10 to 50 or more per cent. in addition to the discount for cash.

It is a curious custom with some firms to give 20 per cent. and 20 per cent. and 5 per cent.; some of these expedients are adopted to confuse the distinction between trade discount and discount for cash to the advantage of the builder in the adjustment of prime costs; but the surveyor in the process of settling accounts, may reasonably reject some of these transparent fictions, and as a rule adopts the principle that anything beyond 2½ per cent. is *trade discount.*

A stipulation in contract conditions that the builder shall produce invoices of materials which are affected by provisional sums, will often show the true price paid, but not always; the receipted monthly statements of the merchant are often better evidence.

Unquestionably the astute builder often gets two profits by the manipulation of invoices.

Use of Priced Bills and Pricing.—Another resource of the surveyor is a collection of old priced bills of quantities; but these do not help him much. Comparison of the sets of quantities for the same building priced by different men will show a diversity for which it is sometimes very difficult to account. The occasion for pricing presents itself in various ways; the preparation of an estimate from a bill of quantities or from drawings and specification; the items of a measured account or of a day account; the position of the man who prices is also varied—he is the surveyor appointed by the architect, in which case he is popularly supposed to administer justice between the building owner and the builder, but is not unfrequently special pleader for the former, or he is estimating clerk to a contractor or a surveyor employed by the builder. In either case pressure is often exerted to induce him to raise every plea in his employer's favour which ingenuity can devise. How far this may be legitimately carried is a question of the morality of advocacy upon which we need not here enter. In the pricing of an estimate, the experienced man has in his mind a store of approximations to the truth which he applies on all occasions. If it were not so, it would be necessary to calculate every price in detail, and it would not be possible in a large business to do the large quantity of work which estimating involves. Preparation for such work is another matter; the student must arrive at every price by careful calculation in detail, and he should beside this be always on the watch for parallel cases and generalisations. When pricing a bill, it will be advisable to write neatly on clean sheets of paper (not rough scraps which are liable to loss or destruction), the calculations used of cost of concrete, brickwork, timber, &c., and the rate of profit, not only as a check upon other calculations which they may affect in the same bill, but for future reference, and if necessity should arise, for the vindication of the judgment. To those familiar with office work it will be unnecessary to speak of the importance of dates. Every document should have written legibly on it the date of its production.

Means of Preserving and Arranging Information.

The general neglect by the student of surveying of the obvious methods of acquiring a thorough knowledge of prices (the most important part of his work) is patent to all experienced persons. A surveyor whose capabilities are only equal to the taking out of quantities is a mere clerk. Unfortunately, many building surveyors know little else. As the examinations of the Surveyors' Institute increase in their stringency, prices will probably take a more important place in them, to the advantage of all concerned. Doubtless, the expert in prices begins with a mind naturally formed for the study, and without that natural faculty probably none become eminent in that direction; but patience and assiduous attention will bring to any mind that adopts the study a familiarity with values far beyond the average, and in their company the confidence of clients or professional colleagues, with their consequent substantial pecuniary reward. Nor will this be all: it will give the student a faculty of generalisation and a power to trace analogies even more valuable than a memory of rates. Important means to this end are a collection of trade lists and the common-place book. The student should obtain trade lists of every raw material or manufactured article used in the building trade. When he receives a list he should write legibly on it the date of its receipt, the trade discount, and the discount for cash. The collection should be carefully revised as changes of prices occur. The smaller lists (not books) he should keep in a guard-book, which is most conveniently secured by a strap around it. It should be paged, indexed, and cross-indexed, thus: "Ashton and Green," slate merchants, should appear in A Ashton, and Green, slate merchants, and in S, slate merchants, Ashton and Green. The bound books or thick pamphlets should each be marked with a number entered in an index, as last described, and kept in portfolios or on a shelf. The student should not merely procure these lists and arrange them; he should make himself familiar with their contents and thoroughly understand them, and he should settle by inquiry any doubts raised by their perusal. The estimator should also keep a book, in which he should enter all the rates of railway carriage. The collection of information as to materials, cost prices, railway rates, etc., is necessary to the man who would achieve distinction in the study of prices, and some method of arrangement for facile access is required. Of one of these methods, the common-place book, it is impossible to overrate the importance. The larger number of those men who have become distinguished in any profession have adopted it. By no other means can such a store of professional knowledge be acquired, nor will it in any other way be so readily accessible. The experience of anyone will be sufficient to recall the recollection of long and tedious search (perhaps of days), for information cursorily read and almost forgotten which might have been by those means found in a few minutes. Much depends upon the system adopted. The leading principle is easy and clear reference. There are two kinds: one, the "index rerum," as it has been called, which contains *references* only to passages of interest. The other, the "common-place book" proper, mainly a collection of the passages themselves. Directions for the management of the "Index Rerum" are thus given by Dr. John Todd, author of 'The Student's Manual': (1) When you meet with any subject of interest, note the subject, the book, the page, and any word distinguishing its qualities. The index should be your constant companion when you read. (2) Make your index according to subjects, as much as possible selecting that word which conveys the best idea of the subject. (3) Put, at the upper left hand corner of the page, a capital letter as A, B, etc.; in the centre, one of the first five vowels, as *a, e, i, o, u*. (4) Place the principal word in the margin under the first letter in that word, and the first vowel in it. For America, I turn to capital A and the vowel *e*, because A is the first letter and *e* the first vowel. A book with faint lines is the best, and it should be ruled as below—

| A | *a* | A | *e* | B | *a* | B | *e* |

Examples of the entries are as follows:—

A	*e*
America.	Supposed to be known in time of Homer. Thomas's 'History of Printing,' vol. ii. p. 20.
Atheism.	Of France, picture of. Schlegel's Lectures, vol. ii. p. 199.

B	*a*
Bradford.	Governor, Notices of. *American Quarterly Review*, vol. ii. p. 497.

B	*o*
Brougham.	Graphic and powerful description of. Post, 'Rhetorical Reader,' p. 248.

The scale of apportionment of pages recommended by Todd is as follows:—

A 3	F 3	K 2	P 6	T 4	W 6
B 6	G 4	L 3	Q 2	U 3	X 2
C 4	H 4	M 6	R 4	V 3	Y 2
D 3	I 3	N 2	S 6		Z 2
E 3	J 2	O 2			

The foregoing will not be found an infallible guide; various circumstances will arise to modify the proportion—as a peculiar direction of study. A common-place book may be constructed in a precisely similar form, except that instead of a mere reference to subject, volume and page, we adopt subject, page, volume, extract, or it may be a mixture of both, or it may include newspaper cuttings and small original memoranda of which you may desire to preserve a record.

Q	u
Quantities.	I advise employers not to let the quantities be taken by anyone who can be made out to be their agent, nor to recognise them in any way except as after mentioned with reference to a schedule of extras. Sir Edmund Beckett, 'A Book on Building,' p. 32.
B	e
Bedrooms.	Decoration of, remarks on. 'House Decoration,' R. and A. Garret, p. 30. This might also appear as below.
D	e
Decoration.	Of bedrooms, remarks on. 'House Decoration,' R. and A. Garret, p. 30.

The reference word should express the leading idea, and should be selected with judgment. Another method of keeping a common-place book has been recommended by Mr. G. A. Sala, and is the method adopted by him. He says: "Take a book, large or small, according to the size of your handwriting, and take care that at the end of the book there shall be plenty of space for an index; begin at the beginning, and make your entries precisely as they occur to you in unordered sequence. But after each entry place a little circle or oval, or parenthesis, thus—( ), and in a portion of these spaces put consecutive numbers. Here is the model for a page:—'The Prince of Wales wore the robes of the Garter at his marriage in St. George's Chapel, Windsor, all the other K.G.'s present wore their robes and collars. Mr. W. P. Frith, R.A., who was to paint a picture of the wedding for the Queen, stood close to the reredos to the right, looking from the organ-loft (1023).' 'Just before the liberation, in 1859, of Lombardy from the domination of Austria, the audiences in the Italian theatres used to give vent to their pent-up patriotism by shouting Viva Verdi; the initiated knew that this meant to signify Viva V (for Victor), E (for Emanuel), R (for Re), D I (for D. Italia) (1024).' Now all you have to do is, immediately you have made your entry, to index it, and if you will only spare the patience and perseverance, to cross-index it. Thus, under letter W you will write Wales, Prince of, married in robes of the Garter (1023), under G, Garter, robes of, worn by P. W. at his marriage (1023). Under F W. P. Frith, R.A., present at marriage of P.O.W. (1023). Thus also Verdi, Victor Emanuel, and Italy will be indexed under their respective letters V and I, and be referable to at 1024. Write the figures in the circumscribed spaces in red ink. The corresponding ones in index may be in black." Of common-place books, as compared with his 'Index Rerum,' Dr. Todd says, alluding to Locke's system: "Neither that nor any other common-place book which I have ever seen will either come into anything like extensive use or be of any essential advantage to the student and man of literary habits; they require too much time and too much labour. Everything that is saved must be copied out in full, and then noted also in the index. Few have the time, and fewer still the patience, to do this." This is not by any means a general opinion. We recall the somewhat hackneyed but pregnant sentence, "Reading maketh a full man, conference a ready man, writing an exact man." For the third reason, irrespective of the value of the store of knowledge amassed, the common-place book is to be strongly recommended. The foregoing are those for the arrangement of general professional information. For prices there is still one other method—the arrangement of the average price-book in trades. The student should have a book for each trade, and enter every price that comes in his way in its respective book. Every element which has affected the cost of work, all the information obtainable as to the working of particular materials, relative values, and weights and measures which are new to you; and in every case date the item and state its derivation, as both source and date are important. The student should also seize upon every opportunity which may offer of observing and making notes of the time and material expended upon certain quantities of work. The most valuable will be those of larger bulk, as from these more reliable results may be deduced for the valuation of smaller quantities. Example : The notes of the production of a thousand yards of concrete will yield a more reasonable average than ten. This practice will, besides enriching his store of information, help the student to foster habits of observation and precision such as would be worth very much more trouble to acquire. The card index, which is becoming deservedly popular, may also be recommended as a very convenient method of classifying and preserving information.

Profit.—Profit may be best defined as a percentage to be added to every item after its complete cost has been computed. The rate of profit must be decided upon in any system of prices, and although the general usage does not countenance it, some estimators hold that it is, in the case of pricing a bill of quantities, best added at the end of the bill.

The question as to what may be considered a fair profit is a somewhat difficult one to settle. The profit placed by builders upon provisional sums varies from 5 to 20 per cent., but it is very questionable whether the majority of builders realise, before sending in a tender, the percentage of profit that its acceptance will ensure; they would be able to say whether the generality of the prices were good or otherwise, but would not be able to fix a definite percentage.

Mr. Lucas, in the course of the discussion upon the paper read by Mr. Brassey before the Royal Institute of British Architects, February 4th, 1878, 'On the Rise of Wages in the Building Trades,' said, " We do not profess to carry out works at less than 10 per cent. profit, and I do not believe any man can do so," and it is certain that only in exceptional cases he does.

Where it devolves upon the surveyor to fix a profit on particular items in which incidental charges like establishment charges are included, 20 per cent. is a fair one. This is little enough when one considers the number of extraneous expenses which this percentage must cover. In pricing an estimate, where all the incidental items have been allowed for in the preliminary bill, 15 per cent. is sufficient.

Ten per cent., besides establishment charges is the least that should be attached, except in a very large work. Sometimes contracts are deliberately taken at a less profit than this; but it is much oftener unwittingly done. It may be safely assumed that work at a less rate than 10 per cent. clear is seldom worth the doing. Undoubtedly work has been taken without profit for such reasons as the keeping together in a period of depression a well organised business and a staff of efficient workmen; but such a case is an exceptional one, requiring exceptional treatment. Profit should be higher on work or material in small quantities. Nails and screws (of which many are inevitably wasted), and small quantities of such things as paint, putty, solder, colours, etc., are examples. The adjustment of profit in cases of variation is usually defined by a condition of the contract that extras or omissions shall be valued at the rates of the original contract or analogous ones; in such case there is no base of argument. In its absence it seems reasonable that the special circumstances should be considered. Several hundred similar doors may be produced by machinery at first, at a much lower rate than one or two later; and these should bear a larger price. On the whole, it may be admitted that the equitable assessment of profit requires mature judgment, and almost every case demands modifications not exactly referable to a fixed rule. A collateral issue is well expressed by Mr. Matheson in his 'Aid Book for Engineers.' He says: "There are certain engineering undertakings in which it is very difficult to frame a schedule which shall apply fairly to additions or omissions. A contractor may have to provide expensive plant, machinery, and apparatus, the extent or cost of which would be but slightly altered if the quantity of work were to be diminished or increased. Therefore, having these fixed expenses already provided for, extra work could be performed at less rates than those in the primary contract, while deductions in price for omissions should also be made at a less rate, because the contractor would only save the material and labour deducted, and not a like proportion of his apparatus and fixed charges."

The size of a work will affect the rate of profit. Work in small quantities should be priced at a higher rate, work in large quantities at a rate proportionally lower; but in the case of a very large contract another element tends to balance this reduction of profit, which is the much smaller number of firms who are able to deal with a large contract, and the consequent lessening of the competition. The profit is usually least upon works offered in open competition.

A conspicuous example of this principle is the large rate of profit uniformly charged by the suburban builder who lives by jobbing. The profit is (and very properly) 40 or 50 per cent.; a lower rate would neither be a sufficient inducement to him to engage in the work, nor would it afford him a decent living. For some small works 50 per cent. would not be sufficient remuneration for a respectable man.

Establishment Charges.—It is best to preserve a distinction between profit and establishment charges; these latter represent the outlay other than labour or material before any profit can be realised, and are most conveniently considered as a percentage to to be added to the prime cost of every item. In a well-drawn bill of quantities, the preliminary bill contains a number of items which, in the opinion of some, might be properly included in that percentage—as sheds, scaffolding, screens, etc.; but as these vary with the requirements of each building, they are probably best dealt with separately. The remaining items for consideration are:—Interest on capital, rent of offices and yard, clerks' salaries, including cost of measuring, or waiting on an appointed measurer, depreciation of plant and machinery. The whole of the prime cost of the building operations for a year being ascertained from the books, exclusive of foregoing items, and a similar calculation of a year's outlay on those items made separately, their comparison will show the proportion and consequent percentage which should be added to every item of prime cost. The judgment and experience of an assessor are often considerably tried to form a reliable opinion in a given case. The conduct of some businesses costs twice as much as some others equally efficient but less pretentious. Sometimes when the rate of percentage is in question the builder will permit the surveyor to inspect his books (in confidence), when he can satisfy himself as to the reasonableness or otherwise of the claim set up. In default of this permission, the builder may be requested to furnish a statement of the detail of his calculation. When no items of the kind are included in a preliminary bill, establishment charges are commonly computed at 5 per cent. Sometimes as much as 7½ per cent. is allowed. Occasionally they are classed in two categories—5 per cent. on work done at the building, 7½ per cent. on work done at the builder's shops. Establishment expenses in some businesses are very great. In the business of a high-class decorator, the submission of sketches (often a great number), artistic and consequently expensive, and the frequent attendance by a principal on the client, add very much to the cost of the work, and are often charged at 15 to 20 per cent.

Another business involving a large percentage for establishment expenses is that of the manufacturer of structural metalwork: sometimes his price has to be accompanied by a key plan of every floor, and details of the larger girders, stanchions, etc., and some of these firms calculate as much as 25 per cent. for their establishment charges.

The necessity of defining these charges before signing a contract is admitted, as, after a schedule of prices has been agreed to and the work done, the contention has been raised that the prices were not intended to include establishment charges. Probably the best way is to mention them in the preamble of a schedule.

The Works Department of the London County Council has adopted the following percentages to cover general and establishment charges.

On Estimated Work.

Add to expenditure on wages and materials charged to a work—

For general charges	1¾ per cent.
For establishment charges	4¾ ,,

On Jobbing Works.

For use and waste of plant	3 per cent.
For general charges	2 ,,
For establishment charges	8 ,,

Supply of Materials to London.

The sources from which materials may be most conveniently procured, and the means of placing them on the site, will always be an important one, and in this connection the extent of the proposed work must materially affect the decision. Taking sand as an example, for a large work it would be procured by the barge-load or truck-load at the cheapest rate; for a small work, the reasonable alternative would be to obtain it from one of those merchants of building materials whose yards are, as a rule, on the banks of the Thames, the canals, or their basins. Similar conditions regulate the supply of stock bricks. Bricks of all kinds are brought to London in large quantities by railway, the merchants having depots at their termini. Whether a truck or barge-load shall be ordered direct from the field, or the bricks bought by the thousand from the London merchant, will depend upon the size of the work. The price of the smaller quantities thus obtained will be about 10 per cent. higher than the larger ones otherwise procured. It will often be possible to save all trouble of barging, and unloading, and yet avoid much increase of cost, if a large order can be given to one of the London merchants. It is obvious that the possibility of the builder's barging for himself must operate to keep down the middleman's profits. Much of the sand and ballast used in London is dredged from the Thames. Not nearly so much is brought from above bridge as from below, although the former is more generally specified. The contractors may either charter a barge, and purchase the sand when dredged, landing it a one of the Thames draw wharfs, where the privilege of unloading will cost him 25s. per day, and he must find his own baskets, barrows, and planks, or he may arrange for a barge-load at a price per yard, including barging, unloading, and filling into carts. Pit sand is also brought by barge. Contractors may, however, generally buy sand (unless he occupies a wharf) as cheaply of the keeper of a draw wharf, who generally keeps a stock, or will have it there to order, and will arrange for its cartage and delivery if desired. From some of the suburbs, as Fulham, Hammersmith, Epping, Walthamstow, etc., pit sand is still carted to London, and in such cases the carts take away rubbish on their return journey. Sand is also brought by truck or barge from Croydon, Redhill, Sevenoaks, West Drayton, Southall, Ealing, Dawley, etc. The supply of stocks and malms for the metropolis is derived from Kent, Essex and Middlesex—Sittingbourne, Teynham, Faversham, Southend, Acton, West Drayton, Ealing, Hayes, and Southall are some of the places. Bricks in small quantity are still made in the more immediate neighbourhood of London, as at Edmonton, Norwood, etc.; but as these must be carted all the way, it will generally prove cheaper to use bricks from more distant places conveyed by rail or barge. Within a few years the use of stocks has been largely superseded by that of Fletton bricks which for hardness and truth of shape are superior to the ordinary stock, but do not take plastering so well. The supply of lime is derived from Dorking, Merstham, Halling, Lewes, and Petersfield. A stock is always kept at the merchant's yard in London, and the price quoted generally includes delivery in loads of two yards, and when the lime is ground the use of sacks. The lias lime used in London, is for the most part brought from Rugby Barrow, Leicestershire, or Lyme Regis, Dorsetshire. Portland cement is brought from Northfleet, Rugby, Arlesley, etc. Most of the manufacturers at these places have London depots, and there is no advantage in purchasing at the works, as it would probably increase the cost. The price generally includes delivery within the usual limits, and use of casks or sacks. The stone used in the metropolis is almost uniformly brought by rail; some Portland and York, however, comes into the Thames. The timber used in London comes to the Surrey, Commercial, East India, and West India Docks.

Labour.—One of the factors in the calculation of values is the cost of labour, of which, in its various phases, only long experience can judge wisely; but even the inexperienced may be certain that sanguine expectation of the result of a day's work will always be disappointed. It is generally admitted that men do much less work per day than they did thirty years ago, when their wages were much lower. Although the trade-unions may be congratulated upon much of their past work, which has notably improved the condition of the workmen, their maxim, "A fair day's work for a fair day's wage," like Talleyrand's "Point de zèle, monsieur," is, in the opinion of all but mechanics, a grave mistake. By it the capable and energetic workman is reduced to the level of the comparatively lazy; but, nevertheless, the testimony from all good judges to the capability of the British workman is no mean tribute to his powers. There are, however, signs of change as to the limiting by the trade-unions of the quantity of the mechanic's daily production, a condition of things of which every Englishman who is not a mechanic is ashamed. The lowest rate of wages is that in the schedules of the Government departments, and as these are tendered upon at a discount of 10 per cent. and upwards, it may be inferred that the labour supplied at such prices cannot be of the highest quality; and when it is remembered that these rates cover establishment charges for which these schedules do not provide by any special clause, the uninitiated are lost in admiration of the ingenuity of the builder, who makes a profit under such inclement conditions. The truth is, however, that there are a number of men content to work at low wages (many of them worth no more) in consideration of the comparative permanence of the employment, and when the periodical contract for a district changes hands, they follow it to the new contractor. As a general rule, the building operative works from 44½ to 50 hours. In London wages are as schedule, p. 470.

There are many men who work at lower rates than the foregoing; but this is a matter with which the surveyor cannot deal: he must adopt the current rates—the consideration is not important, for wages below the market rates always involve poor services. Employers pay workmen's fares when the work is four miles from the shop, or pay for lodgings. Workmen sent from London to the country at the request of employers receive London rates, with an additional allowance for lodgings; the railway fare is paid from London to the works, and, if the workman is discharged by the employer, back to London. Very good workmen are, in addition to the foregoing, paid 1s. per day "country allowance" as it is called. Engineers pay 2s. per day. In London overtime is paid for at the rates shown in the schedule before mentioned. In "tide work" six hours is usually paid for as nine hours, and if more than six hours is available, in a similar proportion; the overtime rate applies to this time also. Working in water is paid at the rate of one-third more than ordinary pay, and the contractor finds water-boots. The War Office schedules provide as follows:—Whenever it may be necessary to employ workmen by the day between the hours of 6 p.m. and 12 p.m., or on Sundays between 6 a.m. and 6 p.m., one-half the contract rates per day will be allowed in addition for such work, and when employed between the hours of 12 p.m. and 6 a.m., or on Sundays between 6 p.m. and 6 a.m., double the contract rates per day will be allowed for such work. In like manner, whenever workmen are ordered to be employed on measured work at tide work, or during those hours of the night before specified or on Sundays, the extra rates for the labour only will be allowed in addition, provided in each case the contractor's claim is supported by a written order signed by the superintending officer, and the work is not specifically provided for in the schedule.

In dealing with labour, whether in estimating or adjustment of accounts, the question arises as to the proportion of labourer's time to be allowed to each mechanic. Day accounts almost invariably demand the same amount of time on a particular item of work for labourer and mechanic, and it requires much judgment and experience of working processes to deal justly with such claims. Observation of the progress of an average building will show a varying proportion at its successive stages; taking an average for the whole period, three labourers to five mechanics is a reasonably close approximation to the truth. The necessities of attendance are various; for works in which there is lifting and supporting, the labourer's constant presence is necessary. In others the labourer takes to the place sufficient material for the piece of work, and goes to it no more, except to remove the unused material. In building brickwork, when the scaffold can be supplied from the ground, one labourer can supply three bricklayers. As the height increases more labourers will be required; but one labourer can generally supply one bricklayer—the exceptions are rare. When men are doing gauged work the progress is slow, and requires very little attendance. A mason fixing stonework will generally require a labourer in constant attendance, and when the stones are heavy, two. The mason working at a banker only requires labourers' attendance for bringing the stone before working, and removing it afterwards. If it is beyond the strength of the mason to turn the stone he is working on, another mason helps him. A slater generally requires constant attendance of a labourer, but often contents himself with a boy. The carpenter's work—roofs, floors, partitions, etc.—generally requires one labourer to three carpenters, and if the timbers are heavy one labourer to two carpenters. One labourer will attend to two joiners as a rule; one labourer will serve two plasterers; one labourer did serve two plumbers, often a boy would do it; now he invariably insists upon a labourer's constant attendance. When considering charges for time in a day account, or in calculating prospective expenditure of time, observe that a workman often has to be specially brought from the contractor's yard to do a certain work, and that the time for his journey should, in justice, be allowed. Capable labourers are often set to pull down brickwork for an alteration by the bricklayer, and bricklayer's time charged for it. Labourers often limewhite and paint, and so save the time of more expensive men. Doubts are not seldom entertained by surveyors and others on charges for time, and their conclusions that it is over-charged are frequently correct; but mere moral certainty is not legal proof—it can only be a matter of opinion, and the *opinions* of experts have not such weight as might be desired. On any disputed point dependent on opinion, evidence of equal force and diametrically opposed may be obtained if only trouble enough be taken to procure it. Contentions as to time are usually met by the builder's production of his books, of which the day accounts are or should be exact copies. The only test of a day account is comparison with the result of a measurement and valuation at current rates; but this is often fallacious. Many surveyors refuse altogether to criticise time charged in day accounts. They maintain that if absolute accuracy and proof is required, the work must be carefully observed by a clerk of works, or by the architect himself. The latter is, in most cases, impossible.

Economy of Labour.

In a large work various labour-saving expedients may be adopted, which, in a smaller one, would be unreasonable. In such as these the advantages of competition lie with those contractors who are constantly engaged upon great works, with the requisite plant for the purpose. The use of overhead travellers, steam engines, and cranes long enough to lift material from the cart or waggon and deliver it directly on the part of building where it is required, are obvious advantages. In London large cranes are now sometimes worked by hydraulic machinery connected with the main of the Hydraulic Power Company. In providing for these, it will be necessary to design the staging and its supports, charging it and the crane as use and waste. The hydraulic machinery may be hired from the Hydraulic Company which supplies the district.

Jobbing.—The question as to what work shall be classed as jobbing is one which often arises. In the case of a contract for a new building, small alterations should, as a rule, be charged day work, the actual time spent on them and the materials used being charged and a reasonable profit added.

Work, although different to the original intention if instructions have been given in time for it to be carried on as a part of the general work, is obviously worth no more than contract rates, and a contract condition usually stipulates for the application of such rates.

Some public departments have two schedules of prices, one for ordinary contract work, the other for jobbing. The School Board for London and the War Department fix the limit for jobbing at about 40*l*.

Approximate Cos of Buildings.

The desirability of some rule by which the approximate cost of a building may be arrived at before it is built has been felt by most persons engaged upon its production. This is a branch of investigation which may resonably be considered in this connection.

An intelligent and experienced surveyor, from the cubic content of a building, and a general description of its materials and finishings, can tell with remarkable accuracy what will be its cost. If the surveyor will adopt the practice of commencing his dimensions with the cubic content of the building for which he is preparing quantities, he will be, in course of time, in possession of some valuable material for comparison of relative costs. These measurements to be of any value must always be done in the same manner, and must comprise the whole content, including the walls. Disregard chimney-stacks, buttresses, and dormers, unless in unusual number. Many architects in stating the price of a building consider that extraordinary decoration and gardening are excluded.

In applying the rates of a previously erected building to the calculations of a new one, consider whether the conditions are similar or not.

Measuring and Valuing at a Price per Cubic Foot.—This is the kind of calculation most frequently adopted for an approximate estimate; the result must comprise the whole cubic content including the walls. Multiply the length of the building by its breadth and the result by the height from the bottom of the footings to half-way up the roof. Disregard buttresses, chimney-stacks and dormers, unless the dormers are in unusual quantity when they should be included. The ordinary cubing is assumed to include drainage of customary extent and character. A different price per foot may sometimes be adopted for the various parts, but not often; it is more convenient to adopt a uniform rate.

For boundary walls and such works an approximate estimate should be made in detail and added to the price arrived at by cubing.

Various causes will operate to modify the price per foot, as very expensive finishings, a considerable proportion of basement, difficulty of conveying materials to the site, an unusual quantity of drainage, etc. In such cases the surveyor commences with his estimate on the general principle above described, and adds the extra cost of such things as last mentioned.

Other elements which vary the price per foot cube, are very small or very large rooms, the utilisation of the whole of the roof space for occupation, etc.

Small buildings cost more than large ones, the finishings being of similar character.

The foregoing are the rules most frequently adopted, but some instructions for architectural competitions prescribe the measurement to be taken to the bottom of the concrete, and to include all excrescences such as chimney shafts, buttresses, turrets, dormers, etc. The result will require the application of a smaller price per foot cube than one produced by the method first mentioned.

In the case of a large collection of buildings of varied character like a hospital, lunatic asylum, or workhouse, the buildings will differ greatly in content and finish, and must be kept in separate categories, so that different rates may be applied: as administrative block, wards, laundry, workshops, chimney shaft, water tower, etc.

Sometimes the addition of a new block of buildings to an existing one may be dealt with in the usual way of cubing, but a higher price than ordinary must be adopted for it, and the collateral work, valued item by item, added.

Occasionally, where an architect fondly hopes to save by avoiding the preparation of quantities, a builder may be found willing to agree to a price for the building based on its cubic content. In such a case the method of cubing must be clearly defined in writing and a clear understanding arrived at as to what extraneous work, as drains, boundary walls, paving, etc. is included. Generally the variations on a contract based on such particulars are difficult to settle and fruitful of disputes, and probably the attempt to save the expense of quantities has been more than counterbalanced by its disadvantages.

The bare price per foot cube of a building is of small value for future use, unless one knows the character of the building; the estimator should preserve with his note of cubic price particulars of the materials and finishings, and the date of its erection. It is obvious that in default of such particulars mere notes of the prices per foot cube of well known buildings of which lists are current are not of much value.

Buildings already built, when the price is not known, may be cubed from the actual work in the same way to arrive at their approximate value. As illustrative of the division of a building into sections for the application of different prices, Mr. Young's preliminary estimate for Glasgow municipal buildings was—basement 9d.; superstructure 1s.; towers 1s. 6d. Mr. Corson's estimate was 1s. throughout.

Professor Kerr, in his 'English Gentleman's House,' estimates by the cost per square; he says in substance as follows:—

The method of calculation is to take the dimension of every room and portion of the house internally, multiplying their relative length by the breadth, then squaring the floor spaces of the passages and stairs in the same manner and adding to the total $\frac{1}{2}$ of the whole for walls and waste. To all the results obtained by estimating the cost of the house by the number of superficial squares, must be added an allowance for incidental expenses. These would be professional services, including law charges, special fittings (grates, mantels and the like), external fences and boundary walls and some internal decoration, and should not be estimated at less than 20 to 25 per cent.

The method suggested for the application of these principles is illustrated by the following table :—

APPROXIMATE COST OF BUILDINGS. Professor Kerr.

Total Outlay required.	The Family Department.							The Servants' Department.							Total Outlay resulting.
	1.	2.	3.	4.	5.	6.	7.	1.	2.	3.	4.	5.	6.	7.	
	Price per cubic foot.	Average height of room.	Price per superficial square.	Corresponding number of squares.	Cost at the prices given.	Number of rooms to correspond.	Average price per room.	Price per cubic foot.	Average height of rooms.	Price per superficial square.	Corresponding number of squares.	Cost at the prices given.	Number of rooms to correspond.	Average price per room.	
£	d.	ft. in.	£		£		£	d.	ft. in.	£		£		£	£
1,250	8	12 0	40	22·30	892	13	68	6	11 3	28	12·90	361	13	28	1,253
2,500	9	12 9	48	37·50	1,800	20	90	6¼	11 7	31½	22·20	700	19	37	2,500
5,000	10	13 9	57¾	63·10	3,639	30	121	7	12 0	35	39·00	1,365	29	47	5,004
10,000	11	15 0	69	106·00	7,314	45	163	7½	12 6	39½	68·00	2,686	43	62	10,000
20,000	12	16 6	83	178·00	14,774	67	220	8¼	13 0	44½	117·70	5,237	65	80	20,011
40,000	13½	18 0	100	298·00	29,800	100	298	9	13 6	50	204·00	10,200	97	105	40,000
80,000	15	19 0	120	501·00	60,120	150	400	10	13 6	56	355·00	19,880	145	137	80,000

As to cubic value Professor Kerr says, a well-built residence is found to cost as follows :—

For the main building or family department, 8d. to 15d. per cubic foot.
For the attached offices or servants' department, 6d. to 10d. ditto.
For stables and farm offices, 4d. to 6d. ditto.

Mr. Wheeler, in his 'Choice of a Dwelling' 1872, recommends calculation by the square, including the walls, and some of his opinions on value are as follows :—

A first-class house in London, not exceeding 1800 feet superficial per floor, that is, 30 feet by 60 feet—

	£	
Principal rooms	50	per square.
Chambers and secondaries..	35	,,
Domestic offices	25	,,

A well-built London or suburban residence, not wholly detached and not exceeding 900 feet superficial each floor, or 22 feet 6 inches by 40 feet—

	£
Principal rooms	40 per square.
Chambers and secondaries	30 ,,
Domestic offices	20 ,,

A London or suburban residence, plainly but honestly built as an investment or for a very moderate scale of living, and not exceeding 900 superficial feet each floor—

	£
Principal rooms	35 per square.
Chambers	25 ,,
Domestic offices	18 ,,

The bare price of a building per foot cube, unless one knows the character of the work, is of little value; notes of the particulars of the nature of materials and finishings are a valuable adjunct to the cubic price.

Builders are sometimes willing to make a contract upon a price per foot cube; if the surveyor is concerned in such an arrangement he should define the method of cubing, settling the question whether dormers, chimney stacks or buttresses should be included in the measurement, and whether the height should be measured from the top or bottom of the concrete.

The cubic price of some well-known buildings is given below; the older ones must be received with caution, first, because we do not know how they were cubed, and second, that many of them are based upon prices of labour below those of this time.

	Per foot cube.	A.D.
	s. d.	
British Museum	1 6	1843
Houses of Parliament	2 6	1843
Royal Exchange	11	1841
Manchester Assize Courts	9½	
Foreign Offices	12¾	1857
Waterlow's Industrial Dwellings	7¼	average
St. Thomas's Hospital	9	
Buildings of the School Board for London, 1870-1880	5½–7¾	Not including boundary walls
Ditto, 1880	6¼	or playgrounds.
Hotel Victoria, Northumberland Avenue	1 6	
Holborn Viaduct Hotel	1 4	
Holborn Town Hall	1 2	
Railway Clearing House, Seymour Street, Euston Square	6¼	
Post Office, corner of Newgate Street	8¼	
Bow Street Police Station	11	

Mr. Street's design for St. Mary's, Edinburgh, was estimated at

12½d. per foot cube.

Mr. Young's design for Glasgow Municipal Buildings was estimated at

Basement	9d. per foot cube.
Superstructure	12 ,, ,,
Towers above general level of buildings	18 ,, ,,

Mr. Corson's design for Glasgow Municipal Buildings was estimated at

12d. per foot cube (throughout).

Messrs. Leeming and Leeming's design for the New Admiralty Buildings was estimated at

Main buildings	12d. per foot cube.
Towers and work above parapets	18 ,, ,,

Baths and Wash-houses.—With the development of the movement for their establishment their cost has greatly increased; the best ought not to exceed 1s. per foot cube; some of the early ones cost 8d. per foot cube.

	d.	
Caledonian Road, Islington	8	About 2d. of this for machinery and boilers.
Hornsey Road, London, N.	8½	
Tibberton Square, Islington	9	

Artisans' Dwellings in London.—These are quoted in various ways: per foot cube, per room, per inmate, per square foot of rent-producing floor space. Blocks in the country have been built as low as 7d. per foot cube. In London, 7d. to 9d. per foot cube. 65l. to 84l. per room.

Underground Conveniences.—These have cost from 2s. 6d. to 4s. per foot cube; about 60l. per w.c., lavatory and urinal.

A few general rates per foot cube are given below; they can only be taken as approximations.

	Per foot cube.
	s. d.
Labourers' cottage	4
The cheapest dwelling houses	5
Good country houses	10

PRICES.

	Per foot cube.	
	s. d.	
The best mansions	1 6	
Stables	8	average.
Very good stables	1 0	
Breweries	5	
Churches	6	have been done at 5d.
Lunatic asylums have been built for	6	not lately.
Warehouses of plain character	5¼–6	
Maltings	2⅞	
Baths of the best and recent London type—		
Building	5¼	
Machinery and apparatus	2	= 7¼
Workhouses	5	

There are some other methods of arriving at the probable cost of a building by comparison—

The price per superficial square (100 feet) for each floor for dwelling-houses. Thus 8d. per foot cube would be commonly equal to 40l. per square, assuming the rooms to be 12 feet high, etc.

The price per stall of stables (without lofts) varying from 25l. to 100l.

The price per bed for hospitals.—Recent buildings show an enormous increase in cost consequent upon the larger cubic content per patient, the general advance of scientific knowledge, increased accommodation for staff, larger proportion of nurses to patients.

Some well-known instances are as follows:—

	£
St. Thomas's Hospital	650
One pavilion without administrative buildings	250
Herbert Hospital, Shooter's Hill, including administrative buildings	320
The most expensive species of hospital is the fever hospital.	
The three new hospitals of the Metropolitan Asylums Board, "Brook," "Park" and "Fountain," will probably cost	400
Salford Sanatorium cost	375
Heathcote Infectious Hospital, Leamington cost, 20 beds	385
Willesden Isolation Hospital cost 50 beds	380
Ruchill (Glasgow) Infectious cost	400
Netley	305

Workhouses, per inmate:—

	£
West London, Holloway	48
St. Pancras Infirmary, Highgate	68
Charlton Union Hospital	50
Constance Road, Camberwell	104
Grove Park, Kent	240

2 H

The price has greatly increased within a few years; they ordinarily cost £150 to £200 per inmate.

Lunatic Asylums, per Inmate.

Claybury	236
Bexley Heath	210
Hanwell (built some years ago)	160

Theatres, per head.—About 1s. per foot cube.

Empire (Glasgow)	13
Brixton	15

Churches, per sitting.—4l. is the least. Churches of the highest class cost from 10l. to 12l.

Breweries at per quarter, i.e. the quantity of malt that can be dealt with at one brewing, including machinery and plant and exclusive of malting or cooperage, the approximate cost is as follows :—

	£	
5 quarters	500	per quarter.
20 quarters	400	,,
40 quarters	350	,,
100 quarters	300	,,

Schools, per scholar.—Schools cost from 6d. to 7d. per foot cube or 4l. to 10l. per scholar.

The London board schools cost for the buildings about 10l. 15s., from 1870 to 1880, the present price is about 8d. per foot cube.

The proportion of carcase to total cost of an ordinary house is about one-third; i.e. *one-third carcase, two-thirds finishings*. The total cost of a building is approximately (a little more than) three times the cost of the brickwork for buildings of the commonest class; but for buildings of average quality five times the value of the brickwork, or four times the bricklayer's bill when there are quantities.

Some estimators profess to have derived from their experience a faculty of fixing factors, varying with the class of building, which if used as a multiplier of the number of rods of brickwork in a building, will produce an amount nearly approximating to the cost of the building in pounds: thus, country houses 75l.; workhouses 70l.; London board schools 100l.

The proportional value of labour to materials and plant, varies with the different trades. The following is a reasonable average:—

	Labour per cent.	Plant and materials per cent.
Excavator	90	10
Drainer	67	33
Bricklayer	75	25
Mason	60	40
Slater	20	80
Carpenter	25	75
Joiner	60	40
Smith	20	80
Plumber	25	75
Plasterer	60	40
Glazier	15	85
Painter	50	50

The following list of wages issued by the London County Council embodies the Trades Union rates of wages in the London District, which is comprised in a circle of 12 miles radius from Charing Cross. The County Council, however, in its contracts stipulates for its adoption within a 20 mile radius:—

LONDON COUNTY COUNCIL LIST OF RATES OF WAGES AND HOURS OF LABOUR.

The rates of wages and hours of labour set forth in this list are paid and observed by the Council in works which are in the nature of construction or manufacture, and which the Council may resolve to carry out without the intervention of a contractor on a site wholly or partially within a radius of 20 miles measured in a straight line from Charing Cross.

Contractors for works in the nature of construction or manufacture to be executed within a radius of 20 miles measured in a straight line from Charing Cross, or on a site partly within and partly outside the radius, will be required to pay wages at rates not less, and to observe hours of labour not greater, than the rates and hours set out in this list. Such rates of wages and hours of labour are inserted in a schedule to and form part of the contract, and penalties will be enforced for any breach thereof, and if the contractor employ any workman or workmen in any trade not included in the Council's list the rates of pay shall not be less, nor the hours of labour more than those recognised by associations of employers and trade unions and in practice obtained in London.

Trades.	Hours of Labour per Week. Winter— 12 weeks after second Monday in November.			
	Six middle weeks.	Until 8 p.m.	8 p.m. until 10 p.m.	After 10 p.m.
BUILDING TRADES— Carpenters Joiners Bricklayers Bricklayers (cutting and setting gauged work) Plasterers Masons Masons (fixing) Masons (granite work) Painters and glaziers Smiths, fitters, gas-fitters, &c.	44	Time and a quarter.	Time and a half.	Double time.
Labourers and navvies	44 {	After 7 p.m. until 8 p.m. time and a half.	Time and a half.	Double time.
Labourers employed on night shifts				
Plumbers Plumbers' mates	41½	.. {	8 p.m. to 11 p.m. time and a half.	11 p.m. to 7 a.m. double time.	1 p.m.
Granite-sett paviors Wood-block paviors Paviors' labourers Timbermen Scaffolders Hot-water engineers Zinc workers Bell-hangers Paper-hangers Painters' labourers French polishers Wheelwrights	48 44	Time and a quarter.	Time and a half.	Double time.

PRICES.

Trades.	Rate of pay per hour.	Hours of Labour per Week.			Week-days (except Saturdays).	
		Summer.	Winter—12 weeks after second Monday in November.			
			Three weeks at beginning and three weeks at end.	Six middle weeks.	Until 8 p.m.	8 p.m. until 10 p.m.
Lath-renders	Piecework Prices as per trade list.					
Machinists employed in working—						
Trying up machines						
Vertical spindle						
Band saw						
Fret saw						
Tenoning machine						
Under and over-hand planing machine						
Over-hand planing machine		50	47	44	Time and a quarter.	Time and a half.
Variety moulding machine	8d. to 1s.					
Frame saw						
Circular saw bench						
Joiner's saw bench						
Cross cut saw						
Universal moulding machine (Elephant)						
Four cutter machine						
Rope feed bench machine						
Rack saw bench						
Mortising machines						
Stablemen	6d.					
ASPHALTE PAVING—	per day.					
Spreaders	6/- to 6/6					
Potmen and labourers	5s.					
CARMEN.						
Employed by the Council—	per week.					
One horse	27s.				6d. per hour after 6 p.m.	
Two horses	30s.		..	..		
Employed by contractors—						
One horse	25s.					
Two horses	27s.					

QUANTITY SURVEYING.

Trades.	Rate of Pay per Hour.	Hours of Labour.	
Electrical Trade—	*d.*		
Wiremen	9¼		*First two hours*—Time quarter.
Jointers	9		*Second two hours*—Time half.
Labourers	6½	54 hours per week.	*After four hours till starti next day*—Double time.
Armature winders	9		
Armature winder helpers or labourers	6½		
Transformer winders	8		
Magnet winders	7		
Engineering Trade—			
Brass finishers			
Coppersmiths			*First two hours*—Time quarter.
Turners			*After first two hours until 6* Time and a half.
Fitters and erectors	8¾ to 9¼		*Sunday work*—Double
Millwrights		54 hours per week	*Night shifts*—Time and a
Smiths			
Pattern makers	9¼		
Borers	8¾ to 9¼		
Slotters and planers			
Hammermen	7		
Labourers (employed in Engineering trade)	6		
Drillers	per day. 5s.		
Barge builders	per hour. 10d.	54 hours per week.	11½d. per hour.
Watermen and lightermen	per day. 6, 7 or 8 a.m., until 6, 7 or 8 p.m., 6s.	12 hours per day.	*After 12 hours' work, up* 1s. *per hour.*
Watermen and lightermen	per night. 8 to midnight, 4s.	4 hours.	
Watermen and lightermen	8 p.m. to 6 a.m., 6s.	10 hours.	
Watermen and lightermen	Sunday. Up to 10 a.m., 4s.		
Watermen and lightermen	After 12 noon, 6s.		
Watermen and lightermen	Whole day, 8s.		
Scotch derrick drivers	per hour. 9d.		
Steam navvy and grab drivers	9d.		
Drivers of steam cranes and travellers	8d.	as arranged	Not less than time and a when engines are wor
Locomotive drivers	8½d.		
Stationary engine drivers	8d.		
Portable engine drivers	7½d.		

The rates to be charged by the contractor are usually settled by a schedule of rates for labour delivered with the form of tender (see Form of Tender). When this has not been done, an addition of 15 per cent. to the above rates is a reasonable one. There are other arrangements for the surveyor to consider in dealing with time.

The valuation of jobbing work will sometimes be affected by working rules of the trades. The London rules are as follows:—

SENT FROM SHOP OR JOB. *Plasterers, Carpenters and Joiners, Stonemasons, Bricklayers, General Smiths.*—Men who are sent from the shop or job, including those engaged in London and sent to the country, shall be allowed as expenses 6s. per day for any distance over six miles from the shop or job, exclusive of travelling expenses, time occupied in travelling and lodging-money.

Plasterers.—Any workman being sent to work over 20 miles and not exceeding 50 miles shall be allowed by his employer return railway fare once per month, between 50 and 120 miles, once in eight weeks, over that distance special arrangement between employer and workman. No payment shall be made for time occupied in travelling. The payment for travelling shall be made on the works on the following pay-day. It is understood that travelling time shall be allowed on completion of job.

Plumbers.—Any skilled workman sent to work over four miles from his employer's workshop shall receive all travelling expenses. If sent over eight miles from his employer's workshop, he shall be entitled to 1s. per day extra, with the usual allowance for lodgings and all travelling expenses. Should there be no accommodation for him to reach his work at 7 a.m. he shall be entitled to 1s. per day, unless he travels in the employer's time, and be paid from 7 a.m.

The payment of wages shall commence at noon, or as soon after that time as practicable, on Saturdays, and be paid on the job, but if otherwise arranged, walking time at the rate of three miles per hour shall be allowed, to get to the pay table by noon.

The cost of work done in the country, if local labour be employed, will usually be from 5 to 15 per cent. less than the same kind of work done in London.

A few local rates are stated on next page.

Approximate Estimate based on Rough Quantities.—Cubing will not suffice in many cases as, for instance, in alterations, roughly taking out the quantities and pricing the items is the only safe

	Liverpool.	Colchester.	Reading.		Liverpool.	Colchester.	Reading.
	d.	d.	d.		d.	d.	d.
Labourer	6	4	5	Joiner	9½	6½ & 7	8
Bricklayer	9½	7	8	Plasterer	9½	8	8½
Mason	9½	7	9	Plumber	9½	8	8
,, Fixer	10	7	9	Glazier	8½	5½	6½
Slater	9½	7	9	Painter	8½	5½	6½
Carpenter	9½	6½ & 7	8				

process. In works of alteration, gutters, rain-water pipes, drainage, water-supply, and the general structure have to be entirely remodelled at great cost; moreover the alterations are not unfrequently confined to such parts of a house as are most expensive, as bathroom, w.c's., etc., which in cubing an entirely new work are balanced by the cheaper parts.

Sometimes the addition of a new block of building to an existing one may be dealt with in the usual way of cubing, but a higher price than ordinary must be adopted for it, and the collateral work valued item by item.

In the absence of information founded on the cubic content of a building similar to the one proposed, an approximate estimate will be the only means of discovering the cost. In such a case avoid going much into detail. The process will be a rough imitation of taking out the quantities.

The cubic content of the building will, however, be a useful check upon an amount arrived at in this manner.

The general principle for taking out quantities is to dissect the work as much as possible, for an approximate estimate to concentrate it into single items. When the building is a very large one, and of great height, it must be remembered that some parts of the work, although of the same material and finish, will be of greater value than others because of their position.

The detail usually taken in the preparation of a bill of quantities, such as rebates, chamfers, beads, mitres, etc., must be included in the rates.

It will be found convenient to prepare approximate estimates on paper specially ruled for the purpose, as in the following example.

Begin with a heading as to a set of dimensions for quantities. Spaces must be left for inserting the prices later.

| 2/ | 26 0
10 0 | 520 0
———
58 yds. | Dig 6" deep and cart, 8d.; cement concrete 6" thick, 2s. 3d.; cement floated face, 1s. 6d.; tiling P.C. 10s.; and laying, 4s. .. .. .. | | 18s. 5d. | £ s. d.
53 8 2 |

Take notes of all the preliminary items and price them in the ordinary way; as many of them will be assessed by a percentage on the total, they may be filled in after the total is computed.

Excavator.—The digging and concrete in trenches may be taken together. The digging, concrete, floated face, pavings and dry rubbish may be taken together; include strutting and planking in the price.

Drains.—Taking the running length of the pipe, keeping each size separate; with this length may be included all bends and junctions, for these about 20 per cent. should be added to the price; include in the description digging and concrete and strutting and planking, if any.

Gullies, grease-traps, connection with sewer must be separately taken.

Inspection pits may often be averaged for size, the notes of one taken out and priced and the result multiplied by the number.

Bricklayer.—The general brickwork must be taken out in the ordinary way, but the price may include rough cutting, rough arches, ends cut and pinned, stock facing; better facings may also be included, if preferred; about 1000 to each rod of brickwork is often adopted.

Moulded strings and cornices must be measured per foot run.

Windows or door openings may often be averaged for size, the work to one taken out in all trades, and then multiplied by the number, dealing with deduction of brickwork, plastering and facing at the same time.

Fireplaces may also be averaged for size, the work to one taken out (everything connected with it in all trades) and multiplied by the number.

In dealing with the brickwork it will save time to abstract and reduce it in the usual way; reducing each item instead of the total is a tedious and cumbrous process.

Damp-proof courses, lime-whiting and a few other items must be measured in the ordinary way.

Mason.—Stone dressings must be taken out, but as stone and all labour brought into one total, and priced at per foot cube, to include everything, and reducing the price to be affixed for the value of a certain cubic proportion of brickwork of which it takes the place.

Take hearths with the fire openings before mentioned, sills with the window openings, steps or thresholds with the door openings.

Stone staircases must be taken in all trades in the usual way, including the balustrades and other adjuncts.

Slater and roof coverings.—Measure the superficial area of the roof surface to include slating (making an allowance in the price for cutting), boarding or battens, felt and rafters, the cubic content of these last obtained from a table as directed in Carpenter.

Ridges may be measured per foot run to include tile-ridge and wooden ridge.

Hips and valleys.—Measure per foot run, including timber *tilting-fillets*, cutting to slating, lead valley or hip or soakers.

Dormers or flèches must be taken out in detail.

The following is an illustration of the measurement of a roof covering.

2/	40 0			$4\frac{1}{2}''\times 2\frac{1}{4}''$ rafters, 15s. 10d.; 1" rough boarding, 19s.; countess slating, 29s. 8d.; felt, 8s. 4d.	72s. 10d.	£	s.	d.
	15 0	1200 0				43	14	0
		———						
		12 sqrs.						

The prices would be computed as follows. Referring to the table of timber (see Laxton), we find against $4\frac{1}{2}'' \times 2\frac{1}{4}''$ rafters 6' 4" cube; adopting 2s. 6d. as the price per foot cube, we have 15s. 10d. per square. For 1-inch rough boarding say 19s., for countess slating say 27s. plus 10 per cent. for cutting, 29s. 8d., for felt say 8s. 4d., making a total of 3l. 12s. 10d.

Slate mason.—The superficial quantity of slab must be measured, and its price may include all labours.

Carpenter.—There are tables in the builder's price books which show the cubic quantity of timber according to scantling in a square of either roofs, floors, or partitions; in these tables the absolute-quantity is arbitrarily increased by the addition of one joist, rafter or quarter in 15 or 16 feet.

For partitions the table is not to be trusted, as the posts are larger than the quarters and the quantity does not include either heads, sills, or braces; an addition of a certain percentage may be made to cover these, and this is perhaps the most practical course; in the preparation of an approximate estimate, 10 per cent. is enough.

Roofs.—The principals and purlins must be measured in the usual way. (See also Slater.)

Partitions.—If the whole surface is measured, the saving of quarters in the openings may be considered equivalent to the extra for posts, braces, heads and sills. Collect the length of partitions of similar height.

				£	s.	d.
150 0						
10 0	1500 0	$4\frac{1}{2}"\times 2\frac{1}{4}"$ quarters, 15s. 10d., L.P.F. & S. partition, both sides, 29s. 7½d.; paper, P.C. 1s. per piece, 4s., both sides; and hang, 3s.	52s. 6d.	39	7	6
	15 sqrs.					

The prices would be computed as follows. Referring to the table of timber (see Laxton), we find for $4\frac{1}{2}"\times 2\frac{1}{4}"$ quarters 6' 4" cube; adopting 2s. 6d. as the price, we have 15s. 10d. per square. For the plastering we have 200 feet = 22⅔ yards at 1s. 4d. = 1l. 9s. 7½d. For the paper, using a 50 feet divisor, we find four pieces per square = 4s. Hanging paper, four pieces at 9d. = 3s. = 2l. 12s. 5½d., say 2l. 12s. 6d.

Floors.—Floors may be treated as follows; include the plan of internal walls and partitions in the superficial area. The voids will be disregarded in consideration of the necessary trimmers, &c.

					£	s.	d.
2/	70 0		9" × 3" joists, 37s. 9d.; sound board-				
	25 0	8500 0	ing, 10s.; pugging, 2s. 6d.; 1¼" Y.B. floor, 27s.; L.P.F. & S. ceiling,				
		85 sqrs.	18s. 6d.; 2ce whiten, 2s. 9d.	98s. 6d.	172	7	6

The prices would be computed as follows. Referring to the table of timber (see Laxton), we find for 9" × 3" joist 16' 2" cube; adopting 2s. 4d. as the price, we have 37s. 9d. Sound boarding say 10s., pugging say 2s. 6d., 1¼ floor say 27s. L.P.F. & S. ceilings say 1s. 8d. per yard = 18s. 6d. square. Twice whiten say 3d. per yard = 2s. 9d. per square; total 4l. 18s. 6d.

Collect the plates and strutting in the usual way.

Ceiling joists should be measured in a similar way with the ceiling and its whitening.

Joiner.—For floors see Carpenter. Skirtings must be measured lineally and the price of mitres and housings included in the price.

Doors, see Bricklayer. Windows, see Bricklayer.

Fittings must generally be measured in the usual way.

For dressers, it is customary to settle a price per foot run, 10s. or 12s.

Staircases will generally require measuring in the usual way, but for approximate estimates the surveyor often settles on a price per step, to include balustrades and other accessories, 10s. to 15s. per step. For instance a complete staircase of 24 steps at the former price would appear as 12*l.*

Plasterer.—The plastering to ceilings and partitions having been taken with the timber, the walls must be measured superficially in the usual way. The junctions of the internal walls and partitions with them may be disregarded and measured in. Deduct the openings in all trades, adopting a similar dimension for all.

The following may illustrate the measurement of one floor 30 by 20 feet and 10 feet high.

| | 100 0
10 0 | 1000 0
—————
111⅑ yds. | R.F. & S. walls, 11*d.*; and paper at 1s. 6*d.* per piece, 3¼*d.*; and hanging, 1¼*d.* .. .. | 1s. 4*d.* | £ s. d.
7 8 2 |

The prices would be computed as follows: plastering say 11*d.* per yard; for the paper, using a 50 feet divisor = 5⅔ yards, and dividing 1s. 6*d.* by it we have per yard 3¼*d.*; for hanging, say 9*d.* per piece, equals 1½*d.* per yard superficial; total 1s. 4*d.*

If the deduction for doors and window has not been dealt with in the bricklayer, it should be made now.

| 10/ | 7 0
3 0 | 210 0 | 1½ brickwork, 1s.; red facing, 3*d.*;
R.F. & S., 1¼*d.* .. .. | 1s. 4¼*d.* | £ s. d.
14 4 5 |

The prices would be computed as follows, brickwork say 13*l.* 12s. per rod = 1s. per foot reduced, red facing say 3*d.*, render float and set, say 11*d.* per yard = 1¼*d.* per foot; total 1s. 4¼*d.*

Cornices measured lineally in the usual way, but let the price per foot include all mitres, stopped ends, bracketing, etc.

Count the centre flowers, if any.

Smith and Founder.—The work to this trade can only be measured in the usual way.

Plumber.—The leadwork must be measured in the usual way, but the superficial price should be computed to cover all extra labours.

Pipes should include all the joints and other labours in the price per lineal foot.

Lead flats may be taken with the carpentry, assuming the rolls as 2" 6" centre to centre.

					£	s.	d.
20 0			Joists 7" × 2½", 3d.; ¾" rough board-				
10 0	200 0		ing and firring, 8d.; 2" rolls, 1d.; 7 lbs. lead, 1s. 11d.; L.P.F. & S. ceiling, 2¼d.; 2ce whiten	2s. 8¾d.	27	5	10

The prices would be computed as follows. Referring to the table of timber, we find for 7" × 2½" joists, 10' 10" cube, about 1¼" cube to each superficial foot, this at 2s. 6d. is 3d. per foot; 1" rough boarding and firrings, say 25s. per square = 3d. per foot. Each superficial foot would require about 4" of roll at 3d. = 1d. The leadwork of each superficial foot of flat would be increased by about 4" for the rolls. 1' 4" of 7-lb. lead weighs 9¼ lbs.; this at say 23s. 4d. per cwt. = 1s. 11d. L.P.F. & S. ceiling at 1s. 8d. per yard = per foot 2¼d.; twice whiten at 3d. per yard = ½d. per foot; total 2s. 8¾d. When one flat is calculated in this manner it will often happen that the price may be applied to other flats of the same building with very slight modification or none.

Sanitary apparatus must be counted and priced from a trade list.

Gasfitter.—This work is not generally measured, the lights are counted and the piping priced at a price per point: 6s. is about the lowest price, 7s. 6d. is a reasonable average.

Count the fittings and price them from a trade list.

Bells.—These, either ordinary or electric, are counted and priced at a price per pull.

Painting.—This may be priced with the joinery or taken out in detail; if we assume four oils as 1s. per yard and add ¼ for edges = about 1¼d., we have 1s. 1¾d. per yard, or 1½d. per foot. Or a percentage may be added to the total of the other trades added together. The amount for plain painting rarely exceeds 2½ per cent. of that total.

Paperhanger.—It is a common practice to paper a house throughout with lining paper if papering is done immediately after the rest of the work is finished, and in such a case methods of dealing with it suggested in the foregoing paragraphs can be applied, but when there is a considerable variety of value in the papers the only practicable way is to measure the various kinds.

Some of the paperhanging manufacturers publish a table of the number of pieces of paper required for a room of a given size, and which may conveniently be used for an approximate estimate, as follows:—

The top line is the measurement round the Walls in Feet, including Doors, Windows, &c.

Height in feet from Skirting to Cornice.	28	32	36	40	44	48	52	56	60	64	68	72	76	80	84	88	92	96	100	
7 to 7½	4	4	5	5	6	6	7	7	8	8	9	9	9	10	10	11	11	12	12	pieces
7½ to 8	4	4	5	5	6	6	7	8	8	9	9	10	10	11	11	12	12	13	13	,,
8 to 8½	4	5	5	6	6	7	7	8	8	9	9	10	10	11	12	12	13	13	14	,,
8½ to 9	4	5	5	6	7	7	8	8	9	9	10	11	11	12	12	13	13	14	14	,,
9 to 9½	4	5	6	6	7	7	8	9	9	10	10	11	12	12	13	13	14	15	15	,,
9½ to 10	5	6	6	7	7	8	9	9	10	10	11	12	12	13	14	14	15	15	16	,,
10 to 10½	5	6	6	7	8	8	9	10	10	11	12	12	13	14	14	15	16	16	17	,,
10½ to 11	5	6	7	7	8	9	9	10	11	11	12	13	13	14	15	16	16	17	18	,,
11 to 11½	5	6	7	8	8	9	10	10	11	12	13	13	14	15	16	16	17	18	18	,,

Glazier.—See other trades.

The pricing of estimates and bills of quantities forms a large part of the ordinary surveyor's work.

The drawings should always be examined before or during the pricing of an estimate, as however well the quantities may have been prepared, a sight of them will generally affect the estimator's conclusions.

It should be observed that when a large quantity of a particular *material* is required, specially low quotations may be obtained from manufacturers. In many cases, as for slating, ironwork, glass, etc., a builder in preparing his estimate sends a copy of the bill of such part of the work to a firm, or several firms, of slaters founders, or glass merchants, and adopts the price furnished by one of them, with the addition of a percentage. The same course is frequently available for the quantity surveyor, and will save him much trouble.

The surveyor, in pricing a bill of quantities, should make himself familiar with the manner in which they have been prepared, concrete being sometimes described as including staging; excavation, as including strutting and planking; brickwork, as including all relieving arches; facings, as including many of the cuttings, as extra on common brickwork, as extra on facing of a commoner kind, etc.; stone, including labour; carpentry and joinery of finished sizes, etc., and it is obvious that such conditions must modify the rates.

In the calculation of the value of a foot superficial of work it will usually be best to compute the value of a large quantity, and from that calculate the smaller, as when the difference in price between a superficial foot of square framed and moulded door 3 feet by 7 feet is required, the whole of the moulding on both sides being collected and a value placed upon it, the result when divided by twenty-one will give the amount per foot superficial to add to the original price.

In pricing bills of quantities the expert estimator does not worry himself with nice distinctions, for instance he will probably use one price per foot cube for all the wooden frames, although the labour upon them may vary, in the result some of the prices may be a little too low but others will be a little too high, and the differences will probably balance each other; he will adopt analogies such as a similar price for rubbed steps and rubbed curbs, etc., and by these means will save himself much trouble and produce a result as safe as by more refinement of practice.

For all work there will be various preliminary charges to allow for. In London there will be district surveyor's fees, settled by the schedules of the Building Act.

Fees for the parochial surveyor for hoarding, connections with sewer, taking up, and making good paving, etc., to be obtained by inquiry at the office of the local board.

Carriage.—Sometimes material is most cheaply carried by a canal route. The charges may be obtained by application to the particular canal company.

A complete map of all the canals and inland navigations is attached to the Report of the Select Committee on Canals, May 1883, vol. 13, Parliamentary Papers.

DOCKS IN LONDON.

The Surrey Commercial Docks is the principal place of entry for the timber and deals, prepared flooring and matched-boarding, pine, oak and pitch-pine from Northern Europe and North America; the West India Docks and Millwall—principally the former—for cedar, mahogany, teak, American oak and walnut, sequoia, whitewood, kauri pine, etc. Much of the material thus landed is conveyed by barge up the Surrey canal for the supply of South London, enters the Regent's canal at Limehouse dock or Bow creek for the supply of North or Central London, or goes by that route into the Grand Junction canal for the supply of the home counties. Large quantities are carried by barge to the Upper Thames, and still larger quantities carted either by the saw-mill proprietors (who saw and deliver), or directly by the purchasers. The dock dues are published in printed schedules; but they only slightly concern the student of prices. The wood supplied to London is usually purchased at the sales held frequently by the merchants, the greater part at Winchester House, Old Broad Street, and the Cannon Street Hotel. When a large quantity of wood is required the builder buys at these sales; when smaller quantities, of a retail timber merchant. At most of the large seaports of the kingdom there are periodical timber sales. Landing rates and saw-mill regulations will be dealt with in the section Carpenter.

THE RIVER THAMES.

It is probable that the wharves of the Thames have been less used for the introduction of materials since the great development of the railway system, the rates for railway carriage being reduced to their lowest point wherever shipping comes into competition with them. Besides this, a coasting vessel of average tonnage carrying a cargo for delivery above London Bridge must transfer it to lighters before it can be deposited at wharves up the river, and this involves additional expense. This consideration does not affect the traffic from the Medway or the estuary of the Thames, which is carried in smaller craft, and consequently large quantities of bricks, sand, cement, lime, ragstone, etc. are brought from Kent and Essex by water. Deals and timber are also conveniently brought by barge from the Surrey, East and West India, and other docks to the river wharves, and as some of these belong to the railway companies, access is thus afforded by their systems to all parts of the country. There is still, however, a large traffic in seaborne materials, as granite, Portland stone, Yorkshire stone, bricks, lime, slates, stoves, ranges and miscellaneous iron castings, iron joists and girders, rain-water pipes and gutters, etc. The drawwharves on the Thames are much used for the shooting of rubbish into barges. Exclusive of loading and unloading, the lighterage from the Surrey or India Docks to any wharf between London Bridge and Westminster costs:—

    25 tons and over, per ton, Deals, 1s. 6d.
    25   „   „   Mahogany, 1s. 9d.

Bricks, Southend to any wharf between London Bridge and Westminster, 4s. per thousand, exclusive of loading and unloading. These charges would include all dues. The Thames Conservancy tolls do not much concern the contractor, as he generally makes his bargain with either merchant or lighterman at a price which includes them.

For the supply of materials for country work, it will be a question which is the dominant town of the district. Timber and other materials are taken from London, until some other port counterbalances the metropolitan advantages, when they are taken from that.

Following the eastern coast northward, we have Colchester Ipswich, Harwich, Yarmouth, Lynn, Grimsby, Hull, Sunderland, Newcastle. On the south coast we have Southampton, Portsmouth, Plymouth. Going north-westward, as soon as Oxford is passed, materials not of local origin would be brought from Gloucester or Bristol. Further to the north-west, we are under the influence of Liverpool, and since the construction of the Manchester Ship Canal Manchester is gradually becoming a new source of supply.

RAILWAY RATES.

A familiarity with railway rates will be useful to the surveyor in the adjustment for claims of carriage, the valuation of materials which are in some cases delivered into trucks at the railway station adjacent to the manufactory, in others at that nearest to the proposed building. In the valuation of country work, as in the conduct of a building distant from London, much of the material must be worked at the contractors' shops in town and conveyed somehow to the building. Except for small parcels sent by passenger train and for return of empties, there is no published scale of charges. As a rule, the rates of the various companies for similar quantities conveyed similar distances in a similar way are very much alike; the exceptions are mainly referable to the acquirement by a large company of a smaller one, whose original Act contained (for some reason now possibly extinct) specially high rates, which are still maintained. Through rates between the stations of one company and another may be best obtained by inquiry of the station master of the station from which the goods are despatched, or at the chief office of its company, or the rate per ton for carriage may be obtained from the manufacturers if the question is asked at the time of making inquiry about prices. As a general rule, the cost of railway carriage does not increase in a regular ratio; the cost of conveyance for 100 miles is much less than twice that for 50 miles. Railway companies convey at two rates: one under which they are liable for damage, called company's risk; the other (a lower one), under which they are not so liable, called owner's risk. It is obviously to the interest of the sender to adopt the latter when the goods are not easily damaged. Goods rates are modified as follows:—A certain set of charges is applied to all consignments not exceeding 500 lb. in weight, regulated by a table used by all companies alike. Under this table nothing less than 28 lb. is charged, and the charges for weights above 28 lb. increase by multiples of 14 lb. This scale applies uniformly according to its rate to all carriage between stations in England, south of Newcastle and Carlisle, and in Wales. A second set of rates is applied to material exceeding 500 lb. and under 2 tons, called class rates, which include collection and delivery within the usual limits (three or four miles) in towns; a third set of rates, for two tons and over, called special class rates, which do not include collection and delivery. It is often cheaper to pay for a consignment of two tons, although it may weigh but little over one. At small country stations nothing is delivered except very small parcels in the immediate neighbourhood. In London, such agencies as the London Parcels Delivery Company, Carter Paterson, and the Metropolitan Railway convey and deliver small parcels at cheap rates, and similar facilities exist in most of the large towns of the kingdom. Small articles, designated by number in a trade list, may consequently be ordered by letter and conveyed in this manner. This may be remembered when a claim is made in a day account for labourer's time for half a day to fetch such a thing as a tap or other small article. When goods are ordered from the manufacturer by the truck load, in stating the position of the work, inquiry should be made as to the station to which he proposes to consign it, and it is advisable to test this information by an independent inquiry of the station master of the nearest railway station to the proposed building, as he can frequently suggest a more convenient station for its delivery. As the Acts of Parliament under which particular lines of railway have been established only enumerate the rates of charges of quite a small variety of articles, there has gradually developed a classification of rates agreed on by the various companies called the Clearing House classification, in many respects unsatisfactory. Some of the railways carry timber at the *measured* ton. One principle of measurement is prevalent in some parts of the country called string measurement, another tape measurement, another caliper measurement; these differ materially one from the other, and all differ from the machine weight. When considering railway charges, the surveyor should insist upon the production of the original railway delivery note, which simplifies the matter when settling the accounts for a building erected or in progress; but in estimating the cost of a prospective work, the surveyor must be familiar with rates and the method of their charge. A map for the measurement of distances will prove useful in the calculation of all kinds of carriage.

Delivery by Manufacturers and Merchants.

The makers of small articles do not usually deliver anything less than a cart-load, but when as much as a load is ordered, they may generally be induced to deliver without extra charge. In ordering a truck or barge load of material, it will be advisable to obtain a price, including carriage and delivery, as the merchant will often, for the sake of obtaining a good order, forego a part at least of the dues involved. The provincial manufacturers' price is generally stated to include delivery into trucks or barges at the nearest available point to their works; but they will quote a price, including delivery, at a station, canal, or river wharf adjacent to the proposed works.

The makers of drain-pipes and similar goods, when a load is ordered, deliver free in London and its suburbs; or, if for the country, at a London railway station; smaller quantities must be sent for and carted by the purchaser, but it does not follow that every single article is separately sent for, as the zealous advocate will sometimes contend. The majority of the country manufacturers of machinery limit their free delivery in London to articles over 5l. in value, and many of them will deliver free at any railway station or shipping port in England. Country manufacturers of cast-iron pipes deliver free in London or any other town where the rate of carriage is equivalent. The saw-mills, in the cases in which wood has involved a fair quantity of sawing, will collect timber and deals at the docks, and deliver to the builder free of charge. The delivery of stone is almost always done by the merchants, who will, however, if desired, quote a price exclusive of carriage; but the builder can rarely cart it so cheaply as the stone merchant.

Packing.— Articles liable to injury are packed in cases, the charge for which is included in the invoice. Packing cases are nearly always charged at exorbitant rates, and in the investigation of p.c. by inexperienced persons, this charge is often allowed, but the general practice of manufacturers is the allowance of two-thirds of the amount charged if returned. The railway companies carry returned empty cases, casks or sacks at a very low rate.

CARTAGE.

Although convenience appears to dictate the separation in this inquiry of the various trades, it is difficult in dealing with cartage to entirely avoid allusion to excavation, to which a trade or separate section will be afterwards devoted. The question of cartage is almost certain to arise in the adjustment of a day account, or a measurement on a schedule of prices; unless it is stipulated that the prices of items shall include it. In the case of a schedule, it is generally understood that the prices include delivery; it is, however, safest to settle the question by a stipulation in its preamble. Cartage in a city like London will cost more than in its suburbs or the country, because of the congested traffic; it should also be remembered that in a hilly neighbourhood the cartage of fewer loads in a day, and consequent greater cost, must be allowed for. A rough engineering axiom is "one shilling a load a mile."

Assuming that a horse, cart and man can in a day cart ten loads each a mile, at 1*s.* per load, we thus have a result of 10*s.* per day. A common valuation of a load on a return journey from an original delivery is one-half the price of the latter. There are a considerable number of men in London who are willing to contract to dig and cart away, and find a shoot at a price per cubic yard, measuring the digging in position before it is dug, as the surveyor does ("hole measured"). The method of measurement should be clearly defined in the agreement. General usage would be a strong argument for this construction, but should not be depended upon. When carting alone is agreed for, the builder arranges with the sub-contractor that his carter shall help to fill the cart. When the material removed is hard brick or stone rubbish, the carter may sometimes obtain more per load by its sale than the cost of the cartage. Sometimes brick or stone rubbish for which removal is provided in the contract is allowed to be used as one of the ingredients of concrete, and in such case the quantity of carting saved may be pleaded as an element of reduction, modified by the cost of breaking. Either in town or country, when the earth is good vegetable soil, it may often be sold for the purpose of making mounds or filling up depressions. In a well-known case in the country an arrangement was made with the freeholder of a field on the opposite side of the road to the contemplated building for the deposit of the earth thereon. A temporary wooden bridge was built across the road, the earth carted over it and a large saving in cartage thus effected. A little trouble in examining a neighbourhood, especially in the country, will often discover a place where the earth will be welcome for filling. In London it will sometimes be cheaper to shoot the earth or rubbish direct into a barge at the nearest draw-wharf, paying the wharfinger's charge for the convenience (usually 2*s.* per load), than to cart it to a land shoot, often in a distant suburb. In the latter case the cost may be reduced by arranging for the cartage of bricks, ballast or sand on the return journey instead of coming back empty.

Current Rates for Cartage, Etc.

Horse, cart and man per day, 10*s.* (10 hours at 1*s.* per hour); ditto in the country, sometimes as low as 7*s.* 6*d.*; two horses, cart and man per day, 19*s.*; chain horse per day 9*s.* Van (to carry

2 tons), horse and man per day, 10s.; timber-carriage, horse and man, 10s.; ditto, two horses, 19s. Stone-truck, one horse and man per day, 10s.; ditto, two horses, 19s.; ditto, three horses, 25s. Furniture van, horse and man per hour, 1s. 6d.; ditto, two horses, 2s. 6d. In the City of London, rubbish is carted, including finding a shoot, at 3s. per load. Soft material, not easy to consolidate, 3s. 6d. per load. In the suburbs, rubbish is carted, including finding a shoot, at 2s. 6d. per load. Permission to shoot rubbish into barge at a river wharf, per load, 2s. In a suburb like Brixton or Camberwell carmen may be found who will dig, cart and shoot at 3s. 6d. per yard (hole measured). When sand is delivered in a London suburb by barge, a local contractor will unload, cart and shoot within two miles for 2s. per yard—sometimes as low as 1s. 9d.; will unload bricks and deliver into carts for 1s. 6d. per thousand; will pay canal dues, unload bricks, load into carts, deliver and stack within two miles at 4s. 9d. per thousand. An approximation to the usual charge of stone merchants for cartage is 5s. per load of 1½ tons within four miles. The charge for delivery of Bath stone (Bath Stone Firms) is 2d. per foot cube within four miles; the minimum charge is 1¼d., increasing at the rate of about ¼d. for each half mile. Cartage of deals from the Surrey Commercial Docks to St. Paul's, or equal distances, 8s. per standard; ditto timber per load, 3s. Some of the distinctions made as to cartage in a schedule of the War Office may usefully illustrate the subject :

		s.	
100	Carting rubbish arising from every description of trades from the premises of the War Department, including filling the carts, or loading, finding a deposit, and unloading, per yard cube, or load	2	0
101	Carting materials, water, rubbish, etc., one furlong or under, including loading and unloading, do.	0	9
102	Ditto for the first load when not more than two loads are ordered to be removed on the same day, do.	1	3
103	Add to items 101 and 102 for every additional furlong ..	0	1¼

Mr. Hurst says in the 'Architectural Surveyor's Handbook':—
" The maximum distance to which earth can be wheeled in barrows economically is 100 yards, in dobbin carts (three-wheeled carts) 300 yards, in ordinary one-horse carts half-a-mile; when the distance is over half-a-mile it will be more economical to use waggons on rails." He has here in his mind very large excava-

PRICES.

tions, like railways, whose design usually involves such a balancing of cutting and embankment as makes it convenient to run the earth in waggons on rails from the former to construct the latter (a process which has enriched the vernacular with the phrase "straight tip"). In the average operations of the builder carting in tumbrels is necessary.

A common price for cartage in the country is 1s. per ton per mile. A loaded cart travels about 3 miles per hour.

The Usual Load for Various Vehicles.

A tumbrel carries ordinary earth $1\frac{1}{4}$ yards, equal to 1 yard before digging; sand, $1\frac{1}{4}$ yards. A builder's cart carries $1\frac{1}{2}$ to 2 tons; do. carries ordinary bricks, 500; do. glazed or Staffordshire blues, 400; do. plain tiles, 1000; do. countess slates, 1000; do. flooring, about 12 squares; do. timber or deals, 50 cubic feet; do. lime, 1 cubic yard (usually conveyed by the merchant in a one-horse van, carrying 2 yards); do. light bulky materials, 80 cubic feet, 72 scaffold-boards. A stone-truck, 3 to 10 tons; railway vans, two-horse, about $2\frac{1}{2}$ tons; railway trucks, 8 to 10 tons; a Thames lighter, 90 to 120 tons; a van and one horse will carry about 35 scaffold-poles, each 25 feet long; a navvy's wheelbarrow about 50 bricks, or 4 hods of mortar; do. earth, about $\frac{1}{10}$th of a yard cube; the average earth waggon, about 50 barrow-loads; a large do., heaped, 3 cubic yards; a small do. do., $2\frac{1}{2}$ cubic yards; a stone-truck takes about 90 feet of freestone.

Rings.

Merchants and manufacturers of various articles used in building have combined to keep the prices up to a certain rate in each case, and issue a tariff to which the majority adhere. The astute builder will nevertheless contrive to get large orders executed at a lower rate by special agreement. This applies to drain pipes, stock bricks, glazed bricks, Bath stone, Portland stone, plate glass, and iron.

Contractions.

Merchants' quotations will often contain contractions which indicate that all charges are paid up to that point:—

F.O.B.	Free on Board
F.O.V.	,, ,, Van
F.O.R.	,, ,, Rail

SCAFFOLDING.

The use of scaffolding and sheds may be looked upon as establishment charges, but they are most conveniently and exactly dealt with in their connection with a particular building. As they are means to the end of obtaining a profit, they had better be treated as a net outlay and no profit added. Some builders calculate cost of scaffolding at so much for each rod of brickwork. Probably under ordinary circumstances it costs 4s. per rod. An exact calculation may be made by taking out the quantities of it, a row of scaffold poles at the usual distance completely encircling the building, ledgers at about 5 feet apart, and the necessary putlogs, boards, ladders and cords, calculate the cost, and on this cost, say, 7½ per cent. per annum for interest on capital and depreciation, and add to it labour, in fixing, and the number of loads carted. When the work is in the country, the number of loads requiring carting and railway carriage must be allowed for. For large halls or churches, interior as well as exterior scaffolding will be required, and the use of this should be separately calculated. The cost of scaffolding will vary according to circumstances. Sometimes the builder may obtain work in the neighbourhood of another building from which the scaffolding has not yet been removed—he can consequently transfer it at a small expense; at other times the cost of carriage is so great that it will better serve his purpose to buy new in the vicinity of the proposed building. It is the frequent practice of the contractor for railways, docks, canals, or other great works of engineering to buy new plant for a particular work, selling it at the close for what it will fetch. When works are delayed for a long period, as sometimes happens, hire of scaffolding will probably be claimed, and properly so. The decay of scaffolding is rapid; a year's delay in the progress of a building leaves the cords useless for another work, and many of the poles badly decayed at their butts. The boards and putlogs would probably have suffered but little. If a definite charge for scaffolding appears in the deposited bill of quantities, the stipulated time for the completion of the building may be compared with the time of delay, and a rate per week, month, or year calculated in proportion, or the scaffolding may be valued, and the value of its hire calculated as a percentage of its first cost; 7½ per cent. per annum would be enough—i.e. 5 per cent. interest on capital, and 2½ per cent. for depreciation; or the charge may be estimated from a list of net charges in London for hire of building plant, which would much exceed a calculation on the lines first suggested, and are rarely allowed by experienced assessors.

Scaffolding represents a larger percentage upon work now than it did before the passing of the Employers' Liability Act.

In cases of alterations and repairs a price must be fixed for scaffolding. The refacing or pointing of a house front or the restoration of a church spire are familiar instances. Some estimators habitually use a price per foot run about 1*l.* for a house of four stories, and make an addition in the same proportion for any height beyond that. For such work as spire restoration it will be necessary to roughly calculate the quantity. Very often substantial beams are laid across the tower, passing out through the belfry windows, and the scaffold carried up from them. It is often not necessary to begin at ground level.

The foregoing deals with scaffolding of the ordinary character. In large buildings, it will be necessary to use scaffolding of squared timber connected by iron dogs and bolts. In such a case it must be designed, and the quantities taken out charging it as use and waste. Twenty feet or thereabouts of the length of a façade will be enough to base the whole calculation upon. Illustrations of scaffolds of this kind are to be found in Spons' 'Dictionary of Engineering' and Seddon's 'Builders' Work.' See also the article on "Scaffold" in the 'Dictionary of the Architectural Society,' and Cresy's 'Encyclopædia of Civil Engineering'; and many interesting examples may be found in the *Transactions* of the Institution of Civil Engineers.

Tarpaulins.—These would generally be used in the construction of temporary roofs, which would involve some arrangement of supports for the tarpaulins. The estimator must make a calculation of the material, pricing it as use and waste, and adding thereto the cost of fixing and removal. The same principle must be adopted for temporary floors. These contingent arrangements are so various that the only course open is clearly to realise the requirements, design something suitable, and price it. The price for the hire of tarpaulins is regulated as follows.

Tarpaulins from 60 *to* 900 *ft. each—per* 10 *yds. superficial or under.*

First Week.		Second Week.		Third Week.		After Third Week.	
day.	week.	day.	week.	day.	week.	day.	week.
s. d.	s. d.	s. d.	s. d.	s. d.	s. d.	s. d.	s. d.
0 1½	0 6	0 1	0 4	0 0½	0 2	0 0¼	0 1

HOARDING.

When the work involves the pulling down of an old building, enough timber and boarding can generally be obtained for use as a hoarding, or the builder has possibly more old material at his yard than he well knows what to do with, and this is available for such purposes. The outlay is then for cartage, fixing and removal only. When there is no old material new must be used, and the only exact way is to realise what will be necessary, and measure and value it.

Some builders add no profit to cost of hoarding, and the dealing with the material as use and waste disposes of that consideration. When there is no stipulation in the contract against advertising, the hoarding may be let to an advertising contractor. The price per foot superficial varies with the locality, ranging between 1*d.* and 7*d.* On some of the larger building estates it is especially prohibited by the freeholder, and often the building owner prevents it by the terms of the contract. Since the passing of the Advertising Stations Rating Act, 1889, many of the vestries and district boards have prohibited advertising on hoardings; others charge a fee of so much per superficial foot of hoarding. Sometimes the advertisement contractor will include in his offer for the advertising privileges the erection and removal of hoarding. It is, however, in many respects, preferable that the builder do it himself. In cases of one builder doing the work of the basement, and another the superstructure, the first builder would erect the hoarding and leave it for the second, in which case the use and waste principle would not apply, and the whole value would be charged.

Some estimators habitually price hoarding at 10*s.* per square and fans at 6*d.* per foot run. The license for hoarding payable to the local authority is commonly about 2*s.* 6*d.* per month.

SHORING, GANTRIES, ETC.

A visit to the building site will be necessary for the valuation of this item. Stock or Blagrove on 'Shoring' will show the student what methods of shoring are best. If the site is cleared, and with old buildings on one or both sides, the shoring will probably be already done, and the question must be determined (if not provided by the specification) as to who is to pay for it, and to whom the timber shall belong when removed. When an old building is to be pulled down, sufficient old timber for the purpose is often available, when labour of fixing and removal, and cost of spikes, wedges, and hoop iron are all that need be considered. If a credit amount has been allowed for the old building, something must be added for use and waste of the old timber. When new timber is involved, its cost must be calculated as use and waste, adding thereto the cost of labour, wedges, spikes and hoop-iron as before. A common contract price in London for use and waste of shoring, including labour, wedges, spikes, hoop-iron, removal and profit, is 1s. 3d. per foot cube. When, from lack of room to store materials, or from necessity to keep the whole footway open, a gantry is required, it must be designed, the quantity of materials and labour calculated and priced, as recommended for shoring. In the country it will often be necessary to inclose the site of the building and a space around it with a post-and-rail fence, to prevent the men from trespassing on the adjacent land. This must be calculated in a similar way; but this will not be estimated for in the absence of an express stipulation. If, however, there is a condition in the contract to preserve the adjoining property or grounds from damage, contractors should take some precautions of the kind.

FOREMAN.

In pricing a bill of quantities or otherwise preparing an estimate, the cost of maintaining a foreman on the works is best kept separate. He will be on the work at the beginning and ending of the building operations, and at both periods there will be very few men to supervise, consequently the general prices are best calculated without foreman. It will be expedient to find out either by calculation or reference to the early part of the bill, or to the conditions attached to the specification, what stipulation there may be as to time of completion. Calculate somewhat more than the prescribed period at about 60s. per week, the average rate of payment. In the country, the foreman would as a rule, be paid lodging money beside. In pricing work the foreman's time in day accounts is generally treated as a charge which the general rate of profit should cover. An allusion to foreman generally appears in a well prepared schedule for day accounts. A clause is inserted which deals with general foreman's time, sometimes as follows:—" The above prices and prices for day work generally to include use of scaffold, and all tackle, tools, tool sharpening, water, and general foreman's time; subordinate foremen to be charged as ordinary workmen. Time for fixing and removing scaffolding will be allowed." It is sometimes advisable to depart from this rule when a foreman has only two or three men to supervise, as in the case of the time for completion being protracted by items of extra work. A good foreman on a building benefits all the parties concerned. He should know the quantity of work every man should do, and the amount of work which all the men together should do in a given time, and can help the work forward to a very great extent by putting men in their right places.

Fire Insurance.

An agreement exists between the leading insurance offices as to rates. The charges for buildings in course of erection are as follows. There are some offices which charge more; but this scale

—	1 month.	3 months.	6 months.	9 months.	12 months.
For each 100l. assured	s. d. 1 0	s. d. 1 3	s. d. 1 9	s. d. 2 0	s. d. 2 0

may reasonably be taken for the pricing of estimates. Where more is claimed, the proper course is the production of the receipt. Insurance is usually postponed until some combustible material is fixed, if it is left to the option of the contractor; but more generally there is a specific stipulation as to time in the conditions of contract.

When the time allotted by contract for the erection of a building is exceeded, it becomes a question who shall pay the further charge for insurance. If the delay is the fault of the contractor, he should pay; if not, the building owner is liable.

WATER.

A bill of quantities should include an item for supply of water, and in any case the estimates must allow for it. When no water company supplies the neighbourhood, water may sometimes be obtained near the surface by digging, and the cost of well curb, digging, and use of tubs for storage must be reckoned. In some cases, when water is near the surface, Norton's tube-wells may be advantageously used. When there is an existing well and pump, the cost of pumping and storage, or it may be necessary to bring it from a lake or river, in which case a butt on wheels is most convenient for the purpose; when one such source of supply is clean and the other dirty, the use of the former will probably be insisted upon. A local carter will probably have a water-cart suited for the purpose, or a street watering-cart may be available, or a liquid manure cart; whichever it may be, its capacity must be ascertained, the number of loads a day, possible or required, and the price per load for cartage. A butt will convey 108 gallons, a liquid-manure cart generally 120 gallons, water carts are made to contain 200, 250, or 300 gallons, water-vans 350, 400, or 450 gallons. The filling with a bucket on the end of a pole is a tedious but common method, and it is preferable to agree for water supply by the load, to hiring cart and man by the day. When there is a local water company, the cost of a building supply may be easily discovered by inquiry. Some of the country corporations supply their jurisdiction with water, and publish their water regulations with their bye-laws. London and its suburbs are supplied by eight water companies, each of which publishes a set of regulations to be observed in its district. These special regulations are similar in character. They generally stipulate that pipes shall be a certain depth below the surface, that the fittings shall be approved by the turncocks before connecting with main; some of them define the size of pipes and describe the cocks. The regulations of the local water company should always be examined before making an estimate, and they should be preserved for future use. Each of the London companies is constituted by a special Act of Parliament, which incorporates the public Acts relating to water companies. The chief concern of the estimator is the schedule of special regulations published by the company (obtainable by any applicant), and the Board of Trade regulations made under the Metropolis Water Act, 1871, which affects all the Metropolitan companies. These are amalgamated by some of the companies with their general regulations; others publish them separately, or copies may be obtained from the Government printers (Eyre and Spottiswoode). Application for a building supply is made generally on a printed form, the different companies requiring different information. The company makes connection and opens ground and makes good, only so far as thus rendered necessary. The list of water companies with their offices is to be found in the 'London Post Office Directory.' The usual charge for making connections and supplying ferule is 5s. (except Lambeth, which see). The cost of a building supply will be made up as follows:—

Water company's charges for connection
— feet of pipe, *use and waste* .. ..
Ball-cock .. ..
Soldered joint .. ..

All the companies require a stop-cock in the permanent supply pipe immediately outside the boundary of the property. The companies will not, as a rule, allow a pipe laid for building supply to be used for the permanent purposes, except by special application; but if a pipe of the necessary size be laid in at first, the application would certainly be granted, and it may be used for the building supply. The charges by the Metropolitan companies for a building supply are as follows. Some of them are based on annual values, some on the cost of the building, others on the work for which water is required, and this latter is the most reasonable. There is a map of the district supplied by each company in Firth's 'Municipal London':—

Chelsea.—¼ per cent. upon estimated cost of building.

East London.—One shilling per rod on brickwork; one penny per cubic yard on concrete.

PRICES.

Kent.

		s.	d.
Not exceeding £100		10	0
100 and not exceeding £150		15	0
150 ,, ,, 200		20	0
200 ,, ,, 250		25	0
300 ,, ,, 350		35	0
350 ,, ,, 400		40	0
400 ,, ,, 500		45	0
500 ,, ,, 600		50	0
600 ,, ,, 700		55	0
700 ,, ,, 800		60	0
800 ,, ,, 900		65	0
900 ,, ,, 1000		70	0
1000 ,, ,, 1100		75	0
1100 ,, ,, 1200		80	0

Above 1200 by special agreement.

New River.

£	each s.	£	each s.	£	each s.
100	10	250	25	400	0
125	13	275	28	450	42
150	15	300	30	500	45
175	18	325	32	600 to £700	50
200	20	350	35	800 .. 900	60
225	23	375	38	1000 .. 1200	70

Above 1200, 5 per cent. additional.

West Middlesex.—5s. per cent. on the estimated cost of building. When engines are used, an extra rate of 1l. 1s. per horse-power per annum will be charged.

Southwark and Vauxhall.—No published scale; rates based on value of building will be given on application.

Grand Junction.—Charges on estimated cost of building:—

£	£	per cent. s. d.	£	£	per cent. s. d.
100 and under	500	8 0	10,000 and under 20,000		4 0
500 ,, ,,	1,000	7 0	20,000 ,, ,, 30,000		3 0
1,000 ,, ,,	3,000	6 0	30,000 and above		2 6
3,000 ,, ,,	10,000	5 0			

Should difference of calculation arise as to the estimated cost, then the charge to be fixed upon the estimated annual value at 1s. per 1l. sterling of such value.

Lambeth.—No published scale; rates based on value of building will be furnished on application.

QUANTITY SURVEYING.

		s.
Charges for connection, including stop-cock, ferule, opening ground, and reinstating, constant supply district only—½ in.		15
¾ in., country district		21
1 in., „ „		30
1¼ in., „ „		35
1½ in., „ „		42

Other sizes by special agreement.

The above charges for opening, reinstating and relaying are for ordinary trenches only; but if in opening a trench tar paving has to be broken and reinstated, an extra charge of 5s. will be made. If asphalt or wood paving, the extra charge will be 10s.

If the required quantity be very large, and the time for erection long, application should be made for supply by meter, and the charge would be at similar rates to the following. It would also be necessary to allow for the use and fixing of a water-meter, or the company would supply a meter at a rental.

Quarterly consumption.	Per 1000 gals. d.
From 25,000 to 50,000 gallons	9
„ 50,000 „ 100,000 „	8
„ 100,000 „ 200,000 „	7
All above 200,000 „	6

If high service, 25 per cent. additional on these rates.

The following are instances of amounts included in estimates by various estimators for temporary plumber's work and storage of water:—

—	Cost of Building.	Amount.	—	Brick-work.	Plaster-ing.	Con-crete.
	£	£		rods	yards	yards
House at Kensington	5200	1	Water Company	52	2470	79
House at Hornsey	2000	4	do.	31	1840	60
House, Oxfordshire	3900	7	Well 75 yards away	55	15 0	177
Alterations to London printing office	2700	10	Water Company	11	539	14

About ⅛ per cent. on the cost of the building is a reasonable average.

The quantity of water required for various kinds of work is as follows:—

	gallons
A cubic yard of concrete, dependent upon the quantity of sand in the ballast, the average quantity is about	25
A rod of brickwork in mortar, one of lime to three sand, about	220
Ditto in cement, one of cement to two of sand	130
A superficial yard of render, float and set	3

CLERK OF WORKS.

This item appears in specifications or bills of quantities somewhat like the following. " Allow for an office for clerk of works, and for the requisite firing, lighting and attendance." The cost of the foregoing will vary considerably; many contractors have movable structures which can be taken from their yard to the site of the building, in which case a percentage on its original cost may be reckoned. These are warmed by standing stoves. Others build a temporary structure with a fireplace, in which case a rough estimate of the cost may be made. Generally, two or three hours a day of a labourer will be expended on attendance. The cost of fuel will depend upon the season of the year.

A common charge in the winter is 6d. per day for fuel, and the rent of the office may be reckoned at about 2s. 6d. per week. The following are sums actually allowed in estimates, but are not valuable guides, as the circumstances vary :—

	Contract.	C.O.W. Office.
	£	£
For an alteration in the country	9000	10
For schools in Surrey	4900	30
„ house at Chelsea	3400	10

Attendance.—A properly written specification or bill of quantities usually contains a clause to the effect that each trade shall attend upon and make good after all other trades; but whether it does so or not, the contractor is bound to complete his work, and under one condition or another of most contracts his liability could be established. In pricing a bill of quantities it must be examined to discover the extent of detail adopted by the surveyor in his treatment of this item. There is a growing tendency with surveyors to enumerate everything in this way that can be clearly defined, and it may be admitted that the less left to speculation in a bill of quantities the better. The expense of attendance depends greatly upon the quality of the management; the estimation of its value must always be regulated by this consideration, and, consequently speculative; only experience and careful observation will enable the estimator to value it. When a sum is provided for the whole of the gas-fitting, hot-water work, or electric lighting, and no mention is made of attendance, it may be assumed that the attendance is included in the amount. In the case of a separate estimate by a specialist, it will always be found that the attendance costs less if he does it than when it is done by the general contractor; for if the latter does it the specialist has no interest in the economy of the labour of his attendant, and he will probably keep a bricklayer or labourer always with him, whether he requires his service or not. An arrangement can always be made (though not without trouble) that he shall supply his own attendance; this course, however, has its drawbacks in the division of responsibility and the inferior quality of the work in making good. Special protection to work done by independent contractors may properly be charged as an extra on the contract if not mentioned in the quantities or specification. Some of the items of attendance enumerated in a bill of quantities by some surveyors and not by others are as follows:—*Bricklayer*, working in conjunction with masons in backing to masonry; cutting holes for pipes, and making good, cutting chases; carpenter, cutting holes for pipes, &c., and making good; plasterer, making good plastering to ends of bearers, etc. Examination of the bill will show this, if the estimator knows what to look for. The uncertainty of estimators as to this item is shown by a few instances as follows. The bill in each case included all the above details:—

Cost of building	£		£ s.
	2000	Dwelling-house	5 = 5 per cent.
,,	3900	,, ...	nothing
,,	5200	,, ...	10 = 4 ,,
..	2700	{ Addition to a printing office .. .. .. }	15 = 11 ,,

Other items of attendance in A—the clearing away all dirt and rubbish, scrubbing floors, and leaving all clean; sometimes with this is described the cleaning of windows, and making good broken glass; in others a clause appears at end of *Glazier*—B: Leave all glass clean and perfect. The following are some instances from priced bills for A :—

	Cost of Building.	Amount.	Percentage on total.
	£	£ s.	s.
Dwelling-house	5300	5 0	2
,, ...	2000	4 0	4
,, ...	3900	2 0	1
Addition to printing office	2700	2 10	2

Probably about 2s. 6d. per cent. would meet most cases. Some instances of item B are as follows :—

Amount of Glazier's bill.	Amount of B.
£ s. d.	£ s. d.
121 0 0	1 0 0
32 0 0	1 10 0
73 0 0	1 0 0
4 0 0	0 5 0

If more exactness be required, the surface of floors and glass may be ascertained from the bill. The work of cleaning them is generally done by boys or labourers at spare times. A reasonable basis for the pricing is that a charwoman at 5s. per day would clean about twelve squares of flooring in a day of ten hours. From these data the value of larger or smaller quantities may be calculated. The quantity of building rubbish will depend upon circumstances. Sometimes joinery and masonry are prepared at building, which would increase the quantity of rubbish. When these are done at the builder's yard, five or six loads of rubbish to each 1000*l.* worth of work is a reasonable estimate. In a work of alteration, that proportion would be largely exceeded.

ITEMS AT CONTRACTOR'S RISK.

In a good set of quantities a definition of the word "allow" is given thus: "Wherever the word 'allow' occurs in these quantities, the cost of the item is at the risk of the contractor."

The zealous architect, or surveyor, in settling variations, will occasionally find in an original estimate an item of this character with a sum attached, for which probably no expenditure of time or material has been incurred, and burns to deduct it intact. He is usually met by argument: "If this item had involved an expenditure of twice the amount included in contract, I should have been forced to do it without extra charge; consequently, the item should not be interfered with." The student will hear many such arguments, which "cut both ways," and may often adopt the principle with advantage in his own defence or for purposes of attack; and in this connection the estimator may be reminded that the fallacy of an argument may often be shown by requesting the claimant to give the details of his calculation.

Excessive Quantities.

There is one other consideration for the estimator which has been a frequent subject of discussion—the extent of excess in quantity that may be expected in a bill. Perhaps excess is not so frequent as it was a few years ago; but it is not rare, even now, to see 5 per cent. taken off at the end of a bill by the estimator for full quantities. No doubt the closeness of the quantities will vary with the temperament of the surveyor—the personal equation, as it has been called—who prepares them; but the general principle now is (whatever it may have been in the past) to give the exact quantity to the best of the surveyor's ability. The author of "Estimating" says, "I would advise estimators to use prices of at least 2½ per cent. higher for works measured at completion and priced from a schedule, than they use for ordinary bills of quantities." Mr. Rickman said, in his paper on "Building Risks," read before the Surveyors' Institution, "Probably it will be near the mark to state that in very careful quantities taken from general drawings only, there is an excess of ½ to 1 per cent. (and there ought not to be more), and that there are various labours taken which a builder tendering will consider either not imperative or included in the general description of the work to the extent of possibly 1 per cent. on the value of the work; these points are commonly discounted by the successful tenderer. In the case of measurement the excess may be considered as occurring to about the same extent, though it is likely to occur in different items from those which are likely to show excess in quantities; moreover all labours, and only those which are executed, are paid for. These points are also discounted in the case of tenders on a schedule." No doubt percentages varying from 5 to 7½ per cent. have been taken off amounts of summary, and it is still the custom with some of the older fashioned builders to do so. Just as some will still omit from their pricing *all* the extra labour in a joiner's bill, or all the labour items of the stone; but a much more rational plan is the pricing every item, and realising its exact rate of profit. It may be safely said that it is most unwise to take off any percentage from quantities or measurement produced by a man who knows his business.

PROVISIONS.

It is commonly the practice in a bill of quantities, or a properly prepared specification, to define the treatment of provisional sums, in respect of packing, carriage, profit and fixing. Provisions will be found either all together at the beginning of the specification or bill of quantities, or at the end of the particular trade to which they may be referred. Sometimes such list has at its head: " On all things for which a sum is provided, allow for packing and carriage." When no specific mention of these occur, they may properly be considered as included in the provisional sum. A clause should settle the question of trade discounts. The following will do it:—

"If contractor desires a profit on any of the following provisional sums, he is to add it to the sum in each case, and such of these amounts as are not used will be deducted *with such profits* from the amount of contract. P.C. or net cost shall mean the net cost after deducting from the list price the trade discount; but not the discount for cash."

In some specifications or quantities a view quite different to that here taken is set forth as follows: "In all cases where letters 'P.C.' are made use of in this specification, they are intended to imply the published catalogue price; and the architect shall be empowered if he thinks proper, to order the articles of any special manufacturer to the full value of sum named."

This is in direct opposition to the admitted interpretation of P.C. and is really a definition of "list price."

Another mode of treatment is provided for thus: "All provisional items shall include 10 per cent. profit for the contractors calculated on the net amount after deducting the trade discount."

In default of a definition clause, it is often maintained that "P.C.," or "prime cost," means trade list price—unreasonably enough; but the contention is often successful. In these times of close competition, many builders put no profit on provisional sums, trusting to their native ingenuity to cajole or delude the architect, or reckoning upon his ignorance of trade usage to get the trade discount. If a contractor wants profit, this is not always safe. On such an item as the following it is generally admitted that a profit will be allowed on the measured items which go to make up that expenditure, and no profit need be added: "Provide for extra works 50*l*." Sometimes a number in trade list and list price is quoted; but a definition of prime cost is none the less necessary, for the purpose of adjustment of accounts. 5 to 10 per cent. is usually added by the builder to sums provided, except that for extra works or other work which will be measured, the price for which will include profit.

THE CHARGES OF VESTRIES AND DISTRICT BOARDS OF WORKS.

The Boroughs, Vestries and District Boards almost uniformly require an application on a printed form in duplicate supplied by them, on which must be drawn a block plan of the building and drains. One copy is retained by the authority; the other returned to the builder with the official permission. In the country the Local Board of Health generally requires a drain plan, and sometimes a fee; but often it does not exact the latter. Reference to the local by-laws will furnish all necessary information, and the by-laws of some municipalities empower them to require the whole set of drawings of a proposed building to be submitted for approval. Whether the building be in town or country, the estimator should include a small charge in his estimate for the preparation of this plan. Nearly all the vestries, district boards and local boards of health have some kind of printed regulations.

Some of the local authorities issue to builders applying for permission to build or drain, a schedule of the rates they charge for the work they do. Most of the Metropolitan authorities supply and lay the drain from the sewer as far as the front walls of forecourt, vault, or house. Some only connect with sewer, and lay two or three lengths of pipe, and they all reinstate the road and footway. In the country the practice varies in all manner of ways. The usual process after application made is for the surveyor to estimate the value of drain and making good paving and road, and to inform the builder, who deposits the amount before commencing the work. If the deposited sum exceeds the expenditure the balance is returned. Generally the charges are based upon a schedule of prices, which is tendered upon periodically. The contractors work under it, being supervised by an officer of the local authority; in other cases the vestry buys its materials and employs its own workmen. In thoroughfares paved with asphalt or wood blocks, the paving is done by special contractors for such work, who generally have a standing contract with the Board. It may be assumed that the charge for work done by a local board or vestry will be about 10 per cent. beyond that it would cost the builder. The average charge for a hoarding license is 2s. 6d. per month. Some vestries charge 2s. 6d. at each renewal, others do not. In London, and in many provincial towns, there is a growing disposition to refuse applications for cellar-flaps, rolling ways, and vaults, and when granted it is under a special agreement which treats them as temporary easements revocable by reasonable notice. The list of Metropolitan Boroughs may be found in the 'London Post-Office Directory.' When the work is in the country, the estimator should invariably apply to the local authority for a copy of the by-laws before he completes his tender. All the information about local authorities and their charges should be carefully preserved in good order for future use. In making an estimate, some considerations will arise on these local powers, which will require treatment differing according to circumstances. When quantities are supplied, the architect and his quantity surveyor are expected to provide for the construction and arrangement required by the local authorities, and all the estimator has to do is to attach prices. If the quantities are a part of the contract, any neglect of this foresight will probably entail extra work, for which the builder will not provide, and which must be paid for by the building owner. When the quantities are supplied, but are not part of the contract, the conditions will (as they might in the first case) probably comprise an obligation to conform to the local Acts; if consideration of them has been neglected by the surveyor, so as to involve extra work by the builder, or when the estimator is preparing an estimate from drawings and a specification in which such a condition appears, he should call attention to anything discordant with that condition and have it rectified, or the builder will possibly have to do such work for nothing. When there is no condition of the kind the estimator may, if he chooses, merely interpret the drawings and specification according to what they express. Often he does choose this course, as the candid man, by drawing attention to or providing for the lapse of the architect, produces a higher amount of tender than another who has studiously ignored everything he was not bound to notice; it is obvious, therefore, that familiarity with local by-laws and regulations is necessary.

DISTRICT SURVEYOR.

In London the district surveyor's fees must be allowed for. The incidence of these is best learned from the London Building Act. The principal points liable to dispute are, What is a building? What is a separate building? Who is liable for fees? and To what extent? The schedules of the Act state the amount of fees. There is an extra fee of 5s. per house under the Metropolis Management and Building Acts Amendment Act, and Bye-laws, which is apt to be forgotten. It is also necessary to observe that when party-walls are cut into, raised, or otherwise added to, the fees for an alteration (half full fee) will be charged for adjoining buildings. A very useful map, showing the boundaries of the districts of the various surveyors, was published by the *Builder* in 1887.

WATCHING AND LIGHTING.

Some watching and lighting must be provided for in almost every estimate. Watching in country places is often dispensed with—in towns it may often be deferred until the carcase work is well advanced. About half the time allotted to the total work is usually enough. 30s. per week is a common payment to a watchman, but often a less sum is paid. Lighting will depend upon the time of the year, and is often not required at all. When overtime is worked the lighting may reasonably be charged as an extra, unless the overtime is worked in order to complete within the contract period.

RAILWAY WAGGONS.

The question of the carrying capacity of railway waggons will sometimes arise during the preparation of an estimate. About 10 tons is the limit of load, as this is the most that a horse can draw on rails in addition to the weight of the truck; and it is hardly necessary to remark that in the course of making up a goods train the loaded waggons are drawn to their position by horses. Waggons are rarely fully loaded. Often a 10-ton waggon will only be loaded with 7 or 8 tons. Very many of the London and North Western waggons are 15 feet 6 inches long 7 feet 8 inches wide, and the sides 20 inches high. The general type of the waggons on the South Eastern railway is 16 feet long, 7 feet 6 inches wide, and the sides 3 feet high. Many of the waggons used for minerals are 15 feet 6 inches long, 7 feet 6 inches wide, with sides only 9 inches high.

From the foregoing particulars and the tables of weights of materials, a complete list of which may be found in Hurst's Handbook, it will be easy to compute the labour that will be involved in unloading, and the consequent necessities of cartage.

EXCAVATOR.

Sand and Ballast.—The prices of the materials used in this trade are a necessary preliminary to the consideration of the cost of the finished work. Pit sand in London delivered, 7s. per yard. In the country, pit sand at the pit usually costs 2s. per yard, and to this must be added the cost of cartage. Thames sand delivered within two miles of a river wharf, 7s. per yard; Thames ballast delivered within two miles of a river wharf, 6s. per yard; for each additional mile beyond two the cost would be 1s. per yard. Good Thames sand by the barge-load can be brought up to a river wharf at 2s. 9d. per yard cube; pit sand at 3s. Pit sand from the Drayton district brought to a canal wharf, Paddington Basin, 3s. 4d. per ton = $\frac{2}{3}$ of a yard = 5s. per yard. To this must be added in either case, unload cart not exceeding two miles and shoot, 1s. 9d. per yard; and to the whole freight probably 100 tons = 67 yards must be added for permission to unload; 10s. for the freight, or 1¾d. per yard in case of canal, and 25s. the freight, or nearly 2d. per yard, for river wharf in such a locality as Thames Street. The charges are highest at the City wharves. But on some parts of the foreshore of the Thames, as at north end of Lambeth Bridge, and the southern shore between Lambeth and Vauxhall, the permission costs only 12s. 6d. per cargo. Carts can be taken to the water's edge, close to the barge, and sand, ballast, or broken stone can be thrown into the cart by a single throw. The time available for work in such a position is limited by the state of the tide, and is not generally more than seven hours in the day; but this inconvenience is met by the use of more men and more carts than ordinary. The man who contracts to unload and deliver will supply baskets, barrows and planks. When a crane and buckets are available the unloading is quicker, but if the builder has to pay for their use not much cheaper. Two buckets of the capacity of a yard each are most convenient, three men filling one bucket during the transit of the other. The average time to fill a yard bucket is eight minutes, but unavoidable delays would increase it to an average of nine minutes; equivalent to 22½ yards per day for each man. Sand brought by the truck-load to a railway station will require one man in the truck to unload, and the driver to assist him. A sufficient number of carts should be used to keep the man in the truck occupied. What this number shall be will depend upon the length of the journey. A truck will contain about 14 yards, but the weight they will carry is limited to about 10 tons—about 7 yards—and will take about two hours to unload, including inevitable delays. When sand is known to exist in the part of the site which will be excavated, it will be (if of sufficiently good quality for use in the work) not only a saving of the purchase-money of sand, but also of carting the earth; the digging being alike in either case. The advantages would be as follows: Sand 7s. per yard cube, carting 2s. 6d. = 9s. 6d. per yard; but frequently the architect who knows his business stipulates that for every yard of sand procured on a site and used in the building, a certain agreed sum shall be allowed. The consequences of failure to settle the question of the disposal of sand are known to professional men by unpleasant experience. A contractor will sometimes come upon an unexpected bed of sand, and will not only obtain enough for his building, but, there being no restraining condition, will dig much more than he requires and sell it. An instructive instance of this kind occurred in a London suburb a few years ago. As a preliminary to the development of a building estate, the freeholder contracted with a road-maker to construct the roads. The whole course of these roads lay over the beds of sand near the surface, and well adapted for building purposes. There being no adverse conditions in his contract, the contractor proceeded to dig it out to a considerable depth and sell it, the excavation being filled with rubbish of all kinds, the permission to shoot being in all cases paid for at per load. A load of sand is a cubic yard and should slightly exceed 21 striked bushels, but it never does. The War Office schedules insist on 22 bushels to the yard.

Ashes of various kinds are used for making ash mortar. The rakings from the furnaces of locomotives are generally most convenient; they vary considerably in price, according to the neighbourhood and consequent demand for them. At a large railway depot like Swindon, where the supply is great and buyers few, they may be bought at 1s. per ton. At most London railway stations they would cost in trucks 4s. 6d. per ton with 6d. per ton reduction if more than 50 tons be bought. Where the work is in the neighbourhood of ironworks, the refuse from the blast furnaces is used. The first cost at the ironworks does not represent the whole outlay, as it requires much grinding in a mortar-mill to make it fit for use. As the ash takes the place of sand, the difference only in value is to be dealt with in calculations of cost.

Black sand from iron-foundries is also used; it costs, delivered into carts at London foundries, 4s. per yard.

Clay for Puddling.—The cost of clay will vary according to the position of the works, the quantity, and the purposes for which the clay is required. In many parts of London, enough for the purpose (it may only be required for puddling round the joints of drain pipes) may be found on the site. In other cases it may be obtained from works in progress near at hand, as in the case of the construction of a section of the Northern Thames Embankment (Mr. Ridley's contract). A very large proportion of the clay required for puddling for the coffer dams was brought from the excavations for the Metropolitan Railway by the contractor for that work, he not only charging nothing for the clay, but paying a sum per load for permission to deposit it. When required in considerable quantity it may be drawn from the nearest brickfield. Clay delivered in the City of London costs 5s. 6d. per yard; in the suburbs 3s. 6d. per cubic yard. In the country it may commonly be bought for 1s. per yard at the field, to which must be added cartage according to distance. The labour of tempering varies considerably; but a labourer should do not less than 4 yards in a day. Rankine says: Tempering, ·3; spreading, ·3 per yard cube. Cost at 7d. per hour, 1s. 4½d. per cubic yard. Constant, ·25.

Portland Cement is sometimes sold by the cental, or trade bushel, of 100 lb. weight, but generally by the ton; will weigh from 90 lb. to 130 lb. per bushel, and costs about 35s. per ton, including the use of sacks, delivered in London within five miles of the work, or at any London railway terminus. The average weight per bushel may be reckoned at 112 lb. There are commonly ten sacks of two striked bushels each to the ton. The weight last mentioned is most frequently specified; but it is rarely the case that anyone takes the trouble to weigh it. In arranging with contractor for its supply the merchant usually agrees to deliver cement of the specified weight and will take the responsibility.

Stone Lime.—The lime merchants supply their own railway trucks, whose capacity varies, carrying 10 yards under 5 tons; 12 yards under 6 tons, 14 yards under 7 tons. For country work it is nearly always necessary to buy lime by the truck-load, and the same arrangement as suggested for sand will be necessary. The driver and a labourer should (including delays) deliver into carts about 4 yards of loose lime in an hour. The usual form in which stone lime is used by the excavator is ground to powder. A yard of ground lime is the result of grinding a cubic yard of stone lime in lumps, and will measure about 16 striked bushels. Is delivered in London at 11s. per yard, in loads of 2 yards; use of two-bushel sacks 1d. each by some merchants, by others nothing.

Lias Lime.—Good ground lias lime costs 25s. per ton delivered in London. The number of bushels in a ton varies (dependent upon the degree of freshness) from 27 to 33 bushels; a common average is 30 bushels. Add for use of bags as before. The uniform charge (included in the foregoing prices) for grinding a yard of lump lime is 1s.

Burnt Ballast.—Sometimes burnt ballast is specified as one of the ingredients of concrete. As such cases occur where the digging is in clay, the requisite quantity is often obtained from the excavation, or is procurable on the site. The digging and depositing of the clay is of equal value whether to be burned or not. A ton of common coal (a mixture of small coal and slack) will burn 12 yards of clay. As the burning has to be done further away from the proposed building than ballast would be shot, filling barrows and wheeling must be reckoned for, and the disposition of the coals and clay will take a labourer about one hour per yard. In the country the wheeling into heap and burning is done as low as 2s. per yard. The whole process is described in Seddon's 'Builders' Work.' Slack is delivered at country stations in loads of 8 or 9 tons; the cost of unloading and delivering into carts and cartage to the site must be reckoned for. The cost of slack at country stations is about 14s. per ton. In the neighbourhood of London the small coal and dust which accumulates in the coal depots of railway stations is used—the price is about 15s. per ton delivered.

Rubbish.—Hard, dry rubbish for filling in often arises in sufficient quantity on the building. If the works are pulling down and alteration, its deposit beneath pavings would be a saving compared with carting away. In town it may often be obtained for nothing at an adjacent building which is being pulled down, in which case so many loads carted would represent the cost; in many cases, however, it would have to be paid for besides. And similarly, the carman who knows where to dispose of hard rubbish will reduce his price for cartage. In the country it will sometimes be necessary to break up bricks or buy gravel to make up the requisite quantity. A common contract price in London for "hard, dry brick or stone rubbish, and filling in and ramming," is 4s. per cubic yard. When it is known that the works will produce enough for the purpose, the cost of basketing or wheeling, and a little breaking and levelling is generally about 1s. 6d. per yard.

Pumping.—The removal of water from the trenches and basement does not often entail much expense, but sometimes it does. The judicious contractor lays his drains as early as he can in the progress of the works, which enables him to dispose of the water as it is baled or pumped. In London, land water seldom troubles the contractor; the ground being thoroughly drained by the deep sewers. In the suburbs of London or in the country the shallower sewers, or their entire absence, give reasonable cause of apprehension. Inquiry in the neighbourhood as to the depth at which water is reached is always necessary so as to know what may be expected; as the architect has generally done this, he has probably taken constructive precautions for the exclusion of water, and these would be adopted by the builder as soon as possible. Except in very wet seasons the rain water causes but little inconvenience if the contractor takes the precaution of cutting shallow channels to prevent the surface water from running into the trenches. It is impossible to give any rule for estimating the cost. Small quantities would be baled, for larger quantities a chain pump would be necessary, and for very large quantities a pulsometer. In most cases the digging a sump towards which the water may drain and be pumped from is a convenient expedient. In any case the trench stage of the work does not last long, and when water is troublesome is shortened to the best of the contractor's ability. Comparison of similar cases and a record of observations will be useful to the estimator, but the cost is generally small.

Digging.—The valuation of excavation is comparatively simple for the estimator of building values, as the work is all of a similar character. The digging for a building in London is frequently sub-let, despite the usual conditions prohibiting sub-contracts, and the cases are rare in which the general contractor can get it done as cheaply in any other way. The common practice is to contract with one man for excavation and carting (including general carting) and the supply of sand and ballast. The professional excavator has engines, cranes and buckets which he brings on the work, and when a basement is required makes an inclined road to bring his carts down to that level. A common price for digging and carting in London is 3*s.* 3*d.* to 3*s.* 6*d.* When the builder does the work with his own men, the points for consideration are—

1. The kind of labour available.
2. The degree of hardness of the ground.

3. The depth from general surface.
4. The distance to deposit, and whether basketed, wheeled, or carted.

1. *The Kind of Labour available.*—Digging can be done best and cheapest by the navvy who has been used to the kind of labour involved in the making of roads, canals, railways, or docks, and when there is a large quantity of digging it is policy to procure such; when the quantity is small, the builder contents himself with the work of the general labourer. A higher rate per hour than that for general labourer is frequently allowed in building accounts for labourers engaged in digging, but it is not always paid by contractors; when allowed, ½d. per hour extra is usual.

Sometimes the builder will let the digging only; it is obvious that the price must be, as a rule, less than the cost by other methods, or the builder would employ his own men. The men who take such sub-contracts work harder and for a longer time each day than those employed by the hour under ordinary conditions. In very large works the steam excavator is used with advantage.

2. *The Degree of Hardness of Ground.*—The estimator should satisfy himself of the nature of the ground by inquiring before pricing. When a building is pulled down, the men who do it leave the basement completely filled with building rubbish. This can be removed entirely with the shovel, although it is frequently measured by the quantity surveyor as digging, and included in the item Excavation to Basement. Such as this and other building rubbish requiring no getting is worth least of any. The classification of digging is simple, and may be reduced to four categories sufficient for all purposes.

A. Building rubbish, sand, loose soil, which requires no actual digging, only shovelling; of these a man will fill into a cart, or throw on to a stage or barrow about 22 yards per day of 10 hours. Instances have been known of a man throwing 25 yards per day. Cost at 7½d. per hour, 3d. per yard cube. Constant ·045.

B. Vegetable soil, as in surface digging, dug with a shovel but not picked. A man will dig and fill into a barrow or cart 14 yards per day. Cost at 7½d. per hour, 5¼d. per yard cube. Constant ·07.

c. Clay, which can be dug with a shovel, but not necessarily picked, though sometimes picked for convenience; a man will dig and fill into a barrow, or throw up 6 feet, seven yards per day at 7½d. per hour, 10¾d. per yard cube. Constant ·143.

D. In all the foregoing cases one man will do all the digging and filling or throwing. Hardest earth, clay, or gravel requiring picking, 6 yards at 7½d. per hour, 12½d. per yard cube. Constant ·167.

When picking is necessary, it is usual to have one man to pick and another to fill. The constant to be applied would be the sum of the constants for each man's labour.

Taking as an instance Mr. Hurst's constants for the excavation of clay, we have—

Excavating only per cubic yard	·100
Throwing with a shovel to a height of 5 feet, per cubic yard	·055
	·155 = 6·45 yards per day.

Digging to Trenches.—Add to the foregoing prices 20 per cent. for work in trenches.

3. *The Depth from General Surface.—Additional Throws.*—The height of throwing is limited to 6 feet. Each additional throw, including staging, 22 yards per day. Cost at 7d. per hour 3¼d. per cubic yard. Constant ·45; staging about 1d.; total 4¼d.

4. *The Distance of Deposit.—Wheeling.*—A run is assumed to be 20 yards, but in the ordinary operations of the builder where the distance is only two or three runs, the wheeler will do it in one stage, using two barrows; one which he wheels, the other which he leaves to be filled during the time spent in going to the deposit and back, a third barrow and a second digger being used when one is not sufficient to keep the wheeler going. In cases of great excavations and longer distances, a more complicated arrangement is necessary. The runs must be carefully planned and the distances apportioned for the wheelers. The number of barrows for each shoveller is always one more than the number of wheelers. Plans of various barrow-runs are to be found in Seddon's 'Builders' Work.'

A filler will fill an ordinary barrow in the time a man takes to wheel it 33 yards, deposit the earth, and return with the barrow.

PRICES. 513

A barrow holds one-tenth of a cubic yard. A man will wheel 20 yards, and deposit in a day of ten hours, about 35 cubic yards of earth. Cost at 7d. per hour, 2d. per cubic yard. As the filler keeps pace with the wheeler add another 2d. making 4d. Constant ·028. Additional runs, 2d.

Filling in and Ramming.—A man will fill in about 22 yards per day, and a rammer must attend on each filler. Cost at 7d. per hour each man, 6¼d. per cubic yard. Constant ·090.

Sometimes in a bill of quantities the excavation is all stated as "part filled in and rammed," or "a small part filled in and rammed." The earth returned should always be kept separate by the surveyor, and should appear in the bill as "earth filled in and rammed," or "dig, fill in, and ram." When it is not separated it must either be measured from the drawings or guessed at from the other items of the bill. The proportion will vary, and will rarely be less than one-third of the whole quantity; assuming the proportion as one-third and adopting the calculation D, we have

		d.
	Digging ..	12¼
Add	20 per cent for trenches	2¼
		15
Add	⅓ yard at 6¼ ..	2¼
		17¼ per yd.

Basketing.—Ordinary earth and clay per cubic yard would increase in digging about one-fourth, and as all earthwork is measured by the bulk before digging, we have just over 26 bushels per yard to dispose of. The basket holds about a bushel, but would be heaped. We may reckon, therefore, the usual 21 bushels as a yard. The case may be treated as analogous to filling barrows and wheeling. A wheelbarrow holds one-tenth of a yard, a basket about $\frac{1}{21}$ of a yard (not quite half so much as a barrow). The difference is, however, enough to compensate for the lifting of the baskets, dealing with the distance as a run. The conveyance would take twice as long, two journeys instead of one. The filling the same as for wheelbarrows. Therefore the cost at 7d. per hour per cubic yard, of filling 2d. as before. The cost at 7d. per hour per cubic yard of basketing 4d., equal twice wheeling; total 6d.

There are a few general considerations which must be re-

2 L

membered in connection with earthwork. That work in trenches, post-holes, wells, or cesspools will cost more than in large surfaces. Small quantities of digging will render it necessary to take a cart from place to place to make up a load at an additional expense. In the cartage of earth or other materials, a good schedule makes a distinction in the price thus:—" Add on the first load when not more than two loads are ordered to be removed on the same day." When earth is wheeled, small quantities differ but little in value per yard cube from larger ones.

Sometimes it is impossible to measure the earth of an excavation before digging; but a proportion can be deducted from the quantity of earth measured while loose after digging, and thus find out what was its bulk before digging, as follows:—

	Increase of Bulk.	Deduction.
Earth and clay	$\frac{1}{4}$	$\frac{1}{5}$
Sand and gravel	$\frac{1}{12}$	$\frac{1}{13}$
Chalk	$\frac{1}{3}$	$\frac{1}{4}$
Rock	$\frac{1}{2}$	$\frac{1}{3}$

Profit.—The author of a valuable series of papers on Estimating says: "A liberal profit should be added to excavating, as there must be a foreman on the site superintending this one trade, and the items are not large." The amount of an excavator's bill is generally small, but keeping the cost of foreman separate is to be preferred for the reasons before mentioned.

Large Earthworks.—In estimating the cost of large earthworks, the proportion of getters, fillers and wheelers requires careful consideration, and in any excavation the estimator must realise the method of doing the work before he can decide upon its cost. The 'Dictionary of Architecture' says, under the heading "Earthwork": "The French military engineers make a distinction between the various descriptions of earth, according to the number of men required to execute the description of work in them. (1) Ground easily moved is called earth of one man—terre à un homme. (2) Ground requiring two men, one to get and one to fill, is called earth of two men—terre à deux hommes. (3) Ground requiring three men, two to get and one to fill, is called earth of three men, —terre à trois hommes."

There is a simplicity in this method of analysis which is, under some circumstances, most convenient. Rankine's 'Civil Engineering' says:—

"The proportion of pickmen to the shovellers in a single rank depends upon the stiffness of the earth.

	Pickmen to one Shoveller.
Loose sand and vegetable soil	0
Compact earth	½
Ordinary clay	½ to 1
Hard clay	1½ to 2

Earth is designated as "earth of one man" if one shoveller can keep one line of wheelers at work. "Earth of a man and a half" if two shovellers and a pickman are needed to keep two lines of wheelers at work. "Earth of two men" if one shoveller and one pickman can keep one line of wheelers at work, and generally, "earth of so many men" according to the number of shovellers and pickmen together, who are required to keep one line of wheelers at work. Let m denote the number, l = horizontal distance the earth has to be wheeled, h = height of ascent if any, then the total number of shovellers, pickmen and wheelers for each line of wheelers will be, approximately—

$$M = m + \frac{l + 6h}{\text{from 100 feet to 120 feet}}.$$

This is most clearly put by Mr. Hurst. He says ('Architectural Surveyors' Handbook'): "As all ground has to be brought nearly to the same state before it can be filled into barrows or carts, the labour of filling may be assumed to be constant for earth of the same class, and the stiffness of the soil or difficulty of excavating it will affect only the number of getters required. Therefore, the proportion which the latter bear to the number of fillers determines the relative amount of labour required to excavate any particular soil. The unit to be adopted is the average quantity of earth which a man can fill into a cart or waggon per day of ten working hours."

"Let the number of fillers be represented by F, the number of getters by G, and the proportion which represents the labour as compared with that on earth which requires no getting by N. Then

$$N = \frac{F \times G}{F}.$$

In this manner excavation can be described as earth of 1, 1½, 2 or more men, according to the difficulty in getting it. The French division of earth into three degrees of hardness, or four at most, is sufficient for all practical purposes."

Slopes.—Formerly the surveyor would, in the measurement of trench-digging, make a certain allowance for sloping sides to trenches This practice is obsolete. When the sides of a trench will not stand vertically without support, strutting and planking is always allowed for.

In pricing bills of excavation, observe whether it is described as to include strutting and planking.

Breaking Stone.— Breaking the larger stones in ballast for concrete requires little attention; but if insisted upon, a labourer could do thoroughly 30 yards in a day. Cost at 7*d.* per hour, nearly 2¼*d.* per cubic yard. Constant ·033.

The breaking of old bricks (sometimes used for concrete): a labourer would break into 2-inch cubes about 4 yards per day of 10 hours. Cost at 7*d.* per hour, 1*s.* 5½*d.* per cubic yard. Constant ·25.

Breaking Kentish rag or limestone. The stone in the sizes generally delivered for rubble walling: a labourer would break into 2-inch cubes 2 yards per day of 10 hours. The stone measured after breaking. Cost at 7*d.* per hour, 2*s.* 11*d.* per cubic yard. Constant ·50.

In Boulnois' 'Municipal Engineers' Handbook' we have the following: " A good stone-breaker will break 2 cubic yards of hard limestone to the ordinary gauge (2½ inch) in a day, and some men more. Hard silicious stone and igneous rocks can only be broken at the rate of 1½ or 1 yard per day. Of Guernsey granite, a man can only break on an average half a cubic yard per day." River gravel, field stones, or flints, which are already of small size, can be broken at the rate of 4 or 5 yards per day. Breaking by hand costs from 1*s.* 3*d.* or 1*s.* 4*d.* to 1*s.* 10*d.* per cubic yard for ordinary silicious rocks and harder limestones, and 2*s.* to 2*s.* 6*d.* for harder silicious or igneous rocks.

In the construction of a building with rubble walls, the fragments produced by the dressing of the stone are available for concrete. For large works, where steam power is convenient, Blake's stone-breaking machine may be used with advantage. The price should always include stacking in regular heaps for convenience of measurement.

Concrete.—The materials for concrete should be delivered as close as possible to their ultimate position, for, as before mentioned, it is the frequent handling of materials that increases the cost of the finished work.

The value of concrete depends upon—first, the amount of void in the aggregate which needs to be filled with the matrix; second, the diminution in bulk of the lime and sand, or cement and sand, as a result of mixing with water.

Stone broken into 2½-in. cubes has voids	..	..	37 per cent.
„ „ 2 in. „ „	..	..	40 „
„ „ 1¼ in. „ „	..	..	42 „
Clean shingle or burnt clay „ „	..	..	33 „
Thames ballast „ „	..	..	15 „

The diminution of sand and lime when made into mortar (matrix), is generally about one-fourth, more or less, and this may be adopted as a reasonable average. For cement and sand, one-sixth may be taken as the average.

We may apply these facts to the calculation of the value of concrete in the following manner.

Taking water supplied in London by meter as 1s. per 1000 gallons, we have less than a farthing for a yard of concrete. Some water companies charge 1d. per yard of concrete; this is a reasonable average cost. Constant for mixing only, ·125 per cubic yard. As a rule the cost of water for the whole building is dealt with in the pricing of the preliminary items.

Concrete composed of one part ground stone lime to six parts ballast:—

	s. d.		£ s. d.
7¼ yds. ballast	6 0	=	2 3 6
1 yd. ground lime	11 0	=	0 11 0
			6) 2 14 6
			0 9 1
Water	—		0 0 1
Labourer .. .. mixing ·125 of	5 10		0 0 8¾
Cost per yard			0 9 10¾

Concrete composed of one part cement to six parts of ballast:—

		s.	d.			£	s.	d.
7¼ yds. ballast		6	0			2	3	6
22 bushels cement		1	10			2	0	4
					6)	4	3	10
						0	13	11½
Water		—				0	0	1
Labourer .. mixing ·125 of 5 10						0	0	8¾
Cost per yard						0	14	9¼

Add to each of the foregoing for filling into trench and ramming:—

		s.	d.		d.
Labourer		·105 of 5	10		7¼

Add to each of the foregoing if filled into barrows and wheeled:—

		s.	d.		d.
Labourer		·028 of 5	10		2

Screening Sand.—Sand is generally selected free from stones, and rarely requires to be screened for mortar-making. When it has large stones in it, screening is necessary. With a screen of the usual size of mesh, a labourer will screen 10 yards in a day of 10 hours. Cost at 7d. per hour, 7d. per cubic yard. Constant ·10. The sand and the screenings will both be usable, the latter for concrete and filling beneath pavings.

Washing Sand.—A man will wash a yard of sand (measured after washing) in three hours. The waste upon the original yard of sand will depend upon the proportion of loam or other soluble elements in it; commonly, a quarter of the original bulk is lost in the washing. The cost of a yard of washed sand would consequently be as follows, dependent, of course, upon the local value:—

		s.	d.	
1¼ yd. of sand delivered (7s.)		8	9	
Labourer, 3 hours at 7d.		1	9	
Cost per cubic yard		10	6	Constant ·30

By inquiry in the neighbourhood or inspection of the site, the experienced surveyor will be able to judge whether or not sand will be found; if it is, and there is no stipulation in the specification that sand shall not be dug, the builder will thus obtain the sand for nothing, and save the carting of so much earth from his trenches as the pit formed by the removal of the sand will contain.

Where there is a large quantity of carting it is often advantageous to obtain an estimate from a local carman, who often brings sand to the works, and on his return carts away earth.

Drains.—The price of glazed stoneware drain-pipes *in London* is regulated by an agreement amongst the vendors, by which the published list always shows the same prices, but the trade discount varies. The list, as published by the merchants, is to be found in the current price books. The merchants will deliver within four or five miles without extra charge if a cart-load is ordered.

Drain-pipes are usually carried by the London merchants in a one-horse van, of size specially adapted to the purpose, 9 feet by 4 feet 9 inches by 2 feet, of capacity as follows:—4-inch pipes, 200; 6-inch pipes, 120; 9-inch pipes, 55; 12-inch pipes, 35; 15-inch pipes, 22; 18-inch pipes, 18.

The average one-horse van is about 9 feet by 4 feet by 2 feet, and will carry about 2 tons. A knowledge of the size of vehicles is useful to the estimator, as he can the more readily reckon the cost of carriage, and he will find it to his advantage to note the quantities of materials carried under various conditions, the sizes of the vans or carts and the number of horses, the number of men occupied on loading and unloading, and the time spent upon such work.

The trade catalogues of the large manufacturers, such as the Bristol Waggon Works Company, contain much useful information as to sizes, carrying power, and price of carts, vans, waggons, etc.

The present trade discount off list prices of town-made pipes is 20 per cent. For country-made pipes, 5 per cent. more than these, and for very large quantities 7½ per cent. more. As the country-made pipes are much more frequently used than any other, whatever may be proposed to the contrary, the examples and the following list are based upon the discount for country pipes, say, 25 per cent.

Large quantities may be bought of the Midland manufacturers at from 7½ to 10 per cent. below the above rates, carriage paid to a London station, or to a country station of equal distance from the manufactory; but this involves carting by the purchaser. Approximate number of pipes to a ton:—

	2 in.	3 in.	4 in.	6 in.	9 in.	12 in.	15 in.	18 in.
Pipes	285	200	150	90	42	27	18	14
Siphons and gullies	..	203	150	68	33	25	..	..
S-traps	..	224	124	62	..	..	..	..
P-traps	..	320	203	80	..	..	..	..

Tested pipes cost generally 25 per cent. more than the foregoing rates, and each length is marked T. For the Stanford jointed pipes the trade discount is 10 per cent. off the published list. The customary measurement and valuation of drains is per foot run, including digging, filling in and ramming; up to 3 or 4 feet deep, it may generally be assumed that no strutting or planking will be necessary, beyond that depth it is often necessary. The extent of it will depend upon the nature of the ground. It is rarely the case that the poling boards require to be close together; they may *generally* be from 4 inches to 6 inches apart. In the neighbourhood of London, the digging, filling-in, and ramming for drain trenches is often priced at 1d. per foot run for each foot in depth.

A bricklayer (10½d.) and labourer (7d. = 1s. 5½d.) will lay and joint and cement in a day of ten hours:

	Per foot.	Constant.	A bushel of cement and sand will joint
4-in. pipes (100 ft.) =	1⅜d.	·010	150 ft.
6-in. „ (66 ft.) =	2¼d.	·015	100 ft.
9-in. „ (44 ft.) =	3¾d.	·022	65 ft.
12-in. „ (33 ft.) =	5¼d.	·030	50 ft.
15-in. „ (26 ft.) =	6¼d.	·038	40 ft.
18-in. „ (22 ft.) =	7¾d.	·045	32 ft.

The labourer's constant attendance is necessary to hand the material, and assist in the adjustment of the pipes.

The smaller works, which usually appear in a bill of quantities for drains, can only be dealt with by resolving them into their elements. The basis of valuation would be the normal value of the larger quantities. The modification of the price per yard cube of concrete or foot of brickwork, as the case may be, would be an increase by a purely arbitrary percentage dependent upon experience, and it should be remembered that, as the particular article increases in number, so the percentage should decrease.

Prices of pipes delivered free in London after deducting trade discount:—

Bore of pipe	2 in.		3 in.		4 in.		6 in.		9 in.		12 in.		15 in.		18 in.	
	s.	*d.*	*s.*	*d.*	*s.*	*d.*	*s.*	*d.*	*s.*	*d.*	*s.*	*d.*	*s.*	*d.*	*s.*	*d.*
Straight pipes, per ft.	0	3¼	0	3¼	0	4½	0	6¾	0	11¼	1	7¼	2	7½	3	11¼
Bends, each	0	10¼	0	10¼	1	1½	1	8¼	2	9¾	4	9¾	7	10½	11	9¾
Single junctions and taper pipes, 2 ft. long, charged in list as 4 ft. straight pipes, each	1	1	1	1	1	6	2	3	3	9	6	5	10	6	15	9
Double junctions, 2 ft. long, charged as 6 ft. of straight pipe, each	1	7½	1	7½	2	3	3	4½	5	7½	9	7½	15	9	23	7½
Siphon traps, long ..	1	6	1	10½	2	7½	4	6	7	6	13	6	..		..	
S and P traps	1	3	1	6	2	3	4	6	7	6	..		..		..	

A few examples will illustrate the method of valuing drains.

Example 1.—4-in. drain jointed in cement and the joints puddled around with clay and 4 ft. excavation.

1.0
2.0
4.0

		s.	*d.*
——8.0 =	$\frac{8}{27}$ cube yard, dig, fill and ram, at 12½*d.*, as before	0	3¾
	4-in drain, per ft. run	0	4½
	Laying 1¾*d.*, cement, ¼*d.*, and clay ¼*d.*	0	2¼
	Cost per foot	0	10½

Example 2.—6-in. drain as before, and 8 ft. excavation.

1.0
2.0
8.0

		s.	*d.*
——16.0	cube digging, about ¾*d.* for each ft. in depth, or $\frac{16}{27}$ at 12½*d.*	0	7½
	6-in. drain, per ft. run	0	6¾
	Laying 2½*d.*, cement ¼*d.*, clay ¼*d.*	0	3

1.0
2.0
6.0

		s.	*d.*
——12.0 =	$\frac{12}{27}$ of a cubic yard, throwing out only, at 3*d.* the lower 6 ft. of earth	0	1½
	Staging (boards laid on the poles), or struts of the strutting and planking) about 5 ft. by 2 ft.—say	0	1
	The real value is the labour of fixing and removal, plus 7½ per cent. per annum on the original cost of the planking.		
	Cost per foot	1	7¾

522 QUANTITY SURVEYING.

The strutting and planking, if necessary, would be valued as hereafter described.

The valuation of bends, junctions, taper pieces, etc., is in accordance with the usual method of measurement, "Extra for" as the length of bend varies, some being 2 feet long, and some less; an average length reasonably applicable to all cases may be taken as 1 foot 6 inches.

The value of extra for bends is as follows. There is no extra cost on the jointing or laying:—

	Extra for 4-in. bend.	d.
Cost price of 4-in. bend..		13¼
Deduct 1 ft. 6 in. of 4-in. pipe		6¾
		6¾

Junctions are of equal length to the pipe upon which they are extra. There is no extra cost on the jointing or laying.

Traps or gullies, which are generally bedded in concrete, would be valued in some such way as follows:—

Doulton's (Lambeth) Fig. 13b yard gully, with cast-iron grating bedded in and including cement concrete 18 in. by 15 in.

```
 2  0                                                              s.   d.
 2  0
 1  3   5  0   Dig and wheel 1/17 cubic yd. at 7¼d.  ..    ..      0   1¼
 ─────
 1  6
 ½  6
 1  3   2 10   Cement concrete.
 ─────
    9
    9
    9   5      Deduct.  Void.
 ─────
        2  5 = nearly 1/11 cubic yd. at 14s. 9d.  ..    ..    ..   1   4
               Trap. List price, 9s., discount 1s. 4d. =..    ..   7   8
               Fixing and connecting. Bricklayer ¾ of an hour
                 at 10½d.  ..   ..   ..   ..   ..   ..   ..   ..  0   8
               Cement ..   ..   ..   ..   ..   ..   ..   ..   ..  0   1
               Labourer, ¼ of an hour at 7d.  ..   ..   ..   ..   0   1¾
               20 per cent. on digging and concrete (1s. 5d.) for
                 small quantities  ..   ..   ..   ..   ..   ..    0   3½
                                                                 ──────
                                                                 10   3½
```

Small cesspools, inspection pits, etc., must be dissected in a similar manner, as in the following example:—

Brick cesspool, 1 ft. 6 in. by 1 ft. 6 in., and 1 ft. 6 in. deep, all in clear, of half-brick sides and brick flat bottom, all built and rendered in cement on lime

PRICES. 523

concrete 6 in. thick, including 2½ in. tooled both sides York cover, 2 ft. 3 in. by 2 ft. 3 in., with tooled edges, and digging and connecting with two drains.

2 9				
2 9				
2 3		17 0	Dig and wheel.	

2 9				
2 9		8 9	Lime concrete.	
6				

2 3				
2 3				
3		1 3	Brick in cement.	

7 6		11 3	Half-brick in cement.	
1 6				
2 3				
2 3		5 1	2½ tooled York cover.	

| 9 0 | | 9 0 | 2½-in. tooled edge. | |

| 1 6 | | | | |
| 1 6 | | 2 3 | render 1 coat in cement. | |

| 6 0 | | | | |
| 1 6 | | 9 0 | | |

| 0 2 | | 0 2 | make good drain-pipe to ½ brickwork in cement. | |

The discount off other pottery varies from 10 to 15 per cent., and is delivered free of charge, if a complete load, within a radius of five or six miles of the manufacturer's premises.

Strutting and Planking.—The excavator's items of a bill of quantities or a schedule are often described to include strutting and planking in some such form as follows:—" Digging straight or curved on plan, except for wells, drains, or pipes, including levelling bottom of trenches or forming same to proper falls, as may be ordered; cutting steps, providing, fixing and removing shoring, close-planking and strutting where required, not exceeding 6 feet in depth." The character and extent of strutting and planking vary greatly with the nature of the soil and the state of the weather. The sides of an excavation in loose sand may require the planks close together and supported by many struts. In clay soil the planks might be a foot apart, and in places would require none. As a familiar instance, a trench 6 feet deep and 4 feet wide in stiff clay would generally have 1-inch or 1½-inch vertical boards about 7 inches or 9 inches wide and about 4 feet long, 12 inches apart, one 9 inches by 3 inches waling piece on each side of trench, and one row of struts about 5 feet apart, generally short pieces of scaffold-pole. Arrangements of poling and planking of trenches are illustrated in Seddon's 'Builders' Work,' and Hurst's 'Tredgold's Carpentry.' If strutting and planking is left in its place, as in some cases of underpinning where it is impossible to withdraw it, the cost of the wood plus the fixing and profit is the proper price, if new material is used; if old material, about a third the price of new is the customary rate.

Strutting and planking to trenches is usually described as follows:—

24 feet run. Strutting and planking to sides of trenches 4 feet wide and 6 feet deep, one side measured for two; a common contract price for this would be 1s. per foot run. The poling boards would be "ends," as they are called, which may be purchased at the timber sales or the wood imported as firewood and sold to the firewood dealers at about 6l. per fathom of 6 feet by 6 feet by 6 feet = 216 cube feet. These ends may be selected of any thickness and of all varieties of length from 3 feet to 6 feet at the same price per fathom. The cartage for four one-horse loads per fathom would be also added, say, 12s., total, 6l. 12s., or, about ⅝d. per foot superficial for 1½ inches. The waling pieces, 9 inches by 3 inches, would cost about 2½d. per foot run. The poles would be scaffold poles, cut to lengths, cost about 1¼d. per foot, including the cutting. The valuation of the above item would be as follows:—

```
  2/ 24  0
      6  0
     ─────    288  0   1¼" boarding delivered at 7s. 4d. per    £   s.   d.
                        square  .. .. .. .. .. ..              1   1   1½
  2/ 2/ 24  0
     ─────     96  0   9" by 3" waling piece, 2½d. per foot    1   0   0
  2/ 10/ 3  9
     ─────     75  0   Struts 3 ft. apart, 1¼d. per foot ..    0   7   9¾
                                                               ─────────
                                                               2   8  11¼
```

Interest on capital at 5 per cent, p.a.
Depreciation at 2½ per cent, p.a.
Say, one month in use at 7½ per cent. p.a. on 2l. 8s. 11¼d.   s.   d.
= 3s. 8d. + 12..   0   4
288 ft. fixing and removal, at 2s. 6d. per square   2   2½
Cartage to yard, 1 load, at   3   0
                                                              ─────────
                     Divide by 144 ft. (one face for two).. .. 5   6¼

                            Cost per foot superficial, about 0 0¼
                            Feet in depth 0 6
                                                                  ───────
                            Cost per foot run 0 3
```
```

PRICES. 525

Strutting and planking to an excavation like a basement would be described by its superficial area. Taking as an example an area 40 feet by 40 feet, and 6 feet deep, a calculation on the same principles would produce the following results:—

						£	s.	d.
	160	0						
	6	0						
		—	960	0	1½" poling boards as before, at 7s. 4d. per square	3	10	4⅜
2/	160	0						
		—	320	0	9" by 3" waling pieces, at 2½d. ..	3	6	8
4/ 9/	4	0	144	0	9 × 3 sole pieces for struts (5 ft. apart) (= 27·0 cube) 2½d. ..	1	10	0
4/ 9/	3	6						
	4½							
	3		11	10	Fir struts.			
4/ 9/	6	6						
	5							
	4		32	6	Ditto.			
			44	4	At 1s. 6d.	3	6	6
						11	13	6¾

	£	s.	d.
Say one month at 7½ per cent. per annum on 11l. 13s. 6¾d. = 17s. 7d. + 12 .. .. =	0	1	5¾
Waste, 5 per cent. on 3l. 6s. 6d.	0	3	4
960 ft. fixing and removing planking, at 2s. 6d. per square	1	4	0
71¼ ft. labour and nails fixing struts, at 4d.	1	3	9
Cartage to yard, two loads, at 3s.	0	6	0
Divided by 960 ft.	2	18	6¾
Cost per foot superficial .. .. ¾d.			

It is rarely the case that more than the foregoing is required. When the earth requires more support than this affords, 3 inch planks are used instead of poling boards, sometimes shod with iron and driven like piles. When still more security is required, square piles are used. The valuation of such work presupposes a knowledge of civil engineering, and is beyond the scope of these Notes.

The section "Quantities and Costs" of the Index of the 'Transactions' of the Institute of Civil Engineers contains a number of references as to cost of coffer dams, piling, concrete in large masses, and similar works. The method of valuation of piling depends upon the mode of measurement, for which see page 46.

Tar Paving.—This is much used for footways in suburbs of the Metropolis and other towns; the price varies from 1s. 9d. to 2s. 6d. The present price paid by the School Board for London for 2½-inch paving of the best quality, made with broken limestone, is 1s. 10d. per yard.

It is almost always done by persons who, like the asphalters and wood-paving contractors, do nothing else, and so possess facilities for the doing of such work, with which the average builder cannot profitably compete.

BRICKLAYER.

Bricks.—In estimating for a work, it will always be necessary to obtain the current price of bricks, as a rainy season will sometimes increase their price from 25 to 30 per cent. In London a contractor may generally make as good a bargain with a London merchant for the delivery of bricks at the site (so saving himself much trouble in arranging transit), as he can by buying at the field. When the builder is apprehensive of a rise in value or failure of supply, he will make an arrangement for the supply of bricks for an agreed term at a fixed, uniform price per thousand. For work in the country, when the quality only is specified, the builder will find out the nearest brick-field capable of producing bricks of that kind, and it will generally cost him less for delivery by the brick-maker than to engage a separate carter.

The manufacturers of the best known bricks in their price list will generally state the weight per thousand of their bricks—useful if contractor pays carriage direct; but they will usually quote a price including railway charges to the railway station nearest to the works. A railway truck will carry about 3000 bricks of the average weight (see also Cartage). A labourer, assisted by the carter, will unload bricks from a cart and stack them, or from railway truck to cart, at the rate of 5000 a day, at 7d. per hour = 1s. 2d. per thousand. The carter's time is included in the cartage rate. A common rate paid for discharging and stacking bricks brought in a barge is 2s. 3d. per thousand. The work is done at per thousand by men who work in gangs.

At Bow Bridge bricks cost 5s. 9d. per thousand for canal dues, lighterage, from Shoebury, unloading, including barrows and planks, loading into carts, and delivery within a mile. Wharfingers in the City will unload and cart within a mile, including use of barrows and planks, for 5s. 6d. per thousand. The alternative is for the builder to find his own barrows and planks, and unload, paying a rate per day or cargo for use of wharf.

At Battersea Bridge (Thames), unloading, use of barrows and planks, cartage, and delivery within two miles costs 4s. 6d. per thousand. At Camberwell Bridge (Grand Surrey Canal), unloading, canal dues, use of barrows and planks, loading into carts, and delivery within two miles costs 5s. 3d. per thousand.

Before pricing a bill of bricklayer's work, it will be necessary to calculate the values of a few of the leading items, from which the other calculations may be deduced in a consistent analogy: such as a rod of brickwork in mortar or cement, a yard of lime mortar, a yard of cement mortar, a yard of lime-and-hair mortar, etc. The estimator must have also at hand the rates of carriage and cartage, the local rates of labour, and the prices of materials such as are to be used. For close investigations of bulk, the reduction of the materials to their cubic quantity will be found more convenient than dealing with bushels and yards. For the majority of calculations, 21 bushels to the cubic yard is quite near enough to the truth for adoption.

Proportion of Bricks to Mortar.—The consideration next in importance, after the price of bricks and carriage, is the number of bricks required for a reduced rod of work. This will depend upon the size of the bricks, the thickness of the mortar joints, the number of flues, the amount of cutting, and the quality of the bricks.

By analogy, the price of bricks for day account may be deduced approximately from the contract price of brickwork, by first deducting the profit from the gross charge per rod. Then deducting the cost of scaffolding, labour and mortar, the remainder will be the amount charged for the calculated number of bricks in a rod.

Waste on Bricks.—The elements which modify the calculated number of bricks per rod are the waste of bricks by breakage or defects. This will vary according to their quality. Some specifications stipulate for the rejection of *all* broken bricks or bats, in the majority of cases they are all packed into the middle of the thicker walls. $2\frac{1}{2}$ per cent. is an ample provision for this contingency.

The waste in the cutting of closers will depend upon the proportion the openings bear to the total quantity of brickwork; generally 2½ per cent. will cover this. This is all the cutting not measured by London surveyors. They measure and bill as rough cutting (and waste implied) all the cutting of squint quoin, birdsmouth, splayed jambs, rakes and skewbacks, measuring the brickwork net.

In the Northern and Midland counties the surveyor measures as follows: "All walls finished with a bevelled upper edge, as to gables, eaves, etc., to be measured to 3 inches above the average height, the lineal dimensions of beam-filling between spar feet and of gable cutting given" (Manchester Society of Architects); from which it will be seen that their description of rough cutting excludes waste.

Saving in Number of Bricks.—The saving of bricks and mortar consequent upon the voids for smoke-flues, is principally confined to dwelling houses. In warehouses the proportion is so small as to be scarcely worth consideration. In domestic work a reasonable average is ·040 per rod. The saving in the space occupied by plates and lintels is about ·014 per rod.

Increase of Mortar.—The increase of mortar for the frogs or depressions in the bricks is about ·040 per rod. In the majority of cases this may be disregarded, as the joints are hardly ever filled.

Proportion of Mortar to Bricks.—The proportion of mortar to bricks charged in day accounts is often unreasonable. Sometimes this may be accounted for by the re-use of bricks, as in the cutting of an opening, or in alterations to it, at others, as in facing an old wall, a large quantity is used in proportion to the bricks. The builder's axiom, 20 bricks can be laid with a hod of cement or mortar, is a safe rule if the joints are not filled; if they are, 18 is nearer the truth.

Mixing Mortar.—The cost of mixing mortar *by hand* is as follows:—

Labourer per cubic yard of mortar, measured after
mixing, and including screening sand, 7 hours   s.   d.
at 7d. 4   1   ·7 constant.

Burnt Clay.—Mr. Burrows (1884) says: "I have turned it out ready for bottoming at a cost of less than 2s. per cubic yard, including digging the clay, clamping, burning, and the necessary fuel. The digging, wheeling and burning cost 1s. 1d. per cubic yard, and a chaldron of breeze costing 9s. burnt from 9 to 12 cubic yards, according to the size of heap."

Lime Mortar.—One of lime to three of sand.

	s.	d.	£	s.	d.
1 yd. of lime	—	—	0	11	0
3 yds. of sand	7	0	1	1	0
4 yds. mixing mortar, labourer ·600 at	5	10	0	14	0
4 yds. water at		2	0	0	8
			2	6	8
Add ¼ for loss of bulk			0	15	6¼
			4)3	2	2¼
Cost per cubic yd.			0	15	6¼

Cement Mortar.—One of cement to three of sand.

	s.	d.	£	s.	d.
22 bushels of cement	1	10	2	0	4
3 yds. sand	7	0	1	1	0
4 yds. mixing, labourer ·600 at	5	10	0	14	0
3 yds. washing sand at ·300 of	5	10	0	5	3
4 yds. water at	0	2	0	0	8
			4	1	3
Add ¼ for loss of bulk			1	0	3¾
			4)5	1	6¾
Cost per cubic yd.			1	5	4¼

Cement Mortar.—One of cement to two of sand.

	s.	d.	£	s.	d.
22 bushels of cement	1	10	2	0	4
2 yds. sand	7	0	0	14	0
Washing 2 yds. of sand at ·300 of	5	10	0	3	6
3 yds. water	0	2	0	0	6
3 yds. mixing, labourer ·600 of	5	10	0	10	6
			3	8	10
Add ¼ for loss of bulk			0	17	2¼
			3)4	6	0¼
Cost per cubic yd.			1	8	8

2 M

Value of a rod of brickwork with ½-inch mortar joints. Mortar, one of lime to three of sand.

	£	s.	d.
4250 bricks at 40s. per thousand	8	10	0
1 yd. of lime at 11s. per yd.	0	11	0
3 yds. of sand at 7s. „	1	1	0
Water	0	1	0
Mixing mortar, labour ·700 of 5s. 10d.	0	4	1
Scaffolding	0	4	0
Bricklayer and labourer 5·00 at 14s. 7d.	3	12	11
Cost per rod	14	4	0
Cost per foot reduced	0	1	0¼

Value of a rod of brickwork in cement with ½-inch mortar joints. Mortar, one of cement to three of sand.

	£	s.	d.
4250 bricks at 40s. per thousand	8	10	0
19 bushels cement 1s. 10d.	1	14	10
57 bushels sand	0	18	1¼
Water	0	1	0
Mixing mortar, labourer ·700 of 5s. 10d.	0	4	1
Scaffolding	0	4	0
Bricklayer and labourer 5·50 at 14s. 7d.	4	0	2¼
Cost per rod	15	12	3
Cost per foot reduced	0	1	1¼

The cost is kept down by leaving many joints unfilled with mortar, by using less than the prescribed proportion of lime or cement, and possibly subletting the labour and supplying bricks below the prescribed quality.

The number of bricks which a bricklayer will lay in a day, will depend upon the character of the work. At the rate of 4½ days and 4250 bricks per rod, we have an average of 945 bricks per day. It is a popular but erroneous belief that a bricklayer lays 1000 bricks per day, he never does. In a building with many openings and consequent cutting and plumbing he rarely exceeds 400.

In the early days of railway construction, in such works as viaducts, large buildings and retaining walls he laid sometimes more than 1000 per day.

The value of erection, use of, and removal of scaffolding has been variously estimated; the price of 4s. per rod is worthy of respectful consideration, if only for its age and its frequent adoption.

It was propounded by Peter Nicholson as early as 1823, and was probably not new then. Seddon says 3s. per rod, which is certainly its maximum value for ordinary work.

Extra only in Cement.—The cost of extra only in cement may be easily obtained by deducting the price of work in mortar from that in cement.

Brickwork in Northern Counties.—The estimator who has to deal with the price of brickwork in the North of England will be surprised by curiously low prices, brickwork sometimes appearing in an estimate as low as 7l. per rod. This is to be accounted for, to some extent, by the difference between the modes of measurement in the Northern Counties and those in the Southern. It will also generally be accompanied by a high price for labour to hollows (forming openings), as it is called, which goes far towards bringing the price up to the value in other localities. This is an item which does not appear in London estimates. It may be conceded that in the want of distinction between brickwork with many openings and large masses with few openings the London bill is weak. For any modification of prices dependent upon these differences, the estimator must examine the drawings, whereas the ideal bill of quantities should be sufficient without any inspection of the drawings.

Hollow Walls.—The labour is little more than equal to that on plain brickwork, and the wall ties are often considered equivalent to the advantage gained by the cavity being measured in the thickness. If haybands or movable boards are used to keep the mortar from dropping into the cavity, their value must be added: 6s. per rod would amply pay for this.

HOLLOW WALLS, *per rod.*—These are measured as solid, the whole thickness including the cavity being measured as brickwork. The saving of brickwork by the cavity will pay for the extra labour in its formation. To the price of ordinary brickwork add for hay bands or boards to protect the cavity.

The following description will illustrate the valuation.

Hollow wall of two thicknesses, 9 inches and 4½ inches respectively, with 2¼ inch cavity, bonded with galvanized wrought-iron wall ties 9 inches long, four to each superficial yard, and weighing 60 lbs. per hundred, and allow for keeping cavity free from droppings or rubbish.

	£	s.	d.
Labour and materials of a rod of brickwork, as before	14	4	0
Haybands	0	6	0
120 wall ties = 72 lb., at 25s. per cwt.	0	16	1
Per rod reduced	15	6	1

The labour on half-brick walls is commonly priced by estimators at from ¼d. to ½d. per foot extra per foot superficial. The extra cost of brickwork circular on plan over 15 feet radius may be taken as per rod, including labour to circular face, 10s. to 15s. according to the radius.

Rough Cutting and Waste.—The rough cutting not usually measured by the surveyor has already been dealt with as modifying the number of bricks per rod reduced, and as it is the London custom to measure brickwork net, it is obvious that waste of bricks as well as the labour of cutting should be allowed for in valuation.

In cutting a superficial foot to a gable raking at an angle of 45° with the horizon, the cutting would waste ½th of the contents of a brick. Assuming the wall to be 1½ bricks thick, and treating the bricks as all headers rising with their joints four courses to a foot, we should have about 2¾ courses or 8¼ bricks affected by a superficial foot of rough cutting, reckoning, therefore, ½th of 8¼ bricks, say 1½ bricks at 40s. per thousand, we have ¾d. representing the cost of the waste. The labour at London rates does not exceed 1d. per superficial foot. Total 1¾d. Constant, ·010 day of a bricklayer.

Joints of Brickwork Struck Fair and Twice Limewhiting.—The striking of the joints would be done as the work was built. The price per rod before mentioned is enough to cover it; but if the value is desired separately, the constant may be taken as ·025 day of bricklayer = at London rate 2½d. per yard.

Limewhiting is generally done by labourers, although ostensibly bricklayer's work. A man should do between 140 and 150 yards in a day of ten hours, including mixing the limewhite, ·007 day of a labourer, about ½d. The ideal limewhite has some tallow or sulphate of zinc in it. The average builder does not indulge in these subtleties; he uses slaked lime and water alone.

A cubic foot of stone lime before slaking will be sufficient for 150 yards of limewhiting first coat, second coat 250 yards. With lime at 11s. per cubic yard, the cost per superficial yard would be as follows:—

	s.	d.
Twice limewhiting, labour	0	1
Lime, use of brushes and water	0	0¼
	0	1¼

Brick-nogging.—This may be priced at the same price as common brickwork; the extra labour of laying between the timbers is quite covered by the saving of bricks, by not deducting the fir, and the saving of labour, as there is no trouble about bonding.

Half-brick Trimmer Arch in Cement.—The value per rod is equal to that of work fair both sides.

	£	s.	d.
Price per rod of brickwork in cement as before calculated	15	12	3
Extra for work fair both sides ·5 days of bricklayer and labourer at 14s. 7d.		7	3¼
272)	15	19	6¼
Per foot reduced .. over	0	1	2
14d. divided by three, gives the value of half-brick, the thickness of the arch	0	0	4¾
Levelling up with fine concrete, generally equal to a layer 3 in. thick all over, ₁/₁₀th of a cubic yd. at say, 10s.	0	0	1
	0	0	5¾

Vaulting.—Usually a segmental arch in two half-brick rings in mortar. The value of this may be taken as equal to plain brickwork with ⅜-inch joints. The trouble of forming the arch is no greater than the cutting of closers in walls; but it is not subject to any of the allowances thereon for plates and flues. The number of bricks required for a rod of solid work is 3921. Adding waste in broken bricks 2½ per cent., we have 98 bricks to add, making a total of 4019.

	£	s.	d.
Say 4020 bricks at 40s.	8	0	10
Labour, mortar, etc., as example	5	14	0
272)	13	14	10
Per foot reduced	0	0	12¼

Two-thirds of 12¼ is 8¼, the cost per foot superficial.

Groin Point.—This requires rather better bricks than the general arching. In a lineal foot about three bricks would be affected. Bricks sufficiently good for the purpose (inferior malm facings would be reasonable) would cost about 12*s.* 6*d.* per thousand more than stocks in London—⅛th of a penny each. A man would cut and rub about 20 feet lineal per day. This, at London rates, gives 5¼*d.* per foot. Constant, ·050 day of bricklayer per yard.

	£	s.	d.
Extra for bricks	0	0	0¼
Labour	0	0	5¼
	0	0	5½

Raking out Mortar Joints and Pointing Soffits of Vaulting with Cement.—The raking out, done by labourers or boys at London rates, is worth, at 7*d.* per hour, about ½*d.* per yard, if done by labourers. Constant ·008. By boys it would cost about ¼*d.*

Pointing with Cement.—Labourer will keep three bricklayers supplied with pointing stuff, and the constant for the bricklayer being taken as 080 day, and labourer ·027 day, we have at London rates, bricklayer, 8½*d.*; labourer, 1¾*d.* A yard of cement mortar will be sufficient to point 225 yards superficial. If made one of cement to two of sand, the cost would be—

		£	s.	d.
1 yd. cement mortar cost	225)	1	8	8
		0	0	1
Raking out ·008 of 5*s.* 10*d.*		0	0	0½
Bricklayer pointing ·080 of 8*s.* 9*d.*		0	0	8½
Labourer ·027 of 5*s.* 10*d.*		0	0	1¾
Cost per yd. sup.		0	0	11¾

The scaffolding had better be separately dealt with.

Damp-proof Courses.—The simplest and most convenient damp-proof course is a mixture of pitch, tar and sand, boiled together and laid hot on the walls. This kind of damp-proof course is generally prepared and laid by the builder's men. For the preparation a cauldron is necessary, which costs, with tools, about 10*l.* Strong iron buckets cost about 3*s.* each.

These cauldrons will boil between 4 and 5 cwt. at a time. The most convenient way of doing this is to buy the mixture in casks of the tar-distillers; cost in London at works, or at a railway station near the works, is 45*s.* per ton, which will cover about

PRICES.

40 square yards, ½-inch thick. It contains sufficient grit for the purpose. The detail of cost in London would be as follows:—

	£	s.	d.
One ton of material	2	5	0
Carriage, say	0	2	0
28 lb. coal-tar as a flux, at 2s. 8d. cwt.	0	0	8
⅕th ton of coals (18s.)	0	3	7
Use of cauldron and buckets, say	0	1	0

(If care be taken to avoid making the material too liquid, deal fillets to confine it to the wall may be dispensed with.)

	s.	d.		
Labourer, 10 hours, breaking up and boiling material, conveying to walls, and assisting to spread	0	7	0 5	10
Spreader labourer, 10 hours	0	7	0 5	10

40) 3 3 11

Cost per yard .. 0 1 7

A common price charged by asphalters for mineral asphalt damp-proof course ½-inch thick laid in London is 2s. per yard superficial.

The contract price for damp-proof course of pitch, tar and sand ½-inch thick is often as low as 2d. per foot.

If for the prepared material we substitute coal-tar, pitch and sand—one of the former to two of the latter—the cost in London would be as follows. The quantity required to cover 40 yards would be as follows:—

	s.	d.
¼ ton coal pitch, at 40s.	10	0
½ yd. sand, at 7s.	3	9
Total of remaining items, as in last calculation	18	11

40) 29 10

Per yard superficial .. 0 10

Asphalt.—Damp-proof courses of special kinds are more conveniently laid by the specialist, who will give prices on a statement of the nature and quantity of the work:—

Seyssel, ¾ in. thick .. .. .. costs 5s. 3d. per yard.
„   ½ in. „   .. .. .. „ 4s. 6d. „

Pavings.—The bed to receive paving is nearly always separately measured in the case of tile paving as a cement-floated face of a given thickness, and it is better thus treated for convenience of valuation. There is in such case only a thin bed of cement to provide for and the jointing. For brick pavings, often laid on gravel, ashes, or brick rubbish, the description includes both the bedding and the jointing.

If the estimator calculates before pricing the items the cost of a cubic yard of cement mortar or lime mortar, he can arrive at a very close approximation to the quantity of such material required for the jointing by calculating it. This laborious process will not be frequently required; but it is an admirable test to apply to the wild statements of workmen—is a ready means of testing their accuracy, and of arriving at a true conclusion.

Paving of Hard Stocks Laid Flat in Sand, including Cutting and Waste other than Raking.

	£	s.	d.
36 bricks at 40s. per thousand	0	1	5¼
Sand 1/27 of a yd. at 7s.	0	0	3
Bricklayer and labourer ·045 of 14s. 7d.	0	0	7⅞
Cost per yd.	0	2	4

Paving of Hard Stocks Laid on Edge in Sand.

	£	s.	d.
53 bricks at 40s. per thousand	0	2	1½
Sand 1/20 of a yd. at 7s.	0	0	4¼
Bricklayer and labourer ·066 of 14s. 7d.	0	0	11½
Cost per yd.	0	3	5¼

Paving of Hard Stocks Laid Flat in Mortar.

	£	s.	d.
36 bricks at 40s. per thousand	0	1	5¼
1/27 of a yd. of mortar at 15s. 6½d.	0	0	7
Bricklayer and labourer ·080 of 14s. 7d.	0	1	2
Cost per yd.	0	3	2¼

Paving of Hard Stocks Laid on Edge in Mortar.

	£	s.	d.
53 bricks at 40s. per thousand	0	2	1½
1/20 of a yd. of mortar at 15s. 6½d.	0	0	9¼
Bricklayer and labourer ·118 of 14s. 7d.	0	1	8½
Cost per yd.	0	4	7¼

Tile Paving.—Pavings of encaustic tiles are usually selected from a manufacturer's pattern-book, the list price being per yard at the manufactory or delivered at the railway station nearest to the manufactory. The manufacturer's charge for packing is commonly, for plain unglazed tiles and glazed thin tiles, 6d. per yard superficial; for encaustic and glazed tiles, 8d. per yard superficial. The trade discount is usually 10 per cent. off the list. Manufacturers will lay encaustic tiles, supplying labour, two colours, 2s. 6d. per yard; small quantities, or intricate patterns in several colours, 3s. per yard; when they supply the cement for bedding and jointing, a charge of another 6d. per yard is made. The carriage will, of course, vary with the distances from the manufactory. The laying of tile-pavements is slow, and the attendance of a labourer is but a small item. He would bring the tiles and cement to the bricklayer at the commencement of his work, occasionally bring him cement, and clear away after him.

If a cement-floated face has not been provided, more cement would be required for bedding. Tesselated pavements, for which a floated bed is always provided, would require $\frac{1}{2}$ a cubic foot of cement per superficial yard.

Paving of 6 in. × 6 in. best Red Staffordshire Quarries Laid and Jointed in Cement and Cleaned off at completion.

	£	s.	d.
36 tiles at 9s. per hundred delivered	0	3	3
Bricklayer ·20 of 8s. 9d.	0	1	9
Labourer ·05 of 5s. 10d.	0	0	3½
$\frac{1}{108}$ of a cubic yd. cement mortar at 25s. 4½d.	0	0	2¾
Cost per yd.	0	5	6¼

Flues, Parget and Core.—Some of the provincial surveyors state the length of smoke-flue, and describe as "forming, pointing, and coring out smoke-flues," and state the sizes if various. This practice might be universally adopted with advantage. The London bill gives no description, but No. — flues, parget and core; 1s. or 1s. 6d. each is generally allowed by the average estimator. If the saving of bricks has been allowed for in the calculation of the value of brickwork per rod, and the estimator is not satisfied with the 1s. 6d. or 1s., which rarely pays for it, he must find out the average length by inspecting the drawings. The forming and labour rendering may be assumed as included in the price per rod for labour on brickwork, and the material used in the rendering one coat may be measured superficially to arrive at the value; thus for a flue 14 inches by 9 inches and 50 feet long:—

```
  3.10
 50.0
 ————   .. .. .. .. ..   191.8 = 22¼ yds. material only,
                                  render one coat in mortar.
```

Cutting and Pinning Edges of Landings with Cement.—It is customary to make all the masonry item include fixing. This item would therefore deal only with the bricklayer's time cutting away the brickwork, and making good. A good foreman will arrange for much of the stonework to be built in as the work proceeds. When it is a case of cutting away, working in conjunction with the mason wastes much time. The bricklayer goes to the place and cuts the chase, and has to return to make good after the stone is in position. A number of such pieces of work are left to be done at once by the workman; but time is inevitably lost, and the price adopted by estimators varies greatly. For one course cut out as to a 3-inch landing, constant ·022 day of a bricklayer per lineal foot; cement for pointing upper joint, ¼d. For two courses cut out as to a 6-inch landing, constant ·044 day of a bricklayer per lineal foot. For a 4-inch landing two courses would generally have to be drawn, the upper course cut and reset in cement. The price should therefore be more than for a 6-inch landing. Thus—

		d.
Chase cutting, bricklayer ·044 at 8s. 9d.		4¾
¼ ft. superficial rough cutting, as before, at 1¾d.		0¼
Cement		0¾
		—
		6

The mason beds landings. Stephenson says ('Estimating'): "Price at 3d. per foot run (for 3-inch landing), including profit; 1d. should be added for every inch of increased thickness of landing."

Level and Prepare Old Wall to receive New Work.—This item varies in quality according to the amount of care exercised in pulling down. It means resetting some of the bricks in the topmost course, and a little more mortar in the first joint before raising on it, and is not calculable. 1d. per foot superficial is a common London rate, and is a reasonable average.

Ends Cut and Pinned.—The sizes are not usually stated, and must be guessed, as time cannot be spared to inquire, and a general average is assumed. A long list of such things generally appears in a London bill of quantities, and the estimator frequently brackets them together and attaches one price. Ends of timbers—this description generally means girders or binders (if joists, it should be stated)—for these, if of large scantling, holes are generally left in the brickwork, and the making good is done after they are fixed. The brickwork, not having been deducted, will generally pay for the material, leaving time only to be dealt with. Taking a timber 11 inches by 9 inches as an illustration, there would be rough cutting on top and two sides and 9 inches into wall = 3 feet 4 inches superficial of rough cutting = $8\frac{1}{4}d.$ for time—say, bricklayer ⅓rd hour; but the work being done in small pieces would involve quite half an hour; consequently—

	d.
Bricklayer, half-hour at $10\frac{1}{4}d.$	$5\frac{1}{4}$
Labourer, half-hour at $7d.$	$3\frac{1}{2}$
	$8\frac{3}{4}$
Cement	2
	$10\frac{3}{4}$

If the holes are cut, another $5\frac{1}{4}d.$ should be added, or bricklayer half an hour; with these girders may be classed all things of similar size. Ends of steps cut and pinned, may be dealt with as last item, according to their assumed sectional area; thus an 11 inch × 6-inch step would be ⅔rds of the last price.

When steps or timbers are built in, the rough cutting, done as the work is carried up, is all that need be allowed for. It is not safe to reckon on less than 9 inches into wall for large timbers, $4\frac{1}{2}$ inches for smaller timbers and steps; a 9-inch by 3 inch joist (two sides) would require $\frac{7}{13}$ of a foot of rough cutting; a step 11 inches by $7\frac{1}{2}$ inches (three sides) $\frac{19}{12}$ of a foot of rough cutting, at London rates less than 1d., and some little delay caused by working in concert with the mason. This would probably waste a quarter of an hour for each step. We should have this result:—

	d.
Cutting	1
Delay	$2\frac{3}{4}$
	$3\frac{3}{4}$

When sand courses are left, the cost may be taken as equal to an item of building-in; the cutting is the same in either case. There would be no waste of time in building-in timbers; a floor of joists, for instance, would be fixed ready for the bricklayers to carry up the work.

Holes for Pipes.—These, it need hardly be said, vary in size, as the pipes vary. The thickness of the walls through which they are cut is usually averaged by the surveyor. A bricklayer should cut a hole through a 9-inch wall in less than five minutes, but as he has to be brought from another part of the building, or has to wait for the plumber's directions, it is not safe to reckon less than a quarter of an hour, $2\frac{3}{4}d.$ at London rates, and $1d.$ would pay for cement. As the thickness of the wall increases, the difficulty of cutting holes is greater; to cut through a two-brick wall should be reckoned three-quarters of an hour, and besides this, many of the holes for pipes are in awkward positions. No labourer's time need be reckoned for the cutting. For the making good there should be allowed a little labourer's time for bringing material for the mechanic's use. The making good, as before remarked, would be done after the whole system of pipes was fixed. The following is approximately the cost:—

Bricklayer, cutting.	Making good.	Labourer.	Material.
Holes in 1-brick wall—			d. d.
$\frac{1}{4}$ hr.	$\frac{1}{4}$ hr.	$\frac{1}{8}$ hr.	1 = 7
Holes in 2-brick wall—			
$\frac{3}{4}$ hr.	$\frac{1}{2}$ hr.	$\frac{3}{8}$ hr.	2 = 17$\frac{1}{2}$

(The average 1$\frac{1}{2}$-brick wall would thus be 12$d.$)

Extra Labour, Cutting and Waste, to Relieving Arches.—Many surveyors make no mention of relieving arches in the bill of quantities. They measure extra only in cement if they are built in cement, and bill it with that item at per rod, and measure the rough cutting including it with the general rough cutting. The work in other respects is not more troublesome than the building of a wall with proper bond. Other surveyors prefer to number the relieving arches, describing them as follows:—No. —. Extra labour, cutting and waste to relieving arches " 4 feet long, 1-brick by 1-brick, averaged." The cutting will be longer than this, as the mean length only has been measured; an addition of 3 inches to the length will generally be sufficient for the extrados. The value of this item will be as follows:—

| 9.8 | | Rough cutting and waste (all round) = 7¼ ft. at | s. | d. |
| .9 | 7.3 | 1¾d. as before.. .. .. | 1 | 0¾ |

Treat segmental arches in the same manner, but observe that the cutting and waste is only to extrados and skewbacks.

Extra Labour and Materials Cutting and Bonding New Brickwork to Old.—In each foot of height one course would be let into the old work 4½ inches, the three others making a straight joint with it. The brickwork would, therefore, amount to one-quarter of a superficial foot 4½ inches thick at the common rates. The cutting out one course of bricks, 1 foot lineal for each foot superficial of work, is on an average:—Constant, days bricklayer, ·022 per foot sup.

		d.
Brickwork 1/12 ft., reduced at 14/4/0, say		1
Extra mortar		0¼
Labour		2¼
		3¼

This calculation forms a base for the pricing of lineal dimensions of the same character, which often appear in bills of quantities as "extra labour and materials, cutting and bonding 1 brick, 1½ brick, etc., wall to old." The one-brick wall would be taken at three-quarters of the foregoing price = 2½d. ; two-brick wall twice the last or one and a half times 3⅓d. = 5d. etc. Some surveyors measure the general brickwork 4½ inches into the old wall, and a lineal dimension of *labour only* cutting and bonding. The former way is preferable. A few of the lineal dimensions commonly measured by the London surveyor are as follows:—

"*Labour, Rough Oversail or Set-back One Course.*"—Many estimators would attach no price to this. It is worth very little. The constant may be taken, ·002 day of bricklayer per foot run, less than ¼d. at London rates.

Rough-cut Chamfer on Brick.—Generally cut with hammer and chisel after the work is built, can be done by a labourer, and affects one brick in one course only when thus described. Constant, ·005 day of labourer per foot run.

Rough-cut Birdsmouth or Squint Quoin.—Equal to the last, but cut by the bricklayer in course of his work. Constant, ·008 day of bricklayer per foot run, ¾d. at London rates.

5-inch *Skewback Cutting.*—Usually measured to trimmer arches, rather more than equal to a superficial foot of rough cutting and waste. Is cut with hammer and chisel after work is built. Constant ·017 day of bricklayer per foot run = at London rates 1¾d.

Cut or Form Chase for Pipe.—It is impossible to tell from the average bill whether these are to be cut or formed. If they are formed, they need not be priced; if they are cut, they would be done by the bricklayer with hammer and chisel. Often it is only a groove in the brickwork for one pipe of small size, at other times it is for a soil pipe. If the chase is assumed as 9 inches wide and 4½ inches deep, we have a girth of 1 foot 6 inches = 1 foot 6 inches superficial of cutting in each, and taking the constant of ·033 of bricklayer per foot superficial, we have, at London rates, 5d. per foot run. As the size, however, depends so much on circumstances, estimators commonly put down 3d. or 4d. per foot run.

Chase, Cut and Parge.—Chase, cut and parge may be taken at the last rate for cutting, adding labour and materials, parging in lime and hair mortar.

Rake out and Point Flashings with Cement.—The raking out is done by a labourer. Constant, ·002 day of a labourer per foot. The pointing by a bricklayer. Constant, ·003 day of a bricklayer per foot. One foot cube of cement will point 200 feet. At London rates, we have the following result:—

	d.
Labourer	0¼
Bricklayer	0½
Cement	0¼
Cost per foot run	1

Commonly priced at 1½d. per foot.

Rake out and Point Stepped Flashings with Cement.—The running length of joint would average one and a half times the length of the raking line as measured. Consequently, the price would be one and a half times the last result = 1½d. Commonly priced at 2d. per foot.

Setting Stoves.—The charge in the average estimate compared with that in a day account is often remarkable.

A common contract price for setting a register stove to a 2 feet 6 inch opening is 5s. A slow-combustion stove to a 2 feet 6 inch opening, 7s. 6d. Range 10s. per foot of opening.

Register stoves to 2 feet 6 inch openings would be set by a—

	s.	d.
Bricklayer, 3 hours at 10¼d.	2	7½
Labourer, 1 hour	0	7
Brickwork, about 2 ft. reduced, at 12½d.	2	1
Cost	5	3½

(A common contract price in London is 6s.)

The setting of slow-combustion stoves, of the Barnard and Bishop type, to 2 feet 6 inch opening:—

	s.	d.
Bricklayer, 5 hours at 10¼d.	4	4½
Labourer, 2 hours at 7d.	1	2
Brickwork, about 2 ft. at 12½d.	2	1
Fireclay	0	6
Cost	8	1½

(A common contract price in London is 8s.)

The setting of close-fire ranges varies considerably in the amount of work required. The average cost is about 8s. per foot run of opening. It is very difficult to say what time a man will spend upon it. The material used in setting stoves is often mere rubbish instead of bricks, and builders reckoning upon this will often price such items at a low rate.

Frames, Bed and Point.—Frames being of very various sizes, an average must be taken. A frame is generally considered as 24 feet superficial, 24 feet to 36 feet large, over that size extra large.

The London surveyor will often, in a window with mullion and transom, reckon four frames; others state the number of feet run. In each (averaged) the last-mentioned practice is better; but the former is most frequent.

Frames are fixed by the carpenter. If properly fixed, they are screeded. The value of bedding and pointing a 24-feet frame is as follows: Taking 3 feet 4 inch by 7 feet 3 inch as producing that dimension, and 2 feet 8 inch by 7 feet as the external opening, the calculation would be as follows:—

```
2/7  3
   10    1¼ yds.  Render one coat in narrow widths  ..   ..
   ─────
   16  8
   ─────  16  8  Pointing with cement  ..   ..   ..   ..   ..
```

There are nearly always a number of smaller frames, tending to reduce the average, and consequently the cost.

A common contract price in London is—for frames, 1s.; large frames, 1s. 6d. each.

Facings.—The variety of facing bricks used in London is considerable; but the enumeration of the various kinds and qualities is not here necessary. They are admirably described in Rivington's 'Notes on Building Construction,' vol. iii. A few items will be sufficient to show the principles commonly adopted in their valuation.

The value of facings will depend upon the kind of bond, as there will be more headers in Old English than Flemish bond. This is, however, sometimes evaded, many half bricks being used as headers instead of whole ones, whole bricks being used only occasionally as bonders.

Other points to be observed in the valuation of items of facing are, whether the work is pointed as it proceeds, or raked out and pointed afterwards; whether it has been measured as "extra on common brickwork," or "extra upon another kind of facing."

The number of facing bricks compared with the rods of brickwork will vary, a detached building commonly requiring a larger proportion than one of a row; but 1000 to each rod is a reasonable average. When a bill of quantities exists, the multiplication by seven of the number of feet superficial of facings therein will give the number.

Red Facings.—Extra on common brickwork for facings of Lawrence's (Bracknell) No. 3 best red facing bricks, raked out and pointed with a neatly struck bevel joint in fine mortar.

	£	s.	d.
Facing bricks per thousand in field	2	17	6
Railway rate to Nine Elms, 2½ tons at 3s. 8d. per ton	0	9	2
Loading into carts	0	1	2
Unloading	0	1	2
Cartage within 3 miles	0	10	0
	3	19	0
Waste, 2½ per cent. on 2l. 17s. 6d.	0	1	5¼
	4	0	5¼
Deduct one thousand stocks delivered	2	0	0
	2	0	5¼

Extra cost of 7 bricks at 2l. 0s. 6d. per thousand = 3½d. per foot sup.

Then for the cost per foot   *d.*

	d.
Extra cost of bricks per foot sup.	3½
Raking out, labourer, ·008 of 5s. 10d.	0½
Bricklayer pointing ·080 of 8s. 9d. per yard	1
Labourer, ·027 of 5s. 10d.	0¼
Cost per foot sup.	5¼

Moulded Courses.—Moulded courses are measured in various ways. Some surveyors present them in the bill as extra on facings, having by this method measured the brickwork in the projection and billing it with the general brickwork, girthing the moulding for facing and billing it with the general facing, the item appearing in the bill in the following form:—

" — feet run. Extra on facings for one course of moulded bricks all headers as string, including setting out and raking, and pointing to match facing."

Other surveyors measure the brickwork in projection, billing it with the general brickwork, and if the facing has been measured over the whole face of wall, deducting it.

The item appears in the bill as follows :—

" — feet run. Extra on brickwork for labour, moulded bricks and facing, to one course of moulded bricks, as string, oversailing and including setting out and raking out and pointing to match facings."

This latter method is probably less liable to misapprehension by the builder.

By either method, if the work is in cement, the extra only in cement goes in with the general item of that work.

The calculation of the value of the last item is as follows:—

	£	s.	d.
1000 moulded bricks, Lawrence's 19a, at per thousand	6	10	0
Carriage, loading and unloading	0	13	4
	7	3	0
Waste, 2½ per cent.	0	3	7
	7	6	7
Deduct 1000 stocks at 40s.	2	0	0
	5	6	7
Extra cost of 2¾ bricks at 106s. 7d. per 1000	0	0	3¾
Exposed face (5-in. girth) 1/12 foot, raking out and pointing as before, at per foot sup., 1¼d.	0	0	0½
Setting out	0	0	0¾
Price per foot run	0	0	5

Mitres (Lawrence's) cost 12l. per 1000, compared with 6l. 10s.; the moulded bricks upon which they are extra = 1½d. each, which represents the whole of the extra cost. All the labour has been included in the price per lineal foot of string-course.

Cut and Rubbed Mouldings.—Cut and rubbed mouldings should be treated in a similar manner for ordinary sections; the constant may be taken as ·215 day of bricklayer per foot superficial of mouldings, if the bricks are of thoroughly good quality, the cost increasing with the inferiority of the bricks. From the foot superficial, the prices of moulding of smaller girth may be calculated. The mitres are generally equal in value to about one superficial foot of moulding.

Glazed Brick Facings.—The treatment of glazed brick facings is similar to that for other kinds of bricks. The printed lists of prices in circulation are subject to a trade discount of 10 to 12 per cent. There is so little difference between first and second quality that the inferior ones are often used when the best are specified. The difference in price is about 2l. per thousand.

A cart will only carry 400 because of their weight. The London merchants charge 10s. per thousand for carting within the usual limit of distance.

The weight is commonly about 4½ tons per thousand. The railway rate to London for the Leeds glazed bricks is 40s. per thousand in 4-ton lots. The Midland Company has of late brought glazed bricks from Kilmarnock (which have successfully competed in London with the better-known sorts) at a rate of 2*l*. 11*s*. per thousand in 4-ton lots.

Salt-glazed bricks are frequently used where economy is important. There are two qualities, the better being about 2*l*. per thousand higher in price.

Picked Stock Facings.—Extra on common brickwork for facings of picked stocks finished with a neatly struck weathered joint as the work proceeds.

These would be selected from the general bricks delivered. Bricks delivered, say 40*s*. per thousand.

Constant, extra for selecting bricks per thousand, ·400 day of a bricklayer's labourer.

Constant for striking bevel joint and fair face, ·005 day of a bricklayer per superficial foot.

The value at London rates would be as follows:—

	d.
Selecting 7 bricks at 2*s*. 4*d*. per thousand	0¼
Striking joints—bricklayer	0½
Per sup. foot	0¾

A common contract price in London is 1*d*. per foot superficial.

Gauged Arches.—Extra on Lawrence's facings for rubbed and gauged segmental arches in Lawrence's No. 10 red rubbers, raked out and pointed to match facings.

Gauged arch is generally reckoned at ten bricks per foot superficial which includes an allowance for waste. When the bricks are very good, eight or nine will be enough, say nine.

	£	s.	d.
Price of rubbers per thousand	4	19	0
Expenses as before	0	12	4
	5	11	4
Cost of nine bricks	0	1	0
Deduct facing bricks 1 foot sup.	0	0	3½
	0	0	8½
Bricklayer, cutting, rubbing and setting ·142 of 8*s*. 9*d*.	0	1	3
Cost per foot sup.	0	1	11½

Oversail.—Labour, oversail or set back one course of facings (brickwork in projection measured with the common brickwork) is worth very little more than labour to rough oversail (which see), say ¾d. at London rates.

Birdsmouth.—Fair cut birdsmouth or squint quoin. Constant ·012 day of bricklayer per foot run, about 1¼d. at London rates.

Skewback Cutting.—Circular and skewback cutting to red facing, as described above. Waste one-sixth of three bricks (see rough cutting), equal one-half of a brick at 40s. 0d. per thousand, about ½d.

	d.
Waste	0½
Bricklayer, ·015 day at 8s. 9d.	1½
Cost per foot run	2

Brick on Edge and Tile Creasing.—Feet run. Extra on common brickwork for coping of brick on edge, and double plain tile creasing, all in cement, to one brick wall. The course on edge is worth no more for labour than common brickwork. The reasonable way to value this is as follows:—

			d.
1 0 6	6 in. 1 brick E.O. in cement 4 in. reduced at 1¼d.		0½
1 0 2 5	2 ft. 5 in. striking joints	9 in. 4½ in. 4½ in. 5½ in. 5½ in. ¼d.	0¾
		29 in.	
1 0 2 0	Labour equal to oversail and *set back* one course on one side of wall .. ¾d. 4½ plain tiles at, say 40s. per thousand Cement fillet at ¼d.		0¾ 2¼ 1
	Cost per foot run		5¼

For the foregoing, the tiles being placed longitudinally, require no cutting. When the walls are thicker, cutting would be necessary. The waste would be small—as nearly all the tile is concealed, 5 per cent. would cover it. The cutting is worth ·003 day of bricklayer per foot superficial.

Terra-cotta.—A great variety of terra-cotta work of stock patterns is to be found illustrated in the catalogues of the brick and pottery manufacturers, and the prices may be obtained from their lists. Estimating for these is simple enough. For terra-cotta, when in large quantities, a separate contract is often made, in which case it is frequently stipulated that the contractor shall assist the manufacturer in setting out, shall unpack, store, and protect it, and be responsible for its damage. No rule of general application can be furnished for valuation of an item of this kind.

In other cases the name of the maker from whom the terra-cotta shall be procured is given, and the general contractor takes all the responsibility.

When the architect draws the details for the piece moulds, as he generally does (or should do) for a large work, the duty of the builder is less troublesome than in those cases where it is left to him. The necessary drawing in the latter case often occupies much of the time of a foreman or a clerk.

Often the modelling of enriched work is done by an artist appointed by the architect, and a sum provided in the contract, and it will be necessary to observe what work thereby falls upon the contractor, as the conveyance to and fro of models, " offering up," etc.

So far as measurement is concerned, the manufacturer of terra-cotta is satisfied by the division of the work into two categories— feet cube of plain and moulded, feet cube enriched, and he frequently puts them together and prices at one uniform rate.

A good bill of quantities states the number of models which the given quantity of terra-cotta will require, and the manufacturer regulates his price accordingly. It is not the custom to deduct the brickwork; the value is therefore the difference between it and terra-cotta.

The manufacturer's price for good terra-cotta in quantity rarely exceeds 5s. per foot cube all round, including the models delivered, and he generally prefers to make his own models. The chambers in terra-cotta, which require filling with cement concrete, commonly amount to three-quarters of the total bulk.

The setting and pointing of terra-cotta may be taken as ·035 day of bricklayer and labourer. The valuation of terra-cotta would be as follows:—

		s.	d.
Terra-cotta		4	6
Bricklayer and labourer, setting and pointing ·035 at 14s. 7d.		0	6
Mortar		0	1
¾ c. ft. cement concrete 6½		0	4¾
Filling, labourer ·016		0	1¼
Rough-cutting and waste on brickwork (rarely exceeding a superficial foot)		0	1¾
		5	8¾
Deduct 1 c. ft of br'ckwork at 14l. 4s. 0d. per rod		0	10¾
Price per foot cube		4	10

As the facings would be deducted in the course of measurement, they are not affected by the last calculation.

In pricing of brickwork, much time may be saved by using 'Laxton's Bricklayer's Tables.'

WALLER.

The use of rubble for walls in London is rare; it is almost entirely confined to Kentish rag, used as a facing only, and is measured and valued as extra on brickwork. In the stone districts the entire wall is built of rubble, or rubble with a thin backing of brickwork.

The London surveyor usually measures rubble by the cubic yard, wherever it may be situated, keeping walls 12 inches thick and under separate, and stating their thickness. The local customs of measurement are various: by the perch of 18 superficial feet reduced to a thickness of 24 inches, the rod of 36 superficial yards reduced to 24 inches thickness. In Glasgow and its neighbourhood by the superficial yard, 24 inches thick; walls exceeding that thickness being reduced to it, and those under 24 inches classed according to their respective thicknesses. In Ireland, by the perch of 21 feet run, stating height and thickness, or by the square perch of 21 feet superficial at a standard thickness of 18 inches. (Seddon.)

The labour upon stone to adapt it for coursed work will vary considerably; some stones as they are raised from the quarry require no labour on their beds, the stratification is so regular; in others the beds are ill defined, and the stone comes out in very irregular shapes. 1½ yards cube of rubble stone as stacked by the quarryman is required for every cubic yard of rubble work if coursed, if uncoursed about 1¼ yard cube. Uncoursed rubble requires about one-third of its total bulk of mortar, coursed rubble about one-fourth. 3s. 6d. to 4s. per ton is a common price at the quarries in the stone districts for rubble stacked ready for carting. Bargate stone (Godalming) costs 3s. 6d. per ton at quarry, roughly dressed for random work, and may be taken as a reasonable average price for the stones near London.

Kentish Rag.—This is almost always sold dressed ready for use. The prices are as follows: Rubble for foundations, 3s. per ton. Hammer-dressed facing in courses of irregular height, or in random courses 10s. 6d. per ton, on trucks or in barges at Maidstone; random-dressed facing, 6s. 6d. per ton; quoins with drafted angle, 1s. per foot run. Railway rate to Bricklayers' Arms Station, 3s. 4d. per ton; rate by barge in the Thames below Westminster, 2s. per ton—about 13½ cubic feet. Unloading and loading into carts will cost 6d. per ton at London rates, and the cartage must be added according to distance. The stone would be tipped from the cart at its destination; allow about 1½ tons to a load. The average bed of squared Kentish ragstone facing is 7 inches, and a ton will cover about 3¼ yards superficial. The random facing has an average bed of 5 inches, and a ton covers 4½ yards superficial. The labour on rubble walling varies so much that nothing short of building a specimen piece to serve as a standard (about a cubic yard) will clearly define the character of work required, and this is frequently done. Oolitic limestone is the easiest to work; the labour on harder sandstones and grits would be about 10 per cent. more.

"Extra on brickwork for facing of Kentish ragstone, hammer-dressed on face, laid random with one through stone 14 inches on bed in each superficial yard, set in ash mortar as described, and finished with a neat bevelled joint as the work proceeds."

	s.	d.
Ragstone in trucks at quarry, per ton	6	6
Railway carriage, do	3	4
Unloading, do	0	6
Cartage (say), do	3	0

13 4 ÷ 40½ ft. = 4¼d. per ft. sup

Add bonders ... 1

5¼

Deduct $\frac{5}{13}$ of a foot cube of brickwork at 14*l*. 4s. per rod ... 4½

Extra cost of stone per foot sup. ... ¾

The value of the item would be as follows:—

	d.
Extra value of stone	0¾
,, ,, ashes	0¼
Mortar extra	0½
Extra labour facing with stone ·010 at 8s. 9d.	1
1 ft. sup. of rough cutting and waste on brickwork (as before)	1¾
Striking joints	0½
Per superficial foot	5

The coursed facing would be treated in a similar manner, varying the quantity of mortar and the constant for labour, and observing the difference of surface covered by a ton of stone.

"*Yards Cube.*—Rubble walling in mortar of local limestone in random courses with one through stone 20 inches on bed to every superficial yard."

Such a description means that every stone shall be squared, but to get this done is difficult. The stones on each internal face and the through stones are squared. The interior parts of the wall are practically random rubble worked up to courses about every 12 or 18 inches of the height. Valuing the work on this assumption, we proceed as follows:—

In a 20-inch wall about one-half of the stone would be squared and the remainder rough. The quantity of material may be taken as the mean between 1¼ and 1½ yards of stone; a ton is often assumed to be equal to a yard in the stack, but about 24 feet is a reasonable average.

	s.	d.
1⅜ yd. cube = 1½ tons at 4s. per ton	6	0
1¼ tons = 1 load carting, say	2	6
Labourer assisting to load	0	6
The mean quantity 7/27 cubic yd. of mortar = 8 cubic ft. of mortar 7d. per cubic foot	4	8
Waller and labourer ·323 at 14s. 7d.	4	8½
Per yd. cube, random rubble	18	4½

The labour in squaring would be included in the next item but one. If, however, the work be squared quite through the wall the other constant ·500 will apply, and the facing item will be for raking out and pointing only.

Yards Cube.—Do. in 14-inch walls.

	s	d.
1½ yd. = 1¾ ton at 4*s*. per ton	6	8
1 load carting	2	6
Labourer assisting to load	0	6
Mortar as before	4	8
Waller and labourer, ·500 at 14*s*. 7*d*.	7	3½
Per yd. cube	21	7½

Feet Superficial.—Extra on rubble walling for local stone facing in roughly dressed courses 4 inches to 6 inches deep, well bonded to the work behind and set in mortar, and the joints raked out and pointed with black ash mortar.

	d.
Extra labour waller ·330 (per yd.) at 8*s*. 9*d*. = 2*s*. 10½*d*. ÷ 9 = ...	4
Raking out (as before) and ashes	0½
Pointing, waller and labourer	1
Per foot superficial	5½

Feet Run.—

	d.
Extra on rubble facing for quoin with drafted angle ·100 at 8*s*. 9*d*.	10½
Deduct 1 ft. 6 in. superficial of facing at 5½*d*.	7¼
Extra per foot run	3¼

MASON.

The principal points to be noted in the valuation of stonework are the cost of the various sizes, the quarry regulations, the distance of the place at which it is delivered, cost of cartage, cost of labour, and the proportion of waste. Where stone brought by sea allow the cost of unloading, and further carriage, by truck or barge.

To take Mansfield stone as an instance:

There is the difference of price between white and red, random blocks, selected blocks, selected blocks sawn to dimensions.

In Ancaster stone the price varies with the thickness of the bed.

In Craigleith and many other stones with the number of cubic feet in the block.

In some cases a saving is effected by getting the stone worked at the quarry, or where a large quantity of similar moulding has to be done, by the use of the stone moulding machine.

Where a large quantity of stone is required, the builder usually obtains a special estimate from the quarry or its agent at a considerable reduction from the usual prices.

The value of templates will be readily obtained by observing what proportion their area bears to a foot superficial or cube.

And generally, if one settles the price of a cubic foot of rubbed, tooled, or sawn stone before beginning to price the items of a bill, there is but little difficulty in fixing the relative values of the items.

The Cost at Quarry.—This does not always represent the whole original value. Sometimes a specification will stipulate for the quarrying of the stone some months before it is used, and if this is not the custom of the quarry, additional expense will be thus incurred, and must be allowed for.

When small quantities of stone are required, the builder will make a sub-contract with a local mason, or obtain it from a stone merchant as near as possible to the proposed building. When a sub-contract is thus made, the production of the mason's invoice is no proof of value, as his charges are often much higher than current rates. The proper course is to adopt the analogy of the contract rates, if a copy of the priced original estimate has been deposited.

When a large quantity of stone is required, and the specification is stringent, the builder will deal directly with the quarry owner, binding him by agreement as to the quantity, rate of supply and price, and generally to carry out the stipulations of the specification to the satisfaction of the architect.

The specification by an architect of stone from a particular quarry will sometimes induce the quarry owner to raise his price for the occasion, and in anticipation of such possibility, it is a wise precaution to obtain from him the price per foot cube before the work is advertised, so that it may be compared later with the price quoted to the builders.

Another consideration is the increased cost of scaffolding beyond that for a brick building, and the necessity for overhead travellers and derrick cranes when much stone is to be used.

When granite is used, coming, as it does, quarry-worked, extra time and trouble is involved in identifying the pieces for particular positions, although this is materially lessened when the blocks are numbered and accompanied by a key-plan. Special storage also is frequently necessary.

In dealing with quarry-worked stone under a special contract, the following stipulations should be made :—The quality shall be equal to a sample of the stone deposited with the architect. The stones shall be fitted together ready for setting, and shall include all necessary joggles and holes for dowels. The stone shall be marked with numbers for identification, shall be accompanied by a key-plan, and shall be packed and delivered at the works at convenient times, stating dates which shall not be exceeded for delivery of stone below various defined levels of the building. The stone merchant shall be responsible for, and make good, damage in transit.

Any departure from recognised methods of measurement, and the extent to which preliminary faces have been measured, should be stated in the bill of quantities.

Preliminary Considerations.—The valuation of stonework is so closely associated with the modes of measurement that a brief reference to them is not unreasonable.

Probably the most exact method, and the one which can most safely be priced, is the measurement of the stone, and all the labour on it, and presenting stone and labour separately in the bill. The experienced estimator will most frequently price the cube stone at a rate per cubic foot to include all labours; but he will look through the list of labours before he does so, generally to see what proportion of moulded work they include. In other cases he will price the labours, and compare the result with the number of cubic feet of stone, when he most frequently finds that the resulting price per cubic foot is beyond anything he can charge with any reasonable hope of obtaining a contract.

Another way which has arisen as a consequence of the uncertainty as to the extent to which preliminary faces have been measured by the surveyor (and this varies very much), is to omit *them* entirely, and make the price per foot cube of stone include all *plain beds and joints.*

" Stone and all labour " admits of two courses: the division into a few categories of stone with similar labour, or separating every stone from another upon which the labour is different. The " stone and all labour " method is inexact at the best, and, when the builder knows his business, supplies numerous opportunities for claims for extras.

In the adjustment of variations, when the work has been billed in this way, the surveyor, if it should be necessary to vindicate a new price, will take out all the labours on a particular piece of work in the bill, as nearly similar as he can find to the new item, and price them at the current rates of stone and labour. If the result is different to the builder's total price for the item he will add or deduct a proportion in the form of a percentage; he will then apply these prices so modified to the details of the new item.

The method of valuing labour on various kinds of stone by comparison with some other well known stone like Portland, tends very much to simplify the subject. Quoting from Laxton's price book (with which, we believe, the system originated), each description of stone comprises a decimal representing the relative labour or cost of plain rubbed work as compared with that upon Portland stone, which is assumed "including sawing and setting to beds and joints per foot superficial at 1·0." This series will be much more valuable when complete; at present many of the stones have no constant attached.

In schedules of prices a similar system is adopted with advantage. After a list of labours on one stone in the schedule, an item appears:—"Allow for Bath stone off price of labour on Portland stone—per cent.," and "Allow for York stone on price of labour on Portland stone—per cent." The trouble of pricing the same set of items for several different stones is thus avoided, and generally a more consistent analogy preserved. When an estimator has much work of the kind to do, he must adopt some such method.

Having settled a price per cubic foot for stone and labour, different pieces of stone on which the labour is similar may be priced at a similar rate. Thus a lineal foot of 12 inches by 6 inches step, tooled and back jointed, 72 inches sectional area, would be worth about the same price as one 9 inches by 8 inches, also 72 inches sectional area.

It is a common practice to price stone and labour in solid steps in this way, and to class with them at a similar price per foot cube such things as curbs and lintels of square section. It is hardly necessary to say that $\frac{1}{4}d.$ per sectional inch equals $3s.$ per cubic foot, $\frac{1}{2}d.$ equals $6s.$, and so on.

Proportion of Beds and Joints.—The proportion of beds and joints to each cubic foot of stone is a question which arises when the stone has been measured to include them; the relative quantity varies with the character of the work, and is generally largest in Gothic stonework.

A reasonable approximation is 1½ superficial feet to each cubic foot of stone in Classic and Classic Renaissance, and 2 feet to each cubic foot in Gothic work.

The item appears in the bill as:—

"Feet. In.——cube. Portland stone, including all sawn faces, plain beds and joints, and hoisting and setting."

The qualities of all the building stones in common use, their weight per cubic foot, and the buildings in which they have been used may be found in Rivington, 'Building Construction,' vol. iii., and cubes of the stones may be seen in the Museum of Practical Geology, Jermyn Street. There is no very definite connection between the crushing weight of stone and the difficulty of its working, and a complete table of information on this head would be a welcome addition to the literature of building. When, however, the estimator has a certain stone to deal with, he can generally obtain information as to its relative hardness from the merchant. In preserving his notes of such inquiries, one uniform standard of comparison, like Portland stone, is desirable.

The difficulty of working seasoned stone, compared with the same kind of stone freshly quarried, is considerable, and varies with the different kinds of stone.

Cost of Carriage.—The quarry owners, whose stone is used in London, all have London agents, and their price includes delivery into vehicles at their depots. They will quote a price for delivery in town, and the average builder cannot do the work at a lower rate. If the builder carts it himself, he will calculate the number of loads per day that a horse, cart and man can convey from the depot to building. In the country, when the quarry is within reasonable distance, the quarrymen will quote a price delivered, and in a hilly district this often varies considerably.

When the stone is worked at the quarry, the charge for railway carriage to the ordinary consignee would be something like 40 per cent. more than for rough stone, and the quarry owner usually having large transactions with the railway company can always make better terms. It should, consequently, be left in his hands.

Labour.—The valuation of labour (as has been before remarked) is materially simplified by a set of constants, and by the adoption of relative percentages, and this is frequently done. The application of the commonly received conclusion that Bath stone is worth 40 per cent. less, and Yorkshire stone 25 per cent. more than that on Portland, is an obvious saving of labour in pricing.

The estimator may either price all his labours at Portland prices, deducting or adding a percentage at the end, or may modify each rate as it occurs. The former is easier.

The enormous variety of stones used in the various parts of the country renders it obviously impossible, within reasonable limits, to deal thoroughly with them all. To a few of the stones used for London work, one may more profitably apply the general principles of valuation.

The table opposite comprises various useful facts.

Machinery.—The advantages which machinery possesses over manual labour are becoming more widely known, and its use is gradually increasing. Stone-working machines are principally used at the quarries where large quantities of stone are worked either ready for setting or so nearly ready as to require but little more labour on its arrival at the building.

Granite is almost always worked and polished at the quarry. York steps and landings, copings and curbs are generally quarry-worked. The turning of Classic balusters is most frequently done at the quarry, and instances are not rare in which the greater proportion of the stonework of a building is quarry-worked. A modification of this arrangement is to have the whole of the stone sawn to sizes at the quarry.

The saving on the cost of labour and carriage for quarry-worked stone is considerable, the cost of the carriage of the waste stone being saved, except for the difference between the railway rate for worked stone and stone in block.

There are few builders whose business will profitably maintain stone-moulding machines; but there are several firms in London and its neighbourhood who work stone by machinery, and will make sub-contracts with the primary contractor for all the stone

PRICE OF STONE AT MERCHANT'S YARD, OR IN TRUCKS, IN LONDON.

Cubic feet per ton.	Content compared with Portland.	Percentage to be added to, or deducted from, Portland price.		Price per foot cube.		Character.	How finished.	Saw.	
16	·6	−40	Ancaster	.. ..	4 to 2 / 4 9 6	} Limestone	Rubbed	Toothed	Random blocks.
Various, average 19, carried by railway as			Bath	Paddington .. .. .. Nine Elms .. .. .. Selected blocks, add Blocks cut to sizes, add	1 1 0 1	} Limestone	Dragged	Toothed	Random blocks.
16	·6	−40			6½ 7½ 1 8				
16	1·20	+20	Bramley Fall	King's Cross or St. Pancras	1 10	Sandstone	Tooled	Sand	Ordinary sizes.
19	·5	−50	Caen Stone	At the Bath stone depôts	1 8	Limestone	Dragged	Toothed	Random blocks.
14½	1·00	··	Cornhill (Red Dumfries)	At merchant's yard	2 5	Sandstone	Rubbed	Sand	Random blocks.
15	1·30	+30	Darley Dale	At St. Pancras	2 1	Sandstone	Rubbed	Sand	Extra price over 80 cubic feet.
13½	1·23	+23	Gazeby	At St. Pancras or King's Cross	2 6	Sandstone	Rubbed or tooled	Sand	Random.
16	1·00	··	Harehill	At St. Pancras or King's Cross	2 6	Sandstone	Rubbed or tooled	Sand	Random.
16	1·23	+23	Parkspring	At St. Pancras or King's Cross	2 8	Sandstone	Rubbed or tooled	Sand	Random.
14½	1·00	··	Portland	Nine Elms or river wharves	2 1	Limestone	Rubbed	Sand	Blocks of 20 ft. average; any average up to 48 ft. cube, 1d. 4 ft. additional or part yond 48 ft. to be at the rate of foot for every 8 ft. additional beyond 48 ft. or part thereof.
15	1·00	··	Mansfield, red and white	St. Pancras	2 3	Sandstone	Rubbed	Sand	Random blocks.
15	1·33	+33	Ditto, yellow (Bolsover)	St. Pancras	2 9	Sandstone	Rubbed	Sand	Random blocks.
15	1·25	+25	Spinkwell	King's Cross or St. Pancras	2 7	Sandstone	Rubbed	Sand	Random.
14½	1·00	··	Chilmark troughbed	Nine Elms	1 8	Limestone	Rubbed	Sand	Random.
16	·60	−40	Scott and general bed	Nine Elms	1 8	Limestone	Rubbed	Sand	Random.
Carried by rail as 16 cub. ft.									
13	1·50	+50	Devonshire granite	Paddington	3 0	..	Axed	Sand	Scabbled to size.
13½	1·60	+60	Aberdeen granite	King's Cross	3 9	..	Axed	Sand	Scabbled to size.

The prices of block stone given in the remarks on their quality and derivation in the price books are net prices, subject only to the discount for cash.

and labour which his building may require. Such a price is generally quoted per foot cube for stone and labour, and the saving on the *labour* by such an arrangement may be taken as about 33⅓ per cent.

Another consideration is the saving of time in production, often an important element of a contract.

The Proportion of Waste.—Some kinds of building stone are raised from the quarry in such irregular shapes that the waste is very great. Others are so brittle that they splinter, and on these the waste is greatest.

Much of the waste incident to the conversion of stone is allowed for in the measurement, stones irregularly shaped when finished being measured of the sizes of the cubes which contain them. The waste is therefore represented by the rough outside pieces of the blocks. Large blocks have a smaller proportion of outside than small blocks, and consequently less waste in conversion.

On most stones which are sawn 5 per cent. is a sufficient allowance for waste. On stones which are divided by coping or splitting the waste will often amount to 25 or 30 per cent.

Many of the quarries supply their stone sawn to sizes at a price higher than the stone in rough blocks as raised from the quarry. The difference on Ancaster is about 9d. per foot cube; on many of the oolites of the Midland counties the difference is 6d.; red Mansfield, 1s. 2d.; Bath stone, 5d.; Doulting, 4d.

Protection.—"Case and protect the stonework from injury, and cover the steps after fixing either with boarding or plain tiles in plaster." Some such clause as the foregoing will appear, either in specification or quantities. About 1 per cent. on the amount of the mason's bill is a reasonable average to adopt as the cost.

Scantling Lengths.—Stones described as in scantling lengths in a bill of quantities are not generally *much* more than 6 feet in length; they would be sawn out of stones containing about 40 feet cube; the difference in cost between stones of 20 feet average and 40 feet average is 5d. for Portland.

A few examples of valuation of stonework are as follows.

Portland Stone.—Feet cube.——Portland stone as described, including all labour, sawing plain beds and joints, hoisting, and setting in fine mortar. For labours see the constants.

	s.	d.
Stone at depôt	2	1
Waste, 5 per cent.	0	1¼
Cartage	0	2
Hoisting and setting ·040, at 15*s*. 2½*d*.	0	7½
Cleaning down, ·045 at 9*s*. 2½*d*.	0	5
Mortar	0	0¾
1½ ft. beds and joints, 6*d*. (·056)	0	9
Per foot cube..	4	2½

When the bill is for stone and labour, the valuation is not quite so simple, especially if the stone is divided into categories. As before described, the stone is often in quite a few divisions, and the estimator can then only compare the stone with that of a similar building of which he has had experience; he must see the drawings and fix a price for stone and labour throughout, and the cost of the stone and labour will vary from 5*s*. to 7*s*.

If the stone is presented in lineal feet, and the labours described, the process of valuation would consist in roughly taking out the quantities of the stone and labour, and pricing them. A few examples would suffice to show the method under such circumstances.

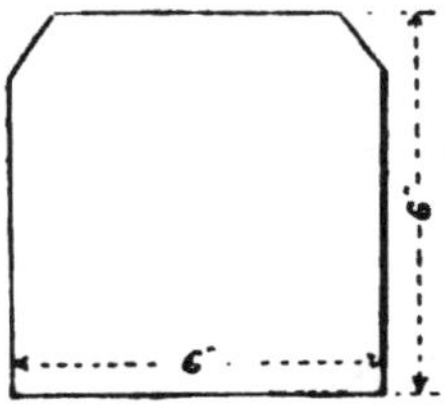

Feet run.——6 in. by 6 in. curb, rubbed on top and two edges, and twice chamfered, 2 in. wide including joints and bedding and jointing in mortar.

The joints are customarily assumed to be 3 ft. apart, consequently we value a length of 3 ft. as follows, adopting the constants before given by which the following rates are calculated.

Dimensions			s.	d.	s.	d.
3 0 6 6	9	Cube stone, including waste, hoisting and setting, and cleaning down ..	3	5½	2	7
3 0 6	1 6	Half bed	0	3	0	4½
3 0 1 6	4 6	Plain face (which includes half-sawing)	0	9½	3	6¾
3 0	6 0	Chamfer 2 in. wide	0	2	1	0
6 6	3	Joints (ends)..	0	6	0	1½
		Cost of 3 ft.			7	7¾
		Cost of 1 ft.			2	6½

Equal to 10*s*. 2*d*. per foot cube.

Feet run.——12 in. by 12 in. sill, stop-splayed, 8 in. and 4 in. wide, rebated for lights and grooved, and bedding and jointing in mortar.

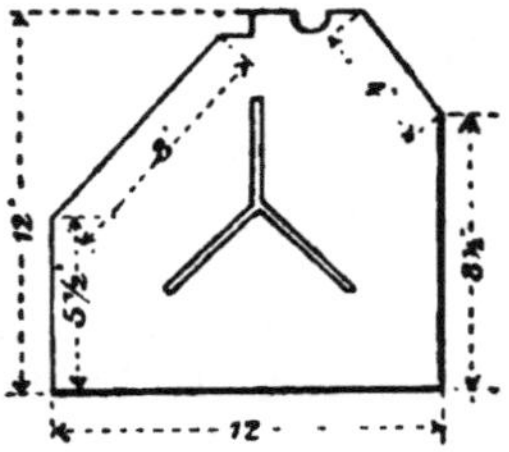

Dimensions		Squared		Description	Waste	s.	d.	s.	d.
3	0								
1	0								
1	0	3	0	Cube stone as before		3	5¼	10	3¾
					ft. in.				
					1 0				
					0 6½				
3	0				0 3½				
1	10	5	6	½ bed and joint	1′ 10″	0	3	1	4½
1	0								
1	0	1	0	Joint		0	6	0	6
					0 8½				
					0 5½				
3	0								
1	2	3	6	Plain face	1 2	0	9½	2	9½
3	0								
	8	2	0						
3	0			Sunk face stopped		1	1	3	3
	4	1	0						
				3 0					
3	0	3	0	Rebate 1 in. girth		0	1	0	3
3	0	3	0	Groove 1½ girth		0	1	0	3
1	1	1	1	Double arris joggle, pebbles and cement		0	1	0	1
				Cost of 3 ft.				18	9½
				Cost of 1 ft.				6	3¼

6*s*. 3¼*d*. per foot cube.

Foot run of 15 in. by 9 in. string moulded 14 in. girth weathered 8 in. wide.

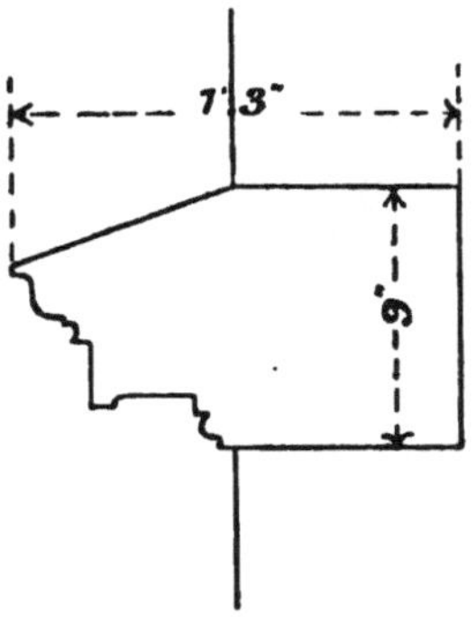

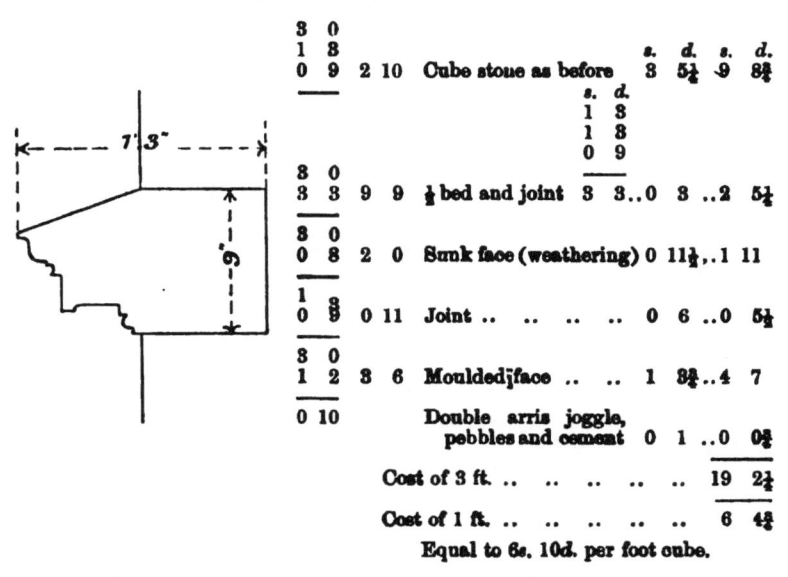

3	0								
1	3					*s.*	*d.*	*s.*	*d.*
0	9	2	10	Cube stone as before		3	5¼	9	8¾
					s. *d.*				
					1 3				
					1 3				
					0 9				
3	0								
3	3	9	9	½ bed and joint	3 3	.. 0	3	.. 2	5¼
3	0								
0	8	2	0	Sunk face (weathering)		0	11½	.. 1	11
1	3								
0	9	0	11	Joint		0	6	.. 0	5½
3	0								
1	2	3	6	Moulded face		1	3¾	.. 4	7
0	10			Double arris joggle, pebbles and cement		0	1	.. 0	0¾
				Cost of 3 ft.				19	2¼
				Cost of 1 ft.				6	4¾

Equal to 6*s*. 10*d*. per foot cube.

It will be seen from the foregoing examples that the price per foot cube decreases as the beds of the stone increase.

Sawn stone, available for hearths, landings, sills, steps and similar purposes, is kept at the London depots at the following prices. To these rates must be consequently added the cost of delivery.

The prices are subject to a discount of 2½ per cent. for cash, and are for random sizes.

Per Foot Superficial.

in.		*s.*	*d.*	in.		*s.*	*d.*
1	.. ..	0	6½	4	.. ..	1	4½
1½	.. ..	0	8	5	.. ..	1	6½
2	.. ..	0	9	6	.. ..	1	9½
2½	.. ..	0	11	7	.. ..	2	0½
3	.. ..	1	1¼	8	.. ..	2	4
3½	.. ..	1	3¼				

Above 12 ft. super., ½*d*. per foot extra.

Exceeding	18	ft. super.	1*d*.	per foot super.	extra.
"	24	"	2*d*.	"	"
"	30	"	3*d*.	"	"
"	36	"	4*d*.	"	"
"	42	"	5*d*.	"	"
"	48	"	6*d*.	"	"

QUANTITY SURVEYING.

Not Exceeding an Average of 4 ft. Random Lengths.

Thickness.	Width.						
	6 in.	7 in.	8 in.	9 in.	10 in.	11 in.	12 in.
	s. d.	s. d.	s. d.	s. d.	s. d.	s. d.	s. d.
3 in.	0 8	0 9	0 10½	0 11½	1 1	1 2	1 3
4 in.	0 9½	0 11	1 1	1 2	1 3½	1 5	1 6
5 in.	1 0	1 1	1 3	1 5	1 6	1 7	1 8
6 in.	1 1	1 3	1 5	1 7	1 9	1 11	2 0
7 in.	..	1 5	1 7	1 9	1 11	2 2	2 3
8 in.	..	..	1 9	2 0	2 2	2 4	2 6
9 in.	..	..	..	2 2	2 4	2 6	2 9
10 in.	..	..	..	..	2 7	2 9	3 0
11 in.	..	..	..	..	..	3 1	3 4
12 in.	..	..	..	..	..	..	3 6

For stones over 4 ft. long, add ½d. per foot to the above rates for each extra foot or part of a foot lineal.

For any of the faces left rough deduct at the rate of 3d. per foot super.

Bath stone and all labour compared with Portland is often priced at 25 per cent. less.

The constant for labour given in the list may be taken as a reasonable average; but the difference between the labour on the softer (as Corsham), and the harder (as Box Ground) is 10 to 12½ per cent.

Yorkshire Stone.—The use of York stone in London is almost entirely confined to pavings, landings, steps, hearths, thresholds, copings, window sills, and templates, and bases for iron bearers, as columns and stancheons. Much the larger portion is worked at the quarry. The greater part of the stone specified as rubbed is despatched from thence with a sawn face, and the rubbing completed at the site of the proposed building.

Templates.—These are often a means of working up pieces of stone which would otherwise be wasted, the alternative being to saw or cope them out of paving. A common contract price is 5s. per foot cube, but rarely less than 1s. is charged for a template however small. They are built in by bricklayer as the work proceeds.

Conversion.—The following are conveniently converted from tooled paving or sawn stone:—Templates, cover stones, hearths, treads and risers, curbs, copings, lintels, thresholds, steps, chimney-pieces. The waste on the conversion of sawn stone will average 5 per cent., on paving 10 per cent.

Yorkshire stone delivered in London, within 4 miles of depot, costs as follows:—If less than a cart-load, carriage would be charged; reckon the load as 1½ ton. The weight per 100 feet will be found in the current price books. The price of paving is usually quoted at per 100 feet superficial:—

	Per foot super. d.
2 in. self-faced paving	5¼
2½ in. " "	6¼
3 in. " "	7½
2 in. tooled "	6¾
2½ in. " "	8
3 in. " "	9
4 in. " "	12
2 in. rubbed "	8¼
2½ in. " "	9
3 in. " "	10¼

For uniform sizes, as for paving in parallel courses, the cost would be about 1½d. per foot more than the above.

	Per foot super. s.	d.
2½ in. tooled landings	0	11
3 in. " "	1	2
4 in. " "	1	5
5 in. " "	1	8
6 in. " "	2	0
7 in. " "	2	3

If over 30 ft. super., add 3d. per foot super.

Sawn York in random sizes per foot superficial:—

	Hare Hill. d.	Park Spring. d.
1 in.	4½	6
1¼ in.	5¼	7
1½ in.	6¼	8
2 in.	8¼	9½
2½ in.	11	11½
3 in.	12½	13½
4 in.	—	20
5 in.	—	25
6 in.	—	29
7 in.	—	33
8 in.	—	38
9 in.	—	43

If cut to sizes one way, add 1d. per superficial foot; two ways, 2d. per superficial foot.

Where the labour appears in a bill of quantities, although the estimator may feel satisfied as to the value put on the cube stone, he should price out the labours in order to compare them with his conclusions; the chief consideration is the proportion of moulded work.

The contractor often saves on the cubic quantity of stone by cutting two stones out of one, which has led to the stipulation in specifications "that all stones shall hold their full width and height square to the back."

Marble Mason.

The marble used in building, if in large quantity, is nearly always the subject of a separate contract, and if not so treated by the architect, the builder sublets it.

Its use for such purposes as counter tops, steps and wall linings is extending; but for chimney-pieces, except in the form of slips in combination with other materials, is decreasing.

Chimney-pieces should be selected by the architect from a manufacturer's stock, and if not, the surveyor should see the invoice at the settlement of accounts.

The independent valuation of marble is rarely required on the part of the surveyor, except for steps, lavatory tops, or counter tops.

Lavatory tops always cost more when supplied by the sanitary specialist, and some surveyors, when preparing quantities, make a point of separating them from the basins and inclosures. Sometimes the specialist, when his particular apparatus is prescribed, refuses to supply it without the top, in which case the builder is involved in a larger expense than he estimated for, and will claim the extra amount.

The following are a few of the current rates for Sicilian veined marble of the best quality at the merchant's yard, to which must be added the cost of delivery and fixing:—

Unpolished Slabs Sawn to Sizes.

		s.	d.
1 in. lavatory or counter tops	per ft. super.	1	9
1¼ in. " "	"	2	3
1½ in. " "	"	3	0
1 in. by 7 in. plain skirting	per ft. run	1	1
Mitres to ditto	each	0	6

		s.	*d.*
Moulded ends	each	0	8
2 in. by 12 in. step, square edges, rubbed but unpolished, not exceeding 6 ft. long	per ft. run	4	0
2 in. by 12 in. ditto, exceeding 6 ft. long ..	"	4	6
Polishing	per ft. super.	0	9
Polishing edges of 1 in.	per ft. run	0	2
" " 1¼ in.	"	0	2¼
" " 1½ in.	"	0	2½
" " 2 in.	"	0	2¾
Polished rounded nosing to 1 in.	"	0	5
" " 1¼ in.	"	0	6¼
" " 1½ in.	"	0	7½
" " 2 in.	"	0	10
Polished moulded edge to 1 in.	"	0	7
" " 1¼ in.	"	0	8
" -- 1½ in.	"	0	9
2 in.	"	1	0
" " 3 in.	"	1	2

	Rounded Edge.		Moulded Edge.	
	s.	*d.*	*s.*	*d.*
Polished quadrant corners, 1 in.	0	4½ ..	0	7
" " 1¼ in.	0	5 ..	0	7½
" " 1½ in.	0	6 ..	0	9
" 2 in.	0	8 ..	1	0
" " 3 in.	0	9½ ..	1	2

Basin holes, with rounded and polished edge, or rebated and polished edges, 1 in.	each	3	0
" " " 1¼ in.	"	3	3
" " " 1½ in.	"	4	0
Holes for taps, 1 in.	"	0	6
" " 1¼ in.	"	0	7
" " 1½ in.	"	0	9
Sinkings for soap	"	1	0

Spandrel steps with moulded nosings, and rebated and splay-rebated joints rubbed but not polished, cost 10*s*. per foot cube at yard.

The valuation of the fixing will be conducted on the principles of the other masonry, if pinned into walls; but this is rarely the case except for steps. A slab is generally laid on a rough deal top prepared to receive it. Skirtings are fixed with copper screws, or brass cups and screws, to plugs in walls.

As polished marble has always to be packed in a case, the expense of it must be allowed for.

The manufacturer's charge for a case is always exorbitant, but as two-thirds of the original charge is allowed if returned, this does not so much matter. The railway companies make a very small charge for their carriage; but this and conveyance to the railway station must be allowed for.

Comparative Values of Marble.—Next to Sicilian marble, St. Anne's is most frequently used. The labour and material may be taken as equal.

For Sienna marble the labour may be taken as equal to, the material about three times the price of, Sicilian.

SLATER.

The work of the slater is frequently sublet by the contractor, though more often in London than in the country. But whether in town or country, it is rarely the case that work is done by the builder's men so cheaply as by subletting. The allowances made for cutting are a compromise between the waste of slates and the extra labour; they usually more than cover the waste of slates.

The practice of allowing superficial quantity for extra labour is the survival of an old custom, which is inexact and inconsistent with the modern surveyor's methods in other trades.

Name.	Size.	Number required to cover a square.		Number of squares covered by 1200.		Weight of 1200.		Number of nails per square.	
		2¼ in. lap.	3 in. lap.	2¼ in. lap.	3 in. lap.	1st qlty.	2nd qlty.	2¼ in. lap.	3 in. lap.
	in. in.					cwt.	cwt.		
Duchesses	24 by 12	112	115	10·75	10·50	60	70	224	230
Marchionesses	22 by 12	124	127	9·75	9·50	55	65	248	254
Ditto small	22 by 11	135	138	8·93	8·70	50	60	270	276
Countesses	20 by 10	165	170	7·28	7·08	40	50	330	340
Ditto	20 by 12	138	142	8·75	8·50	50	60	276	284
Viscountesses	18 by 10	186	192	6·45	6·25	35	45	372	384
Ditto	18 by 9	207	214	5·80	5·62	30	40	414	428
Ladies	16 by 10	214	222	5·62	5·41	30	40	428	444
Ditto	16 by 8	267	277	4·49	4·33	25	30	534	554
Small ladies	14 by 12	209	219	5·75	5·50	32½	40	418	438
Ditto	14 by 7	358	375	3·35	3·20	20	22	716	750
Doubles	12 by 8	379	400	3·17	3·00	18	20	758	800
Ditto	12 by 6	506	534	2·37	2·25	12	14	1012	1068

In London, Welsh slates are those generally used, and are sold by the thousand of 1200 when bought of local dealers; when purchased by the truck load at the quarry, 1260, the 60 (5 per cent.) being an allowance for those likely to be broken.

Most of the slates of the common designations are supplied of a first and second quality, and the very large ones, as Queens and Rags, are sold by the ton.

The Westmorland slates are occasionally used in London, and are sold by the ton. They are prepared in promiscuous sizes, which the slater sorts when laying, using the largest at the eaves and the smallest at the ridge, the gauge and width of slate consequently gradually diminishing from the starting point. A larger allowance for cutting and eaves is often claimed than that on Welsh slating, and it is necessary to state in a bill of quantities what allowance has been made when ton slating is measured. The sorting in sizes to regulate the diminution also slightly increases the cost.

Stone slates are commonly used in various parts of the country. Some of the oolitic limestones are stratified in very thin layers, and are used under this name. They are found, among other places, in Northamptonshire, Staffordshire, Oxfordshire and Gloucestershire. Similar roof coverings are found in Sussex in the neighbourhood of Horsham.

The cartage of slates from local quarries is generally done cheaper by the quarry owner than the builder can do it, and the price agreed should include delivery.

The local allowances for cutting on stone slates are commonly 2 feet for valleys, and 1 foot for eaves.

A constant for their laying may be calculated from that for countesses after averaging the sizes of the slates. To this should be added per square slater and boy ·075 for the extra trouble of sorting to diminishing sizes.

Waste.—When the slates are bought by the truck-load, the 5 per cent. allowed at the quarry is sufficient for waste caused by breaking. If bought of the merchant, a further allowance of 5 per cent. should be made. All the other waste is allowed for by the recognised modes of measurement. The waste of nails will vary with the quality of the supervision, but 5 per cent. is a reasonable average. The following table shows the number of slates and nails required for a square of each kind of slating. The existing tables vary, but when the lap is known, the calculation is easy enough, and any doubt can be readily solved.

Taking countesses, 20 inches by 10 inches with 3-inch lap, as an example, we have 8½-inch gauge, then 8½-inch gauge × 10 inches wide = 85-inch surface; then 100 feet × 144 inches ÷ 85 inches = 170, nearly, to each square. The number of nails is two to each slate, and consequently 340. A simple calculation will show the number of squares which 1200 of the same slates will cover: 85 inches × 1200 inches ÷ 14,400 inches = 7·08 squares. The designation of the slate is not enough to define its size; it varies with different quarries. The size should always be stated in dealing with slates.

The weights are close approximations, but will vary; they are, however, quite near enough to calculate the cost of carriage and cartage.

Nails.—These are commonly of zinc or copper, and are sold by weight. The prices of both vary slightly with the prices of spelter and copper; some of the trade lists quote a price per cwt. beyond the current price of zinc or copper sheet. The price of nails decreases with the increase of length. Sometimes in specifications the nails for slating are described by their length and their weight per thousand, as "1½-inch copper nails, weight 7 lb. per thousand." The following is the weight per thousand of the nails commonly used:—

	1¼ in.	1½ in.	2 in.	
Zinc	3¾ lb.	4¾ lb.	7¼ lb.	per thousand
Copper	5¼ lb.	9½ lb.	11¾ lb	"

The present net price in London for slating nails is as follows:—

	1¼ in.	1½ in.	2 in.
Zinc	26*s.*	24*s.*	23*s.* per cwt.
"	2¾*d.*	2½*d.*	2½*d.* per lb.
Copper	87*s.*	84*s.*	84*s.* per cwt.
"	9¼*d.*	9*d.*	9*d.* per lb.

The current prices of first quality Port Madoc slates delivered in London, are as follows:—

				£	*s.*	*d.*
24 in. by 12 in.	per thousand of 1200		.. .. ..	14	17	0
22 in. by 12 in.	"	"	.. .. ..	13	0	0
22 in. by 11 in.	"	"	.. .. ..	11	14	6
20 in. by 10 in.		"	.. .. ..	9	8	6
18 in. by 10 in.	"	"	.. .. ..	7	13	6
18 in. by 9 in.	"	"	.. .. ..	6	17	6
16 in. by 10 in.	"	"	.. .. ..	6	10	6
16 in. by 8 in.	"	"	.. .. ..	5	2	6

PRICES.

				£	s.	d.
14 in. by 12 in. per thousand of 1200	..	..	..	6	6	6
14 in. by 7 in.	,,	,,	.. .. ..	3	12	0
12 in. by 8 in.	,,	,,	.. .. ..	3	2	6
12 in. by 6 in.	,,	,,	.. .. ..	2	4	6

First quality Bangor slates cost about 10 per cent. more; second quality slates cost about 20 per cent. less than first quality, and are frequently used. It is rarely the case that the specification "best Bangor countess slating" is carried out.

Delivery.—The prices of slate merchants are usually quoted exclusive of delivery. The weights of roofing slates are to be found in the table relating to the various sizes. The weight of slate slab is easily remembered, as 150 feet superficial of 1-inch thick weighs a ton, and other thicknesses are in regular proportion—for instance, 50 feet of 3-inch weighs a ton. (The weight will be necessary for the calculation of cost of cartage; see also section Cartage). When there is a load, the slate merchant can generally deliver it as cheaply as the builder can do it himself, and can often be induced to do so free of charge.

Cost per square.—The cost of slating, if sublet, will vary with the quantity, the larger quantities being done at a less rate per square than the smaller ones.

Labour :—

		d.
Slater per hour	.. ..	11
Boy ,,	.. ..	8¼

Saving is effected by builders by reducing the lap where it can be reduced, using cheaper nails, and only one to each slate instead of two, and not laying double courses where they should be laid.

The usual trade terms are acceptance at three months, or 2½ per cent. discount for cash or monthly account.

The cost of a square of Port Madoc countess slating laid to a 2½-inch lap, with two 1¼-inch copper nails to each slate, would be as follows, assuming that the slates were purchased at the current rate in London, allowing 5 per cent. for waste on slates and nails.

	£	s.	d.
165 + 8 = 173 slates at 9l. 8s. 6d. per 1000 of 1200	1	7	2
330 + 17 = 347 copper nails, 1¾ lb. at 9¼d.	0	1	5¼
Slater and boy, preparing and laying ·205 days at 12s. 1d.		2	5
Cost per square	1	11	0¼

QUANTITY SURVEYING.

With slates smaller or larger than countesses, the constant may be increased or decreased by a simple proportion, which gives a result very near the truth. Taking slates 12 inches by 8 inches as an example, the calculation would be 165 : 379 :: ·200 to the constant required, ·459.

For slates larger than countesses, add ·050 to the constant for countesses, for extra trouble of handling.

For slating to steep roofs, like Mansard or curb roofs, add to the constant ·050. Sub-contractors charge 6d. to 9d. per square extra.

The same method of proportion may be adopted to apportion the labour of preparing slates for laying—that is, sorting them for thickness, marking them, and making the holes for nails, constant for countesses, ·095 days slater and boy.

The calculation for a square of slating, similar to the last, but laid to a 3-inch lap, would be as follows. The slate merchant increases his price by about 1s. per square.

	£	s.	d.
170 + 9 = 179 slates at 9l. 8s. 6d.	1	8	1¼
858 copper nails, 2 lb. at 9¼d. lb.	0	1	6¼
Slater and boy (constant increased by proportion) ·206 days at 12s. 1d.	0	2	5¾
	1	12	1½

Spaced Slating.—In open or spaced slating the spaces vary from 2 to 4 inches. The difference in quantity of slates may be calculated in the same manner as before. The saving of slates is commonly reckoned as one-third.

Taking as an example countess slates 20 inches by 10 inches, with 2½-inch lap and 4-inch spaces, we have for the surface covered by a slate 10 inches + 4 inches × 8¾ inches = 122½ inches; then 100 feet × 144 inches ÷ 122½ inches = 118 slates, nearly. This, compared with the calculation for continuous slating, is $\frac{118}{164}$ about ¾, a saving of one-quarter. The value of a square would be calculated as follows:—

	£	s.	d.
118 + 6 = 124 slates at 9l. 8s. 6d. per thousand of 1200	0	19	5¾
248 nails, weight 1¼ lb. at 9¼d.	0	0	11½
Slater and boy preparing and laying, ·193 at 12s. 1d.	0	2	4
Cost per square	1	2	9¼

PRICES.

The last constant is arrived at by the proportion 165 : 118 : : ·200 to the constant required—i.e. ·143, and we add ·050 = ·198. It is a common assumption that spaced slating is worth about four-fifths of the price of continuous, and circular one-third to a half more.

Circular slating is valued by the application of the foregoing methods. It is always necessary to use smaller slates, and their size must be regulated by the radius of the curve; and this radius should be stated in the bill of quantities. Much of the circular slating in a circular roof to a spire or an apse is ill done. The perpends should radiate regularly from the apex, and this involves the maximum of cutting. This is frequently avoided by the introduction of slates here and there as closers, and the slating is as a consequence more cheaply done. As circular slating most frequently runs up to a point, the best work is that in which several sizes of slates are used, diminishing from eaves to top, and their selection involves increased labour. The following may illustrate the mode of treatment, assuming a third of the space covered with each of the following sizes, and laid to a 2½-inch lap.

Per thousand of 1200.

		£ s. d.		£ s. d.
Ladies	214 at	6 10 6		1 3 3¼
Small ladies	209 at	6 6 6		1 2 0¼
Double	379 at	3 2 6		0 19 9
	3) 802			3 5 0¼
	268			

£ s. d.
268 + 27 = 295 slates at 3*l*. 5*s*. for 802 .. .. .. 1 3 11
(In the foregoing has been added a further 5 per cent. for waste in cutting to graduated sizes).
590 nails, weight 3 lb., at 9¼d. 0 2 3¾
Slater and boy preparing ·170, and laying and cutting 281 = ·451 at 12*s*. 1d. 0 5 5¼

Cost per square 1 11 8¼

All the preceding examples are for slating, exclusive of cutting, and, as before mentioned, these, by custom, are paid for by an allowance of superficial quantity. For valleys, the allowance is, for each lineal foot, 6 inches on each side, or 12 inches in all. The valuation of this may be illustrated by countess slating laid to a 2½-inch lap.

Adopting our calculation for countess slating, 30s. 7d., the allowance would amount to 3¾d. per foot run. Adopting the market price, 25s. 6d., the allowance would amount to 3d. per foot run.

In slating with a 2½-inch lap, the length of cut edge of slates of a single thickness on each lineal foot of raking-line of roof is $3\frac{1}{4} + 12 + 11\frac{1}{4} + 2\frac{1}{2} = 30$ inches. The slate may be taken as half the width of a countess slate, 5 inches :—

	d.
30 in. + 5 in. ÷ 200 in. = ·75 of a slate at 9l. 8s. 6d. per thousand	1¼
Constant for cutting, per foot ; 2¼ ft., slater and boy, ·002 at 12s. 1d.	¾
Per foot run	2

To a valley or hip this calculation would be doubled, and, consequently, 4d. per foot run.

For cutting and double course eaves, the allowance on countess slating at 31s. per square, as before, would be equal to $\frac{10}{12}$ of a foot = 3d. per foot run;

or

	d.
12 in. × 10 in. ÷ 200 = ·60 of a slate at 9l. 8s. 6d. per thousand	1
1 ft. cutting = ·002 at 12s. 1d.	¼
Per foot run	1¼

Slating in Bands.—When slating is arranged in bands of different colour and quality it is most conveniently dealt with by finding out the relative proportion of each kind and valuing them separately. The extra labour of arranging slating in bands would commonly be covered by ·10 on the constant for the particular kind of slate described.

Vertical Slating to Walls.—When the gauge is determined on, the proportional constant for fixing, and the number of slates may be calculated as before, the preparation of the slates would be equal to that for a roof. The fixing is worth more, averaging about 50 per cent. If a gauge of 8¾ inches be adopted and countess slates used, the constant would be calculated thus (see example of countess slating):—

PRICES.

	Slater and Boy.
Preparation	·095
Fixing	·105
Add 50 per cent. on last	·53
Per square	·253

Torching is the term used to describe pointing inside with lime and hair. For the lime and hair, adopting the calculation for the cost of mortar in Bricklayer, 15s. 6½d. per cubic yard, and adding thereto the cost of 14 lb. of hair at 8s. 6d. per cwt. = 1s. 0¾d. = 16s. 6¾d. per yard cube.

	s.	d.
The constant is ·200 day of bricklayer per square at 8s. 9d.	1	9
The constant is ·066 day of labourer per square at 5s. 10d.	0	4½
¼ c.ft. of lime and hair at 16s. 6¾d. per cubic yard	0	4¾
Per square	2	6¼

Slate Damp-proof Courses.—For these the very cheapest slates are used, and consequently the smallest—i.e. 12 inches by 6 inches. These may be bought for 35s. per thousand of 1200 delivered—say 40s.; i.e. about 1d. per foot superficial, and almost every fragment would be used. The calculation would be as follows:—

	d.
Two thicknesses of slate at 1d.	2
2½ per cent. waste on slates for cutting	0¼
Cement 1/10 bushel at 1s. 9d.	0¼
Bricklayer and labourer ·008 at 14s. 7d.	1½
(Mixing cement and labour and laying)	
Cost per foot superficial	4½

Ridges.—Of slate ridges, Thomas's patent is the best, and costs per foot run net at warehouse:—

	s.	d.
1¾ in. roll 5 in. wings	1	10
2 in. roll, 5¼ wings	2	0
2¼ in. roll, 5¾ wings	2	3
2¼ in. roll, 6 in. wings	2	7

If cut to lengths, ¼d. per foot more.
Taking the first price as an illustration.

	s.	d.
1 ft. of ridge	1	10
Carriage	0	0¼
Incidental cuttings	0	0¼
Cement, 1/20 bushel at 1s. 9d.	0	1½
Bricklayer and labourer ·005 at 14s. 7d.	0	1
Oil cement for roll	0	0½
Per foot run	2	1¾

Merchant's Prices.—The preceding calculations will suffice to show that, except by putting himself in the exact position of the slate merchant—i.e. regularly employing slaters, buying his slates at the quarry in large quantities, or quarrying them himself, and conveying them by sea, the average builder cannot get the work done at anything like the price at which the slate merchant does it, and the reason for subletting to him is irresistible.

The fashion of specifying countesses 20 inch by 10 inch tends to keep up their price as compared with other sizes; often as much as 2s. per square may be saved by adopting another size when the 20 inch by 10 inch slate is scarce.

Final Repair.—The cost of final repair will depend upon the quality of the management. Much damage is done by leaving plumbing too late, neglect to remove scaffolding after it should have been removed, and preventible carelessness of the men.

The Slater's bill of an estimate should contain a clause—"Allow for cleaning out gutters, and for leaving roofs perfect and weatherproof at completion of works." The value of the item is usually guessed at—its cost varies with each building. Like other items of its kind, the only way to arrive at a reasonable generalisation would be to keep the account of time and materials in a number of instances, and it is difficult to induce foremen to do so.

With care, 6d. for each square of ordinary slating like Countesses, will cover the cost; on more expensive slating the amount should be increased in proportion to the price per square, and it may be observed that slates are more likely to be broken than tiles.

A few instances of its estimation by different men are as follows:—

			per square. s. d.	£ s. d.
10 squares of slating, countess			8 0	1 10 0
40	„	„	0 6	1 0 0
14	„	Eureka green	1 5	1 0 0
41	„	tiling	0 6	1 0 0
23	„	„	0 4¼	0 10 0
86	„	„	0 2¼	1 0 0

Slate Masonry.—The work of the slate mason is almost always sub-let by the builder. Lavatory tops, with their skirtings, shelves, urinal slabs, and cisterns, are always ordered complete and ready for fixing of the merchant, and the builder fixes.

As the grooving, rubbing, sanding, drilling, etc., are all done by machinery, and can in no other way be so cheaply done, the usual practice of the estimator is to send an extract from the bill of quantities to the merchant, and obtain a special estimate.

In the case of slabs, the price is quoted in the trade lists in two categories; one for slabs as they come in the merchant's stock, and another cut to size. The price is regulated not only by the thickness, but by the width and length. There is also a difference in price between self-faced splitting slate slab, and slate slabs planed two faces edges from saw.

The London prices for slab at merchant's yard are regulated as follows, and the table shows the distinctions of size which are observed:—

PLANED TWO FACES, EDGES FROM SAW.—MEASURED NET.

—	¾ in.		1 in.		1¼ in.		1½ in.		2 in.		2½ in.		3 in.	
	s.	d.	s.	d.	s.	d.	s.	d.	s.	d.	s.	d.	s.	d.
From 3 ft. to 5 ft. long and 1 ft. to 2 ft. 6 in. wide ..	0	10	0	11	1	1½	1	3	1	8	2	3	2	10
From 5 ft. to 7 ft. long and 1 ft. to 3 ft. 6 in. wide ..	1	0	1	1	1	3½	1	6	2	0	2	5	3	0
From 7 ft. to 9 ft. long and 1 ft. to 4 ft. wide	—		1	3	1	5½	1	8	2	2½	2	7	3	2
From 9 ft. to 10 ft. long and 1 ft. to 5 ft. wide	—		—		1	7½	1	10½	2	4½	2	9	3	5½

Slabs cut to angular shapes are measured by the extreme dimensions either way.

Self-faced splitting slabs sawn to size are about 10 per cent less than the foregoing.

The current London net prices for labours by the merchant are as follows:—

—	¾ in.		1 in.		1¼ in.		1½ in.	
	s.	d.	s.	d.	s.	d.	s.	d.
Rubbed edges	0	2	0	2¼	0	2½	0	2¾
Rounded edges	0	2¼	0	2¾	0	3¼	0	3¾
Groove or rebate, per 1 in. girth	0	2	—		—		—	
Rounded corners to 1 in. ..	0	4	—		—		—	
Ditto to 1½ in.	0	5	—		—		—	
Rounded corners with rounded edges, 1 in. ..	0	6	—		—		—	
Ditto 1½ in.	0	7	—		—		—	

Moulded Edges.—

¾ in. and ⅞ in.	1 in.	1½ in.	2 in.
4*d.*	5*d.*	6*d.*	8*d.* per foot run.

Holes drilled or countersunk for screws, per dozen:

	s.	*d.*
Depth of hole ½ in.	0	1
" " ¾ in.	0	1
" " 1 in.	0	1½
" " 1¼ in.	0	1¾
" " 1½ in.	0	2
" " 2 in.	0	2½
" " 3 in.	0	4
Holes for basins in 1¼ in.	2	6
" " 1½ in.	3	0

Sawn Slate Cisterns.—Where special shapes are not important, including delivery within usual limit, including galvanised iron bolts, and fixing complete, hoisting and tackle by purchaser.

	£	*s.*	*d.*
50 gallons 1 in. thick	2	0	0
80 " 1 in. "	2	12	6
100 " 1 in. "	2	17	6
120 " 1¼ in. "	3	6	0
150 " 1¼ in. "	3	18	0
175 " 1½ in. "	4	16	0
200 " 1½ in. "	5	8	0
250 " 1½ in. "	6	2	6
275 " 1½ in. "	6	7	6
300 " 1½ in. "	6	15	0
350 " 1½ in. "	7	7	6
400 " 1¾ in. "	9	0	0
450 " 2 in. "	11	5	0
500 " 2 in. "	12	10	0

Cisterns requiring slabs over 7 feet long, or 2 feet 6 inches wide, or containing less than 20 gallons, are a higher rate. When cisterns are made to specified sizes, they are charged by the foot superficial, and the following is the rate complete, and in all respects as last described.

	s.	*d.*
¾ in. per foot super	1	5¾
1 in. " "	1	7
1¼ in. " "	1	9¼
1½ in. " "	1	11½
1¾ in. " "	2	3
2 in. " "	2	8½
2½ in. " "	2	11½
3 in. " "	3	4

PRICES.

Enamelled Slate.—Long lists of the varieties of enamelling, and which are comparatively useless, will be found in the price books: it is questionable whether anyone who understands the subject ever refers to them. If enamelled chimney-pieces, or enamelled slabs, are required, the architect selects them from a manufacturer's stock, or gets a special estimate. The surveyor, in dealing with them, requires the production of the invoice. The independent valuation of any ornate enamelling is only possible to a valuer possessing technical knowledge beyond the average.

The current price for plain enamelling is $11\frac{1}{2}d.$ per foot super. enamelling, plain edges, as follows:—

$\frac{1}{2}$ in.	$\frac{3}{4}$ in.	1 in.	$1\frac{1}{4}$ in.	$1\frac{1}{2}$ in.	2 in.	3 in.
$2\frac{3}{4}d.$	$3\frac{1}{4}d.$	$3\frac{3}{4}d.$	$4\frac{1}{4}d.$	$5\frac{1}{2}d.$	$6\frac{3}{4}d.$	$8\frac{3}{4}d.$ per ft. run.

TILES.

The roofing tiles used in London are brought from Reading, Maidenhead, Bracknell, Broseley, or Ruabon, and the estimator's price per square usually includes the lathing. In their measurement the same system of allowing superficial quantity for labour and waste obtains as in slater's work.

The current prices are as follows for three localities, which will afford reasonable parallels to any of the districts from which London is supplied.

Lawrence's.—

	Per thousand. s. d.	Weight per thousand. ton.
Red, holed only	34 6	1
Pressed, holed, rubbed and sand faced	42 6	1
Stained brown-red, extra	1 6	—
Fancy tiles (shaped ends)	42 6	1
Hip, valley, plain angle and gable tiles	255 0	1

In trucks at Bracknell.
Railway rate to Nine Elms 3s. 4d. per ton.

Broseley Tiles.—

	Per thousand. s. d.	Weight per thousand.
Pressed plain roofing tiles	40 0	1 ton.
Ornamental ,,	42 6	1 ,,
Gable tiles (tile and half)	80 0	$1\frac{1}{2}$,,
Eaves tiles (7 in. by $6\frac{1}{4}$ in.)	40 0	$\frac{3}{4}$,,
Hip tiles	292 0	$33\frac{1}{2}$ cwt.
Valley tiles	271 0	$33\frac{1}{2}$,,
Angle ,,	250 0	$33\frac{1}{2}$,,

In trucks at nearest railway station.
Railway rate to Paddington, Brentford, or Chelsea basin, 7s. 11d. per ton in 4-ton lots.

The foregoing holed or nibbed at same rates. They generally have both, so that if the nib is broken off, which frequently happens, the tile may be nailed.

Ruabon Tiles.—

	Per thousand.		Weight per thousand.
	s.	d.	
Red, brown, or brindled	45	0	23 cwt.
Ornamental do.	47	6	21 ,,
Plain tile-and-half (10½ in. by 9¾ in.), for gables	90	0	33 ,,
Plain under-eaves tile (7 in. by 6½ in.)	45	0	16 ,,
Valley tiles	312	6	30 ,,
Hip tiles	291	8	30 ,,

In trucks at nearest station to works.
In 4-ton lots, 9s. 6d. per ton. In 2-ton lots, 25s. per ton.
If any of the foregoing are prepared with holes instead of nibs, add 2s. 6d. per thousand.
If blue instead of red, add 5s. per thousand.

Quantity Required per Square.—The tables as to the number of tiles required to cover a square vary considerably, and rarely state whether they express the calculated quantity or include an allowance for waste. The number may be computed in the same manner as directed for slates. The general size is 10½ inches by 6½ inches and ½ inch to ⅝ inch thick. The *calculated* quantity per square is as follows. They are most frequently laid with a 3½-inch lap.

4 in. gauge	2½ in. lap	554 tiles
3½ in. ,,	3½ in. ,,	633 ,,
3 in. ,,	4½ in. ,,	739 ,,

Tile Lathing.—Judging from published tables, considerable uncertainty exists as to the best size of laths for plain tiling. Some say 1¼ inches by ¼ inch, others 1 inch by ¼ inch. For good work they are never specified less than ⅜ inch thick; 1¼ inches by 1 inch is a reasonable size. Their size should always be specified.

They are imported ready sawn in various sizes—generally ¾ inch by 1¼ inches, ¾ inch by 1 inch, ¾ inch by 2 inches, and are sold at the timber sales by the 144 feet run, or they may be bought at the saw-mills, where common stuff is frequently converted into laths for tiles and slates, and sold at per hundred feet run. Anything beyond ¾ inch thick would be obtained from the mills.

PRICES.

Common prices in quantity are as follows:—

	At Timber Sales.		At Sawmills.	
	s. d.		s. d.	
⅞ in. × 1¼ in.	0 6	per 144 ft.	0 8	per 100 ft.
⅞ in. × 1 in.	0 5	,,	0 7	,,
⅞ in. × 2 in.	1 4	,,	1 6	,,

1¼ in. by 1 in. at sawmills would cost 9d. per 100 feet.

			Nails.
The calculated number of feet run of lath and number of nails for 4 in. gauge is 300 ft.			255
Ditto ditto for 3½ in. gauge is 340 ft.			289
Ditto ditto for 3 in. gauge is 400 ft.			340

The gauge of the lath being the same as that of the tiles, in 10 feet of roof slope we should have for 4-inch gauge 30 lines of lath and 30 feet × 10 feet = 300 feet.

With rafters 14 inches from centre to centre we have 8½ by 30, the number of crossings, which gives 255 nails to a 4-inch gauge.

Five per cent. on laths and nails will cover the waste.

The constant for laying tiling with a 3½-inch lap (3½-inch gauge) is ·700 of bricklayer and labourer. Labourer unloading or loading per thousand ·100.

The constant may be modified for the different gauges by a calculation in simple proportion, as recommended for slating.

Two and a half per cent. will cover the waste, exclusive of cutting otherwise allowed for. The value of a square of Broseley tiling 3½ inches gauge, laid with nibs, in London, would be calculated as follows.

A labourer should keep three tilers supplied with material.

	Per thousand.
	s. d.
Broseley tiles in trucks	40 0
Railway rate	7 11
Loading carts ·100, unloading carts ·100 = ·200, labourer at 5s. 10d.	1 2
Cartage, say	4 6
	53 7

	£	s.	d.
633 + 16 = 649 tiles at 53s. 7d.	1	14	9
·700 of bricklayer at 8s. 4d.	0	5	10
·233 of labourer at 5s. 10d.	0	1	4¾
357 ft. 1¼ in. by 1 in. sawn fir laths, 9d., carriage 2d. = 11d. per 100 ft.	0	3	3¼
304 = 1$\frac{1}{16}$ lb. 1¼ in. cut clasp nails (sixpenny), 3½ lb. per thousand, at 12s. cwt.	0	0	1½
Carpenter fixing per square ·40 at 8s. 4d...	0	3	4
	2	8	8

Extra on Common Tiling for Tile and a Half to Verges is measured by the lineal foot. The plain tiles are 10½ inches by 6½ inches, the tile and half 10½ inches by 9¾ inches. The extra-sized tile is used for the alternate courses only. The price per thousand is about twice that of the common tiles. The extra trouble of changing from common tiles to tile and half and fitting to verge (no cutting required) is extra labour, ·008 day of tiler.

The cost of Broseley tile and half is as follows:—

	£	s.	d.
1000 at works	4	0	0
Railway rate on 1½ ton at 7s. 11d.	0	11	10½
Cartage and delivery as before—1½ ton at 11½d. ..	0	1	5¼
	4	13	3¾

Adopting as an example a 4 inches gauge, to half the lineal quantity measured tile and half is substituted for the common tiling; we have, therefore, to each lineal foot 6 inches by 9¾ inches = 5 inches superficial of extra value. Then—

	£	s.	d.	£	s.	d.
568 Broseley tile and half at 4l. 13s. 4d. per thousand			..	2	13	0
568 common tiles at 2l. 13s. 7d. per thousand	1	10	5¼			
Add one-half	0	15	2¾			
				2	5	8
Difference on 1½ squares				0	7	4

Two-thirds of the above difference, 4s. 10d., represents the difference of value per square.

The value per foot run would, therefore, be—

	£	s.	d.
$\frac{5}{12}$ of a foot super. extra value of tiles at 4s. 11d. per square	0	0	1
Extra labour laying and fitting to verge, ·008 day bricklayer at 8s. 4d.	0	0	0¾
Per foot run	0	0	1¾

Extra on Common Tiling for purpose-made Hip and Valley Tiles to Course and Bond with General Tiling and Allow for Cutting and Waste is an item, as it frequently appears in a bill of quantities, and is better measured thus than in two items—i.e. extra for hip and valley tiles and the cutting put into the superficial measurement of the tiling. The difference between the amount of cutting to a valley or hip where lead is used and another where purpose-made tiles are adopted is considerable—in the former case, all the tiles must be cut to a raking line; in the latter, the cutting is only to be discovered by close scrutiny, and there is much less waste, because it is at a right angle with the ridge instead of raking. Hips and valleys are generally measured and billed together, although the mean length of gauge of which they take the place is not the same for hip as valley. The average may be taken as 10 inches wide, and 1½ inches on each edge would more than cover the waste. The width affected would, therefore, be 13 inches in all, equal to the width of two tiles, and for a 4-inch gauge, three courses to the foot run. Three hip or valley tiles would consequently take the place of six common ones. The difference of value of the tiles to each lineal foot would, therefore, be as follows:—

	£	s.	d.
The cost of Broseley hip and valley tiles, per thousand	14	12	0
Railway rate on 1½ ton at 7*s.* 11*d.*	0	11	10½
Cartage and delivery, 1½ ton at 11½*d.*	0	1	5¼
	15	5	3¾

	£	s.	d.
Cost of three hip or valley tiles at 15*l.* 5*s.* 4*d.* per thousand	0	0	10½
Deduct cost of six plain tiles at 2*l.* 13*s.* 7*d.* per thousand	0	0	3¾
Difference in value of tiles per foot run ..	0	0	7¾

	£	s.	d.
Extra cost of tiles per foot run	0	0	7¾
Extra labour—laying, fitting and cutting, ·020 day of bricklayer at 8*s.* 4*d.*	0	0	2
Extra cost per foot run	0	0	9¾

Pegs and Pins.—When tiles have holes instead of nibs, they are secured with wooden or cast-iron pegs. The latter are best and should be about 2 inches long. One thousand weighs 25 lb., and costs at the rate of 9s. 6d. per cwt., and if galvanised, 18s. per cwt. The cost of these may be readily computed by allowing two for each tile. Wooden pegs cost about 1s. 9d. per bushel, and a square requires about a peck.

If tiles are bedded in lime and hair, the extra cost would be calculated as follows:—

		£	s.	d.
$\frac{3}{17}$ of a cubic yard of lime and hair at 16s. 7d. (see Torching, section Slater)		0	1	2¾
Labour, per square: bricklayer ·150 at 8s. 4d.		0	1	3
,, ,, labourer ·050 at 5s. 10d.		0	0	3½
Extra cost per square		0	2	9¼

Tile Ridges.—Of tile ridges the prices and patterns are various. They may be selected by number from a trade list, and such a list will generally give the weight and railway rate per ton for carriage. They are generally made in 12-inch lengths.

If we take as an illustration Lawrence's 12-inch lengths, of which about 150 weigh a ton,—

	£	s.	d.
Say 150 tiles at 6s. per dozen	3	15	0
Railway carriage, 1 ton at 3s. 8d. per ton	0	3	8
Unloading, carting and delivery at building, say	0	5	0
150 ft.)	4	3	8
Per foot	0	0	6¾
Cement, $\frac{1}{10}$ bushel at 1s. 9d.	0	0	1¼
Bricklayer and labourer, ·006 at 14s. 2d.	0	0	1
Cost per foot run	0	0	9

Oak Shingles.—Oak shingles are generally about 12 inches long and 3 inches to 4½ inches wide, and are laid to the same gauge as tiles; about 1000 are commonly reckoned for a square, and the cost, delivered within three or four miles, is 8s. per hundred.

The laying may be taken as equal to the smallest slates or tiles of the same gauge.

The value of the cutting would vary with the angle. Assuming the size of a shingle to be 12 inches by 4 inches, the calculation for the number required to cover a square to a 4-inch gauge would be 4 inches by 4 inches = 16 exposed surface; 100 feet by 144 inches ÷ 16 = 900; 2½ per cent. will cover waste by damage = 23.

There would be, therefore, 923 shingles and 1800 + 90 nails. The calculation for the value of a square would be as follows:—

	£	s.	d.
923 shingles at 8s. per hundred	3	13	10
1890 1¼ in. copper nails, 9/10 lb. at 9¼d.	0	7	7½
·700 carpenter at 8s. 4d.	0	5	10
·283 labourer at 5s. 10d.	0	1	4¼
Cost per square	4	8	7¾

For the double course to eaves the under shingle would be about 5 inches long; say, half a shingle = 1½ shingle per foot run, and three nails = 5¾d., extra labour less than ¼d. = 6d.

The price per square having been computed, the cost for raking cutting may be obtained from it. A width of 3 inches to each out edge of the finished work would pay for the waste. Extra labour, day of carpenter ·010; consequently

	£	s.	d.
¼ ft. at 88s. 8d. per square	0	0	2¾
·010 carpenter at 8s. 4d.	0	0	1
Cost per foot run	0	0	3¾

CARPENTER.

The valuation of carpentry can only be possible to one who has a considerable knowledge of the quality of wood and of its market value, and although a systematic study of qualities will involve much time, its expenditure will be amply repaid.

The best and most recent information as to quality may be found in Rivington's 'Notes on Building Construction' (3rd volume), Hurst's 'Architectural Surveyor's Handbook,' Seddon's 'Builder's Work,' Laslett on 'Timber and Timber Trees,' and a number of timber-trade handbooks published by Rider, of Aldersgate Street, London.

The marks on timber and deals should also be studied; lists of these are given in the builders' price-books.

The trade newspaper is the *Timber Trades Journal*, published by Rider. The student will derive advantage from its diligent study, including the advertisements. The annual reports of the wood-brokers, which are published at the beginning of the year in the *Timber Trades Journal*, and the annual special numbers should also be read. In the same journal is reported after each timber sale by the leading wood-brokers, such as Churchill and Sim, or Foy, Morgan & Co., a complete list of the sold lots, with the price realised for each. Liverpool is the great port of entry for American wood, and the Liverpool dock sales are reported in the same journal. These may with advantage be cut out and pasted in a book for reference. Tabulated analyses of them may also be made.

The student should also get a catalogue of an advertised timber sale and before the date of such sale should go to the docks and examine the lots; he should then attend the sale and mark the catalogue, and he will know better what to observe, and acquire more information if he has previously gone through the course of reading above recommended. It should be observed that the timber sale catalogues adopt the *shipping marks* and the grades of quality indicated by them, as 1st, 2nd, 3rd, &c., quality. In this respect they differ from the timber merchant, who calls 2nd quality 1st, 3rd 2nd, and so on.

Only the larger firms of builders buy their timber at the public sales. The majority purchase of the merchant, and their timber consequently costs them from 5 to 10 per cent. more than the auction prices.

In the analysis of a timber catalogue, a good map of the timber ports will be useful. A map of Scandinavia, published by Rider, shows all the timber ports in the Baltic, Gulf of Bothnia, Gulf of Finland, and the White Sea.

In the *Journal* will be found a number of contractions, which perhaps may embarrass the unaccustomed reader; some of them are as follows:—

F.O.W.—First open water. The earliest time possible for the despatch of timber-ships after the breaking up of the winter ice in the northern ports of Europe.

F.O.B.—Free on board. The seller of the goods puts them on board ship free of all expense to the buyer.

F.A.S.—Free alongside ship. The seller of the goods brings them to the ship's side, and the buyer pays for putting them aboard and the dues and charges for slinging.

C.I.F.—Cost including freight.

Merchants' quotations often contain similar contractions as:—

F.O.V.—Free on van.
F.O.R.—Free on rail, &c.
G.M.B.—Good marketable brands.
G.O.B.—Good ordinary brands.

In comparing reports of timber sales with the catalogues, the lots not mentioned in the report are those for which there was no bidding. The method of description is uniformly in some such manner as follows :—

For timber :—

No. of pieces.	Range of section.	Range of length.	Total contents.	s. d.	Buyer.
15	16¼/18¼	24/29 ft.	772 ft.	59 0	Thompson.

For Deals :—

Brand.	No. of pieces.	Range of length in feet.	Scantling.	Quality.	Price per Petersburg standard.	Buyer.
E.H.B. & Co. 1.	545	15/24	8″ × 9″/12	1st Y.	12*l.* 15*s.*	Smith.

Ports of Entry.—In considering the most convenient point from which timber may be obtained for the construction of a building in the country, the table published annually in the *Timber Trades Journal*, showing the distribution of imports of hewn and sawn timber to the various parts of the United Kingdom, gives a complete list of the shipping ports, and will be found useful by the estimator. It will be necessary to calculate the relative cost of its conveyance from the one or the other. It need hardly be said that the advantages of purchase at a great market like London, Liverpool, Hull, Grimsby, Cardiff, Gloucester, &c., will frequently counterbalance the advantages of shorter distances from other ports of entry.

Timber purchased at dock sales is loaded by the company.

When it is necessary to employ outside labour to load into trucks or other vehicles, the cost is about 2*s.* 6*d.* per Petersburg standard.

The railway rates should, of course, be discovered, as they are by no means uniform. A rough approximation to the cost of carriage of timber is as follows :—

20 miles and under, 4*d.* per load of 50 cub. ft.
30   „   „   3*d.*   „   „
50   „   „   2½*d.*   „   „
75   „   „   2¼*d.*   „   „
100   „   „   2*d.*   „   „

(Deals and battens per standard about three times the above rates, i.e. 20 miles and under, 1*s.* per standard.)

To this, in the case of despatch from a railway station, a small charge for the use of crane and loading must generally be added ; 1*s.* 6*d.* per ton measurement is a common charge.

Limit of average contents of logs of timber without extra price.

	Average.	Limit of Size Obtainable.	
		Length.	Section.
		feet.	in. in.
Sawn and hewn pitch-pine logs ..	65 to 70 ft. cube	80	9 to 19
American yellow pine ditto ..	65 to 70 ft. cube	70	14 „ 26
Swedish fir	27 ft. long	35	8 „ 12
Stettin oak	.. ..	30	10 „ 16
Dantzic oak	.. ..	30	10 „ 16
Dantzic fir	26 ft. long	50	10 „ 20
Memel fir	27 „	35	11 „ 14
Riga fir	27 „	40	11 „ 14
Sawn Oregon pine	40 „	40	16 by 16

(Planks 11 in. and wider any thickness up to 6 in., and length up to 40 ft.)

When these averages are exceeded, the merchant's charge on the whole load of timber is about ¼d. per cubic foot for every foot that it exceeds the average. Thus, if the average limit is 27 feet in length, and the load in question averages 35 feet, 35 feet − 27 feet = 8 feet by ¼d. = 2d. by 50 = 8s. 4d.

Modes of Purchase.—Timber and deals, if bought at a timber sale, must, of course, be carted from the docks. If the builder has the requisite machinery, he can convert it himself; if not (as is most frequently the case), he will arrange with a saw-mill proprietor for its collection from the docks, sawing, and delivery at the works. In this case the dock order, obtained through the timber auctioneer, would be sent to the saw-mills with the instructions for sawing.

In addition to the charge for sawing, a further charge, known as landing-rate, is made on deals and battens fetched from docks.

```
                                           s.  d.
On goods for immediate removal and sawing  3-  9 ⎫ per Petersburg
If piled awaiting orders.. .. .. .. ..     5   0 ⎭ standard.
```

There is no landing-rate on timber.

Dock Charges.—All timber under 9 inches square is landed on the wharves; 9 inches square and over lies in the timber ponds.

Considerable quantities of timber are floated from the dock-ponds into the Thames and the canals. It is delivered at the lock-gates of the docks by the company's men, the charge for this being an importer's charge.

Much of the baulk timber is conveyed by barges from the dock-ponds to the rivers or canals, and by those routes to its destination.

The importer pays the dock rent on timber and deals sold at auction, usually for a period of four weeks; after that period, a rental of 1¼d. per load per week for timber, and 3d. per standard for deals is charged by the dock company, and is payable by the purchaser.

Mahogany, 1¾d. per ton weight per week; teak in log, 1d. per load of 50 feet cube per week; teak in plank, 1¼d.

An extra charge of 6d. per standard is made by the dock company on any delivery of deals less than a standard, and is payable by purchaser.

When the purchaser of a float of timber desires to remove less than half a float, it is subject to an extra charge of 1s. 6d. per load, paid by purchaser.

Timber delivered by order at the dock gates must be taken away within two tides, or is charged 2d. per load per tide until removed.

For timber loaded into barges the dock company charges 1s. per load cranage, paid by purchaser.

For timber loaded on to timber carriages or other vehicles the dock company charges 1s. 6d. per load for cranage, paid by purchaser.

Barges remaining in dock more than 12 hours after obtaining a pass to go out are subject to a charge of 5s. each for every week or part of a week.

The charge by lightermen for rafting timber or towing it in raft from the docks to Westminster is 9d. per load in quantities of not less than 10 loads.

The charge for conveyance the same distance in barges is, timber 9d. per load in quantities of not less than 10 loads; deals and battens, 2s. 6d. per standard in quantities of not less than 5 standards.

Cartage.—The cartage per standard of deals from Commercial Docks to the City, exclusive of loading and unloading, is 10s. The loading is done by the deal porters employed by the dock company, and is a part of the dock charge paid by the importer.

The cartage of timber from the Commercial Docks to the City is 4s. per load of 50 cubic feet. As timber exceeding 35 feet in length may not be carried through the City except between the hours of 7 p.m. and 10 a.m. (Metropolitan Streets Act, 1867) a charge of about 2s. more per load is made by the carters in such cases.

Purchase of Timber from Timber Merchants.—If timber is not bought at auction, it would be bought at per load of the timber merchant, who would probably be also a proprietor of saw-mills. Another way commonly adopted by estimators is to send a timber merchant or saw-mill proprietor a copy of a carpenter's specifications, and contract with him to supply the timber, sawn to scantlings, for the whole of the requirements of the building at one uniform price. This is sometimes done at as low a rate as 1s. 6d. per foot cube.

The time of year slightly affects the price of timber at auction; but the practical estimator takes but little heed of it, except to delay his buying as long as possible. The import is certainly stopped during the time the timber ports of Northern Europe are frozen; but the keen competition for trade, and the large stocks kept by many of the merchants, keep the prices down.

Timber, by what Measure :—

Fir, American pine, greenheart timber is sold by the load of 50 ft. cube, sometimes caliper and sometimes string measure.
Oak, birch, ash, elm ditto 50 ft. ditto ditto.
Teak ditto 50 ft. cube or by the foot superficial, according to thickness, ditto ditto.
Oregon pine, ditto ditto ditto.
Pitch pine, ditto ditto ditto.

At sales of English timber the wood is sold in lots of so many trees as they stand, and the purchaser fells them.

Deals, planks and battens, and pitch-pine deals up to 6 in. thick, by St. Petersburg standard hundred.
Flooring by the reputed square.
Matched and beaded or grooved boarding ditto.
Beads, mouldings, skirtings and weather boards by the 100 ft. run.
Battens for slates or tiles by the 144 ft. run.
Thick American birch planks at per foot cube.
Crown deck deals at per 40 ft. run of 8 in.
Cedar and mahogany at per superficial foot of 1 in. thick.
American walnut at per cubic foot or foot superficial, according to thickness, sometimes caliper, sometimes string measure.
Ditto whitewood ditto ditto ditto.
Ditto hickory ditto ditto ditto.
Ditto quartered oak thicknesses above 1 in. at per cubic foot; thinner, at per superficial foot of 1 in.

Sequoia (Californian redwood), at per cubic foot.
Wainscot at per cubic foot, at per 18 ft. cube.
Kauri pine at per load, or per cubic foot, or foot superficial of 1 in.
Plasterer's laths at per bundle of 300 ft., 360 ft., or 500 ft. run.
Mahogany ends at per ton weight.
Rose-wood, satin-wood, ebony ditto ditto.

English Oak.—Oak of good quality and large size sells by auction in the log before felling at about 2s. per cubic foot. Fair average quality, about 1s. 6d.

The merchant will sell it in hewn logs at about 70s. per load of 50 cubic feet; sawn plank at 150s. per load of 50 cubic feet.

Sash sills and sawn scantlings of similar size can be obtained for 3s. 6d. per foot cube of the merchants.

Dry and well-seasoned oak of good figure and large size sometimes costs as much as 6s. per cubic foot.

Elm.—The value is approximately half that of oak.

Elm of good quality in logs of 12 by 12 inches fetches before felling about 10d. per foot cube.

The merchant will sell it in hewn logs at about 55s. per load of 50 cubic feet. Sawn planks or scantling 2s. 6d. per foot cube.

Ash.—Ash in log sells by auction in log before felling at 1s. to 1s. 6d. per foot cube.

Ash is but rarely used by the builder. It is approximately twice the value of oak.

The merchant will sell it in hewn logs at about 8l. per load of 50 cubic feet; scantling 3s. 6d. to 4s. 6d. per foot cube.

Removal.—As the removal of English-grown timber is often expensive, it is obvious that it will fetch a larger price if it lies convenient to a good road. It rarely costs less than 2d. per cubic foot for loading and carriage four miles; 3d. is a reasonable average.

Railway companies charge 1s. 6d. per ton of 40 feet for loading oak, ash, or beech on their trucks. Other woods per ton of 50 feet.

Conversion of English Oak.—The method of conversion, and consequently the size of the trees, materially affects the price per foot cube. The cheapest form is a post converted from a single trunk sawn square for the greater part of its length, and the butt left rough, to be afterwards buried in the ground. This is good enough for gate posts, and is the method usually adopted to produce them; the pith is consequently left in the wood. The next production in order of cheapness is sash sills and fence rails as they can be cut in various ways, and from their small size the pith and sap are easily avoided.

Broad planks or large scantlings for better purposes than the foregoing are invariably cut "on the quarter," and consequently must be sawn out of much larger trees, with a corresponding increase of cost.

Long lengths also increase cost.

Waste in Conversion.—The practice of specifying sawn timber for the carcasing of buildings is still frequent. A clause in common use is as follows: "The fir timber is to be of the best description from Memel, Riga, or Dantzic, sawn die-square, free from sap, shakes, large loose or dead knots, and all other defects, and sawn into scantlings immediately after signature of contract."

Probably in nineteen cases out of twenty this stipulation is met by the use of planks, deals, or battens of poor quality, whose use is tolerated, and their defects almost ignored.

Better and more business-like conditions will come some day, when the architect becomes a better judge of the values of wood, with the consequent familiarity with brands and what they mean. He will then realise whether he wants timber sawn uniformly out of balks, or of sizes imported as deals, battens, or planks; he will specify his timber by its trade mark or importer's name, and see that he gets what he specifies.

But the use of deals instead of log timber is always a saving in sawing, and if they are of good quality they make a better building, principally for the reason that they are nearly always better seasoned than any balk timber procurable, however stringent the stipulations of the specification in that respect may be. A large variety of scantlings, ready sawn, is now imported as:—

"	"	"	"	"	"	"	"	"	"
2	4	2	8	2¼	6¼	3	8	4 × 7
2	4½	2	9	2¼ × 7	3 × 9	4 × 9
2	5	2	10	2¼ × 8	3 × 10	4 × 10
2 × 5½	2 × 11	2¼	9	3	11	4 × 11
2	6	2¼	5¼	3	7	3	12	4 × 12
2	7	2¼	6					

Seddon's remarks on this subject state the case admirably. He says, in 'Builders' Work':—

"Planks, deals and battens are largely used for cutting into small scantlings, such as 3″ × 2″, 2½″, 3½″, 4½″, &c., which are, if out from good deals, better than those cut from whole timbers, though the contrary is the most generally received opinion. In proof of this, as has been already pointed out, the export of planks, deals, and battens is almost entirely confined to the more northern districts, which produce the closest-grained timber; while the coarser and more open grained timber, grown in the southern districts of Livonia and Prussia, is exported in balk, and though superior in strength when used in large beams, is unsuitable for cutting up into small scantlings, since the large scale of the knots and other defects would tell too much in small sections, where they would bear an undue proportion to the sound part of the wood."

One does occasionally see in a specification the deals or battens designated by a shipping mark; but not unfrequently it is one which has disappeared from the market for years. The book published by Rider, 'Shipping Marks on Sawn and Planed Wood,' contains all the shipping marks used on timber and deals, &c., of European growth, as also charts of all the shipping ports, and the names of the shippers and their agents. This book is annually revised. Careful watching of the reported prices at the timber sales, and comparison of the prices obtained for goods with certain marks, which should be referred to in the book, will afford the student much valuable information as to quality and price.

American timber and deals are often imported without brands. Observations of the prices at the trade sales, and the port of consignment, will soon show the student the best kinds and the names of their shippers.

For the judgment of value of timber after conversion, the student must depend upon a knowledge of quality, derived from observation of its natural characteristics. The marks will have disappeared from the timber; but in the case of deals used for timbering, they can often be seen on the exposed ends.

Waste in Sawing.—The waste upon balk timber, if sawn all round to remove the outside, is rarely less than 25 per cent.; but there are few buildings where the original outside is not made to do for one side and one edge of sawn timbers. There is a still further waste on sawing into scantlings of the average size. The waste upon the saw-cuts is commonly disregarded, as the timbers are passed if they are not diminished more than $\frac{1}{16}$ inch for each sawn side; but when the specification prescribes finished sizes, it

2 q

should be allowed for, and the waste would be equal to another 5 per cent.

The waste on cross-cutting is only equal to another 2½ per cent., assuming that the conversion is done with care and judgment.

Waste by Slabbing or Sawing " Die-square."—To illustrate the question of waste in slabbing, we may take the following example: A log of Dantzic timber 32 ft. × 15 in. × 15 in. = 50 ft. cube. If we assume 1 inch as the least thickness, including waste of the saw-out, which can be removed from each face, we have the following dimensions:—

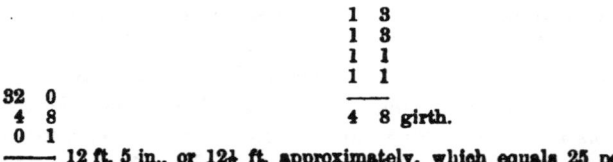

```
                         1  3
                         1  3
                         1  1
                         1  1
                        ------
    32  0                 4  8  girth.
     4  8
     0  1
    ------
           12 ft. 5 in., or 12¼ ft. approximately, which equals 25 per cent. of the
           whole.
```

And in this connection a note in Laxton's 'Builders' Price-Book' may be quoted, which more that corroborates the above conclusion: "The waste in conversion of fir timber when slabbed all round (comparing the net quantity obtained with the contents of the balks cut up) is probably much greater than is generally supposed. A careful attention to the subject has proved it to be quite 30 per cent., based upon an ordinary fair average of scantlings; and at a prime cost in docks of 80s. per load, the net timber cut out cost 2s. 9½d. per foot cube. Of course, an allowance may be made for value of slabs, &c.; but, practically, they are more or less waste only to the builder."

Waste by Sawing into Scantlings.—The specification should show the architect's intention as to sawing. Some specifications stipulate for finished sizes of timber by some such clause as the following: "The whole of the specified or figured dimensions of timbers shall be the finished sizes when fixed in the building." In such a case an allowance must be made for the waste produced by the saw-cuts. If the builder converts the timber at his own works, he usually does it with a circular saw, and the waste is greater than in mill-sawing, where the frame-saw is almost uniformly used for the purpose. The waste on each saw-out by frame-sawing will average about $\frac{1}{32}$ inch; by the circular saw, about $\frac{1}{8}$. The waste will, however, vary with the quality and condition of the wood, and $\frac{1}{8}$ inch may be adopted as a reasonable average.

Adopting the average of 288 feet of sawing per load, we have the calculations 288 feet × 1 foot + $\frac{1}{8}$ inch = 3 feet cube for waste per load in sawing, exclusive of the waste in slabbing, elsewhere dealt with.

A clause relating to sawing and waste, used in the War Office schedules, is as follows: "The net lengths, breadths, thicknesses and girths ordered to be supplied, and no allowance to be made beyond those demanded. One-eighth of an inch to be allowed on the several denominated thicknesses of superficial measured work, if wrought on both sides—viz. for 1 inch it must measure $\frac{7}{8}$ inch, for $1\frac{1}{4}$ inch it must measure $1\frac{1}{8}$ inch, and so on for the various thicknesses; but if planed only on one side it must measure full $\frac{7}{8}$, or $1\frac{1}{8}$ inch, as the case may be."

See also the paragraph relating to allowance for sawing—Section I. *Carpenter and Joiner*—"General Statement of Methods recommended by the (Manchester) Society (of Architects) to be used in taking Quantities and Measuring up Works."

Quality of Deals and Battens.—Deals and battens for carcasing are frequently used of poor quality, even in good buildings, unseasoned, shaky, waney, sappy, or knotty, costing often as little as 7*l.* per Petersburg standard.

The architect should not tolerate any quality inferior to shippers' mixed Swedish, a common price for which at the dock sales is about 10*l.* per standard, a price which will be adopted in the following calculations. Battens would commonly be about 2*l.* per standard less.

Difference of Value of Labour and Materials expressed by a Percentage.—Examination of price books or priced estimates, with this principle in view, will show to what a great extent a series of prices of a certain different wood is modified by the method of adding or deducting a certain percentage to or from the price of work in fir or deal, and this course may reasonably be adopted either for the valuation of material or material and labour.

This is also the usual treatment of rates in schedules of prices; but care should be taken to make the prices bear a consistent relation, as, if the price of deal be too low or too high, a percentage would be added to the deal price different to what it would be if the deal price were a fair one.

Labour on oak timbers is generally assumed to be twice that on deal—i.e. 100 per cent. Labour on oak carcasing, one-third more than fir—i.e. 33 per cent. Labour and materials to roofs, floors and quarter partitions is often assumed to be equal to roofs and quarter partitions, being almost universally taken as equal. Planing on hard woods, one-third more than fir. Approximately, labour to curved work is worth 50 per cent. more than to straight.

Mode of Measurement.—The valuation of timber is affected by the mode of measurement.

The dock custom as to measurement of squared timber is so close to the actual content as to be unimportant; in the cases of round or eight-sided timber, which are sometimes measured quarter-girth "string" measurement, and sometimes "caliper" measurement, the difference between it and the actual content is considerable. The former method gives less than the actual content, and is consequently in purchaser's favour; the latter more, and so is against him.

The dock mode of measuring mahogany gives the purchaser at dock sales an excess of about 25 per cent. over the actual contents.

Tables.—The custom of selling at these dock auctions by the St. Petersburg standard of 120—12 feet 11 inches by 1½ inches—has made necessary for the comparison of prices a series of tables, of which the one showing the relative value of a standard hundred of deals, battens, or St. Petersburg hundred, and a given price for a load of timber is probably the most useful. Thus:—

Petersburg standard hundred, 12 ft. 1½" × 11", 165 ft. cube.	12 ft. 3" × 9"	12 ft. 3" × 7"	12 ft. 2½" × 7"	Timber, per load.
8*l.* 10*s.*	13*l.* 18*s.* 2*d.*	10*l.* 14*s.* 1*d.*	9*l.* 0*s.* 4*d.*	2*l.* 11*s.* 6*d.*

It is useful to remember that 1*l.* per St. Petersburg standard is approximately 1*s.* per square for 1 inch. One shilling per foot cube is 8*l.* 5*s.* per St. Petersburg standard: that 2*d.* each is 1*l.* per standard; and that a St. Petersburg standard is 165 feet cube. A variety of these tables is to be found in the current price-books, and they save much calculation.

At the auctions, the boarding and flooring is sold at the "customary square," which is not invariably 100 feet superficial.

in.		ft. run.		ft. in. super.	in.		ft. run.		ft. in. super.
9	=	140	=	105 0	6	=	200	=	100 0
8	=	160	=	106 8	5¾	=	210	=	100 8
7½	=	170	=	106 3	5½	=	220	=	100 10
7	=	180	=	105 0	5¼	=	230	=	100 8
6¾	=	185	=	104 1	5	=	240	=	100 0
6½	=	190	=	102 11	4½	=	270	=	101 3
6¼	=	195	=	101 7	4	=	300	=	100 0

Imported Sizes.—The London contractor, when he receives a bill of quantities, as a rule looks at the carpenter's specification to see whether any of the prescribed scantlings are to be obtained of imported ready-sawn sizes, and settles his price accordingly; and for a similar reason the contractors of the North of England prefer timber stated in a bill of quantities as running lengths of each particular scantling, with the labour on each. As:—

500 ft. run, 4" × 2½", wrought all round.
500 ft. „ 4½" × 3" ditto, splayed on edge, etc.

***Mill Charges for Sawing*:—**

***Fir timber*:—**

	s.	d.
Fir timber under 12 in. square, 3 cuts to the load of 50 ft. cube	6	6
Ditto 12 in. and over, 4 cuts ditto	—	
Timber sawing, per 100 ft. super.	4	0
Cross cuts, each	0	4
Cutting 4 in. arris rail, per 100 ft. run	2	0
Ditto 5 in. ditto, ditto	2	3
Fir scantlings, 6 in. and under, per foot run	0	0¼
Ditto, above 6 in., ditto	0	0½
Cartage, 1s. per mile per load of 50 ft. cube.		

***Hard woods*:—**

—	Inches deep.	Under 1¼ 100 ft. super.		1¼ in. 100 ft. super.		Planks, over 1¼ in.
		s.	d.	s.	d.	
Mahogany, Honduras	under 24	6	3	7	3	27s. per ton of 40 ft. cube.
„ „	„ 36	7	6	8	6	
„ „	„ 42	9	6	10	6	
„ Spanish	„ 24	7	6	8	6	
Cedar, Honduras and Cuba	„ 24	5	6	6	6	
„ Paraguay	„ 24	6	3	7	3	
„ Ak and Padouk	„ 24	8	0	9	0	
Sabicu	„ 24	12	0	13	0	
Yellow pine	„ 24	5	0	6	0	
Pitch pine and Kauri pine	„ 24	6	0	7	0	

	Inches deep.	Under 1¼ in. 100 ft. super.	1¼ in. 100 ft. super.	Planks over 1¼ in.
		s. d.	s. d.	
Wainscot	under 24	6 0	7 0	
Birch and maple	„ 24	6 6	7 6	27s. per
Hornbeam, foreign	„ 24	7 0	8 0	ton of
American ash, whitewood and satin walnut	„ 24	6 0	7 0	40 ft. cube.
American oak, elm, hickory and black walnut	„ 24	7 0	8 0	

Log ends, cut by horizontal saw, 1s. 6d. per 100 ft. super. extra.
(The prices for the above woods rise 1s. for every 6 in. extra depth, excepting mahogany and cedar.)
Planks and flitches other than whitewood and Sequoia planks, same price as logs.
Whitewood planks, 4 × 12 and under, 4s. per 100 ft. super.; above, same price as logs.
Sequoia, 4 × 12 and under, 4s. per 100 ft. super.; above, and under 30, 5s.
(1 cut in plank charged as two cuts.)

		100 ft. super.	Planks 1¼ in.	Planks over 1¼ in.
		s. d.	s. d.	
English oak, beech, sycamore, plane tree, English elm, ash, chestnut and lime-tree ..	all depths	7 0	8 0	27s. per ton of 40 ft. cube
English hornbeam		8 4	9 4	

Cleating and canvassing boards, 1d. per end.
Cross cuts, under 14 in. 6d., above, 9d. each.
Rosewood, all kinds, inch and under, 12s. 6d. per 100 ft. super.
    „   „   above, 16s. 8d.   „   „

	s.	d.
Flatting (sawing through thinnest way of boards) up to 4 in. per 100 ft. run	3	0
Ditto 4 in. to 8 in. ditto	4	6
Over 8 in. ditto	7	0
Breaking cuts (dividing balks into boards or planks) 30 in. deep (nothing less charged), per foot super. ..	0	2
Ditto 36 in. ditto ditto	0	2½
Ditto 42 in. ditto ditto	0	3

Cross cuts, mahogany 9d. each, ash 6d. each.
Cartage charged on 7 cuts and under, at 7s. 6d. per ton of 40 ft.

The foregoing charges are subject to a discount of 20 per cent.
The foregoing prices for sawing and preparing include collection from docks by barge and delivery after sawing within three miles of mills, except the extra charges for cartage and landing rate. All measurements centre out.

PRICES.

Price for Sawing per dozen cuts:—

Lengths.	Battens.	Deals.	Planks.	Lengths.	Battens.	Deals.	Planks.
	s. d.	s. d.	s. d.		s. d.	s. d.	s. d.
6	1 4	1 6	2 0	19	3 6	4 9	5 11
7	1 6	1 9	2 3	20	3 9	5 0	6 3
8	1 8	2 0	2 6	21	4 0	5 3	6 8
9	1 10	2 3	2 9	22	4 3	5 6	7 0
10	2 0	2 6	3 0	23	4 6	5 9	7 6
11	2 2	2 9	3 3	24	4 9	6 0	8 0
12	2 3	3 0	3 6	25	5 0	6 3	8 6
13	2 4	3 3	3 10	26	5 3	6 6	9 0
14	2 6	3 6	4 3	27	5 6	6 9	9 6
15	2 8	3 9	4 8	28	5 9	7 0	10 0
16	2 10	4 0	5 0	29	6 0	7 3	10 6
17	3 0	4 3	5 3	30	6 3	7 6	11 0
18	3 3	4 6	5 6				

Feather-edged cutting one-third extra per dozen cuts.
Pitch-pine planks one-third more per dozen cuts.
Arrising up to 3 in. per 100 ft. run, 1s. 3d.
Flatting battens, deals and planks 3 in. and under, 1s. per 100 ft. run; 4 in. ditto, 1s. 4d. per 100 ft. run.

(The above subject to a discount of 25 per cent.)

Yellow pine planks over 11 in. wide, 3s. 6d. per 100 ft. super.
Extra cartage on 1 cut in flatting, 2d. per 12 ft. 3 in. by 9 in.; 2 cuts, 1½d. per 12 ft. 3 in. by 9 in.
Extra cartage on 1 cut in deeping, 1½d. per 12 ft. 3 in. by 9 in.

(The above prices are net.)

Mill sawing is always to be preferred to hand sawing, because of its superior precision, and only costs about half the price of the latter.

Observe that dry timber takes longer to saw than that freshly imported.

Seddon says: "The cost of sawing dry, seasoned timber being about one-fourth more than new. The value of sawing on African oak, teak and mahogany is about two-and-a-half times that on fir, and on oak, elm, ash and beech about two-thirds as much again as on fir."

Nicholson says: "Sawing mahogany is worth three times fir; oak, ash, elm and beech twice fir."

Flooring :—

	1¼ in. and under.			1½ in.	
	s.	d.		s.	d.
Sawing and planing	2	3	..	2	6
Sawing, planing and grooving	3	0	..	3	3
Preparing imported boards, same prices.					
Planing boards (when sawing charged separately)	1	6	..	1	9
Grooving prepared boards at yard	1	6	..	1	9
„ boards from docks ..	1	9	..	2	0

Stacking, 3d. per square extra.
Stacking and sticking, 9d. per square extra.
Rent charged on all goods stacked for a longer period than six months at the Surrey Commercial Dock Company's rates.

The foregoing prices subject to a discount of 5 per cent.

Matchboarding, &c., per square :—

	1¼ in. and under.			1½ in.	
	s.	d.		s.	d.
Sawing, planing and plain matching ..	3	3	..	3	6
Sawing, planing, matching, and beading or jointing	3	9	..	4	0
Ditto ditto prepared both sides	5	0	..	5	6

Net prices of desiccating (or drying) per square. Extra on prices for sawing and preparing.

	1 in. and under.	1¼ in.	1½ in.	
In quantities not less than	200	160	140	Squares.
	s. d.	s. d.	s. d.	
Boards sawn or prepared at the same mills	1 10	2 3	2 7	per square.
Collecting boards from docks per barge, landing and stacking in drying-room, and delivery within three miles	2 2	2 7	3 0	ditto.
Time required for drying will average about	4	5	6	weeks.

The foregoing saw-mill charges, although ostensibly the fixed charges throughout the trade, are very often modified by special agreement, by which landing rate or extra cartage, or both, are dispensed with.

A few of the dock rates above mentioned are as follows :—

	Rent per week.	
	s.	d.
Deals, planks, battens and ends, 3 in. and under (per St. Petersburg standard)	0	3
Deals, planks, battens and ends, 4 in. and upwards, and scantling 5 in. and under (ditto)	0	3
Deck deals, yellow pine, pitch pine deals and boards (ditto)	0	4½
Fir timber and balk under 9 in. square, per load ..	0	1½
Hardwood timber and plank	0	1

We will base our analyses on wood of good quality, as it will be quite easy to alter the prices of material by a percentage if we know the conditions which modify the price.

Nails.—The nails generally used for carpenter's work are steel "cut clasp," and for ordinary work the following may be adopted on calculations:—

Thickness of wood—	in. $\frac{1}{2}$	in. $\frac{3}{4}$	in. 1	in. $1\frac{1}{4}$	in. $1\frac{1}{2}$	in. $1\frac{3}{4}$	in. 2	in. $2\frac{1}{4}$	in. $2\frac{1}{2}$	in. $2\frac{3}{4}$	in. 3
Length of nails—	$1\frac{1}{4}$	$1\frac{1}{2}$	2	$2\frac{1}{2}$	3	$3\frac{1}{2}$	$3\frac{1}{2}$	4	$4\frac{1}{2}$	$4\frac{1}{2}$	5
Weight per thousand in pounds—	3	$3\frac{1}{2}$	8	12	20	25	25	40	50	50	67
Price per cwt. of merchants—	14/9	13/9	12/9	12/	11/9	11/9	11/9	11/9	11/9	11/9	11/9

German nails cost about 1*s.* 6*d.* per cwt. more than the foregoing. Much of the waste of nails and screws is avoided under the supervision of a careful foreman; but 5 per cent. is the very least that occurs, and in the calculation of the number of nails or screws for a piece of work that percentage should be allowed. Nails, when there is no cross-strain, may be about 9 inches apart, otherwise about 6 inches apart.

For the estimation of the number of nails required for a particular piece of work, it is, of course, necessary to know the whole process of fixing. Those for floors will depend upon the distance apart of the joists; for roof boarding, on the distance apart of the rafters, jamb linings and skirtings of the backings, etc.

After the prices for material, the value of the ordinary labours should be settled, when the analogy of similar labours may be easily drawn, and the conclusion arrived at as to whether the labour in a given case is worth less or more than the item with which it is compared.

Wrought Timbers.—The surveyor treats wrought timbers in a variety of ways and does not always measure as much wrought face as he should do. As an instance, it will be found that a wrought plate, the front only of which is exposed, will in many cases require planing on its two returns if the result is to be satisfactory. The various ways of dealing with wrought face are as follows:—Measuring the timbers of the size specified, and measuring the surfaces planed by the superficial foot, and describing as wrought face on fir, in which case the timbers will finish less than the specified size.

When there is a stipulation that the specified sizes of the timbers shall be the *finished* sizes when fixed in the building, the wrought timbers are measured as ¼-inch larger each way than specified, to allow for loss in planing; thus 4 inches by 3 inches each way should appear in the dimensions as 4¼ inches by 3¼ inches, the surfaces planed being measured by the superficial foot, and described (as before) as wrought face on fir, in such case the timbers finishing the specified size.

A modification of this last treatment is to measure the timbers of the sizes specified, and to measure the surfaces planed by the superficial foot, describing it as *wrought face on fir, including waste.*

Another way is to measure the timbers ¼-inch larger each way than specified to allow for loss in planing, and describe them as fir wrought and framed. This last way is not just to the builder, as he often attaches a lower price to the item when treated in this manner.

It is therefore necessary to observe which of these courses has been adopted by the measurer.

The planing of boards is nearly always done at saw-mills; timbers generally by the builder, who sometimes has a planing machine. The machine planing will always require further smoothing by the builder's men.

The cost of planing of boards by machinery is to be found in the saw-mill prices already given.

Conversion.—The quantity of sawing necessary to convert timber to the sizes adapted to the ordinary purposes of a building will vary somewhat with the scale of the structure and its details of construction.

The scrutiny necessary to arrive at a *precise* result is impossible without an expenditure of labour on each occasion of making an estimate far beyond that the subject deserves. An *average* percentage is necessary. The following table shows the results of an analysis of the quantities of three commonplace examples, marked A, B and C.

As the conversion of timber into scantlings involves sawing all around, it is obvious that each of the saw-cuts produces two surfaces, equal to *half* sawing on each. For example, a lineal foot of 4 inches

× 3 inches would require 1 foot by 1 foot 2 inches = 1 foot 2 inches half-sawing.

	Total Cubic Quantity of Timber.	Super. ft. of Half Sawing.	Super. ft. of Whole Sawing.	No. of Loads of Timber.	Feet of Sawing per Load.
	feet.				feet.
A.—Country house ..	1533	19,259	9,630	30·66	314
B.—Country house ..	1755	20,186	10,093	35·10	288
C.—School - hall and classrooms ..	5216	55,198	27,599	104·32	264

In example A the rooms were rather small, the timbers of small span, and consequently comparatively light; in example B, the timbers were of larger span, and consequently heavier; in example C the hall was about 80 feet by 40 feet, and the classrooms larger than any of the rooms in A and B. It will be seen that the proportion of sawing slightly decreases with the increase in the bulk of the timbers.

The average of these calculations gives a result of about 288 feet of sawing per load.

A further allowance must be made of half-sawing all around the original log, and adopting the former example illustrating loss in slabbing as a reasonable average size we have 32 feet by 4 feet 8 inches = 149 feet 4 inches superficial of half-sawing, or 74 feet 8 inches superficial of whole sawing; 288 feet + 75 feet = 363 feet superficial per load of 50 cubic feet.

If we examine the same three sets of quantities in order to compare the cost of the timber of imported sizes with that sawn out of balk we shall find that the material divides itself into three categories.

First.—Scantlings which must be sawn out of timber.

Second.—Scantlings which can be obtained from deals, battens, or planks, as 4½ by 3 inches, 4 by 3 inches, 4½ by 2½ inches, 4½ by 3½ inches, 5½ by 3 inches, 4 by 4 inches, 5 by 4 inches, &c.

Third.— Scantlings of imported sizes requiring no sawing.

The result of this analysis is shown in the following table.

The average quantity of sawing *per load* is consequently reduced to 145 feet—i.e. 436 feet ÷ 3.

	Total cubic quantity of Fir.	Categories of Material.			Sawing.			Sawing feet per load.
		First.	Second.	Third.	Calculated by proportion at 383 ft. per load. First.	Measured and calculated from dimensions. Second.	Third.	
Example A	1,533	189	776	568	1,372	2,180	None	115
" B	1,755	410	661	684	2,977	1,727	None	133
" C	5,216	2,358	2,117	741	17,119	2,750	None	188
	8,504	2,957	3,554	1,993		6,657		436

Before pricing an estimate or measured bill, it will be necessary to decide upon the price per cubic foot of timber and deals, upon which the prices will be based. The value of the various common labours must also be settled.

Peter Nicholson, in the "Practical Builder's Perpetual Price Book,' calculates the value of timber as follows. It may be usefully compared with a more modern calculation. All the sawing at that date was done by hand; now it is all done by the steam saw.

	£	s.	d.
Prime cost of 50 cub. ft. of timber at 2s. 8d.	6	13	4
Average value of sawing at 3d. per foot cube ..	0	12	6
Average cost of carting one mile, 1d. per foot.. ..	0	4	2
Allowance 6 ft. for waste in sawing at 2s. 8d.	0	16	0
	8	6	0

Or 3s. 3¼d. per foot cube.

It was an old fashion, still recommended by the price-books, to add to the price of timber, at the docks or merchant's yard, 1l. per load for sawing and cartage.

Gwilt gives :—

	£	s.	d.
Prime cost of a load of fir	4	10	0
Suppose the cartage (dependent on distance)	0	5	0
Sawing into necessary scantlings	0	10	0
	5	5	0
Waste in converting equal to 5 ft. at $2\frac{1}{10}$s. per foot, the load being 105s.	0	10	6
	5	15	6

Or 2s. 3¼d. per foot cube.

the dock timber-ponds, there to lie for a further period until it is sold and sawn, and forthwith carried to the building. The ordinary Baltic timber would take something like six months to season, and this would represent not less than five per cent. per annum on its cost; consequently it, as a rule, gets but little seasoning, the more especially because it is concealed in the structure.

American yellow pine floated deals are often very wet.

Baltic deals and battens, being sawn before shipping and delivered on the dock wharves, are, of course, drier; but these always require natural or artificial seasoning. The better class of builder stacks his deals and battens and flooring a considerable time before use, and probably has a drying-room for drying deal after it is sawn to sizes. But all builders have not the requisite capital which this involves.

If thoroughly good work is required, another 5 per cent. at least should be added to the price of the material.

Labour.—The customary definitions of fir in London bills of quantities are unsatisfactory, although they have the sanction of long usage. Fir in ground joists is usually a separate item. But all the other fir in floors is described in one item as fir framed in floors. In the majority of cases it means only bridging joists; but an arrangement of binders and bridging joists (the girders being kept separate) would appear in the same category. It is, however, to be noted that the simplest floors require some trimming and consequent framing, which in London practice is not separately stated.

In Manchester and the Northern and Midland counties the fir is presented in one total, as "1500 feet cube fir converted"—that is, sawn to sizes—and separate items are given for the labour on it, as "two squares labour and nails to plates and ground joists"; "thirty-one squares labour and nails to naked flooring, with joists notched and framed to beams." The practice is explained in the "General Statement of methods, recommended by the Manchester Society of Architects, to be used in taking quantities and measuring up works."

Some of the schedules of prices used by the public departments make the distinction "timber fixed but not framed," defined thus :—

"The prices in the foregoing table include spikes, nails, oak treenails, and all workmanship and labour in preparing and connecting timbers together by lapping, notching, bevel, or bird's-mouth cuttings, including boring for bolts, and also in halving or dovetailing; to bond timbers, plates, lintels, sleepers, floor and ceiling joists, hip and common rafters, purlins and girders, and all labour, tackle, and machinery to hoist, raise, fit, place and fix the same, exclusive of digging when necessary."

Timber framed and fixed is defined as follows:—" The word framed being understood to comprehend all modes of connecting by mortise and tenon, and dovetailing and housings where necessary, and in curved work to include oak keys to joints. Boring for bolts, &c., if required, is also included in these prices."

The labour described as fir framed in roofs seldom varies much; the fir framed in roof-trusses is, or should be, always a separate item. The description "fir framed in quarter partitions" is a liberal one, considering the way in which they are usually put together. The theory is that all the vertical timbers should be tenoned to the heads and sills. It is but seldom the case that anything beyond the posts are tenoned.

In "fir framed in trussed partitions," the whole of the trussed portion is really framed, but the quarters are not. This custom as to quarter-partitions is so prevalent, that it is a question whether the tenoning of the quarters could be insisted on without an express stipulation of the specification to that effect.

Sub-contractors for carpentry will contract for labour to the whole of the timbering of a building at one uniform price per foot cube, sometimes as low as 7d. all round.

The adoption of an average quantity of sawing, and a given proportion of deals to timber, which may apply to all items, very much simplifies calculations.

Thus, the valuation of fir framed in double framed floor would be as follows, adopting the average values of wood and sawing as before :—

	s.	d.
1 ft. cube of fir	1	7¼
Labour.—Fir framed in double-framed floor (·090 days of carpenter at 8s. 9d.)	0	9¼
Cost per foot cube	2	4¾

Thicknesses.—The thicknesses used in the carpentry of a building are not generally much criticised, and are consequently most frequently of very poor quality. They consist principally of sound boarding, roof boarding, gutter boarding, grounds and fillets.

The rough boarding is imported ready sawn ¾ inch, 1 inch and 1¼ inch thick; beyond that thickness it is usually sawn out of deals or battens of inferior quality.

Not infrequently fillets are either imported deal slating battens, or are sawn out of them.

The common price in the docks for rough boarding is as follows:—

		s.	d.
¾ in.	per square	6	0
1 in.	„	8	0
1¼ in.	„	10	0

Common flooring is sometimes used when rough boarding is described.

For other thicknesses a table may be constructed as follows:—

Assuming the dock price to be 7*l*. per Petersburg standard, and taking 12 feet as the average length, and the scantling 3 by 9 inches, we have in a Petersburg standard seventy-four pieces 12 feet long, 9 by 3 inches (nearly). The calculation would be as follows:—

	£	s.	d.
Cost of deals per standard	7	0	0
Landing rate	0	3	9
Collection by saw-mills proprietors, and delivering included in charge for sawing, if in reasonable quantity			
Unloading at building	0	1	6
Waste 10 per cent.	0	14	0
	7	19	3
Say	8	0	0

The list price for sawing 12 feet 9 inches by 3 inches is 3*s*. per dozen cuts, less 25 per cent. = 2*s*. 3*d*. Where the following is one cut only, there is a charge for extra cartage 1½*d*. per deal = 9*s*. 1*d*. per standard, or 1½*d*. per cut = 1*s*. 6*d*. per dozen.

The prices of similar thicknesses arrived at in two ways, it will be seen by the column of figures on the right and left of the description, are identical.

PRICES.

ANALYSIS OF COST OF A ST. PETERSBURG STANDARD OF ROUGH DEAL SAWN TO VARIOUS THICKNESSES, ASSUMED TO CONTAIN 74 TIMES 12 FT. × 9″ × 3″ = 666 FT. OF 3″.

	£ s. d.	
¼″ = 11 cuts in 3″ × 9″ (producing 12 thicknesses of ¼″) = 814 cuts = 67⅚ doz. at 2s. 3d.	8 0 0 7 12 7½	Deal. Sawing.
(feet) 7992) 15 12 7½		
½d.		

½″ = 5 cuts in 3″ × 9″ (producing 6 thicknesses of ½″) = 370 cuts = 30⅚ doz. at 2s. 3d.	8 0 0 3 9 4½	Deal. Sawing.
3996) 11 9 4½		
¾d.		

¾″ = 3 cuts in 3″ × 9″ (producing 4 thicknesses of ¾″) = 222 cuts = 18½ doz. at 2s. 3d.	8 0 0 2 1 7½	Deal. Sawing.
2664) 10 1 7½		
⅞d.		

1″ = 2 cuts in 3″ × 9″ (producing 3 thicknesses of 1″) = 148 cuts = 12⅓ doz. at 2s. 3d.	8 0 0 1 7 9	Deal. Sawing.
1998) 9 7 9		
1⅛d.		

s. d.
6 8   Deal.
6 11¼ Sawing

1¼″ = 2 cuts in 3″ × 9″ (producing 2 thicknesses of 1¼″ and 1 of ½″) = 12½ doz. at 2s. 3d. *Note.*—One-quarter of the total of the foregoing sawing is apportioned to the ½, leaving three-quarters for the calculation, and the deal reduced ⅛.	6 13 4 1 0 9¾	Deal. Sawing.
1332) 7 14 1¾		
1⅜d.		

1½″ = 1 cut in 3″ × 9″ (producing 2 thicknesses of 1½″) = 74 cuts = 6⅙ doz. at 3s. 9d.	8 0 0 1 3 1½	Deal. Sawing.
1332) 9 3 1½		
1⅜d.		

2 R

£. s. d.			£. s. d.	
3 6 8	Deal.	1¾" = 1 cut in 3" × 9" (producing one	4 13 4	Deal.
0 11 6¾	Sawing.	thickness of 1¾", and one of 1¼),	0 11 6¾	Sawing
666)3 18 2¾		74 cuts = 6⅙ doz. at 3s. 9d.	666)5 4 10¾	
		Note.—One-half of the total of the foregoing sawing is apportioned to the 1¼", and the deal reduced ₁/₁₂.		
1⅜d.			1⅞d.	
2 13 4	Deal.	2" = 1 cut in 3" × 9" (producing one	5 6 8	Deal.
0 11 6¾	Sawing.	thickness of 2", and one of 1"),	0 11 6¾	Sawing
666)3 4 10¾		74 cuts = 6⅙ doz. at 3s. 9d.	666)5 18 2¾	
		Note.—One-half of the foregoing sawing is apportioned to the inch, and the deal reduced ⅓.		
1½d.			2⅛d.	
2 0 0	Deal.	2¼" = 1 cut in 3" × 9" (producing one	6 0 0	Deal.
0 11 6¾	Sawing.	thickness of 2¼" and one of ¾"),	0 11 6¾	Sawing
666)2 11 6¾		74 cuts = 6⅙ doz. at 3s. 9d.	666)6 11 6¾	
		Note.—One-half of the foregoing sawing is apportioned to the ¾", and the deal reduced ¼.		
⅞d.			2⅜d.	
1 6 8	Deal.	2½" = 1 cut in 3" × 9" (producing one	6 13 4	Deal
0 11 6¾	Sawing.	thickness of 2½", and one of ½"),	0 11 6¾	Sawing.
666)1 18 2¾		74 cuts = 6⅙ doz. at 3s. 9d.	666)7 4 10¾	
		Note.—One-half of the foregoing sawing is apportioned to the ½", and the deal reduced ⅙.		
¾d.			2⅛d.	
0 13 4	Deal.	2¾" = 1 cut in 3" × 9" (producing one	7 6 8	Deal.
0 11 6¾	Sawing.	thickness of 2¾", and one of ¼"),	0 11 6¾	Sawing
666)1 4 10¾		74 cuts = 6⅙ doz. at 3s. 9d.	666)7 18 2¾	
		Note.—One-half of the foregoing sawing is apportioned to the ¼", and the deal reduced ₁/₁₂.		
½d.			2⅛d.	
		3" No sawing.	666)8 0 0	Deal.
			3d.	

In the price for deal in day accounts waste may be disregarded. The contractor charges the gross quantity.

The variety of size of rough fillets used in the carpentry of a building and the small difference of their value indicates the reasonableness of referring the value of each size to a certain *range* of tabulated sizes to which one rate shall be attached. This is the ordinary practice in the production of a schedule for measured work, and as their size is rarely specified, and if specified, not rigidly enforced, the contractor is able to utilise material which would otherwise be wasted.

The estimator will find it an advantage to have such tables ready to his hand, something like the following. One of fillets sawn out of common deals or battens, thus:—

	½ in.	¾ in.	1 in.	1¼ in.	1½ in.	2 in.
2 in. wide and under rough, per ft. run	⅛*d.*	⅛*d.*	⅛*d.*	¼*d.*	¼*d.*	¼*d.*
2¼ in. to 3 in. wide rough ditto	¼*d.*	¼*d.*	¼*d.*	½*d.*	½*d.*	½*d.*

etc. The other of imported sizes slate or tile battens, thus:—

½ in. by 1 in.	⅝ in. by 1½ in.	½ in. by 2 in.
$\frac{1}{16}$*d.*	⅛*d.*	⅛*d.*

Cost per foot run at saw-mills.

The principle of the construction of the foregoing table is as follows:—For convenience of calculation we may take 100 feet superficial of deal of the thickness required, say 100 feet by 1 foot. For the first line 2 inches wide and under.

Adopting the former price of 8*l.* per standard, and the table of values of thicknesses based thereon, and assuming the ***average width*** of the fillets as 1 inch, the calculation would be as under:—

		£	s.	d.
100 ft. super. of ½ in. deal, at ¾*d.* per foot		0	6	3
Add 5 per cent. for further waste and breakage ..		0	0	3¾
11 cuts = 1100 ft. run, at 9*d.* per 100		0	8	3
Lineal feet produced	1200)	0	14	9¾
Per foot run		0	0	0⅛
100 ft. super. of 2 in. deal, at 2⅛*d.* per foot		0	17	8½
Add 5 per cent.		0	0	10⅝
11 cuts = 1100 ft. at 9*d.*		0	8	3
	1200)	1	6	10⅛
Per foot run		0	0	0¼

For the second line, 2¼ inches to 3 inches wide, assuming the average width as 2½ inches, the calculation would be as under:—

		£	s.	d.
100 ft. super. of ½ in. deal, at ¾d. per foot		0	6	3
Add 5 per cent.		0	0	3¾
4 cuts = 400 ft. at 9d.		0	3	0
Lineal feet produced	500)	0	9	6¾
Per foot run		0	0	0¼
100 ft. super. of 2 in. deal, at 2⅛d.		0	17	8½
Add 5 per cent.		0	0	10⅝
4 cuts 400 ft., at 9d.		0	3	0
Lineal feet produced	500)	1	1	7⅛
Per foot run		0	0	0½

Small timbers under 3 inches square commonly appear in a bill of quantities in lineal, and timbers 2 inches thick and under, in superficial dimensions, the theory being that they should be separately stated, because of the larger proportion of sawing to the cubic foot—½d. per foot cube generally more than covers the extra cost. The estimator commonly prices such items at the same rate per foot as the timber adjacent, and if an average of sawing has been adopted, as before suggested, the procedure is the more reasonable one.

The following are a few examples of the application of the constants. The labourer's time need not be considered, except in the case of specially large timbers.

	s.	d.
Fir in ground joists	2	2¼
Labour (carpenter) fitting and fixing, ·050 at 8s. 9d. ..	0	5¼
Nails	0	0¼
Cost per foot cube	2	7¾
Fir and Deals in ground joists	1	7¼
Labour (carpenter) fitting and fixing, ·050 at 8s. 9d. ..	0	5¼
Nails	0	0¼
	2	0¾

For the foregoing either the 2s. 2¼d. or the 1s. 7¼d. may be adopted as the price per foot cube, according to the judgment of the estimator and the terms of the specification.

	s.	d.
Fir framed in floors	2	2¼
Labour (carpenter) fitting an fixing, ·066 at 8s. 9d. ..	0	7
Nails	0	0¼
	2	9½

PRICES.

	s.	d.
Fir framed in roof trusses, exclusive of hoisting	2	2¼
Labour (carpenter) fitting and fixing, ·128 at 8s. 9d.	1	1
One-third of a labourer's time = ·041 at 5s. 10d.	0	3
	3	6¼

The pricing of an item in this form is but a guess. Fir wrought and framed in roofs would commonly comprise rafters and purlins. Assuming the rafters as 4½ inches by 2½ inches (a common size), about 13 feet lineal would be required to each cubic foot; 13 feet × 11½ inches = 12½ feet superficial × ·010 × 105 = 13d. per foot cube. Assuming purlins to be 8 inches × 5 inches, about 4 feet lineal would be required to each cubic foot; 4 feet × 1 foot 9 inches = 7 feet superficial × ·010 × 105d. = 6½d. per foot cube, which would tend to reduce the average price somewhat. The proportion of purlins to common rafters in an ordinary roof is as one to three, consequently—

	d.
Purlin	6½
Common rafters 3 × 13d.	39
4) 45½	
11¼ average.	

If we take a king-post roof truss of ordinary scantlings to a 20 feet span, we find that it contains about 11 cubic feet of fir, and that the wrought surface is about 90 feet superficial, equal about 8 feet of planing to each cubic foot of fir. As the spans of the trusses increase, the timbers are of larger scantling, and the proportion of planing is lessened; 8 feet is, however, not an unreasonable average to adopt, therefore 8 × ·010 × 105d. = 8½d. per foot cube, and adding this amount to the last example, we have 3s. 6¼d. + 8½d. = 4s. 2¾d. per foot cube.

If of finished sizes add 1¼d. per foot cube, as shown in a former item.

If the hoisting is included in the description of roof-trusses it will be necessary to adopt a price per foot cube for uniform application, although the trouble of hoisting will vary with the circumstances of the contract: 10d. per cube foot is the price commonly adopted.

Sometimes the labour of hoisting is saved by framing the roof-trusses on a platform at the level of the wall-plates.

The fairest way of presenting the item of hoisting in a bill of quantities is to separately state it thus:—

"No. 5. Hoisting and fixing roof trusses, 30 feet by 10 feet, and 40 feet from ground to ridge."

Fir framed in circular ribs to roof trusses occurs in a bill of quantities in diverse ways, as " fir framed in circular ribs to roof trusses," " fir framed and wrought in circular ribs to roof trusses," " fir framed and wrought in circular ribs to roof trusses, including grooved and rebated joints and dowelling with oak dowels."

The most lucid way is to present them as first described, and take out all the labours on them. Unless this is done the pricing must be uncertain.

In either case the timber has been measured of the superficial area sufficient to contain the various component pieces of the rib, so that no further waste need be allowed for, and the price for the timber in previous items may be adopted. The calculation would be as follows:—

	s.	d.
Fir framed in circular ribs to roof trusses	2	2¼
Labour, carpenter, ·123 at 105d.	1	1
Labourer, ·041 at 70d.	0	2¾
Per foot cube	3	6

Fir *wrought* and framed in circular ribs to roof trusses:—

	s.	d.
Fir	2	2¼
Carpenter as last	1	1
Labourer	0	2¾
Planing	0	8¼
Per foot cube	4	2¼

Fir framed in circular ribs to roof trusses, including sunk edges and grooved and rebated joints and dowelling with oak dowels:—

	s.	d.
Fir	2	2¼
Carpenter ·250 at 105d.	2	2¼
Labourer ·041 at 70d.	0	2¾
Oak dowels	0	1
Per foot cube	4	8¼

When such ribs as the foregoing are obtained out of timber of

PRICES.

unusual depth, it is, or should be, stated in a bill of quantities and the price per load for extra size being decided on, the calculation would be as follows, adopting the method before stated. supposing the increase of cost to be 1*l*. per load:—

	s.	*d.*
Extra cost of fir	20	0
25 per cent. waste for slabbing	5	0
5 per cent. waste for cross-cutting	1	0
50) 26	0
Per foot cube	0	6¼

As before remarked, the extra sawing on small timbers increases the value a little, and there is more labour in fixing. Ceiling joists may be taken as an instance of this, as follows: "500 feet run 3½ inches by 2¼ inches framed ceiling joists." Although these would undoubtedly be obtained out of deals, if an average of the timber and deals in a building has been adopted for estimating purposes as before suggested, it is obvious that the *average* price should be applied—

	s.	*d.*		*s.*	*d.*
Fir	2	2¼	or	1	7¼
Carpenter ·100 at 105*d.*	0	10		0	10
Nails	0	0¼		0	0¼
	3	0½		2	5½

Centring.—This is an item the price of which varies considerably in different estimates, as does its cost also, according to the circumstances of the work. Much of it is made of old material, and whether new or old, when not much cut, is capable of future use. In the use of flat-boarded centring for large areas, an intelligent labourer takes the place of a carpenter in laying the boarding and renders important help in fixing the struts. He furnishes the same assistance, but in a less proportion, in the preparation of centring to vaulting.

The simplest kind of centring would be described as follows:—

<small>squares. ft.</small>
"100   50 supl.   Flat-boarded centring to concrete floors and horsing to 12 ft. story."

An apartment 20 feet by 20 feet, as sketch A, the area of which

QUANTITY SURVEYING.

is four squares, will illustrate the method of investigation, if new material be used according to the above description.

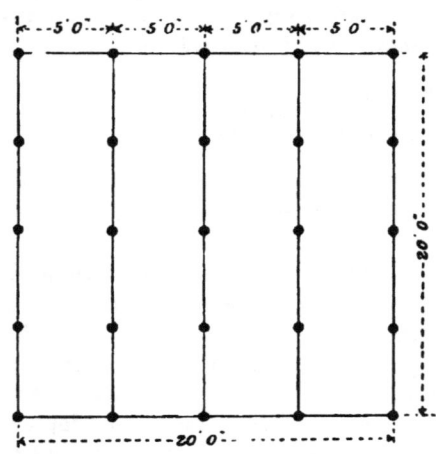

```
           |<--5'0"-->|<--5'0"-->|<--5'0"-->|<--5'0"-->|
```

(diagram of scaffold grid 20'0" × 20'0")

				£	s.	d.
4/20 0	80	0	Sill on floor.			
3/30 0	60	0	Cross-sills ditto.			
5/20 0	100	0	Heads.			
25/12 0	300	0	Struts.			
	540	0				
	4					
	3		45 ft. cube fir at 1s.	2	5	0
20 0						
20 0						
	400	0	1 in. rough boarding at 8s. per square at docks	1	12	0
			⅕th of a standard cartage at 10s.	0	2	0
			4)	3	19	0
					19	9
			Use and waste in conversion 10 per cent. =	0	2	0
			Percentage on outlay 5 p.c. P.A.			
			Depreciation .. .. 5 p.c. P.A.			
			Say 2 months (⅙th of a year) at .. .. 10 p.c. on 19s. 9d. =	0	0	4
				0	2	4

PRICES. 617

Then :—

	£	s.	d.
Material ..	0	2	4
Carpenter, ·60 at 105d.	0	5	3
Labourer, ·60 at 70d.	0	3	6
50 2 in. nails, weight say ¼ lb., at 1¾d.	0	0	0¾
Cost per square	0	11	1¾

Ordinary contract prices for such work vary from 11s. 6d. to 15s. per square.

Centring to Vaults.—Analysis of such an item may be conducted in a manner similar to the last. Adopting, as an example, a vault of section as sketch B, and 20 feet long, the calculation and description would be as follows :—

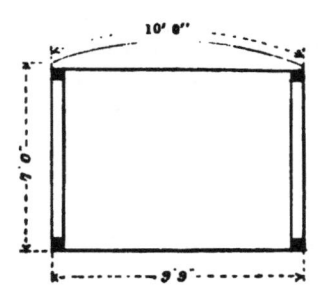

```
         sqrs.  ft.  in.
                      super. centring to vault and horsing to
         20    0       7 ft. story
         10    0
         ——  200   0    1 in. rough boarding as before, at 10s. 0d.
                      square  .. .. .. .. .. .. ..   20  0
                              4/20  0  =  80  0
                              10/ 6  6  =  65  0
                                          ———
         145   0                           145  0
                4
                3    12   1 Fir in struts at 1s.  .. .. .. ..  12  1
         ———
         5/9   9
               11
         ——   44    8  1¼ in. rough deal at 1⅜d. .. .. ..  5  1½
                                                 Ribs ————
                                                    2)37  2½

                 Per square .. .. .. .. .. ..  18  7¼
```

	s.	d.
10 per cent. waste on 18s. 7½d.	1	10½
Two months at 10 per cent. per annum Use and waste .. .. }	0	3¾
Nails	0	1
Carpenter, 1·00 at 105d.	8	9
Labourer, ·50 at 70d.	2	11
Per square	13	11¼

Ordinary contract prices for this work vary from 12s. to 15s. per square.

Centring to Openings (per foot superficial).—To analyse this price take an opening, the width of which must be assumed—say 3 feet 6 inches—and as the surveyor measures by the foot superficial any soffit the width of which exceeds 9 inches, the centre would require two ribs.

Centring for ordinary openings is rarely used a second time; if it is, the cost of cartage to another work, and of its alteration, more than counterbalances the value of the material, consequently the whole value of the material should be used for the calculation of cost.

						d.
2/3	6 5	2	11	1 in rough Ribs. deal at 1⅛d.		3⅜
23/1	0	23	0	1 in. by 1 in. lagging, 1 in. apart, ⅛d.		2⅞
						6¼

This centre would be 3 ft. 9 in. × 1 ft. = 3 ft. 9 in. super. for 6¼d., or 1¾d. per foot super.

	d.
Then, per foot super. material	1¾
Labours (·030 at 105d.)	3¼
Nails	0¼
Cost per foot super.	5¼

Centring to 9-inch soffits would require two ribs, as last, and the lagging would be less. The application of the last process is obvious.

Turning piece measured per foot run by the surveyor would require only a single rib and no lagging. One of the ribs in the last calculation is a reasonable example.

	d.
Material	½
Labour (carpenter, ·011 at 105d.)	1¼
Per foot run	1¾

Large centres should appear as numbers in a bill of quantities and described in something like the following manner:—

No. centres to Gothic pointed arch 8 ft. 2 in. span, 5 ft. rise, 13 ft. 6 in. around, and 12 in. soffit.

The estimator can then roughly draw them, take out the quantities, and price the items.

Arches for gauged work are usually close-boarded instead of lagged, and are consequently worth a little more for material; the labour is about equal to that on the lagged centres.

Deal cradling, fitted to iron, per foot superficial.—This is formed of rough fillets, about 2 inches by 2 inches, nailed together and about 12 inches apart.

	s.	d.
1 ft. 0 in. run, 2 in. by 2 in. fillet (see table, Rough Fillets)	0	0¼
Labour, ·028 at 8s. 9d.	0	2¾
Nails	0	0¼
Cost per foot super.	0	3¼

Deal bracketing for cornices, per foot superficial.—The labour on this work varies, and is dependent on the profile of the cornice. The material should not be less than 1½ inches thick, nor should the brackets be more than 12 inches from centre to centre. The profile of each bracket is a rough approximation to that of the cornice. A common proportion is one-third on the wall and two-thirds on the ceiling. A fillet at top and bottom of brackets is generally necessary, but these are separately measured. Two brackets like those of sketch would be cut out of a piece of deal 11 inches by 6 inches, and there would be one to each foot.

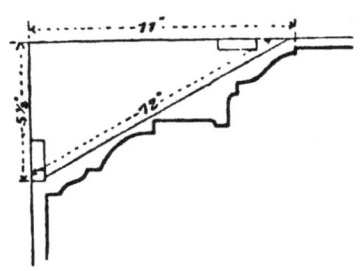

	s.	d.
3 in. super. of 1½ in. deal at 1⅝d.	0	0¼
Waste 10 per cent. say	0	0¼
Labour ·024 at 8s. 9d.	0	2¾
Nails	0	0¼
Cost per foot super.	0	3¼

Angle brackets to bracketing, 12-inch girth.—Two brackets would be sawn out of 16 inches by 6 inches.

	s.	d.
4 in. super. of 1½ in. at 1⅝*d*.	0	0¼
10 per cent. waste	0	0¼
Labour ·048 at 8*s*. 9*d*.	0	5
Nails	0	0¼
Cost each	0	5¾

The foregoing are mere triangular pieces of deal, and are the most frequent shape; but if the cornices are much undercut they would be notched and roughly shaped, and although the above is the theoretical value, the ordinary estimator prices them somewhat higher to be safe, about 50 per cent. more.

" . . . . *¾-inch by 2-inch slating battens spaced for countess slating*," per square.—¾-inch by 2 inch is an imported size, and may be bought for 1*s*. 6*d*. per hundred feet at saw-mills. The ordinary gauge of countess slates is 8½ inches, and this would be the distance apart of the battens (from centre to centre).

	s.	d.
150 ft. of ¾ in. by 2 in. battens, at 1*s*. 6*d*. per hundred ..	2	3
5 per cent. for waste	0	1¼
135 nails plus 5 per cent. = 142 nails 1½ in., weight ½ lb. at 1½*d*.	0	0¾
Labour ·17 at 8*s*. 9*d*.	1	5¾
Cost per square	3	10¾

The prices for battens to slating of smaller or larger size than countesses may be analysed by drawing the rafters and battens to scale, measuring the battens, and counting the nails.

" . . . . *Inodorous felt, and allow for laps and nailing with clout nails*," per square.—A common contract price is 1*d*. per foot superficial, or 8*s*. 4*d*. per square. The ordinary merchant's price is 1*d*. per square foot, and in quantity they will allow 25 to 40 per cent. off; for this calculation, let us say 33 per cent. off. The felt is 32 inches wide; the laps are about 1 inch; the nails ⅞-inch clout, about 6 inches apart, and should weigh 2⅛ lb. per thousand

	s.	d.
100 ft. super. of felt	5	7¾
Waste 5 per cent.	0	3⅜
Laps, $\frac{1}{32}$nd of 5*s*. 7¾*d*.	0	2⅛
Labour, ·18 at 8*s*. 9*d*.	1	7
150 nails at 1*s*. per thousand	0	1¾
Cost per square	7	10½

Boarding for Slating or Tiling.—1-*inch rough boarding for slating, 7 inches widths.*—In such items as this, waste consequent upon *irregular* cuttings, and for cuttings around openings like skylights or dormers, is allowed in the measurements; but a further waste occurs in adapting to a building. The width most frequently used is 7 inches.

	s.	d.
1 in. rough boarding per square at docks	8	0
Landing rate ₁/₄₀th of a standard at 3s. 9d.	0	2¼
Cartage, ₁/₂₀th of a standard at 10s.	0	6
Further waste, 10 per cent. of 8s. 8¼d.	0	10½
2½ lb. 2 in. nails (about 300, which includes 5 per cent. for waste) at 1¼d.	0	3¼
Unloading	0	1
Laying ·33 at 8s. 9d.	2	10¼
Cost per square	12	9½

1-*inch rough boarding traversed for lead and firring to falls.*—

	s.	d.
1-in. rough boarding at docks	8	0
Landing rate	0	2¼
Cartage	0	6
Laying	2	7¼
Traversing ·15 at 8s. 9d.	1	3¼
Waste 10 per cent.	0	10½
Unloading	0	1
2½ lb. nails at 1¼d.	0	3¼
Firrings, as under	9	3¼
Cost per square	£1 3	1¼

Firrings.—At the ordinary distance apart there would commonly be 9 joists in a width of 10 feet, and for a square they would be 10 feet in length; there would, consequently, be 90 lineal feet of firring to each square. In a long flat with the drips disposed across its length, the firring at the higher level would be deep, but flats are generally arranged to fall in the direction of the shorter of their two dimensions, and the firring is consequently shallower. Surveyors usually include the firring with the description of the boarding, and under ordinary conditions this does fairly well. When the firrings are very deep, pricing would be more exact if they were measured and billed separately, stating the average depth. 2-inch deal is commonly used.

Sometimes when firrings are specially deep the surveyor bills them and the boarding separately, and they appear as *squares superficial deep firrings to flats.*—Taking as an example a flat 10 feet wide, with a fall of 1½ inches in 10 feet, the firring pieces would probably be 1 inch in height at one end, and 2½ inches at the other, and as two would be sawn out of a piece 3½ inches deep, the average depth may be taken as the material used—i.e. 1¾ inches deep, we should consequently have 90 feet run 1¾ inches by 2 inches; but, unlike ordinary fillets, the sawing would be done by hand instead of machine.

Referring to the table before given for value of deal per superficial foot, we find 2⅛d. for 2 inches = 1¾ inches by 2 inches would consequently be about ⅐th or ⅜d. The nails would average 3½ inches long, and about 126 + 5 per cent. = 133; weight about 3¼ lb. would be required. The calculation for cost would be as follows:—

	s.	d.
90 ft. run 1¾ in. by 2 in. at ⅜d.	2	9¾
3¼ lb. nails at 1¼d.	0	4
Cutting out, fitting, and fixing, ·70 at 8s. 9d.	6	1½
Cost per square	9	3¼

¾-inch Sound-boarding and deal fillets (the joists not deducted) per square:—

	s.	d.
¾-in. rough boarding at docks, per square	6	0
Landing rate $\frac{1}{27}$th standard at 3s. 9d.	0	1¾
Cartage $\frac{1}{27}$ standard at 10s.	0	4½
Further waste, covered by the excess of measurement (joist measured in)		
180 ft. run, 1 in. by 1¼ in. rough fillet at ¼d. (see table of Fillets)	3	9
Fitting and fixing boarding and fillets ·80 at 8s. 9d. ..	7	0
2 lb. nails at 1¼	0	2½
Cost per square	17	5¾

Such an item as the foregoing is often priced below this rate, sometimes as low as 10s. per square, as the work is rarely criticised, and not seldom is of old or refuse material which would otherwise be burned or thrown away.

1-*inch Gutter boards and bearers, per foot superficial.*—The bearers are often described as framed. This is but rarely done.

For this the commonest rough boarding is used. The ordinary gutter increases about 1 inch in width for each inch of rise, but this is allowed in the measurement. The waste for raking, cutting on each edge, must be allowed for; about one-sixth of a foot is sufficient.

	s.	d.
1⅛ ft. of 1 in. rough boarding at 8s. 9d. square	0	1¼
Labour to boarding and bearers, ·030 at 8s. 9d.	0	3¼
Nails	0	0¼
Bearers, say 3 in. by 2 in. by 20 = 1⁰⁄₁₂ in. cube at 1s. 7¼d. per foot	0	1½
Cost per foot..	0	6¼

The woodwork of dormer cheeks sometimes appears in a bill of quantities in two places, thus:—" feet superficial 1-inch rough boarding in small quantities and spandrel shapes in dormer cheeks, measured net"; and the timber, " feet cube fir in small quantities in dormer cheeks." In other bills it is put together thus: " feet superficial 1-inch rough boarding and small fir-framed quarters in small quantities and spandrel shapes in dormer cheeks, measured net." In either case the product of the length of the raking edges multiplied by 3 inches would be added to make up the total of the boarding.

The value of the item as last described is as follows:—

	s.	d.
1 ft. super. 1 in. rough boarding at 8s. 9d., as before	0	1¼
1 ft. run 4 in. by 2¼ in. = ₉⁄₁₂ in. cube at 1s. 7¼d.	0	1¼
Labour (equal to gutter-boards and bearers), ·030 at 8s. 9d.	0	3¼
Nails	0	0¼
Cost per foot super.	0	6

The above labour is an analogy, but may be analysed as follows:—

	s.	
Laying the boarding, per square, about twice as much as in large surface—i.e. ·66 at 8s. 9d. = per foot (nearly)	0	0¾
Labour to the timber per foot cube, about thrice the value of partitions per foot cube, and ₉⁄₁₂ in. cube, ·30 at 8s. 9d.	0	2
Cost of labour per foot super.	0	2¾

The valuation of items measured lineally is mainly based upon the tables previously given. Being commonly of small scantling, the table of fillets will be found useful. The respective sectional area is another useful element of comparison. Thus, 3s. per cubic foot is ¼*d.* per inch of sectional area; 4 inches by 3 inches = 12 inches at 3*s.* per cubic foot = 3*d.* per foot run; 1*s.* 6*d.* per foot cube is ⅛*d.* per inch of sectional area. A similar scantling would be consequently 1½*d.* per foot run. This principle is applied in the valuation of small timbers, mouldings, solid frames, moulded architraves, &c.

Rolls for lead are commonly bought ready for fixing at a moulding mill.

The prices of ordinary lineal labours have been already given. A few examples of treatment of lineal items of labour and materials are as follows:—

Plugging.—Plugs are commonly driven into the brickwork about 18 inches apart. 1*d.* per foot run is an ordinary contract price. The value of the material is so small that it is not worth consideration. Constant ·088 at 8*s.* 9*d.* = ¾*d.*

½-inch by 4½-inch deal pads and building-in as wood bricks, per foot run.—These are usually 9 inches long. A common contract price is 1*d.* per foot.

	s.	d.
½ ft. super. ½-in. rough deal at ½*d.*	0	0¼
Building-in	0	0½
Cost per foot run	0	0¾

2-inch cross-rebated drip, per foot run.

	s.	d.
1 ft. 2 in. by 1 in. rough fillet	0	0⅛
1 ft. labour, rebate cross grain ·009 at 8*s.* 9*d.*	0	1
Fixing and nails	0	0½
Cost per foot run	0	1⅝

2-inch deal roll for lead, per foot run, costs 1*d.* per foot run at saw-mills—

	s.	d.
1 ft. of roll	0	1
Cutting to lengths, and nailing, and nails and waste ..	0	0¾
Cost per foot run	0	1¾

2-inch deal roll birdsmouthed, per foot run—

	s.	d.
1 ft. of roll	0	1
Labour to birdsmouth, ·003 at 8*s.* 9*d.*	0	0¼
Cutting to lengths, nailing, nails, and waste	0	0¾
Cost per foot run	0	2

Mitres to ditto.—Estimators commonly adopt some fraction of the price of a foot run as the value of mitres. Three-quarters of a foot is reasonable for a 2-inch roll = 1½*d.*

4 inches by 2 inches rough fillet nailed, per foot run.—The fact that this item is of small scantling need not be considered, as an average of sawing has been adopted which is applicable throughout.

	s.	*d.*
1 ft. of 4 in. by 2 in. fillet at 1*s.* 7½*d.* per foot cube ..	0	1⅛
Two 3½ in. nails	0	0⅛
Fixing ·003	0	0¼
Cost per foot run	0	1½

4 inches by 2 *inches feather-edged springer, per foot run.*—

	s.	*d.*
1 ft. of 4 in. by 2 in. springer (two out of 4 in. by 2 in.) half value of last	0	0⅝
Extra sawing	0	0¼
Nails and fixing as last	0	0⅜
Cost per foot run	0	1¼

Eaves fillets of similar size are equal in value to last.

3½ *by* 2¼ *inches fir-framed ceiling joists, per foot run.*—

	s.	*d.*
1 ft. of 3½ in. by 2¼ in. = 8 sectional inches at 1*s.* 7½*d.*	0	1⅛
Labour, ⅜ of an inch cube at ·100 per cubic foot at 8*s.* 9*d.*	0	0½
Nails	0	0⅛
Cost per foot run	0	1¾

2 *inches by* 2 *inches herring-bone strutting to* 9-*inch joists, per foot run*—

	s.	*d.*
2 ft. 2 in. by 2 in. rough fillet (see table) at ¼*d.*	0	0½
Labour, ·014 at 8*s.* 9*d.*..	0	1¼
Nails	0	0⅛
Cost per foot run	0	1⅞

The waste in cross-cutting is more than covered by the practice of measuring the joists in.

A few examples of items which appear in estimates as numbers are as follows:—

Short lengths of 2-*inch cross-rebated drip, each.*—A similar item appears in the foregoing lineal example: but when so measured the estimator may reasonably assume that it is in fairly long lengths. Short lengths are usually drips in gutters, and their length may be taken as averaging 1 foot 3 inches.

	s.	d.
1¼ ft. of 2 in. by 1 in. rough fillet	0	0¼
1¼ ft. labour to rebate, cross grain, ·009 at 8s. 9d. ..	0	1¼
Fixing and nails	0	0½
100 per cent., add for short length	0	2
Cost each	0	4

Short lengths of 2-*inch deal roll for lead, each.*—These bear the same relative value to the lineal item as the last: 1¾*d.* by 2½ = 4½*d.* each.

Fir-framed bridging pieces 4 *inches by* 3 *inches, and about* 15 *inches long, framed at each end between joists, each*—

	s.	d.
5 in. super. of 3 in. deal at 3*d.*	0	1¼
2 dovetail halving, ·030 each at 8s. 9d.	0	6¼
Cost each	0	7½

Deal sprockets, two out of 18 *inches by* 4 *inches by* 2¼ *inches, each*—

	s.	d.
3 in. super. 2¼ in. deal at 2⅜*d.*	0	0¾
19 in. run of sawing to 2¼	0	1
Cutting to lengths, nailing, nails, and waste	0	1
Cost each	0	2¾

Filleting soffits of trimmers for lathing.—The foregoing item presupposes the removal of the centring to trimmer, whereas it is cheaper to leave it in.

One way of doing the work is to fix fillets extending from trimming joists to trimming joists. The size of the arch has to be assumed, ordinarily 1 foot 6 inches wide, and 4 feet 6 inches long. Two fillets or ceiling joists are sufficient.

	s.	d.
2/ 4 ft. 6 in. = 9 ft. of 3 in. by 2 in. rough fillet ½*d.* ..	0	4½
Fixing 9 ft. at ½*d.*	0	4½
Nails	0	0¼
Cost each	0	9¼

1½-*inch deal dovetailed cesspools,* 9 *inches by* 9 *inches by* 6 *inches, holed and fitted, each.*—A common contract price is 2*s.* 6*d.*

```
0 10 in.
0 10    0  8 in.
─────
3  8
   6    1 10
```
                                                                    s.   d.
            2 6 of 1¼ in. deal at 1⅜d. 0 3¼
2 0 dovetailed angle at 6d. 1 0
      ─────
            1/ dished hole 0 3
Fixing and fitting 0 6
                                                                    ─────
                        Cost each 2 0¼

Fixing only, ironwork.—In official schedules the fixing of straps, bolts, etc., to wooden structures is priced per lb., in the case of bolts the price per lb. decreasing as the weight increases. Flitches, kingheads, shoes, &c., at per cwt. See Constants.

Some estimators price the fixing of bolts 12 inches and under so much each, over 12 inches a certain additional price per foot run.

Fixing to oak, teak and pitch-pine is worth about 50 per cent. more than to fir.

JOINER.

In the pricing of joinery it should be observed whether or not it is to be of finished sizes, whether the mouldings are worked on solid or worked to detail, as the increase of cost is considerable in cases where mouldings cannot be obtained from a manufacturer's stock.

Take care to preserve a reasonable analogy between the prices. In framings, for instance, start with a certain value for square framing, and add to that the various further labours upon the piece of work in question. Estimators sometimes disregard all the extra labours when pricing a bill, putting such price on the superficial items as will cover them.

The broad distinction between carpentry and joinery is that the former is generally framed and prepared at the building, while joinery is prepared at the contractor's workshops and brought to the building ready for fitting and fixing. Its cost would, therefore, properly involve a larger percentage for establishment charges than other portions of the building; but this, save in exceptional circumstances, it is not the custom to consider, one uniform rate being applied to all the items of an estimate, as before recommended.

The material for carpentry, as before remarked, is often of very common quality, possibly because it is nearly all concealed; but the wood used for joinery must be fairly good, as it receives a higher finish, is constantly in view, and its defects more easily judged. Moreover, the warping and cracking of material of poor quality and without seasoning would be at once apparent. It need hardly be said that the architect who defers his examination of the joinery of a building until its delivery on the work puts himself at a disadvantage. The artistic stopping and priming sometimes done before it leaves the workshops makes it very difficult to discover defects.

In the production of joinery the builder derives more help from machinery than in any other branch of his trade. Sawing, planing, mortising and tenoning, when there is sufficient work to justify the use of machinery, are all done by its means. The mouldings are "stuck" by the moulding-machine, while the profiles of shaped pieces of wood are cut by the band-saw with great speed and accuracy.

When the builder has not a sufficient business to justify the extensive use of machinery, he contents himself with a circular saw and a general joiner, or some such machine, and obtains long lengths of mouldings, door-frames, beads, grounds, parts of sashes and sash-frames, &c. from a moulding mill, which would, on directions given, supply both labour and material for items of this kind.

There are several firms in London and its vicinity who make sub-contracts with the original contractor for the whole of the joinery of a building, most of whom produce work of very good quality, and often at less price than it could be done in the contractor's own shop.

The preliminaries to the valuation of joinery must be a list of constants of labour and a table of cost of deal in thicknesses, similar to that given in the trade of carpenter.

Much of the wood used for joinery is brought into the market machine-planed, and only requires smoothing; while for the commoner parts of joinery work, prepared floorings can often be conveniently used at a saving of cost.

Properly seasoned wood is often difficult to get, and always commands a high price, consequently those building firms who have the reputation for good joinery keep a large stock of deals and battens in stock, so as to insure well-seasoned wood for their work, and when the wood is well seasoned it is worth more by the interest on the outlay for the time it has been kept.

The preamble of a bill of good work will stipulate that the deals for joinery shall be the best St. Petersburg or Onega yellow. The panels, the best bright St. John's or Quebec pine. The St. Petersburg best would be marked with a crown over the word "Gromoff," the name of the shipper; or "C. & Co.," for Clarke & Co.; or a crown between the letters "P. and B.," the mark of the shippers, A. P. Belaieff; or "P. B. S. & Co.," the mark of the shippers, Peter Balaieff's Successors & Co. The best St. Petersburg deals generally cost, at the dock sales, about 13*l.* per St. Petersburg standard.

The pine would be yellow. The best is known as 1st bright pine deals. Gilmour or Booth are shippers of the best; they commonly cost, at the dock sales, about 20*l.* per St. Petersburg standard. The use of pine for panels is almost universal, as its width enables the tradesman to dispense with a glued joint, which, if deals were used, would generally be necessary. Its fine grain and facile working render it most convenient for mouldings, for which it is much used by the moulding makers. For panels of work circular on plan, "Basswood" (American lime) is most useful, and produces good work. It need hardly be said that a great quantity of deal and pine is sold at the dock sales at much below the above prices, and that it is all converted into joinery for some destination. Nothing but mature experience and naturally good judgment will enable the estimator to assess the price of the material.

The above price will, however, be adopted for the following calculations, and the total price for deal will consequently be as follows:—

	£	s.	d.
Joiners' deals at docks, per St. Petersburg standard	13	0	0
Landing rates	0	3	9
Unloading	0	1	6
Cartage by sawmill proprietors, as before.			
Two years' compound interest at 5 per cent. per annum on the total of the first three items, 13*l.* 5*s.* 3*d.*, two years being assumed as the time required for thorough seasoning	1	7	3
Waste in sawing and conversion, 10 per cent. on 14*l.* 12*s.* 6*d.*	1	9	3
Cost per standard	16	1	9

In ordinary framed work, as doors and partitions, the cost of the pine used would but slightly exceed that of good joiner's deals, and the difference has been disregarded,

Analysis of cost of a St. Petersburg standard of rough deal sawn to various thicknesses :—

Assumed to contain 74 times 12 feet × 9 × 3 inches = 666 super. feet of 3 inches, equal 166 feet 6 inches cube, a little in excess of the actual quantity (165 feet cube).

	£	s.	d.	
¼" = 11 cuts in 3" × 9" (producing 12 thicknesses of ¼") = 814 cuts = 67⅚ doz. at 2*s.* 3*d.* ..	16	1	9	deal.
	7	12	7½	sawing.
7992 ft.)	23	14	4½	
	¾*d.* per ft.			
½" = 5 cuts in 3" × 9" (producing 6 thicknesses of ½") = 370 cuts = 30⅚ doz. at 2*s.* 3*d.*	16	1	9	deal.
	3	9	4½	sawing.
3996 ft.)	19	11	1½	
	1¼*d.* per ft.			
¾" = 3 cuts in 3" × 9" (producing 4 thicknesses of ¾") = 222 cuts = 18½ doz. at 2*s.* 3*d.*	16	1	9	deal.
	2	1	7½	sawing.
2664 ft.)	18	3	4½	
	1⅝*d.* per ft.			
1" = 2 cuts in 3" × 9" (producing 3 thicknesses of 1") = 148 cuts = 12⅓ doz. at 2*s.* 3*d.*	16	1	9	deal.
	1	7	9	sawing.
1998 ft.)	17	9	6	
	2⅛*d.* per ft.			
1¼" = 2 cuts in 3" × 9" (producing 2 thicknesses of 1¼" and 1 of ½") = 12⅓ doz. at 2*s.* 3*d.*	13	8	1½	deal.
	1	0	9¾	sawing.
1332 ft.)	14	8	11¼	
	2⅝*d.* per ft.			

Note.—One-quarter of the total of the sawing last mentioned is allotted to the ½", leaving three-quarters (the remainder) for the calculation. The deal is reduced by one-sixth.

	£	s.	d.	
1½" = 1 cut in 3" × 9" (producing 2 thicknesses of 1½") = 74 cuts = 6⅙ doz. at 3*s.* 9*d.*	16	1	9	deal.
	1	3	1½	sawing.
1332 ft.)	17	4	10½	
	3⅛*d.* per ft.			

	£	s.	d.	
1¾" = 1 cut in 3" × 9" (producing 1 thickness of 1¾" and 1 thickness of 1¼") = 74 cuts = 6⅙ doz. at 3*s.* 9*d.*	9	7	8¼	deal.
	0	11	6¾	sawing.
666 ft.)	9	19	3	
	3⅝*d.* per ft.			

Note.—One-half of the total of the sawing last mentioned is allotted to the 1¼", and the deal reduced by $\frac{5}{12}$.

	£	s.	d.	
2" = 1 cut in 3" × 9" (producing 1 thickness of 2" and 1 of 1") = 74 cuts = 6⅙ doz. at 3*s.* 9*d.*...	10	14	6	deal.
	0	11	6¾	sawing.
666 ft.)	11	6	0¾	
	4⅛*d.* per ft.			

Note.—One-half of the total of the sawing last mentioned is allotted to the 1" and the deal reduced by ⅓.

	£	s.	d.	
2¼" = 1 cut in 3" × 9" (producing 1 thickness of 2¼" and 1 of ¾") = 74 cuts = 6⅙ doz. at 3*s.* 9*d.*...	12	1	3¾	deal.
	0	11	6¾	sawing.
666 ft.)	12	12	10½	
	4⅝*d.* per ft.			

Note.—One-half of the sawing last mentioned is allotted to the ¾", and the deal reduced by ¼.

	£	s.	d.	
2½" = 1 cut in 3" × 9" (producing 1 thickness of 2½" and 1 of ½") = 74 cuts = 6⅙ doz. at 3*s.* 9*d.*	13	8	1½	deal.
	0	11	6¾	sawing.
666 ft.)	13	19	8¼	
	5*d.* per ft.			

Note.—One-half of the sawing last mentioned is allotted to the ½", and the deal reduced by ⅙.

	£	s.	d.	
2¾" = 1 cut in 3" × 9" (producing 1 thickness of 2¾" and 1 of ¼") = 74 cuts = 6⅛ doz. at 3s. 9d.	14	14	11¼	deal.
	0	11	6¾	sawing.
666 ft.)	15	6	6	
	5½d. per ft.			

Note.—One-half of the sawing last mentioned is allotted to the ¼", and the deal reduced by $\frac{1}{12}$.

3". No sawing .. .. 666 ft.)	16	1	9	deal.
	5¾d. per ft.			

Some estimators, in addition to the table of value of rough deal, prepare a table of the various thicknesses with ordinary labours upon them like the following, adopting the prices of deal in the last table given. Such a table is found in the majority of good schedules of prices, and is not only easily applied, but is a good means of comparison with prices otherwise calculated.

Cost of deal in thicknesses, based on the prices before calculated.—

	s.	d.
½" rough deal, exclusive of fixing	0	1¼
Ditto, including rails and fixing	0	2
Ditto, edges shot	0	2¼
Ditto, wrought one side and one edge	0	3¼
Ditto, wrought two sides and two edges	0	4
Ditto, wrought one side, and framed, clamped and dovetailed	0	6
Ditto, wrought both sides	0	6¾
Add to either if ploughed and tongued	0	0¾
Ditto if cross-tongued	0	1

Complete the table by using the various thicknesses, as ¾", 1", 1¼", &c.

A table of fillets at the same rate for deal (16*l.* 1*s.* 9*d.* per standard) will also be necessary. The principle of its construction will be the same as before described in the section Carpenter, and the result as follows :—

Fillets.—

—	½ in.	¾ in.	1 in.	1¼ in.	1½ in.	2 in.
	d.	*d.*	*d.*	*d.*	*d.*	*d.*
2 in. wide and under, per foot run, rough	¼	¼	¼	⅜	⅜	½
Ditto, wrought	½	½	½	¾	¾	1
2½ in. to 3 in. wide	⅜	½	½	⅝	¾	1
Ditto, wrought	¾	1	1	1	1½	1½
Add if fixed with screws	¼	¼	¼	½	½	½

The smallness of difference in value between rough and wrought makes any nicety of distinction in their valuation impossible and unnecessary.

The following difference between straight and circular fillets is adopted in ordinary schedules:—

Add to fillets, "if bent circular, one-fourth the foregoing rates"; "if cut circular on both edges, once the foregoing rates"; if cut circular on one edge, half the foregoing rates."

Screws.—Nails have been considered in a previous section of this series. Screws are much more used in joinery than in carpentry, and their value may reasonably be mentioned here.

Their variety is considerable, and those used by the carpenter or joiner are sometimes by them called "wood-screws," possibly to distinguish them from those used for metal. The old shape of screw (with a blunt point) has been almost entirely superseded by Nettlefold's patent pointed screw, which is pointed like a gimlet, and acts in the same way.

The screw merchants publish a screw list, the prices in which are invariable, but subject to a discount, sometimes, but not often, changed. This discount is now 60 per cent. off iron, and 50 per cent. off brass, and for specially large quantities another 2½ per cent.; but this latter does not concern us, as it is practically a cash discount.

The numbers indicate the diameter of the screw, the first number being the smallest in each series, except ¼ and ⅜-inch, which include screws of still smaller diameter than No. 1, designated 0 and 00, and ½-inch designated 0.

The stereotyped list is here appended:—

Iron wood-screws, per gross:—

⅜, ½, and ⅝ inch.		¾-inch.		⅞-inch.		⅞ and 1-inch.		1¼-inch.	
No.	s. d.	No.	s. d.	No.	s. d.	No.	s. d.	No.	s. d.
1	0 8¼	2	0 9¼	3	0 10	4	1 0½	5	1 3½
2	0 8¼	3	0 9¾	4	0 10	5	1 1	6	1 4
3	0 8¼	4	0 9½	5	0 11	6	1 1¼	7	1 5
4	0 8¼	5	0 10	6	1 0	7	1 2¼	8	1 5¼
5	0 9	6	0 11	7	1 1	8	1 3	9	1 6¼
6	0 9½	7	1 0	8	1 1½	9	1 3¼	10	1 7¼
7	0 11	8	1 1	9	1 2½	10	1 4¼	11	1 9
8	1 0	9	1 2	10	1 3	11	1 6¼	12	1 11
9	1 1	10	1 3	11	1 4	12	1 7½	13	2 1
10	1 2	11	1 4	12	1 5¼	13	1 10	14	2 4
11	1 3	12	1 5	13	1 8	14	2 0	15	2 8
12	1 4	13	1 8	14	1 10	15	2 3	16	3 0
13	1 8	14	1 10	15	2 2	16	2 7	17	3 3
14	1 10	15	2 2	16	2 6	17	2 10	18	3 7
15	2 2	16	2 6	17	2 10	18	3 2	19	4 0
16	2 6	18	3 2	18	3 2	19	3 7	20	4 5
				19	3 4	20	4 0	21	4 9
				20	3 7	21	4 5	22	5 2
				22	4 9	22	4 9	23	5 6
				24	5 6	24	5 6	24	5 10
				26	6 2	26	6 2	26	6 2
				28	8 0	28	8 0	28	8 0

1½-inch.		1¾-inch.		2-inch.		2¼-inch.		2½-inch.	
No.	s. d.	No.	s. d.	No.	s. d.	No.	s. d.	No.	s. d.
6	1 6	7	1 10½	8	2 0	9	2 5	10	2 8
7	1 6¼	8	1 11	9	2 1	10	2 6	11	3 0
8	1 7¼	9	1 11½	10	2 2	11	2 8	12	3 4
9	1 8¼	10	2 0	11	2 4	12	3 0	13	3 8
10	1 9	11	2 2	12	2 8	13	3 4	14	4 0
11	1 11	12	2 4	13	3 0	14	3 8	15	4 5
12	2 1	13	2 8	14	3 4	15	4 0	16	4 9
13	2 4	14	3 0	15	3 8	16	4 5	17	5 2
14	2 8	15	3 4	16	4 0	17	4 9	18	5 6
15	3 0	16	3 8	17	4 5	18	5 2	19	5 11
16	3 4	17	4 0	18	4 9	19	5 6	20	6 4
17	3 8	18	4 5	19	5 2	20	5 11	21	6 10
18	4 0	19	4 9	20	5 6	21	6 4	22	7 4
19	4 5	20	5 2	21	5 11	22	6 10	23	8 0
20	4 9	21	5 6	22	6 4	23	7 6	24	8 8
21	5 2	22	5 11	23	6 9	24	8 2	25	9 8
22	5 6	23	6 4	24	7 1	26	9 6	26	10 6
23	5 10	24	6 8	26	8 6	28	12 6	28	13 0
24	6 2	26	8 0	28	12 0	30	16 0	30	18 0
26	7 6	28	10 0	30	14 0	32	20 0	32	22 0
28	8 6	30	12 0	32	18 0	34	25 0	34	26 0
30	10 6	32	15 0	34	22 0	36	30 0	36	32 0
				36	27 0	38	35 0	38	38 0
				38	32 0	40	40 0	40	42 0
				40	38 0				

PRICES.

2¼-inch.		3-inch.		3½-inch.		4-inch.		4½-inch.				
No.	s.	d.	No.	s.	d.	No.	s.	d.	No.	s.	d.	No.	s.	d.
11	3	4	12	4	0	14	5	10	16	8	1	14	10	0
12	3	8	13	4	5	15	6	6	17	9	0	16	10	0
13	4	0	14	4	9	16	7	2	18	9	10	18	11	6
14	4	5	15	5	2	17	8	0	19	10	8	19	12	6
15	4	9	16	5	10	18	9	0	20	11	6	20	13	6
16	5	2	17	6	6	19	9	8	21	12	4	21	14	6
17	5	9	18	7	3	20	10	8	22	13	0	22	15	6
18	6	6	19	7	11	21	11	10	23	13	8	23	16	6
19	7	0	20	8	7	22	12	4	24	14	4	24	17	6
20	7	6	21	9	4	23	13	0	25	15	6	25	18	6
21	8	0	22	10	2	24	13	8	26	16	6	26	19	6
22	8	8	23	11	0	25	14	8	27	17	6	27	21	0
23	9	6	24	11	8	26	15	6	28	18	6	28	22	6
24	10	4	25	12	6	28	17	6	30	24	0	29	24	9
26	12	0	26	13	6	30	22	0	32	32	0	30	27	0
28	14	0	28	15	0	32	30	0	34	37	0	32	32	0
30	20	0	30	20	0	34	34	0	36	42	0	34	40	0
32	25	0	32	26	0	36	40	0	38	48	0	36	45	0
34	30	0	34	32	0	38	45	0	40	56	0	38	50	0
36	35	0	36	38	0	40	50	0				40	60	0
38	40	0	38	42	0									
40	45	0	40	46	0									

5-inch.		5½-inch.		6-inch.		7-inch.		8-inch.				
No.	s.	d.	No.	s.	d.	No.	s.	d.	No.	s.	d.	No.	s.	d.
18	13	6	18	16	0	20	20	0	20	23	9	20	25	0
19	14	6	20	18	0	21	21	3	22	25	0	22	27	6
20	15	6	22	20	0	22	22	6	24	27	6	24	30	0
21	16	9	23	21	0	23	23	9	26	34	0	26	38	0
22	18	0	24	22	0	24	25	0	28	40	0	28	45	0
23	19	0	26	25	0	26	30	0	30	45	0	30	50	0
24	20	0	28	30	0	28	35	0	32	50	0	32	60	0
25	21	0	30	35	0	30	40	0						
26	22	0	32	40	0	32	45	0				9-inch.		
28	26	0	34	45	0	34	50	0				No.	s.	d.
30	30	0	36	50	0	36	55	0				26	42	0
32	35	0	38	60	0	38	65	0				28	50	0
34	40	0	40	70	0	40	75	0				30	60	0
36	45	0										32	70	0
38	55	0												
40	65	0												

Bright round heads charged one size higher than flat; japanned round heads or flat heads two sizes higher; lath-screws charged 4d. extra per gross net.

Brass wood-screws, per gross :—

½, ⅝, and ¾-inch.		⅞-inch.		1-inch.		1⅛ and 1 inch.		1¼-inch.	
No.	s. d.	No.	s. d.	No.	s. d.	No.	s. d.	No.	s. d.
4	1 6	4	1 8	4	1 10	6	2 9	6	3 3
5	1 7	5	1 10	5	2 0	7	3 0	7	3 7
6	1 9	6	2 0	6	2 3	8	3 4	8	4 0
7	2 0	7	2 3	7	2 6	9	3 8	9	4 5
8	2 3	8	2 5	8	2 8	10	4 1	10	4 11
9	2 6	9	2 8	9	3 0	11	4 6	11	5 5
10	2 9	10	3 0	10	3 4	12	5 0	12	6 0
11	3 4	11	3 4	11	3 10	13	5 7	13	6 8
12	3 8	12	3 8	12	4 4	14	6 3	14	7 5
		13	4 6	13	4 10	15	7 0	15	8 3
		14	5 9	14	5 9	16	7 10	16	9 2
				15	7 0	17	8 11	17	10 4
				16	7 10	18	10 0	18	11 6
				17	8 11	19	11 2	19	12 9
						20	12 6	20	14 2
						22	17 3	22	17 3

1⅜-inch.		1½-inch.		2-inch.		2¼-inch.		2½-inch.	
No.	s. d.	No.	s. d.	No.	s. d.	No.	s. d.	No.	s. d.
6	4 0	7	5 3	8	6 3	8	7 0	10	10 0
7	4 4	8	5 6	9	6 9	9	7 6	11	10 6
8	4 8	9	5 11	10	7 5	10	8 3	12	11 0
9	5 2	10	6 7	11	8 2	11	9 1	13	12 1
10	5 9	11	7 3	12	9 0	12	10 0	14	13 3
11	6 4	12	8 0	13	9 11	13	11 0	15	14 6
12	7 0	13	8 10	14	10 11	14	12 1	16	15 10
13	7 9	14	9 9	15	12 0	15	13 3	17	17 5
14	8 7	15	10 9	16	13 2	16	14 6	18	19 0
15	9 6	16	11 10	17	14 7	17	16 0	19	20 8
16	10 6	17	13 2	18	16 0	18	17 6	20	22 6
17	11 9	18	14 6	19	17 6	19	19 1	22	26 5
18	13 0	19	15 11	20	19 2	20	20 10	24	30 8
19	14 4	20	17 6	22	22 9	22	24 7	26	36 0
20	15 10	22	20 11	24	26 8	24	28 8	28	44 0
22	19 1	24	24 8	26	32 0	26	34 0	30	48 0
24	22 8	26	30 0	28	38 0	28	40 0	32	54 0
26	28 0	28	36 0	30	44 0	30	46 0		
28	34 0	30	42 0	32	50 0	32	52 0		

PRICES.

2½-inch.			3-inch.			3½-inch.			4-inch.			4½-inch.		
No.	s.	d.	No.	s.	d.	No.	s.	d.	No.	s.	d.	No.	s.	d.
12	13	0	12	14	6	12	16	6	14	22	6	14	27	0
13	13	9	13	15	0	13	17	3	15	23	6	16	30	0
14	14	5	14	15	7	14	18	0	16	25	0	18	34	0
15	15	9	15	17	0	15	20	0	17	27	0	20	37	0
16	17	2	16	18	6	16	22	0	18	29	0	22	48	0
17	18	10	17	20	3	17	24	0	19	31	6	24	50	0
18	20	6	18	22	0	18	26	0	20	33	6	26	55	0
19	22	3	19	23	10	19	28	0	22	39	0	28	65	0
20	24	2	20	25	10	20	30	0	24	45	0	30	75	0
22	28	3	22	30	1	22	35	0	26	50	0	32	90	0
24	32	8	24	34	8	24	40	0	28	60	0			
26	38	0	26	40	0	26	45	0	30	70	0			
28	45	0	28	50	0	28	55	0	32	85	0			
30	52	0	30	60	0	30	65	0						
32	60	0	32	70	0	32	77	6						

5-inch.			5½-inch.			6-inch.			6½-inch.		
No.	s.	d.	No.	s.	d.	No.	s.	d.	No.	s.	d.
16	34	6	18	46	0	20	60	0	22	75	0
18	37	0	20	50	0	22	65	0	24	80	0
20	40	6	22	55	0	24	70	0	26	90	0
22	47	0	24	60	0	26	80	0			
24	55	0	26	70	0	28	96	0			
26	60	0	28	84	0	30	120	0			
28	72	0	30	100	0	32	150	0			
30	86	0	32	130	0						
32	100	0									

Round heads are charged one size higher than flat heads.
Galvanising ½d. per lb.   Tinning 3d. per lb.

Ideas as to the proper length of screw to be used for a given thickness of wood will vary with different men; but the following table shows some reasonable conclusions. The maximum length ever used is equal to twice the thickness of the wood.

Screws for deal or other soft woods.

Thickness of wood—
½ ¾ 1 1¼ 1½ 1¾ 2 2¼ 2½ 2¾ 3 in.

Length of screw—
1 1¼ 1¾ 2¼ 2¾ 3 4 4 4 5 5 in.

List number of screws—
8 9 10 11 12 13 14 14 14 18 18

For hard wood, the screw would be brass, and two or three numbers of the list earlier in each case.

For 3-inch butts about 1½-inch, No. 9, and for 4-inch butts 1½-inch, No. 12, are commonly used.

In day accounts the numbers are never stated, and often not the lengths. When the lengths are not given, the valuation can be little more than a guess, founded upon the nature of the item for which they are charged. When the lengths are mentioned, the numbers may be adopted from the foregoing table.

When charged in day accounts, the contractor expects to get, and obtains when he can, 3*d.* per inch *per dozen* for iron screws and 4*d.* for brass. It must be remembered that the waste of nails and screws on some buildings is considerable; but with good management 5 per cent. ought to cover it.

Best Scotch glue costs 65*s.* per cwt.; glass paper costs 14*s.* per ream.

Analysis.—The cumbrous processes of analysis such as follow are illustrations of the methods of obtaining prices; but it will be obvious to the reader that the practised estimator will rarely require these. Reference to priced bills of quantities of past work, records of the cost of particular items in the builder's prime cost book, his private memoranda or recollections of similar things, will save him the time and trouble of such tedious work.

The constants for labour only, for labour "from bench," and for fixing, respectively, have been kept separate; as the fixing of joinery often involves bearers, backings or plugging, which are more conveniently valued separately according to the requirements of each case.

For each item requiring valuation, the cost will, therefore, be the value of the labour and fixing plus the value of the material.

As a preliminary to pricing it is necessary to settle the value of the fixing only of various kinds of joinery and of the incidental labours of common occurrence. See Constants.

The application of the constants may be illustrated by the analysis of a *few* ordinary items of joinery. An exhaustive selection from the enormous variety of joinery in modern use would unduly increase the bulk of this section without an adequate advantage. It is, however, hoped that they will prove sufficient to explain the principles which should be observed.

Skirtings.—Large quantities of the ordinary kinds of skirtings are imported ready worked from Sweden, or are prepared in a considerable variety of stock patterns by the moulding mills. These are usually sold by the hundred feet lineal, and are subject to a discount of from 5 to 30 per cent. off list prices. The ordinary discount may be taken as an average of 15 per cent. including waste. If a cartload of skirtings or mouldings is purchased, it would be delivered free within ordinary limits. Smaller quantities must be fetched.

1 in. by 9 in. *torus skirting*, imported—

	s.	*d.*
1 ft. of 1 in. by 9 in. torus at 22*s.* 6*d.* per hundred ..	0	2¾
Waste 15 per cent. equal to discount		
Cleaning up	0	0¼
Fixing, ·018 at 8*s.* 9*d.*	0	1½
Cost per foot run	0	4½

Skirting in two pieces, rebated together, the upper 1 inch by 4½ inches, moulded 3 inch girth; the lower 1¼ inch by 7 inches, hollow moulded and rebated on lower edge. The upper piece would be wrought both sides, the lower, one side.

			s.	*d.*
1 ft.	0 in.	run of 1 in. by 4½ in. deal at 2⅛*d.* super.	0	0⅞
1	0	„ of 1¼ in. by 7 in. deal at 2⅝*d.* super.	0	1½
1	0	„ labour moulding, 3 in. girth, at ·072 ft. super. of 8*s.* 9*d.*	0	1¾
1	0	„ labour moulding, 1 in. girth, at ·012 ft. of 8*s.* 9*d.*	0	1¼
1	4	super. planing, at ·010 of 8*s.* 9*d.*	0	1⅜
1	0	run fixing at ·015 of 8*s.* 9*d.*	0	1⅝
1	0	„ rebated edge to 1¼ in., ·003 of 8*s.* 9*d.*	0	0¼
1	0	„ rebated edge to 1 in., ·003 of 8*s.* 9*d.*	0	0¼
1	0	„ fixing rough fillet on floor, ·005 of 8*s.* 9*d.*	0	0½
1	0	fillet, 1¼ in. by 1 in.	0	0⅜
		Cost per foot run	0	9¾

A few conclusions adopted by estimators which are near the truth and useful may be here given.

Mitres are generally estimated at some proportion of the value of a foot of the skirting, and this is probably the most convenient way of dealing with them. They are usually tongued as well as mitred. Estimators price them at from one-half to one-and-a-half a foot of skirting. The price of *a foot* of skirting may be adopted as a reasonable average.

Fitted Ends.—The price of three-quarters of a foot run may be adopted as a reasonable average.

Housings.—The price of one foot of skirting may be adopted as a reasonable average.

> Skirting, circular on plan, twice straight.
> Ditto, ramped, twice straight.
> Ditto, circular quick sweep, thrice straight.
> Ditto, wreathed, five times straight.
> Ditto, raking, one-half more than straight.

Floors.—

		s.	d.
1¼ in. yellow batten floor, at *Docks*		13	0
Landing rate ₁⁄₂₀ of a standard, at 3s. 9d. per standard		0	1¾
Unloading, ₁⁄₂₀ ditto., at 2s. 6d.		0	1
Cartage, say, ₁⁄₂₀ of a load at 5s. per load		0	5
15 per cent. for waste in conversion and defects for the four foregoing items		2	0½
5½ lb. of 2⅜ floor brads, at 10s. 6d. cwt.		0	6¼
Labour, laying, ·440 at 8s. 9d.		3	10¼
Cost per square		20	0¾

Casements.—The valuation of casements would be as follows:— Taking, as an example, a 2-inch ovolo moulded casement, 2 feet 6 inches by 5 feet 6 inches in a single square (13 feet 9 inches superficial), hung with butts:—

```
ft. in.   ft. in.
 2  6
 0  4½    0 11    2 in. deal.
                  Bottom rail.
13  6
 0  2¼    2  6    2 in. deal.  Top rail and stiles.
          ─────
           3  5                                              s.  d.
 3  6    super. of 2 in. deal at 4½d.    ..   ..   ..   ..   1   2¼
13  9       „   labour at ·037 of 8s. 9d.  ..  ..   ..   ..  4   5¾
13  9       „   labour hanging, ·016 of 8s. 9d.  ..  ..   ..  1  11
                 Glue and glass paper    ..   ..   ..   ..   0   0½
                                                            ─────────
                              13 ft. 9 in.)  7   7½
                 Cost per foot super.    ..   ..   ..   ..   0   6¾
```

This price may be analysed in another manner:—

```
 5  0
 5  0  ft. in.                                                          s.  d.
 2  2  3  6 super. of 2 in. deal as before at 4½d.  ..    ..    ..   1  2¼
 2  2  3  0   ,,     planing at ·010 of 8s. 9d.     ..    ..    ..   0  7½
    —— 14  4 run,   labour to moulding at ·012 of 8s. 9d.  ..    ..  1  6
       14  4   ,,   labour to rebate at ·003 of 8s. 9d.  ..    ..    0  4½
           13  9 super., putting together and cleaning up, at ·016
              of 8s. 9d.  ..  ..  ..  ..  ..  ..  ..  ..  ..  ..   1 11
                   Glue and glass-paper   ..   ..   ..   ..   ..   0  0¼
                   Labour hanging, as before  ..   ..   ..   ..   1 11
                                                      13 ft. 9 in. ) 7  7

                   Cost per foot super.   ..    ..    ..    ..    0  6¾
```

When casements, skylights, sashes and frames are not specially described or drawn to detail, their various parts can be obtained at moulding mills, at some of which a great variety of sections are kept in stock, or the complete sashes and frames will be made to order from such sections. Sashes and frames are also very frequently sublet. The following are some of the current generalisations about casements, sashes, fanlights, &c.

That small sashes, fanlights, or casements 12 feet superficial and under, are worth 20 per cent. more than larger ones. That casements hung on butts or centres are worth 1d. per foot superficial beyond those which are fixed.

Sashes and Frames.—The valuation of sashes and frames requires some nicety of judgment. The variety of size or thickness of the parts of the frames, the quality of the lines and pulleys, the amount of the labour, and the adoption of several kinds of wood in the same sash and frame effectually test the surveyor's knowledge and skill.

Ordinary sashes and frames are rarely drawn to detail—are very often described but superficially in specifications, and only cursorily looked at when supplied. Even when specifically described, the thicknesses may vary if not watched—1 inch pulley stiles instead of 1¼ inch; ⅞ inch inside and outside linings, instead of 1 inch; thinner sills, &c.

An ordinary description, such as the following, will serve to illustrate the valuation of sashes and frames.

Deal-cased frames of 1 inch inside and outside linings, 1¼ inch pulley-stiles, oak double-sunk and weathered sills, and 2-inch deal moulded sashes in single squares, double-hung with best flax lines, best brass axle-pulleys and iron weights, and grooved to receive finishings. Assuming the example to be 4 feet by 7 feet = 28 feet superficial.

					s.	d.
	ft. in.					
	3 3					
	6 10			2 in. sashes, and hanging, same price as already calculated for casements, 6¾*d.*	12	6
		22 3				
				7½ yds. sash line, at 1*d.*	0	7½
				4 sash weights (at 10 lb.) 40 lb. at ½*d.* per lb.	1	8
				4 brass pulleys, at 10*d.*	3	4
				4 ft. 2 in. run, 6 in. by 3 in. oak sill sawn to shape, 4*s.* per ft. cube, delivered 6*d.*	2	1
	4 2					
	1 0			4 ft. 2 in. superficial planing on oak, ·020 of 8*s.* 9*d.* = 2*d.*	0	8¾
	4 2			run, rebate, 4 in. in girth, on oak, ·012 of 8*s.* 9*d.* = 1¼*d.*	0	5¼
		4 1				
		5½	1 10	2 in. deal, *head* 1 ft. 10 in. supl. at 4⅛*d.*	0	7⅝
	4 0					
	7 0	18 0				
	7 0	0 4½	6 9	1 in. deal, *outside lining*		
2/		7 0				
		0 4½	5 3	1 in. deal, *inside lining*		
			12 0	12 ft. supl. 1 in. deal at 2⅛*d.*	2	1½
2/		6 10				
		0 5	5 8	1¼ in. deal, *pulley style*, 2⅝*d.*	1	3
2/		6 8				
		0 5½	6 1	¾ in. deal, *back lining*		
		0 2½				
2/		6 7	2 9	½ in. deal, *parting slip*		
			8 10	8 ft. 10 in. supl. ¾ in. deal, 1¼*d.*	0	11
2/		0 8		13 ft. 4 in. parting bead at 2*s.* 1*d.* per 100 ft. plus 10 per cent. waste, 2*s.* 3½*d.*	0	3½
		3 3	3 3			
2/		6 7	13 2	16 ft. 6 in. inside bead at 2*s.* 8½*d.* per 100, plus 10 per cent. waste, 2*s.* 11⅝*d.*	0	6¼
			16 5			
		4 2	4 2	Groove in oak, 4 ft. 2 in. groove in oak, ·006 of 8*s.* 9*d.*	0	2½
		4 0	4 0			
2/		6 8	13 4			
2/2/		6 7	26 4	Pulley-stile and back lining		

PRICES.

2/	6 8	13 4	Outside lining.					
		57 0	57 ft. 0 in., groove in deal at ·003 of			*s.*	*d.*	
		8*s.* 9*d.*	1	6
3/2·11 =	8 9	½/2/ 6	8 6 8 Blocking out to lengths, 1*d.*	0	6½
4/3·1					
	12 4	21 1	Labour to moulding, ·012, 8*s.* 9*d.*	2	2¼
	21 1	21 1	Labour to rebate, ·003, 8*s.* 9*d.* ..	0	6¼
			Glue	0	1
			Glass-paper	0	1
	28 0		Supl., putting together and cleaning up, at ·020 of 8*s.* 9*d.*	4	10¾
2/	3 3	6 6	Labour splayed edge to 1¼ in.⎫		
	3 3	3 3	⎬ ·006 of ⎫	0	6
		9 9	Labour splayed edge to 2 in. ⎭ 8*s.* 9*d.* ⎭		
	3 3		Labour to groove ·003 of 8*s.* 9*d.* … ..	0	1
2/			Planing. ⎫		
	1 10		⎪		
	1 0	3 8	⎪		
2/			⎪		
	12 0		⎬		
	1 0	24 0	·010 of 8*s.* 9*d.* = 1*d.*	2	9¼
			⎪ 28)40	6¼
	5 8		⎪		
	1 0	5 8	Cost per foot super. exclusive of fixing	1	5¼
		33 4	⎭		

The following are a few of the current generalisations about sashes and frames:—

That sashes and frames under 12 ft. superficial each are worth 20 per cent. more than those of larger size.
That the difference between single and double hanging is 1*d.* per foot superficial.
That circular on plan flat sweep is worth 1½ times the price of straight.
That circular on plan quick sweep is worth twice the price of straight.

Doors.—The ordinary door has top and frieze rails, stiles, and muntins out of 4½ inches wide, middle or lock rail out of 9 inches wide, bottom rail out of 9 inches wide, panels out of ½-inch or ¾-inch—seldom less than the latter. Square framing, therefore, with parts of these proportions must be the starting-point in the valuation of all framed work, and the value of any labour or material beyond this, as moulding on one or both sides, should be valued for the whole door, and its cost added to the total price of the original square framing. The result thus obtained divided by the superficial quantity will give the price per foot superficial.

2 T 2

Thus, in a four-panel door 7 feet by 3 feet, we have 21 feet superficial, and assuming a price of 10*d.* per foot, its value would be 17*s.* 6*d.* The quantity of moulding on one side of such a door would be about 29 feet at ¾*d.* per foot = 1*s.* 9¾*d.* + 17*s.* 6*d.* = 19*s.* 3¾*d.* ÷ 21 feet = 11*d.* per foot superficial.

The thickness of a piece of joinery is ordinarily described as that of the wood it is out of: 1¼-inch deal wrought both sides would be about 1⅛-inch thick. If finished sizes are specified, then the piece of work must *finish* 1¼-inch thick, and would be produced from 1½-inch deal—a difference as shown by the last table of deal in thicknesses of ½*d.* per foot superficial.

The following example illustrates the valuation of panelled doors:—

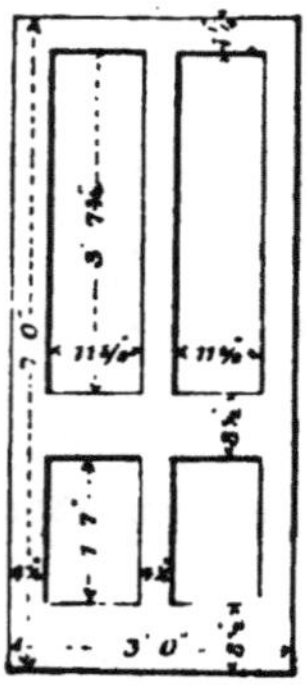

1¼-inch four-panel square door, 3 feet by 7 feet. The dimensions are for the original sizes of the deal:—

Bottom and Middle Rails.		
Top rail ..	3	0
Stiles	7	0
	7	0
Muntin ..	5	10
Horns .. .	0	4
	23	2

2/3	0						
	9	4	6	1¼ in. deal.			
23	2						
	4½	8	8	1¼ in. deal.			
		13	2	13 ft. 2 in. of 1¼ in. deal at 2⅝*d.*		2	10½
2/1	1						
1	8	3	7	¾ in. deal			
2/1	1						
3	9	8	2	11 ft. 9 in. of ¾ in. deal at 1⅝*d.*		1	7
		11	9				

	s.	d.
Glue	0	3½
Glass paper	0	1½
21)	4	10½
Cost of material per foot	0	2¾
Labour making ·042, at 8*s.* 9*d.*	0	4¼
Labour fitting and hanging, ·008 at 8*s.* 9*d.*	0	¾
Cost per foot super.	0	7¾

The value of moulding one side is arrived at as follows:—

PRICES. 645

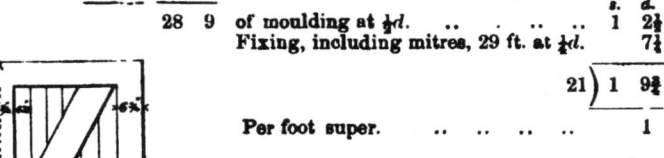

```
8/    0 11¾   7 10
4/    1  7    6  4
4/    3  7¾   14 7
            ─────
            28 9   of moulding at ½d.  ..  ..  ..  ..   1  2¼
                   Fixing, including mitres, 29 ft. at ¼d.   7½
                                                    ────────
                                                 21) 1  9¾
                   Per foot super.   ..  ..  ..  ..        1
```

The following example illustrates the valuation of framed and braced doors:—

The sizes of the parts of framed and braced doors are not described in specifications so often as they might be, nor are they criticised after they are made.

The more substantial ones have the bottom and middle rails out of plank, and the stiles, top rails and braces of battens. In other cases, they have the bottom and middle rails of deals, the stiles, top rails and braces of half-deals.

2-inch framed and braced doors would be valued as follows:—

```
Bottom rail       8  3
                  0 11   3  0
Stiles and  ⎧ 3  3
   rails    ⎨ 7  6
            ⎩ 7  6
                 ───
                 18  8
                  0  7  10  7
                        ─────                              s.  d.
                            13  7   2 in. deal at 4½d.     4  8½
Braces      ⎧ 3  3
            ⎨ 3  4
            ⎩ 4  2
                 ───
                 10  9
                  0  7   6  3
                  2  3
                  5  8  12  9
                        ─────
                            19  0   1 in. deal at 2¼d.    3  4¼
Glue    ..   ..   ..   ..   ..   ..   ..   ..   ..   ..   0  3½
Nails   ..   ..   ..   ..   ..   ..   ..   ..   ..   ..   0  3¼
Glass-paper ..   ..   ..   ..   ..   ..   ..   ..   ..    0  1½
                                               ────────────────
                                          24 ft. 6 in.)   8  9½
            Cost of material per foot   ..   ..   ..   ..  0  4¼
Labour making, ·060 at 8s. 9d.   ..   ..   ..   ..   ..   0  6½
Labour fitting and hanging, ·010 at 8s. 9d.  ..   ..   ..  0  1
            Cost per foot super.   ..   ..   ..   ..   ..  0 11¾
```

The value of 2½-inch framed and braced doors would be found in a similar manner.

If hung folding, labourer's assistance will be required, one-half the joiner's constant for hanging, priced at labourer's rate. Thus—

	s.	d.
Joiner ·020 at 8s. 9d.	0	2
Labourer ·010 at 5s. 10d.	0	0¾
Cost per foot super. of hanging	0	2¾

The following are a few of the current generalisations about doors:—

That moulding adds to the price of square framed 1d. per foot superficial per side.

That finished sizes add 1d. per foot superficial to the value of framing.

That deal selected for staining is worth 1d. per foot superficial more than that intended to be painted. Others add from 5 to 10 per cent. on the whole of such joinery to cover extra labour and greater waste of material in selection.

That doors with segmental heads, although measured square, are worth 25 per cent. more than square.

Doors with three-centred heads 75 per cent. more than square.

Doors with semicircular heads, 50 per cent. more than square.

Doors circular on plan to flat sweep twice straight.

Doors circular on plan to quick sweep thrice straight.

That a door prepared for glass is worth 1d. per foot superficial over ordinary.

That doors hung folding are worth 1d. per foot superficial more than single doors.

That bolection moulding is worth 2d. per foot superficial for each side.

That dwarf doors are worth 20 per cent. more than those of ordinary size.

That loose beads add 1d. per foot superficial to the value.

The following are a few of the current generalisations about framings:—

That a piece of framing under 4 ft. is worth about 25 per cent. more than the rate for a piece of ordinary size.

That pieces of framing measuring 1 ft. superficial and under are worth about twice the rate per foot of ordinary sizes.

That partitions of spandrel shape although measured net are worth about 20 per cent. more than rectangular ones.

That framings circular on plan are worth twice the price of straight. Some estimators apply a different principle, dependent upon the degree of curvature—thus: For each ⅛th of an inch rise on a chord line of a foot, add ⅛th of the original price of straight.

Stairs.—The following example illustrates the valuation of stairs:—

1¼ in. treads with rounded nosings and 1 in. risers, all rebated and grooved together, glued, blocked, and bracketed on, and including strong fir carriages and prepared for close strings. Taking one step 4 ft. long as an example, 10 in. "going" and 7 in. rise:—

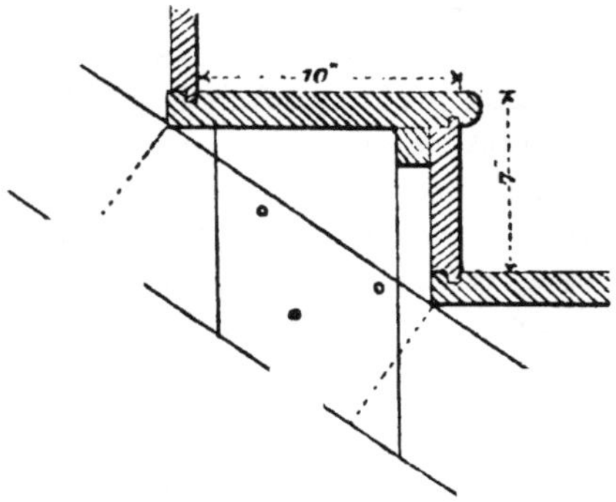

					s.	d.
4	0					
1	0	4	0	1¼ in. deal at 2⅝*d*.	0	10½
4	0					
0	8	2	8	1 in. deal at 2⅛*d*.	0	5¾
4	0	4	0	Cross-tongued joint at ·045 of 8*s*. 9*d*.	0	6¼
2/4	0	8	0	Labour to groove at ·003 of 8*s*. 9*d*.	0	2½
2/4	0	8	0	Labour rebated edge to 1 in. at ·003 of 8*s*. 9*d*.	0	2½
4	0	4	0	Labour rounded edge to 1¼ in. at ·006 of 8*s*. 9*d*.	0	2½
1	6	1	6	1¼ in. by 1¼ in. deal blocking cut to lengths, at 1*d*.	0	1½
1	1					
0	5½	0	6	2½ in. *rough* deal carriage, at 2⅝*d*.	0	1¼
1	1			1¼ in. *rough* deal bracket, at 1⅜*d*.	0	1½
				3 8		
0	7	0	8	2 8		
6	4			6 4		
1	0	6	4	Planing, at ·010 of 8*s*. 9*d*.	0	6¾
				6 ft. 8 in.)	3	5
				per foot	0	6
				Glue and nails	0	1
				Fixing, except housing and wedging, at ·045 of 8*s*. 9*d*.	0	4¾
				Cost per foot super.	0	11¾

Housing and wedging is always separately priced.

Frames.—Frames for doors, casements, and for similar purposes are ordinarily valued by the foot cube, and in an estimate of any size a price per foot cube is adopted which will cover the *average* value, the small differences of labour on each being disregarded.

The price, including profit and fixing, will vary from 4s. 6d. to 6s., according to the estimator's judgment. If we apply the latter price to a 4 by 3-inch rebated and beaded frame, 4 by 3 inches = 12 sectional inches = 1 inch cube = $\frac{1}{12}$th of a foot cube = 6d. per foot run; or we can calculate by farthings:—3s. per foot cube is $\frac{1}{4}$d. per inch of section; 6s. per foot is $\frac{1}{2}$d.; 9s. is $\frac{3}{4}$d., &c.; in this case 12 times $\frac{1}{2}$d. is 6d. as before.

An analysis of the value of the same frame may be made as follows:—Taking 16l. 1s. 9d. per standard as a base, we have for the deal nearly 1s. 11½d. per foot cube. This, for facile calculation, may be called 2s.

	d.
$\frac{1}{12}$th of a foot cube at 2s.	2
$\frac{1}{12}$th labour, ·032 at 8s. 9d.	3¼
Cost	5¼ } per foot run, or 5s. 3d. per foot cube.

The following are a few of the current generalisations about door-frames:—

That segmental heads to frames are worth about twice straight.

That semicircular or Gothic heads to frames are worth about two-and-a-half times straight.

That transomes, being in shorter lengths, are worth 10 per cent. more than frames.

Mouldings.—If mouldings are not specified or described as "to detail," stock mouldings only would be allowed for; but these would only be obtainable either of the character of architraves or those cornices of comparatively thin material, intended for fixing anglewise, or "sprung," as it is called. The prices of these may be obtained from a trade list of a moulding manufacturer. To the price thus found would be added the cost of fixing and mitres.

When mouldings are to detail they are usually referred by the estimator to a price per foot cube, according to their size, the smallest being of the highest price.

Some estimators adopt the following scale, which includes fixing and profit:—

		s.	d.
2 in. by 2 in. and under	per foot cube	12	0
2 in. by 2 in. to 4 in. by 3 in.	,, ,,	7	6
Over 4 in. by 3 in.	,, ,,	6	0

For the value of mitres to mouldings, the estimator usually adopts a proportion of the price per foot run, as 1 foot for ordinary mitres, 2 feet for irregular mitres, &c.

Sometimes a percentage, as 15 per cent. on the price per foot cube.

A large moulding of plain character presents a reasonable analogy to a door frame.

The application of the foregoing constants and values of deal produces a result somewhat similar to the above. Assuming the girth of the mouldings to be twice the width of the deal, we have for the first category 2 by 2 inches as follows:—

1 0	run of 2 in. by 2 in. wrought fillet	.. ..	..	0 1
1 0	4 in. supl. of moulding at ·072 of 8s. 9d.	..	..	0 2¼
0 4				
1 0	run fixing at ·005 of 8s. 9d.	.. ..	..	0 0½

$$\text{Cost per foot run} \quad .. \quad .. \quad .. \quad 0 \ 4 \brace \text{Nails less than ¼d.} = 12s. \ 0d. \text{ per foot cube.}$$

For the second category, 4 by 3 inches:—

1 0	4 in. supl. 3 in. deal at 6d.	.. ..	..	0 2
0 4				
1 0				
1 2	1·2 supl. planing at ·010 of 8s. 9d.	..	..	0 1¼
1 0	3 in. supl. moulding at ·072 of 8s. 9d.	..	..	0 5
0 8				

$$\text{Cost per foot run} \quad .. \quad .. \quad 0 \ 8¼ \brace \text{Nails less than ¼d.} = 8s. \ 3d. \text{ per foot cube.}$$

When a price for moulding is inclusive of mitres, it should be increased from 10 to 20 per cent., according to their frequency, which may be guessed at from the description of the purpose for which the moulding is required.

Attendances.—Beyond the general item for "each trade to attend on all others," and which is generally interpreted as attendances which cannot be specifically described, it is the custom for surveyors to describe attendances of joiner upon plumber to every w.c. apparatus, sink, cistern and bath. Such things often require none worth mentioning, and when they do their position and necessities vary so much in different buildings, that no estimator takes the trouble to investigate it, and he usually brackets the items together in the bill of quantities, and prices them all alike, commonly at about 2s. 6d. each, and often less.

Yellow Pine.—Yellow pine is the shippers' name for this wood. Mr. Hurst (see 'Architectural Surveyor's Handbook') calls it white pine. Yellow pine deals are shipped at Quebec, Montreal, Botwoodville, Three Rivers, etc. Generally the Quebec are the best.

They are classed as bright, dry-floated and floated, and there are first, second, third and fourth qualities of each. Little grey specks of long shape, parallel to the grain of the wood, are a characteristic feature of yellow pine. When these are conspicuous the wood should be rejected.

Ironfounders' patterns are usually made of good yellow pine, and the better qualities of *bright* pine are used for specially good joinery and fittings, and as it can be obtained as wide as 30 inches, it is specially useful for large panels and similar purposes.

As before remarked, large quantities of pine, which cost no more than good deal, are used for joinery. The price of yellow pine increases with the width and length of the deals. When less than 12 feet long, or 11 inches wide, the price falls rapidly.

Exceptionally good bright pine of long lengths and wide widths will fetch, at the dock sales, sometimes as much as 28*l*. per St. Petersburg standard; but as comparatively short lengths and moderate widths are often as convenient as long and broad "deals," a reasonable average would be 20*l*. per standard, as before mentioned.

For the best work, the average price of best pine should be applied, and a scale of similar construction to those previously given for deal is as follows:—

	£	s.	d.
Average price of first bright pine deal at docks, per St. Petersburg standard	20	0	0
Landing rate	0	3	9
Unloading	0	1	6
Cartage by sawmill proprietors, as before.			
Two years' compound interest at 5 per cent. per annum on the total of the first three items 20*l*. 5s. 8d., two years being assumed as the time required for thorough seasoning	2	1	6½
Waste in sawing and conversion, 10 per cent. on 22*l*. 6s. 9½d.	2	4	8
	24	11	5½

PRICES. 651

The *net* price for sawing yellow pine planks over 11" wide is 3s. 6d. per 1000 ft. super.

Assuming (as in the tables of deals) a St. Petersburg standard to be 666' super. of 3", the price of the various thicknesses will be as follows:—

$\frac{1}{4}$" = 11 cuts in 3" (producing 12 thicknesses of $\frac{1}{4}$") = 7326' of sawing = 73$\frac{1}{4}$ hundred at 3s. 6d. per hundred

	£	s.	d.	
	24	11	6	pine.
	12	16	4$\frac{1}{2}$	sawing.
7992 ft.)	37	7	10$\frac{1}{2}$	
	1$\frac{1}{8}$d. per ft.			

$\frac{1}{2}$" = 5 cuts in 3" (producing 6 thicknesses of $\frac{1}{2}$" = 3330' of sawing = 33$\frac{3}{10}$ hundred at 3s. 6d. per hundred

	24	11	6	pine.
	5	16	6$\frac{1}{2}$	sawing.
3996 ft.)	30	8	0$\frac{1}{2}$	
	1$\frac{3}{4}$d. per ft.			

$\frac{3}{4}$" = 3 cuts in 3" (producing 4 thicknesses of $\frac{3}{4}$" = 1998' of sawing = 20 hundred at 3s. 6d. per hundred

	24	11	6	pine,
	3	10	0	sawing.
2664 ft.)	28	1	6	
	2$\frac{1}{2}$d. per ft.			

1" = 2 cuts in 3" (producing 3 thicknesses of 1" = 1332' of sawing = 13$\frac{3}{10}$ hundred at 3s. 6d. per hundred

	24	11	6	pine.
	2	6	6$\frac{1}{2}$	sawing.
1998 ft.)	26	18	0$\frac{1}{2}$	
	3$\frac{1}{4}$d. per ft.			

1$\frac{1}{4}$" = 2 cuts in 3" (producing 2 thicknesses of 1$\frac{1}{4}$" and 1 of $\frac{1}{2}$") = 1332' of sawing = 13$\frac{3}{10}$ hundred at 3s. 6d. per hundred

	20	9	7	pine.
	1	14	11	sawing.
1332 ft.)	22	4	6	
	4d. per ft.			

Note.—One-quarter of the total of the sawing last mentioned is allotted to the $\frac{1}{2}$", leaving three-quarters (the remainder) for the calculation; the pine is reduced by $\frac{1}{6}$.

	£	s.	d.	
1½" = 1 cut in 3" (producing 2 thicknesses of 1½" = 666' of sawing = $6\frac{7}{10}$ hundred at 3s. 6d. per hundred	24	11	6	pine.
	1	3	5½	sawing.
1332 ft.)	25	14	11½	
	4⅝d. per ft.			

1¾" = 1 cut in 3" (producing 1 thickness of 1¾" and 1 thickness of 1¼") = 666' of sawing = $6\frac{7}{10}$ hundred at 3s. 6d. per hundred	14	6	8½	pine.
	0	11	8¾	sawing.
666 ft.)	14	18	5¼	
	5⅜d. per ft.			

Note.—One-half of the total of the sawing last mentioned is allotted to the 1¼", and the pine reduced by $\frac{5}{12}$.

2" = 1 cut in 3" (producing 1 thickness of 2", and 1 thickness of 1") = 666' of sawing = $6\frac{7}{10}$ hundred at 3s. 6d. per hundred	16	7	8	pine.
	0	11	8¾	sawing.
666 ft.)	16	19	4¾	
	6d. per ft.			

Note.—One-half of the total of the sawing last mentioned is allotted to the 1", and the pine reduced by ⅓.

2¼" = 1 cut in 3" (producing 1 thickness of 2¼" and 1 thickness of ¾") = 666' of sawing = $6\frac{7}{10}$ hundred at 3s. 6d. per hundred	18	8	7½	pine.
	0	11	8¾	sawing.
666 ft.)	19	0	4¼	
	6⅞d. per ft.			

Note.—One-half of the sawing last mentioned is allotted to the ¾", and the pine reduced by ¼.

2½" = 1 cut in 3" (producing 1 thickness of 2½" and 1 thickness of ½") = 666' of sawing = $6\frac{7}{10}$ hundred at 3s. 6d. per hundred	20	9	7	pine.
	0	11	8¾	sawing
666 ft.)	21	1	3¾	
	7⅝d. per ft.			

Note.—One-half of the sawing last mentioned is allotted to the ½", and the pine reduced by ⅙.

$2\tfrac{3}{4}''= 1$ cut in $3''$ (producing 1 thickness of $2\tfrac{3}{4}''$ and 1 thickness of $\tfrac{1}{4}''$) = 666' of sawing = $6\tfrac{7}{10}$ hundred at 3s. 6d. per hundred

	£	s.	d.
	22	10	6¼ pine.
	0	11	8¾ sawing.
666 ft.)	23	2	3¼

8½d. per ft.

Note.—One-half of the sawing last mentioned is allotted to the ¼", and the pine reduced by $\tfrac{1}{12}$.

$3''$ no sawing .. .. 666 ft.) 24 11 6 pine.

8⅞d. per ft.

A table of thicknesses of yellow pine, with the ordinary labours on them, will be found useful for reference, and may be constructed in a similar manner to that given after the table of rough deal in thicknesses.

The labour on yellow pine is rather less than on yellow deal, but the difference is hardly enough to vary constants, which can only be taken as averages.

Hardwoods.—Work in pitch-pine, oak, mahogany, teak, etc., may be either valued by preparing a table of various thicknesses of the wood, and pricing the material and labour as suggested for deal; or the price of the work in deal may be adopted as a base, and increased by a percentage. This must of necessity depend for its consistency on the price of the deal work, as, if the price for the deal work is low, the percentage must be high, thus:—In one recent schedule, work in wainscot was charged two-and-a-half times the price of deal, but the deal prices were high. In another schedule, work in wainscot was charged three-and-a-quarter times the price of deal, but the deal prices were low; in each case including profit.

Pitch-Pine.— Pitch-pine is imported from Darien, Mobile, Pascagoula, Pensacola, etc. It comes into the market as hewn logs, sawn logs, and planks. Joinery is most frequently out of plank. It varies greatly in quality and in difficulty of working: some of it is so hard and "curly" that it must be finished with a scraper instead of the plane to obtain even a decent finish. Technically it is classed with "hardwoods."

Although the freight is higher, it has within the last few years successfully competed with Baltic fir.

Sawing at the mills is charged at one-third more than the rate per dozen cuts in deals.

The labour on pitch-pine will be least on wood which most closely resembles deal. That which is most resinous and figured—the latter the distinguishing characteristic of the best wood—will require the most labour. Old and dry pitch-pine is sometimes harder to work than seasoned English oak. The builder judiciously avoids very old pitch-pine. The average labour on pitch-pine may be taken generally as about 50 per cent. more than the value of that on deal. Some will cost more and some less.

The average price of pitch-pine plank is about the same as that of good joiners' deals. Adopting the items of the last deal calculation and increasing the price for sawing by one-third, a table may be constructed as follows:—

	£	s.	d.
½" deal	16	1	9
Add sawing	7	12	7½
Add ⅓ of last item for extra cost of sawing	2	10	10½
7992 ft.)	26	5	3

¾d. per ft.

	£	s.	d.
¼" deal	16	1	9
Add sawing	3	9	4½
Add ⅓ of last	1	3	1½
3996 ft.)	20	14	3

1¼d. per ft.

	£	s.	d.
¾" deal	16	1	9
Add sawing	2	1	7½
Add ⅓ of last	0	13	10½
2664 ft.)	18	17	3

1¾d. per ft.

	£	s.	d.
1" *deal*	16	1	9
Add sawing	1	7	9
Add ⅓ of last	0	9	3
1988 ft.)	17	18	9
	2⅛*d*. per ft.		

	£	s.	d.
1¼" *deal*	13	8	1½
Add sawing	1	0	9¾
Add ⅓ of last	0	6	11¼
1332 ft.)	14	15	10½
	2⅝*d*. per ft.		

	£	s.	d.
1½" *deal*	16	1	9
Add sawing	1	3	1½
Add ⅓ of last	0	7	8½
1332 ft.)	17	12	7
	3⅛*d*. per ft.		

	£	s.	d.
1¾" *deal*	9	7	8¼
Add sawing	0	11	6¾
Add ⅓ of last	0	3	10¼
666 ft.)	10	3	1¼
	3⅝*d*. per ft.		

	£	s.	d.
2" *deal*	10	14	6
Add sawing	0	11	6¾
Add ⅓ of last	0	3	10¼
666 ft.)	11	9	11
	4⅛*d*. per ft.		

	£	s.	d.
2¼" *deal*	12	1	3¾
Add sawing	0	11	6¾
Add ⅓ of last	0	3	10¼
666 ft.)	12	16	8¾
	4⅝*d*. per ft.		

	£	s.	d.
2¼" deal	13	8	1¼
Add sawing	0	11	6¾
Add ⅓ of last	0	3	10¼
666 ft.)	14	3	6¼

5¼d. per ft.

	£	s.	d.
2¾" deal	14	14	11¼
Add sawing	0	11	6¾
Add ⅓ of last	0	3	10¼
666 ft.)	15	10	4¼

5⅝d. per ft.

| 3" deal (no sawing) | 666 ft.) | 16 | 1 | 9 |

5¾d. per ft.

A table of thicknesses of pitch-pine, with the ordinary labours on them, as given after the table of rough deal in thicknesses, may be constructed in a similar manner, and will be found useful for reference.

It will be seen that the price of the material is not appreciably increased by the cost of sawing. As, however, pitch-pine is almost invariably varnished or French polished, the care requisite to protect it and keep it clean has to be considered. About 1d. per foot superficial will cover the additional cost.

The following are a few of the current generalisations as to the differences of value of pitch-pine and deal, including profit:—

That the labour on pitch-pine floors is 25 per cent. more.
The labour and material in pitch-pine jamb-linings is 25 per cent. more.
The labour and material in wall-strings and skirtings is 25 per cent. more.
The labour and material in door-frames is 20 per cent. more.
The labour and material in w.c. seats and risers is 25 per cent. more.
The labour and material in doors and framings is 25 per cent. more.
The labour and material in newels is 25 per cent. more.
The labour and material in handrails is 25 per cent. more.
That the relative values of pitch-pine labour and materials compared with wainscot, is one-half to two-thirds of wainscot value.

Issues as to relative values are often confused by including painting, staining and varnishing, and French polishing; they will be found easier to estimate if the calculation is confined to the woodwork, the finishing being compared separately.

The comparative value of deal and pitch-pine as represented by a percentage may be illustrated by two simple items.

If we refer to the illustration on a former page of a skirting in two pieces, we find the material valued at $2\frac{3}{4}d.$, the labour at $7d. = 9\frac{3}{4}d.$ total. If we increase the labour constant by 50 per cent., we have the following :—$2\frac{3}{4}d. + 10d. = 12\frac{3}{4}d.$, which, compared with $9\frac{3}{4}d.$, represents an increase of about 30 per cent. on the price of the same work in deal.

In the application of the same process to the calculation of the value of 1¼-inch square framed doors, we find the material valued at $2\frac{3}{4}d.$, the labour $5\frac{1}{4}d. = 8d.$ total. If we increase the labour constant by 50 per cent., we have the following :—$2\frac{3}{4}d. + 7\frac{3}{4}d. = 10\frac{1}{2}d.$, which compared with $8d.$, represents an increase of 33 per cent.

Oak.—Well-seasoned, well-grown English is the strongest oak procurable, and is still commonly used for fences, sash-sills and other external work; but its hardness, when sufficiently seasoned, and its tendency to crack and warp have led to an increased use of foreign oak in its stead.

For internal joinery, good foreign oak is to be preferred, as it works easier, and does not warp or split so readily as English. Baltic oak has been much used, known as Riga, Dantzic, Stettin, or Memel. Riga wainscot makes good work, but is scarce, and Austrian or Hungarian is now most plentiful in the market. Much of this is shipped from Fiume or Trieste, in the Adriatic, in exceptionally large logs or wide planks of very fine quality.

Sometimes staves are used. In the specification of the Houses of Parliament occurs the following clause :—" The wainscot to be used in the joiners' work is assumed to be from the best Crown Riga wainscot in the logs, and from pipe staves of the best quality in equal proportions, to be prepared for use by steaming or otherwise." Riga staves are so dear that they are now imported but rarely. Memel and Stettin are of good quality, but are not much used by builders, as stuff of larger size is found to cut up to greater advantage. Really good staves are now very expensive.

The use of American oak is steadily increasing.

Oak and Honduras mahogany joinery are generally assumed to be of equal value, but the former does not work so easily as average mahogany, and there is undoubtedly more waste. The labour on oak is about twice the value of that upon deal. When deal work is priced judiciously, three times its price is a reasonable rate for oak-work labour and material, including profit.

The waste on Riga or Hungarian oak in conversion may be taken as 10 per cent. more than on deal. This was estimated for the tables of deal as 10 per cent.—i.e. 5 per cent. waste for sawing and 5 per cent. conversion. The oak will, therefore, be 10 per cent., and 10 per cent. = 20 per cent. in all.

The price of oak is quoted by the retail timber-dealer by the inch in thickness, the other thicknesses above 1 inch increasing in uniform proportion. The London timber merchants' present price for Austrian or Hungarian oak of good quality, well-seasoned and of good figure (being cut on the quarter), in thicknesses, is as follows:—

$\frac{1}{2}$ in.	$\frac{5}{8}$ in.	$\frac{3}{4}$ in.	1 in.	1¼ in.	1½ in.	1¾ in.	2 in.
s. d.	s. d.	s. d.	s. d.	s. d.	s. d.	s: d.	s. d.
0 5	0 6	0 7	0 8	0 10	1 0	1 2	1 4

Add 10 per cent. for conversion :—

0 0½	0 0½	0 0¾	0 0¾	0 1¼	0 1¼	0 1¼	0 1¼
0 5½	0 6¼	0 7¾	0 8¾	0 11	1 1¼	1 3¼	1 5¼

per ft. super.

Riga wainscot would cost about ½d. per inch more.

It is questionable whether the builder can buy wainscot in the log or plank at the docks, and convert it so as to save much on the foregoing prices; certainly he rarely does so. We shall, therefore, adopt the above rates in the following examples.

The extra care in preparation and protection on work in hardwood is worth an additional 10 per cent.

Adopting, for the purpose of illustration and comparison, the same items as before, we have, for the deal skirting in two pieces —material 2¾d., labour 7d. = 9¾d. total. Excluding the deal fillet and its fixing, we have 8¾d. remainder. If we increase the labours by 100 per cent., and adopt the oak prices in the table of Austrian oak, the calculation will be as follows:—

ft.	in.			s.	d.
1	0	run	of 1 in. by 4½ in. oak, at 8¾*d.*	0	3¼
1	0	,,	of 1½ in. by 7 in. oak, at 11*d.*	0	6⅜
1	0	,,	labour to moulding 3 in. girth at ·144 super. of 8*s.* 9*d.*	0	3¾
1	0	,,	labour to moulding 1 in. girth, at ·024 of 8*s.* 9*d.*	0	2½
1	4	super.	of planing, at ·020 of 8*s.* 9*d.*	0	2¾
1	0	run	fixing, at ·030 of 8*s.* 9*d.*	0	3
1	0	,,	rebated edge to 1½ in., ·006 of 8*s.* 9*d.*...	0	0½
1	0	,,	rebated edge to 1 in., ·006 of 8*s.* 9*d.*	0	0½
				1	10¾
			Add 10 per cent for extra care and protection	0	2¼
			Cost per foot run	2	1

Three times 8¾*d.* is 2*s.* 2¼*d.*

Applying the same process to the calculation of the value of 1¼-inch square-framed doors, we find the material valued at 2¾*d.*, the labour 5*d.* = 7¾*d.* total. If we increase the labour constants by 100 per cent., and adopt the oak prices in the table of Austrian oak, the calculation will be as follows :—

						s.	d.
2/	3	0					
	0	9	4	6			
	23	2					
	0	4½	8	8			
			13	2	1¼ in. oak at 11*d.*	12	0¾
2/	1	1					
	1	8	3	7			
2/	1	1					
	3	9	8	2			
			11	9	¾-in. oak at 7¾*d.*	7	7
					Glue	0	3½
					Glass paper	0	1½
					21)	20	0¾
					Cost of material per foot super.	0	11½
					Labour making ·084 at 8*s.* 9*d.*	0	8¾
					,, fitting and hanging ·016 at 8*s.* 9*d.*	0	1¾
						1	10
					Add 10 per cent. for extra care and protection	0	2¼
					Cost per foot super.	2	0¼

Three times 7¾*d.* is 1*s.* 11¼*d.*

Oak of any growth, either English or foreign, is not to be thoroughly depended on. It will crack in the most unexpected cases. Consequently, whenever possible, work is made up in pieces. A familiar instance is the treatment of door-frames, which is nearly always adopted for good work. The door-frame in the illustration is made up of pieces of wainscot cross-tongued and glued to a deal core. The whole of the pieces must be thoroughly seasoned, or it will be no more trustworthy than the solid oak.

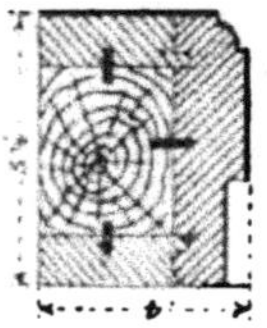

Analysis of the cost of such a frame may be instructive. Adopting a lineal foot for the purpose, and assuming the annexed sketch as the finished size, the material must be taken of such thickness as will allow for working.

					s.	d.
	1 0 0 3¾	4	3	deal at 5¾d.	0	2
2/	1 0 0 3½	0	7	1½ oak at 11d.	0	6½
	1 0 0 5¾	0	6	2" oak at 1s. 5½d.	0	8¾
3/	1 0	3	0	oak, cross-tongue, at ½d.	0	1½
3/	1 0	3	0	groove in oak, ·006 of 8s. 9d.	0	1⅞
3/	1 0	3	0	groove in deal, ·003 of 8s. 9d.	0	1
2/	1 0	2	0	rebated edge to 1½ in. oak, ·006 of 8s. 9d.	0	1¼
2/	1 0	2	0	groove in oak, ·006 of 8s. 9d.	0	1¼
	1 0	1	0	labour to rebate 2½ in. girth in oak, at ·008 of 8s. 9d.	0	0⅞
	1 0	1	0	labour to moulding 1½ in. girth in oak at ·024 of 8s. 9d.	0	2½
	1 0 1 1	1	1	planing on deal, ·010 of 8s. 9d.	0	1
				Glue	0	1
2/	1 0 9½	1	"			
	1 0 1 3½	1	4			
		2	11	planing on oak (the whole surface of each piece) ·020 of 8s. 9d.	0	6¼
				Labour, putting together, and framing ·040 of 8s. 9d.	0	4¼
				Cost per ft. run	3	3¾
				dd 10 per cent for extra care and protection	0	4
					3	7¾

PRICES.   661

A fir solid frame of the same size would cost, at 6s. per foot cube, 11d. Three times 11d. is 2s. 9d.

Sometimes, however, the use of English oak for internal joinery is insisted on, and this is most conveniently purchased in the form of planks or scantlings from the dealers in English timber. This is the course adopted by the various waggon manufacturers (who are the largest consumers), as only a timber dealer can advantageously dispose of the various sizes and lengths which arise in the conversion of timber from the log.

Let us assume freshly-sawn planks to cost 7l. 10s. per load delivered, and that it requires three years' seasoning—some specifications prescribe five years' seasoning; but the insistance on such a clause is extremely difficult, and oak seasoned so long is rarely obtained. English oak twists so much in the process, that the work will sometimes, for that reason alone, require wood $\frac{1}{4}$-inch thicker to produce a specified thickness.

The waste on the following is put at 10 per cent. more than deal.

The calculation would be as follows, assuming the thickness of the plank as 3 inches.

	£	s.	d.
Oak plank, delivered, per load	7	10	0
5 per cent. compound interest for three years	1	3	7¾
Waste, 20 per cent. on the two foregoing items for sawing and conversion	1	14	8¾
Unloading	0	0	6
	10	8	10½

The load of 50 ft. cube would contain 200 super. ft. of 3 in.

$\frac{1}{2}''$ = 5 cuts in 3" (producing 6 thicknesses = 1200')
$\frac{1}{2}''$) = 1000' of sawing at 5s. 8d. per hundred } 10 8 10½ oak.
                                                          2 16 8 sawing.

1200) 13. 5 6¼

2⅝d. per ft.

$\frac{3}{4}''$ = 3 cuts in 3" (producing 4 thicknesses = 800' of
$\frac{3}{4}''$) = 600' of sawing at 5s. 8d. per hundred } 10 8 10½ oak.
                                                          1 14 0 sawing.

800) 12 2 10½

3⅝d. per ft.

QUANTITY SURVEYING.

	£	s.	d.	
1" = 2 cuts in 3" (producing 3 thicknesses = 600' of 1") = 400' of sawing at 5s. 8d. per hundred	10	8	10½	oak.
	1	2	8	sawing.

600) 11 11 6½

4⅝d. per ft.

1¼" = 2 cuts in 3" (producing 2 thicknesses of 1¼" = 400' of 1¼", and 1 of ½") = 400' of sawing at 5s. 8d. per hundred	8	14	0¾	oak.
	0	17	0	sawing.

400) 9 11 0¾

5¾d. per ft.

Note. — One-quarter of the total of the foregoing sawing is apportioned to the ½", leaving three-quarters for the calculation, and the oak is reduced by ⅛.

1½" = 1 cut in 3" (producing 2 thicknesses of 1½" = 400' of 1½") = 200' of sawing at 5s. 8d. per hundred	10	8	10¼	oak.
	0	11	4	sawing.

400) 11 0 2¼

6⅝d. per ft.

1¾" = 1 cut in 3" (producing 1 thickness of 1¾" = 200' of 1¾", and 1 of 1¼") = 200' of sawing at 5s. 8d. per hundred	6	1	10¼	oak.
	0	5	8	sawing.

200) 6 7 6¼

7⅝d. per ft.

Note — One-half of the total of the foregoing sawing is apportioned to the 1¼' and the oak reduced by 5⁄12.

2" = 1 cut in 3" (producing 1 thickness of 2" = 200' of 2", and 1 of 1") = 200' of sawing at 5s. 8d. per hundred	6	19	3	oak.
	0	5	8	sawing.

200) 7 4 11

8¾d. per ft.

Note. — One-half of the total of the foregoing sawing is apportioned to the inch, and the oak reduced by ⅛.

PRICES.

$2\frac{1}{4}''$ = 1 cut in 3″ (producing 1 thickness of $2\frac{1}{4}''$ = 200′ of $2\frac{1}{4}''$, and 1 of $\frac{3}{4}''$ = 200′ of sawing at 5s. 8d. per hundred .. ..} £ s. d.
7 16 8 oak.
0 5 8 sawing.

200) 8 2 4
$9\frac{3}{4}d$. per ft.

Note.—One-half of the total of the foregoing sawing is apportioned to the $\frac{3}{4}''$, and the oak reduced by $\frac{1}{4}$.

$2\frac{1}{2}''$ = 1 cut in 3″ (producing 1 thickness of $2\frac{1}{2}''$ = 200′ of $2\frac{1}{2}''$ and 1 of $\frac{1}{2}''$) = 200′ of sawing at 5s. 8d. per hundred .. ..} 8 14 $0\frac{3}{4}$ oak.
0 5 8 sawing.

200) 8 19 $8\frac{3}{4}$
$10\frac{3}{4}d$. per ft.

Note.—One-half of the total of the foregoing sawing is apportioned to the $\frac{1}{2}''$, and the oak reduced by $\frac{1}{6}$.

$2\frac{3}{4}''$ = 1 cut in 3″ (producing 1 thickness of $2\frac{3}{4}''$ = 200′ of $2\frac{3}{4}''$ and 1 of $\frac{1}{4}''$) = 200′ of sawing at 5s. 8d. per hundred .. ..} 9 11 $5\frac{3}{4}$ oak.
0 5 8 sawing.

200) 9 17 $1\frac{3}{4}$
$11\frac{3}{4}d$. per ft.

Note.—One-half of the total of the foregoing sawing is apportioned to the $\frac{1}{4}''$, and the oak reduced by $\frac{1}{12}$.

3″, no sawing .. .. 200) 10 8 $10\frac{1}{4}$
$12\frac{1}{2}d$. per ft.

The above table is based upon the assumption that the oak is for quite ordinary uses, and plank for these purposes is cut right through the tree. But when the oak is required for good joinery it is cut "on the quarter," as for such work there must be "figure." The angular method of cutting causes much more waste; there should consequently be added to each of the above prices 25 per cent., which will about cover the extra cost.

Mahogany.—Only a few years ago mahogany was the principal wood used for internal joinery of the best quality—sometimes solid, sometimes veneered. This is not now the case; pitch-pine varnished or French polished, wainscot, teak, and American walnut have, to a great extent, taken its place.

For w.c., bath, and lavatory fittings, and hand-rails, and as a backing for veneers of more expensive wood, it cannot be surpassed, and will probably continue to be preferred for these purposes.

Much of the common mahogany is little better than cedar, which wood is sometimes substituted for mahogany. This, and the common qualities of mahogany, are stained before polishing, to give them a richer hue and to delude the unwary.

Cuba, St. Domingo, and Honduras were formerly the chief sources of the supply of mahogany; but it is now obtained from many other countries. Yucatan, Panama, Mexico, and Africa send large quantities.

The quality of mahogany varies greatly, and the best figured wood is nearly always cut into veneers. This will sometimes fetch at auction 5s. or more per foot of inch, while the price of ordinary qualities may be under 4d.

In London and Liverpool mahogany in the log is sold by auction calliper measure, and about 70 per cent. of the cubic content only is charged for, as it is a custom of the trade to allow the difference, about 30 per cent., for the waste assumed to accrue in the process of cutting into thicknesses. The amount of advantage derived by the timber merchant from this allowance will, of course, vary according to the condition of the log. A log measuring 200 feet superficial calliper measure usually gives from 135 to 150 feet sale measure. The following average prices per foot superficial of inch at the dock sales of 1894 may be compared with the timber-merchant's prices for the thicknesses; but these latter include the cost of sawing, while the former do not. It must also be remembered that the merchant's purchases are in larger quantities than it would suit the average builder to buy.

London:—	s.	d.
Honduras	0	4
Tabasco	0	4⅝
Mexican	0	3¼
Cuba	0	3⅝
Panama	0	3
African	0	3⅞

Averages are, however, deceiving, for the larger proportion of the mahogany imported is common, and its consequent low price reduces the average. The freights to Liverpool are lower, and the buyers there have a trifling consequent advantage over those in London. The prices per foot superficial of inch at the public sales of 1894 were as follows:—

Liverpool:—	s.	d.		s.	d.
Honduras	0	3½	to	0	6
Tabasco	0	3½	to	0	6
Mexican	0	3	to	0	4
Cuba	0	3½	to	0	6
Panama	0	3¼	to	0	5
African	0	3¼	to	0	5

Honduras (in the north of England and Liverpool sometimes called bay-wood) is imported in logs 24 inches to 48 inches square and 12 feet to 14 feet in length. Sometimes planks 5 feet wide are imported, but very seldom. It is distinguished from Spanish mahogany by its absence of figure, by its even grain, and grey specks instead of white, as in the former. It is mostly shipped from Belize.

Panama is similar to Honduras, but is misshapen and in short lengths, and so badly prepared that it is somewhat unpopular. Is shipped from Colon and Darien.

Mexican is also similar to Honduras. Was imported in logs 15 inches to 36 inches square and 20 feet to 30 feet in length, but of late has been mostly of small sizes. Shipped from Minatitlan, Tecolutla, Tlacotalpam, &c.

Tabasco is, next to Spanish, the best mahogany brought into the market. Is imported in logs 15 inches to 36 inches and 20 feet to 30 feet in length. The better qualities have good figure and are very often used when Spanish is specified. Shipped from Chiltepec, Laguna, Tonala, Santa Ana, &c.

Cuba (Spanish) is the best mahogany imported, and fetches the largest price. Is imported in logs 12 inches to 24 inches square and 20 feet to 30 feet in length, but the logs get smaller year by year. It is distinguishable by small white specks. Is shipped from Gibara, Nuevitas, St. Jago, &c.

St. Domingo (*Hayti*) is as good or better than Cuba mahogany, but always smaller, and but little comes into the market now. Is sometimes to be had 20 inches to 26 inches square, and about 10 feet in length.

African.—This mahogany is of recent introduction, but is good and well prepared. It value is rising. Is imported in logs 12 inches to 36 inches square, and some as much as 36 feet in length; is shipped from Axim, Assinee, &c. The Liverpool import in 1888 was 88,499 feet; in 1893, 4,983,357 feet. The waste in conversion of mahogany is 5 per cent. less than on wainscot—i.e. 7½ per cent. sawing and 7½ per cent. conversion = 15 per cent.

The London timber merchants' present price for *Honduras* mahogany is as follows. *Honduras*:

½ in.	¾ in.	1 in.	1¼ in.	1½ in.	1¾ in.	2 in.
s. d.	s. d.	s. d.	s. d.	s. d.	s. d.	s. d.
0 5	0 7	0 8	0 10	1 0	1 2	1 4
Add 7½ per cent. for conversion:—						
0 0⅜	0 0½	0 0⅝	0 0¾	0 0⅞	0 1	0 1¼
0 5⅜	0 7½	0 8⅝	0 10¾	1 0⅞	1 3	1 5¼

Good Tabasco costs 20 per cent. more than Honduras, as follows:—

½ in.	¾ in.	1 in.	1¼ in.	1½ in.	1¾ in.	2 in.
s. d.	s. d.	s. d.	s. d.	s. d.	s. d.	s. d.
0 6	0 8⅜	0 9⅝	1 0	1 2⅜	1 4¾	1 7⅛
Add 7½ per cent. for conversion:—						
0 0½	0 0⅝	0 0¾	0 0⅞	0 1⅛	0 1¼	0 1⅜
0 6½	0 9	0 10⅜	1 0⅞	1 3½	1 6	1 8½

Good Spanish (*Cuba*) costs 50 per cent. more than Honduras, as follows:—

½ in.	¾ in.	1 in.	1¼ in.	1½ in.	1¾ in.	2 in.
s. d.	s. d.	s. d.	s. d.	s. d.	s. d.	s. d.
0 7½	0 10½	1 0	1 3	1 6	1 9	2 0
Add 7½ per cent. for conversion:—						
0 0⅝	0 0¾	0 0⅞	0 1⅛	0 1¼	0 1⅜	0 1¾
0 8⅛	0 11¼	1 0⅞	1 4⅛	1 7¼	1 10⅜	2 1¾

The *labour* on *Honduras* mahogany is twice that on deal; on *Spanish* mahogany, thrice that on deal.

Approximately, labour and materials for work in Honduras mahogany is *thrice* deal; Spanish mahogany, *four* times deal. The different proportion of labour to material in the various kinds ot work will vary the comparative value; but it is absolutely necessary in the case of pricing in large quantities to adopt some generalisation of uniform application.

Adopting, for the purpose of illustration and comparison, the same items as before, we have, for the deal skirting in two pieces, material 2¾*d.*, labour 7*d.*, total 9¾*d.* Excluding the deal fillet and its fixing, we have 8¾*d.* remainder. If we increase the labour constants for deal by 100 per cent., and adopt the prices for *Honduras* mahogany in the table, the calculation will be as follows:—

ft.	in.			*s.*	*d.*
1	0	run	of 1 in. by 4½ in. mahogany at 8⅝*d.*	0	3⅜
1	0	„	of 1¼ in. by 7 in. mahogany at 10¾*d.*	0	6¼
1	0	„	labour to moulding, 3 in. girth, at ·144 ft. supl. of 8*s.* 9*d.*	0	3¾
1	0	„	of labour to moulding, 1 in. girth, at ·024 of 8*s.* 9*d.*	0	2½
1	4	super.	planing, at ·020 of 8*s.* 9*d.*	0	2¾
1	0	run	fixing, at ·030 of 8*s.* 9*d.*	0	3
1	0	„	rebated edge to 1¼ in., ·006 of 8*s.* 9*d.*	0	0½
1	0	„	rebated edge to 1 in., ·006 of 8*s.* 9*d.*	0	0½
				1	10⅝
			Add 10 per cent. for extra care and protection ..	0	2¼
			Cost per foot run	2	1

Three times 8¾*d.* is 2*s.* 2¼*d.*

Applying the same process to the calculation of the value of 1¼-inch square framed doors, we find the material valued at 2¾*d.*, the labour 5*d.* = 7¾*d.* total. If we increase the labour constants by 100 per cent., and adopt the prices for Honduras mahogany in the table, the calculation will be as follows:—

						s.	*d.*
2/	3	0					
		9					
	28	2					
		4½	8	8			
			13	2	supl. 1¼ in. mahogany at 10¾*d.*	11	9½
2/	1	1					
	1	8	3	7			
2/	1	1					
	3	9	8	2			
			11	9	supl. ⅜ in. mahogany at 7½*d.*	7	4
					Carried forward	19	1½

	s.	d.
Brought forward	19	1½
Glass paper	0	1½
Glue	0	3½
21)	19	6½
Cost of material per foot supl.	0	11⅛
Labour making, ·084 at 8s. 9d.	0	8⅞
Labour fitting and hanging, ·016 at 8s. 9d.	0	1⅝
	1	9⅝
Add 10 per cent. for extra care and protection	0	2¼
Cost per foot supl.	2	0

Three times 7¾d. = 1s. 11¼d.

Applying the same process to the calculation of the value of 2-inch moulded casement in a former article, we have for deal 6⅝d. If we increase the labour constants by 100 per cent., and adopt the prices for Honduras mahogany in the table, the calculation will be as follows:—

2	6		
	4½		11
13	6		
	2½	2	6
		3	5

Say—

ft.	in.		s.	d.
3	6	super. 2 in. mahogany, at 1s. 5¼d.	5	0⅜
7	0	„ labour planing, at ·020 of 8s. 9d.	1	2¾
14	4	run labour moulding, at ·024 of 8s. 9d.	3	0
14	4	„ „ rebate, at ·006 of 8s. 9d.	0	9
13	9	super. „ putting together and cleaning up, at ·032 of 8s. 9d.	3	10¼
13	9	„ „ hanging, ·032 of 8s. 9d.	3	10¼
		Glue and glass-paper	0	0⅝
		13 ft. 9 in.)	17	9¼
		Per foot super.	1	3
		Add 10 per cent.	0	1½
		2½ times 6⅝d. is 1s. 4½d.	1	4½

Skirting as before, but in Spanish mahogany (Cuba). Constant three times deal.

ft.	in.			s.	d.
1	0	run	of 1 in. by 4½ in. mahogany, at 1*s*. 0⅞*d*.	0	5
1	0	„	of 1¼ in. by 7 in. „ 1*s*. 4⅛*d*.	0	9⅜
1	0	„	labour to moulding 3 in. girth, at ·216 supl. of 8*s*. 9*d*.	0	5½
1	0	„	labour to moulding 1 in. girth, at ·036 of 8*s*. 9*d*.	0	3¾
1	4	super.	planing, at ·030 of 8*s*. 9*d*.	0	4
1	0	run	fixing, at ·045 of 8*s*. 9*d*.	0	4¾
1	0	„	rebated edge to 1¼ in., at ·009 of 8*s*. 9*d*.	0	1
1	0	„	rebated edge to 1 in., at ·009 of 8*s*. 9*d*.	0	1
				2	10⅜
			Add 10 per cent. for extra care and protection ..	0	3⅝
			Cost per foot run.	3	2

Four times 8¾*d*. is 2*s*. 11*d*.

1¼ in. four-panel square door as before, but in Spanish mahogany.

						£	s.	d.
2/	3	0						
		9						
	23	2						
		4½	8	8				
			13	2	1¼ in. mahogany at 1*s*. 4⅛*d*.	0	17	8¼
2/	1	1						
	1	8	3	7				
2/	1	1						
	3	9	8	2				
			11	9	⅝ in. mahogany at 11½*d*.	0	11	0
					Glass-paper	0	0	1⅛
					Glue	0	0	3¼
					21)	1	9	1½
					Cost of material per foot super.	0	1	4⅝
					Labour making, ·126 of 8*s*. 9*d*...	0	1	1¼
					Fitting and hanging, ·024 at 8*s*. 9*d*.	0	0	2½
						0	2	8⅜
					Add 10 per cent. for extra care and protection ..	0	0	3¼
					Cost per foot super.	0	2	11⅝

4½ times 7¾*d*. is 2*s*. 11*d*.

There are a few current generalisations about the relative value of handrails which, as they are so frequently made of mahogany, may be reasonably mentioned here:—

That ramped handrail is worth twice straight.
That level circular ditto is worth two and a half times straight.
That wreathed ditto is worth four times straight.

The comparative value of deal and mahogany handrails may be illustrated by a 4½ by 3 inch moulded handrail, assuming a joint and handrail screw every 10 feet in length.

				s.	*d.*
1 0 0 4½	0 5	super.	3 in. deal, at 5¾*d*.	0	2½
1 0 1 3	1 3	„	planing, at ·010 of 8*s*. 9*d*.	0	1¼
1 0 1 0	1 0	„	moulding, at ·072 of 8*s*. 9*d*.	0	7½
			1/10th of labour to joint, and fixing handrail screw, ·085 at 8*s*. 9*d*.	0	1
			1/10th of handrail screw at 2*d*.	0	0¼
		run	labour fixing, at ·015 of 8*s*. 9*d*.	0	1½
			Cost per foot run	1	2

The adjuncts of fixing, as housings, &c., would be separately stated in a bill of quantities.

				s.	*d.*
1 0 0 4½	0 5	super.	3 in. Honduras mahogany at 2*s*.	0	10
1 0 1 3	1 3	„	planing, at ·020 of 8*s*. 9*d*.	0	2½
1 0 1 0	1 0	„	moulding, at ·144 of 8*s*. 9*d*.	1	3
			1/10th of labour to joint and fixing handrail screw ·170 at 8*s*. 9*d*.	0	1¾
			1/10th of handrail screw at 2*d*.	0	0¼
		run	labour fixing, at ·030 of 8*s*. 9*d*.	0	3
				2	8½
			Add 10 per cent. for extra care	0	3¼
			Cost per foot run	2	11¾

2½ times 1*s*. 2*d*. is 2*s*. 11*d*.

Teak.—The use of teak in buildings has within a few years much increased, and for many purposes it is a valuable wood. No timber bears the weather better; it is consequently frequently used for the exposed fittings of ships, such as decks, seats, gratings to hatchways, &c. Vast quantities have, of late years, been used as backing for the armour-plating of war ships, and this demand has increased the price. The lessons taught by the naval war between China and Japan will probably induce a greater use of iron instead of wood, and the demand for naval purposes will decrease.

For sills of sash-frames, for dormer fronts where the wood is to remain uncovered, for hospital floors, and for good internal fittings, it cannot be surpassed.

Malabar produced the best teak, but it is now seldom seen in European markets. The chief supply is drawn from Burmah or Siam, and is shipped from Rangoon, Moulmein, and Bangkok.

It is imported in squared logs from 15 feet to 40 feet in length, and from 12 inches to 27 inches square; also in planks of various widths and thicknesses.

The so-called African teak is not teak, and is an inferior wood of poor quality. Teak, whether log or plank, is sold at the timber sales at per load of 50 cubic feet.

The labour on teak is equal to that on Honduras mahogany (which see). The waste may be taken as about 5 per cent. more than deal—i.e. $7\frac{1}{2}$ per cent. for sawing and $7\frac{1}{2}$ per cent. for conversion = 15 per cent. in all.

The price of good teak is quoted by the retail timber merchant by the inch in thickness: the other thicknesses above inch increasing in uniform proportion. The rates are as follows:—

$\frac{1}{4}$ in.	$\frac{1}{2}$ in.	$\frac{3}{4}$ in.	1 in.	$1\frac{1}{4}$ in.	$1\frac{1}{2}$ in.	$1\frac{3}{4}$ in.	2 in.
s. d.	s. d.	s. d.	s. d.	s. d.	s. d.	s. d.	s. d.
0 $5\frac{1}{4}$	0 $6\frac{3}{4}$	0 $7\frac{3}{4}$	0 9	0 $11\frac{1}{4}$	1 $1\frac{1}{2}$	1 $3\frac{1}{2}$	1 6

Add $7\frac{1}{2}$ per cent. for conversion:—

0 $0\frac{3}{8}$	0 $0\frac{1}{2}$	0 $0\frac{5}{8}$	0 $0\frac{5}{8}$	0 $0\frac{7}{8}$	0 1	0 $1\frac{1}{8}$	0 $1\frac{3}{8}$
0 $6\frac{1}{8}$	0 $7\frac{1}{4}$	0 $8\frac{3}{8}$	0 $9\frac{3}{8}$	1 $0\frac{1}{8}$	1 $2\frac{1}{2}$	1 $4\frac{5}{8}$	1 $7\frac{3}{8}$

Adopting for the purpose of illustration and comparison the same items as before, we have for the deal skirting in two pieces, material $2\frac{3}{4}d.$, labour $7d. = 9\frac{3}{4}d.$ total. Excluding the deal fillet and its fixing, we have $8\frac{1}{4}d.$ remainder. If we increase the labour constants for deal by 100 per cent., and adopt the teak prices in the table of teak, the calculation will be as follows:—

ft.	in.			s.	d.
1	0	run	of 1 in. by 4½ in. teak, at 9⅝d.	0	3⅝
1	0	,,	of 1¼ in. by 7 in. teak, at 12⅛d.	0	7⅛
1	0	,,	labour to moulding 3 in. girth, at ·144 super. of 8s. 9d.	0	3¾
1	0	,,	labour to moulding 1 in. girth, at ·024 of 8s. 9d.	0	2¼
1	4	super.	of planing, at ·020 of 8s. 9d.	0	2¾
1	0	run	fixing, ·030 of 8s. 9d.	0	3
1	0	,,	rebated edge to 1¼ in., ·006 of 8s. 9d.	0	0½
1	0	,,	rebated edge to 1 in., ·006 of 8s. 9d.	0	0½
				1	11¾
			Add 10 per cent. for extra care and protection	0	2½
			Cost per foot run	2	2¼

Three and a-quarter times 8¾d. is 2s. 4½d.

Applying the same process to the calculation of the value of 1¼-inch square framed doors, we find the material valued at 2¾d., the labour 5d. = 7¾d. total. If we increase the labour constants for deal 100 per cent., and adopt the teak prices in the table of teak, the calculation will be as follows:—

						s.	d.
2/	3	0					
		9					
	23	2					
		4½	8	8			
			13	2	1¼ in. teak at 12⅛d.	13	3⅝
2/	1	1					
	1	8	3	7			
2/	1	1					
	3	9	8	2			
			11	9	⅝ in. teak at 8¾d.	8	2⅜
					Glue	0	3¼
					Glass-paper	0	1½
					21)	21	11
					Cost of material per foot super.	1	0½
					Labour making, ·084 of 8s. 9d.	0	8¾
					Labour fitting and hanging, ·016 of 8s. 9d.	0	1¾
						1	11
					Add 10 per cent. for extra care and protection	0	2¼
					Cost per foot super.	2	1¼

Three and a-quarter times 7¾d. equal 1s. 11½d.

American Walnut.—American walnut work is of equal value to wainscot, both as to labour and materials. The labour on some is exceptionally troublesome, but as a general rule wainscot is a fair parallel. Its use for internal fittings is increasing. It is imported in logs, planks and boards of various thicknesses. The leading brand is <K>. The greater part is shipped from Baltimore or Quebec.

American White Wood.—American white wood or Canary white wood is shipped from Baltimore or New York in logs, planks and boards. The better qualities are very finely grained and of good figure. It takes polish well. It is frequently used for shop fittings, and when stained and French polished, bears a fair resemblance to mahogany. The labour is not more expensive, nor the waste greater, than on deal. The timber merchants' prices are quoted by the inch in thickness, the prices for the other thicknesses above inch increasing in uniform proportion.

	½ in.	¾ in.	1 in.	1¼ in.	1½ in.	1¾ in.	2 in.
	d.	*d.*	*d.*	*d.*	*d.*	*d.*	*d.*
	2	3	3½	4⅜	5¼	6⅛	7
Add 5 per cent. for conversion :—	0⅛	0⅛	0⅛	0¼	0¼	0⅜	0⅜
	2⅛	3⅛	3⅝	4⅝	5½	6½	7⅜

Adopting, for the purpose of illustration aud comparison, the same items as before, we have for the deal skirting in two pieces, material 2¾*d.*, labour 7*d.* = 9¾*d.* total. Excluding the deal fillet and its fixing, we have 8¾*d.* remainder. Adopting the labour constants as on deal, and adding 10 per cent. for extra care and protection, the calculation will be as follows :—

ft.	in.			s.	d.
1	0	run	of 1 in. by 4½ in. white wood at 3⅝*d.* per ft. super.	0	1⅝
1	0	„	of 1¼ in. by 7 in. white wood at 4⅜*d.* per ft. super.	0	2⅝
1	0	„	labour to moulding 3 in. girth at ·072 super. of 8*s.* 9*d.*	0	1¾
1	0	„	labour to moulding 1 in. girth at ·012 of 8*s.* 9*d.*	0	1¼
1	4	super.	planing, at ·010 of 8*s.* 9*d.*	0	1⅛
1	0	run	fixing, at ·015 of 8*s.* 9*d.*	0	1⅝
1	0	„	rebated edge to 1¼ in., at ·003 of 8*s.* 9*d.*	0	0¼
1	0	„	rebated edge to 1 in., at ·003 of 8*s.* 9*d.*	0	0¼
				0	10½
			Add 10 per cent. for extra care and protection	0	1
			Cost per foot run	0	11½

One and a-third times 8¾*d.* is 11⅓*d.*

Applying the same process to the calculation of the value of 1¼-inch square-framed doors. We find the material valued at 2¾*d.*, the labour 5*d.* = 7¾*d.* total. Adopting the deal constants for labour, the calculation will be as follows:—

						s.	*d.*
2/	3	0					
		9	4	-			
	23	2					
		4½	8	8			
			13	2	1¼ in. white wood at 4⅝*d.*	5	1
2/	1	1					
	1	8	8	7			
2/	1	1					
	8	9	8	2			
			11	9	¾ in. white wood at 3⅛*d.*	3	0¾
	Glue				.. ..	0	3¼
	Glass-paper				.. ..	0	1½
					21)	8	6¾
					Cost of material per foot super.	0	4⅞
					Labour making, ·042 of 8*s.* 9*d.*	0	4⅜
					Labour fitting and hanging, ·008 of 8*s.* 9*d.*	0	0¾
						0	10⅛
					Add 10 per cent. for extra care and protection	0	1
					Cost per foot super.	0	11⅛

One and one-third times 7¾*d.* is 10½*d.*

Sequoia.—Sequoia, known also as Californian red wood, is imported generally in planks of various thicknesses, 2 inches and upward, and sometimes as wide as 3 feet 6 inches. The shipping port is San Francisco. Some of this wood has a pleasing figure; it is used for internal fittings, French polished. The labour and waste is about equal to that on deal.

The timber merchants' prices are quoted by the inch in thickness, the prices for the other thicknesses above inch increasing in uniform proportion:—

	½ in.	¾ in.	1 in.	1¼ in.	1½ in.	1¾ in.	2 in.
	d.	*d.*	*d.*	*d.*	*d.*	*d.*	*d.*
	2¼	3¼	4	5	6	7	8
Add 5 per cent. for conversion:—							
	0⅛	0⅛	0⅛	0¼	0¼	0⅜	0⅜
	2⅜	3⅜	4⅛	5¼	6¼	7⅜	8⅜

Adopting, for the purpose of illustration and comparison, the same items as before, we have for the deal skirting in two pieces, material 2¾*d.*, labour 7*d.* = 9¾*d.* total. Excluding the deal fillet and its fixing, we have 8¾*d.* remainder. Adopting the labour constants as in deal, and adding 10 per cent. for extra care and protection, the calculation will be as follows:—

ft.	in.			s.	d.
1	0	run	of 1 in. by 4½ in. red wood, at 4½*d.* per foot super.	0	1½
1	0	„	of 1¼ in. by 7 in. red wood, at 5¼*d.* per foot super.	0	3
1	0	„	labour to moulding, 8 in. girth, at ·072 super. of 8*s.* 9*d.*	0	1¾
1	0	„	labour to moulding, 1 in. girth, at ·012 of 8*s.* 9*d.*	0	1¼
1	4	super.	planing, at ·010 of 8*s.* 9*d.*	0	1⅛
1	0	run	fixing, at ·015 of 8*s.* 9*d.*	0	1⅛
1	0	„	rebated edge to 1¼ in., ·003 of 8*s* 9*d.*	0	0¼
1	0	„	„ „ 1 „ „	0	0¼
				0	11
			Add 10 per cent. for extra care and protection	0	1
			Cost per foot run	1	0

One and one-third times 8¾*d.* is 11⅓*d.*

Applying the same process to the calculation of the value of 1¼-inch square-framed doors, we find the material valued at 2⅝*d.*, the labour 5*d.* = 7⅝*d.* total. Adopting the deal constants for labour, the calculation will be as follows:—

						s.	d.
2/	3	0					
		9	4				
	28	2					
		4½	8	8			
			13	2	of 1¼ in. red wood at 5¼*d.*	5	9⅛
2/	1	1					
	1	8	3	7			
2/	1	1					
	3	9	8	2			
			11	9	¾ in. red wood at 3⅜*d.*	3	3⅝
					Glue	0	3½
					Glass-paper	0	1½
					21)	9	5¾
					Cost of material per foot super.	0	5⅜
					Labour making, ·042 at 8*s.* 9*d.*	0	4½
					Labour fitting and hanging, ·008 at 8*s.* 9*d.*	0	0¾
						0	10⅝
					Add 10 per cent. for extra care and protection	0	1
					Cost per foot super.	0	11⅝

One and one-third times 7⅝*d.* is 10⅛*d.*

Elm.—Elm is very rarely used for building work. Occasionally it is seen in stables as wall-lining or boarding to stall divisions. It has no particular merit to recommend it for such purposes. In water it is said to be exceptionally durable.

The labour and waste is equal to that on wainscot. It can be bought of country timber merchants, in sawn planks or scantlings, at the rate of 2*s.* 6*d.* per foot cube. In fair quantities the various thicknesses could be obtained at similar proportional rates. Coffin plank, which is usually of fair quality, is quoted by the timber merchants at per 100 feet of inch, commonly about 17*s.* in the summer, and about 18*s.* in the winter—i.e. 2*d.* and 2¼*d.* per inch; 2*s.* 6*d.* per foot cube is equal to 2½*d.* per inch. If this price be uniformly adopted, the difference will be sufficient to cover any extra sawing, and the prices for various thicknesses would be as follows:—

¾ in.	1 in.	1¼ in.	1½ in.	1¾ in.	2 in.	2½ in.	2¾ in.	3 in.
d.	*d.*	*d.*	*d.*	*d.*	*d.*	*d.*	*d.*	*d.*
1⅞	2½	3⅛	3¾	4⅜	5	6¼	6⅞	7½

The comparison of labour and materials may be investigated in the same manner as recommended for the other kinds of wood.

Ironmongery. — Builders' ironmongery varies very much in quality, and consequently in price. Those manufacturers who have a high reputation for their wares—such as, among others, Chubb, Hobbs, and Tucker and Reeves, of London, and Gibbons, of Wolverhampton, for locks; Baldwin, of Birmingham, for butts, &c. —are able to command high prices; but much of the ironmongery in use is of very poor quality and low priced, and in the case of brasswork the weight of the articles is reduced to the lowest possible limit. If bought by retail, as it often is in the case of small jobbing work, the price is high; for the profits of a retail ironmonger, selling (as he does) articles in small quantities, must, of necessity, bear a large proportion to the original cost.

The builder who has a large business buys his ironmongery in large quantities of the wholesale manufacturer, or he opens an account on special terms with a local ironmonger, who reduces his prices in consideration of the quantity required.

The trade lists of the large houses which deal in builders' ironmongery—such, for instance, as Tonks, of Birmingham, or Pfiel

Stedall, of London—are indispensable to the estimator, and the various trade discounts on the various kinds must be known.

Articles of ironmongery are frequently inadequately described in specifications; butts, of which there are a great variety, being frequently described by their length only. All kinds of ironmongery are conveniently designated by numbers from a trade list.

Knowing the net prices of the articles, the only other considerations will be fixing and screws, and profit.

The constants for fixing to deal will serve as a guide for *similar* articles not mentioned therein. The fixing of butts is properly a part of the hanging of a door; but, although this view is correct, the prices in most bills of quantities will be found to include something for fixing.

It is a common practice with estimators to charge a certain number of pence per inch for bolts and fixing and profit—as 2*d.* for iron bolts, 3*d.* for brass, 2*s.* for the fixing of a mortise lock, &c. Some such ready rules will be found convenient.

For fixing ironmongery to hardwoods, add 25 per cent. on the price of fixing to deal.

Prices of rim locks generally include the furniture; for mortise locks they do not.

RELACQUERING.—Satisfactory repainting of good work requires the removal and refixing of the whole of the ironmongery except hinges and locks; but the lock furniture would be removed also. Often, in such a case, the brass lock furniture and general brasswork would be relacquered. Relacquering should be done with judgment, as it is not an uncommon thing to relacquer old brasswork at a cost beyond the value of new articles of a similar kind.

Relacquering involves, besides the lacquerer's charge, taking off, labelling, conveying to and from the lacquerer's, and refixing.

As a rule, the value of taking off is about two-thirds of the value of original fixing.

The following list gives the prices commonly charged to builders by the wholesale people for relacquering. They will serve as a guide for the pricing of similar articles not mentioned therein:—

		s.	d.
Butts up to 4 in., per pair		0	3
Flush bolts, per inch		0	0¼
Rim lock furniture, including striking plate		0	4¼
Mortise lock furniture ,, ,,		0	6¼
Turnbuckles		0	1¼
Striking plates		0	1½
Escutcheons		0	1
Brass barrel bolts, per inch		0	0¾
Sash fasteners		0	2¼
Casement fasteners		0	3
Sash lifts		0	1½
Letter plates		0	6
Bell levers		0	4

Where the price of wood is calculated for day account, waste need not be considered; where for finished work $\frac{1}{16}$ is a fair average allowance; where finished sizes are specified ¼ should be allowed.

IRONMONGERY.

The trade discount off ironmongery varies from 15 to 30 per cent.

Screws, 70 per cent. off list prices.

Stock mouldings, often as much as 40 per cent. off list prices.

PLUMBER.

The surveyor should observe how the labours in this trade have been measured; some surveyors making many more items of extra labour than others, which will of course make a difference in the prices.

As lead is now rolled in a great variety of thicknesses, the surveyor will, in measuring, find it good policy to test the weight, when he will possibly meet with the curious phenomenon of 3¼ lbs. and 4½ lbs. lead where he had least reason to suspect its use. Weights of pipes should also be tested, "middling" sometimes being put in place of "strong," and composition pipes instead of lead.

The trade discount off plumbers' brasswork is from 10 to 15 per cent.

Discount 2½ per cent. for cash.

Discount off sanitary goods, as w.c.'s and lavatories, 10 per cent.

The allowance for waste on old lead varies from 4 lbs. to 6 lbs. per cwt.; but 4 lbs. is that most generally adopted.

Valuation of Plumbing.—The valuation of plumbing requires a thorough knowledge of the operations of the mechanic: nevertheless, much may be obtained from books. The subject has within the last few years been exhaustively treated in such books as Hellyer's 'The Plumber and Sanitary Houses,' and others of a similar character.

The trade lists of Tylor and Sons, Bolding, Doulton and others are instructive, but the book knowledge should be supplemented by careful observation of the actual work.

The basis of a price will, of course, be the cost of labour and material.

Price of Lead.—The price of milled sheet lead has fluctuated much during recent years, and the market price should be watched. Most of the daily papers publish market reports, and under the heading "Metals" will be found the price of pig lead per ton. The merchant's price of sheet lead is usually about 20s. per ton more than the pig. Lead pipe is sold at 10s. per cwt. above the price of sheet lead, and soil pipe 2s. 6d. per cwt. above the price of sheet lead; drawn square soil pipe 5s. 6d. per cwt. above the price of sheet. The merchants will supply sheet lead cut to sizes for 6d. per cwt. above the price of sheet lead.

Subletting Plumbing.—The larger contractors employ the mechanics direct, but a large proportion of the builders sublet their plumbing. The quality of the labour varies greatly, and not infrequently the weight of lead prescribed is not supplied.

Plumbing as a Provision.—Architects who wish for specially good work include a provisional sum for all the plumbing, based upon the estimate of a practical plumbing firm. The general contractor in such a case has very little to do with the work, except cutting away and making good, and sometimes even this is done by the plumber.

This arrangement is met by a clause in the provisions: " Provide for plumbing, £—." " The whole of the plumber's work will be done by . . ." " The general contractor shall supply all scaffolding and ladders and shall cut away and make good."

Some architects prefer that the sub-contractor shall do his own cutting away and making good.

Plumbing as a Separate Contract.—Unquestionably the best plumbing is done by the men who do nothing else, but their prices are much higher, and there is good reason for this: not only is the work better done, but often the specialist writes the specification and affords an amount of attendance and advice which must somehow be paid for.

When the architect writes a workmanlike specification, he can get competitive prices from a selection of good men, and in such a case the difference between their prices and those of the general contractor is not so large.

Sub-contract for Labour.—Often the building contractor buys all the materials and contracts with a working plumber to lay all the external lead at a price per cwt. including every labour, and to fix all the lead pipes at a price per cwt. inclusive or exclusive of solder and firing.

A common price for laying external lead work by a working plumber is 4s. cwt.; it is sometimes done as low as 3s. 6d. For fixing lead pipe 10s. per cwt. including firing, wall-hooks and running joints.

Delivery of Lead.—The London merchant's price for sheet lead and pipe includes delivery free in London and its suburbs if there is a cart load. When the material is for a country building the price would include delivery at a London railway station; to this price must be added the railway carriage and the cartage thence to the building.

Old Lead.—When an old building is demolished, the old lead is often a valuable item of material. In London and its suburbs an arrangement may be made with the lead merchant to exchange the old lead for new, an allowance of 2s. 6d. and 5 lbs. per cwt. from the weight of the old in payment for the new; the removal from the old roof, and the loading, is the contractor's work, and must be allowed for. The merchant on delivering the new lead will take the old on his return journey.

The distinctions observed in a bill of quantities of plumbing by the competent Quantity-Surveyor pretty clearly mark the differences of price.

Weighing old Lead.—Sometimes a bill of quantities for a work of alteration contains a clause: "The old lead will be weighed as it is removed from the roofs, and the exact quantity will be charged at the settlement of the accounts. The contractor to supply weights and scales and to weigh the lead."

Distinctions.—The price is always fixed at per cwt. "Gutters, flats and flashings" is one category; "Stepped flashings" another, and various extra labours on these, as "bossed ends to rolls," "extra labour and solder to cesspools, dressing lead to mouldings, and any dressing of unusual character, welted lap, close copper nailing, bedding edges of flashing in white lead, lead wedging, dressing lead to secret gutter," etc.

Labourers or "Mates."—The theory is that every plumber shall have a mate in constant attendance; but although there are various works of the plumber which cannot be done without an attendant, there is much that can be and is done without assistance. In dressing lead to flats the mate does a good proportion of the work, and if his service is used to the best advantage he can do a good share of the proper work of the plumber.

In day accounts identical time is always charged for plumber and mate, and in the majority of schedules of prices the two appear in one sentence, thus: "plumber and mate per hour . . . ."

Fire.—In a general way fire is an insignificant item of expense. The pieces of wood which arise from the conversion of the timber are commonly sufficient.

Work in Position.—So much of the work of the plumber is done "in position," that the time spent upon it varies greatly, and even when the work is straightforward and easy of access, the different degrees of skill or industry in the mechanic often affect the results to an important extent.

Contract rates.—The rates for plumbing in a builder's estimate are nearly always too low to be fairly remunerative. Rates for similar work in the estimate of a plumbing specialist are invariably much higher, probably referable to his greater experience of the cost of the work.

Materials.—The present price of plumbers' materials is as follows:—

	s.	d.	
Sheet lead	14	0	per cwt.
ditto, cut to sizes	14	6	,,
Lead pipe	14	6	,,
Soil pipe	16	6	,,
Drawn square soil pipe	19	6	,

	s.	d.	
Tin, in ingots	1	4	per lb.
Solder, in ½ cwt. casts, 68s. per cwt. .. .. =	0	7½	„
Copper nails	1	2	„

Valuation of pipes.—Pipes of the weight required by the water companies including "running" joints.

	d.
1 ft. run ½" pipe, weight 2 lb. per foot (made in 15 ft. lengths), at 14s. 6d. per cwt.	3
Plumber and mate, ·014 of 15s.	2¼
1/15 of a solder joint, solder only (5½d.)	0¼
Wall hooks	0½
Cost per foot	6

	s.	d.
1 ft. run of 1¼" pipe, weight 5⅛ lb. (made in 12 ft. lengths), per foot, at 14s. 6d. per cwt.	0	8
Plumber and mate, ·056 of 15s.	0	6½
1/12 of a solder joint, solder only (10¾d.)	0	1
Wall hooks	0	0¼
Cost per foot	1	3¾

Weight of Pipes as required by the Water Companies.—The following is the average weight of lead pipes as required by the water companies.

Inch.	Lbs. per yd.	Inch.	Lbs. per yd.
⅜	5	1	12
½	6	1¼	16
⅝	7½	1½	18
¾	9	2	24

Weight of Soil Pipes as required by the London County Council.—The following is the weight of soil pipe required by the London County Council regulations.

Inches.		Lbs.	
3½	not less than	67	per 10 ft. length.
4	„	76	„ „
5	„	94	„ „
6	„	112	„ „

Solder.—Solder may be bought of the lead merchants in ½ cwt. casts or in bars by the pound, but the majority of plumbers make it themselves, buying tin by the ingot and using the cuttings of lead arising from the ordinary work of the plumber, melted and mixed with the tin. The quantity of solder required for wiped joints for the various sizes of pipes is as follows :—

The larger pipes 3 inches to 6 inches would be soil pipes and the weight given is for the ordinary wiped joint, but these are seldom required, as the running joints are very often block joints and for these the solder required is much less in quantity.

½ in. pipe	¾ lb.	2½ in. pipe	2¾ lb.
¾ "	1 "	3 "	2 "
1 "	1¼ "	4 "	2½ "
1¼ "	1½ "	5 "	3 "
1½ "	1¾ "	6 "	3½ "
2 "	2¼ "		

cwts.	qrs.	lbs.		£	s.	d.
			Lead and laying in gutters, flats and flashings			

	s.	d.
1 cwt. of lead	14	0
Plumber and labourer, ·233 of 15s.	3	6
Cost per cwt.	17	6

There are cuttings from the lead, but these involve no loss, as in a general builder's business they can be melted and used either for solder, or the running of iron in mortises. No solder would be used except in cesspools.

cwts.	qrs.	lbs.		£	s.	d.
			Lead and laying in stepped flashings ..			

	s.	d.
1 cwt. of lead	14	0
Plumber and labourer, ·300 of 15s.	4	6
	18	6
Deduct cuttings about $\frac{1}{16}$	1	2
Cost per cwt.	17	4

Stepped flashings are customarily measured square, and 12 inches wide. The cuttings are available for solder and running as before mentioned; if not, they would be sold as old lead with the usual allowance for tare.

Soakers.—For soakers, the lead would be cut up into the sizes required, and ordinarily, as the slating or tiling of a building is laid to a uniform gauge, they would be nearly all alike. The charge by the merchant for lead cut to sizes is 6d. per cwt. beyond the price of sheet lead. The general contractor can cut it as cheaply as the merchant. After cutting they would be bent and then given to the tiler or slater to lay them with his tiles or slates.

When they are used for ton slating, or stone slating, the gauge of the courses would vary from eaves to ridge, and there would consequently be varied sizes of soakers; one dimension would however be constant, and there would not be probably more than five or six varieties.

cwts.	qrs.	lbs.	Lead in soakers and fixing by tiler	£	s.	d.

		s.	d.
1 cwt. of lead		14	0
Cutting		0	6
Bending, plumber, 1 hour		0	11
Laying by slater, 1 hour		0	10
Cost per cwt.		16	3

In the case of a varied gauge, the time of the plumber would probably be about twice that in the last example.

Dressing lead to mouldings.

cwts.	qrs.	lbs.	Lead and labour in dressing to mouldings	£	s.	d.

	£	s.	d.
1 cwt. of lead	0	19	0
Plumber and labourer, 1·00 of 15s.	0	15	0
Cost per cwt.	1	14	0

Cesspools.

		No.	–	Extra labour and solder to cesspools	£	s.	d.

No. 1 extra labour and solder to cesspools. (Commonly 10″ × 10″ × 6″ in clear.

	s.	d.
Plumber and labourer, dressing and soldering ·160 of 15s.	2	4¾
3 lbs. solder in angles, 7¼d.	1	9¼
Nails	0	0¼
Cost of each	4	2¾

The lead would be a rectangular sheet with a rectangular piece cut out at each angle, and, after dressing the angles, soldered up.

Soldered dots.

| | | No. | – | Soldered dots and brass screws .. .. .. | | £ | s. | d. |

No. 1 soldered dot and brass screw.

	s.	d.
1 lb. solder	0	7¼
Plumber and labourer, ·033 of 15s.	0	6
Brass screw	–	
Cost of each	1	1¼

Soldered angle.

| | | ft. | in. | | | | | | £ | s. | d. |
| | | – | – | run. | Soldered angle .. .. | | | | | | |

	s.	d.
1 lb. of solder	0	7¼
Plumber, ·050 of 9s. 2d.	0	5½
Cost per foot	1	0¾

Soldered angle as to sinks, cisterns and cesspools.—The angle to a sink of moderate size takes less than that to a large cistern. In the former the lead would be shaved about ¾ inch each way, and the solder of moderate size; in the cistern the lead would be shaved 1 inch each way and the solder heavier.

Angle to sinks takes about 1 lb. of solder, to cisterns 1¼ lb.

Soldering on tacks.

	d.
½ lb. of solder	3¼
Labourer, plumber and mate, ·030 of 15s.	5½
Cost per foot	9

INTERNAL PLUMBING.

Terms "Strong" and "Middling."—Some specifications stipulate that all supply pipes shall be "strong," all wastes "middling," lead. pipe. Strong and middling are not exact terms, and to be definite should form part of a table of weights.

Sedden's 'Builder's Work' gives the following table, but it is purely arbitrary as to the classification.

Inside diameter.	Light.	Middle.	Stout.	Inside diameter.	Light.	Middle.	Stout.
in.	Per yard in lbs.	Per yard in lbs.	Per yard in lbs.	in.	Per yard in lbs.	Per yard in lbs.	Per yard in lbs.
⅜	2½	3½	4½	2	20	25	30
½	3½	4½	5½	2¼	25	30	35
⅝	4½	5½	6½	2½	30	35	40
¾	6	7½	9	3	32	42	52
⅞	7	8½	10	3½	46	51	56
1	8	10	12	4	52	63	74
1¼	11	13	15	4½	60	70	80
1⅜	13	15½	18	5	73	–	84
1½	14	16½	19	6	78	–	92
1¾	15	18	22				

Water Company's Regulations.—Sometimes pipes are specified to be of the weight prescribed by the regulations of the local water company.

A collection of sets of regulations of the various water companies of the district where the estimator's work chiefly lies will be found useful.

Plumber's Brass-work and Sanitary Apparatus.—Valves, unions, boiler-screws, traps, caps and screws, thimbles, and all the various water-closet apparatus, baths, lavatories, etc. are illustrated in such lists as Bolding's, Doulton's, Tylor and Sons', etc. The list prices are all subject to a trade discount of 10 to 15 per cent.

Fixing of Taps, Apparatus, etc.—In the case of taps and brass work, there is little to allow for in the way of fixing beyond the joints, but taps often require adjustment and brass work manipulation, which takes time.

Closet apparatus is often in several pieces, which have to be fitted together, or they have to be taken to pieces and put together again before they are in working order, and for these operations allowance should be made. Moreover, they have to be carried from one part of the building to another and fitted into their position.

Baths have to be hoisted and fitted into their places and into their cradles; lavatories fitted to their tops, etc.

Solder Joints.—The work of the mechanic varies considerably. ome will leave much more solder on joints than others, others will waste solder by splashing it about. When the plumber can do so, he makes the joints on the bench. Often the position of a joint is very inconvenient; at other times water cannot be entirely shut off, and a joint may take three or four times as long as it would under ordinary circumstances.

The joints in pipes, other than branch joints, will vary in their distances, according to the plan of the building and the position of the apparatus. The adoption of the trade lengths of pipes will, however, give a fair average.

⅜-inch to 1-inch pipe is supplied in lengths of 5 yards. There would consequently be a joint every 15 feet.

1¼-inch to 2-inch pipe is supplied in lengths of 4 yards. There would consequently be a joint every 12 feet.

The foregoing is also supplied in coils of 12 to 20 yards. The use of extra lengths such as these without a joint is rare, and may be disregarded.

"Branch joints" or "extra soldered joints" are measured at any junction of one pipe with another one, also to brasswork or apparatus.

Soil pipes being supplied in 10-feet lengths, there must be a joint at every 10 feet. In passing through roofs or walls, joints are often unavoidable, although there may be no bend, or a very slight one. In quantities, these should be measured as being caused by the bend.

In valuing, allow one joint in 10 feet. The weight of the tacks and the labour and solder to them must also be considered.

Proportions of Lead and Tin in Solder.—The most frequent proportions of tin and lead in plumber's solder are one of tin to two of lead.

Price for Fixing Pipes.—A working plumber will fix all the pipes of a building, including solder, wall-hooks and running joints, but exclusive of extra soldered joints (branch joints), sometimes as low as 10s. per cwt.

Wall-Hooks.—Wall-hooks are sold by the cwt. at about 21s. per cwt., and vary from 2½ inches to 6 inches in length. They are rarely used shorter than 3 inches. Of the various lengths the following numbers weigh about 1 lb.: 3-inch, 11; 4-inch, 7; 5-inch, 5; 6-inch, 4; they may be assumed to be at an average distance apart of 24 inches.

Valuation of Soldered Joints.—

	s.	d.
One ½″ soldered joint—		
Solder, ¾ lbs. at 7¼*d*.	0	5½
Plumber and mate, ·050 of 15*s*.	0	9
Fire	0	0¼
Cost of each	1	2¾
One 1″ soldered joint—		
Solder, 1¼ lbs. at 7¼*d*.	0	9
Plumber and mate, ·066 of 15*s*.	0	11⅞
Fire	0	0¼
Cost of each	1	9
One 1½″ soldered joint—		
Solder, 1½ lbs. at 7¼*d*.	0	10⅞
Plumber and mate, ·075 of 15*s*.	1	1½
Fire	0	0¼
Cost of each	2	0½
One 2″ soldered joint—		
Solder, 2¼ lbs. at 7¼*d*.	1	4¼
Plumber and mate, ·100 of 15*s*.	1	6
Fire	0	0½
Cost of each	2	10¾
One 4″ soldered joint—		
Solder, 2½ lbs. at 7¼*d*.	1	6¼
Plumber and mate, ·175 of 15*s*.	2	7½
Fire	0	0½
Cost of each	4	2¼

Bends.—Bends in pipes not exceeding 1 inch diameter are made by the plumber as he fixes the pipes, and involve so little labour that they are not worth separate consideration.

The bends on the larger pipes depend upon the weight of the pipes and their angle. The constants must be received as averages for the various kinds.

	Cost of each. s.	d
One 1½″ bend—		
Plumber and mate, ·075 of 15*s*.	1	1½
One 2″ bend—		
Plumber and mate, ·100 of 15*s*.	1	6
One 3″ bend—		
Plumber and mate, ·200 of 15*s*.	3	0
One 4″ bend—		
Plumber and mate, ·300 of 15*s*.	[illegible]	[illegible]

One 36" length of 4" soil pipe, weight 76 lbs. per 10 ft. length, with two bends and two soldered joints—

	s.	d.
3' 0" of 4" soil pipe, 23 lbs. at 16s. 6d. per cwt.	3	4¾
Two 4" bends, at 4s. 6d. (see Bends)	9	0
Two 4" soldered joints (see Joints), at 5s. 10½d.	11	9
Cost of each	24	1¾

Lead Soil Pipes.—When properly measured the branch joints would be separately stated. The running joints will occur every 10 feet, or thereabouts, as the pipe is made in 10 feet lengths.

Lead soil pipe, weight 76 lbs. per 10 foot length, with a tack 10 inches by 9 inches of 6-lb. lead soldered to either side of the pipe alternately and three to each length. The tacks fixed with three 4-inch wall-hooks to each.

	s.	d.
10 ft. of pipe, weight 76 lbs., at 16s. 6d. per cwt.	11	2¼
One 4" soldered joint (as before)	4	2¼
1' 6" supl. of 6-lb. lead in tacks, cut to sizes, 9 lbs. at 14s. cwt.	1	1½
2' 6" run of soldering, at 9d.	1	10½
Fixing, plumber and mate, ·056 of 15s.	0	10
Nine 4" wall hooks, 1½ lbs. at 2¼d.	0	3
	10) 19	5½
Cost per foot run	1	11¼

Galvanized Iron Soil Pipes.—These are sold of various qualities and weights. It is inexpedient to use any of less weight than those prescribed by the London County Council regulations.

The present cost price, with or without ears, is as follows:—

	2¼ in.	3 in.	3½ in.	4 in.	4½ in.	5 in.	6 in.	
Price	1/10	2/-	2/3¼	2/4	3/2	3/5	4/-	per yard.
Weight in lbs.	33	40	48	54	62	69	84	per length.

For coating with Dr. Angus Smith's solution, add 10 per cent.

For galvanizing, add on weight when galvanized, 7s. 3d. per cwt.

These pipes should be jointed with molten lead, caulked with oakum, and "set up" when cold.

Value of 4" pipes as follows:—

	s.	d.
6 ft. of 4" pipe	4	8
Galvanizing ¼ cwt., at 7s. 3d.	3	7½
Nails and plugs	0	2
Fixing, including plugging, plumber and mate, ·100 of 15s.	1	6
One joint, ·020 of 15s.	0	3¼
1¼ lbs. lead, at 1¼d.	0	2
Oakum and fire	0	0¼
6)	10	5¼
Cost per foot run	1	9

Protection of Pipes.—In many buildings the whole of the supply pipes are protected; in others, those supply pipes which are outside the building only. The methods are various. Pipes out of the ground are swathed in hair-felt, or "matted-felt," tied on with string or copper wire. Supply pipes beneath the ground are laid in a wooden box, filled up with dry fine sand. Pipes inside the building are swathed in hair-felt, or matted-felt, tied on as before, or the chase is filled in with "slag-wool" (silicate cotton), packed to a definite consistency.

The net price of 32 oz. felt, in sheets 34" × 20" is 1¼d. per sq. ft.
The net price of matted felt in strips 24 ft. long, 4" wide, is 2d. per sq. ft. (about ¾d. per foot run).
The net price of loose silicate cotton is 8s. per cwt.

Cover the pipes normally charged with water with 32-oz. hair-felt, bound on with string.

	d.
1 ft. run strip of felt, say 4" wide	0½
Cutting into strips and binding on	1½
String	0¼
Cost per foot run	2¼

Cover the pipes normally charged with water with Croggon's matted-felt, 1 inch thick, wound *spirally* around the pipe, and secured with copper wire.

	d.
1 ft. run of matted felt 4" wide	0¾
Winding around pipe and securing the junctions of the strips with wire	1¼
Wire	0¾
Cost per foot run	2¾

PRICES.

Pack chase for pipe 9 inches by 5 inches with slag-wool, rammed to a consistency of 12 lbs. per cubic foot.

		d.
⅓ of a cubic foot at 12 lbs. = 4 lbs. at 8s. per cwt.	..	3½
Packing ⅓ of ·083 of 15s.		5
Cost per foot run		8½

Slag-wool, moderately compressed, will weigh about 10 lbs. per foot cube. When packed in a chase, enclosing several pipes, the work is often troublesome.

Lay the water pipes, which are in the ground, in 1-inch deal rough box, 8 inches by 3 inches clear, nailed together, tar the wood inside and out, and fill the box with fine dry sand.

1·0			d.
1·8	1' 8" supl. of 1" deal nailed 2d., labour and material	..	3¼
—	3' 0" supl. of tarring, 2¼d. per yard		¾
	⅟₁₂th of a foot cube of dry fine sand }		¼
	Sand 8s. per yard, screening 7d.		
	Cost per foot run		4¼

Domestic Hot-water Supply.—The common practice is to measure the pipes to include all their adjuncts as follows:—

ft.	in.			£	s.	d.
		run	¾" best wrought-iron welded steam tubing, jointed in red lead cement, including all tees, bends, elbows, connectors, etc. and fixing complete			

The manufacturers generally adopt a printed list, in which the prices never vary, but are modified by a discount, which rises or falls with the price of iron. The present discounts off the list are as follows:—

	Per cent.
Black	60
Steam	50
Galvanized steam	37½

Steam pipe is stamped "steam" on the sockets and connections; ordinary barrel painted is sometimes supplied for hot-water work by dishonest persons. The normal list is published in all the price books.

It is a common custom with contractors in day accounts, to

charge small quantities at the list price and the time for fixing, and this is a reasonable practice.

	d.
Boy per hour	4
Fitter per hour	10

Percentage for Fittings to Pipes.—In pricing hot-water pipes it is usual to add a percentage on the value of the pipe for the tees, bends, elbows, etc. These vary in number, according to the plan of the building and the position of the apparatus. About 50 per cent. of the net cost of the pipe is a reasonable average. An exact calculation is impossible in preparing quantities, as the course of the pipes must be assumed. In measuring from the actual work they may be counted if it is preferred; but this is rarely done.

Valuation of Hot-water Pipes.—

	d.
1 foot of ¾" steam pipe	2¼
Fixing ·015 of 11s. 8d.	2
Hooks and red lead	¼
Tees, bends, etc., 50 per cent. of 2¼d.	1⅛
Cost per foot run	5¾

	d.
1 foot of ⅞" steam pipe	3
Fixing ·018 of 11s. 8d.	2⅛
Hooks and red lead	¼
Tees, bends, etc., 50 per cent. of 3d.	1½
Cost per foot run	7¼

	d.
1 foot of 1" steam pipe	4¼
Fixing ·020 of 11s. 8d.	2¾
Hooks and red lead	¼
Tees, bends, etc., 50 per cent. of 4¼d.	2¼
Cost per foot run	9½

	d.
1 foot of 1¼" steam pipe	6
Fixing ·020 of 11s. 8d.	2¾
Hooks and red lead	¼
Tees, bends, etc., 50 per cent. of 6d.	3
Cost per foot run	12

Heating Apparatus.—The heating of a building is rarely done by the general contractor. The systems are so various that the work is nearly always done by a specialist, and a sum of money included in the general contract as a provision.

The treatment of this provisional sum should be observed : and whether any profit has been allowed for the general contractor. If not, he should add it, also who is to do the cutting away, attendance and making good (sometimes the specialist does it himself).

Boiler Setting.—If there is a boiler to set, the labour and materials should be charged day account.

Sometimes the specialist sets it himself, and the general contractor supplies the materials.

None of the above mentioned work of the general contractor will pay at the normal prices of work, because of the inevitable delays and general waste of time.

Cast-iron Heating Pipes.—Usually the pipes will be cast-iron socketed pipes with rust-cement joints and occasional expansion joints. They are fully illustrated in the catalogues of Bailey, Pegg, of Bankside, and the General Iron-foundry Co., Upper Thames Street.

ZINC WORKER.

Sub-letting.—Zinc work is nearly always sub-let by contractors, except when quite a small quantity.

Eaves gutters and rain-water pipes.—Eaves gutters and rain-water pipes are commonly bought from a zinc worker, and fixed by the general contractor.

Zinc Work done Cheapest by Specialist.—Zinc work can always be supplied by the specialist at a lower price than the general contractor can do it by his own men.

The present rate of wages is—

	d.
Zinc worker	9¼
Labourer	7

Price of Zinc.—The present price of zinc is—

	£	s.	d.
Vieille-Montagne, per ton	26	0	0
Silesian, per ton	25	10	0

The difference in cost between the two is less than ¼d. per foot superficial, and the Vieille-Montagne is decidedly the better.

Valuation of zinc.—A square foot of V.-M. No. 16 gauge zinc work laid with square roll caps, solid stopped ends, holding-down clips and patent saddle plates laid without solder, would be as follows:—

	d.
1 ft. of zinc	4¼
Plumber and labourer, ·015 of 13s. 9d.	2¼
	6¾

The specialist would do this work for about 6d.

Ornamental Zinc Work.—There are several English price lists in which ornamental articles of zinc are illustrated, such as dormer fronts, crestings, finials, etc., but the French lists present a much greater variety, the use of zinc in France being so much more general than in Great Britain.

Copper.—The lower price of copper during recent years has increased its use in building. It has always been popular for the covering of domes and cupolas, now it is frequently used for the covering of flats.

The ordinary plumber not being used to the material, if he is able to lay it at all, takes an unduly long time over it. Consequently copper work is nearly always sub-let. The specialist is able to buy it in large quantities, and can employ men who are used to copper work. Roofs of small cupolas and turrets will often cost 50 per cent. more for a covering of copper than of lead, and besides this the travelling expenses of a coppersmith are incurred.

Flats may be covered 16 oz. copper nearly as cheaply as with 6 lb. lead.

The present prime cost price for 16 oz. copper laid with welted caps, ends and saddles is 11¾d. per foot supl. of copper used.
Ditto, with standing up welts without rolls, 11¼d.

The price of copper in the ingot is at present about 53l. per ton. Best copper sheets about 20 per cent, more, say 64l. per ton.

There is another quality called "best selected" about 54l. per ton, which is commonly used = 9¾d. per lb.

	d.
1 ft. 16-oz. copper	9¾
Laying	3
	12¾

Comparison of the cost of covering of a small flat 10 feet by 5 feet in lead with 2 rolls and copper, lead at the price previously given, copper at Ewart and Sons' current rate, may perhaps be useful.

						£	s.	d.
	10	0	5	0				
3/8	2	0	0	6				
2/6	1	0	0	2				
	13	0	5	8				
	13	0						
	5	8						
			73	8	6 lbs. lead in flat = 4 cwt., at 17s. 6d...	3	10	0
	6/				Bossed ends to rolls, at 5d.	0	2	6
	10	8			Close copper nailing, at 3d.	0	2	6
						3	15	0
	10	0	5	0				
3/7	1	9	0	6	turn up			
			0	3	drip			
3/7	1	0	0	2	turn down			
	12	9	5	11				
	12	9						
	5	11						
			75	5	16 oz. copper			
6/	0	4						
	0	3						
			0	6	Ditto saddle clips			
3/	0	3						
	0	3	0	2				
			76	1	at 11¾d.	3	14	5
	10	0			Close copper nailing, at 3d.	0	2	6
						3	16	11

PLASTERER.

The chief differences in the quality of plastering are produced by the use of a deficient quantity of lime, inferior sand, inferior quality and thickness of laths, deficiency of hair (quantity and quality), saving of nails by overlapping the laths, etc.

Uncertainty of Cost.—The scarcity of plasterers, partly referable to the discouragement of apprenticeship by their trade union, the difficulty of management of them as mechanics, and the much smaller quantity of work they do compared with that done in former years, has raised prices in this trade, and made any anticipation of ultimate cost very uncertain. As a consequence contractors protect themselves by a generally increased set of rates.

Sub-letting.—Plastering in past years was frequently sub-let by contractors, and many of the sub-contractors did the work remarkably well. The plasterers' trade union, by repeated strikes, has made this arrangement difficult if not impossible. The unions do not, however, forbid lathing by the lath-render, and as they do nothing else they can do it at less cost than the general contractor's men.

Labourers.—The arrangements for the supply of materials to the mechanic and the necessary attendance upon him have changed within recent years. Formerly the operative plasterer was served by a hawk-boy, who handed up the stuff ready mixed to the men on the scaffold, a labourer having previously prepared it. The boy has almost disappeared. As a rule the plasterers are supplied by a labourer, who, having previously mixed the stuff, brings it into the room where the plasterers are working, keeps it tempered and hands it to each workman as he requires it. In ordinary cases one labourer will supply two or three plasterers.

Materials delivered, the prices of cement including use of sacks or casks:—

	s.	d.	
Lime	11	0	per yard.
Sand	8	0	,,
Portland cement	39	0	per ton.
Fine Keene's cement	72	6	,,
Coarse ditto	50	0	,,
Selenitic lime	29	0	,,
Single laths	30	0	per load = 30 bundles
Lath and half	45	0	,, ,,
Hair	7	0	per cwt.
Coarse plaster	24	6	per ton.
Fine ditto	27	6	,, = 1s. 1¼ per bushel.
⅞" cut steel lath nails	10	1	per cwt.
1" ditto	10	1	,,

Render—

		d.
1/9 yard of lime, at 11s.	..	1¾
1/54 yard of sand, at 8s.	..	2
2¼ oz. hair, at ¾d. per lb.	..	0¼
Plasterer, ·013 of 9s. 2d.	..	1¼
Labourer, ·007 of 5s. 10d.	..	0½
		6

Render and set—

	d.
$\frac{1}{54}$ yard of lime, at 11*s.*	$2\frac{1}{2}$
$\frac{1}{54}$ yard of sand, at 8*s.*	2
8 oz. hair, at $\frac{3}{4}$*d.* per lb.	$0\frac{1}{4}$
Plasterer, ·023 of 9*s.* 2*d.*	$2\frac{1}{2}$
Labourer, ·012 of 5*s.* 10*d.*	1
Cost per yard	$8\frac{1}{4}$

Render, float and set—

	s.	d.
$\frac{1}{45}$ yard of lime, at 11*s.*	0	3
$\frac{1}{35}$ yard of sand, at 8*s.*	0	$2\frac{3}{4}$
$8\frac{1}{2}$ oz. hair, at $\frac{3}{4}$*d.* per lb.	0	$0\frac{1}{4}$
Plasterer, ·038 of 9*s.* 2*d.*	0	$4\frac{1}{4}$
Labourer, ·019 of 5*s.* 10*d.*	0	$1\frac{1}{4}$
	0	$11\frac{1}{2}$

Lathing with lath and half—

$\frac{1}{4}$ bundle of laths, at 1*s.* 6*d.*	$4\frac{1}{2}$
Plasterer, ·020 of 9*s.* 2*d.*	$2\frac{1}{4}$
Labourer, ·010 of 5*s.* 10*d.*	$0\frac{3}{4}$
$\frac{1}{3}$ lb. nails, at $1\frac{1}{2}$*d.*	$0\frac{1}{2}$
	8

Lathing is sometimes done by lath-renders, the ordinary price being $5\frac{1}{2}$*d.* per yard for single lathing, $7\frac{1}{2}$*d.* per yard for lath and half.

Lath and half plaster float and set partitions.

	s.	d.
Lathing as before		8
$\frac{1}{45}$ yard of lime, at 11*s.*		3
$\frac{1}{35}$ yard of sand, at 8*s.*		$2\frac{3}{4}$
4 oz. of hair, $\frac{3}{4}$*d.* per lb.		$\frac{1}{4}$
Plasterer, ·042 at 9*s.* 2*d.*		$4\frac{1}{2}$
Labourer, ·019 of 5*s.* 10*d.*		$1\frac{1}{4}$
	1	$7\frac{3}{4}$

Lath plaster float and set ceilings.

	s.	d.
$\frac{1}{4}$ bundle of laths, at 1*s.* 6*d.*		$4\frac{1}{2}$
Plasterer, ·020 of 9*s.* 2*d.*		$2\frac{1}{4}$
Labourer, ·010 of 5*s.* 10*d.*		$\frac{3}{4}$
$\frac{1}{45}$ yard of lime, at 11*s.*		3
$\frac{1}{35}$ yard of sand, at 8*s.*		$2\frac{3}{4}$
4 oz. hair, $\frac{3}{4}$*d.* per lb.		$\frac{1}{4}$
$\frac{1}{3}$ lb. nails, at $1\frac{1}{2}$*d.*		$\frac{1}{2}$
Plasterer, ·046 at 9*s.* 2*d.*		5
Labourer, ·019 at 5*s.* 10*d.*		$1\frac{1}{4}$
	1	$8\frac{1}{4}$

Lath, plaster, float and set, *flueing* ceilings.

	d.
$\frac{3}{8}$ bundle of laths, at 1s. 6*d*.	7
Plasterer, ·020 of 9s. 2*d*.	$2\frac{1}{4}$
Labourer, ·010 of 5s. 10*d*.	$\frac{3}{4}$
$\frac{1}{45}$ yard of lime, at 11*s*.	3
$\frac{1}{36}$ yard of sand, at 8*s*.	$2\frac{5}{8}$
4 oz. hair, at $\frac{3}{4}$*d*. per lb.	$\frac{1}{4}$
Plasterer, ·069 of 9s. 2*d*.	$7\frac{1}{2}$
Labourer, ·019 of 5s. 10*d*.	$1\frac{1}{4}$
$\frac{1}{2}$ lb. nails, at $1\frac{1}{2}$*d*.	$\frac{1}{2}$
	2 $1\frac{1}{2}$

Selenitic Plastering.—Its recommendations are: its superior hardness, in which it nearly approaches Portland cement; the greater quantity of sand that may be used, compared with ordinary lime; the greater rapidity with which one coat may follow another; the saving of one coat on brickwork.

For selenitic plastering, first coat on brickwork: two bushels of selenitic lime to ten or twelve bushels of clean sharp sand.

First and second coat for lathwork: two bushels of selenitic lime, six or eight bushels of clean sharp sand; two hods of well-haired ordinary lime putty.

Third or setting coat: two bushels of carefully sifted selenitic lime, two hods of ordinary lime putty, three bushels of fine washed sand.

Its value is generally considered to be equal to plastering with ordinary lime.

Rough Cast.—Rough cast is specified in a great variety of ways. If in lime, the valuation would be based on the cost of render, float and set; if in cement, on plain face. If the work is in panels between timbers, it would be technically narrow widths, and should be so valued. A little advantage to the contractor is derived from the practice of measuring over the timbers and not deducting them.

Applying this reasoning to the following description, we have as follows:

Rough cast made with two parts clean, sharp, washed shingle and sand and one of cement between timbers—

	s.	d.
Portland cement (see plain face)	1	11½
Add for narrow widths, 20 per cent.	0	4¾
Cost per yard	2	4

The advantage derived from the measurement of the timbers in the superficial quantity will contribute to reduce the cost of the narrow widths. Probably, in most cases one-third of the 4¾d. may be safely deducted.

Fibrous Plaster.—When it is necessary to paint or paper on walls or ceilings promptly, fibrous plaster slabs may be used; they may be obtained with a finished surface, in which case the stopping of the joints is all that is required for completion, or they may be had with a rough surface which requires a coat of plaster or Keene's cement.

The following are two ordinary descriptions:—

A. "Cover the quarter partitions and ceilings with approved well-seasoned fibrous plaster slabs, made with a smoothly finished surface, screwed to the timbers with 1¼-inch galvanized iron screws and carefully stopped with Keene's cement;

or,

B. "Cover the quarter partitions and ceilings with approved well-seasoned fibrous plaster slabs of ordinary surface, screwed to the timbers with 1¼-inch galvanized iron screws and carefully stopped and finished with a thin coat of Keene's cement trowelled."

Either kind, rough or smooth, may be nailed with galvanized iron nails driven into the joints of the brickwork, or fixed with galvanized iron screws to 1-inch by 3-inch deal grounds plugged to the walls.

The present price for the ordinary surface slabs is 10d. to 1s. 3d. per yard; for the slabs with finished surface, 2s. 6d. to 4s. 6d. The ordinary size of the slabs is 3 feet 6 inches by 2 feet 6 inches.

A.

	s.	d.
Fibrous plaster slab	2	6
Waste, 5 per cent.		1½
1 doz. screws		1¼
Fixing, carpenter and labourer, ·024 of 14s. 7d.		4¼
Stopping material		1½
Plasterer, ·015 of 9s. 2d.		1¾
Cost per yard	3	4½

B.

	£.	d.
Fibrous plaster slab	1	0
Waste, 5 per cent.	0	0¼
1 doz. screws	0	1¼
Fixing, carpenter and labourer, ·024 of 14s. 7d.	0	4¼
1/10 bushel Keene's cement, at 2s. 9d.	0	3¼
1/10 bushel washed sand, at 5½d.	0	0¾
Plasterer, ·013 at 9s. 2d.	0	1¼
Labourer, ·007 at 5s. 10d.	0	0½
Cost per yard	2	0¼

Mouldings.—Mouldings over 12 inches girth are usually measured and valued by the superficial foot. Those not exceeding 12 inches girth by the foot run.

Mitres, stopped ends, returned and mitred ends, etc., are usually priced by estimators at some proportion of a lineal foot of the moulding. The following are some of the ordinary practices:—

A mitre at the value of a lineal foot of the moulding.

A stopped end at the value of a lineal foot of the moulding (some estimators half a foot).

A returned and mitred end at the value of three-quarters of a lineal foot (some as much as two feet lineal).

An irregular mitre at the value of one and a half lineal foot of the moulding.

The estimator usually calculates the price of cornices from a rate per inch of girth, as a 12-inch cornice at ¾d. per inch = 9d. per lineal foot.

The smaller mouldings are worth more per foot than the larger ones, the price per foot increasing as the mouldings decrease in girth.

Enrichments.—Plaster enrichments in great variety of pattern and girth, also modillions, trusses, etc., may be bought ready cast. Those commonly used in the classic entablatures are readily obtainable. When they are specially designed, they must be modelled and cast, and the cost is very much greater. The general contractor cannot produce enrichments as cheaply as he can purchase them; he therefore buys them of a plasterer's modeller, in which case the value will be the prime cost of the enrichment plus the fixing and the mitres.

Composition enrichments, made of glue, whiting, resin and oil are also made by special makers, but they are more frequently used in conjunction with woodwork.

Carton-pierre is often used instead of plaster for cornices and enrichments, in which case, a price per foot run is usually stated in a specification or bill of quantities. They are best and most economically fixed by the maker; the most practical course is to send the particulars to a maker of such things, get his price, and add fixing and profit.

Plaster of Paris.—This work, if plain, is usually in narrow widths as in friezes, coves, fascias, etc.; if moulded, in cornices or strings; if cast, in enrichments, trusses, etc.

The usual proportion of material used in fine plaster work is one of plaster of Paris to one of putty.

Putty.—The putty is one part lime to one part sand. About one-fourth of each may be allowed for waste.

Value of a cubic yard of putty.

	s.	d.
1¼ yard of lime at 11s. 0d.	13	9
1¼ yard of sand at 8s.	10	0
2) 23	9
11	10¼
Labourer, sifting and running lime, screening sand and mixing putty, ·100 of 5s. 10d.	0	11
Cost per yard	12	9½

Plaster Moulded Cornices.—These are dubbed out with coarse and finished with fine plaster.

Mitres, stopped ends, returned and mitred ends, etc. are usually priced by estimators at some proportion of a lineal foot of the moulding. The following are some of the ordinary practices:—

A mitre at the value of a lineal foot of the moulding.

A stopped end at the value of a lineal foot of the moulding (some estimators half a foot.)

A returned and mitred end at the value of three-quarters of a lineal foot. (Some estimators allow as much as two feet lineal.)

An irregular mitre at the value of one and a half lineal foot of moulding.

The smaller mouldings are worth more per foot than the

larger ones, the price per foot increasing as the mouldings decrease in girth.

Moulds.—In the ordinary bill of Quantities the whole of the cornices not exceeding 12 inches girth are in one category, all above 12 inches girth in another; although there may be, and in a large building usually are, a considerable variety of girths and sections, and every variety requires a different zinc mould, an additional expense which the estimator does not always consider.

Value of plaster moulded cornice per yard including preparation of moulds—12 inches girth and over:—

	s.	*d.*
½ bushel of fine plaster at 1*s.* 1½*d.*	0	6¾
½ bushel of putty (see detail of putty) at 7½*d.*	0	3¾
Plasterer, ·500 of 9*s.* 2*d.*	4	7
Labourer, ·250 of 5*s.* 10*d.*	1	5½
9)	6	11
Cost per foot		9¼

The value of plaster moulded cornice under 12 inches girth is approximately 20 per cent. more than the last.

The value of circular cornices is about 25 per cent. more than straight.

Enrichments.—Plaster enrichments of stock patterns will cost for labour and material for each inch in girth per foot run:—straight 1½*d.*, circular 50 per cent. more.

Soffits and friezes over 12 inches wide, per foot superficial:—

	s.	*d.*
⅜ bushel of putty, at 7½*d.*	0	2¾
⅜ bushel of fine plaster, at 1*s.* 1½*d.*	0	5
Plasterer, ·250 of 9*s.* 2*d.*	2	3½
Labourer, ·125 of 5*s.* 10*d.*	0	8¾
Cost per yard .. 9)	3	8
Cost per foot		5

The value of plaster soffit or frieze under 12 inches wide is about 20 per cent. more than the last.

The value of plaster soffit or frieze circular is about 25 per cent. more than straight.

Portland Cement.—In good work the material should be mixed in the proportion of one part of cement to two parts of washed sand.

Render float and trowel=plain face

	s.	*d.*
¼ bushel of Portland cement, at 39*s.* ton	0	6
½ bushel of washed sand, at 10*s.* yard	0	2¾
Plasterer, ·100 of 9*s.* 2*d.*	0	11
Labourer, ·050 of 5*s.* 10*d.*	0	3½
	1	11¼

If in narrow widths, add 20 per cent.

Moulding exceeding 12 inches girth per foot superficial, including preparing zinc mould:—

	s.	*d.*
3/10 bushel of Portland cement, at 39*s.* per ton	0	7¼
⅗ bushel of washed sand, at 10*s.* per yard	0	3¼
Plasterer, ·750 of 9*s.* 2*d.*	6	10½
Labourer, ·875 of 5*s.* 10*d.*	2	2¼
Cost per yard .. 9)	9	11¼
Cost per foot	1	1¼

Mouldings under 12 inches girth and over 6 inches girth may be valued at 20 per cent. more than the last.

Mouldings 6 inches girth and under 50 per cent. more those exceeding 12 inches girth.

Mouldings circular may be valued at 25 per cent. more than the straight.

Skirtings.—The value of skirting is most conveniently deduced from that of "plain face."

A Portland cement square skirting 9 inches high, would be valued as follows:—

	s.	*d.*
"Plain face," per yard	1	11¼
Add 20 per cent. for narrow widths	0	4¾
Cost per yard supl.	2	4
Cost per foot	0	3⅛
¾ of 3⅛*d.*	0	2⅜
Arris, ·007 of 9*s.* 2*d.* (plasterer)	0	0¾
Labourer, ·004 of 5*s.* 10*d.*	0	0¼
Cost per foot run	0	3⅜

If circular, add 25 per cent.

Estimators commonly price each mitre at the value of a lineal foot of skirting; stopped ends, half a lineal foot.

Parian and Keene's Cement.—This work is, by some, done on a backing of coarse Parian or Keene's, by others on a backing of Portland. The backing of Portland is likely to be the most durable.

These cements are more difficult to manipulate than those in plaster of Paris, and depraved plasterers are prone to mix plaster of Paris with the cement to make it work more freely.

Parian and Keene's cement are of equal value.

Much of this work is done by contract at the same price as Portland. Other estimators price it at from 33 per cent. to 50 per cent. more.

Mouldings are worth (approximately) 20 per cent. more than in Portland cement.

Plain face on backing of Portland.

	s.	*d.*
Backing—		
⅛ bushel of Portland cement, at 39*s.* per ton	0	4¾
⅜ bushel washed sand, at 10*s.* per yard	0	2¼
Finish—		
⅛ bushel of fine Keene's, at 72*s.* 6*d.* per ton	0	6½
⅛ bushel of washed sand, at 10*s.* per yard	0	1¼
Plasterer, ·100 of 9*s.* 2*d.*	0	11
Labourer, ·050 of 5*s.* 10*d.*	0	3½
Cost per yard	2	5¼

Moulding per foot superficial, including preparing zinc moulds:

	s.	*d.*
⅓ bushel fine Keene's cement, at 72*s.* 6*d.* per ton	1	4½
⅓ bushel of washed sand, at 10*s.* per yard	0	2¾
Plasterer, ·900 of 9*s.* 2*d.*	8	3
Labourer, ·450 of 5*s.* 10*d.*	2	7½
Cost per yard .. .. 9)	12	5¾
Cost per foot	1	4¾

Founder and Smith.

The quality of iron and steel, and consequently the price, vary greatly.

Observe whether rain-water pipes and gutters have been measured net or by the yard, and regulate the price accordingly.

For large quantities of ironwork special prices may be obtained.

Note the way in which iron joists, girders, etc. have been measured, whether the hoisting is included or separately stated.

Iron cisterns are increased in value if not of stock sizes.

The trade discount off stoves and ranges is usually 20 to 25 per cent.

The basis of all pricing of smith and founder's work must be the weight of the article, and this ascertained, the comparative values of the labour on each are easily adjusted.

The value of wrought ironwork may often be deduced by comparing its proportional value with some article of which the value is well known, like straps to roof trusses.

The importance of iron and steel in the construction of buildings has greatly increased within a comparatively few years, and its manufacture has largely developed. It is probable that in Great Britain the limit has been reached despite all our natural advantages in the possession of coal and iron. For this the trade unions will be mainly responsible; their short-sighted policy in restricting the quantity of labour to be done in a day by a workman, the discouragement of apprenticeships and other disreputable vagaries, has already sapped that predominance which England has so long maintained in the markets of the world, and she must inevitably succumb to the better industrial conditions which obtain in America and Germany.

The basis of price throughout the trade is the weight; and before pricing ironwork the market price must be ascertained, as it fluctuates greatly.

The current price of iron and steel is quoted weekly in 'Iron' and 'The Builder.'

For the various qualities of iron and steel and the extra rates on iron, see "Building Construction," vol. iii. (Longmans).

Rolled Joists.—Much of the iron or steel in rolled joists is imported from Belgium and is about 25 per cent. less in price than English.

Structural steelwork, as columns, stanchions, rolled joists, girders, and roof-trusses, are always supplied by an iron merchant, sometimes fixed by the merchant, at other times by the general contractor.

The trade lists such as Dorman and Long, Measures, Moreland, etc., give illustrations of the ordinary sections of rolled iron and steel. Lindsay gives illustrations of steel stanchions and columns bolted up in sections. Dorman and Long illustrate the various connections.

The ordinary limit of length is 30 feet and the depth 10 inches.

Joists above 30 feet in length, 1s. 6d. per cwt. extra, above 10 inches, in depth 6d. per cwt. extra. If over 30 feet long, 2s. per cwt. extra. If cut to exact length, 6s. per ton extra.

If ordered in good time the normal price will be charged. If obtained from stock (where they are kept in lengths of even feet) about 1s. per cwt. extra. This condition, if they are required to exact lengths, will involve cutting, the joist being charged in even feet, and the cut extra.

Cambering of rolled joists will cost 8s. per ton.

Cutting ends to bevel up to 12 inches deep 1s. 6d. per end. Ends notched 1s. 6d. Holes 3d. The simplest holes in position 1s. each; in difficult positions sometimes as much as 2s. 6d.

Compound girders 2s. 6d. per cwt. extra.

In the case of a smith and founder's bill, the common practice is to send it to an iron merchant to price.

If the quantity required is large, the price per ton will be less than for a small quantity.

The iron merchant in stating a price usually attaches some sort of conditions as follows:—

> Time of delivery—From stock.
> Where delivered—On vans, Bloomsbury.
> Terms of payment—Cash, less 2½ per cent.
> Subject to reply—In three days.
> The extras on girders are as follows: 1s. 6d. per foot per ton for lengths over 36 ft., and 6s. per ton if cut to exact lengths. Where no margin is stated on order, we cut to an inch under or over without extra charge.

Any work quoted for upon the girders, such as punching holes, notching out ends, etc., is subject to approval of drawing, and where tests are required, it is understood they are made at our works and are final.

The time for delivery named is subject to the usual Strikes and Accidents clause, or any other unavoidable delay.

Fixing.—The fixing of structural ironwork is worth (approximately) 20s. a ton.

The fixing of ironwork is done by various mechanics.

The bricklayer fixes air-bricks, gratings, hoop-iron, many of the rolled joists, flue-plates, soot-doors, stoves, ranges, coppers, chimney-bars, safe and party-wall doors, covers to inspection pits, etc.

The carpenter fixes bolts, straps, heads and shoes to trusses, flitches, and the ironwork in connection with carpentry. He also fixes eaves-gutters and rain-water pipes.

Cleats and Holes.—The lengths of angle-iron with the holes and bolts used for connecting rolled joists to others transverse to them are always supplied by the merchant who supplies the joists and girders. No general contractor has equal facilities for the work.

Materials.—The following are the current prices of structural steelwork painted two coats at the works and delivered in London. Unloading, getting into building, hoisting, and fixing by general contractor are charged extra on the following rates.

	£ s. d.	
Steel joists, ordinary section	8 10 0	per ton.
Combination girders of steel joists and bottom plates	12 10 0	,,
Riveted stanchions of steel joists, with top and bottom plates as caps and bases	14 0 0	,,
Roof trusses, bolted and riveted together, of L or T section, rafters, L struts, and rods varying from ⅜" to 1¼"	25 0 0	,,
If these are unpainted, 2s. 6d. a ton less.		
Builder's bolts, of various sizes, with heads, nuts and washers, delivered in London	0 18 0	per cwt.

Galvanizing is commonly charged as a percentage, and is greater on small articles than on large. Approximately the distinction is between articles not exceeding 28 lbs. each and articles over 28 lbs. each.

Some of the iron merchants state rates per cwt. for specific items. The following are some of the present rates:—

	Per cwt.			Per cwt.
	s.	d.		s. d.
Air bricks	9	1¼	Manhole covers, under 1 cwt. each	7 8¼
Boilers	9	5¼	Ditto, 1 to 2 cwt. each	7 0¼
Dust-bins	12	7¾	Rain-water pipes	7 2¼
Furnace pans	4	11½	Ditto connections	8 9¾
Eaves gutters	7	10	Soil pipes	6 6¾
Ditto connections	9	5¼	Wall ties	9 9¼
Ditto brackets	10	10¾		

Galvanizing increases the weight of iron articles by about 10 per cent.

Iron Merchants' Prices.—

	s.	d.	
Ends of joists up to 12" bevelled	1	6	each.
Ditto notched	1	6	„
Riveted L-cleats 9" long	2	0	„
Ditto 15" long	3	0	„

CAST IRON: *Structural Work.*—Structural cast iron, as columns, stanchions, etc. is often very elaborately described as to the quality of the iron, that the castings shall be from the second melting, and the like.

The contractor usually sends the bill of quantities to the ironfounder and adopts his price with the addition of a percentage for profit. If the ironfounder does not fix the work, an amount must also be added for fixing.

Patterns.—The patterns are nearly always made by the ironfounder, and his price includes them.

Sometimes there is a stipulation in the specification that the patterns shall be submitted before casting. In such a case the contractor must either have the patterns made by his own men or arrange with the founder to submit them.

Stanchions and Columns.—Cast-iron stanchions have been superseded to a great extent by wrought-iron or steel ones made up of rolled joists with plates bolted and cleated on, as caps and bases.

For cast-iron columns steel or rolled iron bolted up in sections are frequently substituted.

Sections of rolled stanchions and columns are illustrated in most of the iron merchants' trade lists.

General Articles in Cast Iron.—Nearly all the articles of cast iron used in building are supplied by the ironfounder; the general contractor fixes and paints only.

PRICES.

Stoves and ranges are usually selected from a trade list, and are generally subject to a large trade discount off the list prices.

Gratings, eaves-gutters, and rain-water pipes, iron chimney-pieces, baluster-panels, iron staircases, iron casements and sashes, are but a few items of the bewildering variety which is illustrated in the trade catalogue of a large firm.

ft.	in.	run	4" cast-iron rain-water pipe, with ears cast on, and fixed with wrought-iron nails to plugs in brickwork	£	s.	d.

	s.	d.
1' 0" run 4" rain-water pipe	0	6
Nails and plugs	0	0¼
Fixing, including plugging, ·030 of 8s. 9d.	0	3
Cost per foot	0	9¼

Shoes—

	s.	d.
1 shoe	1	1¼
Fixing, ·030 of 8s. 9d.	0	3
Cost of each	1	4¼

Although these are measured "extra only," they nearly always involve some cutting and waste of the pipe, the value of which does not more than balance the cost of the labour.

Plinth bends, 2¼" projection—

	s.	d.
1 plinth bend	1	7
Fixing, ·050 of 8s. 9d.	0	5¼
Cost of each	2	0¼

Although measured "extra only," these nearly always involve some cutting and waste of the pipe.

Awkward plinth bends and swan necks are sometimes made in lead and fitted to the iron pipes; the necessary pattern for iron and special casting, often very expensive, is thus avoided.

Swan necks, 15" projection—

	s.	d.		s.	d.
Swan neck ..	3	2¼			
Ddt. 2 ft. of pipe	1	0			
				2	2¼
Fixing, ·350 of 8s. 9d.				0	5¼
Cost of each				2	7½

Heads—

		s.	d.
1 head		1	5¼
Nails		0	0¼
Fixing, ·040 of 8s. 9d.		0	4¼
Cost of each		1	10

| | | No. | 6 | Cast-iron king heads, weight 112 lbs. each, including pattern and fixing by carpenter | £ | s. | d. |

Iron-founders will always give a price per cwt. including patterns. In this case a pattern casting about 7s. 6d. would be 1s. 3d. for each cwt.

	s.	d.
1 king head, weight 112 lbs., at 8s. per cwt., including pattern	8	0
Fixing, ·300 of 8s. 9d.	2	7½
Cost of each	10	7½

| | | No. | | Cast-iron ornamental newels, P.C. 5s. each, and fixing by smith | £ | s. | d. |

The mortises and lead and running would be part of the mason's work.

	s.	d.
Newel, P.C. at warehouse	5	0
Smith and labourer, ·050 of 14s. 2d.	0	8½
	5	8½

| cwts. | qrs. | lbs. | | No. hollow columns, and hoisting and fixing at various levels | £ | s. | d. |

	s.	d.
Founder's price per cwt., including pattern	8	6
Unloading, hoisting and fixing 1/30 of ·270 of 5s. 10d.	0	9¼
1/30 of ·600 of 8s. 4d.	0	3
Cost per cwt.	9	6½

Materials.—

	£	s.	d.
Columns and stanchions of ordinary pattern, including pattern, not including fixing	8	10	0 per ton.

—	2″	2½″	3″	3½″	4″	4½″	5″	6″	
Rain-water pipes—									
6 feet lengths ..	-/10½	-/11½	1/1¾	1/5	1/5½	2/1	2/4¾	3/0¼	per yard run.
***Heads,* flat or angle**	-/9¾	1/0½	1/2	1/4½	1/5½	2/3	3/-	4/10	each.
Shoes	-/6½	-/7¾	-/9½	-/11¾	1/1½	1/7½	1/10¾	2/10½	„
Boots	-/9½	-/9¾	-/11½	1/1½	1/7½	1/11½	2/8½	3/2½	„
Plinth bends—									
2½″ projection ..	-/9½	-/9¾	-/11½	1/1¼	1/7	1/11½	2/6¾	3/0¼	„
3″ „ ..	-/9¼	-/9¾	-/11½	1/1½	1/7	1/11½	2/6¾	3/0½	„
4½″ „ ..	-/10¾	-/11¾	1/2¼	1/5½	1/11½	2/4¾	3/0¼	3/8¾	„
6″ „ ..	-/10¾	-/11¾	1/2¼	1/5½	1/11½	2/4¾	3/0¼	3/8¾	„
Elbows—									
Square or obtuse	-/7	-/9½	-/10½	1/2¼	1/5½	2/1	2/10½	3/7¾	„
Branch pipes, single	-/10¾	1/1½	1/3	1/7½	2/1	3/1	3/4½	4/3¾	„
Ditto, double ..	1/2¼	1/8	2/0¾	2/6¾	2/10½	3/10	4/7¾	6/2½	„
Offsets—									
3″ projection ..	-/8½	-/9¾	-/11½	1/1½	1/7	1/11½	2/6¾	4/3¾	„
4½″ and 6″ ditto	-/10¾	-/11¾	1/2¼	1/5½	1/10¾	2/4¾	3/0¼	3/8¾	„
9″ projection ..	1/2¼	1/5	1/7¼	1/11½	2/0¾	3/1	3/7¾	4/7¾	„
12″ „ ..	1/5½	1/8	1/10	2/0¾	2/5½	3/6½	4/0¾	5/1½	„
15″ „ ..	1/7¼	1/10½	2/3	2/6¾	3/2¼	4/5¾	4/7½	5/7¼	„
18″ „ ..	1/10½	2/1¾	2/4¾	2/9½	3/6¼	4/5¾	5/6¾	6/1	„
21″ „ ..	2/1¾	2/5½	2/9½	3/2¼	4/3½	6/2¾	7/0½	8/5¾	„
24″ „ ..	2/5½	2/10½	3/3¾	3/8¾	4/11½	7/2¼	8/1½	9/7¼	„
30″ „ ..	3/0¼	3/6½	4/0¾	5/1½	6/4¾	8/7¾	9/11	11/8¼	„

—	3″	3½″	4″	4½″	5″	6″	
Half-round gutters—							
6 feet lengths	-/7½	-/7½	-/7½	-/8¾	-/10½	1/1½	per yard.
Angles and nozzles ..	-/6¼	-/6¼	-/7½	-/9½	-/10½	1/1½	each.
Stop ends	2/3	2/3	2/3	2/8½	3/2¼	4/3¾	dozen.
Union clips	2/8½	2/8½	3/2¼	3/10	4/9¾	6/4¾	„

—	3½″	4″	4½″	5″	6″	
Ogee gutters—						
6 feet lengths	-/9¾	-/10¼	1/0¾	1/2½	1/7	per yard.
Nozzles and square or obtuse angles	-/11¾	1/0¾	1/2¼	1/5	1/8¾	each.
Lion heads	3/2¼	3/2¼	3/2¼	3/2¼	3/2¼	per dozen.
Stop ends	2/4¾	2/4¾	2/10½	3/10	4/9¾	„
Union clips	5/9	5/9	7/8¼	7/8¼	9/7¼	„
Gutter bolts and nuts ..	..	..	..	..	1/10	per gross.

Gutter brackets—	Black.	Galvanized.	
Wrought rafter	1/8¾	..	per dozen.
Driving	1/8¾	2/4¾	„

A few items of valuation are as follows:—

Eaves-gutters and Rain-water Pipes.—These vary a good deal in quality and weight; the better class of iron-founders supply heavier and better finished goods, and have two qualities, "ordinary" and heavy.

	ft.	in.			£	s.	d.
			run	4" cast-iron ogee gutter, bolted and jointed with red lead cement, and fixed with screws to fascia, measured net..			

All surveyors, although they say measured net, invariably measure rain-water pipes and gutters liberally, consequently waste or laps may be disregarded.

	s.	d.
1′ 0″ run of 4″ gutter	0	3¼
Red lead, bolts and screws	0	0¾
Fixing, ·030 of 8s. 9d.	0	3
Cost per foot run	0	7½

Stopped ends—

Stopped end	0	2½
Red lead and bolts	0	1
Fixing, ·030 of 8s. 9d.	0	3
Cost of each	0	6½

Outlets with nozzles—

Outlet	1	0¾
Red lead and bolts	0	2
Fixing, ·040 of 8s. 9d.	0	4¼
Cost of each	1	7

	ft.	in.			£	s.	d.
			run	4" cast-iron half round gutter, bolted and jointed in red lead cement, fixed with and including wrought-iron driving brackets			

	s.	d.
1′ 0″ run 4″ gutter	0	2¼
Red lead and bolts and screws	0	0¾
½ of two brackets, at 1¾d. each	0	0½
Fixing gutter and brackets, ·030 of 8s. 9d.	0	3
Cost per foot run	0	6¾

Stopped ends—

	s.	d.
Stopped end	0	2¼
Red lead and bolts	0	1
Fixing, ·030 of 8s. 9d.	0	3
Cost of each	0	6¼

Outlets with nozzles—

	s.	d.
Outlet	0	7¼
Red lead and bolts	0	2
Fixing, ·040 of 8s. 9d.	0	4¼
Cost of each	1	1¾

Angles—

	s.	d.
Angle	0	7¼
Red lead and bolts	0	2
Fixing, ·060 of 8s. 9d.	0	6¼
Cost of each	1	3¾

Although angles are measured as "extra only" on the gutter, they usually involve cutting and waste on the gutter, either at their junction with it or further along its line.

WROUGHT IRON: *Wages.*—The rate of wages is—

Smith .. .. .. 10d.   Labourer .. .. .. 7d.

The proportion of labour to material is about 25 per cent. labour, 75 per cent. material.

For this work the iron is often elaborately specified. What is generally supplied for wrought-iron articles is ordinary or merchant bar, or at the most Staffordshire "Best." Such things as chimney and bearing bars are produced from the former; work which requires more forging, as hinges, core rails, hand rails, etc., from the latter.

Materials.—

	s.	d.	
Merchant bar costs	10	0	per cwt.
Marked bars cost	12	0	,,

The ordinary sizes of rod, round or square, are ½ inch to 3 inches; anything beyond this is charged extra at various rates.

The ordinary size of bar is from 1 inch to 6 inches wide, and ¼ inch to 1 inch thick; anything beyond this is charged extra at various rates. For these extras see 'Notes on Building Construction,' vol. iii. (Longmans).

QUANTITY SURVEYING.

It is obvious that the various stock sizes and sections of bars, rods and rails, if judiciously selected, will save much labour of the smith. These various sections are all illustrated in 'Building Construction,' vol. iii. (Longmans).

Many contractors keep no smith, and consequently buy wrought-ironwork from a local smith in London. There are iron-founders and smiths who can supply builder's ironwork much cheaper than the general contractor can do it in his own shops.

cwts.	qrs.	lbs.	In cambered and caulked chimney bars, and fixing by bricklayer	£	s.	d.
					s.	d.
			1 cwt. of bar iron		10	0
			Smith and labourer, ·45 of 14s. 2d.		6	4½
			Bricklayer, fixing		0	8
			Cost per cwt.		17	0½

Local blacksmiths will supply these very cheaply, as they can use old bars which have come into their possession. The observer will be able to trace the past history of the iron; often old wheel tyres and similar things are turned into chimney bars. A working smith with a boy to blow the fire would dispense with a labourer, or blow for himself.

cwts	qrs.	lbs.	In straps, including perforations and fixing by carpenter	£	s.	d.
				£	s.	d.
			1 cwt. of bar	0	12	0
			Smith and labourer, 1·00 of 14s. 2d.	0	14	2
			Carpenter, fixing, ·448 of 8s. 9d.	0	4	1
			Cost per cwt.	1	10	3

The same remarks about smith and boy as above apply to this item.

cwts.	qrs.	lbs.	In framed guard-bars, and fixing by bricklayer	£	s.	d.
				£	s.	d.
			1 cwt. of rails and round bars	0	12	0
			Smith and labourer, ·920 of 14s. 2d.	0	13	0
			Bricklayer, fixing, and labourer, ·010 of 16s. 7d. ..	0	1	5½
			Cost per cwt.	1	6	5½

PRICES.

| cwts. | qrs. | lbs. | In framed balusters, and fixing by smith .. | £ | s. | d. |

The mason cuts the holes and runs them with lead.

		£	s.	d.
1 cwt. of bar		0	12	0
Smith and labourer, ·900 of 14s. 2d.		0	12	9
Cost per cwt.		1	4	9

| cwts. | qrs. | lbs. | In 2" × ⅜" half round handrail, riveted to balusters and newels, and fixing by smith | £ | s. | d. |

	£	s.	d.
1 cwt. of handrail	0	13	0
Smith and labourer, 1·40 of 14s. 2d.	0	19	10
Cost per cwt.	1	12	10

| | | No. | Extra for angles .. ..
No. 1 labour to angle .. .. | £ | s. | d. |

	s.	d.
Smith and labourer, ·150 of 14s. 2d. .. (Cost of each)	2	1½

| | | No. | Extra for scrolls .. ..
No. 1 labour to scroll .. .. | £ | s. | d. |

	s.	d.
Smith and labourer, ·180 of 14s. 2d. .. (Cost of each)	2	6½

| | | No. | 1 | Extra for short wreath .. ..
Labour to short wreath .. .. | £ | s. | d. |

	s.	d.
Smith and labourer, ·170 of 14s. 2d. .. (Cost of each)	2	5

| cwts. | qrs. | lbs. | In framed grating of ½" × 1½" frame, and ⅜" × 1½" bars (fixing by mason) .. .. | £ | s. | d. |

	s.	d.
1 cwt. of iron	12	0
Smith and labourer, 1·13 of 14s. 2d.	16	0
Cost per cwt.	28	0

Ornamental Wrought Ironwork.—This is nearly always the subject of a provision in an estimate; observe whether the price includes fixing and carriage; if not, they must be added.

The ironworker should supply a drawing of what he intends to supply, and state a price before he does it.

Bolts.—These are bought from the iron merchants at about 18s. per cwt. for large quantities. Smaller quantities by the gross or dozen. The weights of the various sizes may be obtained from Hurst's 'Architectual Surveyors' Handbook.'

No.	¾" bolts, average 9" long, not exceeding 12" long, with heads, nuts and washers, and fixing by carpenter	£	s.	d.

		d.
1 ¾" bolt, 9" long, weight 1¾ lbs., at 18s. per cwt.		3½
Fixing, ·038 of 8s. 9d. (carpenter)		4
Cost of each		7½

No.	¾" bolts, average 15" long, all over 12" long, with heads, nuts and washers, and fixing by carpenter	£	s.	d.

		d.
1 ¾" bolt, 15" long, weight 2¼ lbs., at 18s. per cwt.		5
Fixing, ·040 of 8s. 9d. (carpenter)		4¼
Cost of each		9¼

Some contractors price their bolts and fixing at per inch, ¾d. for ½ inch, 1d. for ¾ inch.

Some estimators price the fixing of all bolts, not exceeding 12 inches in length, at 3d.; over 12 inches in length, 6d.; other bolts, not exceeding 12 inches in length, 2½d.; over that length, per foot run, 2d. Fixing bolts in hard wood is generally considered to be worth 50 per cent. more than in fir.

GAS-FITTER.

The discount off list prices of gas fittings is usually about 25 per cent.

Observe in pricing whether the pipe has been measured to

include tees, bends and connections, whether it includes cutting away and making good, whether it includes painting the pipes.

Sub-letting.—Gas-fitting is very often sub-let by the general contractor.

Mode of Measurement.—The common practice is to measure the pipes to include all their adjuncts as follows:

ft.	in.			£	s.	d.
		run	½" wrought-iron welded tubing, including all tees, bends, elbows, connectors, etc., jointing in red lead cement, and fixing complete			

Pipe List.—The manufacturers of tubing adopt a printed list in which the prices never vary, but are modified by a discount which rises or falls with the price of iron.

The present discounts off the list are as follows:

	Per cent.
Gas tubes and fittings	60
„ „ galvanized	47½
Water tubes	55
„ galvanized	42½
Steam tubes	50
„ galvanized	37½

Gas at per Point.—Often a contractor is content to do the work at an agreed price per point, generally 6s. or 7s. a point for domestic work.

Gas-fitter, per hour	10d.
Boy	4d.

Proportion of Labour to Material.—The proportion of labour to material is approximately 20 per cent. labour, 80 per cent. materials.

Iron and Composition Tubing.—In all except the commonest work, wrought-iron welded tubing is used, and in very good work every pipe is tested to a considerable pressure before fixing (200 lbs. to the square inch).

In common work composition pipe is largely used.

Percentage for Fittings.—In pricing gas-fitter's work it is usual to add a percentage on the value of the pipe for the tees, bends, elbows, etc.; about 50 per cent. of the net cost of the pipe is a reasonable average.

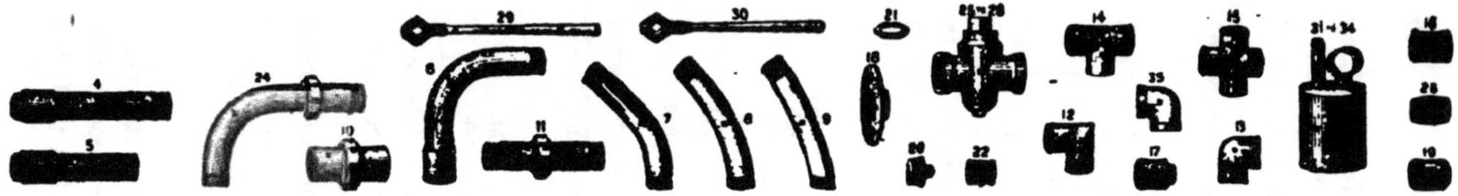

	Internal diameter, in inches	⅛ & ¼	⅜	½	¾	1	1¼	1½	1¾	2	2¼	2½	2¾	3	3½	4	4½	5	5½	6
1	Tubes, 2 to 14 ft. long, per ft.	–/3	–/3½	–/4½	–/6	–/8½	1/–	1/3	1/6	1/9	2/6	3/–	3/3	3/6	4/6	5/6	6/–	6/9	7/9	8/9
2	Pieces, 12 to 22½ in. long, each	–/8	–/9	–/11	1/3	1/9	2/6	3/3	3/11	4/7	6/9	8/6	9/6	10/6	13/2	16/–	21/–	24/6	29/6	35/–
3	" 3 to 11½ " "	–/5	–/5½	–/7	–/10	1/1	1/6	2/–	2/5	2/10	4/3	5/6	6/3	7/–	9/1	11/–	15/3	18/3	22/3	26/6
4	Long Screws, 12 to 23½ " "	–/9	–/10	1/–	1/4	1/10	2/8	3/6	4/3	5/–	7/6	9/7	11/3	12/–	15/3	17/–	22/6	27/–	32/6	38/6
5	" 3 to 11½ " "	–/6	–/6½	–/8	–/10	1/2	1/8	2/3	2/9	3/3	5/–	6/4	7/6	8/3	10/9	12/3	17/–	20/3	24/6	29/3
6	Bends .. .. "	–/7	–/8	–/10	1/–	1/6	2/6	3/–	4/–	5/–	8/6	12/–	15/–	18/–	25/–	32/6	80/–	105/–	135/–	150/–
8, 9	Springs, not socketed .. .. "	–/5	–/6	–/7	–/9	1/1½	1/11½	2/3½	3/1	3/11	6/9	9/6	12/–	14/6	20/–	26/6	70/–	93/–	120/–	132/–
, 11	Socket or pipe unions .. .. "	2/–	2/6	3/–	4/–	5/6	6/9	8/–	9/–	10/–	15/–	17/6	20/–	22/6	27/6	35/–	48/–	66/–	84/–	105/–
2	Elbows, square .. .. "	–/8½	–/9	–/10	1/–	1/4	1/10	2/5	3/–	3/10	6/3	9/–	11/6	14/–	22/–	28/–	75/–	95/–	120/–	150/–
3	" round .. .. "	–/9½	–/11	1/–	1/3	1/6	2/1	2/7	3/6	4/6	6/9	10/–	13/–	17/–	25/–	32/–	75/–	95/–	120/–	150/–
4	Tees .. .. "	–/9	–/10	–/11	1/2	1/5	2/–	2/6	3/2	4/3	6/6	9/6	12/6	16/6	24/–	30/–	78/–	98/–	125/–	155/–
5	Crosses .. .. "	1/4	1/6	1/11	2/4	3/–	4/–	4/10	6/–	7/9	14/–	21/4	28/–	40/–	56/–	66/8	175/–	220/–	280/–	350/–
6	Sockets, plain .. .. "	–/2	–/2	–/3	–/3½	–/4½	–/6½	–/8½	–/11	1/1	1/9	2/6	3/–	3/6	5/–	6/–	10/–	12/–	15/–	18/–
7	" diminished .. .. .. "	–/3	–/4	–/5	–/6	–/7	–/9	–/11	1/2	1/4	2/3	3/3	4/–	5/–	7/–	9/–	25/–	35/–	45/–	55/–
8	Flanges .. .. "	–/9	–/10	1/–	1/2	1/4	1/9	2/–	2/3	2/9	4/–	5/–	6/9	8/6	10/–	11/6	10/–	18/–	28/–	27/–
9	Caps .. .. "	–/3½	–/3½	–/5	–/6	–/8	1/–	1/8	1/7	2/	3/–	4/4	5/3	6/–	9/9	10/6	25/	30/	37/6	45/–

20	Plugs .. .. "	-/3	-/3	-/5	-/6	-/8	-/10	1/-	1/3	2/-	2/6	3/6	4/9	7/-	10/-	22/-	30/-	39/-	48/
21	Backnuts .. .. "	-/2	-/2	-/3	-/3¾	-/5	-/6	-/8	-/10	1/1	1/9	2/3	3/-	3/6	4/6	5/6	14/-	18/-	21/-	26/
22	Nipples .. .. "	-/2	-/2	-/3	-/3¾	-/4	-/6	-/8	-/10	1/-	1/9	2/3	3/-	3/6	4/6	5/6	14/-	18/-	21/-	26/
23	Barrel nipples .. .. .. "	-/5	-/5	-/7¼	-/9	-/10	1/3	1/8	2/1	2/6	4/4	5/8	7/6	8/9	11/3	13/9	28/-	36/-	42/-	52/
24	Union bends .. .. .. "	2/6	3/-	8/9	5/-	6/3	8/6	10/-	11/6	13/6	21/-	27/-	32/-	37/-	49/-	58/-	78/-	100/-	135/-	160/
25	Main cocks .. .. .. "	2/3	2/9	3/6	4/6	6/6	8/6	11/-	14/-	18/-	27/-	40/-	50/-	60/-	85/-	110/-	210/-	270/-	360/-	420/
26	" with brass plugs "	..	7/-	8/9	11/6	16/-	21/6	28/-	35/-	45/-	68/-	100/-	112/-	125/-	212/-	250/-	360/-	450/-	570/-	600/
27	Round way cocks .. .. "	3/6	4/-	5/6	7/6	10/-	13/-	17/6	22/-	38/-	60/-	65/-	75/-	120/-	160/-	..	..	..	..
28	" " with brass plugs "	10/6	12/-	16/6	22/6	30/-	39/-	52/6	66/-	114/-	180/-	195/-	225/-	360/-	480/-	..	..	..	..
29	Cock spanners, wrought .. "	1/6	1/8	2/2	2/8	3/2	3/6	4/-	4/3	4/9	6/-	7/6	9/-	12/-	14/-	16/-	19/-	22/-	24/
30	" " malleable cast.. "	1/-	1/4	1/8	2/-	2/4	3/-	3/6	4/-	4/9	6/-	7/6	9/-	12/-	14/-	16/-	19/-	22/-	24/
31	Syphon boxes, 1 quart .. .. "	..	22/9	23/-	23/4	24/-	24/6	25/2	26/3	28/6	..	..	..	..	..	..	..	..	..
32	" " 2 quarts .. .. "	..	..	27/-	27/4	28/-	28/6	29/2	30/3	32/6	35/6	38/6	42/6	..	..	..	..	..	..
33	" " 3 quarts .. .. "	..	..	32/-	32/4	33/-	33/6	34/2	35/3	37/6	40/6	43/6	47/6	55/6	61/6	..	..	..	..
34	" " 4 quarts .. .. "	..	..	38/-	38/4	39/-	39/6	40/2	41/3	43/6	46/6	49/6	53/6	61/6	67/6	..	..	..	..
35	Malleable cast round elbows .. "	-/6¼	-/7	-/8	-/10	1/2	1/9	2/3	3/-	3/6	5/6	9/-	12/-	15/-	30/-	40/-	..	..	..	..

Tubes, screwed and socketed, supplied in short random lengths of from 2 feet to 6 feet inclusive, or in exact lengths of any length, are charged 2½ per cent. less gross discount.

Tubes, not screwed, supplied in random lengths of any length, are charged at the current discount with an *extra allowance* of 2½ per cent off the net.

Tubes, not screwed, cut to exact lengths of any length, are charged at the *same discount* as tubes screwed and socketed in random lengths.

Tubes, screwed and s n t *without sockets,* are charged at the current discount with an *extra allowance* of 1¼ per cent. off the net.

All short lengths under 2 feet to be considered pieces.

Tubes of intermediate diameters are charged at the list price of the next larger sizes.

Springs, if *socketed*, will be charged extra for the sockets at list prices.

Diminished sockets reduced by more than 1 inch, and bull-headed and bottle-headed tees are charged at the list price of the next larger sizes.

Orders not amounting to 2l. net value will be sent carriage forward.

Lad.—The gas-fitter, as a rule, only requires the assistance of a lad at about 4*d.* per hour = 3*s.* 4*d.* for 10 hours.

Pipe Hooks.—These are bought by the gross. They cost as follows:—

½″	¾″	1″	1¼″	1½″	
1/11½	2/7	3/4	4/3	5/-	per gross.

Fittings in Small Quantity.—When small alterations are made to gas-fittings or small additions to the gas system after the original scheme is complete, it is usual to charge the time and materials as day account, and with justice, such work will perhaps include quite small quantities of pipe and a few fittings, as bends, elbows, springs, tees, etc. List price should be allowed for these.

Separate Contract for Gas-Fitting.—The whole of the gas pipes and fittings are often the subject of a separate contract, which may include the cutting away and making good, in which case, the general contractor has nothing to do with the work.

Attendance on Gas-Fitter.—Attendance, cutting away and making good after gas-fitter will generally be covered by 5 per cent. on the amount of the gas-fitter's bill.

Provision for Fittings.—For gas-fittings a sum of money is nearly always provided in a contract, and they are selected from a trade list; in such a case, carriage, profit and fixing must be added to the net cost.

Trade Discounts.—The prices for the large variety of gas-fitters' requisites may be obtained from the trade lists. Most of these list prices are subject to 25 per cent. discount.

Valuation of Gas-Pipes.

	d.
1 foot of ¾″ gas pipe	2½
Fixing, ·018 of 11*s.* 8*d.*	2½
Hooks and white lead	0¼
Tees, bends, etc., 50 per cent. of 2½*d.*	1¼
Cost per foot run	6½

	d.
1 foot of 1″ gas pipe	3½
Fixing, ·020 of 11*s.* 8*d.*	2¾
Hooks and white lead	0¼
Tees, bends, etc., 50 per cent. of 3½*d.*	1¾
Cost per foot run	8¼

PRICES.

		d.
1 foot of 1¼" gas pipe	.. ..	4¾
Fixing, ·020 of 11s. 8d.	.. ..	2¾
Hooks and white lead	.. ..	0¼
Tees, bends, etc., 50 per cent. of 4¾d.	.. ..	2¾
Cost per foot run	.. ..	10¼

Contractors will often agree to do the work at a price per point, i.e. to lay on the pipe in each case to the place where the bracket or pendant is to be fixed.

BELL-HANGER.

Bells are usually valued at a price per pull.

Average Price per Pull.—Sometimes it will happen that a bell and its adjuncts have to be measured and valued in detail; but as a bell in a building is sometimes near to and sometimes distant from the pull, an average price is generally satisfactory.

The work is often sub-let by the contractor at so much per pull. A common price with a plain pull is 10s., general contractor doing the cutting away and making good.

Electric Bells.—Electric bells are usually sub-let by the contractor, or the architect makes a separate contract for them with an electrician. A common price, including a plain push, is 15s. per push, the general contractor doing the cutting away and making good.

Attendance.—About 20 per cent. on the amount of the bell-hanging will usually cover the cutting away, attendance and making good.

Trade Discounts.—Bell-pulls, cranks, bells, etc. are illustrated in many of the wholesale ironmongery catalogues, and the discounts vary from 10 to 20 per cent. off the list prices.

GLAZIER.

Observe whether or not the measurement of plate-glass is to include beads.

On large quantities of glass there is a trade discount of 20 per cent. for cash.

The usual plate-glass tariffs are subject to a trade discount of 25 per cent. The low prices in this trade are sometimes to be

8 A

accounted for by the substitution of sheet-glass of weight or quality inferior to that specified.

Glazing by Glass Merchant.—The common practice, by a general contractor, is to send sashes to the glass merchant, who glazes them at his works at an agreed price. The merchant carries them both ways and includes the value of carriage in the price per foot which he charges for the glass. Most merchants will send to the works and do the glazing there at about 1d. per foot for labour, if a large quantity.

In the case of a bill of quantities for glazing, the contractor will send it to a glass merchant for a price for sending for the sashes, glazing, and returning them.

Glazing by Contractor.—The glass which must be glazed on the building, as in roofs, lanterns, doors, and the like, is delivered there by the merchant and is glazed by the general contractor's own men.

Lead Lights.—Lead lights of ordinary character are fixed in the sashes by the contractor. Lead lights in stone windows are usually fixed by the maker, who either takes his own dimensions, or makes the lights to templates supplied by the general contractor.

The manufacturers of lead lights charge all lead lights 12 inches wide and under at 12 inches, and all lights 12 inches by 12 inches and under as 12 inches by 12 inches. Irregular shapes are measured as square, and charged as if they were square, but, although this is the custom, the manufacturer will commonly make a special arrangement at a lower price than this if the quantity of square and irregular is kept separate and so submitted to him.

Lead lights are always bought ready-made by the general contractor; he never makes them. The simplest kinds glazed with 21 oz. sheet, may be bought for 1s. per foot; if leaded in geometrical patterns, 2s.

Manufacturers who supply lead lights will fix, including saddle-bars, packing, and the use of cases, at 6d. per superficial foot.

British Plate.—British polished plate glass manufacturers issue a uniform printed tariff, which is subject to a trade discount which varies at different times, but is usually 25 per cent. Special terms may often be obtained when the quantity is large. The tariff is published in the price books.

Prices of Glass.—Common glass may be had at very low prices; for instance, foreign sheet may be obtained in boxes of sheets cut to sizes varying from 12 inches by 10 inches to 24 inches by 18 inches, all sheets of one size in each box, 15 oz. 4ths, 11*s.* per 100 feet; thirds, 12*s.* 6*d.* per 100 feet; 21 oz. fourths, 15*s.* per 100 feet; thirds, 16*s.* 6*d.* per 100 feet.

The glass is usually described in a specification as best: it is rarely better than seconds.

The present prices of glass in crates delivered in London are as follows:—

			d.
English	15 oz. sheet,	seconds	3¼
”	15 oz. ”	thirds	2¾
”	15 oz. ”	fourths	2¼
”	21 oz. ”	seconds	4¼
”	21 oz. ”	thirds	3½
”	21 oz. ”	fourths	3
”	26 oz. ”	seconds	5¼
”	26 oz. ”	thirds	4½
”	26 oz. ”	fourths	4
”	32 oz. ”	seconds	6½
”	32 oz. ”	thirds	5½
”	32 oz. ”	fourths	5
”	15 oz. fluted sheet		3¼
”	21 oz. ditto		4¼
”	⅛″ Hartley's rolled plate		3
”	$\frac{3}{16}$″ ditto		3½
”	¼″ ditto		4

Obscured sheet 1*d.* per foot more than clear.
Matted sheet 1½*d.* per foot more than clear.
10 per cent. in the cost of the glass will cover cutting and waste and putty.
Circular cutting and risk to sheet glass 1½*d.* per foot run.

On next page is shown the tariff for British polished plate glass delivered, the prices reduced by the 25 per cent. trade discount.

There is a further 2½ per cent. for cash.

Classification of Size of Glass.—Although most bills of quantities separate the sheet glass into items, as "not exceeding 2 feet to 3 feet," etc., the contractor commonly brackets them together and attaches one rate for the whole. Up to 6 or 8 feet superficial, the distinction is not worth consideration, the more especially that the larger sizes usually bear but a small proportion to the total quantity.

Broken Glass.—If the supervision is good, ½ per cent. on the amount of the glazier's bill should cover this.

NET PRICE OF BRITISH POLISHED PLATE GLASS PER FOOT SUPERFICIAL.

In plates containing not above	Ordinary Glazing.	Best Glazing.	Silvering quality.	In plates containing not above	Ordinary Glazing.	Best Glazing.	Silvering quality.
ft. supl.	s. d.	s. d.	s. d.	ft. supl.	s. d.	s. d.	s. d.
1	1 3	1 4½	1 6	30	1 9¾	2 0¾	2 5¼
2	1 3¾	1 5¼	1 6¾	35	1 9¾	2 0¾	2 5¼
3	1 4½	1 6	1 8¼	40	1 9¾	2 0¾	2 5¼
4	1 5¼	1 6¾	1 9¾	45	1 9¾	2 0¾	2 6
5	1 6	1 7½	1 10½	50	1 9¾	2 0¾	2 6
6	1 6¾	1 8¼	1 11¼	55	1 10½	2 1½	2 6¾
7	1 7½	1 9	2 0	60	1 10½	2 1½	2 6¾
8	1 7½	1 9	2 0	65	1 10½	2 1½	2 6¾
10	1 8¼	1 9¾	2 0¾	70	1 10½	2 1½	2 7½
12	1 8¼	1 9¾	2 1½	75	1 10½	2 1½	2 9¾
14	1 9	1 11¼	2 3	80	1 11¼	2 3	3 0
16	1 9	1 11¼	2 3	85	1 11¼	2 3¾	3 2¼
18	1 9	1 11¼	2 3¾	90	1 11¼	2 4½	3 4½
20	1 9	1 11¼	2 3¾	95	2 0	2 6	3 6¾
25	1 9	1 11¼	2 3¾	100	2 0	2 7½	3 9

Valuation of Sheet Glass.—

	d.
1 foot 21 oz. seconds, sheet	4¼
10 per cent. waste and putty	½
·012 of 7s. 1d.	1
Cost per foot supl.	5¾

Valuation of Plate Glass.—Fixed with beads, elsewhere valued.

	s.	d.
1 foot supl, British polished plate, not exceeding 20 feet in a square at 1s. 9d.	1	9
·012 of 7s. 1d.		1
Cost per foot supl.	1	10

	s.	d.
1 foot supl. British polished plate, not exceeding 100 feet in a square at 2s.	2	0
2 man at ·012 of 7s. 1d.		2
	2	2

PAINTER.

In this trade inferior oils and white-lead, and the use of other materials instead of white-lead, the dilution of varnish with turpentine, boiled oil, etc., are means by which the cost of work is reduced.

Tenders for Painting.—Competitive tenders for painting vary perhaps more than those for any other trade. The quality of the materials used is one reason, but a stronger one is the amount of labour contemplated.

Labour on Painted Work.—The finished results of merely plain painting afford some striking contrasts to experienced eyes. Although the same number of coats may be applied in a bad example as in a good one, the latter may have 50 per cent. more labour bestowed upon it.

Materials in Day Accounts.—These are often supplied in great variety and in small quantities. The prices may be obtained from a merchant's trade list, and should have a profit of 25 per cent. added to the P.C. The supply of materials in small quantities often involves much trouble.

The proportion of the cost of painting to the whole work of a building contract, is commonly about $\frac{1}{40}$, or 2½ per cent.

Materials.—The present price of good materials is as follows:—

		s.	d.	d.	
White lead	per cwt.	27	0	= 3	per lb.
Red lead	,,	24	6	= 2¼	,,
Linseed oil	per gallon	3	0	= 4½	per pint.
Boiled oil	,,	3	4	= 5	,,
Turpentine	,,	2	10	= 4¼	,,
Putty	per cwt.	9	0	= 1	per lb.
Patent driers	,,	28	0	= 3	,,
Litharge	,,	28	0	= 3	,,

		s.	d.	s.	d.	
Best elastic copal varnish	per gallon	18	0	= 2	3	per pint.
Knotting	,,	10	0	= 1	3	,,
Size	per cwt.	20	0	= 0	2¼	per lb.
Pumice stone	,,	35	0	= 0	3¾	,,
Glass paper	per ream	13	4	= 0	8	per quire.

The above mentioned materials may be bought for much lower

prices, and the inferior qualities are frequently used with impunity; sufficient knowledge to criticise them is not common to professional men.

White lead.—The white lead, which is the base of good paint, is usually specified to be the "best old white lead." Common white lead is much adulterated.

Linseed Oil.—The oil is frequently adulterated, or is used when too new. It may be either raw or boiled.

Driers.—To all paint is added a small proportion of driers, in the mixing. Litharge is that most frequently used.

Varnish.—The varnish most frequently used is copal. It varies greatly in quality and price, and the way to ensure varnish of good quality is to specify the maker and the price.

Painting on Items Measured Lineally: Skirtings.—Of these, the leading item is skirting, and the various heights are not usually distinguished. In pricing them, it is reasonable to assume the whole as 12 inches wide. Each running foot will consequently be 1 foot supl., i.e. $\frac{1}{9}$ of a yard, and that proportion of the price of a foot supl. of plain painting may be adopted as the price. As much of the skirting will not be as much as 12 inches wide, this will pay for the cutting in.

Bar or Rail.—These are such things as bar balusters. They may be priced at half the rate of skirtings, if cut in the same price as skirting.

Items Numbered.—In the case of items numbered, like chimney-pieces, the superficial area should be considered, and the price per yard for plain painting applied.

Painting on Iron.—This involves some scraping and rubbing down. On the other hand, there is no knotting or stopping. It may be priced at the same rate as on wood.

Observe that eaves-gutters and rain-water pipe heads will be painted inside and out.

For rain-water pipes, the same price as for skirting may be adopted. For eaves-gutters, twice as much.

Iron balusters, if measured superficially on both sides, may be priced at the rate per yard for superficial painting for each side. This is a common practice.

In calculating the value of painting, it is most convenient to deal with a fairly large quantity. The following is a calculation for 50 yards superficial of knot, prime, stop and three oils, technically four oils.

Knotting, stopping and rubbing down—

	d.	
⅜ pint of knotting, at 1*s*. 8*d*.	7½	
2 lbs. putty, at 1*d*.	2	
½ quire of glass-paper, at 8*d*.	4	
¼ lb. pumice stone, at 3¾*d*.	1	
50)	14½	
	0⅜	per yd.
·010 of 7*s*. 1*d*.	0⅞	
Cost per yard	1¼	

Priming—

	s.	d.	
8 lbs. white lead, at 3*d*.	2	0	
¼ lb. red lead, at 2½*d*.	0	0⅝	
¼ lb. driers, at 3*d*.	0	0¾	
3½ pints linseed oil, at 4½*d*.	1	3¾	
50)	3	5⅛	
	0	0¾	per yd.
·015 of 7*s*. 1*d*.	0	1¼	
Cost per yard	0	2	

Second coat—

6 lbs. of white lead, at 3*d*.	1	6
⅛ lb. driers, at 3*d*.	0	0⅜
1¼ pint linseed oil, at 4½*d*.	0	5⅝
¼ pint turpentine, at 4½*d*.	0	1
50)	2	1⅛
	0	0½
·014 of 7*s*. 1*d*.	0	1¼
Cost per yard	0	1¾

Third and following coats—

5¾ lbs. white lead, at 3*d*.	1	5¼
⅛ lb. driers, at 3*d*.	0	0⅜
1 pint linseed oil, at 4½*d*.	0	4½
½ pint turpentine, at 4½*d*.	0	2¼
50)	2	0⅜
	0	0½
·014 of 7*s*. 1*d*.	0	1¼
Cost per yard	0	1¾

QUANTITY SURVEYING.

Collection—		s.	d.
knot, stop and rub down	.. ..	0	1¼
Prime	.. ..	0	2
Second coat ..	.. ..	0	1¾
Third „	.. ..	0	1¾
Fourth „	.. ..	0	1¾
Cost per yard of knot, prime, stop and three oils		0	8¼
Add 10 per cent. profit		0	0¾
		0	9¼

The contract price for four oils varies from 10d. to 1s. per yard.

Sash Squares.—It is the custom of surveyors to count both sides of a sash for squares. Thus a sash with twelve squares would appear in an estimate as two dozen squares. In the painting of a sash, the quantity of material for its superficial area on one side is very little different to that on a plain surface. Some estimators measure each side of a sash and frame or a skylight, and price it at the ordinary rate for superficial painting.

In the foregoing calculation, the cost of materials of four oils is:

$$\tfrac{1}{4}d. + \tfrac{3}{4}d. + \tfrac{1}{2}d. + \tfrac{1}{2}d. + \tfrac{1}{2}d. = 2\tfrac{1}{4}d. \text{ per yard superficial.}$$

24 feet superficial is considered an ordinary sash frame, and this, if divided into twelve squares, would be a fair type of a dozen ordinary squares.

24 ft. supl. = 2⅔ yards superficial.

			s.	d.
	2⅔ yds., at 2¼d. for materials		0	7¼
Constants per dozen	·025 ·040 ·035 ·035 ·035			
	·170 of 7s. 1d.		1	3
	Cost per dozen		1	10¼

Estimators usually refer the price of a dozen squares to the price per yard of plain painting. Their conclusions will be found to vary from 1½ to 2½ yards.

Large squares 50 per cent. beyond ordinary.

Sash Frames.—An ordinary sash-frame is one not exceeding 24 feet superficial.

Estimators price these at from 1¼ to 2½ yards of plain work. About 1¾ yards is a reasonable average.

A large frame, which is one not exceeding 36 feet, at the same rate would be 2⅝ yards.

Sheets.—Some surveyors when a sash is in a single square describe it as a sheet. It is an inexact term, as the sizes of the sashes vary. Twice the price of ordinary squares will generally pay for it.

Painting on Stone or Cement.—Painting on stone or cement, both being more absorbent and involving a little more trouble, are usually priced at 10 per cent. more than painting on woodwork.

Graining.—This is measured as an after coat, several coats of ordinary paint having been previously applied.

Graining is very rarely done by an ordinary painter. The contractor sublets it to a man who, as a rule, does nothing else. In quality and cost, it varies very much.

Combing costs 4d. to 5d. per yard		
Oak and maple	9d.	„
Walnut	1s.	„

These are all common.

Good graining, 2s. 6d. „ And this is about as good as appears in ordinary work; but for very good work much more than this is paid.

Varnishing.—Varnish varies greatly in quality. A gallon will cover 50 yards of surface. The second coat takes rather less than the first. Some schedules, after stating a price for the first, insert a note, "deduct ⅓ for each additional coat."

50 yards of grain, size and *twice* varnish, would be valued as follows:—

	£	s.	d.
50 yds. grain, at 2s. 6d.	6	5	0
1⅔ gallon varnish, at 18s.	1	10	0
⅛ cwt. of size, at 20s.	0	2	6
50)	7	17	6
Cost of materials per yard	0	3	1

Constants, per yard.
Sizing ·010
Varnishing ·014
„ ·014

·038 of 7s. 1d.	0	0	3¼
Cost per yard	0	3	4¼

Staining and Varnishing.—Stain for wood may be either oil or water stain; the latter is that most frequently used. Stephens' is about the best and it is sold by the gallon, and for good work is thus described:—

"The joinery of dining-room shall be carefully stopped and stained with Stephens' (191 Aldersgate Street, E.C.) stain to a tint to be approved, laid on with two brushes, one to apply it, the other to lay it off with, sized and twice varnished with copal varnish, as described."

Light oak stain costs 4*s.* per gallon; medium oak stain 6*s.* All the rest, 8*s.* per gallon. A gallon covers about 100 superficial yards. 100 yards of stain, size and twice varnish would be valued as follows:—

	£	s.	d.
1 gallon stain	0	8	0
¼ cwt. size, at 20*s.*	0	5	0
4 lbs. putty, at 1*d.*	0	0	4
8½ gallons varnish, at 18*s.*	3	0	0
	100) 3	13	4
Cost of materials per yard	0	0	8¾

Constants, per yard.

Stopping	·005
Sizing	·010
Varnishing	·014
"	·014
	·043 of 7*s.* 1*d.*	0	0	3¼
Cost per yard	0	1	0¼

Painting on Old Work.—The amount of labour involved by the paint being worn away or blistered, depends very much upon the condition of the work to be painted. It must be examined and a proportion of the whole, as one-third, one-fourth, or as the case may be, must be fixed and allowed as extra. It is commonly described in some such way as follows:—

"Thoroughly wash and clean all the work previously painted, rub down, stop and paint three oils. Any part of the painted work which is blistered shall be burnt off bare to the wood and shall receive, in addition to the coats previously described, two coats of oil colour. Any paint-work where worn off shall also receive two additional coats."

		d.
Wash, ·018 of 7s. 1d.		1¼
Three oils, as before		5¼
		6½

	s.	d.
Say one-quarter of the whole requires burning off.		
Burn off, ·141 of 7s. 1d.	1	0
Face up, ·015 of 7s. 1d.	0	1¼
Two extra coats, at 1½d. (as before)	0	3½
4)	1	4¾
	0	4¼	4¼
Cost per yard		11

Conventional Conclusions.—There are a few generally received rates which may be mentioned.

Flatting, 1d. per yard superficial, extra on last coat. Thus if four oils are charged 10d., three oils and flat would be 11d.

Painting in parti-colours, 1d. per yard superficial, extra beyond ordinary painting.

Picking out mouldings of doors and framings in one tint, ½d. per foot run.

Sashes and frames are sometimes measured as though they were plain surfaces and charged by the yard at the ordinary rate.

Work done from ladders is commonly charged at 1½d. per yard superficial extra.

Stippling is commonly charged about 2d. per yard superficial, extra on the ordinary painting. Thus, if four oils are charged 10d. four oils with the last coat stippled would be 1s.

PAPER-HANGER.

Pattern-Books.—Wall papers are almost invariably specified at a price per piece and are selected from a pattern-book in which each pattern has the gross price per piece marked on its back.

Trade Discounts.—The trade discount is ordinarily one-third of the marked price or 33⅓ per cent. Sometimes it is as much as 55 per cent. Some of the firms which produce the more artistic wall-papers give no trade discount.

It is, therefore, best to specify at P.C. so much per piece, or, if the name of the maker is prescribed, the list price may be specified.

Preparation of Walls.—The preparation of new walls for paper is usually of the most perfunctory kind. For good work the walls should be carefully rubbed down, stopped and sized, or rubbed down, clearcolled with white lead and sized.

Allowance for Waste.—The measurer allows one piece in seven for waste.

Sub-contracts.—The general contractor rarely keeps a paper-hanger: he employs a man who does nothing else, and agrees for the hanging and paste at a price per piece. The charge varies, the price increasing with the increased value of the paper, as the more expensive papers demand great care in the hanging.

The cost price of hanging common papers is about 8d. per piece; satin papers 1s.; papers in gold and colours 1s. 3d. Paperhanger, per hour, 9d.

Materials.—Lining paper from 2½d. to 1s. per piece.

	s.	d.
Size	20	0 per cwt.
Paper varnish	14	0 per gallon.

The order of wall-papers is—

       1. Machine printed;   2. Hand printed;

and the price varies greatly.

In cases where colour is of importance, the makers will prepare special lengths of the same pattern as samples in various tones or hues, so that they may be hung up on the walls for comparison; this of course adds greatly to the expense.

The work of the paper-hanger is charged by the piece.

CONSTANTS OF LABOUR.

Constants are the most convenient form for the preservation and application of the observation of results of labour. When their importance is fully recognised, we may hope to see an extended and complete series, which will to a great extent supersede the current price books.

In the words of Gwilt, it is manifest that, if the average time of executing each species of work were known, no difficulty could exist in fixing uniform rates of charge for it.

In Gauthey's 'Traité de la Construction des Ponts,' 1809, there are constants in all trades. Rankine gives some of these in his 'Manual of Civil Engineering,' and Cresy, in his 'Encyclopædia of Engineering,' translates them complete.

They appear, so far as they go, to have been very carefully considered, but the changed circumstances of the artificers' work, the lapse of time, difference of nationality, and the inferior quality of the French workman (forcibly illustrated by Thomas Brassey's experience during his construction of French Railways), detract considerably from their value, and they are consequently hardly worth the trouble of translation.

Other French engineers, Geniey, Claudel and others, have pursued the study later, but only to a limited extent.

Peter Nicholson, in his 'Dictionary of Architecture,' gives a series for the trades of carpenter and joiner, characterised by the care which most of his work evinces.

Gwilt, in his 'Encyclopædia of Architecture' repeats, them and adds a few in the mason's trade.

Dobson, in his 'Guide to Measuring and Valuing Artificer's Work,' gives constants in all trades.

Rankine gives a few in his 'Useful Rules and Tables for Engineers.'

For Mr. Hurst, in his 'Architectural Surveyors' Handbook,' has been reserved the distinction of presenting a series in all trades, which are admitted (by everyone qualified to judge) to be by far the most valuable contribution of recent years to the literature of estimating. The careful thought and long experience which they embody will be apparent to all who may have occasion to use them.

Dobson's constants are described as including profits, and he does not say what the rate of profit is—a return to the worst characteristic of the price-books, and an apparent failure to grasp their essential idea—the recording of cost.

The mode of application of constants is so well known that a few words will be sufficient.

A constant represents in days and decimal parts of a day the time which a certain unit of work as a yard, a foot, a rod, will require for its performance. The amount of the workman's wages for ten hours being multiplied by such constant, the result will be the cost of the unit as a yard, foot or rod.

An example is as follows:—

The wages of a joiner for 10 hours' work being .. ·105d.
Constant for 1¾" four-panel square-framed door, including hanging .. .. ·077 of a day.
8·085d. per ft. supl.

If it should be necessary to analyse a constant to discover the quantity of work per day of ten hours to which it is equivalent, the division of one day by the constant will produce the result. The following is an instance.

Days of a Labourer.
Excavating vegetable soil, per cubic yard .. .. .. ·045

Day.
·045) 1·00 (22 yards

The student of prices will find it instructive to collect all the constants of labour he can find, arranging them in a book in parallel columns, for comparison, thus:—

          Rankine. Hurst. Dobson.
Excavate vegetable soil, per cubic yard .. ·050 ·045 ·088

It is also a good practice to put acquired information into constant form. Thus, three men excavate 45 cubic yards of clay of a certain character in ten hours.

45 yds. ÷ 3 men = 15 yds. per man.
Yds. Day. Constant.
15) 1·00 (·066

The observation of one man's work is of little use in an inquiry of this kind. The industry and capability of men vary so much that constants can only be derived from the *average* results of the labour of a number of men.

The value of the older existing constants like Nicholson's is affected by the smaller quantity of work done in an hour by a man now, than at the time they were produced.

For the purposes of these constants, the day is assumed to be of the length of ten hours.

EXCAVATOR.

Excavator, per cubic yard—   Days of Labourer.

Tempering and spreading clay	·25
Digging building-rubbish, sand and loose soil which only requires shovelling	·045
Digging vegetable soil, as in surface digging	·070
Digging clay with a shovel, not necessarily picked, but sometimes picked for convenience	·143
Digging hardest earth-clay or rubble, requiring picking	·167
Add for each additional throw	·045
Add for each run when wheeled	·028
Forming contour of road bed, per yard supl.	·014
Spreading road metal 6" thick (deposited in heaps at a convenient distance, per yard supl.	·033
Picking up macadamised road to a depth of 4", per yard supl.	·025

Laying granite setts, including gravelling the bed and grouting—

                                                                       Pavior and Labourer.

Sets 5" deep and under, per yard	·09
Sets 5" to 7", ditto	·10
Sets 7" to 9", ditto	·11

Drains.—   Days of Bricklayer and Labourer.

Laying and jointing 4" pipes, per foot	·010
" 6" "	·015
9" "	·022
12" "	·030
" 15' "	·038
" " 18" "	·045

Concrete.—

Mixing, wheeling, depositing and ramming, per cubic yard	·23
Mixing mortar, including screening sand, ditto	·7

BRICKLAYER.

Brickwork in mortar, per rod	5·00
Ditto in cement	5·50
Ditto in cement and underpinning	7·95
Ditto in walls, fair both sides	5·50
Paving of hard stocks, flat in sand, per yard	·045
Ditto on edge	·066
Ditto flat in mortar	·080

                                                     Days of
                                              Bricklayer.   Labourer.

Joints of brickwork struck fair, per yard	·025	
Raking out joints, per yard		·008
Ditto cement, per yard		·040
Pointing with cement, per yard	·080	·027

QUANTITY SURVEYING.

	Days of Bricklayer.	Labourer.
Colouring or limewhiting, per yard	·007	
Cutting and bonding new brickwork to old, per foot supl.	·022	
Striking bevel joint and fair face, per foot supl.	·005	
Selecting bricks, per thousand		·400
Groin point on vaulting, per yard	·083	
Rough cut birdsmouth and squint quoins, per foot run	·008	
5" skewback cutting, per foot run	·017	
Cut and pin 3" landing, per foot run	·022	
Ditto 6" ditto, per foot run	·044	
Rake out and point flashings, per foot run	·003	·002
Ditto stepped, per foot run	·005	·003
Cutting and rubbing mouldings, per foot supl.	·215	

Laying tiles.—

	Days of Bricklayer.
Laying plain quarries, per yard	·20
Ditto encaustic tiles, two colours	·30
Ditto ditto, three colours	·40
Labourer on either of the above	·05

Terra-cotta.—

	Bricklayer and Labourer.	Labourer.
Setting and pointing, per foot cube	·035	
Filling hollow blocks with concrete, per foot cube		·016

WALLER.

	Days Waller and Labourer.
Building random rubble wall in mortar, per yard cube	·323
Building coursed rubble wall in mortar, per yard cube	·500
Building coursed rubble wall in cement, per yard cube	·555

	Days of Waller.
Preparing hammer-dressed stone as delivered for random facing, per yard supl.	·200
Ditto for coursed facing	·330
Ditto hammer-dressed quoins coursed, including drafted angle, per foot run	·100
Rough cutting to rakes and skewbacks, per foot supl.	·015
Extra labour facing brickwork with coursed facing, per foot supl.	·007
Ditto random facing	·010

TILER.

	Days of Bricklayer.	Labourer.
Lay plain tiling 3" gauge	·720	·240
,, ,, 3½" ,,	·700	·233
,, ,, 4" ,,	·680	·226
Loading or unloading tiles		·100

MASON.

Portland Stone.—To work done in position, add 50 per cent. to the following rates:—

	Days of Mason.		Days of Mason.
Half sawing	·028	Sunk face	·111
Beds and joints	·056	Sunk face, stopped	·125
Sunk joint	·063	Ditto, circular	·140
Circular joint	·063	Moulded work, rubbed	·150
Circular sunk joint	·063	Ditto, stopped	·166
Plain face tooled	·075	Ditto, circular	·200
Plain face rubbed	·090	Rough sunk work	·080
Circular face	·111	Cleaning down	·045

Days of Mason and Labourer.

Hoisting and setting up to 40 feet from ground, per foot cube ·040

Days of Labourer.

Extra hoisting, 40 to 80 feet ,, ·025

Days of Mason.

Beads and mouldings up to 2" girth, per foot run	·032
Ditto, circular	·048
Mouldings over 2" and under 6" girth, per foot supl.	·180
Ditto, stopped	·200
Ditto, circular	·240
Grooves for lead lights or tongue	·010
Ditto, circular	·015
Rebates not exceeding 3" girth (over that girth may be measured as sunk work)	·030
Ditto, circular	·045

Internal.—

Mitres to mouldings, beads, etc., not exceeding 2" girth, each	·020
Ditto, exceeding 2" girth, per foot run	·100

External.—

Mitres to mouldings, beads, etc., not exceeding 2" girth, each	·015
Ditto, exceeding 2" girth, per foot run	·062
Mortises for balusters	·033
Running ¾ lb. lead	1¾d.
Labour	¾d.
Fuel	¼d.
	2¾d.

Mortises for newels approximately 50 per cent. more than last.

Perforations.—

Drilling or jumping holes through stone per foot run } 1" 1½" 2" 2¼" 3" 3¼" diameter.
·08 ·09 ·100 ·123 ·166 ·220 days of mason

	Days of Mason.
Above the foregoing diameter for every inch in depth, per foot supl.	·230

For perforations to stone in position add 50 per cent. on the above rates.

	Days of Mason.
Perforations square or circular in stone for traps, flues, coal-plates, etc., the edges rebated and tooled (circular perforations measured square) for areas *not exceeding* 1 ft., per foot supl.	·066
Ditto for areas exceeding 1 ft.	·044

Yorkshire stone.—

Edges, Coped or Sawn.

	Days of Mason.
2" thick, per foot run	·016
2¼" „ „	·020
3" „ „	·025
4" „ „	·033

For circular edge, add 50 per cent. to the above rates.

SLATER.

	Slater and Boy.
Preparing and laying Duchess slating, per square ..	1·65
Ditto Countesses ditto, ditto	2·00
Ditto Ladies ditto, ditto	2·65
Ditto Doubles ditto, ditto..	4·00

CARPENTER.

	Days of Carpenter.
Labour, fitting and fixing plates, lintels and timbers of similar character—such as curbs. (The bricklayer beds them, and there is a customary allowance of brickwork, to pay for the labour and mortar)	·040
Ditto if dovetail-halved and halved at angles	·050
Ditto fir in ground joists	·050
Ditto fir framed in floors (bridging joists and trimmers for hearths)..	·066
Ditto ditto framed in double-framed floors	·090
Ditto ditto in quarter-partitions (only the posts tenoned)	·090
Ditto ditto in quarter-partitions, all the parts tenoned	·100
Ditto ditto ditto and trussed	·123
Ditto ditto in roofs	·090
Ditto ditto in roof trusses, exclusive of hoisting ..	·123
Ditto fir in flitches, fitted together for bolting and fixing, not including fixing bolts	·066
Ditto framed in ceiling joists..	·100
Wrought face on fir..	·010
Ditto, circular	·015

If done by machinery and *finished* by the carpenter, it would cost ⅜d. (about) for straight planing.

PRICES.

The approximate cost of planing timbers by machinery is generally assumed by estimators as ½d. per foot superficial, and for hard woods about ¼d. more.

It is the custom to assume, in the valuation of joinery, that wrought one side includes one edge; wrought both sides, two edges wrought.

The labours on oak, elm, ash, and mahogany, for ordinary calculations, may be assumed as of equal value, and twice that on fir. The following constants are for ordinary labours on fir:—

	Days of Carpenter.
Chamfers 1" wide and under, straight	·002
Ditto, cross-grain	·003
Ditto, circular	·003
Beads 1" wide and under, straight	·003
Ditto, circular	·006
Ditto, cross-grain	·005
Staff beads, 1" wide and under, straight	·009
Ditto, circular	·015
Ditto, cross-grain	·012
Chamfer above 1" and not exceeding 2" wide, straight	·006
Ditto, ditto cross-grain	·009
Ditto ditto, circular	·009

(If stopped, including plain square stops, increase the constant by one-half.)

Cutting other than straight, excluding waste, 2" thick and under, raking	·006
Ditto, ditto, circular	·009
Ditto, over 2" thick, raking	·030
Ditto, ditto, circular	·045
Rounded edges on 2" and under, straight	·006
Ditto, ditto, circular	·009
Ditto, ditto, cross-grain	·009
Rounded edges over 2", and not exceeding 6", straight	·018
Ditto, ditto, circular	·025
Ditto, ditto, cross-grain	·025
Plough groove, per foot run	·003
Groove 2" girth and under, ditto	·009
Ditto, circular, ditto	·012
Rebate, not exceeding 2" girth, ditto	·003
Ditto, cross-grain, ditto	·009
Ditto, circular, ditto	·012
Moulding, 2" girth and under, ditto	·012
Ditto, cross-grain, ditto	·018
Ditto, circular, ditto	·024
Moulding, over 2" girth, per foot supl.	·072
Ditto, cross-grain, ditto	·108
Ditto, circular, ditto	·144

If any above are stopped, add 50 per cent.

QUANTITY SURVEYING.

	Days of Carpenter.
Notching or scribing 1" and 1¼", per foot run	·009
Ditto 2", ditto	·014
Circular ribs to roof-trusses, including sunk edges and grooved and rebated joints and dowelling with oak dowels per foot cube	·250

Fixing only—

	Straight.	Circular, if the boards can be bent.
¾" rough boarding to roofs, per square ..	·30	·45
1" ditto, ditto	·33	·50
1¼" ditto, ditto	·38	·57
Preparing and fixing firrings to flats, ditto..	·70	..
Slating battens to roofs for Countess slating, ditto	·17	
Sound boarding and fillets	·80	
Inodorous felt to roofs	·15	
Gutter boards and bearers, per foot supl. ..	·030	
Rolls for lead, per foot run	·009	
Ditto and birdsmouthing, ditto	·012	..

	Days of Carpenter.
Eaves fillet, per foot run	·006
Cross-rebated drips, ditto	·010
The average length of "short lengths of cross-rebated drip" is about 15".	
Herring-bone strutting to 9" joists, per foot run ..	·014
Ditto 9" to 12" joists, ditto	·017
Rough fillet, ditto	·003
1¼" dove-tailed cesspools, 9" × 9" × 6" holed and fitted	·125

Preparing and fixing —

	Days of Carpenter.	Days of Labourer.
Centering to vaults, per square	1·00	·50
Flat boarded centering and horsing to concrete floors, ditto	·60	·60
Centering to trimmer arches, per ft. supl.	·030	..
Ditto to openings, ditto	·030	
Ditto 4½" soffits, per foot run..	·011	
Ditto 9" ditto, ditto	·024	..
Extra for groin point	·036	..
Bracketing for cornices, per foot supl.	·024	..
Cradling fitted to iron, ditto	·028	..
Traversing roof boarding, per square ..	·15	..

Fixing only wrought ironwork to fir.—

	Days of Carpenter.
Ties and straps, per lb.	·008
Flitches, per cwt.	·260
Bolts with heads, nuts and washers under 1 lb., per lb.	·027
Ditto, 1 lb. and under 2 lbs., ditto	·022
Ditto, 2 lbs. and under 4 lbs., ditto	·016
Ditto, 4 lbs. and under 8 lbs., ditto	·011
Ditto, 8 lbs. and upwards, ditto	·008

Fixing only cast ironwork.—

Heads and shoes to roof trusses, per cwt.	·300

JOINER.

	Days of Joiner.
Driving of screws, per inch, per dozen	·024
Add if in hard wood	·010

Labour only.—

	Days of Carpenter.
Chamfers 1" wide and under, straight	·002
Ditto, cross-grain	·003
Ditto, circular	·003
Beads 1" wide and under, straight	·003
Ditto, cross-grain	·005
Ditto, circular	·006
Staff-beads 1" wide and under, straight	·009
Ditto, cross-grain	·012
Ditto, circular	·015
Chamfers above 1" and not exceeding 2" wide, straight	·006
Ditto, cross-grain	·009
Ditto, circular	·009
If stopped, including plain square stops, increase the constant by one-half.	
Cutting other than straight, 2" thick and under, raking	·006
Ditto, circular	·009
Ditto, over 2" thick, raking	·030
Ditto, circular	·045
Rounded edges on 2" and under, straight	·006
Ditto, circular	·009
Ditto, cross-grain	·009
Ditto over 2" and not exceeding 6", straight	·018
Ditto, circular	·025
Ditto, cross-grain	·025
Plough-groove 2" girth and under, straight	·003
Ditto, cross-grain	·009
Ditto, circular	·012
Cross-tongued joint up to 2" thick	·015
Tongued angles	·015
Rebate not exceeding 2" girth	·003
Ditto, cross-grain	·009
Ditto, circular	·012
Moulding 2" girth and under, straight	·012
Ditto, cross-grain	·018
Ditto, circular	·024

	Per foot supl.
Moulding over 2" girth, straight	·072
Ditto, cross-grain	·108
Ditto, circular	·144

If any of the above are stopped, increase the constant by one-half.

	Per foot run.
Notching or scribing 1" and 1¼"	·009
Ditto 2"	·014
Plugging	·008

	Days of Carpenter.
	Per foot supl.
Planing, straight	·010
Ditto, circular	·015
Rebating	·048
Ditto, cross-grain	·072
Ditto, circular	·095

Fixing only.—

	Per foot run.
Fillets	·005
Grounds for skirtings and similar purposes	·005
Ditto framed as to doors and windows	·008

When plugged add the constant for plugging in the foregoing list.

Fascias or *skirtings*, exclusive of grounds and backings, 6″ and under	·010
Ditto, 6″ to 9″	·013
Ditto, 9″ to 11″	·015
Ditto in two pieces, 12″ high in all	·021

Floors Laid and Cleaned Off.—

	Per square.
1″ yellow batten, straight joint with splayed headings	·400
1¼″ ditto	·440
1½″ ditto	·485
1″ tongued and grooved, or filistered or rebated with loose tongues	·535
1¼″ ditto	·590
1½″ ditto	·645
Add to either of the foregoing if punched, puttied and traversed	·250

Skylights, Sashes and Frames.—

	Per foot supl.
Fixing fanlights or skylights	·010
Hanging casements, 1½″ or 2″	·016
Fixing deal-cased frames and sashes, or casements and frames	·007

Doors.—

Hanging 1¼″ and 1½″ doors	·008
Ditto 2″ and 2½″ ditto	·010
Ditto, folding, 1¼″ and 1½″ doors	·016
Ditto, 2″ and 2½″ ditto	·020

	Days of Joiner.
2″ framed and braced, filled in with ploughed, tongued and beaded boarding, in 4½″ widths, per foot supl.	·069
2½″ ditto, ditto	·073
Add if boarding V-jointed both sides	·016

A common price for hanging a door is 1s. 6d. in speculating work. The men will hang them (piece work) at 1s. each. In all cases the fixing of doors involves and includes the fixing of the hinges.

Framings.—

	Per foot supl.
Framed partitions	·012
Framed spandrels	·025
Window backs	·016

Thicknesses.—

Jamb linings	·012
Ditto 6" wide and under	·020
Window linings	·010
Ditto boards and bearers	·016
Shelves and bearers	·020
w.c. flaps and frames, fixing and hanging	·016
Skylight curbs	·016
Pipe casing and grounds	·030
$\frac{3}{4}$" matched and beaded boarding, in $4\frac{1}{4}$" widths, per square	·60

Staircases.—

Treads and risers and bearers	·045
Wall string	·020
Outer string	·010

	Each.
Housing treads and risers, rounded nosings	·050
Ditto, moulded nosings	·060
Housing winders, rounded nosings	·075
Ditto, moulded nosings	·090
Housing ends of moulded handrail	·040

	Per foot run.
Newels	·025
Handrail	·006
Architraves up to 3" wide, including mitres	·012

Labour on Casements from Bench.—

	Days of Joiner. Per foot supl.
$1\frac{3}{4}$" ovolo moulded casements, single squares	·032
Ditto, circular on plan	·095
Ditto, circular quick sweep	·127
Ditto, add for small squares	·032
Ditto, semi-heads separately measured and measured square	·095
2" ovolo moulded casements, single squares	·037
Ditto, circular on plan	·111
Ditto, circular quick sweep	·147
Ditto, add for small squares	·037
Ditto, semi-heads separately measured and measured square	·111

Subject to the saving remarks as to limit of size, the foregoing constants may be applied to fixed sashes, fanlights, and skylights.

Observe that the words "from Bench" mean that fixing or hanging is not included in the constant.

QUANTITY SURVEYING.

Labour to Panelled doors and Framings from Bench.—

	Days of Joiner. Per foot supl.
1¼" square framed, 2-panel	·036
1¼" ditto, 4 ditto	·042
1¼" ditto, 5 ditto	·045
1¼" ditto, 6 ditto	·048
1½" ditto, 2 ditto	·036
1½" ditto, 4 ditto	·042
1½" ditto, 5 ditto	·045
1½" ditto, 6 ditto	·048
2 " ditto, 2 ditto	·042
2 " ditto, 4 ditto	·048
2 " ditto, 5 ditto	·050
2 " ditto, 6 ditto	·053
2¼" ditto, 2 ditto	·049
2¼" ditto, 4 ditto	·054
2¼" ditto, 5 ditto	·057
2¼" ditto, 6 ditto	·059
2½" ditto, 2 ditto	·050
2½" ditto, 4 ditto	·056
2½" ditto, 5 ditto	·059
2½" ditto, 6 ditto	·060

The above constants may be adopted for framing, observing that the distribution of panels is similar.

Labour on Sashes and Frames from Bench.—

	Days of Joiner. Per foot supl.
Deal-cased frames, and 1½" sashes, single hung	·066
Ditto, ditto, double hung	·078
Ditto, circular plan, flat sweep	·117
Ditto, ditto, quick sweep	·156
Deal-cased frames, and 2" sashes, single hung	·078
Ditto, ditto, double hung	·090
Ditto, circular plan, flat sweep	·140
Ditto, ditto, quick sweep	·185

Labour to Thicknesses from Bench.—

1" shelves wrought both sides	·020
1" window linings, rebated one edge and tongued at angles	·028
1¼" window boards with rounded nosing	·016
1" mortise and mitre-clamped flap and beaded frame	·080
1½" jamb linings, double rebated and tongued at angles	·045
1¼" landing, cross-tongued	·025
1¼" wrought and framed outer string	·048
1½" ditto	·053
1¼" wrought cut and mitred outer string	·060
1½" ditto	·065
1¼" treads with rounded nosings and 1" risers, glued, blocked and bracketed	·060
1¼" wrought plain wall string	·036
1½" ditto	·042

PRICES.

	Days of Joiner. Per foot cube.
Wrought rebated and beaded door frames, labour and fixing, 12 sectional inches and under	·032
Ditto 24 ditto	·029
Ditto 36 ditto	·026

For transomes increase the constant by 10 per cent.

Fixing only, Ironmongery.—

	Days of Joiner.	Screws. d.
2½" butts, per pair	·022	1
3" ,, ,,	·022	1
3½" ,, ,,	·027	1¼
4" ,, ,,	·033	2
4½" ,, ,,	·033	2
5" ,, ,,	·037	2¼
15" cross garnets	·053	1⅜
3" to 6" tower bolt	·043	1
9" to 12" tower bolt	·065	1
Flush bolts, per inch	·008	0¼
Espagnolette bolts, per inch	·006	0¼
3" cupboard locks	·085	1
3" cut ditto	·137	1
Rim locks	·095	1
Mortise locks	·252	1
Kaye's locks	·286	1
Rim dead locks	·095	1
8" drawback locks	·137	1
10" ditto	·170	1
W.C. latch	·137	1
Night latch	·085	1
Door knobs	·033	0
Knockers	·126	0
Door chains	·085	1
Letter-box and plate	·252	1½
Padlock, hasp, and staple	·095	1
Thumb-latch	·095	1
Mortise lock furniture, per set	·033	0½
Sash lifts	·043	0½
Ditto flush	·065	0½
Sash pulleys	·025	0½
3" sash centres, per set	·095	1
Casement fastenings	·043	0½
Sash fastenings	·033	0½
Casement stays	·033	1
Patent water bars, per inch	·006	0¼
1½ knobs	·027	0
Turnbuckles	·033	0
Button and plate	·022	0½
Cabin hooks 6" to 9"	·033	1

PLUMBER.

	Plumber and Mate.
Lead in gutters, flats and flashings, per cwt.	·233
,, stepped flashings, per cwt.	·300
,, safes and sinks, per cwt.	·300
,, cisterns, per cwt.	·300

QUANTITY SURVEYING.

	Plumber and Mate.
Lead in covering to draining boards, per cwt.	1·08
" covering and dressing to mouldings, per cwt.	1·00
" taking up old lead and depositing on works, per cwt.	·066
Extra labour, dressing lead to secret gutter, per foot run	·011
" lead wedging to flashings, per foot run	·005
" lead wedging to stepped flashings, per foot run	·008
" welt	·010
" four-way intersections, each	·055
" bossed ends to rolls, each	·028
" to cesspools, each	·160
" to soldered dots, each	·033
" burning in lead to stone, per foot run	·017
" close copper nailing, per foot run	·013
" soldered angle, plumber only	·050
" soldering tacks to soil pipes	·030

Solder Joints.—

	Plumber and Mate.		Plumber and Mate.
½" each	·050	2½" each	·125
¾" "	·058	3" "	·150
1" "	·066	4" "	·175
1¼" "	·075	5" "	·200
1½" "	·088	6" "	·250
2" "	·100		

Fixing pipes, including labour to running joints.—

	Plumber and Mate.		Plumber and Mate.
½" per foot run	·014	2½" per foot run	·035
¾" "	·021	3" "	·042
1" "	·028	3½" "	·049
1¼" "	·038	4" "	·056
1½" "	·042	5" "	·065
2" "	·056	6" "	·076

Bends.—

	Plumber and Mate.		Plumber and Mate.
½" each } In the price of pipe		2½" each	·150
¾" "		3" "	·200
1" "		4" "	·300
1¼" "	·075	5" "	·500
1½" "	·100	6" "	·500
2" "	·100		

	Days of Plumber and Mate.
Fixing ¾" ball valve with rod and ball	·022
1" ditto	·028

PRICES.

Drilling Iron Cisterns for Unions.—

	Days of Plumber and Mate.		Days of Plumber and Mate.
½″	·050	1¼″	·078
¾″	·060	1½″	·090
1″	·067	2″	·100

Fixing Unions for Lead Pipes, exclusive of soldered joints.—

	Days of Plumber and Mate.		Days of Plumber and Mate.
½″	·006	1¼″	·012
¾″	·006	1½″	·012
1″	·006	2″	·012

Fixing Apparatus, exclusive of joints.—

Pedestal closet	·170
Valve closet	·350
Copper bath	·170
Porcelain bath	·170

The cost of hoisting and getting in of cisterns varies greatly, depending on position.

| Packing chases with slag wool, per foot cube | ·083 |

Hot-water Pipes.—

Fixing only pipe, including tees, bends, connections, etc.

	Days of Fitter and Lad.
½″ pipe, per foot run	·015
¾″ ditto, ditto	·018
1″ ditto, ditto	·020
1¼″ ditto, ditto	·020
1½″ ditto, ditto	·025
2″ ditto, ditto	·030

ZINCWORKER.

	Plumber and Labourer.
Laying zinc in flats, gutters and flashings, per foot	·015
Ditto, stepped flashings	·009
Extra labour to cesspools	·150

PLASTERER.

	Days of Plasterer.	Days of Labourer.
Render, per yard	·013	·007
Ditto, circular, ditto	·017	·007
Render and set, ditto	·023	·012
Ditto, circular, ditto	·029	·012
Render, float and set, ditto	·038	·019
Ditto, circular, ditto	·048	·019

	Days of Plasterer.	Days of Labourer.
Sifting and running lime, screening sand and mixing putty, per cubic yard	..	·100
Lath, plaster, float and set partitions	·042	·019
Ditto, circular	·053	·019
Extra on surfaces for finishing trowelled stucco	·018	..
Ditto, circular	·023	..
Lath, plaster, float and set ceilings	·046	·023
Ditto ditto flueing	·069	·019
Fine plaster cornices and mouldings (exclusive of mitres), per yard supl.	·500	·250
Ditto, circular	·625	·312
Fine plaster cornices above 6″ and not exceeding 12″ girth, per foot run	·034	·017
Mitres	·034	·017
Quirks, per foot run	·007	..
Ditto, circular	·011	..
Portland cement, plain face, per yard	·100	·050
Ditto, in narrow widths, ditto	·120	·060
Ditto, moulding, ditto	·750	·375
Ditto, mouldings above 6″ and not exceeding 12″ girth, per foot run	·050	·025
Mitres	·050	·025
Portland cement mouldings not exceeding 6″ girth, per foot run	·061	·031
Mitres	·061	·031
Portland cement, jointed as stone..	·105	·053
Ditto, plain face, circular, per yard	·115	·057
Arris, per foot run	·007	..

Parian or Keene's Cement.—

Plain face, per yard supl.	·100	·050
Ditto in narrow widths, ditto..	·120	·060
Moulding, ditto	·900	·450
Mouldings over 6″ and not exceeding 9″ girth, per foot run..	·060	·030
Ditto not exceeding 6″ girth, ditto	·072	·036
Plain face, circular	·115	·057

FOUNDER AND SMITH.

	Days of Labourer.	Smith.
Unloading, getting in, hoisting and fixing rolled joists, columns and stanchions, per ton	2·70	·600
Ditto large girders	3·50	·800

The foregoing are only approximate; the time spent on getting in and hoisting varies with the character of the building.

	Smith and Labourer.
Ends of joists forged to fit to the flange and feathering of girder	·150

PRICES.

Fixing only.—

	Days of Carpenter.
Eaves gutter, per foot	·030
Stopped ends, each	·030
Outlets	·040
Angles	·060
Rain-water pipe	·030
Plinth bends	·050
Swan necks up to 15" projection	·050
Shoes	·030
Heads	·040 cwt.
Ties and straps, per lb.	·004 ·448
Flitches, per cwt.	·260
Bolts, with heads, nuts and washers, under 1 lb. weight, per lb.	·027
Ditto, 1 lb. and under 2 lb., ditto	·022
Ditto, 2 lb. „ 4 lb., ditto	·016
Ditto, 4 lb. „ 8 lb., ditto	·011
Ditto, 8 lb. and upwards	·008
Heads and shoes to roof trusses, per cwt.	·300

Wrought Iron.—

	Smith and Labourer.
Chimney bars, per cwt. (fixed by Bricklayer)	·450
Framed guard bars	·920
Framed balusters and fixing	·900
Oval or half-round handrail and fixing	1·40
Ditto, wreathed	2·70
Gratings, framed (fixing by Mason)	·113
Roof straps and perforations (fixing by Carpenter)	·100
Hook and eye hinges (fixing by Carpenter)	2·45
Core rail, framed and fixed	1·30
Ditto, wreathed	2·00
Extra on handrail for wreaths	·170
Ditto scrolls	·180
Ditto, angles	·150
Fixing newels, each	·050

Observe that the above constants may be applied to work of similar labour.

Work in position is much more costly than work done before it is fixed.

GASFITTER.

Fixing pipes, including fixing of tees, bends, angles and connections:

	Days of Fitter and Lad.
⅜" per foot run	·015
½" ditto	·015
¾" ditto	·018
1" ditto	·020
1¼" ditto	·020
1½" ditto	·025
2" ditto	·030
Fixing gas-brackets with nipple and rose, each	·200

GLAZIER.

	Days of Glazier.
Glazing sheet glass in new sashes, per foot supl.	·012
Ditto in old ditto, ditto	·025
Hacking out glass, including painting putty one coat, per foot supl.	·030
Glazing plate glass	·015

When the squares of plate glass are very large two or three men will be required, and the price should be two or three times the constant.

Cleaning windows inside and out..	·005

The quantity of glass to be cleaned can be seen by referring to the glazier's bill if there is a bill of quantities.

PAINTER.

	Days of Painter.
Priming, per yard	·015
Knotting, stopping and rubbing down, per yard..	·010
Second and after coats, each	·014
First coat on iron, including preparation, per yard	·020
Second coat on iron, per yard	·015
Sash squares, knot stop and rub down, per dozen	·025
Sash squares, first coat, per dozen..	·040
Second and after coats, each per dozen	·035
Large sash squares, stop and rub down, per dozen	·037
Large sash squares, first coat, per dozen	·060
Second and after coats, each	·050
Frames, knot stop and rub down, each	·018
Ditto, priming one side, each	·026
Ditto, second and after coats one side, each	·025
Large frames, knot stop and rub down	·028
Ditto, priming on one side, each	·040
Ditto, second and after coats one side, each	·038
Extra large frames, knot stop and rub down	·036
Ditto, priming one side, each..	·055
Ditto, second and after coats one side, each	·050
Washing old paint, per yard	·018
Burning off old paint	·141
Facing up after burning off	·015

In common work it is customary with many to allow one-third of the total value of painted work for labour, and in difficult and ornamental work, two-thirds of the total value for labour.

	Days of Painter.
Sizing, per yard	·010
Varnishing, per yard	·014
Staining	·018

PAPERHANGER.

	Days of Paperhanger.
Stripping walls, per piece	·044
Rub down, clearcolle and size walls, per yard	·020
Hanging common papers, per piece	·088
Ditto, satin papers, per piece	·134
Ditto, best papers, per piece	·167
Ditto ceiling papers extra only	·033
Ditto Japanese papers, per 12 yards supl.	·330
Twice sizing and varnishing paper, per piece	·204
Lining marble paper to imitate blocks, per piece	·033
Hanging marble paper in blocks, per piece	·266

DEFICIENT QUANTITIES.

The adjustment of variations, in cases where quantities have been supplied, frequently raises questions of quantity, often unreasonable.

Some builders institute a rough check upon the quantities by comparing the amount of material delivered at the building, as bricks, stone, timber, lead, etc., with the item in the bill of quantities. Naturally all their allegations turn upon deficiencies; but the quantities must be taken as a whole, and the surveyor should be allowed the benefit of anything which may have been measured in excess.

In another place, the question of liability to the contractor is dealt with.

In claims of this kind, it will generally be found on inquiry that the demand is based upon invoices, and that questions of waste have not been considered. Moreover, the delivery of certain materials at a building is not absolute proof of their use there. Part may have been taken away again.

A few of the checks adopted by the contractor are as follows:—

Comparsion of the carter's bills with the items of the excavator's bill.

The quantity of ballast, sand and lime, delivered with the concrete and brickwork.

Digging.—As earth after digging increases in bulk and the surveyor measures it *in situ*, there will be a discrepancy between the digging and the carting. In sewer work, this is very possibly conspicuous, and the surveyor protects himself by stating in his bill the width he has allowed for the various trenches. See preamble to a sewers bill.

The carting involved by a given building is not all for excavation, some of it will be the carting of rubbish, and in cases of alterations, this will form a large proportion. The bricks, stone and timber may also be carted by a hired carter, and there may be no means of distinguishing the various materials that formed the load; although it is a common custom with hired carters to state their load on their periodical account.

Bricks.—The number of bricks delivered on a building is sometimes used as an argument for alleged deficiency of brickwork. There is always a percentage of waste: the commoner the bricks the larger is the proportion. Keen supervision will also involve the rejection of a good many.

In the comparison of the number of bricks delivered with the bill, brickwork will probably have been divided in such bill into a number of categories, and these must be collected for the purpose of comparison.

Stone.—When stone is worked at the building, there will be waste, and its percentage will depend upon the kind of stone, and the detail of the stonework. The proportion of waste in conversion varies considerably. See Waste, in section on Prices, page 560. It will vary from 5 to 25 per cent.

Rubble Walling.—The waste on rubble stone will depend materially on the finish of the work.

If the rubble is uncoursed, 20 per cent. will cover the waste; if coursed, 33 per cent.

Slater and Tiler.—As slating is nearly always sublet, the builder who alleges deficiency, produces the slater's account. Discrepancies are often referable to the different allowances for cuttings made by the slate-merchant and the surveyor, and these may be compared.

If the builder buys the slates and does his own slating, the quarry-owner will deliver 1260 slates for the 1200, the 60 being an allowance (5 per cent.) for possible damage in transit.

Carpenter.—As there is always waste in the adaptation of timber to the purposes of the building, the number of loads bought at the docks, or at a timber sale, largely exceeds the cubic quantity in the finished work, sometimes by as much as 30 per cent.

If the builder contracts with a timber merchant for timber sawn to scantlings at a price per foot cube, there is still a proportion of waste in cutting to lengths and fitting, certainly not less than 5 per cent.

When all the timber is wrought, the waste is of course increased. A comparison of the boarding with the slating will show that it is commonly 15 to 20 per cent. less in quantity than the slating.

Joiner.—Flooring. If this is very well specified and critically supervised, a large proportion is rejected for knots, sap, etc., sometimes 50 per cent.

Plumber.—Claims for extra lead are frequent.

Invoices are produced showing the weight of lead delivered at the building, and in many cases the proofs of its delivery are unquestionable. Assuming that the intentions of the contract have been carried out, there is always a percentage of waste which has been taken away as cuttings, varying with the nature of the work from 5 to 15 per cent.: 5 per cent. is a common average.

But a frequent explanation is found in the fact that the lead has been stolen, and this always happens if the watching is lax.

Another common reason is to be found in the widening of flashings and secret gutters, increase of laps, etc., especially when the laying only of the lead is let to a sub-contractor at a price per cwt. He lays as much as he can.

Founder and Smith.—The invoices of the ironfounder often show a greater weight of iron than that taken in the quantities: sometimes the fault of the founder in making his castings thicker than ordered.

Invoices of rolled joists show a greater weight consequent upon rolling margin.

Gasfitting.—Gas and hot-water pipes delivered in a building have to be cut, and this is a source of waste.

General Considerations.—The foregoing are a few of the sources of difference between quantities and material delivered; they are sometimes useful in considering claims for deficiencies, and may save the trouble of measuring an item in dispute, but often there is no sufficient proof short of remeasuring the item in question.

PRICING OF DAY ACCOUNTS.

In pricing day accounts, surveyors must remember that the greater part of materials have been supplied in small quantities and are consequently worth more than in large. Nails, screws, small quantities of painters' colours, single articles, etc., are usually worth a price equal to that charged by the general retail trades.

This treatment will apply if the work properly measurable has been excluded from the day account.

When making an estimate for a builder, the estimator may, if pressed for time, price it on the abstract and carry the amounts to a summary; the bill may never be wanted, but can be produced whenever desired.

It is not usual to abstract the notes, they can be priced on the dimension paper, putting one column of dimensions only on each folio and leaving the other blank for pricing.

Record of the proportion of the cost of work in each trade may advantageously be kept by the surveyor in some such form as follows. Examples from actual practice are given in the following table.

Place.	Name of Contractor.	Cubic contents.	Lowest tender.	Per foot cube.	Provisions.	Excavating and draining.	Bricklayer.	Mason.	Slater and Slate Mason.	Carpenter.	Joiner.	Gasfitter.	Plasterer.	Founder and Smith.	Plumber.	Glazier.	Painter.	Bells.	Preliminary.	Surveyor and Litho grapher.
House at Oxford.		93,318	£ 3837	d. 9¾	670	101	800	496	155	407	314		177	263	191	73	45	4	47	89*
House at Kensington		111,654	4764	10¼	399	144	1223	555	61	331	882	18	276	222	240	91	110	29	51	127†

REMARKS.—* Brick walls faced with coursed rubble, roof of stone slates, stone windows with mullions and transomes, iron casements, deal joinery. Three miles from a railway station. Roads "uphill" all the way.
† Blanchard's red facings, Portland stone strings and cornices, about half the slating green Westmorland, joinery deal, electric bells.

The estimator will find himself but imperfectly prepared for his work, if he has not previously acquired the art of taking off quantities with reasonable facility; such practice is an admirable training of that analytical faculty which good pricing requires, and he should accustom himself to value in accordance with the best system of measurement he may know. The best systems were framed and adopted with special view to that end.

Notes of the number of rods of brickwork, compared with the cubic content of the building, may be preserved with advantage. A few notes of this kind, from a recent paper read by Mr. Rickman, may be useful:—

	Content in cubic ft	Rods of reduced work.
A university building of monumental character	1,010,000	1 to 150
A country house	160,000	1 „ 1,600
A London club house	2,270,000	1 „ 1,849
A set of insurance office buildings	527,000	1 „ 1,850
A hospital	3,250,000	1 „ 1,184
A country house	392,000	1 „ 1,912
Ditto	300,000	1 „ 1,962
A paper warehouse	400,000	1 „ 2,090
A set of residential chambers	386,000	1 „ 2,100
A hotel	4,000,000	1 „ 2,296

Such an investigation might be pursued as to the proportion of beds and joints, and other labours to a given quantity of stonework, according to its character, cubic feet of timber, yards of plastering, etc.

Records of such information as the above are of use in the scrutiny of a bill of quantities before sending it out. A surveyor will frequently be surprised by the quantities in his bill, and will need some such data to reassure him.

ALLOWANCE FOR SMALL QUANTITIES.

Work in small quantities is worth more than in large quantities. Builders in pricing and estimating usually increase the price by 20 to 25 per cent., and this probably covers the extra cost as far as a general rule can, as if applied throughout the bill, one error will balance another; but it is better to take each case on its own merits, and the percentage may reasonably vary from 10 to 50 per cent. It requires considerable experience to decide what this percentage shall be, and when it shall be added.

MATERIAL IN DAY ACCOUNTS.

The argument for an increased price for all the material charged in daywork account is often a fallacy; in many cases it costs no more to supply material in small quantities than in large, increased cost is after all the only reason for allowing a higher rate.

METHOD OF BUILDING.

Intimately connected with the question of the cost of a building, is the method adopted for its erection; the relative advantages of various methods are in the order following, the cheapest and best first:—A contract on quantities, a contract without quantities (on drawings and specifications), on a schedule of prices, purchase by the building owner of materials, and employment of workmen supervised by his clerk of works.

ARBITRATION.

In arbitrations on the value of building, the work of the Quantity-Surveyor is indispensable, and his service and evidence usually the most important.

An arbitration on building values may occur either during the progress of the building or after its completion; the latter time is doubtless the more convenient. In either case if the Quantity-Surveyor who prepared the original quantities for the building is an expert in prices, he will probably be the principal witness.

A few considerations as to his functions may therefore prove useful.

The evidence of the Quantity-Surveyor is most frequently required on the following questions:—

The value of extra work on a contract.

As to whether or not work for which a claim has been made is extra on a contract.

The value of a completed work as affected by the workmanship.

The value of the opposing interests in provisional sums in a contract.

The value of work done when a contract is incomplete either by the abandonment of the building by the contractor, or his ejectment.

The amount of damage sustained by the building owner by the abandonment of a building contract by the builder.

The loss of profit claimed by builder after ejectment.

The value of party walls.

The quality of bills of quantities.

The justice of the fees charged by a Quantity-Surveyor for his work.

All the foregoing are fruitful sources of dispute.

With the inauguration of an arbitration the Quantity-Surveyor has little to do.

The appointment of the arbitrator lies with the building owner, the architect and the contractor, and their action at the beginning, judicious or otherwise, may make all the difference between success and defeat.

If the arbitrator is judiciously selected the advantages of arbitration compared with ordinary litigation are conspicuous.

The arbitration clause commonly stipulates that the arbitrator shall be an architect, but when the issue is the quantity and value of artificers' work (and it is on these questions that disputes most frequently arise), the Quantity-Surveyor is the more capable judge, and in many cases the ostensible decision of the architect arbitrator is really that of his Quantity-Surveyor.

Many of the modern Quantity-Surveyors have had a proper architectural training and are also expert Quantity-Surveyors. Such a combination of experience is calculated to produce a valuable arbitrator.

If the parties to a building dispute begin by resort to the ordinary process of law, they often involve themselves in great and futile expense, for in matters of intricate account, the judge will either refer the case to the official referee or he may appoint an arbitrator whose only recommendation is a knowledge of law, and in either case the expense of the preliminary process is wasted.

The knowledge of the official referees of building operations and values appears to be superficial; of this fact their frequent and artless production of Laxton's Price Book in the course of an examination, and their astounding decisions, is sufficient proof.

The reasonable course is the selection by the parties of a real expert on the matter in dispute who has besides a knowledge of the law of evidence. Such men are fairly numerous in the professions of either architect or Quantity-Surveyor.

A nomination by the President of the Royal Institute of British Architects, or of the Surveyors' Institution, is not always a warrant of the possession of the essential qualifications by the nominee, but the right man for the work may be discovered by the litigants with very little trouble.

When an Architect or Quantity Surveyor is the arbitrator and sits with a legal assessor, it is commonly the case that the more exact training of the lawyer enables him to master the judgment of the arbitrator, and the lawyer is the real arbitrator. It is probably better for either Architect or Quantity Surveyor to sit alone and refer to his solicitor privately for advice on legal questions.

The common practice of arbitrators of receiving an account from each of the parties differently arranged and differently priced without remonstrance, is to be deprecated; as they start from different premises comparison and criticism are extremely difficult. Counsel on each side (often comparatively ignorant of building operations) pick out items which have no parallel in the account of the opposite side and argue them to the confusion of everyone concerned.

There is small doubt that the insistence by the arbitrator upon the remodelling of the statements of claim and the production of counter statements on similar lines by both parties would produce a great saving of valuable time and expense, would much simplify their consideration, and would be productive of more reasonable decisions than those commonly given.

The essential conditions to produce a rational result, and to eliminate the main causes of confusion, are as follows:

A measurement by the surveyors of the respective parties at the same time, in company, and on the same lines.

A bill produced from such measurement, with items in the same order, the items being placed in juxta position, the differences would be obvious, and the arguments for and against easily understood by all.

As the arbitrator can not only decide as to the evidence he will admit, but may prescribe the form in which it shall be presented; it is probable that a judicious representation to the arbitrator at the first hearing of these considerations would, in the majority of cases, ensure his order for the production of such documents.

The whole of the evidence of the Quantity Surveyor must be corroborated, and if he is the leading witness he will, as a rule, be allowed to select the persons to do it. He should choose expert and experienced men and such as are reputed successful.

Builders are often called to give evidence of value, but Architects or Surveyors are to be preferred. Nevertheless, a genuine tender, especially if competitive, as in the case of completing work unfinished, is often valuable evidence.

In the case of specialities as heating, electrical and mechanical engineering, etc., the evidence of specialists will be necessary.

The production of the builder's business books will supply evidence of a certain value if the builder is making a claim, but if the building owner is resisting one, he is often in a better position if they are not produced.

There are a few precautions which the witness should observe.

To measure accurately and have the work examined by another person.

To ensure the careful checking, including clerical errors, of all documents furnished to the arbitrator.

To personally examine all the work done by assistants.

To date every document.

To observe neatness and precision in the measuring books. These small precautions impress the arbitrator to an extent which the inexperienced often fail to realize.

A series of trifling errors (although the main contention may be perfectly true) afford opportunities for opposing Counsel to impugn the whole of the evidence of the man, who is responsible for them, and to cast doubts upon work and evidence which is for the most part correct and praiseworthy.

The law of arbitration is beyond the scope of this work. Russell on Arbitration and Award, and Hudson's Building Contracts will admirably furnish all the requisite legal information.

CHAPTER IX.

THE LAW AS IT AFFECTS QUANTITY SURVEYORS.

THE mixture of common law and custom of the trade has tended to a complication of the relations between surveyor, builder, and building owner, and the want of familiarity of the average solicitor with building contracts renders it incumbent upon the majority of quantity surveyors to acquire a knowledge of those general principles of law which specially affect their professional duties and responsibilities.

And this fact has been so generally admitted that suggestions have been made for the establishment of a court specially for the settlement of questions affecting building contracts and cognate subjects. The portentous combination thus indicated of lawyer and surveyor in one person does not at present appear practicable, but is none the less desirable.

The custom of the trade bears a large share in defining the legal construction of any contract which may affect the quantity surveyor, and such custom is assumed by law to become (in the absence of any condition to the contrary) a part of the contract; but a custom to be good must be certain, and it must be clearly proved to exist and to be general and notorious. Nevertheless, any trade custom may be set aside by special arrangement.

The much argued question, Shall the quantities form a part of the contract? is one which materially affects the quantity surveyor. Sir Edmund Beckett says (in 'A Book on Building'), "There is another intermediate element or document which in modern times has assumed great importance, though the employer generally knows nothing of it, *nor (be it remembered) is it any part of the contract legally, and nothing but confusion arises from recognising it as such*, though it may sometimes be referred to for information — that is, the bill of quantities of every kind of work throughout the building." And, again, "I advise employers not to let the quantities be taken by any one who can be made out to be their agent, nor to recognise them in any way, except as after mentioned, with reference to a schedule of prices for extras." Others assert that, if the quantities form a part of the contract, the amount to be paid by the building owner is always uncertain until the building is completed, and that such arrangement takes away, to a large extent, the feeling of responsibility which a surveyor has for the accuracy of his work.

The strongest argument for the adoption of the quantities by the building owner is the justice of that course. The employer should obtain neither more nor less work than is included in the quantities, and it is believed that this would be the desire of the majority of employers. The assumption of the loss of feeling of responsibility by the surveyor is unfounded, the surveyor's responsibility is of deeper origin. The solicitude as to error on the part of the quantity surveyor is also, to a great extent, mistaken, for when a properly qualified surveyor prepares quantities, he makes remarkably few mistakes; the mistakes in quantities are usually made by persons who have but small right to be called quantity surveyors, and have not been specially trained for the work. It may also be remarked that claims for deficiencies in quantities are sometimes paid by the quantity surveyor, but they are very much more frequently included in "additions" at the adjustment of accounts, and the surveyor would, in such a case, take special trouble to look for items of possible excess as a set-off. And, although when the quantities are not a part of the contract the builder may legally refuse to allow them to be consulted (taking his stand upon what plans and specification alone indicate), yet in the majority of cases they are adopted as the basis of settlement.

There is thus considerable show of reason in the contention that the quantities should form a part of the contract, for they are in the majority of cases practically so.

The making the quantities a part of the contract would tend to abolish the absurd practice of an architect taking out quantities and disclaiming responsibility for their accuracy, which has resulted in the anomaly of the contractor being under the necessity of going through all the dimensions and checking the quantities before signing the contract (usually impossible) or taking them as correct—the architect often receiving for such work as this an equal percentage to that charged by surveyors who take the responsibility of accuracy.

The making the quantities a part of the contract, and the elimination of "sporting items" as they are called, will also reduce prices. Uncertainty induces estimators to add sums to cover possible contingencies.

There is a strong feeling among surveyors, that when an architect supplies quantities for his own works they should invariably form a part of the contract. But it is also the belief of the majority that a general practice of adopting the quantities as a part of the contract, while relieving surveyors to some extent from the responsibility for accuracy, would inevitably induce a decline in the scale of their remuneration, and encourage the supply of inaccurate quantities by irresponsible persons.

When are quantities a part of the contract?

A wholesome fear exists in the minds of architects as to allusions in the specification to quantities, and although such allusions are best avoided, except as to the stipulation that they shall be used as a schedule for pricing items of variation it is very doubtful whether casual reference to them is sufficient to incorporate them with the contract. It is unquestionable that when it is *intended* that they should form a part of it, there is express reference to the fact either in the conditions or the contract, or both. In the contract of the National Association of Master Builders these clauses are used, the intention being to make the quantities a part of the contract.

"In the construction of these presents when the contract will admit of it, the term contractor shall mean the said   the term proprietor shall mean the said   , the term architect shall mean   , or other the architect for the time being employed by the proprietor to superintend the erection and completion of the works, and the term works shall mean all the works, acts, matters and things specified and described in the specification, plans and other drawings and detailed bills of quantities supplied hereinafter mentioned, and also such other works, matters and things as are hereby contracted to be done and performed by the contractor."

Also,

"The proprietor shall pay to the contractor for the full and perfect completion of this contract, the sum of £   ; but if the architect shall direct any addition to, or omission of, or variation from the works, the value of such addition, omission or variation shall be added to or deducted from the said sum of £   , as provided in clause 8, as the case may be, and if there should be found to be any error in the detailed bills of quantities supplied, such error shall be rectified and an addition be made to the contractor or deducted from him as the case may be in respect of such error."

Some surveyors use a clause like the following in the preliminary part of the bills of quantities, but in default of its embodiment in the contract it is questionable whether it has the legal effect desired.

"The bills of quantities will form the basis of the contract, and duplicate bills will be supplied to the contractor whose tender is accepted to fill in the prices of his original estimate, and extras and omissions will be valued at these rates." That the clause must be specific and without option is sufficiently proved by the clauses in Stevenson *v.* Watson and Young *v.* Blake, upon which those actions were based.

As Mr. Hudson says to the builder, "Unless the *contract* provides that you shall only perform such work as is described in the quantities, they become so much additional description which increases your liabilities instead of reducing them."

The new conditions of the Royal Institute of British Architects do not admit the quantities as a part of the contract.

A properly drawn contract in which the quantities are adopted as the basis of the contract, will operate against that legal principle that a contract to do a whole work involves the liability of the contractor to complete it, whatever shortcomings there may be in the drawings and specification.

It was generally supposed, until the decision in Priestley and Gurney *v.* Stone, that when a contract is proceeded with, and the builder discovers loss consequent upon deficient quantities, the surveyor is liable for the amount of such loss.

The case of Priestley and another *v.* Stone, 4 T.L.R. 730, is shortly as follows:—Tuke wishing to build a church, employed Kelly as architect; Kelly instructed Stone, quantity surveyor, to prepare quantities; Stone did so, and sent the lithographed copies to Kelly, who issued them. Priestley and Gurney tendered thereon. The tender, after being reduced, was accepted. The drawings were in pencil when the quantities were prepared; Kelly finished the drawings and added others. In finishing the drawings he materially altered them. Priestley alleged that the quantities were deficient, and sued Stone for the value of that deficiency.

Stone says, in his answer to interrogatories:—"The plan marked B was never before me at all. When I prepared the said quantities, the plans C, D, E and F were before me, but since they left my hands they have been altered in several particulars." Mr. Justice Stephen says, in commenting on the evidence of the architect:—"This shows that the design is altered in a material particular after the quantities were taken out; that alteration in respect of the plan would run through the whole estimate"; and "he (the architect) said he had to alter everything." At some later date the architect wrote the specification. He included several things not taken in bills of quantities, elaborated the doors, and practically treated the quantities as non-existent. Stone was not consulted about the specification, and was not informed of the alterations in the plans. The architect advertised for tenders, quantities to be obtained from him on payment. Priestley and Gurney's first tender was 3750*l.*; this was more than Tuke wished to spend. Kelly, Priestley and Gurney then revised the work without Stone's assistance, and the original tender was reduced by 576*l.* Priestley and Gurney state that, in answer to their inquiries, the architect said that the drawings and specification were those from which the quantities had been prepared

The architect inserted a clause in the contract to which much reference was made during the hearing. This Clause XI. says, "The contractor will be allowed the opportunity, and must previously to signing the contract examine the drawings and specification and compare them, if requested, upon the site with his estimate, so as to satisfy himself that his price includes all the works and materials shown or described in the specification and drawings which constitute the contract, as no departure or alteration will be allowed to be made from or in his tender in the event of any error therein being discovered, however arising." As to this clause Mr. Justice Stephen said:—"This clause shows that the building owner and his architect, and more particularly the architect, were well aware that they might be liable for the estimate on which the matter was based, and that they determined that they would not be liable but would throw the liability upon the builder himself by the express terms of Clause XI. I can hardly conceive how there could be another intention than that which appears to me to be plain common sense in itself, that the contract is between the owner and the contractor, and that the owner will not be liable for any alterations or anything there may be in any want of correctness however arising which has taken place in the early stages of the matter."

In plaintiffs' statement of claim they alleged that they had suffered damage from defendant's negligence and breach of duty as a quantity surveyor, of, and for, the plaintiffs; that defendant, in consideration that the plaintiffs would agree to pay him his stated charges if their tender was accepted, supplied for the use of plaintiffs a bill of quantities, and defendant represented that these quantities would be sufficient for building the church according to plans and specification; that it was the duty of defendant to use ordinary care and skill in the preparation of the quantities, as he well knew that the tenders would be based on the quantities, and that if they were insufficient, loss would be occasioned to the persons tendering; that defendant warranted the accuracy of the quantities; that plaintiffs, relying on defendant having used ordinary care and skill, and on his warranty, based their tender on the said bills of quantities and entered into a contract to build the church according to said plans and specification for the amount of their tender, which contract they performed; that defendant did not use ordinary care and skill in the preparation of the said quantities, and did not correctly set forth detailed particulars of all works and materials required for the building according to plans and specification. The particulars of loss were set forth in an accompanying statement, and plaintiffs claimed 689*l*. Stone, in his statement of defence, denied that he was employed by plaintiffs as quantity surveyor. He stated that there was no privity of contract between plaintiffs and defendant, and defendant did not owe the plaintiffs any duty; that he had not been guilty of negligence or breach of duty, that the bill of quantities was not supplied by defendant to plaintiffs, but to the architect, and it was not supplied in consideration that plaintiffs would agree to pay defendant therefor, but in consideration that the building owner would cause him to be paid the amount of his charges, which were fixed by an agreement made with the architect; he denied the warranty. He states that the plaintiffs did not rely on any representation or warranty of defendant, but did rely either on their own investigation or upon the representation of the architect ; that he did use ordinary care and skill; that he was instructed by the architect to prepare a bill of quantities according to certain plans supplied by the architect, and according to certain verbal instructions given by the architect; that the architect did not supply defendant with any specification, and did not prepare one until after the bill of quantities was finished; that the bill of quantities correctly set forth the particulars of all work and materials required according to the said plans and instructions; that after defendant had delivered the bill of quantities to the architect, several alterations were made in the works of the intended building and in the said plans without defendant being consulted thereon ; that the total amount of work and materials in the bill of quantities is not less than the total amount shown to be necessary by the said plans, &c., handed to defendant by the architect.

In his answers to interrogatories, Stone refers to the alterations in the drawings and the absence of a specification. He explains the omission of certain stone cores to cornices and iron bond as arising from express instructions by the architect to omit them. He denies that plaintiffs arrived at the amount for which they contracted by simply pricing out the particulars appearing in the bill of quantities, and says that before they contracted to do the work they reduced their estimate by 576l.; but in what way the reduction took place, or in what way the quantities were affected by that reduction, he cannot say, as he was not consulted in the matter. He entered into a detailed explanation of many alleged omissions as arising from the alterations in the drawings or from entries in the specification in excess of the verbal instructions given to him. He admits certain trifling omissions, valued in all at about 12l., but claims that these are more than set off by other items taken in excess. At the hearing in January 1888, before Mr. Justice Stephen the plaintiffs' case was endeavoured to be established on the ground of an alleged custom of the trade by which privity of contract between plaintiffs and defendant was sought to be established, and also on the ground of inaccurate statement of facts, i.e. short quantities. It was agreed that the question of liability only should be settled by the Judge, and if he found, on the question of liability, in favour of the plaintiffs, the details of the alleged deficiencies were to be referred to arbitration. Plaintiffs' counsel were heard fully on the custom sought to be established, and this custom was set forth as follows :—

"The employment is originally by the building owner; the quantity surveyor upon that employment does certain things. As soon as the tender of the successful builder is accepted, then the liability which the building owner was under to pay for these quantities shifts to the builder. The builder becomes liable, and then there is a corresponding liability on the part of the surveyor to be answerable (to the builder) for any negligence in the quantities."

This is the custom which the learned counsel for the plaintiffs, Mr. Meadows White, Q.C., said in a paper read before the R.I.B.A. in May 1888, in referring to this case, failed to be established even upon the evidence of the first witness called for the plaintiffs. Having heard plaintiffs' counsel and witness, without hearing the defence, Mr. Justice Stephen gave judgment. His Lordship said :—"There must be a non-suit; there are substantially two points (1) that there is a contract, (2) that there was misrepresentation. If the plaintiffs are under a misrepresentation or have anything to complain of in regard to the quantities which they acted upon, they must go against either the building owner or the architect who made these representations. The evidence about custom on the whole case appears to me not only unsatisfactory and vague in itself, but, as far as it went, was rather in favour of the defendant than in favour of the plaintiffs. Clause XI. is intended obviously to put an end to the question. Suppose it were not there: the man who made the estimate (the builder) would say, I know nothing about your quantity surveyor. You have represented to me that these quantities are correct. I made my estimate from these quantities; I have been deceived; you must pay me damages because you induced me by false representations to make an improvident tender." (His Lordship then read and commented on Clause XI., for which see *ante*.) "The quantity surveyor, if he did anything

wrong, would be liable to his architect and the building owner for any damage he has caused to them. Apart from that I cannot see that there is any further liability unless such liability can be proved." (For his Lordship's remarks as to the alteration of the drawings by the architect see *ante*.) His Lordship further said:—
"If the architect has made serious alterations after the estimate is drawn up, it shows to my mind clearly that the representations made by it might very well be the representations of the architect and not the representations of the quantity surveyor, because they apply to different things. He (the architect) said he had to alter everything. It is impossible to say that the representation made to the contractor is the same representation which Stone authorised the architect to make to anybody, and therefore I say it is the architect's representation and not Stone's representation, and the only representation from Stone is a representation to the architect for which, if inaccurate, or if the architect is damaged by it, he may take his remedy. Non-suit with costs." Judgment was then given on application, without comment, for Stone, with costs.

The plaintiffs appealed. The appeal was heard in July 1888 before the Master of the Rolls, Lord Justice Lindley and Lord Justice Bowen. The builders' case was argued. Their counsel relied on custom and misrepresentation. As to custom, it was that set forth at length in noticing the hearing in the Court below. He sought to prove privity by custom and by that only. It was admitted that this was the first case in which this custom had been endeavoured to be established in a superior Court. There was no reported case which could be found in which the liability of a quantity surveyor to a builder for deficient quantities had been established. This drew from the Master of the Rolls the remark—"You are seeking to prove custom. You say you prove a well-known custom, and nobody has ever heard of a case." After much pressure the learned counsel had to admit that as to custom his case was gone. He then relied on "the holding out these things as correct." He said :—"I ask you to say that the relationship existing between the surveyor and the builder is such that the quantity surveyor if negligent is liable to the builder." He cited various cases bearing collaterally on the question, some of which were read at length, but he was met at all points by the Court—" There is no fraud here, no allegations of fraud. You

should have said, 'I am going to show not only that there were mistakes, but that they were so reckless and careless as to amount to fraud.' The surveyor may have been negligent as against the architect, but the question is whether he owed any duty to all the world of builders." Their Lordships gave judgment, dismissing the appeal without hearing the counsel for the defence. The Master of the Rolls said in his judgment:—" This is a contract between the architect and the man who takes out the quantities; the architect employs that man, and he has to pay that man; therefore the contract, if any, is between the architect and that man, or between whoever is the principal of the architect, and that is the only contract. There is no evidence of negligence in the way of taking out the quantities—none whatever. It is true that negligence is evidence of fraud, that we are all agreed upon; some people call it fraud and some evidence of fraud, but there was none here, because it was arranged that the question of how great the errors were, or what the errors were, was to be left until after the liability was decided by the Judge; then those who agreed to that, if they wanted to rely upon the amount of error as showing gross negligence and reckless statement, ought to have presented that, because that is the question upon which liability depends. There is no evidence upon which the learned Judge could be asked to find anything against the defendant." Lord Justice Lindley said—" The action was brought upon two grounds, firstly that there was some sort of contract between plaintiffs and defendant. That breaks down. There is no contract between the two, and the alleged custom which was relied upon as establishing such a contract was not proved. If the custom had been proved I do not know whether it would be good for anything." His Lordship then referred to the case of Bradburn v. Foley. This was an agricultural case in which some custom was proved as between an outgoing tenant and the landlord, by which the incoming tenant, although not a party to the arrangement between the others, was to be bound. It was referred to during the argument before the judgment, with a remark, 'That custom would bring two people together who had never been together at all, and such a custom must be bad." His lordship continued:—
"This custom has not been established as a fact, and if it were, that authority (Bradburn v. Foley) goes to show that custom is

3 D

nothing and there is no contract. The second ground was that Stone made a negligent representation to plaintiffs in order that plaintiffs might act upon it, and that that being the case, the representation having been made and the plaintiffs having suffered damage, they are entitled to maintain an action in order to prove that suggestion. Look at the facts. The quantity surveyor is employed by the architect and the architect acts as agent for the building owner; at all events a quantity surveyor is not employed by a builder, although it appears to be customary that the successful builder shall pay the surveyor. These quantities are supposed to be made with reference to the instructions given at the time, and suppose that after quantities were taken out they desired to change the design, and yet the architect laid the quantities before the builders with the design, how is this to be turned into a representation by the surveyor to the builder that the quantities taken out are applicable to the changed state of things? Under the circumstances there is absolutely no proof of want of possible care, or anything that should make the action stand." Lord Justice Bowen said:—" This action is without any precedent. In this case there are two fatal difficulties, (1) there is no privity between the builder and the surveyor; and (2) there can be no duty out of which the supposed liability arises, because whatever duty there is must be a duty arising out of the relations between the two, and if there is no privity between the two, there really is no responsibility. However, it has been suggested that the surveyor made a statement which is a misstatement, knowing it is to be handed on to a third person, and therefore is liable to the third person provided the case comes within the doctrine of false representation, as in the case of Peek v. Derry. But all the surveyor does is to represent that these are his quantities, which he believes to be accurate. As to that being false representation, how can you say that any amount of negligence on his part in taking it out, unless it amounted to recklessness, any amount of negligence, unless it was so gross as to raise a suspicion of fraud, in which case it would be evidence of fraud, makes it fraudulent? But mere negligence would not make the representations fraudulent, evidence of negligence in the quantities is not enough to render him liable towards the builder, because there is no privity, no duty between them, and in the absence of misrepresentation there is no liability."

THE LAW.

The rights of the quantity surveyor, in the absence of express agreement, are as follows :—

The architect having disclosed building owner's name (that name on the drawings and specification being sufficient), and having been instructed to obtain tenders, is, by that instruction, empowered to have the quantities prepared. If the work goes on, the contractor is liable to the surveyor for his charges by virtue of custom of the trade. If the work does not go on, the building owner is liable; settled by decision in the case of Moon *v.* Guardians of Witney Union.

3 Bingham, New Cases, page 814, which is shortly as follows :—

Kempthorne, architect to the guardians, prepared plans and specification, and instructed Moon to prepare quantities. The drawings and specification were deposited with the clerk to the guardians, with a copy of instructions for the builders, which ran thus :—" The builders desirous of contracting for the erection of the Witney workhouse are informed that the quantities of the works are now being taken out for their use, and will be ready by the 28th inst. Builders requiring a copy of the same are requested to leave their names, with the sum of 2*l.* 2*s.*, at Mr. Kempthorne's office or at Mr. Leake's, Clerk to the Union, Witney, before 26th inst. The successful competitor will have to defray the expenses of taking out the quantities, the charge for which will be stated at the foot of the bill of quantities when delivered." A dispute arose between Kempthorne and the guardians, and they refused to go on with the work. Kempthorne sent in his bill with Moon's bill appended. They paid Kempthorne's bill, and refused to pay Moon's charges. They had never heard of Moon till the account was delivered. Moon proved the usage for architects to employ surveyors to prepare quantities, and that the successful contractor included the charge in the amount of his tender. The jury gave a verdict for the plaintiff.

A *rule nisi* was obtained to set aside this verdict and enter a non-suit, or to have a new trial, on the grounds that no privity had been established between the plaintiff and defendants, that the usage was not binding on defendants, and that the usage had not been sufficiently proved. The judges held that the defendants were liable.

Lord Chief Justice Tindal said :—

"The question was whether Kempthorne, as agent for the defendants, had any authority to bind them in a contract with plaintiff."

"The jury found that there was a usage in the trade for architects or builders to have their quantities made out by surveyors. Then the defendants themselves had an intimation that such was the practice, for Kempthorne wrote to Leake, the defendants' attorney, and Leake gave out that the successful competitor should defray the expenses of taking out the quantities. If this was to be so, what was to be the result if by the defendants' fault there was no successful competitor, because there was no competitor at all? In such a case the defendants must be liable for the amount of a charge which they have authorised their builder to incur."

"Then, when, upon their refusal to proceed, Kempthorne sent in his account, comprehending the charge for the surveyor, they had express notice of the existence of the charge. They came to an agreement with Kempthorne in respect of his own demand, to pay a sum which could not comprehend the plaintiff's charge, and their conduct upon that occasion was a recognition of the claim which they did not settle or object to."

"It is contended, on behalf of the defendants, that a contract cannot shift so as to have two different parties liable to the plaintiff at the same time; that may be so in some cases, but the difficulty is got over here by considering this a conditional contract, a contract under which it was arranged that the expenses of making out the quantities should be paid by the successful competitor, if any, but if by the act of defendants there should be no competitor, then that the work which was done by their authority should be paid for by them."

Gwyther v. Gaze is a case of similar bearing. Queen's Bench Nisi Prius, reported in the 'Times,' 8th February, 1875.

Gaze wished to build a warehouse, and instructed Sayward, an architect, to prepare plans. Sayward instructed Gwyther to prepare quantities. Three tenders were sent in upon the quantities, none of which were accepted. Afterwards the building was proceeded with on Sayward's plans, the height of the building being reduced. The builder employed was not one of those who first tendered, and was paid as the work proceeded. Gwyther afterwards sent in his account to the building owner, charging 2½ per cent. on the lowest of the three tenders. Mr. Justice Quain thought 2½ per cent., where no tender was accepted, an unreasonable charge.

Defendant asserted that the architect had only a limited authority, that he was told that the defendant had engaged a builder, so that there would be no necessity to take out the quantities to enable builders to tender. His Lordship left it to the jury to say whether the architect's authority was not limited in the way stated by the defendant, and told them if it was not limited the defendant was liable to pay the plaintiff a reasonable remuneration for taking out the quantities. The jury returned a verdict for plaintiff. It was agreed that his Lordship should assess the damages, which he did at the rate of 1½ per cent. on the lowest of the three tenders.

Mr. Justice Quain's artless remark may be noted. As soon as a quantity surveyor issues his bill his work is done, and the amount of the charge should not be reduced.

Waghorn *v.* Wimbledon Local Board of Health, Queen's Bench Nisi Prius, reported in the 'Times,' June 4th, 1877, introduced a somewhat different element.

The defendants are by statute the burial board as well as local board for the district of Wimbledon.

Defendants, at a meeting of the board, passed a resolution which embodied instructions to their surveyor, Rowell, to prepare plans and specification and procure tenders for the erection of a cemetery chapel.

Rowell prepared the drawings and specification and instructed Waghorn, quantity surveyor, to prepare quantities; these were sent to the builders in the usual way, and builders tendered upon them. The estimate being beyond what the board desired to spend, none of the tenders were accepted, and they refused to pay Waghorn's charges. The defendants pleaded that they never authorised Rowell to employ plaintiff.

His Lordship ruled that as they had instructed him to procure tenders, and as tenders could not be made without quantities, they had impliedly authorised him to get the quantities taken out.

It was then submitted that the defendants, being a corporation, could only contract under seal, and there being no contract under seal here, the plaintiff could not recover. His Lordship ruled that as the defendants had, by resolution, impliedly authorised Rowell to get the quantities taken out, and had the benefit of the work which had been done, their objection was not tenable.

Judgment was accordingly entered for the plaintiff for the amount claimed.

Taylor *v.* Hall, Irish Law Reports 4, Common Law, 467 (1869–70), illustrates another point which arises not infrequently.

Walsh, building owner, wished to make alterations to his house, and communicated with Hall, a builder, who prepared a plan and estimate. Walsh, being dissatisfied therewith, instructed Fogarty, architect, to prepare drawings and specification, and obtain tenders. Fogarty instructed Taylor, quantity surveyor, to prepare quantities; these were sent to builders, one of whom was Hall. Hall's tender was the lowest, but greatly in excess of the amount Walsh desired to spend. Taylor thereupon prepared a bill of reductions and sent it to Hall, who modified his estimate; the amount of this reduced estimate was also too large, and was not accepted. At an interview, shortly after this, Taylor cautioned Hall that if he obtained the work on any modified plan, he would still be liable for the charges. Hall answered that he would feel himself liable and would apprise Mr. Walsh of it.

Walsh then employed Hall to carry out a modification of his original plan under a contract by which it was agreed between the owner and Hall that the latter should not be liable for Taylor's charges.

Taylor thereupon brought an action against Hall for the charges, maintaining that the plan adopted was a mere modification of Fogarty's reduced plan.

Taylor's witnesses stated that there was the strongest similarity between the work as carried out and Fogarty's reduced plan.

Hall's foreman stated that the works were carried out according to a reduction of Hall's original plan, and that Taylor's calculations were not used to produce the accepted tender.

Chief Justice Monahan said that, according to the usage of the trade, if the owner discarded the architect's plan he was liable to the surveyor, the architect having the express or implied authority of the owner to employ him; that the builder who carried out the new plan, though he might have tendered under the old plan, was not liable if he did not use the surveyor's calculations; and that the contract here not being any modification or continuation of the original contract which was in negotiation with defendant, but an entirely new, independent contract by the present defendant in his own right, he was not liable, the uncontradicted evidence of his foreman, who acted for him, being that the plaintiff's calculations were not used.

Held that Taylor could not recover.

In this last case the agreement made by Walsh with Hall did not affect the decision, nor did Walsh by that agreement make himself liable, for he was already liable, and the action should have been brought against him.

In cases of measuring extras and omissions, the builder is liable for the charges, and on the same principles, the charges being usually added to the builder's account.

All the foregoing cases derived their force from custom of the trade. The following cases depend upon special arrangements which nullified the effect of custom.

If an architect instructs a surveyor to prepare quantities, not disclosing his client's name, or acting without his client's authority to obtain tenders, or having made an arrangement with his client to prepare drawings, specification and quantities for a given sum, he thereby makes himself personally liable for the charges, and the architect may so act as to give strong reasons for the assumption that he has made a personal contract by his acts, as, for instance, paying money on account of the work.

A case in point is as follows:

Burnell v. Ellis, Common Pleas, a short report of which may be found in the 'Times,' 25th June, 1866, and the 'Builder,' 30th June, 1866, the substance of which is as follows:—

Ellis, architect, was employed to prepare plans and specification for an hotel and stables proposed to be erected by a company. He employed Burnell, surveyor, to prepare quantities. The estimates being higher than was expected, the intention of erecting the buildings was abandoned. Plaintiff said his charges were 100*l.*, but as the work did not go on, and defendant did not expect to get all he was entitled to, he asked plaintiff to reduce his charge; this he did, agreeing to take 50*l.* The evidence showed that defendant had paid plaintiff 10*l.* 10*s.*, which plaintiff asserted was on account, and which defendant said was a loan, and further, that the plaintiff was engaged to do the work by Hall, a builder, and that it was understood that he was to look to the company for payment.

The jury found a verdict for the plaintiff.

Another case bearing upon this part of the question is—

Richardson and Waghorn v. Beales and others. Common Pleas, Nisi Prius. Reported in the 'Times,' June 29th, 1867.

Beales and other gentlemen formed themselves into a committee to establish a club, to be called the New Reform Club, and obtained designs in limited competition for the building. Porter, architect, was the successful competitor. Porter instructed Richardson and Waghorn, quantity surveyors, to prepare quantities. Tenders were obtained. The amount being too large, the intention of building was abandoned. Richardson and Waghorn applied to the committee for their charges, and payment being refused, brought their action.

The architect, on being examined, stated that he had received a communication from the committee informing him that his design was accepted, and desiring him to get in tenders provided he did not pledge the liability of the committee in so doing.

The Lord Chief Justice expressed a strong opinion that this letter put plaintiffs out of court, and that the architect was the proper person to be sued, but said he would leave the evidence to the jury.

Upon this plaintiff's counsel elected to be non-suited.

The following case was an action against a building owner for deficient quantities.

Scrivener and another v. Pask, Law Reports 1, Common Pleas, 715.

The circumstances were as follows:—

Pask, desiring to build a house, employed Paice to prepare plans and specification and obtain tenders. Paice prepared quantities (thus placing himself in the position of quantity surveyor). Scrivener tendered on these quantities. His tender was accepted, and he paid Paice his charges. The quantities proved deficient, and Scrivener sued Pask for the amount of the deficiency.

Justice Blackburn said:—

"To entitle the plaintiffs to recover they must make out three things: That Paice was defendant's agent; that Paice was guilty of fraud or misrepresentation; and that the defendant knew of it and sanctioned it. There is no evidence here of either of these things. *If there has been misconduct upon the part of Paice the plaintiffs have their remedy against him.*"

Primâ facie, an architect who supplies quantities for his own works is in a like position with the independent quantity surveyor, but the question is nearly always complicated by some provision of the contract between building owner and builder.

The quantity surveyor is bound to furnish a copy of dimensions to the person who pays him, but the copying must be paid for. But if the surveyor refuses, there is no way of compelling him to supply them but by bringing action, when they would be obtained by the usual notice to produce.

The responsibilities of the quantity surveyor in cases of express agreement are dependent upon such agreements, and are consequently various.

When the building owner arranges to pay the quantity surveyor's charges, the quantity surveyor is liable to the building owner for the value of any error.

Sometimes the building owner stipulates that the surveyor shall also be liable to the builder for inaccuracies.

An architect who prepares quantities under a similar arrangement places himself in the position of the independent quantity surveyor.

A surveyor acting as an appraiser, as in valuing dilapidations, should have an appraiser's license.

Mr. ARTHUR CATES, in the course of a debate upon the paper "Quantities and Quantity Practice," at the Surveyors' Institution (April 12, 1880), read a series of propositions which very clearly expressed the then generally received conclusions as to the custom and practice of quantity surveyors. Most of the clauses are based on legal decisions and are still admitted.

1. That, when tenders are required from a number of builders for the erection of a building, it is essential that bills of quantities should be prepared, in order that each of them may make his tender on identically the same basis.

2. That, in the absence of special instructions to the contrary, it is the duty of the architect to make the necessary arrangements for the providing of such bills of quantities.

3. That, for this purpose, the architect is the agent of his client, and the client is bound by his acts, whether cognisant or not of them or of the custom.

Moore *v.* Guardians Witney Union; Richardson and Waghorn *v.* Beales and others; Gwyther *v.* Gaze.

4. That, until a tender is accepted, the client is liable to the surveyor so appointed by the architect for the amount of his commission, and the expenses incurred.

5. That, on the acceptance of a *bonâ fide* tender, the liability to the surveyor shifts from the client to the builder; the surveyor accepts the builder as responsible to him, and his right of claim against the client ceases.

Young *v.* Smith; North *v.* Bassett, see also conditions of contract R.I.B.A., clause 14.

6. That if the work is abandoned before a tender is accepted, the client pays the surveyor; if after, the accepted builder pays, and has his remedy against the client.

7. That, as a matter of convenience, it is the custom that the architect should include, and it is an obligation on the architect so to include, in his first certificate such reasonable charges and expenses of the surveyor; but the liability of the builder is not affected by this practice, and commences immediately on the acceptance of his tender.

8. That a surveyor employed directly and solely by the builder, without the intervention or concurrence of the architect, has no claim against the client, and must look for payment to those who employed him, and on whose instructions he acted.

9. That there are circumstances under which an architect may make himself personally liable to a surveyor, but they are of infrequent occurrence, and are not likely to arise with architects of any standing or repute.

Richardson and Waghorn *v.* Beales and others.

10. That the surveyor is liable to the builder for proved inaccuracies or deficiences in the quantities, and it is an obligation on him to prepare his quantities with the utmost care and accuracy, that the client may not suffer by excess or the builder by want therein, and to fulfil his important duties with the strictest honour and integrity.

Priestley *v.* Stone.

Although this responsibility for accuracy may be admitted, it is not to be supposed that the currently professed readiness to pay for deficiencies is to be taken quite literally; it is a question of degree. Many surveyors who in such a case pay for a deficiency of a small amount, are not to be depended upon to pay a very large one. They would, as is most frequent, endeavour to make terms with the building owner or his architect. It must be remembered that although a given item may have been omitted from the quantities, the building owner should not in justice accept it without paying for it, as he will do if the quantity surveyor pays. If the building owner, however, should prove unkind, the builder should show that he sustains loss on the *whole* contract from the surveyor's fault; he should allow as set-off to his claim any excess that may be discovered in the quantities, and he certainly should be satisfied with the *cost* price of the work in question.

The case of Priestley and Stone appears to dispose of the principles of any legal obligation to the builder for deficient quantities, except in cases of gross negligence or fraud—allegations very difficult to prove.

But the moral obligation to the builder has been generally admitted by quantity surveyors, and is not altered by this decision. Various suggestions have since been made with the view of defining more clearly the obligations of the surveyor.

Mr. Hudson suggests several ways (Professional Notes, Surveyors' Institution, vol. iii., p. 105):—

The builder in order to protect himself has four courses open,

(*a*) To employ his own quantity surveyor, so that the quantity surveyor shall owe him a duty, or

(*b*) To take a proper guarantee from the quantity surveyor with the building owner's consent, supported by some consideration (but this arrangement would not relieve the quantity surveyor from liability to the building owner or architect who employed him), or

(*c*) To enter into a conditional agreement with building owner and quantity surveyor, that in the event of the builder's tender being accepted, the quantity surveyor shall be paid by him and be considered as employed by him instead of by the building owner, so far as it is necessary to render the quantity surveyor liable to him for any inaccuracies in the quantities, or

(d) To refuse to tender unless a quantity surveyor whom he can trust is employed, and then take the risk of inaccuracies.

Mr. J. Hayward Studwick (Professional Notes, Surveyors' Institution, vol. iii. p. 107).

"That the quantity surveyor is responsible for the general accuracy of his signed bills of quantities (where privity exists) scarcely admits of doubt, and I think he is responsible equally to the builder for deficiency and to the building owner for excess, privity being ensured by the retainer in the one case, and by the surveyor himself sending to the builder his signed bills of quantities in the other.

The question is, how can this be best effected so as to satisfy the reasonable requirements both of the building owner and of the builder.

The Surveyors' Institution is a corporate body, and it seems to me that it should by a resolution in council declare that the names of surveyors who undertake to become responsible for their signed bills of quantities could be enrolled at the Institution, the precise terms of the responsibility undertaken to be legally expressed in clear and definite language and signed by the applicants, the sole condition of enrolment to be that the surveyor should have been in practice for a prescribed period.

It may be said that the enrolment suggested would not add to the legal liability for negligence to which surveyors and others are already subject; but the effect would be to indicate by the surveyor's own act and in a marked manner the *bona fides* of the person so enrolled, and to show their willingness to submit themselves to such existing liability rather than resort to technicalities to repudiate it."

The folly of accepting quantities prepared by the architect to the building without making them a part of the contract, is forcibly shown by Stevenson *v.* Watson and Young *v.* Blake.

In both these cases the builders assumed that the quantities were a part of the contract.

The responsibilities of the independent quantity surveyor are pretty clearly defined, those of the architect who prepares quantities for his own works are not so clear.

The facts of the case first mentioned are briefly as follows:—

Watson, architect, prepared plans and specification and supplied quantities for the public hall at Nottingham; Stevenson, builder, tendered for the work, and his tender was accepted. The contract was as follows :—

"The said Richard Stevenson and Field Weston agree to erect and build for the said company, upon a certain piece of land situate in North Circus Street, in the town of Nottingham, the temperance hall, according to the drawings, general conditions of contract, and bills of quantities, now produced and signed by the parties hereto, and intended to form parts of the agreement, and shall and will finish and complete the said temperance hall in such manner and of such materials, and within such time as is provided by the said general conditions of contract and bills of quantities, and according to the said drawings; and, further, that they, the said Richard Stevenson and Field Weston, will well and truly observe and perform all and every the said conditions and stipulations contained in the said general conditions of contract on the part of the contractors required to be observed and performed; and in consideration thereof the said company to pay unto the said Richard Stevenson and Field Weston the sum of 13,560*l.*, in the manner set forth in the said general conditions of contract, and in other respects to perform and keep the conditions and stipulations of the said general conditions of contract, so far as the same on their part is or ought to be performed and kept."

The conditions of contract contained the following clauses :—

"The general conditions of contract for artificers' works required to be done in the erection and completion of a new hall for the Nottingham Temperance Hall Company (Limited), Nottingham. Fothergill Watson, architect, Clinton street, Nottingham, Jan. 1874.—The architect is at all times to have access to the works, which are to be entirely under his control and his clerk of the works. The architect may order any additions to or deductions from the contract without in any way vitiating the contract, and the amount of such additions to or deductions from the contract shall be ascertained by the architect in the same manner as the quantities have been measured and at the same rate as they have been priced at.

"The contractor and the directors will be bound to leave all questions or matters of dispute which may arise during the progress of the works or in the settlement of the account to the architect, whose decision shall be final and binding upon all parties.

"The contractor will be paid on the certificate of the architect."

The bill of quantities embodied the following note:—

"Note.—These quantities will, with the drawings and general conditions, form the basis of the contract. Should there be more or less measure than is here given, there will respectively be an addition to or a deduction from the contract.

"All measurements to be made in the same manner as the quantities have been taken, and all additions and deductions to be priced out at the same rate by the architect."

As the work proceeded there were many omissions and additions. The quantities proved deficient, the great discrepancy being in the brickwork and the stonework; the builder's surveyors estimated the deficiency on the brickwork alone at upwards of 400*l*. These variations the builder requested the architect to measure or have measured at convenient times, as before the striking of the scaffolds, &c., but without effect. As to the production of the bill delivered, the builder states:—"At various stages I had the building measured off by two competent surveyors, with the assistance of myself and staff, with the utmost accuracy, and priced out and measured the same way as the bill of quantities, item by item. Every care was taken to have the account correct, as I could see there would be a difficulty with the architect, and when the dimensions were squared up and the extra measurement found, I had the building measured over again. The measurements were gone over and tested by me and the foremen of the different branches of the trade."

It is obvious from this statement that due care was observed in the investigation.

Stevenson, on the completion of the works, made a claim amounting to 1616*l*. 6*s*. 7*d*., of which between 500*l*. and 600*l*. was for errors of deficiency in the original quantities, and the remainder a balance due on the contract.

The architect, without calling for any explanation of this account, certified for a balance of 251*l*. 14*s*. 4*d*., and refused to give any explanation. Thereupon Stevenson brought his action for 1364*l*. 12*s*. 3*d*. and interest. The claim asserted that—"The defendant did not use due care and skill in ascertaining the amounts to be paid by the company to the plaintiff under the said contract, but, in ascertaining the net balance due to the plaintiff, neglected and refused to ascertain, and did not ascertain, the amount of the said additions to and deductions from the contract in the same manner as the quantities had been measured, and at the same rate as they had been priced out, or that there was more measure in the said descriptions of works than was given in the bill of quantities by making measurements in the same manner as the quantities had been taken, and neglected and refused to price out, and did not price out, the excess at the same rate, and make the stipulated additions to the contract in respect thereof, according to the terms of the contract, nor did he use due care and skill to ascertain, in the manner provided by the contract, what was, in fact, the net balance payable to the plaintiff by the company in respect of the works executed, for which the defendant was entitled to his certificate; but the defendant, knowingly or negligently, certified as aforesaid for a much less sum than was, in fact, the net balance payable to the plaintiff in respect of the works executed.

"Upon the receipt of the said certificate the plaintiff requested the defendant to inform him of the data upon which the same was based, but he refused to furnish the plaintiff with them, or to give him any information on the subject. The plaintiff thereupon requested the defendant to reconsider the said certificate, and offered to point out to him the said errors in the bill of quantities, and to give him any explanation he might require of the said accounts; but the defendant refused to reconsider the said certificate, and to allow the plaintiff to point out to him the said errors in the bill of quantities, or to explain the said account, or to hear any objection whatever on the part of the plaintiff to the said certificate."

The case was heard by Lord Chief Justice Coleridge and Justice Denman. Lord Coleridge said the case was one of considerable importance, and but for the intimation that had been given that whichever way the Court decided the matter would be afterwards reviewed by the Court of Appeal, he should have taken time before giving his judgment. The decision was in substance as follows :—

That the architect was in the position of an arbitrator, and therefore not liable for want of care or skill.

An action will not lie against an architect for not using due care and skill in ascertaining the amounts to be paid by a builder's employer to the builder under a contract which provides that the builder is to be paid on the certificate of the architect; that all matters of dispute are to be left to the architect's decision; that he may order any additions to or deductions from the contract; and that the amounts of such additions or deductions shall be ascertained by him at a certain fixed rate, the functions of the architect under the contract being not merely clerkly, but requiring the exercise of a judgment or opinion.

An allegation that the architect *knowingly or negligently* certified for a much less sum than was due does not disclose a cause of action, as it does not amount to a charge of *fraud*. The architect is not bound upon the application of one of the parties to reconsider his certificate, or to give reasons for it.

Leave was given to amend the claim on payment of costs, equivalent to permission to appeal against the decision. The decision was not appealed against for the following reasons:—The builder was not a rich man; he had been kept waiting for the money for his last certificate for six months; the expenses of the case had been very heavy; and the adverse judgment resulted in his bankruptcy.

Young *v.* Blake. High Court of Justice, Queen's Bench Division, November 22nd, 1887, is a case of similar bearing to Stevenson *v.* Watson.

Blake employed Farrell and Edmonds as architects for a proposed house. They prepared quantities for the work, supplied them to the builders whose tender was accepted, and were paid in the usual way.

The quantities were found to be deficient. The builder brought action against Blake and the architects for loss sustained thereby. In the judgment which follows, it is held that the quantities were not warranted correct, that the quantities were not a part of the contract, that the allegation of negligence was not sufficient cause of action.

Mr. Justice Denman's judgment was as follows:—

This was an action brought by Messrs. Young, builders, against Mr. Blake, the head-master of the Sherborne Grammar School, and against Messrs. Farrell and Edmonds, architects and surveyors. It was an action of a very peculiar description, and upon a basis which was somewhat unusual, very unusual I should any, and it was agreed, as I understand, between the parties that the question which was to be raised before me here, and really it was the only question which could be reasonably raised before me, should be the question of the respective liability of Mr. Blake on the one hand, or the architects on the other. The two actions, though mixed up in the sense of being with reference to the same transaction, are totally distinct and different causes of action. I think the best way of dealing with them will be first to deal with the action against Mr. Blake, and then to deal with the action against Messrs. Farrell and Edmonds.

It is not necessary to state the facts in very great detail, but I will state the facts so far as I find them upon the evidence, and so far as they bear on the question which is for me, viz. the liability of the parties. Young and Company were builders, residing and carrying on their business at Salisbury, and the work to be done was to be done at Sherborne, in Dorsetshire, about one hour and a quarter's journey by fast train from the one place to the other. Farrell and Edmonds were architects and also surveyors. In 1884 Farrell and Edmonds had been employed by Mr. Blake to prepare plans at all events, and specifications probably, for a new house, and other matters connected with the school. It appears from the evidence that Farrell and Edmonds had, in consequence of their employment by Mr. Blake, the other defendant, informed him that it would be necessary that quantities should be taken out. A conversation had taken place between him and them in which he showed very great ignorance of the usages of architects and surveyors. He asked them certain questions as to who would have to pay for it, and who would do it; he had heard of such persons as quantity surveyors, but they informed him that they were quantity surveyors as well as architects, and they did, for the purpose of getting a tender for the then contemplated school houses and buildings, take out certain quantities and prepare plans and specifications. At that time I find, in accordance with Mr. Blake's evidence, that they had somewhat led him to suppose that the building which was to be built, or he had rather intimated to them that the building was to be erected, was not to be a building which would cost above 3000*l*., but it appears that they did, in their plans and specifications or in their contemplated building upon which they drew up their quantities, devise a building which would cost a great deal more. The consequence was that he declined to have anything to do with that building, the tenders were not accepted, and the whole thing went off. There is still a question pending between the parties as to whether in those circumstances the sum which it might be said to have cost Farrell and Edmonds to take out the quantities should be paid to them for the trouble they had in taking out quantities. I do not in any way intend by what I say to decide that question, it may be that the one or the other side may be right in that contention. On the one hand Mr. Blake contends that it was owing to their fault with the instructions that they had, that they did not effectively bring about a tender for the thing that he intended, by reason of the devising of a very much more expensive thing; they, on the other hand, contending that as he did not choose to take the work, they acting for him in the matter, they would be entitled to remuneration for taking out those quantities. The only importance, therefore, of that original transaction is with reference to what might be gathered as the terms upon which those two parties dealt with one another, and what authority the one gave to the other upon a subsequent transaction which took place, and which is the one upon which the present dispute arises. Now the present dispute began at a later date, and the correspondence I think very clearly shows the course of events so far as it bears upon the present action. It would appear that on the 1st of December, 1884, after the other transaction had altogether gone off, Messrs. Farrell and Edmonds being architects, and being no doubt at that time authorised by Mr. Blake to set to work to obtain tenders, did write to Messrs. Young and Co., the plaintiffs, and in fact to a good many other builders, a letter in which they asked them whether they would be willing to tender for the erection of so and so, " quantities to be supplied, an early answer will oblige." Then on the next day Messrs. Young write to say that they would be quite willing to tender, and then at a time which is not exactly fixed, but between that date and the 18th (as clearly appears by a letter from the plaintiffs referring to the "bill of quantities"), the bill of quantities was sent which is impeached in the present action. Farrell and Edmonds say, " We have pleasure in sending you herewith a bill of quantities, tenders to be sent to us," and so on. Now the bill of quantities itself is an important document: it is headed "Estimate for master's residence, boarding house and preparatory school, to be built at Sherborne for W. Blake, Esq.—FARRELL and EDMONDS, architects." Then comes the quantities "excavation"—so many yards for this, so many yards for that, in fact an ordinary bill of quantities. At that time, of course, it was not priced, that would have to be done by the persons before they tendered, and the bill as I now have it has in pencil the figures that were put in by the plaintiffs, Messrs. Young, and put in with a view of seeing what the cost at which they could do the building was to be. On the 18th December there was a letter referring to that bill of quantities. On the 19th there was a small matter which the architect spoke about as being an overlooking of a certain sum, and certain arrangements were made between them and the builders. On the 23rd came the acceptance of the tender, and Messrs. Young, on the 23rd of December, write, "We will undertake to do the work comprised in the bill of quantities for the erection of master's residence, &c., at Sherborne, for W. Blake, Esq., for the sum of 3800*l*." Now a question arises as to the exact date at which an interview, and the only interview, took place between the plaintiffs and Mr. Farrell before the contract was actually signed. I do not think it is necessary to determine (and I do not myself feel absolutely certain) what that date was; on the one hand it is sworn by Messrs. Young that it was at some date before the estimate was actually accepted, before the tender was actually made for 3800*l*., that they had the bill of quantities before them. On the other hand there was some doubt raised about that—as to the only time that they had seen the specification and the plan, because there is a letter from Mr. Farrell to Messrs. Young, preparing for a meeting at a certain time, namely, half-past two on the Monday, which is after the 29th of December or the 27th of December, and as there was only one meeting, it would look as though that contemplated the meeting which actually took place, but I do not think it is very important in the view I take of the case. I only allude to it because it must not be taken that my decision depends on it in any way. Eventually the builders, the plaintiffs, set to work. There is a good deal of correspondence, which I have read, and which was referred to and read at length before me, as to small matters about which special arrangements had to be made; certain plans to be supplied are got by the architects themselves, and there was a question as to delay in beginning, and delay in the work, and so on, but all these matters seem to me to have been disposed of practically, and to have been wiped away by the fact (and it is really an admitted fact in the case upon the evidence) that at a certain date, namely, by September 1886, at all events, there had been a payment in full for the work done, certain arrangements being made, and certain gives and takes having taken place, that there had been payment for all the work done, and all the matters arising upon the contract, subject only to the question which is raised in this case, that is to say, whether there is any liability on the part either of Mr. Blake or of Mr. Farrell in respect of deficient quantities, because that is the complaint. Assuming the quantities to be correct, granted that there has been payment for all that has been done in the way of work and extras, the contention is that upon the events which have occurred and upon the discovery by the plaintiffs at a subsequent date, or rather by August 1886, that does give the plaintiffs as against Mr. Blake, and secondly as against Mr. Farrell, a right of action. First as regards Mr. Blake, I must look at the claim against him. The claim against Mr. Blake is shaped in two or three different ways. First it states that Mr. Blake, in or about December 1884, warranted and represented to the plaintiff that the work to be done and the materials provided therein were full and accurately stated in a certain bill of quantities, in writing, which he delivered to the plaintiffs. It is necessary to observe that the relations between Mr. Blake and the plaintiffs are those contained in the agreement of a date subsequent to December, namely, some day in January 1886, that is the date of the actual contract between the parties, and it seems to me upon the true construction of that contract coupled with all the evidence that there is in the case as regards Blake, that it is quite clear that Mr. Blake never did at any moment warrant the accuracy of the quantities. With regard to the contract, it is to this effect. It is between Messrs. Young and Blake—the architect is no party to the contract. It recites that he wishes to build, and that he has appointed Messrs. Farrell & Edmonds architects and surveyors, and that they are hereinafter called " the said architects " —they are to be architects for the purpose. Then it recites that plans and sections, elevations and specifications have been prepared and approved of, and then it recites the fact that, for the sum of 3800*l.* this building was to be erected. Then it recites that plans and specifications have been signed by the contractors and by the architects, and then in paragraph 6, which is the paragraph bearing most upon the bill of quantities, there is the following provision:—" The bills of quantities supplied by the architects are believed to be correct, but should any error or misstatement be found therein, either in favour of or against the contractor, it shall be lawful and in the power of the architects to measure any or all of the works contained and described in the bill of quantities, and to adjust the same in accordance with the prices therein contained and whereon the said tender was based, and the work undertaken, and the said contractor shall, for this purpose, produce on demand to the architects the said bill of quantities." Now the case both against Mr. Blake and against Mr. Farrell depends very much indeed, so far as the liability here is concerned, upon the real meaning of that clause, and I may at once therefore state what I understand it to be. It begins by the assertion that the bills of quantities supplied to the architects are believed to be correct. That certainly seems to me to dispose altogether of the statement that Blake warranted the quantities to be correct. The parties are parties to a contract which merely states that they are believed to be correct. Then it goes on to state, " Should any error or misstatement be found therein in favour of or against the contractors, it shall be lawful and in the power of the architect to measure any or all of the works contained or described in the bill of quantities." There I will stop for the present. Now what is the meaning of that? It appears to me that it is a discretionary power given to the architects, making it lawful and in their power to measure any or all of the works—" any," during the progress of the works, or " all "as soon as they are finished—to measure any of the works in the bill of quantities if any error or misstatement is found therein. Now I do not think that that gives the power, upon the mere assertion of the builders themselves that there is an error or misstatement, to compel and to call upon the architect as a matter of right to remeasure the work. I think that would too unreasonable a power to suppose to be given to the builder in a contract of this kind, because it really would involve a most expensive and a most damaging operation to the works at the mere option of a person who undertakes to do them properly, and undertakes by this agreement to do certain works. I think, therefore, that the meaning of that must be that the architect is to be left in the discretion to remeasure if he finds there is reasonable ground to think that there is an error or misstatement of the works, and that then he may remeasure them and he may adjust them, as the words follow on—"and to adjust the same in accordance with the prices therein contained"—then he may adjust them as between the parties, and the difference, of course, would be allowed to the party, and the gains would be allowed as against the party who loses by that remeasurement. The architects had a right in order that that may be efficiently done to ask for the priced bill of quantities from the builder, in order that he might do that work. That being so, it appears to me that the architect stands in a *quasi*-judicial position between the parties, he is architect not merely as a person who is employed as the agent of the building owner for all purposes, nor is he a person who is employed by the builder in any sense so as to be liable to him as a person at his will and pleasure to be ordered to do anything because the builder is dissatisfied. Now, if that be true, it seems to me to dispose of the case, not only as against Blake, but as against the architects themselves, because if that be the true position, then the architects here are made by the parties, persons in a *quasi*-judicial position—they are persons who, unless they are guilty of fraud or misconduct of any kind beyond mere allegations of negligence, are not to be harassed with actions against themselves, nor are the people whose architects they are as well as the architects of the other party to be harassed with actions merely because the architect in his discretion may refuse, on the demand of the builders, to remeasure any of the works. That principle seems to me to be the principle of a case which counsel called my attention to yesterday, which is reported in the 4th Common Pleas Division. It is a case of Stevenson *v.* Watson, in which when a tender is made upon a statement that the quantities had been negligently certified for a much less sum than was the net balance, an attempt was made to fix a liability in consequence of that, and the decision was, "That the functions of the architect in ascertaining the amount due to the plaintiff were not merely ministerial, but such as required the exercise of professional judgment, opinion, and skill, and that he therefore, occupied the position of an arbitrator, against whom, no fraud or collusion being alleged, the action would not lie." That was a decision of the Chief Justice and myself, and I do not see any reason to doubt that it was right. Several cases were relied upon for that principle; it was very well argued on the other side by the present Mr. Justice Cave. I do not think it went to any Court of Appeal, and therefore, at present, I must take it to be good law. Another case cited, Scrivener *v.* Pask, though the facts are not identical with the present case, also seems to me to go to a considerable length in favour of the defendants in this case—both of them. Scrivener *v.* Pask seems to me only to amount to this, that where there is nothing more than the ordinary employment of a quantity surveyor, the quantity surveyor being paid out of the first receipts by the builder of a building owner who must have quantities taken out in one sense, that is to say, who will not get tenders unless the quantities are taken out—where there is that simple case, there is nothing in that case merely from the fact that the building owner orders things which involve the probability of quantities being taken out, to fix him with a liability to pay for the quantities so taken out which are taken out by a quantity surveyor ordered by the architects, or by the architect himself. It appears to me that that case at least goes so far as this, that unless there is something binding the owner, some understanding between the parties to be gathered from correspondence or words making himself liable for the taking out of the quantities, that he is not so liable.

Now, as regards Mr. Blake, I can see nothing. There is no agreement such as that which is set out in the second paragraph of the statement of claim—namely, that it was agreed that the bill of quantities should be treated as the basis of the contract, and that if they were found to be incorrect, the plaintiffs should be paid by the defendant for the work and materials done and provided. That seems to indicate, if found incorrect by any body who may persuade the builder that he has had a bad bargain, and that the quantities have been too small. I do not think that that is the meaning of the contract. The meaning of the contract is that the architect is to be responsible for deciding between the parties, and if he decides honestly, that the parties should have no right to complain of anybody. That really disposes of the case as against Mr. Blake.

Now then, as regards the architects themselves. I think the same observation almost entirely disposes of the case against them, because if the relations had been that which I have stated—I think it is between the architects and the parties—then I think that they were in the position of persons trusted with a judgment, and, unless it were a dishonest judgment, a mere inaccuracy would not render them liable. They are not parties to the contract in any way; they do not sign the contract, and they are only sought to be made liable by the supposition—first, that there is a custom that they should be liable under such circumstances, which, I think, is entirely unavailing in this case because the clause is a very peculiar one in its language and in its relations in every way. That will not avail them, and next it was alleged that they were liable upon a certain special undertaking on their part to be liable, which was to be gathered from language which was said to be used by them in an interview between them and Messrs. Young & Company. Now, I have carefully considered the language which has been sworn to on both sides, and, though I do not know that I can say that I disbelieve the statements on the part of Messrs. Young —the words "the quantities are full" may have been used, I think it is admitted almost by one of the defendants—Mr. Edmonds, I think, admits that those words may have been used—I do not think that that can be, looking at the relation between the parties and at the contract which they knew of, and which both parties were fully aware of, construed into a warranty. There is no evidence of warranty against them. Then is there evidence of a warranty independently of that? There may be such a thing, and if the plaintiffs' evidence is believed, there would be evidence, perhaps, to go to a jury on it, if it stood uncontradicted; but we have evidence on the other side, and the evidence on the other side is very distinct and positive to the effect that the question of guarantee was actually raised, and the question was put, "Do you guarantee the quantities?" and here I believe the evidence of those witnesses, that when the expression, "guarantee" was put, they positively repudiated any such notion. I forget the exact expression in which it was repudiated, but it was as strong as it could be—" nothing of the kind," or "certainly not." I think that that is the probability of the case—they would have been very foolish, and it would have been very unlikely that they should have guaranteed the quantities, when they had the discretion in them by the contract itself, which said that the quantities were believed to be correct, but that if any error or misstatement should be found, then it was lawful for them and in their power to measure any or all of the works, and so on. This is no action for not doing anything under the contract, it is an action founded upon, in the first instance, warranty, and in the next place upon an allegation of negligence. Now, the allegation of warranty is not made out, but disproved. The allegation of negligence, to my mind, is one which does not give a cause of action upon such a contract as this. It puts the architect in the position of persons trusted by both parties, and if a man with his eyes open chooses to enter into such a contract giving the architect that power, he cannot turn round, and by a mere allegation of negligence say, "I have discovered that there was an error in the measurements, and I have asked you to act under the contract and you won't do it in your discretion, but I will now sue you because you negligently took out those quantities." I do not think that is the relation between the parties at all. The architect is not put in that position so as to be the servant of either party in that sense, but he is in a different position altogether; and, therefore, I think the action fails against Farrell and Edmonds as well as against Blake, and I must, therefore, give judgment for both defendants, with costs.

The legality of the custom by which an architect employs a quantity surveyor to adjust the variations on a contract, the quantity surveyor's charges being added to the amount arrived at by his calculations, and paid out of the next certificate, is held by some to be questionable; the custom is almost universal in works of any size, and is greatly to the building owner's benefit. The decided cases are few. Neither Birdseye v. Dover Harbour Commissioners ('Times,' 14th April, 1881), nor Plimsaul v. Lord Kilmorey (1884 1, 'Times' L.R. 48) are ordinary cases, and although it has been asserted that it is one of the obligations of the architect to adjust the accounts without charge beyond the ordinary 5 per cent., the probabilities are strongly in favour of the success of the contention that the custom is, in ordinary cases, so extensive and reasonable, as to be a binding one.

The most important decision relating to the deposit of a priced bill of quantities is Warburton v. Llandudno Urban District Council, reported 'Building News' December 15, 1899.

Warburton contracted with the Council of Llandudno to erect municipal buildings.

One of the conditions of the contract was that the builder should furnish a "verified sealed copy of the original estimate." Defendant's architects contended that the verification should be by the surveyor who prepared the quantities, plaintiff that verification by the agent of the architects would compromise his interest, and that so long as the bill was verified, any competent person was elegible for the purpose. Plaintiff having sent a copy verified by a accountant, and refusing to allow the architect's nominee to verify, they ejected him from the site and terminated the contract.

Thereupon the builder brought action for loss of profit and for a sum for damage to his reputation. The decision of Mr. Justice Cozens Hardy was (shortly) as follows.

"The Council took up a position which he had been entirely unable to appreciate. They said that clause 3 of the contract meant that the plaintiff must not only furnish a verified or certified copy of his original estimate which was sealed, but that he must furnish a copy which was verified by one individual and one individual only, namely the Quantity Surveyor, who had prepared the bill of quantities. He entirely failed to see any justification for such a contention. All he had to decide was, whether the Llandudno Council were right in refusing to allow the plaintiff to go on with the work, and making a fresh contract on the ground that he had failed to furnish the architect with a verified sealed copy of the original estimate as provided in the contract. He thought the Council were wholly in the wrong, and that the plaintiff absolutely and entirely complied with every obligation under clause 3 when he sent the sealed copy of the original estimate verified by Mr. Lees the accountant. He therefore must make the declaration that the plaintiff asked, and that on the true construction of the agreement, the defendants were not entitled to have the sealed copy verified by the defendant's Architect or Quantity Surveyor. That being so it followed that the defendants has been guilty of a breach of their contract with the plaintiff. He did not find a trace of any intention by the plaintiff to insist on breaking any terms of the contract. On all these grounds he thought the defendants could not be justified in what they did, and he thought the plaintiff was entitled to damages. Damages were claimed under three heads. First, the plaintiff claimed loss of profit on the contract work of 15 per cent., which worked out at upwards of 200*l*. On on the whole, having regard to the evidence, he thought it would be right to order the defendants to pay 1000*l*. for damages for breach of contract. Then the plaintiff claimed 78*l*. for building materials which had been left by him on the site and had been sold by auction on the order of the defendants who had tendered the amount they realized to the plaintiff. As the plaintiff had been requested to remove these materials and had failed to do so, he was not disposed to give damages beyond what the defendants had offered; namely the sum realized at the auction sale. The third claim of the plaintiff was for 500*l*. on the ground of damage to his reputation in consequence of the fact of his having lost the contract being referred to in certain trade and local papers. He was bound to say he could not see any justification for this claim, especially as it was not a case in which there was any allegation of misconduct on the part of the plaintiff. The only damages would be the 1000*l*. he had mentioned, and there would be an order for the defendants to pay the 13*l*. 5*s*. which they had undertaken to pay for the materials sold by auction. The defendants must pay the costs of the action.

The obvious remedy for a difficulty of the foregoing kind is a condition comprising the following stipulations.

The deposit of the priced bill of quantities before the signing of the contract (preferably with the tender). Its examination and comparison with the copy intended for sealing before signing the contract. A statement as to who shall examine it, and that the sealing shall be after the examination.

It has been thought reasonable to quote the decisions in the foregoing cases at length, as they supply instructive parallels to similar cases which are of frequent occurrence. A short list of decided cases, and the principles they establish, follows.

BURNELL *v*. ELLIS, *Times*, June 25, 1866; *Builder*, June 30, 1866. Architect personally liable for quantities.

FORD *v*. BEMROSE (on appeal), *Builder*, March 22, 1902. No warranty by building owner of the accuracy of quantities supplied nor liability for their inaccuracy.

QUANTITY SURVEYING.

GWYTHER v. GAZE, *Times*, February 8, 1875. Building owner liable for quantities not used.

GORDON v. BLACKBURNE, *Builder*, February 1, 1879. Architect personally liable for quantities.

INEBY v. PAWSON & BRAILSFORD, *Builder*, July 2, 1887. Building owner liable for the measurement of variations.

MELLOR v. BRITTON, 28 *Times Law Reports*, 465 (1900). Contractor liable for quantity surveyor's charges, although not in receipt of first instalment of contract sum.

MOON v. WITNEY GUARDIANS (3 *Bingham New Cases*, page 814, 1837). Building owner liable for surveyor's charges for quantities.

McCONNELL v. KILGALLEN (2 *Irish Common Law Reports*, 119, 1878). Quantity surveyor's charges due on acceptance of tender.

NORTH v. BASSETT (1 Q. B. Div. 333, 1892). When tender accepted liability for quantity surveyor's charges shifts from building owner to builder by admitted custom, and builder must pay them.

PLIMSAUL v. LORD KILMOREY (1 *Times Law Reports*, 48, 1884). Architect has no implied authority from building owner to employ a quantity surveyor to measure variations.

PRIESTLEY & GURNEY v. STONE (4 *Times Law Reports*, 730). Quantity surveyor not liable for errors in quantities.

RICHARDSON & WAGHORN v. BEALES AND OTHERS (*Times*, June 29, 1867). Building owners not liable for quantities, because of express agreement.

SCRIVENER v. PASK (*Law Reports*, 1 C. P. 715, 1866). Architect supplied inaccurate quantities taken out by himself. Contractor has no claim against employer.

STEVENSON v. WATSON, 4 C. P. D. 148 (1879), 48 L. J. C. P. 318 (1879). Architect preparing his own quantities (inaccurate) not liable to contractor.

WAGHORN v. WIMBLEDON LOCAL BOARD (*Times*, June 4, 1877). Architect instructed to obtain tenders, has implied authority to employ quantity surveyor.

YOUNG v. BLAKE (Q.B. Div. November 22, 1887). Architects prepared and supplied inaccurate quantities, neither building owner nor architects liable.

WARBURTON v. LLANDUDNO URBAN DISTRICT COUNCIL (Chancery Division), *Building News*, December 15, 1899. Ordinary condition of contract as to verification of the prices of original estimate, does not warrant insistence upon the verification by the architect or his nominee.

CHAPTER X.

ORDER OF TAKING OFF IF THE OLD METHOD BE ADOPTED.

It will be seen that the trades are separated to accord with the finished bills. As the only recommendation of this course is the facility of collection into single items, and consequent saving of labour in squaring and abstracting, the surveyor should see that this advantage is not lost, and should therefore keep additions and deductions respectively as much as possible together, so as to get as few items on his abstract as may be, generally taking all the additions of an item of work first, and then all the deductions.

Some surveyors who adopt this method commence by heading sheets of dimension paper with the titles of each trade, so that when they come to an item of a trade other than that upon which they are engaged, and which may possibly be forgotten if left to be taken with the trade to which it belongs, they put it at once on its proper sheet, as when measuring a roof truss the templates are at once written on the sheet headed "Mason," and the bolts and straps on the sheet headed "Founder and Smith."

Many surveyors begin every trade with the lowest floor, working upwards, acting on the assumption that any necessary expedient of construction can be thus introduced with less trouble than in working from the top of the building downwards. This argument will only apply so far as the carcass is concerned, but whether working upwards or downwards the same order should be maintained by the surveyor in all cases, and with all the trades.

In taking off the work by trades, some persons prefer to commence with the joinery, as they maintain that they thus carry on the work and acquire a familiarity with the drawings at the same time.

A common modification of this method is in the case of openings. The surveyor commences with the joinery and completes the work, both additions and deductions, to the openings in all trades, except when the fronts of the building are of stone or cement, in which case he leaves the masonry or cement work of the openings to be taken with the work to front; on this system openings without joinery are taken with the general brickwork.

Excavator and Bricklayer.

Surface excavation.
Basement trenches and concrete. First for external, then for internal walls, keeping them separate.
Strutting and planking.
Footings.
Brickwork complete for basement, up to a certain level, as ground floor or top of plinth.
Cuttings and extra labours, basement floor.
Deductions, windows, doorways, recesses.
Fender walls.
Sleeper walls.
Trimmer arches.

Window frames, bed and point.
Door frames, bed and point.
Flues, parget and core.
Arches, and their cuttings internal and external.
Vaulting.
Pavings.
Extra only in cement to brickwork.
Damp-proof course.
Bricklayer's work to each floor in the same manner and order.
Chimney shafts.
Chimney moulds or flue divisions.
Facings.
Brick copings and extra labours on facings.
Drains, cesspools and inspection pits.

The foregoing order for a large work; if a small one, take the whole of the additions of brickwork first, then the omissions.

Mason.

The external masonry, beginning at bottom of building.
The deductions.
The internal masonry, beginning at bottom of building.
The deductions.

Tiler, Slater and Slate-Mason.

Tiling or slating.
Cuttings.
Deductions.
Hips.
Ridges.
Slate-mason's work, internal and external.

Carpenter.

Timbering and carpentry of floors, commencing with lowest floor.
Ditto partitions.
Lintels and wood bricks, floor by floor.
Timber in roofs.
Plates.
Trusses.
Purlins.
Ridges.
Hips and valleys, and rolls.
Rafters.
Trimmers.
Battens or boarding and their cuttings.
Eaves, boards or fillets.
Tilting fillets.
Gutter boards, cesspools, &c.

TAKING OFF BY TRADES.

Ceiling joists.
Sundries which are not included in any of the above sections, as bracketting, centering, &c.

Joiner and Ironmonger.

Commence with the lowest story and proceed in the following order:—
Floors throughout the building.
Skirtings and their grounds and dados.
Skylights.
Windows complete, except glass and painting.
Shutters.
Doors complete.
Fittings.
Staircases.
Take the ironmongery with the joinery to which it belongs.

Plumber.

Lead to roofs and external work.
Internal plumbing, cisterns, baths, w.c. apparatus, &c.

Plasterer.

First internal, then external plastering, the deductions following each.
For the former measure floor by floor in the following order. The deductions of each floor to follow the additions of each kind of plastering.
Ceilings.
Walls.
Cornices.
Partitions.
Centre flowers.

Work in Parian, Keene's, or other cement throughout the building, beginning with lowest story.
For the external work begin at bottom of building and work upwards.

Founder and Smith.

Take the ironwork with the trade in which it occurs, or as follows:
Cast-iron work.
Wrought-iron work, not taken with carpentry.
Beginning with lowest floor and working upwards.

Gasfitter.

Pipes.
Fittings.

Bellhanger.

Bells.
Pulls.

Glazier.

Measure each kind of glass throughout the building, beginning with lowest story and working upwards.

Paperhanger.

The dimensions may be obtained from dimensions of plastering, and will consequently follow their order.

Painter.

Refer to the dimensions or the bill. See remarks "Modes of Measurement, Painter."

Reid's 'Young Surveyor's Preceptor' is an instructive illustration of the foregoing system.

CHAPTER XI.

NORTHERN PRACTICE.

The "General Statement of the Methods Recommended by the Society to be used in Taking Quantities and Measuring-up Work," published by the Manchester Society of Architects, appears to have been prepared with a view to facilitating the Northern system of obtaining separate tenders for each trade.

It is maintained in some parts of the country that the small number of men who are able to take the entire responsibility of a large work has rendered such arrangements necessary, but whenever practicable it is beyond question the better plan to obtain tenders for the entire work.

The paper above referred to is as follows:—

GENERAL STATEMENT OF THE METHODS RECOMMENDED BY THE SOCIETY TO BE USED IN TAKING QUANTITIES AND MEASURING-UP WORKS.

October 1866. *Revised January* 1878.
Further Revision, July 1886.

In taking quantities it will always be desirable to bear the character of the works in mind, and so to measure and describe them as to give parties estimating the clearest idea, and at the same time in the most concise manner, of their cost and character.

In measuring-up work already executed, it is of course only necessary to ascertain on what principle of measurement the prices have been determined, and then proceed accordingly.

GENERALLY.

Fees to Corporation, hoarding, propping sides of foundations or walls, and use of water for the different trades, to be mentioned in the trade that has to provide them, or under the head of charges to be borne by the general contractor. (If the right of advertising on hoarding is to be reserved by the proprietor, it should be so stated).

Each trade to provide its own scaffolding, unless specially mentioned.

Each trade to provide its own mortar.

Protecting masonry with slabs, &c., to be put by preference in carpenter's work.

Expense of watchman, making good damaged work, &c., to be inserted as occasion may arise.

Each trade to provide for the expense of the attendance on the other tradesmen, and for cutting holes and making good, &c.

When the general contractor is expected to be responsible for damage by fire, an item for insurance to be put in the quantities—by preference in the carpenter's work.

Any provision of sums of money or values for articles, shall denote the net cash to be paid by the builder, exclusive of his profit and the cost of materials and labour required for fixing, which, however, should be fully stated.

Clerk of works' office and custody of drawings to be provided for.

EXCAVATOR.

1. Stripping surface soil . . . . deep, and wheeling in heaps . . . . yards run, to be measured in superficial yards.

2. Excavating to be given in cubic yards, measured to 3 inches beyond the outer edge of footings; with extra beyond for batter of sides of excavation, depending on depth and nature of soil. State whether material is to be left on the ground or carted away; and if wheeled, state the distance. Extra price for wheeling every 20 yards run additional. The excavating for each successive depth to be kept separate, and the nature of the material to be excavated to be mentioned if possible.

3. Excavating to trenches for walls, &c., to be measured in cubic yards, and to be kept separate from the above, as also for underpinning or any special work that has to be executed separately.

4. Filling in and ramming to foundations to be given as an item, or if thought desirable, to be done by day work.

5. Clearing away rubbish from time to time to be given as an item.

6. Filling in or forming foundation for paving and flagging to be taken in superficial yards, and state depth and material; or add this to the item of paving or flagging in mason's work.

7. Drains to be given in lineal yards with description thereof, to include cutting trenches and laying; describe method of jointing and filling in; and state average depth of each kind, and when practicable, mention whether in rock or other kind of ground.

8. Junctions, bends, &c., to be counted extra beyond the length of drain.

9. Cutting trenches and filling into gas and water pipes (see Plumber), to be given in lineal yards.

10. Wells, cesspools, eyes, &c., to be given with proper particulars, as the work may require.

11. Keeping foundations clear of water beyond ordinary rainfall, and for propping to streets or adjoining buildings, to be given as items.

BRICKSETTER.

1. Give description of materials and mortar, and quality of work. The work, unless otherwise mentioned, to be reduced to one brick thick, and called "brick-length walling," in yards super. If sand, gravel, or water on the spot is not intended to be used, state so. If there be much work of half a brick only, it is desirable to mention it. Where a building is lofty it is desirable to divide the work into stages vertically.

2. To obviate any misunderstanding as to so-called trade usages with regard to other materials, as stone, &c., built in, it is proposed to measure the net quantity of brickwork to be executed; deducting entirely all labour and materials in openings having more than 100 square feet "*face*" measure, and deducting materials only (leaving "hollows" for labour) on the following:—

- a. All other openings than the above, the shape they are actually executed, provided they are openings in the walls and built above with the same materials.
- b. All sills, strings, cornices, &c., and other masonry or dressings built in, and being 6 or more inches high. The "hollows" thereon being assumed to pay for the labour in providing proper bed therefor, filling up thereto, and pointing up.

ѹ. Fireplace openings from under side hearth, and all flues, to be deducted as "hollows," and the lineal dimensions of flues (with size, if various) to be given for extra labour forming, pointing and coring out.

3. Joists and beam ends, wall plates, door frames, band and gudgeon stones, codge stones, beam stones of ordinary dimensions, not to be deducted if built in with the work.

4. All walls finished with a bevelled upper edge, as to gables, &c., to be measured to 3 inches above the average height; and the lineal dimensions of "beam filling" between spar feet and of "gable cutting" given.

5. Any work intended to be whitewashed, to be measured "superficial" over all openings for pointing.

6. All splay cutting (with width), bevelled arrises, moulded arrises, bands of fancy work (with description), cutting to any shape not square, for ramps, hood moulds, &c. &c., double or other shaped reveals to be given lineal.

7. Stops to chamfers, and other single points requiring special labour, to be counted.

8. All brickwork in projecting bands, cornices, &c., to be measured as walling, the labour as above in No. 6.

9. Facing to be measured super. for "extra price over common;" and the net quantity executed only to be given, after deducting all strings, sills, &c. &c. All openings less than 100 feet to be deducted and kept separate as "hollows for extra labour over common work," to pay for the labour, plumbing and setting out. Reveals to be measured for facing separately.

10. Cavity walling to be measured the actual thickness of the bricks, as for instance, to be measured at 1½ brick or 2 bricks, &c., as the case may be; and the superficial dimensions of the wall measured across all openings under 100 square feet, as "extra labour and extra materials for bond, forming cavity walls." The nature of the ties to be mentioned.

11. Work in cement, or other material than the general run, to be taken for "extra price over mortar," and the description of the material given.

12. Backing up to ashlar walling to be kept separate from the ordinary walling.

13. Wrought-iron bond, if used, to be given in lineal feet.

14. Common relieving arches to be counted with average span, and separated into various thicknesses of wall, and number of rims in depth, and this to cover all extra labour and materials.

15. Arches in facing to be either counted, with spans, depth, and soffit given, and this to include all extra labour and materials, including skewbacks, and cutting super-imposed work to fit the rake; or else measured super. as executed, for "extra labour and materials over facing;" girthing the face and soffit net, and measuring separately the skewbacks, and cutting super-imposed work to fit the rake, when the shape of the arches requires it.

16. Ordinary arching to fire-proof floors, &c., to be measured stuff and work in square yards, with description and thickness, girthing along the line of average thickness, and measure the lineal feet of skewbacks with description. Groins to be measured lineal. Allow for cutting and fitting to ironwork.

17. Trimmer arches to be counted, with average size, for "stuff and work," including backing-up, and state whether solid or concrete backing is intended.

18. Backing to arches that have been measured, as in No. 16, to be taken super. with average thickness and description of material.

19. Damp-proof courses to be given lineal with widths.

20. Risers for slop-stones, steps, flag shelves, &c., to be counted for "stuff and work," or if measured as walling, to be kept separate and so stated.

21. Wine bin divisions measured super. with thickness, and kept separate or with the above.

22. Eyes, air-grids, chimney pots, setting grates, or other special items, to be counted, with description, and the mode of executing the work.

23. Covering walls to protect them on special occasions and manner of doing it, to be added to the bricksetter's contract.

24. Items: To cleaning down, making good, pointing, &c., at finish.

25. It would be well to call the contractor's attention to the fact that his price must include making good any damage done by frost.

26. Concrete in foundations to be measured by the cube yard.

MASON.

1. Plain ashlar walling and parpoint walling to be measured super. net work, for "stuff and work," giving the kind of face, and average sizes, and bed of the stones; allowance to be made where square blocks have to be cut and to suit shaped openings, ramps, &c., and defining the number and dimensions of the through stones. Reveals to be measured extra. Rubble walling up to 18 inches in thickness to be measured superficial and described; above that thickness to be measured in the cube and described.

2. Other masonry to be measured as hereafter mentioned, keeping the stone and labour separate, or when conducing to the better understanding of the work, as in strings, &c., to be measured in lineal feet, "materials and labour," with sufficient particulars to enable the number of joints to be included; count the fair ends, quoins, &c.

3. Work in chimneys, or cornices, &c., requiring special appliances for hoisting, or at considerable heights, or of extra dimensions, to be kept separate, so far as the material, hoisting and setting are concerned; the labour, when done on the ground, may be thrown in with the rest.

4. Under the head "stone" must be included the labour in hoisting and setting, and state approximately the quantity to be set above 40 feet from the ground.

5. Special appliances for hoisting, such as travelling cranes, &c., to be specially mentioned.

6. In measuring the cubic feet of stone for other than the work previously mentioned, 1 inch each way beyond the net dimensions of each block, when worked, to be added.

7. In measuring the labour, the necessary operations of the workman to be followed. The beds and joints of each block to be measured, and kept under that head; and it will conduce to more easy pricing of the work when these can be given in lineal feet, with the average width, particularly for strings, cornices, architraves, jambs, &c.

8. The work exposed to sight to be classified under its different heads, mentioning whether bosted, tooled or polished, and giving plain work, sunk, moulded, sunk and moulded (this when the straightedge will not work the block from a mould applied at each

end), weathered, sunk and weathered, and the various kinds of fancy surfaces, and clearly distinguishing from straight work that which is raking, circular, circular on circular, &c.

9. Raised or sunk panels will require that face of the block to be first measured for plain work (for setting out on), then the face work, and the lineal feet of margin of its particular kind.

10. The points from which to girth moulded work will be best ascertained by a consideration of the manual process followed in its execution; in some strings it will thus have to be girthed from wall above to wall below; in other instances, as cornices, where the top is measured for sunk and weathered work, the moulded work will girth from the nose only.

11. All mitres to be counted, with the girth of the mould, &c., they belong to, and state whether internal or external.

12. Throats to be clearly given, either separately in lineal feet, where the soffit has been measured as a "bed," or, in other instances, girthed in with the moulded work.

13. The back of masonry will not generally require any notice, except where it shows through a wall that is not plastered, and in some quoins that bed more than the thickness of the walls, where the sinking must be taken into account. Tooling or rubbing backs of architraves, mullions, &c., to bed frames against, to be measured lineal, and all checking out for the same purposes to be measured in the same way.

14. Holes for flues, timbers, &c., to be counted.

15. Rough sinking down as a preparation for the carver to be given superficially, girthing round the cap, truss, &c. &c.

16. All carving to be clearly set forth with reference to the drawings or special marginal sketches; running ornaments to be given lineal with the girth; but caps, bosses, trusses, modillions, paterae, &c., numbered. State if carver is to find his own scaffolding.

17. Ordinary window sills to be numbered, with dimensions.

18. Others than these to be taken cube, and the labour taken out as before, and in addition the seats counted for jambs, mullions, &c.

19. As tracery will generally be of a description between the work of an ordinary mason and that of a carver, the most satisfactory way would appear to be to measure one face over all the

work as plain work for setting out on, and then number each piece of tracery with its dimensions and reference to the drawings or special marginal sketch for the remainder of the labour, but measuring separately any groove or rebate for glazing.

20. Mullions and other work with little material, as compared with the labour, to be measured lineal, "stuff and work," with particulars or sketch, and give general indication of the lengths.

21. Columns to be girthed for circular work (with or without entasis, as the case may be), and flutes measured lineal with sketch. and number of stops counted; if in extra lengths, keep both materials and labour under a separate head.

22. Rustics and other channelled work to be measured lineal (after the surface and beds have been fully measured) with sketch.

23. Cramps and dowels to be counted, with average size or weight, if metal, and letting in and running; and state whether mason is to find lead.

24. Copings, where worked out of flags or thin material, to be measured lineal, and net as fixed, with proper description of each face and mode of jointing. All knees, apex and foot stones to be counted with sketch or particulars of labour thereon. Any perforations for flues, &c., to be counted.

25. Flagging to be measured net, in square yards, with proper description of materials, average and minimum size of flags, method of jointing and laying; and whether mason to provide bed and mortar, and if so, describe same. All exposed nosings to be measured lineal; and any perforations or notchings out to fit special corners or other objects, to be counted or measured, and the portion so notched out of any flag included in the gross measurement.

26. Flags, if required to be above 12 feet super. to be kept under the head of "landings."

27. Any special mode of jointing flags or landings, as lap or joggle joint, to be measured once along the joint.

28. Hearths to be given in superficial feet, and if the fore and back hearth be in one piece, state "large sizes," and count the notchings for jambs.

29. Paving to be measured net, with description of sets and method of laying, and whether mason finds the bed or not.

30. Keep each kind of tiling to floors separate, and state whether and what kind of bed is to be provided.

31. Band and gudgeon stones, codge stones, beam and pillar stones to be counted, with dimensions and particulars of work thereon, and whether mason has to let in and run any iron work therein, or to find lead therefor.

32. Steps, where practicable, are best counted, with dimensions and particulars; solid steps may be taken lineal, with average length, and count the number for pinning in, and also the worked ends. All letting in of balusters, newels, &c., to be clearly given, and joggle or notched joints measured.

33. Landings, half spaces, &c., to be either counted with description or taken super. with all joints, worked edges, and soffits measured.

34. Letting in of grids to drains, coal places, areas, &c., to be counted, and state whether lead to be found, and if the stone or curb has to be rebated, state so.

35. Area and other curbs to be taken lineal, and if cramped, so stated, with average lengths of the stone.

36. Dubbing out with flags, for cornices, &c., for the plasterer, to be taken lineal, with width and thickness, and any special labour.

37. Slop-stones, &c., to be either counted with particulars or measured super. with the labour; the sinking to be girthed bottom and sides each way; mention the hole for grid, and whether to be fixed by the mason.

38. Mason's work generally requires very minute subdivision in measuring, and a knowledge of the method of working stone is essential to the proper performance of that duty. Each necessary operation of the workman should be taken into account, although it may appear that the same surface (as in panelled or enriched work, &c.) has to be measured more than once for different descriptions of work.

39. Clean down and leave all perfect at completion, as an item.

CARPENTER AND JOINER.

1. Unless a special provision be made that timber and joiner's work must finish net to the dimensions given (the waste being calculated then in the price), it must be understood that all work will follow the original marking or "pricking" for sawing, thus each sawn face would reduce the scantling by nearly $\frac{1}{16}$-inch or

half the width of sawcut, and each wrought face would entail a further reduction of about $\frac{1}{16}$ inch. A 12 × 6-inch scantling would thus measure $11\frac{7}{8}$ × $5\frac{7}{8}$ inches full, and a 2-inch door would finish $1\frac{3}{4}$-inch full.

2. Labour framing and nails to be measured super. in square yards, for floors, roofs and ceiling joists; keeping the different descriptions, whether for single-joisted or framed, separate, and defining the mode of framing, and taking the dimensions over the extremities of the timbers, or the timber may be taken cube and to include the labour.

3. Labour framing principals to be measured the length of span, with wallhold, and separating the different kinds; the same with framing trussed partitions.

4. Ordinary studded partitions, and filling in to the above, to be taken super. in yards, for "stuff and work," with scantlings and distances apart.

5. Labour framing hips and valleys, in lineal feet.

6. Gutters lineal, or if wide, super. and state if with bearers; count cesspools, with dimensions.

7. The timber to be taken cube when sawn to scantling, keeping, as far as practicable, large scantlings, as beams, purlins, large joists, &c., separate from small, as spars, plates, &c. Beams are best given in lineal dimensions, and when over 35 feet long should be kept separate.

8. When conducing to the better understanding of the work, however, large timbers, or those with special labour, to be measured lineal and properly described.

9. Deals, planks, and battens used for joists, &c., to be taken superficial or cube.

10. Planing or other special labour to timbers, measured super. or lineal in feet, and stops to beads or chamfers counted.

11. All timbers to be measured nett lengths as fixed, and trimmings for hearths, wells, skylights, &c., counted.

12. Angle beads, staff beads and tilting fillets, in lineal feet.

13. Bolts, dogs, straps, &c., of iron, to be counted, keeping joint bolts separate from others.

14. Nogs and templates to be counted, or kept separate under a cube or lineal dimension.

15. Snowboards, in lineal feet, with width and description.

16. Ridge and other rolls, and hip and valley boarding, in lineal feet, with width.

17. Centering, in square yards, except centres for doors and window openings, &c., which are to be counted, with span and width. In extensive fire-proof works, a small quantity of centering will often suffice if "taking down and refixing" be given, together with the necessary staying.

18. Cornices and face boards in lineal feet, with description and particulars of bearers, &c., and count mitres to the former.

19. Ordinary flooring in square yards net; mitred margins to be counted.

20. Pugging to floors, measure across the timbers and state description, whether on slabs and fillets, or laths, and if filled in, describe.

21. Skirtings in lineal feet as fixed, with dimensions and description, and state whether to include grounds; where above 7-inch, count all mitres and ramps, and labour housing to architraves, chimneypieces, &c.; if tongued to floor boards, state so.

22. Door casings and frames in lineal feet (allowing length for tenons, &c.), and state whether framed, and number of rebates; count frames for dowelling.

23. Doors, gates, &c., net size in super. feet, allowing for rebate in folding doors, keeping each description separate, and state if flush-beaded at meeting edges, or for double margins, &c.; hanging to be counted and described, locks and other fittings at a price each and labour fixing; bolts to dimensions, and ditto.

24. Architraves in lineal feet net measure, and state if with grounds, and, except for single moulds up to 3-inch, count all mitres, or special adaptations; blocks to be counted in pairs, including all labour fitting architraves thereto.

25. Window sashes and frames to be measured full size of frames; if not square headed, state so, or else count the heads for extra price, with description; state if extra strength be required in any part of frame, or any particular way of working weatherings, sinkings, &c.

26. Casement sheets and frames to be measured separately and as above, with scantlings.

27. State if holdfasts or other particular mode of fixing frames be required.

28. Skylights to be measured full size. Any portion of a sheet made to open, to be measured again in addition, if it is a separate piece of framing, and any grooving, weather fillets, &c., accounted for.

29. Window backs and elbows, and soffits, and linings, worked one side only, to be kept separate from shutters, and state if to have grounds; if linings are under 9 inches wide, take lineal, with the width. Plinths and capping and flush beads to be measured lineal. Hanging sheets and casements to be counted, with particulars of pulleys, cords, weights, hinges, fastenings, &c. Hanging shutters and backlaps ditto. Window bottoms lineal, and count returned ends.

30. Bracketting for cornices in lineal feet, with girth or sketch of brackets, and distances apart.

31. Coves, super. feet, for stuff and work.

32. Cradling round beams, super. for stuff and work.

33. Bridging for floors in lineal feet, and state whether slab or herring-bone bridging, also whether tie rod or hoop iron is to be used.

34. Wall boarding and dados in super. feet, and give lineal feet of grounds, and also fitting to architraves, &c.

35. All casings to under side beams, gutters, cisterns, &c., in super. feet, with description, or else lineal, with girth and number of beads.

36. Pipe casing, ditto, and state if to be fitted with screws for taking down at pleasure.

37. The different items of water-closets to be kept separate, and state if fitted with screws, &c., to take down; seat and bearers, fall and frame, riser, all super.; skirting, super. or lineal according to requirements; capping separate. Holes cutting, falls hanging, hinges, paper boxes, &c., and attendance on plumber to be counted.

38. For bath framing take riser and skirting as for water-closets; bearers for curb lineal: curb lineal, with average width, or super.

39. French polishing to be, where practicable, measured separately in superficial feet.

40. Fixtures require careful measurement in detail; skeleton fronts for drawers and small cupboard fronts separate from the fronts themselves; bearers, false bottoms, drawers fitting with

stops, hanging doors, knobs and other fastenings, divisions, guides, &c., all to be taken into account; shelves with widths and bearers or brackets, also hook rails, all lineal.

41. Stairs in ordinary cases to be counted, with dimensions, and state whether returned nosings and cut string boards, notch boards, and number of carriages; measure hand rail, and balusters and newels separate; and casing and nosing, hand rails and balusters along landings; count ramps, scrolls, curtail ends, and circular corners to wells; give spandril framing separate from square framed work; state French polishing; landings to be taken super. in feet including bearers, and nosings, &c., at the edge, to be taken lineal.

42. All other items of ironmongery to be counted, with particulars, or price and labour fixing in addition.

43. All circular work throughout to be kept separate from straight work.

44. Enter up reserved amounts, provisions of materials or cash &c., and clearly state if such are to include contractor's profit or to be deducted in full.

45. The carpenter generally undertakes the fixing of the ironwork, and in many instances it might be desirable to put the fixing and staying during erection in his quantities, giving the weight of cast- and wrought-iron beams, &c., and counting the bolts, rods, &c.

46. Insurance, if to be provided for by the carpenter, to be entered.

PLASTERER AND PAINTER.

1. State description of materials, and keep the work of each kind separate.

2. Plastering on walls to be measured from the floor upwards, or from the point where each description of work commences.

3. Where cornices are lathed on brackets, measure ceiling and walls to the edge of the brackets only.

4. Where cornices are not bracketted, measure the ceiling full size of room, and the walls up to ceiling; all in super. yards.

5. Deduct all openings 100 square feet and over; deduct materials and add labour (hollows) for net sizes of doors, windows, fire-places, and other openings under 100 feet super.

6. Where ceilings are panelled and coffered, or coved, girth round all portions that are lathed, keeping circular work separate.

7. Ceilings plastered between spars, &c., to be measured across the spars and purlins, and even then kept separate and described as such.

8. All work run with a mould to be measured lineal on the wall and the girth given, as cornices, rustics, strings, architraves, soffits, quirks, &c.; count all mitres, with the girth of mould they belong to; count mitres in panelled work.

9. All cornices, &c., lathed on brackets to be kept separate, and described as such.

10. All cast work to be counted, except running enrichments.

11. Enriched members to be measured lineal, with girth.

12. Modelling of enrichments to be, if special, so stated, and the models to be the property of the architect.

13. Ceilings or walls covered with panels, formed by small moulds, to be measured super. with illustration or drawing, for "extra price over plain work"; larger panelling or special decorative features to be measured in detail.

14. Angles to pilasters, &c., if specially formed in cement or otherwise, to be so measured, lineal and extra to plastering.

15. Door and window frames, bedding and pointing, to be counted, and state material to be used; also flushing to inside of frames after fixing, or behind casings, window backs, or other work to be given.

16. Making good generally, and after plumber, gasfitter, bellhanger, &c., and chimneypieces, as an item, stating numbers.

17. Colouring and whitewashing walls, &c., to be in super. yards, measuring over all openings under 100 super. feet; if the work has to be pointed by a plasterer, state so.

18. Painting to include stopping and knotting, and to be given in square yards. Priming to be separate, if on work painted before being fixed. Painting to be girthed round all exposed surfaces, except as below.

19. Balusters, if ordinary square, and grids, gates, and other metal work painted on both sides, with bars about 5 inches to 6 inches apart, to be measured one surface only; if closer or slightly ornamental, $1\frac{1}{2}$ surfaces; and for very close or very ornamental work, 2 to $2\frac{1}{2}$ surfaces.

20. Windows to be measured each surface over full size of opening for painting frame and sheets, or else the frames counted,

and the sheets, if large squares, counted; but if in small squares (as old-fashioned crown-glazing), then count the squares instead of the sheets.

21. Fancy or ornamental painting to be measured in detail, with lengths of mouldings picked out, gilt, &c. All work in particolours to be kept separate from plain work.

PLUMBER AND GLAZIER.

1. The lead to be reduced to weight, and the different kinds of work kept separate, as gutters, flashings, valleys and ridges, and flats; the work requiring solder, as cisterns and cesspools, dressing over finials, and other fancy work; allow proper lap, as specified, to flashings, drips and rolls; the net quantity of lead, as fixed, only to be measured.

2. All water-supply, service, waste, or other pipes to be given in lineal feet, with the weight per foot of thickness; the dimensions stated to mean *in all cases, whether specified or not, the clear internal bore.* The price to include all soldering and forming joints, wall-hooks and fixing.

3. No allowance to be measured for sockets and joints in iron piping, down-spouts, &c.; but state whether flanges or holdfasts are to be included.

4. If pipes are to be laid in trenches in the ground, state so, and how, and whether plumber is to do the excavator's work.

5. Count all traps, and also all bends and shoes to down-spouts or soil pipes, spitters from cesspools to spouts, rain-water heads, taps, plugs, overflows, wastes with plug and washer, water-closet and bath-fitting (the pipes thereto and therefrom measured with the other piping), wash-basins, urinals, hot-water cisterns, and other special fittings, and in all such instances give a clear, unmistakable description of what is required, or the price, exclusive of fixing.

6. Iron gutters and down-pipes to be measured in lineal yards; no allowance for joints, but elbows, stopped ends, &c., to be counted and described.

7. The different descriptions of glass, with thickness or weight per foot, to be given in superficial feet, assorting each into different average sizes, and keep bent sheets separate. Curved or other

special edges to be measured lineal, the glass being first measured the size such special shape has to be cut from. Special descriptions of work, such as lead lights, &c., to be described.

8. Pointing to flashings to be measured lineal and described.

9. Making good and leaving all perfect at conclusion of works, as an item.

IRONFOUNDER.

1. The most suitable method is to reduce each description of work to weight, keeping columns separate from beams, small castings from large, and intricate ones, as railing, grids, &c., separate from plain ones; the cost to include pattern making; any fancy work to be specially mentioned, and the castings from each such pattern kept separate from others; the metal to be taken at 40 lbs. per foot super. 1 inch thick.

2. State whether price to include fixing. (*Vide* No. 45 in Carpenter.)

3. Special labour, as turning columns, coupling boxes, &c., to be given in detail.

4. State whether beams are to be tested, and if at contractor's expense and risk.

5. Bolts and other small fittings to be counted and described, with the labour necessary in preparing for and fixing them.

6. State whether lead or other material is to be found for running lugs, &c., and indicate the number of them.

7. Long bolts, tie-rods, &c., to be measured lineal, with allowance for head and nut, and count the number of nuts, and screwing, and washers.

8. Swing sheets, gates, &c., and their fittings, hangings, and fastenings to be fully described and counted, both for materials and labour.

9. In measuring wrought-iron beams and frame-work, ascertain the weight of metal in plates and other shapes, adding the rivets and bolts, with the labour separate on any particular forgings or cuttings.

10. All wrought-iron, rolled or built girders to be fully described and given in lineal lengths, with section.

11. If painted before fixing, to be measured in full.

SLATER AND TILER.

1. State size and description of slates or tiles, nails and battens, and whether and how pointed underneath, and amount of lap.

2. The usage varying with respect to allowances at eaves for double course, and hips and valleys for waste, the Manchester Society of Architects purposes to measure slating net as finished, and to give the length of eaves for extra price of double course, and the length of each bevelled edge at hips and valleys, &c., for "single bevel cutting and waste," and also the bevel cutting where so done to land-gutters, &c. All openings of 100 feet to be deducted entirely, and any others, but allowing for labour as "hollows," below that amount down to 6 feet super., below which no deductions to be made. Any special cutting, as close hips, &c., to be separately mentioned and described.

3. In tile roofs, hips and valleys to be measured lineal, and fitting included.

4. Ridge tiling to be fully described and given in lineal yards, and state how to be bedded and pointed.

5. Pointing to overhanging eaves or gables, in lineal feet.

6. Sweeping and cleaning out gutters, leaving all clean and perfect.

7. Circular slating, fancy courses, slating to spires, or other special work to be kept separate, and fully described.

SUNDRIES.

Floor or wall tiling, paper hanging, cooking and heating apparatus, bellhanging and gasfitting, are generally matters of separate arrangement with the tradesmen. If requisite to include the two latter, bellhanging may be given at so many bells with 1 pull, and so many with 2, the furniture being described and counted; gasfitting at so much per position, exclusive of meter, for piping, and brass bits, or else measured in detail as for water piping.

In the foregoing paper there are a few items worthy of remark.

Generally.—These clauses are framed to meet the arrangement of separate tenders for each trade.

Excavator, Clause 2.—Excavation measured 3 inches beyond outer edge of footings, instead of 6 inches, as in London practice.

Bricksetter, Clause 1.—The reduction of the brickwork to superficial yards of 9 inches thick.

Clause 2.—The deductions of openings. (The clear external aperture only.) The way the work is usually treated being to make an item in feet superficial of "labour to openings," but this only applies to apertures not exceeding 100 feet superficial each.

Allowance of "labour to openings" to sills, strings, cornices, &c

Measuring lineal dimensions of flues for extra labour forming and pointing.

Clause 4.—The allowance of 3 inches for gable cutting is not recognised in London practice, but is measured as raking cutting to facing, which deals with 4½ inches inwards from face of wall. The remainder of thickness of wall being measured as rough cutting.

Clause 5.—Measuring over openings for pointing. Only the net quantity measured. London practice.

Clause 9.—"Hollows for extra labour over common work" unknown in London practice.

Clause 10.—Cavity walling measured in London, including the cavity, and no allowance made for "extra labour and extra materials for bond forming cavity walls."

Clause 17.—Numbering trimmer arches—measured by the foot superficial in London practice.

Mason, Clause 5.—Unusual in London practice, the various heights of hoisting being stated, the means are left to contractor's choice.

Clause 6.—Allowance of 1 inch each way, unknown in London practice.

Carpenter, Clauses 1, 2 *and* 3.—Labour and nails never measured for bills of quantities in London practice.

Clauses 4 and 5.—Unknown in London practice.

Clause 7.—These distinctions only observed in London practice when timbers are unusually small or unusually large.

Clause 11.—Trimmings never counted in London practice.

Clause 16.—Hip and valley boarding measured superficial in London.

Clause 19.—Mitred margin usually included in description of floor, and the latter measured in feet, in London practice.

Clause 23.—" Hanging to be counted "; the hanging is included with the door in London practice though not mentioned.

Clause 24.—Grounds are always separated in London practice.

Clause 41.—" Stairs (steps?) in ordinary cases to be counted ; " always measured per foot superficial in London practice.

Plasterer and Painter, Clauses 5, 17.—All openings deducted and no item of " labour to openings " taken in London practice.

Clause 19.—Only the actual surface painted if measured in London practice.

Clause 20.—All counted in London practice.

Plumber and Glazier, Clause 1.—Not so many distinctions observed in the labours on sheet lead in London practice.

Clause 2.—Branch joints always taken in London practice.

Slater, Clause 2.—The allowance for cutting to eaves is well defined, and labour to hollows is unknown in London practice.

A few instances of items as they would appear in a bill to accord with the foregoing system are as follows:—

Yards	ft.			£	s.	d.
11,200	-	supl.	Brickwork in mortar reduced to one brick length			
	278	„	Labour only to openings or "hollows" reduced to one brick length			
149	-	„	Extra for forming ¾" cavity in thickness of wall and filling with White's Hygeian Rock Composition			
„	No.	2	One story flues parget and core			
„	„	2	Two story ditto			
„	„	25	Making good tiling around ventilating trunks, 12" × 12"			
„	„	2	Ditto skylights, 4' 0" × 6' 0"			
	1800	cube.	Fir converted .. .. Sometimes labour and nails do not appear, and in such case a heading is made thus, and the number billed in order of scantlings with description.			
			Work in Roofs and Floors.			
	100	run	4½" × 3" plates			
	100	„	7" × 2½" floor joists			
	100	„	5½" × 3" ditto, &c.			
Squares 2		supl.	Labour and nails to plates and ground joists			
31		„	Ditto to naked flooring with joists notched and framed to beams			

Squares	ft.			£	s.	d.
33		supl.	Ditto, with joists, plates, and beams part fitted and fixed to iron			
19	50	„	Labour and all materials in 3″ framed and braced partition, with 4½″ × 3″ heads and sills, 3″ × 2½″ quarters and braces, and 3″ × 2″ interties..			
			No. 5, extra to forming doorways.			
36		„	Labour and all materials in 3″ × 2½″ ceiling joists, 15″ from centre to centre, and framed to roof timbers			
80		„	Labour and all materials to 3″ × 2½″ rafters, 15″ from centre to centre, and labour and nails only to roofs, with purlins, plates, ridge and collars			
	No.	3	Labour and nails in framing, hoisting and fixing truss 24′ 0″ × 8′ 0″ and the ridge 30 ft. from ground			
	„	2	Extra for trimming joists for trap-door ..			
	„	5	Ditto, roof for chimney-stack			
	„	2	Ditto, large skylight			
	„	2	Ditto, joists for stairs 11′ 0″ × 3′ 0″ on plan			
Yards						
190	–	supl.	Render float and skim			
180	–	„	Lath plaster, float and skim			

In the North builders prefer to have timbers stated in lineal dimensions with the particular labour stated.

CHAPTER XII.

EXAMPLES OF COLLECTIONS.

A COLLECTION of excavation, brickwork, &c., see "Examples of Taking off."

In the case of a collection for a provision, show its detail thus:—

Stoves and Chimneypieces, Exclusive of Setting.
For setting and fixing see Col. 10.

	Stove.	Chy.-piece.
Dining-room	£10	£15
Drawing-room	10	20
Library, &c. ,..	4	4
	£24	£39

Provide for three stoves and three chimneypieces, exclusive of fixing, 63*l*.

A Collection of damp proof courses.

	1 B.	1½ B.	2 B.
Column 3	140 0	20 0	
		9 0	
4	18 3	54 0	29 6
			6 0
6	20 0	10 2	10 2
7	9 9	18 6	8 6
	188 0	111 8	54 2

188 0
   9

111 8
  1 2

54 2
 1 6

Damp proof course ½" thick of best Seyssel asphalt.

COLLECTIONS. 821

Collect timbers in the following manner:—

Possibly several sizes of joists may occur on one floor, 9″ × 3″, 7″ × 2½″, &c. Commence with the 7″ × 2½″ thus:—

					Trimmer	Ex. for T.	H.B.S.	R.Y.C.
	ft.	in.	ft.	in.	ft. in.	ft. in.	ft. in.	
Bed-room 3.—	6/10	0 =	60	0				
	3/ 8	0 =	24	0				
	2/ 2	0 =	4	0	4 9	20 0	10 0	2
Bed-room 4.—	6/11	0 =	66	0				
	3/ 9	0 =	27	0				
	2/ 2	0 =	4	0	4 9	22 0	10 0	2
			185	0	9 6	42 0	20 0	4

```
185  0
     7
     2½
  ─────       22  6   Fir frd. in floors.              Joists.
  9  6
     7
     3½
  ─────        1  7   Add.                             Trimmers.
 42  0
     7
     1
  ─────        2  1   Add.                             Ex. for trimmers.
 20  0
              20  0   2″ × 1″ herring-bone strutting spiked to 7″
                        joists.
   4/          ▲      Rough York corbels 9″ × 9″ × 3″ and
                        building in.
```

Proceed to collect the other sizes in a similar manner.

Where there are a number of quarter partitions, a collection in some such form as follows is a considerable saving of labour, and each dimension may nevertheless be easily identified.

The letter placed against each dimension will be found an assistance, as H for head, S for sill, &c.

	Head, Sills, Posts.	Interties and braces.	Quarters.
South of bath-room.	4/11 0 = 44 0 P 2/ 8 0 = 16 0 H & S	8 0 I 9 0 B	2/11 0 = 22 0 2/ 4 0 = 8 0
North of bed-room 9.	4/11 0 = 44 0 P 2/ 9 0 = 18 0 H & S	9 0 I 10 0 B	3/11 0 = 33 0 4 0
	122 0	36 0	67 0

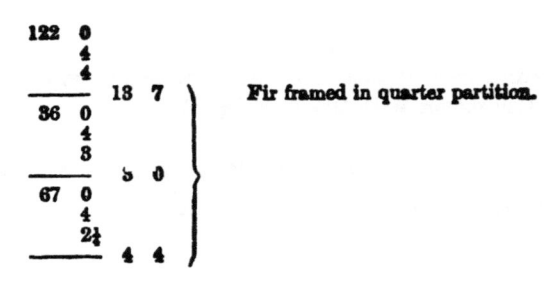

```
122  0
       4
       4
  ─────    13  7  ⎫
   36  0           ⎬  Fir framed in quarter partition.
       4           ⎪
       3           ⎪
  ─────     5  0  ⎪
   67  0           ⎪
       4           ⎪
       2¼          ⎪
  ─────     4  4  ⎭
```

A Collection of Skirtings.

See the plan, Illustration No. 42.

In good work, make the distinction for skirting with grounds plugged or fixed with wall hooks.

SKIRTINGS.

	Skirting.	Mitres.	Irreg. M.	F.E.	H.
Drawing-room.—	5 6				
	5 6				
	18 9				
	15 1				
	25 9	6	4	2	2
Dining-room.—	16 8				
	14 11				
2/1 11 =	3 10				
	15 11				
	14 11	8		2	10
	136 10	14	4	4	12
136 10					
─────	136 10	Deal moulded skirting in three pieces, the lower part of 1¼″ torus moulded 6″ high tongued to floor; the middle part of 1″, the moulding 4½″ girth, all tongued together; 13″ high in all, including groove in pitch-pine floor and the necessary grounds and backings, as sketch.
14/	14	Tongued and mitred angles.
4/	4	Irregular ditto.
4/	4	Fitted ends.
12/	12	Housings.

Some surveyors measure the skirtings across the openings, deducting a length when measuring the joinery to such opening. They maintain that this is the more convenient course in view of possible omissions.

In a large building, doors of uniform size and finish, but in

COLLECTIONS.

walls of various thickness, may sometimes with advantage be collected as follows:—

LININGS.

	6"	11"	15"	R.F.S.	L.P.F.S.
Room 1.	1	1		2	2
" 2.	1			2	2
" 3.	1		1	2	2
" 4.		1	2	6	
	3	2	3	10	6

```
        8/  3 0
            7 0
              ——    168  0   2" four-panel mo. b.s. door.
        8/ 1/  8           Pairs 3½" W. I. butts.
        8/ 1/  8           6" mortise lock and brass furniture.
                                                              7 0
                                                              7 0
                                                              3 0
                                                                5
                                                             ————
                                                             17 5
        3/ 17 5
              6
             ——    26  2   1¼" double rebd. jamb linings tongd. at
                                 angles.
        2/ 17 5
             11
             ——    31 11 ⎫
        3/ 17 5           ⎬  1¼" do., cross tongd.
            1 3           ⎭
             ——    65  4
                                                   17 5
                                          4/3" =    1 0
                                                  ————
                                                   18 5
     8/ 2/ 18 5
             ——   294  8   3" x 1" framed and splayd. grounds.
                                                   18 5
                                          4/2" =      8
                                                  ————
                                                   19 1
     8/ 2/ 19 1
             ——   305  4   2" x 1¼" moulded architrave and mitres.
       10/  3 6
            7 3
             ——   253  9   Ddt. R. F. and S. Walls,
                                and
                             Paper at 2s.
        6/  3 6
            7 3
             ——   152  3   Ddt. L. P. F. and S. partns.,
                                and
                             Paper at 2s.
```

QUANTITY SURVEYING.

A Collection of Plastering and Papering of Walls.

Select the rooms which are of the same height.

	R. F. & S.	L. P. F. & S.	Angle.	Do. Splay.
Bed-room 3.—2/14	0 = 28 0			
2/12	0 = 24 0		2	
Bed-room 4.—2/14	0 = 28 0	12 0		
	12 0		2	1
	92 0	12 0	4	1

```
    92  0
    10  0
         ──── 920  0   R. F. & S. walls
                         and
                       Paper at 2s. per piece.
    12  0
    10  0
         ──── 120  0   L. P. F. & S. partns.
                         and
                       Paper at 2s. per piece.
 4/  9  6
         ────  38  0   Keene's cement angle and two 2" returns.
     9  6
         ────   9  6   K. C. splay 3" wide and two 2" returns.
```

Then proceed with those of differing heights in a similar manner. Sometimes in attics a limited range of heights will frequently recur, and may be collected as follows:—

	R. F. & S.			Arris.	L. P. F. & S.			Arris.
	3' 0"	4' 0"	7' 0"		3' 0"	4' 0"	7' 0"	
Room 1.—4 0	4 0	5 0	7 0 7 0	4 0	5 0	6 0	6 0 6 0
Room 2.—5 0	4 0	4 0	4 0 5 0	6 0	5 0	4 0	4 0
	9 0	8 0	9 0	23 0	10 0	10 0	10 0	16 0

```
 9  0
 3  0
 ──── 27  0 ⎫  R. F. & S. walls.
 8  0       ⎪
 4  0       ⎪
 ──── 32  0 ⎬
 9  0       ⎪
 7  0       ⎪
 ──── 63  0 ⎭
```

COLLECTIONS.

```
10  0
 3  0
─────   30  0 ⎞
10  0         ⎟   L. P. F. & S. partns.
 4  0         ⎟
─────   40  0 ⎟
10  0         ⎟
 7  0         ⎠
─────   70  0
                                                    23  0
                                                    16  0
                                                    ─────
                                                    39  0
39  0
─────   39  0   K. C. arris and two 2" returns.
```

A Collection of Cornices.

	Mitres.	Irreg. do.
Bed-room 3.—52 0	8	
Bed-room 4.—52 0	8	1
104 0	16	1

```
104  0
─────   104  0   P. Mo. Cornice or P. P. C., 6 in. girth
                                and
                         Twice distemper.
16/  16   Mitres 6 in. cornice.
 1/   1   Ditto, irregular.
```

Then collect other similar heights in the same manner.

A Collection of Flashings.

It will generally be expedient to collect all the general flashings to a roof in one collection, omitting those to chimneys.

	S. F.	Apron.	Flashg.
Eastern and western gable	2/20 0 = 40 0		3 0
South of dormitory		10 0	
S.E. angle of main roof		4 6		15 0
	4 6	10 0	21 0

```
21  0
 5
─────    8  9   5 lb. lead
                       Flashings.
44  6
 1  0
─────   44  6   5 lb. lead
                       Steppd. flashing.
```

826   QUANTITY SURVEYING.

```
10  0
 1  0
-----
       10  0    5 lb. lead
                      Apron.                                      10  0
                                                                  21  0
                                                                  -----
                                                                  31  0
31  0
-----
       31  0    R. O. & P. flashing with Ct.
                        and
                Lead wedging.
44  6
-----
       44  6    R. O. & P. S. F. with Ct.
                        and
                Lead wedging.
```

A Collection of Labour and Materials connected with
Chimney-gutters, Flashings, &c.

	Flashg.	Apron.	S. F.	Gutter.	T. F.	Roll.
Stack over B. R. 4	2 5	2 5	4 9	2 5	2 5	1
			4 9			
S. W. stack	2 5	2 5	6 0	2 5	2 5	1
			6 0			
S. E. stack		2 5	9 6			
		2 5	9 6			
	4 10	9 8	40 6	4 10	4 10	2

```
 4 10
    5
 -----
         2  0    5 lb. lead
                      Flashg.
 9  8
 1  0
 -----
         9  8    5 lb. lead
                      Apron.
40  6
 1  0
 -----
        40  6    5 lb. lead
                      Steppd. flashg.
 4 10
 -----
         4 10    1 in. deal gutter boards and bearers 6 in. wide average.
  2/
 4 10    2       Short lengths of 1¼" deal roll.
 -----
         4 10    Deal tilting fillet.
```

              Width   = 6 in.   4 10
              Turns up = 6 in.   Ends 4/9 in. = 3 0
                         ----   Roll 2/6 in. = 1 0
                         9 in.   -----
                          21   8 10

8 10			
1 9			
——	15 6	6 lb. lead gutter.	
2/ 2/ 4		Bossed end to rolls.	

Flashings 4 10
Apron   9 8
         ————
         14 6

14 6		
—— 14 6	R. O. & P. flashings with Ct.	
	and	
	Lead wedging.	
40 6		
—— 40 6	R. O. & P. S. F.	
	and	
	L. W.	

A Collection of Rain-water Pipes.

RAIN-WATER PIPES.

	R.W.P.	Heads & C. W. Covers.	Shoes.	Plinth bends.	Swan necks 4½ projn.	Cutting bk., string, &c., 6 in. high.	Cutting string 3 in. high.
North of dining-room	24 0	1	1	1		2	3
N. E. of ditto	30 0	1	1	1	1	3	2
East of ditto	30 0	1	1				2
	30 0	1	1		1		2
	114 0	4	4	2	2	5	9

114 0		
—— 114 0	4 in. C. I. R. W. P. with ears cast on and fixing with rose-headed nails to and including oak plugs in brick-work.	
4/ 4	Heads	
	and	
	Strong copper wire covers to R. W. P. heads.	
4/ 4	Shoes.	
2/ 2	Plinth bends, extra for.	
2/ 2	Swan necks 4½ in. projn., extra for.	
5/ 5	Labour cutting brick-moulded string 6 in. high for passage of R. W. P. & M. G.	
9/ 9	Ditto 3 in. high.	

Collect eaves gutters in a similar manner.

A Collection of Bells.

	Bells.	Lever pulls.	Ceiling pulls.	Sunk plate pulls.
Bed-room 3	1		2	
" 4	1		2	
Room 6	1	1		
Dining-room	1	2		
Drawing-room	1	2		
Landing (call bell)	1	1		
Front entrance	1			1
Servants' ditto	1			1
	8	6	4	2

8/   8   Bells with six lever pulls, four ceiling pulls, and two plate pulls.
(A sum elsewhere provided for the pulls.)

Provide for six lever pulls, four ceiling pulls, and two sunk plate pulls and fixing .. .. £

CHAPTER XIII.

EXAMPLES OF TAKING OFF.

Earthwork.

THE following illustrate a few simple instances of earthwork, such as are of common occurrence.

The illustration shows the slope of a hill, upon which a level plateau is to be formed to receive a house. A B C D is the original surface, A E F D the required new surface. The levels at A B C D being taken on the ground and reduced, and three sections drawn, A E B, G H I, D F C, the booking of the dimensions would be as follows, based on the well-known formula (Prismoidal):—

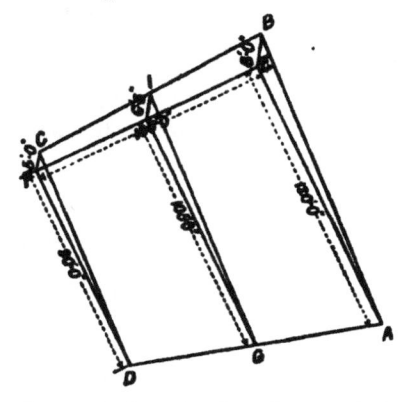

$$\left.\begin{array}{l} A \\ A^1 \end{array}\right\} = \text{area of ends sections}$$

$$a = \text{area of middle section}$$

$$L = \text{length}$$

$$\text{Solidity} = \frac{L(A + 4a + A^1)}{6}$$

830   *QUANTITY SURVEYING.*

The distance of the deposit would depend upon the required destination of the earth.

⅜/½/	120	0			
	8	0			
	100	0			
			8,000 0	Dig to general surface, wheel 2 runs, deposit and level.	End *A.E.B.*
⅜/4/½/	105	0			
	6	6			
	100	0			
			22,750 0		*Middle G.H.I.*
⅜/½/	90	0			
	5	0			
	100	0	3,750 0		End *D.F.G.*
			34,500 0		

= 1278 cubic yards.

Sometimes it may be necessary to calculate the quantity of digging over an area of irregular level. In such a case the ground should be staked out in triangles as the surface varies, and the more frequent the changes of level, the smaller must the triangles be, so that the enclosed areas shall be as nearly level as may be; a level should then be taken at each angle of each triangle. A plan will then be made, the triangles plotted on it, and the levels marked thereon and the new level settled.

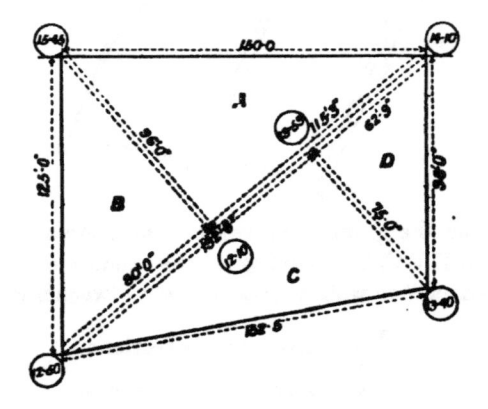

In the example, a datum has been assumed 10 feet below the new level. The new level being 10·00, the original surface levels

EXAMPLES OF TAKING OFF. 831

will be subject to a deduction of 10·00 before use in the calculation.

					4·70
					2·10
					5·45
				3)	12·25
½/	96 0				4·8 = 4' 10"
	115 3				
	4 10	26,738	0	Dig general surface, wheel three runs and deposit in heaps as directed. Triangle A.	
½/	80 0				2·50
	96 0				2·10
	3 4	12,800	0	Triangle B.	5·45
				3)	10·05
					3·35 = 3' 4"
½/	75 0				3·40
	132 6				2·50
	3 2	15,734	5	Triangle C.	3·50
				3)	9·40
					3·13 = 3' 2"
½/	75 0				3·40
	62 9				3·50
	3 10	9,020	4	Triangle D.	4·70
		64,292	9		3) 11·60
		= 2381 yds. 10 ft.			3·86 = 3' 10"

In the case of an excavation for a road 500 feet long, levels should be taken along the centre of the proposed line at *regular* intervals, marked on section A B C D E F, and at these points sections should be taken and drawn; intermediate sections must be drawn, adopting the mean depth between A B, B C, D E, etc., respectively. Then calculate the end sections, A and F; to these add twice the quantity of each of the other sections, B, C, D and E, and add to these four times the quantity of the intermediate sections, *a, b, c, d* and *e*, multiply the total by one-sixth of the length of one of the intervals, and the result will be the quantity of digging.

QUANTITY SURVEYING.

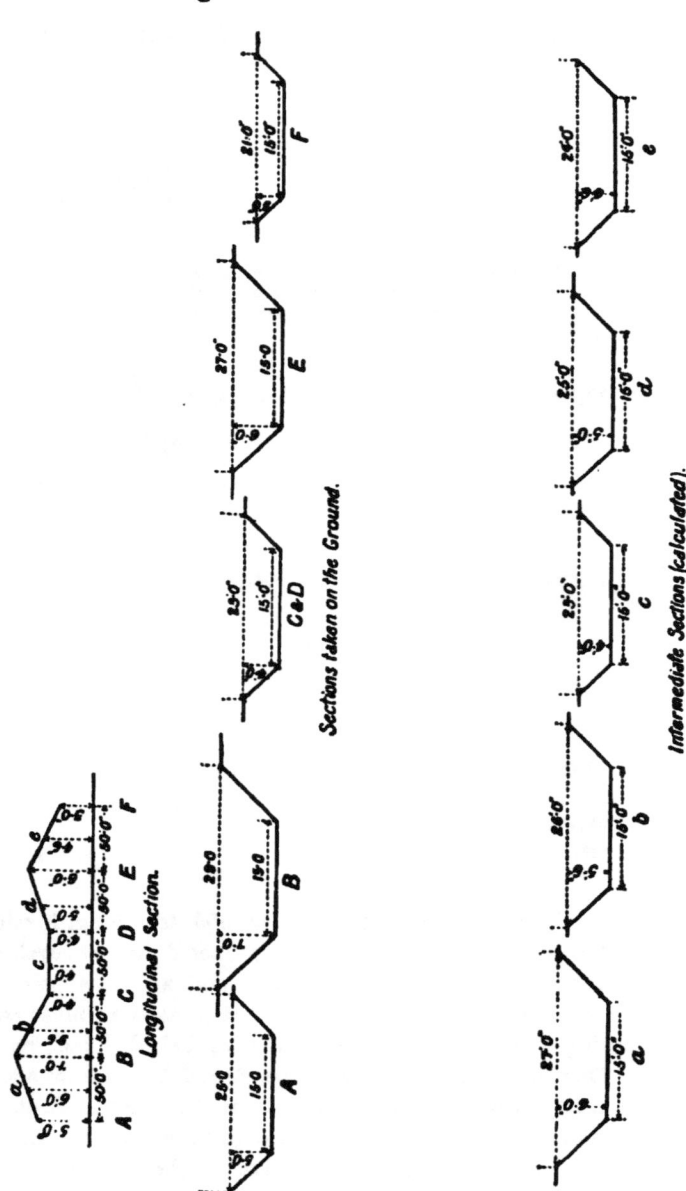

EXAMPLES OF TAKING OFF. 833

```
                                                      25  0
                                                      15  0
             20  0                                   ─────────
              5  0                                  2) 40  0
            ──────  100  0  Section A.                20  0

                                                      29  0
                                                      15  0
      2/     22  0                                   ─────────
              7  0                                  2) 44  0
            ──────  308  0  Section B.                22  0

                                                      15  0
                                                      23  0
 2/  2/      19  0                                   ─────────
              4  0                                  2) 38  0
            ──────  304  0  Section C and D.          19  0

                                                      15  0
                                                      27  0
      2/     21  0                                   ─────────
              6  0                                  2) 42  0
            ──────  252  0  Section E.                21  0

                                                      15  0
                                                      21  0
             18  0                                   ─────────
              3  0                                  2) 36  0
            ──────   54  0  Section F.                18  0

                                                      15  0
                                                      27  0
      4/     21  0                                   ─────────
              6  0                                  2) 42  0
            ──────  504  0  Section a.                21  0

                                                      15  0
                                                      26  0
      4/     20  6                                   ─────────
              5  6                                  2) 41  0
            ──────  451  0  Section b.                20  0

                                                      15  0
                                                      23  0
      4/     19  0                                   ─────────
              4  0                                  2) 38  0
            ──────  304  0  Section c.                19  0
```

3 H

834   QUANTITY SURVEYING.

```
                                              15  0
                                              25  0
                                              ─────
   4/   20  0                              2) 40  0
         5  0                                 ─────
        ────    400  0  Section d.             20  0


                                              15  0
                                              24  0
                                              ─────
   4/   19  6                              2) 39  0
         4  6                                 ─────
        ────    351  0  Section e.             19  6

                3,028 0  Total area of the sections.
                                                         ┌ length of
                                           6) 50  0    ──┤
                                              ─────      └ interval.
        3028 0                                   8  4
           8 4
        ──────  25,233 4  Dig, wheel or cart within half-
                             a mile and deposit.
                = 935 cubic yards.
```

The illustration shows a cutting for a road, 1230 feet in length. Levels of ground would be taken at A, B, C, D, and sections drawn at those points. A mean section must be drawn and constructed on the principle of the average depth between A and B, 400 feet, and similar slopes at sides, adopting the Prismoidal formula, as before.

The dimensions for the digging from A to B, 400 feet, would be as follows. B to C and C to D would be calculated in a similar way.

```
                                              130  0
                                               30  0
                                              ─────
                                           2) 160  0
                                              ─────
   ⅓/   400  0                                 80  0
         80  0
         20  0
        ────    106,666 0  Digging to cutting for roadway, varying from
                             50 to 20 feet, from original surface, and
                             wheeling or carting to form embankment,
                             average 200 yards distant.
                                                         End section A.
                                              205  0
                                               30  0
                                              ─────
                                           2) 235  0
                                              ─────
                                              117  6
   ⅓/ 4/  400  0
          117  6
           35  0
          ────   1,966,668  0
                                                       Middle section a.
```

EXAMPLES OF TAKING OFF.

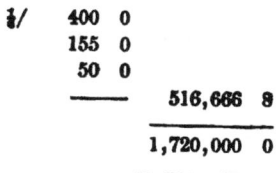

```
                                           280  0
                                            30  0
                                         ─────────
                                        2) 310  0
⅓/   400  0                                155  0
     155  0
      50  0
     ─────
             516,666  8                 End section B.
           ─────────────
           1,720,000  0

        = 63,704 cubic yards.
```

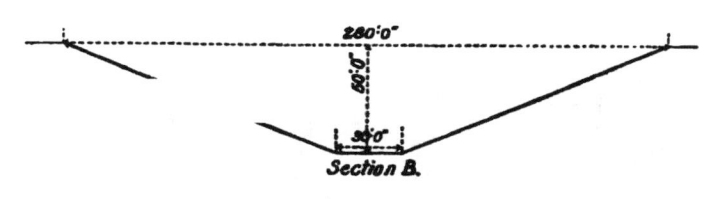

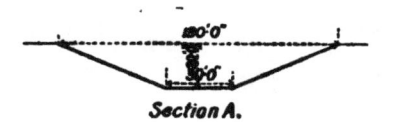

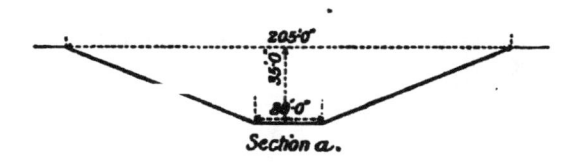

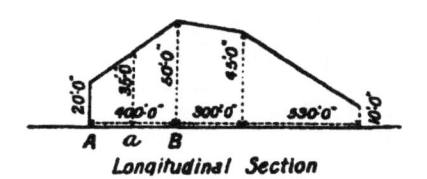

Section B.

Section A.

Section a.

Longitudinal Section

A a B

3 H 2

836   *QUANTITY SURVEYING.*

In the case of a site of irregular surface and large area, upon which it is intended to erect a number of buildings, it will be advantageous to survey and plot all the section lines, and to mark the levels in their proper position on the block plan. The relation of the levels of the buildings to those of the site will thus be readily seen.

Excavation and Brickwork.

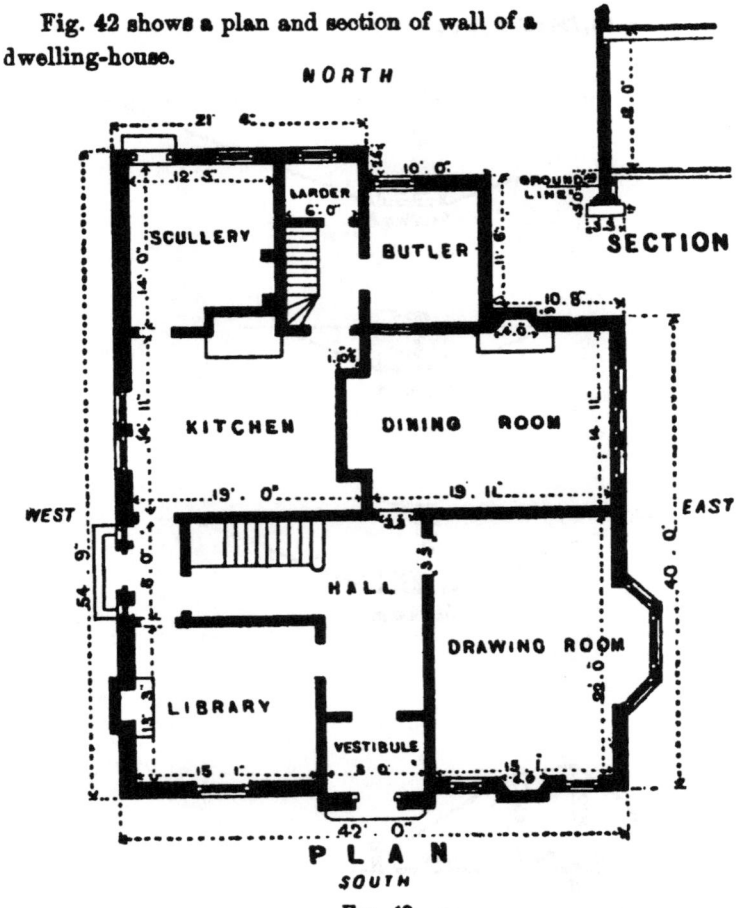

Fig. 42 shows a plan and section of wall of a dwelling-house.

Fig. 42.

EXAMPLES OF TAKING OFF. 837

		40 0	42 0
	Beyond base } of wall	2 0	2 0
		2 0	2 0
46 0		——	——
44 0		44 0	46 0
——	2024 0	Excavation 12 in. deep to surface and wheeling and depositing where directed on the site at an average distance of two runs, and including separating vegetable soil.

(Main Block.)

			21 4
		9	10 0
		11 6	2 0
		2 6	2 0
		——	——
		14 9	35 4
35 4			
14 9			
——	521 2	Add.	

Offices

External Walls.

Assumed to be carried up uniformly to first floor level. Measure each wall to its extremity. If this course is always adopted, the surveyor will never be in doubt as to how far he has measured.

In the case of a rectangular building of 1½ brickwork and 50′ × 30′ external dimensions, the front and back wall would be taken from outside to outside, and the end walls in clear, thus:—

50 0	
50 0	
27 8	
27 8	
——	
155 4	

South of drawing-room	17 0
South of porch	9 6
South of library	17 0
Western wall	53 7
Northern wall	20 2
East of larder	2 6
North of butler	10 0
East of ditto	12 3
North of dining-room	10 8
Eastern wall	37 8
	——
	190 4

190 4
3 3
1 0
—— 618 7 Excavation to surface trenches, wheeling and depositing as before,
and
Concrete.

838 QUANTITY SURVEYING.

Note—Things intended to be separately abstracted should begin on a fresh line, as shown in the foregoing item.

						2	0
				Ddt. surface digging		1	0
						1	0

190	4				
3	3				
1	0				
		618	7	Excavation to surface trenches P. F. I. and B., the remainder wheeled and deposited as before.	

		2	B.
		2¼	B.
		3	B.
	3)	7¼	
		2¼	Bk. av.

190	4				
	9				
		142	9	2¼ B.	

Footings.

	12	0
	1	3
	1	3
	14	6

190	4				
14	6				
		2759	10	1¼ B.	

From foots to first floor level.

Internal Walls.

Assumed to be carried down to the same level as external walls with 12 inches of concrete and two courses of footings.

South of larder	..	..	..	..	..	7	6
South of scullery and butler	..	..		..	29	0
South of kitchen and dining-room	..	..	89	8
North of library..	..	..	..	..	15	10
North of vestibule	..	..	..	..	9	6
East of scullery..	..	..	..	..	13	3
West of butler. *Colld.*	..	..	..	10	9
West of dining-room. *Colld.*	..	..	18	8
East of western entrance	..	..	..	8	0
East of library	..	..	..	..	12	6
West of drawing-room	..	..	..	21	3
					185	11

EXAMPLES OF TAKING OFF.

```
185 11
  2  6
  1  0
 ─────   464 10   Excavation to surface trenches, wheel and deposit as before,
                      and
                    Concrete.
185 11
  2  6
  1  0
 ─────   464 10   Excavation P. F. L and R., &c. a. b.
185 11
     6
 ─────    93  0   1¾ B.
```
Foots.
```
185 11
 14  9
 ─────  2742  3   1 B.
```
To first floor level.

The collection forming the basis of the foregoing dimensions has been made on the principle of measuring the horizontal lines first and afterwards the vertical ones, and does fairly well for a small work; but in the case of a large building this would confuse the majority, and such a plan as the following is better: assume that the work is divided into two sections by the wall south of kitchen and dining-room, measure this wall first, and proceed as before; measuring next the walls north of the latter and then those south of it.

South of kitchen and dining-room	39 8
South of scullery and butler	29 0
South of larder	7 6
East of scullery	13 3
West of butler	10 9
West of dining-room	18 8
North of library	15 10
North of vestibule	9 6
East of western entrance	8 0
East of library	12 6
West of drawing-room	21 3
	185 11

The bay window is assumed to go no higher than the ground floor, and is best taken immediately after the deduction of opening leading into it. In a case where a bay goes up several stories, its foundations and brickwork would be collected with the general collection.

For chimney breasts and stacks see "Order of Taking off."

840   *QUANTITY SURVEYING.*

Deduction of Openings 3′ × 6′ in clear in 1 bk. and 1½ bk. Walls respectively.

A.   **B.**

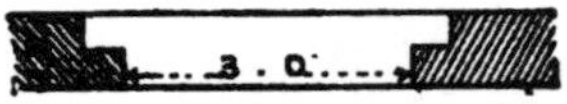

Fig. 43.   Fig. 44.

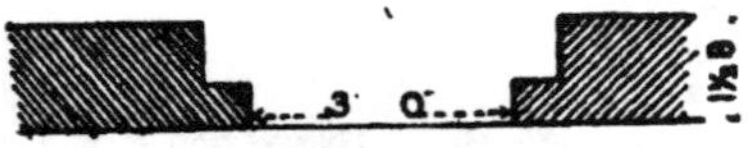

```
                    6  0     6  0     3  0
        half reveal    1½       3       4½  half the reveals.
                    ─────   ─────   ─────
                    6  1½    6  3     3  4½
A.    3  4
      6  1
      ────     20  3    Ddt. 1 bk., or in two dimensions as B.
B.    3  0
      6  0
      ────     18  0    Ddt. ½ bk.
      3  9
      6  3
      ────     23  5    Ddt. 1 bk.
```

Deduction of a Window Opening in a Hollow Wall.

The following is a deduction of a square-headed window opening 3′ 0″ × 7′ 0″ externally, with 4½″ reveal in a hollow wall of two thicknesses of 9″ and 4½″ respectively, with 2¼″ cavity (15¾″ in all). The solid work on each side of opening would be alternately 4½″ and 9″, average 7″ to add to width, two courses below the oak sill and about three above the lintel, equal 21″; i.e. five courses of brickwork 15″; height of soffit of lintel above external soffit 3″; lintel 3″; total 21″ to be added to the height.

```
                  Width.         Height.
                  3  0           7  0
                     7           1  9
                     7           ────
                  ────           8  9
                  4  2
      4  2
      8  9
      ────  36  6   Ddt. 16″ hollow wall as described,
                    and
                    Add 1½ B. solid.
```

EXAMPLES OF TAKING OFF. 841

	3 0			
	7 0			
	——	21 0		Ddt. ½ B.
	3 9			
	7 3			
	——	27 2		Ddt. 1¼ B.
	3 0			
	——	3 0		Centering 4½" flat soffit.
	3 0			⎫ Extra on common brickwork for gauged arch in second
	5			⎬ malms set in putty.
	——	1 3		⎭
	3 6			
	1 0			
	——	3 6		
	3			
	1 0			
	——	3 6		Ddt. facing as described.
2/	1 1			
	——	2 2		Skewback cutting to facing.
	3 0			
	7 0			
	——	21 0		Ddt. facing as described.

	3 0
	7 0
	7 0
	——
	17 0

17 0			
5			
——	7 1		Facing.

Reveals.

3 5			
——	3 5		9" × 4" York window sill, rubbed, sunk, weathered and throated.
	2/	2	Fair ends,
			and
			M. G. facings to ends of window sill.
	1/	1	Frame B. & P.
	8/	8	Wright's fixing blocks, 9" × 4½" × 3".
4 6			
11			
3			
——	1 0		Fir lintel.
	1/	1	Extra labour, cutting and waste to relieving arch 5' 0" × 1¼ B. × 1 B.

	4 6
	9
	9
	——
	6 0

6 0			
2 0			
——	12 0		5 lbs. lead, and building in as gutters over lintels.

QUANTITY SURVEYING.

FACINGS.

See also remarks in section on "Billing."

A Window with Brick Dressings (Fig. 45).

The deduction of brickwork having been taken with general brickwork. Assumed depth of reveal 4½ inches,
General facings, picked stocks; dressings, red bricks.

$$\begin{array}{r} 3\ 0 \\ 7\frac{3}{4} \\ 7\frac{3}{4} \\ \hline 4\ 3\frac{1}{2} \end{array} \quad \begin{array}{r} 6\frac{3}{4} \\ 9 \\ \hline 2)15\frac{3}{4} \end{array} \bigg\} \text{Length of Quoins.}$$

Average 7¾

4 3			
6 0			
——	27 0	Deduct facings of picked stocks, as described.	

Average of face 7¾
Reveal 4½
———
12¼

2/ 1 1
    6 0
  —— 14 0 Facings of red bricks, as described, finished with a neatly struck joint.

Jambs

2/ 4 average.
   1 0
  ——   8 Add
               and
            Ddt. picked stock facings at skewbacks.

3 1 intrados.
4 2 extrados.
———
2)7 3
———
3 7½ mean.

3 8
1 0
———
3 8   3 8 Ddt. picked stock facings for arch.
1 0
———
3 8 ⎫ 3 8 ⎫ Rubbed and gauged arch in red bricks, set in cement,
3 1 ⎬   ⎬ raked out and pointed to match facings.
  5 ⎭ 1 3 ⎭   Face and soffit.

Skewback 1 0
     „   1 0
Extrados 4 2
————
6 2

EXAMPLES OF TAKING OFF. 843

		6 2			
	2/	6 3	6	2	Circular and skewback cutting to facings.
		——	12	6	Labour cut and rubbed moulding, 4-in. girth on facings.

Jambs.

	3 5			
	——	3	5	Ditto, circular.

Arch.

		2/			
	2/	7 0	2		Mitres.
		——	14	0	Extra labour to bonding quoins.

Jambs.

or, instead of last item, write in bricklayer's bill—

"All the red facings to include any extra labour for bonding with the general facings."

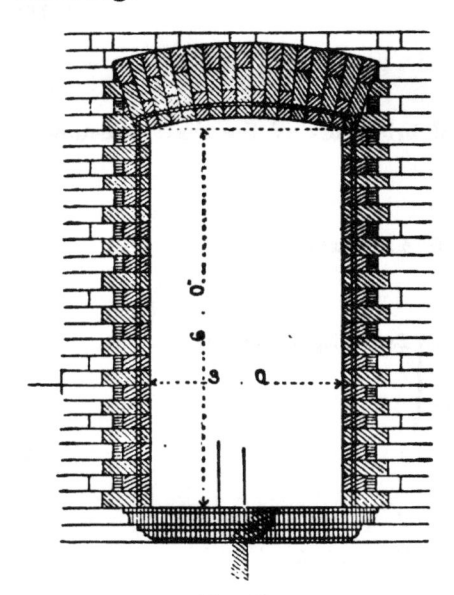

FIG. 45.

3 6			
6			
——	1	9	½ Bk.

Oversail for sill average.

3 6			
6			
——	1	9	Ddt. facings of picked stocks.
4 0			
11			
——	3	8	Red facings top of sill.

844 QUANTITY SURVEYING.

```
  4  0
    10
  ─────   3  4   Red facings, front of ditto (girth).
  4  0
  ─────   4  0   Extra labour on facings for two courses of moulded bricks,
                 as sill, including setting out.
  2/      2      Returned and mitred ends.
```

See also modes of measurement, "Facings."

Facings of a Chimney Stack (Fig. 46).

All the mouldings to be cut and rubbed.

```
                          Girth of shaft   13   2
                          Ditto cap        16   8
                                          ─────────
                                          2) 29  10
                                          ─────────
                          Mean girth       14  11
```

The general brickwork of the shaft measured with ordinary brickwork.

```
      14  11
       1   6
      ─────   22  5   ½ B (averaged).
                                                              Oversail for cap.
      12   2
           5
      ─────   5  1 ⎫ Red facing as described.     2/   3  9 =  7   6
  3/ ½/  9        ⎬                               4/   1  2 =  4   8
           5      ⎪                                             ─────────
      ─────     11 ⎭                                           12   2
                                                              Top.
                                                  2/   5  8 = 10   6
                                                  2/   3  1 =  6   2
                                                               ─────────
      16   8                                                  16   8
           6
      ─────   8  4   Facings a. b.
                                                              Above cornice.
          6/    6    Extra labour on facings for forming pilasters triangular on
                     plan 9 in. wide 5 in. projection and 6 in. high, including
                     cutting to chimney shafts.
      14  11
       1   6
      ─────   22  5   Facings a. b.                           Cornice
      14  11
      ─────   14 11   Extra labour on facings for moulding 18 in. girth in short
                      lengths to chimney shafts, including setting out.
         10/    10    Mitres to 18 in. moulding.
```

EXAMPLES OF TAKING OFF. 845

6/	2/	12	Irregular ditto.
6/½/		9	
	10	11	
—	24	7	½ Bk.

Projection for pilasters.

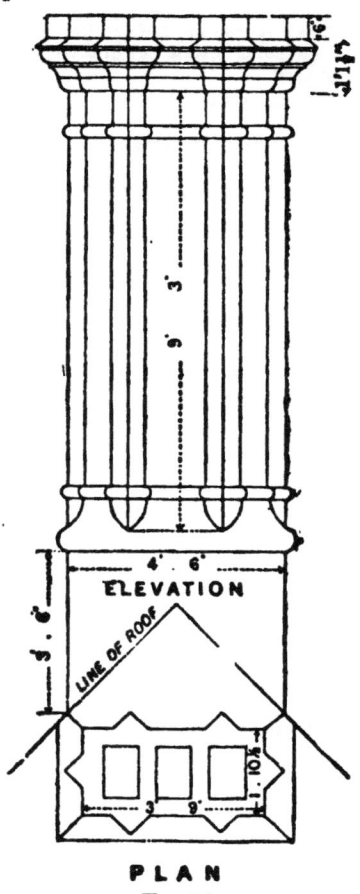

ELEVATION

PLAN

Fig. 46.

	2/	4	4 = 8	8	Necking		9
	2/	2	3 = 4	6	Shaft	7	6
			—		Two }		5
	Girth of shaft		13	2	mouldgs. }		5
						9	1

13 2
9 1
—— 119 7 Facings a, b.

QUANTITY SURVEYING.

2/	13 2				
	3				
	——	6	7	½ Bk.	

Projn. of mouldgs.

2/	13 2			
	——	26	4	Extra labour on facings for moulding one course high, including setting out in short lengths to chimney shafts.
2/	10/	20		Mitres.
2/	12/	24		Irregular ditto.
6/	9 3			
	——	55	6	Extra labour on facings for forming pilasters 9 in. wide 5 in. projn. a. b.
6/	1/	6		Stopped ends to ditto on moulded face.

2/ 4 10 = 9 8
2/ 2 7 = 5 2
————
14 10

14 10
  6
—— 7 5 ½ Bk.

Projn. for base moulding.
3 9
3 9
1 10½
1 10½
4 10
4 10
3 0
3 0
————
2) 26 11
————
13 5½

13 6				
——	13	6	Lab. to moulding 16 in. girth.	
	4/	4	Mitres.	

Base moulding.

13 6			
1 4			
——	16	11	Facings as described.

Base moulding.

6/½/	9			
	9			
	——	1	8	Ddt. facings.

Stoppings of pilasters.

6/2/½/	6			
	9			
	——	2	3	Facings as described.

Sides of ditto.

3 6
Below surface of roof 3
————
3 9

2/	2 8			
	3 9			
	——	20	0	Add.

Case.

EXAMPLES OF TAKING OFF. 847

				Mean 2 3
				3
2/	4 6			——
	2 6			2 6
	——	22 6	Add.	
				Ditto.

In the measurement of the carcass of the building, the shaft would be measured without its projections, leaving them to be taken with the facings as in the foregoing dimensions.

Facings of a Chimney Shaft (Fig. 46).
Alternative Method.

	14 11			
	1 6			
	——	22 5	½ B.	
				Oversail for cap.
	12 2			
	5			
	——	5 1	Red facing as described.	
C/½/	9			Top.
	5			
	——	11		
	16 8			
	6			
	——	8 4	Facings a. b.	
				Above cornice.
6/	6	Extra labour on facings for forming pilasters, triangular on plan 9" wide 5" projection and 6" high, including cutting to chimney shaft.
	14 11			
	——	14 11	Extra on common brickwork for facings and cut and rubbed moulding 18" girth in short lengths to chimney shaft, including setting out.
	10/	10	Mitres to 18" moulding.
6/	2/	12	Irregular ditto.
6/½/	9			
	10 11			
	——	24 7	½ Bk.	

				Projection for pilasters.
				Necking 9
				Shaft 7 6
	13 2			——
	8 3			8 3
	——	108 8	Facings a. b.
2/	13 2			
	3			
	——	6 7	½ Bk.	
				Projn. of mouldings.
2/	13 2			
	——	26 4	Extra on common brickwork for labour, moulded bricks and facing for 1 course moulded bricks in short lengths as string, including setting out to chimney shaft.
2/	10/	20	Mitres.

QUANTITY SURVEYING.

2/ 12/ 6/ 9 3	24		Ditto, irregular.
	55	6	Extra labour on facings for forming pilasters triangular on plan 9" wide 5" projection, including cutting.
6/ 1/ 14 10 6	6		Stopped ends to ditto on moulded face.
	7	5	½ Bk.

Projn. for base moulding.

13 6	13	6	Extra on common brickwork for facings and cut and rubbed moulding 16" girth, including setting out.
4/ 6/½/ 9 9	4		Mitres.
	1	8	Ddt. facings.
6/2/½/ 6 9			
	2	3	Facings as described.
2/ 2 8 3 9	20	0	Add.
2/ 4 6 2 6	22	6	Add.

An Inspection Pit.

The following work in S. Q. to inspection pits.

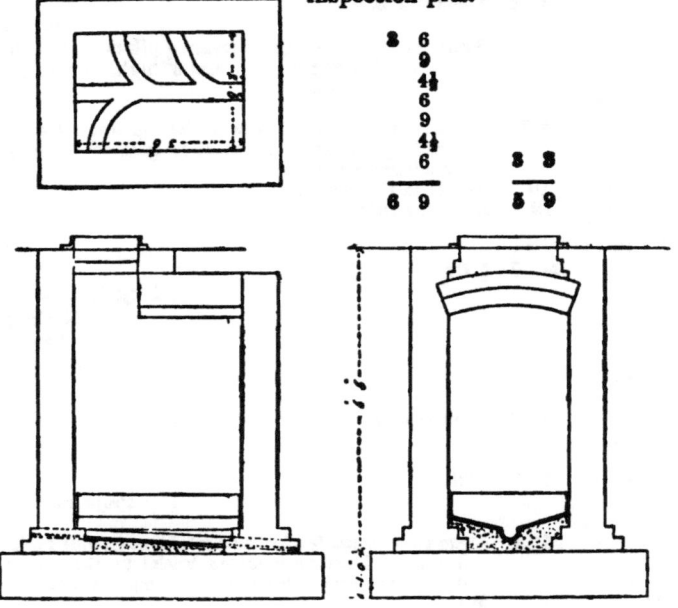

```
  2  6
     9
     4½
     6
     9
     4½
     6
  ─────       2  3
  6  9        5  9
```

Fig. 47.

EXAMPLES OF TAKING OFF. 849

```
  6  9
  5  9
  1  0
 ———      38 10   Cement concrete as described.
  6  9
  5  9
  7  5
 ———     287 10   Dig and cart, extreme depth 7' 6".
```

```
                                                   6  9
                                                   6  9
                                                   4  0
                                                   4  0
                                                  ————
                                                  21  6
```

```
 21  6
    11
  6  5
 ———     126  6   Ddt.
                   and
                  Add Dig, fill and ram.
```

```
                                           2/  5  0 = 10  0
                                           2/  2  6 =  5  0
                                                      ————
                                                      15  0
```

```
 15  0
     6
 ———       7  6   1½ B. in Ct.
```
Footings.

```
 15  0
  5 11
 ———      88  9   1 B. in Ct.
```
Up to surface.
```
                                                   1  5
                                                   1  5
                                                  ————
                                                   2 10
```

```
  2 10
     6
 ———       1  5   Ddt. 1 B. in Ct.
```
Above crown of arch.

```
  2  6
  2  2
 ———       5  5   Centering to vault,
                   and
                  Rake out and point soffit of vaulting.
        2  2
        2  9
 ———       6  0   Vaulting in two half-brick rings in cement.
 2/  2  2
       10
 ———       3  7   Ro. cutting.
```
Skewbacks

```
  2 10
     9
 ———       2  2   Add.
```
Back.

3 I

850 *QUANTITY SURVEYING.*

```
 2/  1  4
        6
       ——    1  4    ½ B. in Ct.
                                                              Oversail
2/ 2/  1  4
       ——    5  4    Labour rough oversail 1 course.
     1/      1       Bolding's (Grosvenor Works, Davies Street, W.) cast-iron
                     air-tight cover, with flange and india-rubber packing.
                     Size of opening 24" × 16", and fixing and bedding in
                     cement.
     1/      1       Making up bottom of manhole 3' 6" × 2' 6" with Portland
                     cement concrete average 8" thick to falls, and rendering
                     with Portland cement trowelled, including making good
                     to channels.
     4  0
     ——    4  0      Winser's (Buckingham Palace Road) 4" white glazed
                     channel pipes, and bedding and jointing with cement.
     3/      3       Extra on 4" straight channel pipe for 4" junction.
     3/      3       4" long channel bends, and bedding and jointing with
                     cement.
     5/      5       M. G. drain to 1 brick.
    12  0
    ——    12  0      Portland cement trowelled skirting 8" high (averaged).
     4/      4       Mitres.
    12  0
     3  9
    ——    45  0      Joints of brickwork struck fair.
                                                              1  4
                                                              1  4
                                                              2  6
                                                              2  6
                                                              ———
                                                              7  8
     7  8
     1  5
    ——   10 10       Joints of brickwork struck fair.
     1/      1       Strutting and planking to hole 6' 9" × 5' 9" and 7' 6" deep.
```

A Stone Pier Cap in Two Stones.

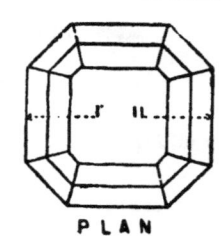

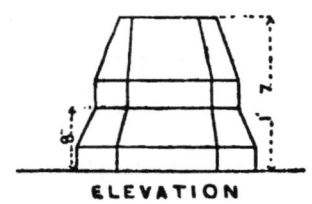

PLAN   ELEVATION

Fig. 48.

```
     1 11
     1 11
        8
    ——     2  5     Box Ground stone.
```
                                                    Pier cap.
                                                    Lower stone

EXAMPLES OF TAKING OFF.

```
        1 11
        1 11
        ——   3  8   Bed.
                                        Top and bottom.
        1 11
           8
        ——   1  3   Joint.
                                        Two sides.
  4/    1  0
           3
        ——   1  0   P. F.
                                        Sides.
  4/       8
           8
        ——   1  9   S. F.
                              (The whole height at angles.)
```

In the measurement of labour, small dimensions like the foregoing must be liberally treated.

```
                                  4/  1  0 = 4  0
                                  4/     8 = 2  8
                                           ——————
                                             6  8
        6  8
           5
        ——   1  9   S. F.
                                        Weathg.
  8/       7
        ——   4  8   Mitre to splay.
        1  6
        1  6
          11
        ——   2  1   B. G. stone.
                                        Upper stone.
                                    Bed   1  6
                                    Joint    11
                                           ——————
                                             2  5

        2  5
        1  6
        ——   3  8   B. and J.
  4/      10
           3
        ——     10   P. F.
                                        Sides.
  4/       5
          11
        ——   1  6   S. F.
                              (The whole height at angles.)
                                  4/    10 = 3  4
                                  4/     5 = 1 8
                                           ——————
                                             5  0
                                        3  1  2
```

852   QUANTITY SURVEYING.

```
        5 0
          8
        ───   3 4   S. F.
         11
         11
        ───  10     P. F.
                     and
                    Ddt. half bed.
8/      8   5 4    Mitre to splay.                          Top
       ───
  2/   2   2       8" × 1" × 1", slate dowels, mortises and cement.
```

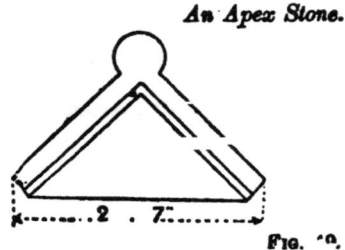

An Apex Stone.

Fig. 0.

```
        2 7
        1 2
        1 9
        ───  5 3   Box Ground stone.
        2 7
        1 2
        ───  3 0   Bed.
2/ ½/   2 7
        1 9
        ───  4 6   P. F.                       Front and back.
2/      1 2
          3
        ───    7   Sunk joint.
2/      1 2
        1 7
        ───  3 8   S. F.
                    and
                   R. S.
2/ ½/   2 3
        1 2
        ───  2 8   Sunk work stopped.
2/ 2/   1 7                                    (Back and front.)
        ───  6 4   Sunk and stopped margin, 2 in. wide.
2/ 2/   1 9
        ───  7 0   Moulding 3 in. girth stopped.
                                               (Throat and chamfer.)
          2/   2   Mitres to 3 in. moulding.
        1 2
        1 2
        ───  1 4   Mo. face.
                                               Roll
```

A Stone String Course in a Brick Wall.

Exposed for a length of 50 feet, and running into wall 6 inches at one end. The stones alternately 4½ and 9 inches into wall.

Fig. 50.

```
                                                        4½
                                                         9
                                                       ─────
   50  0                                            2) 18½
       6                                               ─────
  ─────                                                  6¾
   50  6              Ddt. facings as described.   Projn. 4
       11                                              ─────
        6                                               10¾
  ─────  23  2     Box Ground stone,                   ─────
                     and                              String.
                   Ddt. one half brickwork.

   50  6
       11
  ─────  46  4     Bed.
17/    11
        6
  ─────   7  9     Joint.
   50  6
        6
  ─────  25  3     ½ joint.
                                                        Back.
```

NOTE.—The back is often left rough, in which case no labour need be taken thereon, or a superficial dimension may be taken and described as back.

```
   50  0
    1  0
  ─────  50  0     Mo F.
```

NOTE.—This includes the weathering. Small weatherings are worth as much and are often measured with mouldings.

```
    2  6
    1  0
  ─────   2  6     Ddt. Mo. F.
                     and
                   Add Mo. F. stopped.
                                                     End stone.
    1/   1         Stopped end to mo., 12 in. girth.
     6
     6
  ─────        5   ½ joint.
```

854   *QUANTITY SURVEYING.*

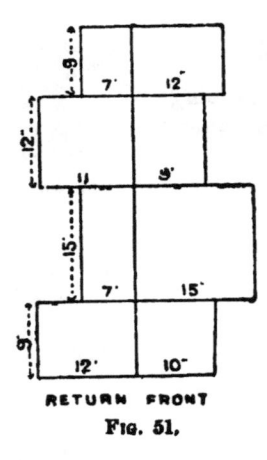

Stone Quoins in a Brick Wall.

Usually averaged, this will sometimes produce a rather larger quantity than squaring each stone, but is close enough for all practical purposes; the long and short dimensions must be respectively added together for the average.

```
         7     12
         9     11
         7     15
        10     12
        ---    ---
     4) 33   4) 50
        ---    ---
         8¼    12½
```

Fig. 51.

```
   1  1
      9
   8  9
   ---   3  1   Box Ground stone,
                   and
                Ddt. brickwork.

4/ 1  1
      9
   ---   3  5   Bed.


   1 10
   3  9
   ---   6 11   ½ joint
                   and
                P. F.
```

Quoins in four stones.

```
   1  1
         9
   ---
   1 10
```

Faces and Backs.

Stone Quoins in a Brick Wall. *Alternative treatment.*

If the measurement of quoins is to be absolutely correct, some proportion of the various sizes must be settled before measuring; thus, "the sizes of the quoins shall be as follows: one third 12" × 7" × 12"; one third 15" × 7" × 9"; one third 9" × 9" × 15"." The running length of quoin, i.e. height of vertical angle, will then be measured and divided by 3, which will give the length for quoin of each size, as in the following example:—

EXAMPLES OF TAKING OFF. 855

$$\frac{3)\ 100\ \ 0}{33\ \ 4}\qquad \begin{array}{r}12''\\7''\\\hline 19''\end{array}$$

1 7 33 4 ————	52	9	Ddt. facing.
1 0 7 33 4 ————	19	5	Stone, and Ddt. brickwork.
1 7 33 4 ————	52	9	½ joint (or back) and P. F.
33/ 1 0 7 ————	19	3	Bed.
1 10 33 4 ————	61	1	Ddt facing.
1 3 7 33 4 ————	24	4	Stone, and Ddt. brickwork.
1 10 33 4 ————	61	1	½ joint and P. F.
44/ 1 3 7 ————	32	1	Bed.
1 6 33 4 ————	50	0	Ddt. facing.
9 9 33 4 ————	18	9	Stone and Ddt. brickwork.
1 6 33 4 ————	50	0	½ joint and P. F.
27/ 9 9 ————	15	2	Bed.

A Stone Balustrade and Entablature.

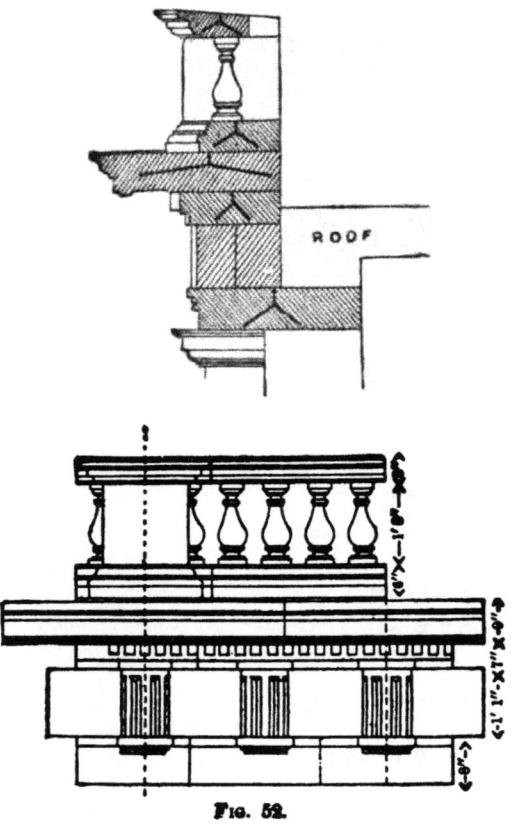

Fig. 52.

```
    5  9
    2  7
       8
      ――  9 11   Portland stone.
                                    Architrave in three stones.
    5  9
    2  7
      ―― 14 10   Bed.
3/  2  7
       8
      ――  5  2   Joint.
3/  1  2
      ――     3   Double V joggle pebbles and cement.
```

EXAMPLES OF TAKING OFF. 857

	5 9			
	8			
	——	3 10		P. F.
	5 9			Back.
	7			
	——	3 4		S. F. stopped.
	5 9			Front.
	——	5 9		Narrow margin.
				Face of fillet.

$$2/ \quad 1 \quad 1 = 2 \quad 2$$
$$2/ \qquad\quad 7 = 1 \quad 2$$
$$\overline{\qquad 3 \quad 4}$$

3 4			
——	3 4		Ddt.
			and
			Add narrow margin, sunk and stopped.
6/ 1/	6		½ in. length of ditto, with one external and one internal mitre to each.
			Return of fillet.
5 9			
——	5 9		Narrow sunk margin.
			Under side of fillet.
3 4			
——	3 4		Ddt.
			and
			Add stopped.
5 9			
——			Narrow margin.
			Upper face of fillet.
3/	3		9 in. length of narrow sunk margin with two ½ in. returns, two mitres, and two stopped ends to each.
3/	3		Labour to guttæ 9 in. long, 1 in. high, ½ in. projection.
6 9			
1 5			
1 1			
——	10 4		Portland stone.
			Frieze in two stones.
6 9			
1 5			
——	9 7		Bed.
2/ 1 5			
1 1			
——	3 1		Joint.
6 9			
1 1			
——	7 4		½ joint
			and
			P. F.
			(Back and front.)
2/ 1 1			
——	3 3		Joggle and cement a. b.

QUANTITY SURVEYING.

```
 4/  1  2
     1  1
     ───   5  1    S. F.
                   and
                   Ddt. P. F.
                                                         Between triglyphs.
 6/  1  1
     ───        6  6   Narrow sunk margin.
                                                         Return of triglyphs.
3/ 2/ 1/    6         12 in. lengths of arris groove ½ in. deep 1½ in. wide, stopped,
                      and with one splayed stop to each.
                                                         Channels of triglyphs.
3/ 2/ 1/    6         12 in. lengths of stopped chamfer 1 in. wide, with one
                      splayed stop to each.
                                                         Outer edges of triglyphs.
     5  9
     1  8
        7
     ───   5  7    Portland stone.
                                                         Bed mould in three stones.
     5  9
     1  8
     ───   9  7    Bed.
3/   1  8
        7
     ───   2 11   Joint.
 4/    10
     ───   3  4   Joggle and Ct. a. b.
     5  9
        7
     ───   3  4   P. F.
                                                                        Rack.
     5  9
       10
     ───   4 10   Mo. F.
                                                                           7
                                                                          13
                                                                          13
                                                                           7
                                                                          ──
                                                                          40
     3  4
     ───   3  4   Labour to stopped sinking 2 in. wide and ¾ in. deep in
                  short lengths.
                                                         To form upper fillet of triglyphs.
         6/      6    ¼ in. lengths of ditto, with one external and one internal
                      mitre to each.
                                                         Returns of do.
     5  9
     ───   5  9   Labour to dentil course 4 in. high, the blocks 2 in. wide,
                  8½ in. high and 1 in. apart.
     7  6
     2 10
        9
     ─── 15 11   Portland stone.
                                                         Cornice in two stones.
```

EXAMPLES OF TAKING OFF. 859

```
        7  6
        2 10
        ─── 21  3   Bed.
    2/  2 10
           9
        ───     4  3   Joint.
        7  6
        1  8
        ───    12  6   Sunk bed.                       Under side.
        7  6
           8
        ───     5  0   P. F.                           Back.
        7  6
        1  3
        ───     9  5   S. F.                           Weathering.
        5  8
           5
        ───     2  4   P. F.
                       and
                       Ddt. ½ bed.
                                                       Top.
                                                       1  4
                                                       1  4
                                                       ─────
                                                       2  8

        7  6
        1 10
        ─── 13  9   Mo. F.
    3/  2  0
        ───     6  0   Joggle and Ct. a. b.
        2  9
        1  4
           6
        ───     1 10   Portland stone.
                                              Base of balustrade.
                                                       2  9
                                                          6
                                                       ─────
                                                       3  3
        3  3
        1  4
        ───     4  4   Bed and joint.
        2  9
        ───     2  9   2½ in. plain margin.             Front.
        2  9
           6
        ───     1  5   P. F.                           Back.
        2  9
        1  0
        ───     2  9   P. F.
                       and
                       Ddt. ½ Bed.                     Top.
```

QUANTITY SURVEYING.

```
      2   9
      ─────    2   9   Mo. 6 in. girth.
      2   2
      1   8
          6
          ───    1  10   Portland stone.
                                                Base to pedestal.
                                                         2   2
                                                             6
                                                        ─────
                                                         2   8


      2   8
      1   8
      ─────    4   5   Bed and joint.
      2   2
          6
          ───    1   1   P. F.                       Back.
      1  10
          ───    1  10   2½ in. margin.              Front.
                                                         1½
                                                         4½
                                                        ───
                                                          6

  2/      6
          ───    1   0   2½ in. ditto, sunk.         Returns.
      1  10
          ───    1  10   Moulding 6 in. girth.       Front.
  2/      6
          ───    1   0   Ditto, stopped in short lengths.  Returns.
  2/  2/          4     Mitres to 6" mo.
  2/      5
      1   0
          ───          10   P. F.
                             and
                           Ddt. ½ Bed.              Top.

  3/      9
          ───    2   3   Joggle and Ct. s. b.
  4/      6
          6
      1   5
          ───    1   5   Portland stone.            Balusters.

  4/      6
          6
          ───    1   0   Bed.
 4/ 4/    6
      1   5
          ───   11   4   P. F.                      Preparatory faces.
```

EXAMPLES OF TAKING OFF. 861

```
    4/  1  1
        1  5
           ─── 6  2     Mo. F. circular continuous to balusters.
   or   4/     4        Turning and labour to stone balusters, 6" × 6" and 17"
                          high, instead of last.
    4/  2/     8        3' × 1" × 1" slate dowels, mortises, and cement.
```

Or the balusters may be taken as follows:—

```
        4/     4        Portland stone and labour to moulded balusters 6" × 6"
                          and 17" high.
        1 11
        1  5
        1  5
        ─── 3 10        Portland stone.                    Die of pedestal.
                                                           Bed   1 11
                                                           Joint 1  5
                                                                 ───
                                                                 8  4
        3  4
        1  5
        ─── 4  9        Bed and joint.                     (Top, bottom and sides.)
    2/  1  4
        1  5
        ─── 3  9        P. F.                              ...ont and back.
  2/ 2/    6
        1  5
        ─── 2 10        S. F. stopped.                     Sides.
     2/    6
        1  5
        ─── 1  5        Plain face.
                          and
                        Ddt. ½ Joint.
  2/ 2/    3
        1  5
        ─── 1  5        S. F.                              Prepy. face for half balusters.
     2/    7
        1  5
        ─── 1  8        Mo. F. circular stopped.           To balusters.
  2/ 2/ 1  5
        ─── 5  8        Stopped end to Mo.
        2  8
        1  1
           5
        ─── 1  2        Portland stone.                    Capping.
        2  8
        1  1
        ─── 2 11        Bed.
        1  1
           5
        ───    5        Joint.
```

862 QUANTITY SURVEYING.

	2 8					
	4		11	P. F.		Back.
	2 8					
	8		1 9	P. F.		
				and		
				Ddt. ½ bed.		Soffit.
	2 8		2 8	Mo. 6 in. girth.		
						Front.
	2 8					
	1 1		2 11	S. F.		
						Top.
	2 1					
	1 9					
	5					
			1 6	Portland stone.		Capping to pedestal.
	2 1					
	1 9		3 8	Bed.		
	1 9					
	5					
			9	Joint.		
	2 1					
	4					
			8	P. F.		
				and		
				Ddt. ½ bed.		Front part of top.
	2 1					
	1 5					
			2 11	S. F.		Remr. of top
2/	5					
	8		7	P. F.		
				and		
				Ddt. ½ bed.		Soffit.
	2 1					
	4					
			8	P. F.		Back.
	2 1		2 1	Moulding 6 in. girth.		
						Front
2/	5		10	Do. 6 in. girth stopped in short lengths.		Returns.
2/	2/	4		Mitres.		
2/	7	1 2		Joggle and Ct. a. b.		

EXAMPLES OF TAKING OFF. 86₃

A Stone Arch.

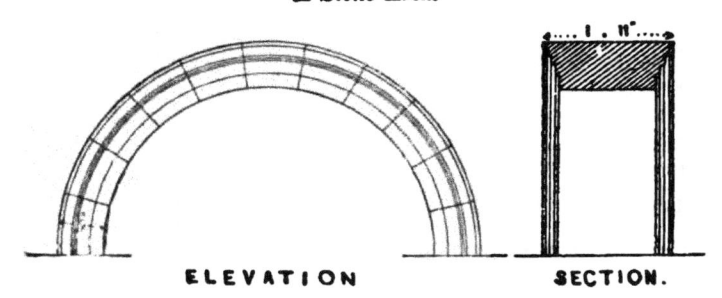

ELEVATION   **SECTION.**

Fig 53.

                                                        Extrados = 10 5
                                                        Intrados = 8 3

        10 5
         1 11
          10
         ———— 16 8   Portland stone.   arch in eleven stones

11/   9
     1 11
     ———— 15 10   Bed.
11/   9
     1 11
     ———— 15 10   Sunk bed.
    ·10 5
     1 11
     ———— 20 0   Circular joint.

                                                     Extrados.
        8 3
        1 7
        ———— 13 1   S. F. circular.   Soffit.
                                                       10 5
                                                        8 3
                                                      2)18 8
2/ 9 4
   1 1
   ———— 20 3   Moulded F. circular.   9 4

                                                      Faces of arch.

864 QUANTITY SURVEYING.

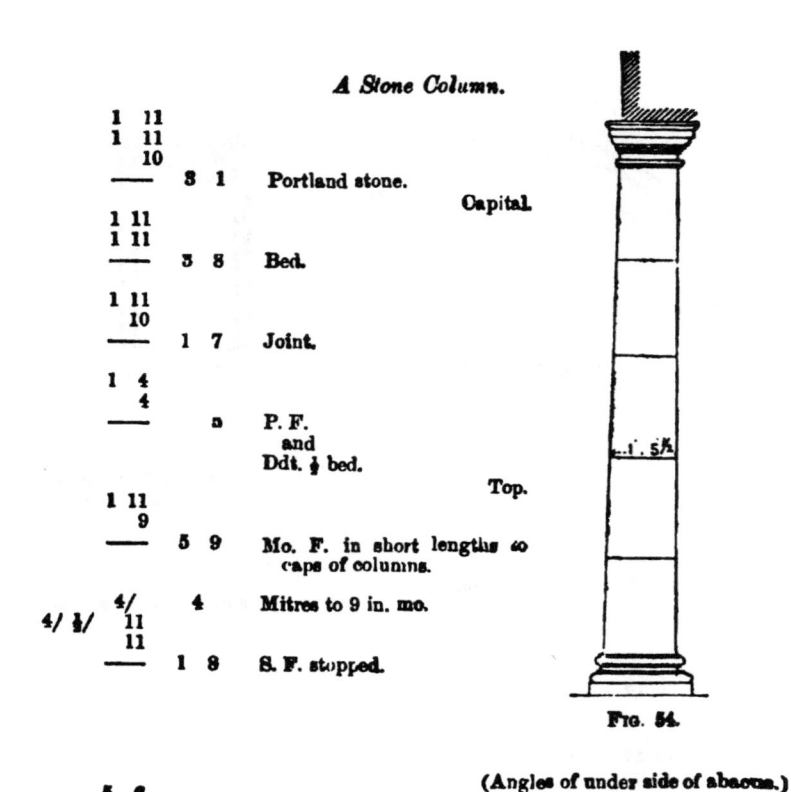

A Stone Column.

```
   1 11
   1 11
     10
   ─────  3  1   Portland stone.
                                   Capital.
   1 11
   1 11
   ─────  3  8   Bed.

   1 11
     10
   ─────  1  7   Joint.

   1  4
      4
   ─────     a   P. F.
                 and
                 Ddt. ½ bed.
                                   Top.
   1 11
      9
   ─────  5  9   Mo. F. in short lengths to
                 caps of columns.

       4/
4/ ½/  11     4   Mitres to 9 in. mo.
       11
   ─────  1  8   S. F. stopped.
```

FIG. 54.

(Angles of under side of abacus.)

```
   5  6
      7
   ─────  3  3   Circular sunk joint.
                                   Prepy. faces.
   4  8
     11
   ─────  4  3   Mo. F. circular continuous to cap of column.

                              Diar. ⎰ 1  8
                                    ⎱ 1  3
                                    ─────
                                      2 11
                                    ─────
                                      1  5½ av.
   1  6
   1  6
  10  4
   ───── 23  3   Portland stone.
                                   Shaft in five stones (average size).
```

EXAMPLES OF TAKING OFF. 865

```
    5/  1  6
        1  6
            ——— 11  3   Bed.
    2/  1  6
       10  4
            ——— 31  0   Joint.
        4  7
       10  4
            ——— 47  4   Circular S.F.
```
In shaft of column.

If with entasis call it circular S.F., to swelled and diminished shafts.

```
        2  1
        2  1
       10
            ———  3  7   Portland stone.
```
Base.

```
                                              2  1
                                             10
                                             ———
                                              2 11

        2 11
        2  1
            ———  6  1   Bed and joint.
    4/  2  1
       10
            ———  6 11   P.F.
```
The whole height.

```
        6  7
           7
            ———  3 10   Circular sunk joint.
                                              2  1
                                              1  7
                                             ———
                                              3  8
```
av. diar. 1 10

```
        5 10
          11
            ———  5  4   Mo. F. circular, continuous to bases.
```
To bases.

```
  4/ ½/  1  0
         1  0
            ———  2  0   S.F. stopped.
```
Top of plinth.

```
    6/  2/  12   3" × 1" × 1" slate dowels, mortises, and Ct.
```

3 K

866   QUANTITY SURVEYING.

Gothic Windows of Stone.—It is generally convenient to first deduct the walling clear of the stonework of the jambs and head for the thickness occupied by the stonework. The deductions for the other stonework, as sill, jambs, head, etc., may be written with the stonework as "Ddt. ¾ rubble," or such proportion as may be decided. The internal opening will be deducted in the usual way.

When measuring stonework of tracery, examine the jointing and amend it if necessary. As each piece of stone is measured, a pencil tick on each of the pieces of stone will be helpful, and preserve the measurer from taking the same piece twice.

In the section Northern Practice the local treatment of traceried windows is also described. The builders of that part of the kingdom in many cases prefer to have such things as sills, jambs, mullions, cornices and the like stated in the bill at per foot run, stating the size, and giving a sketch, thus:

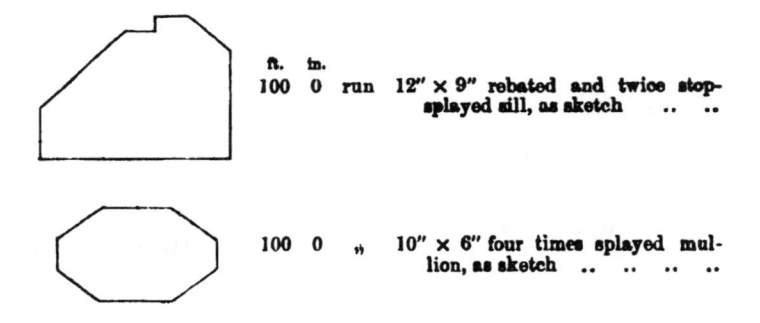

```
              ft.  in.
             100   0   run   12" × 9" rebated and twice stop-
                            splayed sill, as sketch     ..   ..

             100   0    „    10" × 6" four times splayed mul-
                            lion, as sketch     ..   ..   ..   ..
```

The grooves for lead lights and for condensation will be separately measured.

                    A Traceried Window (Fig. 55).

```
     6  8
     1  4
     1  6
     ——    13  4    Box Ground stone
                       and
                    Ddt. rubble.
                                               Sill in two stones
```

EXAMPLES OF TAKING OFF.

Fig 55.

QUANTITY SURVEYING.

```
        6  8
        1  4
       ─────  8 11   Bed.
   2/   1  4
        1  6
       ─────  4  0   Joint.
        1  6
       ─────  1  6   Double V-joggle pebbles and cement.
```
                                                            Back 1 3
                                                            Front   7

```
        6  8
        1 10
       ─────  12  8  P. F.
   2/      8
          11
       ─────   1  3  Add.
```
                                                            At ends of front.

```
        4  5
           6
       ─────  2  3   P. F.
                     and
                     Ddt. ½ bed.
```
                                                                Top.

```
        5  4
        1  2
       ─────  6  2   S. F. stopped
                     and
                     Rough sunk.
```
                                                               Front weathering.

NOTE.—The length of the foregoing being measured from A to A gives sufficient to include from A to B, as also the sunk work to stool of mullion.

```
        4  9
           4
       ─────  1  7   Sunk F. stopped.
   2/   1  4
       ─────  2  8 ⎫  Labour mitre to splay.
   6/      5       ⎬
       ─────  2  6 ⎭
```
                                                               Splay inside.

                                                                  1 9
                                                                  1 7
                                                                  1 3
                                                                  1 10
                                                                  1 3
                                                                  1 7
                                                                  1 11
                                                                  1 9
                                                              8) 12 11
                                                                  1 7½ av

EXAMPLES OF TAKING OFF. 869

```
    2/  1  8
        1  4
        6  0
           ——— 26  8   B. G. stone
                         and
                       Ddt. rubble.
                                              Jambs in eight stones.
 2/ 8/  1  8
        1  4
           ——— 35  7   Beds.                           1  4
                                                          8
                                                       ————
                                                       2  0
    2/  2  0
        6  0
           ——— 24  0   ½ joint.                         Backs.
                                                 Front  1  3
                                                 Back      4
                                                       ————
                                                       1  7
     |  1  7
        6  0
           ——— 19  0   P. F.
    2/     6
        6  0
           ———  6  0   P. F.
    2/    10
        6  0
           ——— 10  0   S. F.
                         and
                       Rough sunk F.
                                              Outer chamfers.
    2/     4
        6  0
           ———  4  0   S. F.
                                              Inner chamfers.
           5
        1  0
        6  0
           ———  2  6   B. G. stone scantling.
                                              Mullion.
    2/  1  0
           5
           ———   10    Bed.
                                                       1  0
                                                       1  0
                                                          5
                                                          5
                                                       ————
                                                       2 10
        2 10
        6  0
           ——— 17  0   P. F.
    4/     4
        ·6  0
           ———  8  0   S. F.
    2/ 2/  4          3″ × 1″ × 1″ slate dowels, mortises, and Ot.
```

870   QUANTITY SURVEYING.

The stone out of which each piece of tracery is obtained is indicated by dotted lines on the illustration.

```
2/  1  1
    1  4
    2  8
    ─────
          7  8⎫  B. G. stone                    Lowest stone of arch.
2/  1  9   ⎪                                              ①
    1  4   ⎪
    2  0   ⎪
    ─────  ⎪
          9  4                                            ②
2/  1 11   ⎪
    1  4   ⎪
    1  8   ⎪
    ─────  ⎪
          8  6                                            ③
2/     8   ⎪
    1  4   ⎪
    1  2   ⎪
    ─────  ⎪
          2  1                                            ④
2/     8   ⎪
    1  4   ⎪
    1  7   ⎪
    ─────  ⎪
          2 10                                            ⑤
    2  0   ⎪
    1  0   ⎪
    2  3   ⎪
    ─────  ⎪
          4  6                                            ⑥
2/  2  1   ⎪
    1  0   ⎪
    1  1   ⎪
    ─────  ⎪
          4  6⎭                                           ⑦
2/ ½/ 5 10
       11 4
       ─────
          70 1  P. F.                          Front and back.
```

The foregoing dimension is that of a triangle, the apex of which is marked a and is assumed to be equal to the surface of the head. If the surveyor is doubtful of his judgment in this manner he may inscribe a triangle in the head and calculate the segments of the circle forming the remainder, as sketch.

```
                                           14
                                            8
                                            8
                                            8
                                            8
                                           11
                                           11
                                            8
                                            8
                                            8
                                          ─────
              Fig. 56.                    10)92
                                          ─────
                                            9¼
```

EXAMPLES OF TAKING OFF 871

```
2/ 10/  10
        1  4
        ──── 13  4   Sk. joints.
```
                                    Collected and averaged.
                                    Stones 1, 2, 3, 4, 5.
                                        2/   5 = 10
                                        2/ 11 = 22
                                        2/   7 = 14
                                                  7
                                             ─────
                                             7) 53
                                             ─────
                                                 7½

```
2/ 7/   8
        1  0
        ────  9  4   Sunk joints.
```
                                    To tracery.
```
        1  0
           5
        ────  ·5     Joint.
```
                                    Over mullion.
```
2/  8   9
    1   4
    ────  23  4      Circular joint.
```
                                    Back of arch.
```
    14/               Slate dowels, mortises, and Ct. s. b.
```
                                    Width   = 4 11
                                    Depth of sinking{   4
                                                        4
                                                    ─────
                                                      5 7

```
½/  5   7
    9   4
    ────  26  1      S. F.
                    and
                    Ro. sunk F.
```
                                    Front of tracery.
```
2/  8   2
        6
    ────   8  2      S. F. circular.
```
                                    Outer splay of arch.
```
    3   6
    2   6
    ────   8  9      Ddt. S. F. stopped.
```
The lower part of tracery where the thinner stone occurs.

       1   1   Labour triangular perforation 5" × 5" extreme through
                    12 in. stone.
                                    Over mullion.
                                    Head of light   4   2
                                    Eye next arch 2 11
                                    ½ triangle at top 3   2
                                    ½ eye over mullion   8
                                                                ─────
                                                                 10 11

QUANTITY SURVEYING.

2/ 10 11
    1 0
——— 21 10   S. F. circular part stopped in tracery.
                                                   12 in stone
14/ 1 0
    ——— 14 0   Mitre to splay.
                                                        8 3
                                                            3
                                                            3
                                                           —————
                                                           8 9

2/ (..3 . 3..) 22 0   S. F. stopped.
                                                  Faces of cusped circle.

       6/   6   Labour to triangular perfs, 3" × 3" extreme, through
                           6 in. stone.
                                                               Eyes.
6/ 8/   6
      ——— 9 0   Mitre to splay.
                                             Eyes 6/   9 = 4 6
                                             Cusps 6/ 2 1 = 12 6
                                                                   ————
                                                                    17 0

      17 0
        6
    ——— 8 6   S. F. circular in tracery.
12/   6
    ——— 6 0   Mitre to splay.
                                         Points of cusps.
                         Heads of lights   2/ 4 8 = 9 4
                                                                 3 10
                      Eye over mullion 2/ 5 4 = 10 8
                                                               11 0
                                                                 8 6
                                                                 ————
                                                                13 4

2/ 43 4
    ——— 86 8   Circular splay, 4 in. wide in tracery, part stopped.
                                                                 Inside and outside.
2/ 14/ 6
    ——— 14 0   Mitre to splay.
                                                    Average of long and short.
2/ 6/ 2 0
    ——— 24 0⎫
                   Circular splay as last, but 2 in. wide.
2/ 6/ 2 3   Eyes in foiled circle.
    ——— 27 0⎭
                                                                 Foils.
2/ 6/ 3/   36   Mitre to 2 in. splay.
                                                                Eyes.
2/ 2 0
    ——— 4 0   Stopped rebate 5 in. girth, including stops.
                                                                  On sill

EXAMPLES OF TAKING OFF. 873

	2/	2 0		
		———	4 0	Stopped groove 1 in. wide and ½ in. deep, incg. stops.
				For condensation.
	2/	1/	2	¼ in. perforations 9 in. long, as eject.
2/ 2/	6 1			
		———	24 4	Groove for lead lights
				and
				Pointing both sides with mastic cement.
	2/	4 2		
		———	8 4⎫	Ditto, circular in tracery.
		1 5		
		———	1 5	
	2/	3 0		
		———	6 0	
	6/	2 1		
		———	12 6	
	6/	8		
		———	4 0	
		6 3		
		———	6 3⎭	
2/ 4/	2 1			
		———	16 8	½″ × ½″ galvanised iron saddle bar.
2/ 4/	2/	16		Mortises for saddle bars and cement.
	1/	1		Ring 18 in. diar. of saddle bar as before, with six points.
	6/	6		Mortises and cement a. b.
	2/	2 1		
		6 1		
		———	25 4	Stout lead quarry lights, secured with strong copper bands to saddle bars, and glazing with rolled cathedral glass in varied tints (to be approved), with border of plain white glass.
	2/	2 1		
		1 10		
		———	7 8⎫	Ditto, in cusped heads, and tracery measured square.
		6		
		5		
		———	8	
	2/	7		
		1 3		
		———	1 6	
	6/	8		
		4		
		———	6	
		1 6		
		1 6		
		———	2 3	
	6/	11		
		11		
		———	5 1	
		2 2		
		1 10		
		———	4 0⎭	

If the stone in a traceried head is all of one thickness, the stone may be taken in one dimension, unless the tracery is unusually open; the quantity of stone resulting from this method

QUANTITY SURVEYING.

will be sufficiently near for all practical purposes. To take the foregoing as an example, and assuming the thickness to be 12 inches, the dimensions would be as follows:—

```
½/   5 10
     1  0
    11  4
    ——— 33  1   B. G. stone in tracery.
```

A Quarter Partition.

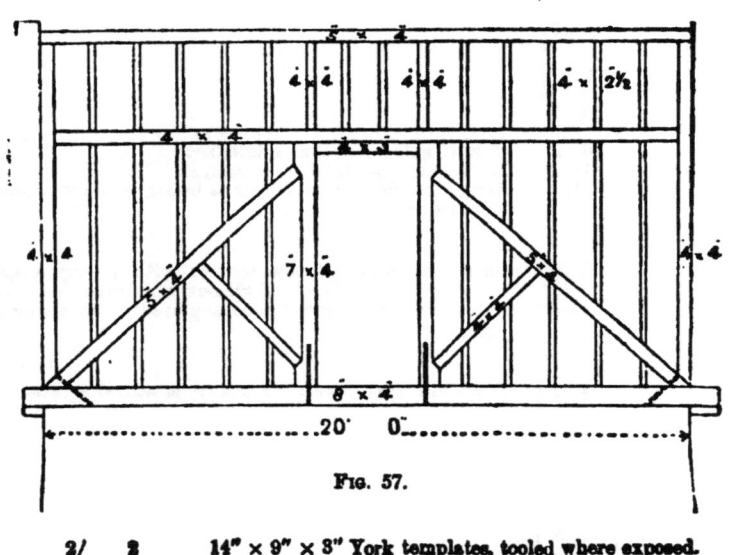

FIG. 57.

```
2/   2   14" × 9" × 3" York templates, tooled where exposed.
                                    Partition south of B.R. 4.
2/   2   Ends of timbers cut and pinned.
                                              20  0
                                                  9
                                                  9
                                              ———
                                              21  6

21  6
 8
 4
———   4  9   Fir framed in trussed partition.
                                        Tie beam.
                                   Head  21  6
                                 Queens {  8  6
                                           8  6
                                          ———
                                          38  6
```

EXAMPLES OF TAKING OFF. 875

```
     38  6
         7
         4
        ——   7  6   Add.
2/   10  9
         5
         4
        ——   3  0   Add.                          Principals.
2/    4  9
         4
         4
        ——   1  1   Add.
                                                  Struts.
      3  5
         4
         3
        ——      4   Add.
                                                  Straining piece.
2/    1/   2   ⅝-in. W. I. bolt 17 in. long, with head, nut and washer,
                 and fixing by carpenter.
                                                  Heel bolts.

                                                       2  0
                                                       2  0
                                                          4
                                                      ——
                                                       4  4
2/    4  4
        ——   8  8   2″ × ¼″ W. I. strap, incg. perforations, and fixing by
                      carpenter.
2/    1/   2   Sets of W. I. gibs and keys.
                                       Posts 2/  11  6 = 23  0
                                       Intertie          20  0
                                       Posts           { 3  6
                                                       { 3  6
                                                         ——
                                                         50  0
     50  0
         4
         4
        ——   5  7   Fir frd. a. b.
                                           10/  11  6 = 115  0
                                            2/   3  6     7  0
                                                         ——
                                                         122  0
    122  0
         4
         2½
        ——   3  6   Add.
                                                  Quarters.
```

A Circular Rib to a Roof Truss.

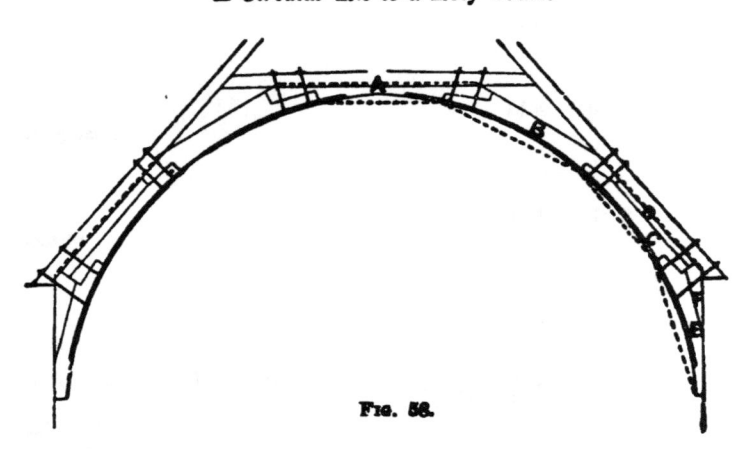

FIG. 56.

		ft	in		
	14	8			
	1	2			
	—	16	8	4-in. fir framd and wrought in circular rib.	Piece A.
2/	12	3			
	1	9			
	—	42	11	4-in. ditto out of timber 21 in. deep.	B.
2/	9	3			
	1	7			
	—	29	4	4-in. ditto out of ditto 19 in. deep.	C.
2/	7	4			
		7			
	—	8	7	4-in. fir as first.	D.
2/	9	6			
	2	1			
	—	39	7	4-in. ditto out of 25 in. deep.	F.
2/	5	6			
	1	3			
	—	13	9	4-in. fir as first.	F.

Describing the timber as wrought gives an excess on the planing, but does not usually increase the builder's price. The word wrought can however be omitted, and the net quantity of the planing measured, if that course be preferred.

EXAMPLES OF TAKING OFF. 877

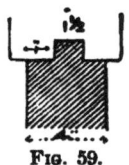

			1½	12	6
			1	9	9
			1	9	9
			3½	32	0

	32 0		
	———	32 0	Labour stopped groove 3½ in. girth in fir, incg. stops. (Edge of rib grooved into collar and principal rafter.)

Fig. 59.

2/	32 0		
	———	64 0	Labour to rebate 2¼ girth.

		Between A and B = 3 10
		„ B and C = 3 6
		„ C and E = 4 0
		11 4

2/	11 4		
	———	22 8	Grooved and rebated joint and dowelling with oak dowels.
2/	7 4		
	———	14 8	Between C and D.
2/	5 3		
	———	10 6	Between E and F.

	2 4
	2 10
	2 7
	2 6
	3 1
	3 9
	6) 17 1
	2 10

2/	6/	12	⅜-in. W. I. bolts, average 2 ft. 10 in. long with head, nut and washer, and fixing by carpenter.
2/	6/	12	Labour heads of bolts let into fir and pelletted or covered.
	58 6		
	———	58 6	Labour circular sunk and sawn edge to 4-in. fir.
2/ 2/	25 3		
	———	101 0	Labour to stopd. chamfer 2 in. wide circular.
2/ 2/	2/	8	Splayed stops.

Cupolas.

The surface of a hemisphere is best written as the result of a calculation. Such a measurement sometimes occurs in the roof of a turret, like the following example :—

⌒ <·4' 0"·> The rule for a sphere is $3·1416 \, d^2$. d = diameter = $50·265$ ft.
One half of this is $25·13 = 25$ ft. 2 in. nearly.

```
25  2
 1  0
 ——     25  2    ¾" boarding in small quantities ¦bent circular to
                 hemispherical roof of cupola, measured net, and
                 allow for all cutting and waste.
```

Doors.

Doors of various heights and widths may often be averaged with advantage where one of the dimensions is the same in each case; thus a number of doors 7 feet high and of various widths as follows :—

EXAMPLES OF TAKING OFF. 879

```
                    Bedroom No. 2    2/  2  6 = 5  0
                    Housemaid's closet      = 3  0
                    Bedroom No. 4    2/  2  9 = 5  6
                        „    No. 5   2/  2 10 = 5  8
                                         ─────────
                                          7) 19  2
                                         ─────────
                                          Av. 2  9
```

```
7/  2  9
    7  0
   ─────  134  9    2-in. four-panel sqr. doors.
```

(4 / Grain & 2ce Varnish)

```
                                              7  0
                                              7  0
                                              2  9
                                     4/1¼     5
                                            ──────
                                             17  2
                                     2/1"     2 horns.
                                            ──────
                                             17  4
```

```
7/  17  4
        6
   ─────  60  8    1¼ in. double rebated jamb linings,
                       tongued at angles.
```

(4 / Grain & 2ce Varnish.)

```
                                             17  2
                                     4/3" =   1  0
                                            ──────
                                             18  2
```

```
7/ 2/ 18  2
   ─────── 254  4   3" × 1¼" moulded architrave
                              and
                    3" × 1" framed and splayed
                        grounds plugged.
```

(4 / Grain & 2ce Varnish.)

Often the difference in length between grounds and architraves is so small that it is assumed to be the same, and they are taken together, as in the foregoing dimensions.

Staircases.

The space shown on a general drawing as occupied by a staircase is from outside to outside of strings.

As the measurement of wreathed lines will often be requisite in dealing with handrails, strings, &c., and may perhaps be

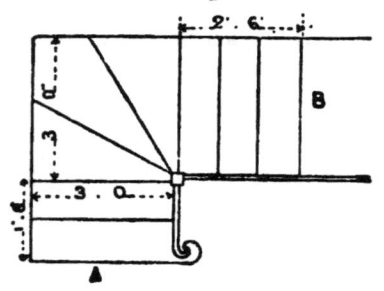

FIG. 60.

puzzling to inexperienced persons, see the following example:—
Measure the length on plan (say 3 feet) and find the rise (say 2 feet), set out 3 feet as the base of a triangle, and 2 feet as the perpendicular, and the hypothenuse will be the required length.

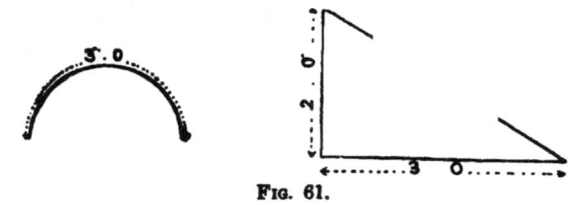

FIG. 61.

Although the length of the treads minus the strings would be a trifle less than 3 feet, as each housing would be less than ¾ of an inch, it is usual to take the whole length as follows:—

Assuming the height from A to B to be 4 feet 6 inches.
Each riser will be 6 inches.

Treads	{	1 8
		2 6
Risers	{	1 0
		2 0
Nosing 6/1″		6
		7 . 8

EXAMPLES OF TAKING OFF.

3 0			
7 8			
——	23 0	1¼-in. deal treads, with rounded nosings and 1-in. risers glued, blocked and bracketted, all tongued together on and including two strong fir carriages, and prepared for cut strings.	
			Fliers.
3 0			
3 0			
——	9 0	1¼-in. ditto, cross-tongued winders, measured net.	

		Treads.
		3 0
Riser	6″	3 6
Nosing	1″	8 6
7		10 0

10 0		
7		
——	5 10	Add.

Risers.

1/	1	Extra labour and materials curtail end to bottom step and veneered front to riser, the step 8 ft. 6 in. long.
3 0		
——	3 0	4″ × 1½″ rounded nosing and tonguing to edge of floor.

At top.

Or the fliers may be measured by single steps, thus :—

Tread	10
Riser	6
Nosing	1
	17

5/	3 0		
	1 5		
	——	21 3	
	3 0		
	7		
	——	1 9	

Topmost riser.

The way first described usually makes fewer dimensions.

6/	6	Housings of steps.
8/	8	Ditto winders.
4/	4	Mitred and returned nosings to steps.
2/	2	Ends of treads notched and fitted to newel.
3/	3	Ditto winders.

Square-headed Sashes and Frames in 14-inch Wall.

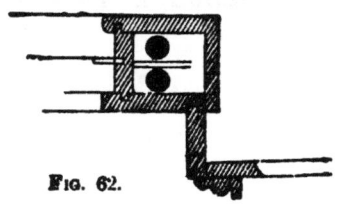

FIG. 62.

QUANTITY SURVEYING.

Begin with the size of the external opening "3' 0" × 6' 0"."

                4¼
                4¼ 3
                3 9 6 3

3 9
6 3
―― 23 5 Deal-cased frames O.S. and weathd. sills, 1-in. inside and outside linings, 1¼ pulley stiles, ¾-in. back linings, and 2-in. moulded sashes double hung with best No. . . . flax lines, best brass axle pulleys, and iron weights, and the frame grooved all round for finishings.

4/ 4 Extra for moulded horns to 2-in. sashes.
3 9
―― 3 9 1¼" × ¼" galvd. W. I. tongue in white-lead.
        and
     Groove in oak,
        and
     Groove in York stone.
1/ 1 Strong brass spring sash-fastening.
2/ 2 Brass sunk sash lifts.
2 10
5 6
―― 15 7 B.P.P.

                 2 squares.
          Length of 3 9
         outer edge 6 3
         of frame. 6 3
                16 3

16 3
 5
―― 6 9 1-in. rebated linings, tongued at angles.
                16 3
                 8

16 11
―― 16 11 2" × 1¼" moulded architrave and mitres.
               16 11
               16 3
                1 0

17 3
―― 17 3 3" × 1" framed and splayed grounds.
               17 3
                3 9
                 2¼
                 2¼
                4 2

 6
4 2
―― 4 1 1½-in. rounded and rebated window board and bearers.
2/ 2 Notched, returned and mitred ends.
          Height of frame 6 3
         Width of one ground 3
                 6 6

         Width of frame 3 9
          and grounds.   6
                 4 3

EXAMPLES OF TAKING OFF. 883

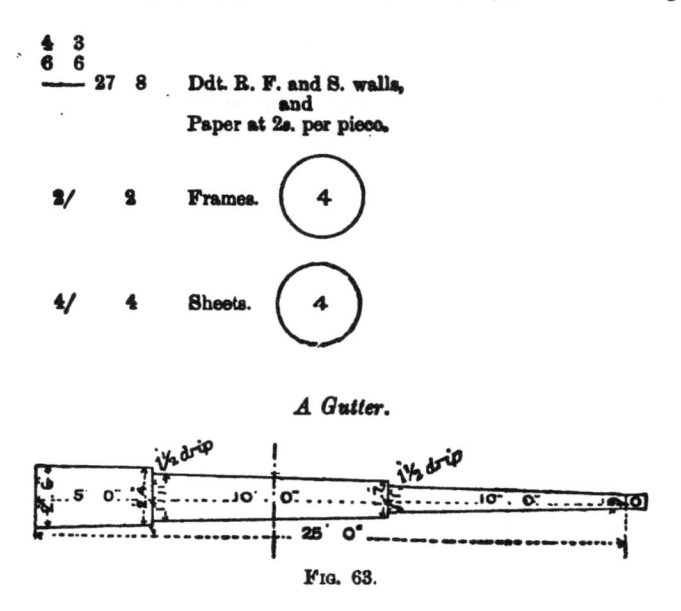

```
  4  3
  6  6
  ───  27  8     Ddt. R. F. and S. walls,
                         and
                 Paper at 2s. per piece.


  2/    2    Frames.   ( 4 )


  4/    4    Sheets.   ( 4 )
```

A Gutter.

FIG. 63.

The width of gutters must always be calculated, as they are frequently incorrectly drawn; when the distances apart of the drips are equal, long lengths may be taken together, and the width may be averaged, but not otherwise, for when the drips are at unequal distances the average will not be a true one, and considerable error will result; nevertheless, in a large number of gutters of similar character the probability is that errors will neutralise each other.

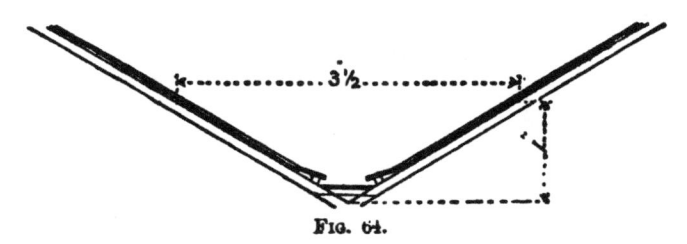

FIG. 64.

First find out the increase of width for every inch of rise (see dotted lines on diagram), by drawing a section of the roof slope to a scale sufficiently large to be depended on. In this instance the

3 L 2

QUANTITY SURVEYING.

rise is 1¼ inch in 10 feet, and the increase of width for each inch of rise 3½ inches.

Observe that lead is often turned up as much as 12″ under the roof covering (measured from the sole of the gutter).

```
                                                            9
                                                       1    1
                                                      ─────────
                                                      2) 1 10
                                                      ─────────
                                                            11
```

```
A.   10  1  ⎫
         11 ⎪
     ───── 9 3 │  1-in. deal gutter-boards and framed bearers.
B.   10  1  ⎪                                           1  7
      1  9  ⎪                                           1 11
     ───── 17 8                                       ──────
C.    5  1  ⎪                                         2) 3  6
      2  5  ⎪                                         ──────
     ───── 12 3 ⎭                                        1  9

                                                         2  4
                                                         2  6
                                                       ──────
                                                       2) 4 10
                                                       ──────
                                                          2  5

     2/   2       Short lengths of 1¼-in. cross rebated drip.
     1/   1       1¼-in. deal dovetailed cesspool 9″ × 9″ × 6″, all in clear,
                      holed and fitted.
```

```
                          10  0              Sole of gutter  11
                 At drip     3                  Turn up  ⎱    9
                 At cesspool 2                           ⎰    9
                          ──────                              ──
                          10  5                               29
```

```
A.  10  5  ⎫
     2  5  ⎪
    ───── 25 2 │  6-lb. lead gutter.
B.  10  6  ⎪
     3  3  ⎪
    ───── 34 2 ⎭
```

```
                                    10  0        1  9
                           At drip {  3          9
                                      3          9
                                    ──────       ──────
                                    10  6        3  3

                                     5  0        2  5
                           Turn up    9          9
                           At drip    3          9
                                    ──────       ──────
                                     6  0        3 11
```

EXAMPLES OF TAKING OFF. 885

```
C.   6  0
     3 11
     ──── 23  6   }   6-lb. lead gutter.
       10
       10
       ────      8                           Cesspool bottom.
                                                       6
     3  1                                              1    3  0
        8                                              1       1
     ──── 2  1                                        ───    ────
                                                       8    3  1
                                                          Ditto sides.

     1/    1    Extra labour and solder to cesspool.
     1/    1    24 in. length of socket pipe out of 7-lb. lead all bent, one
                end tafted and soldered to cesspool.
     1/    1    Stout copper wire cover to cesspool.
     1/    1    Perforation in 1½ Bk. wall for socket pipe, and making
                good.
```

Measuring from the last Illustration in the manner used when the Drips are at equal Distances.

```
                  Wider end   2  6              25  0
                  Narrower end   9   Passings 2/1   2
                              ──────             ─────
                                3  8              25  2
                              ──────
                                1  7½

     25  2
      1  8
     ───── 41 10   1″ gutter board and bearers.
```

It will be seen, by comparison with the former dimensions, that the result is too great.

Alternative for Cesspool.

Cesspools are often of one rectangular piece of lead with the corners cut out.

```
                                         3     9
                                         6     6
                                         9     9
                                         6     6
                                         6     9
                                        ───   ───
                                         30    39

     2  6
     3  3
    ─────        6 lbs. lead.
                                              Cesspool.
```

A Balustrade and Cornice in Portland Cement and Brick.

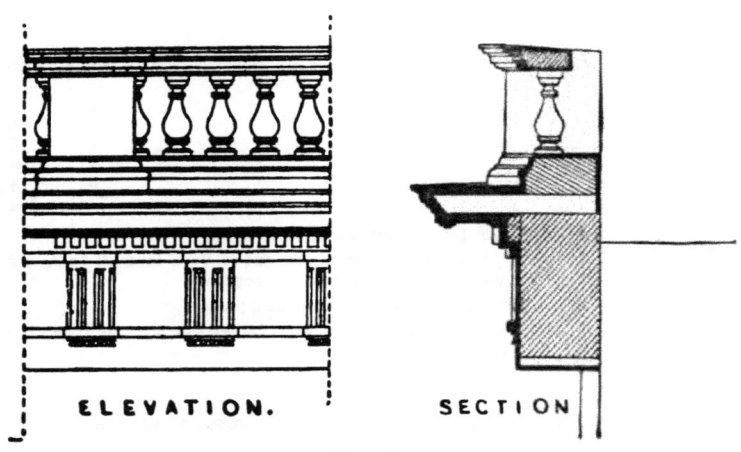

FIG. 65.

4 9 / 3 0	14 8	1½ Bk. in Ct.		Frieze and cornice.
1 8 / 6	10	¼ Bk. in Ct.		Projection base of pedestal.
1 2 / 1 9	2 1	1¾ Bk. in Ct.		Pedestal.
3 7 / 3	11	1 Bk. in Ct.		Capping.
1 11 / 3	6	¼ Bk. in Ct.		Projection for pedestal.
4 9 / 6	2 5	¼ Bk. in Ct.		Projection for bed mould of cornice.
4 9 / 2 5	11 6	4-in. rough York core, and bedding and jointing in Ct. Cornice.		

EXAMPLES OF TAKING OFF. 887

		4 9			
			4 9	Roughly splayed edge to 4-in. York	
2/		4 9			
			9 6 ⎫	Ro. splay one course of bkwk.	
2/	2/	5		⎬	
			1 8 ⎭		

Capping and base of balustrade, &c.

	4 9			
	1 2			
		5 7	8-in. Ro. York core, and bedding and jointing in Ct.	

Soffit of architrave.

All in Narrow Widths.

	1 4		
	1 9		
		2 4	P. F. in Portland Ct. on brick.

Back of pedestal.

	8 5		
	4		
		1 2	P. F.

Back of capping.

	4 9		
	1 6		
		7 2	P. F.

Back of base.

2/	4		
	4		
		5	P. F

Returns of pedestal at top.

	2 2		
	5		
		11	Weathering, including dubbing.

Top of pedestal.

	1 4		
	1 5		
		1 11	Add.

Ditto.

Ddt. 4 9
    2 2
―――
    2 7

	2 7		
	1 1		
		2 10	Add.

Top of capping.

	4 9		
		4 9	Moulding 6-in. girth.

Front.

2/	5		
		10	Ditto, in short lengths.

Returns.

| | | 4/ | 4 | Mitres to 6 in. moulding. |

QUANTITY SURVEYING.

```
                                          1  5
                                          1  5
                                          1  4
                                          ----
                                          4  2
```

```
      4  2
      1  5
      ----
           5 11    P. F.
                                    Front and returns of pedestal.
      4  9
      ----
           4  9    Moulding 6 in. g. in S. L. a. b.
                                    Base of balustrade.
2/    5
      ----
           10      Ditto, in short lengths.
                                    Returns.
      4/   4       Mitres to 6-in. moulding.
                                          4  9
                                             4½
                                             4½
                                          ----
                                          5  6
```

```
      5  6
         8
      ----
           1  5⎫   P. F. a. b. ⎫
      3  5       ⎬              ⎬  Base.
      1  1       ⎪              ⎭
      ----
           3  8 ⎬
      3  5       ⎪
         8       ⎪
      ----       ⎪
           2  3⎭
                                    Soffit of capping.
2/    4  9
      ----                          Capping.
           9  6   Arris part in short lengths.
2/       4
      ----
              8
2/       4½
      ----
              9
2/    3  5
      ----                          Base.
           6 10
2/    1  3
      ----                          Vertical to pedestal.
           3  4
2/    1  5
      ----
           2 10
2/       4
      ----
              8
```

NOTE.—When, as in the above case, a distinction is necessary, but the difference of value for short lengths is small, say " part in short lengths."

```
      4  9
      1  7
      ----
           7  6    Weathering and dubbing a. b.
                                    Cornice.
```

EXAMPLES OF TAKING OFF.   889

	1 10			
	4			
	───	7	Ddt. do.	
				For base of pedestal.
	4 9			
	───	4 9	Arris.	
	4 9			
	2 6			
	───	11 11	Moulding.	

Cornice measured from top of frieze to edge of weathering.

	4 9		
	───	4 9	Dentil course of blocks 2 in. wide, 8½ in. high, 1 in. apart, 2 in. projection.
	4 9		
	1 2		
	───	5 7	P. F. a. b.

Frieze.

2/	2	Cast Doric triglyphs, 10 in. wide, 18 in. high, and 1 in. projection, and including modelling, fixing and m. g.
1/	1	Half do. and do.

```
                                          1 0
                                          1 0
                                            7
                                          ───
                                          2 7
```

	2 7		
	───	2 7	Moulding 8 in. girth in short lengths.

Fillet between lower part of triglyphs.

| 10/ | 10 | Mitres. |

```
                                            6
                                          1 3
                                          ───
                                          1 9
```

	4 9		
	1 9		
	───	8 4	P. F. a. b.

Fascia and Soffit.

	4 9		
	───	4 9	Arris.

4/	4	Cast moulded balusters, 6 in. extreme diameter, and 17 in. high, including modelling and fixing and m. g.
2/	2	Half do., and fixing against pedestal and m. g.

An Archway in Keene's Cement (Fig. 66).

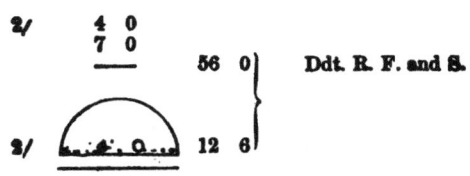

2/	4 0		
	7 0		
	───	56 0	Ddt. R. F. and S.
2/		12 6	

QUANTITY SURVEYING.

2/	6 3/9			
		—	9 5	Keene's Ct. P. F. in N. W.
				(The width usually allowed is equal to thickness of wall.)
	6 4/9			
		—	4 9	Do. Circular.
2/	9			
		—	1 6	Junction of circular and straight.
2/	6 0			
		—	24 0	K. C. mo. 5 in. girth
				and
				Rough chamfer on bk.
				(The latter only when a chamfer is required.)

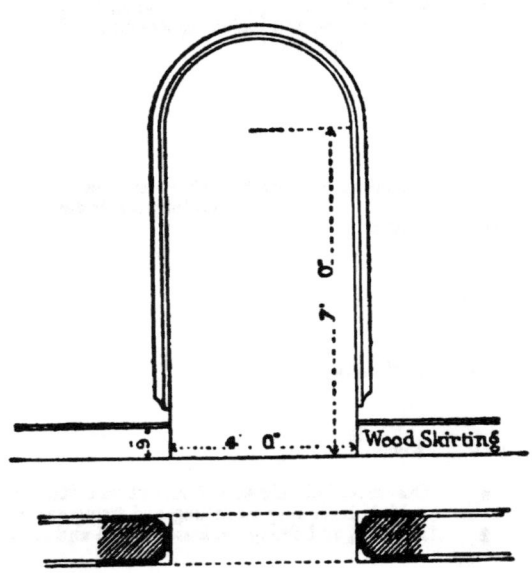

Fig. 66.

2/	7 1			
		—	14 2	Do. circular,
				and
				Rough chamfer on brick.
2/	2/		4	Junction of circular and straight mo. in K. C.
2/	2/		4	Moulded stops to 5 in. moulding.

EXAMPLES OF TAKING OFF.

A Cast-iron Column and its Base.

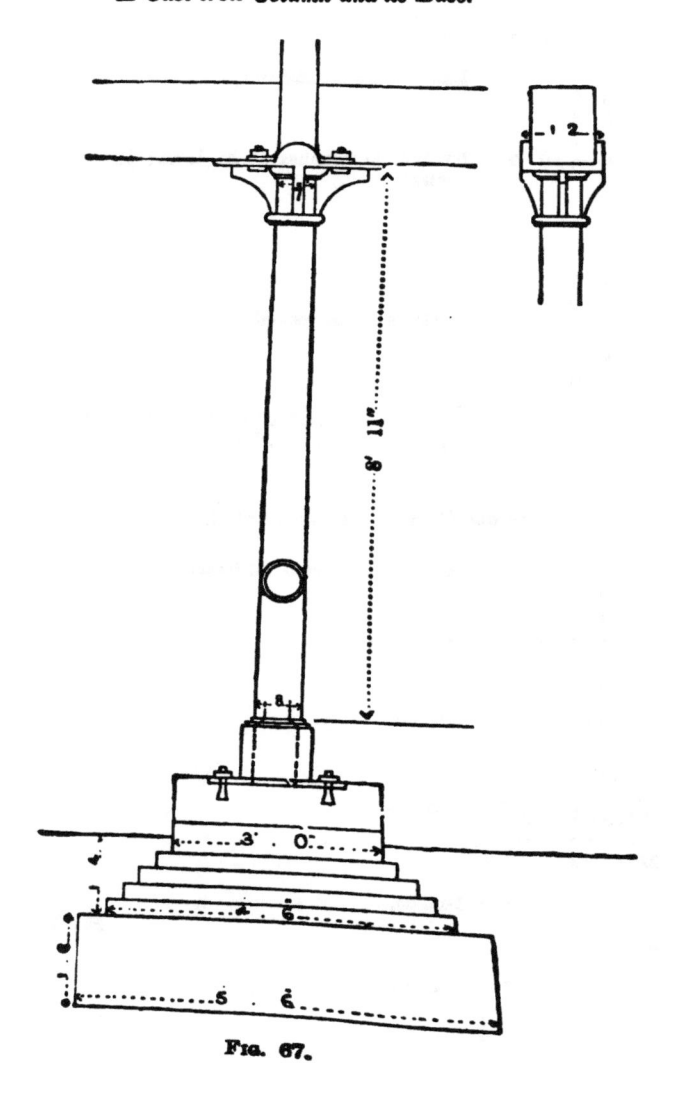

Fig. 67.

```
  5  6
  5  6
  1  6
  ─── 45  5   Concrete as desd.,
                  and
              Excavn. and carting away.
  5  6
  5  6
  1  4
  ─── 40  4   Excavn. to trenches P. F. L and R. and remr. carted
              away.
```

		4¼ B.
		6 B.
		───
		2) 10¼
		5¼
		Foots.

```
  4  0
  1  0
  ───  4  0   5¼ bk. (avd.) in cement.
  3  0
     6
  ───  1  6   4 B.
  1/     1    Tooled York base 3' 0" × 3' 0" × 9", with sinking for base
              plate 24" × 24" × 1" deep.
```

C. I. in one Column, and fixing at Ground Floor Level.

And add 2½ per cent. for featherings.

```
  2  0
  2  0
  ───  4  0   1¼-in C. I.                              Base.

  2  6
  1  2
  ───  2 11   1-in. do.                                Cap.

2/  ⌒8"⌒  4   1-in. do.                               Flanges

2/    5
      9
      ───  5  1-in. do.                        Brackets averaged.

2/    4
      9
      ───  6  1-in. do.                                Ditto.
```

```
                                                        11
                                                         7
                                                        ──
                                                       2) 18
                                                        ──
                                                       9 diar.
```

EXAMPLES OF TAKING OFF. 893

```
(Circum-    2  5
 ference.)     2
            ─────
               5     2-in. C. I.
                                               Moulding of cap.
(Circum-    2  5
 ference.)    1½
            ─────
               4     2-in. C. I.
                                               Moulding of necking.
                                        Diar. at bottom =   8
                                        Do.  at top    =   7
                                                         ────
                                                       2) 15
                                                         ────
  8 11                                                     7½
 ────      8 11   C. I. hollow column 5½ in. intl. diar.
                     (averaged) 1 in. metal.
```

Last item abstracted as taken, weight per foot run to be obtained from published tables. See Laxton, Cast-iron Cylinders, Columns and Pipes.

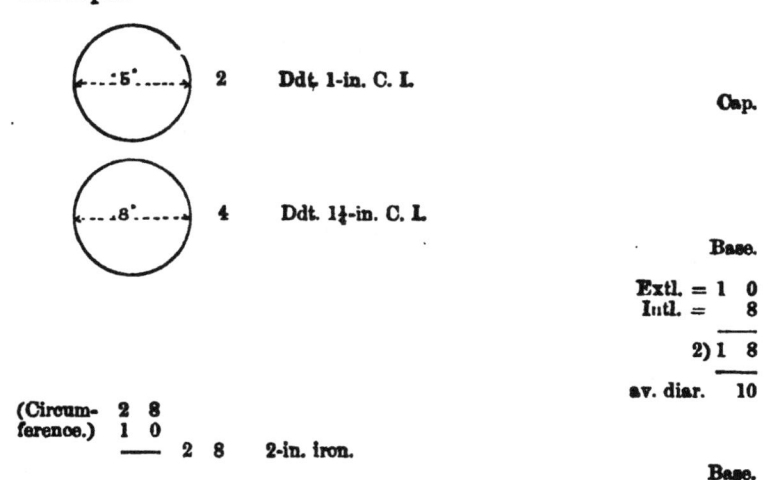

```
  ○ 5"       2     Ddt. 1-in. C. I.
                                                    Cap.

  ○ 8"       4     Ddt. 1¼-in. C. I.
                                                    Base.
                                            Extl. = 1  0
                                            Intl. =    8
                                                   ──────
                                                 2) 1  8
                                                   ──────
                                            av. diar.  10
(Circum-    2  8
 ference.)  1  0
            ─────
               2  8    2-in. iron.
                                                    Base.
```

The extra thickness of iron in base is introduced for the sake of variety, the true principle in cast columns is metal of as nearly as possible the same thickness throughout.

```
           4/   4    1-in. W. I. bolts 7 in. long, one end with lewis, the
                        other with nut.
                                                    Base.
```

No washers required to bolts for ironwork.

4/	4		Mortises in York for lewis bolts, and lead and running.
4/	4		Holes rimed out in 1¼-in. O. I.
4/	4		½-in. W. I. bolts 4 in. long, with head and nut.

Cap.

4/	4		Holes rimed out in 1-in. iron.
1/	1		Pattern to hollow diminished column, 7½ in. average diameter, with moulded cap, four brackets, and moulded base, 10 ft. 2 in. high in all.

A Wrought-iron Box Girder 25 feet in length.

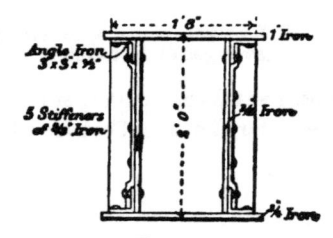

Fig. 68.

2/	2		York templates tooled where exposed, 27″ × 12″ × 4″.
2/	2		Stout pads 24″ × 12″ of boiler felt, 16 oz. to the sheet.
2/	2		Ends of large iron girders, cut and pinned.
25 0 1 8 ————	41 8		1-in. W. I. in riveted box girder, with angle irons and stiffeners, and hoisting and fixing 15 ft. from ground level, including holes and rivets.

Top flange.

25 0 1 8 ————	41 8		⅜-in. ditto.

Bottom ditto.

2/ 25 0 1 10¼ ————	93 9		½-in. ditto.

Webs.

4/ 25 0 6 ————	50 0		½-in. ditto.

Angle irons.

2/ 5/ 1 11 ————	19 2		Add parallel T iron, weight 7·5 lbs. per foot run.

Stiffeners.

Add 5 per cent. for rivets.

EXAMPLES OF TAKING OFF. 895

NOTE.—If it should be necessary to take out the rivets separately, proceed as follows, very rarely done :—

    Rivets ¾ in. diameter for the preceding plate girder (or girders), and driving, as follows.

    Rivet, including holes, through one thickness of ½-in. and one of ½-in. iron.
                    Angle irons.

    Ditto, one thickness of ½-in. and one of 1-in.
                    Ditto.

    Ditto, one thickness of ¼-in. and one of ⅜-in.
                    Ditto.

    Ditto, two thicknesses of ¼" iron.
                    Stiffeners.

Painting.

For dimensions see examples for measuring doors and linings, p. 493.

```
2/  134  9
      1  3
    ───────
          336 11  ⎫
                  ⎬   and
                  ⎪   Grain and 2ce varnish.
     60  8        ⎪
      1  2        ⎪
    ───────       ⎪
           70  9  ⎪
    254  4        ⎪
         6        ⎪
    ───────       ⎪
          127  2  ⎭
```

Or if abstracted directly from dimensions of joiner they are marked as shown ; see dimensions of doors, "Illustration of Averaging," p. 493.

CHAPTER XIV.

THE PRESENT SYSTEM OF ESTIMATING.

It may be interesting to note the manner in which the practice of measuring has developed into the existing system, the which, if pursued to its legitimate conclusions, will be the perfection of analysis and synthesis.

In a work of this character an exhaustive account of the process would be out of place, a few of the leading facts will be sufficient for our purpose. The student to whom the inquiry may chance to be interesting, will find in Rymer, Dugdale, Britton, the Archives of the Universities, and the Record Office, interesting facts relating to the history of estimating in medieval times.

In the case of large works erected in ancient times, as churches, abbeys, colleges and cathedrals, there is little doubt the process adopted by the ruling body was to obtain from its own domain the stone, timber, &c., it produced, purchasing such other material as could not otherwise be obtained. The labour, so far as the masonry was concerned, was executed for the most part, during the twelfth, thirteenth, fourteenth and fifteenth centuries, by the fraternity called Freemasons, who, so far as can be known, were paid by time. The other artificers were usually paid by time, the rate per day being often regulated by Act of Parliament. Occasional instances are to be found, and they were probably frequent, wherein small contracts were made with an individual workman for particular items of work, such as single windows, doors, &c.

The next advance was to the system of contracts by separate tradesmen for the work in each trade. This method still prevails to a large extent in the Northern counties. And in all parts of the country where a builder is ostensibly a general contractor, it is frequently the case that he is merely representative of an association of the several tradesmen who will do the work.

The division of responsibility is an evil, and the various trades are apt to spoil each other's work.

"In the seventeenth century the existence of 'resident surveyors,' as at Wollaton and Holkham, rendered the condition of master builder precarious, if even it were possible, as a monopolist; but in the eighteenth century there was a change of practice on the part of the clients, who either preferred to have the work measured and valued (sometimes on a schedule of prices) or else employed an architect, who, as in the case of Jupp at the East India House in 1799, made his design and submitted an estimate upon which he obtained advances of money, making with his tradesmen his contracts, which with the receipts he produced at the termination of the works to his client, and it was customary for the latter to make a present beyond the commission if the works were executed within the estimate. The Irish architects so late as 1803 practised in this manner.

"Sir W. Chambers was a contractor in the erection of Park's Head at Roehampton, 1767, and his contracts for Pepper Harrow, 1775-6, still exist.

"As the building trade was rapidly falling into few hands, capitalists and others embarked in it, especially about 1815, and then the system of competitive contracts, with all its stratagems, came into full force."—'Dictionary of the Architectural Society.'

The practice of erecting a building and having it afterwards measured and valued by a surveyor, either with or without a schedule of prices previously deposited, was prevalent about the beginning of this century. When it was decided that the work should be measured and valued, the prices for a few leading items only in each trade were arranged, such as brickwork per rod for the bricklayer, timber per foot cube for the carpenter, and the value of the remaining items of the measurement were left to be settled by the measurers at completion, the tradesmen appointing one measurer and the architect or building owner the other. The great cost of work done under these circumstances has led to its gradual disuse. The client seldom knew beforehand what amount the work was likely to cost; moreover the uncertainty as to modes of measurement, and want of definition in the preliminary arrangements, was a frequent source of litigation. The late Mr. Edward Blore, in the erection of Worsley Hall, and a building for the Bridgewater Trust,

3 M

1839, employed a clerk of works who bought the materials and engaged and paid the workmen. A course of this kind is probably the most expensive one that it is possible to adopt. The desire of the building owner to know beforehand the amount that a work would cost, led naturally to the system of contracts by one master builder, who either prepared his own quantities or paid a surveyor for their preparation; thus each builder was put to expense which was often considerable. The increase of competition, which rendered this trouble and expense so often futile, produced the arrangement by which the builders, upon being asked to tender, met and appointed a surveyor to prepare quantities, arranging that his charges should be paid by the successful competitor. This arrangement has been gradually superseded by the present system, under which a quantity surveyor is appointed by the architect, or in the case of unusually large works, one surveyor by the architect and another by the builders, in which case it is popularly supposed that both surveyors take off the dimensions.

The erection of a building under a schedule of prices is another way which has its advocates, but is gradually falling into disuse. There are sometimes circumstances which render any other course impossible; the objection to it is that the work always costs more, however well the schedule of prices may have been drawn, and if ill drawn, very much more.

This practice bears some resemblance to that of the modern French architect. The method adopted for the erection of the Grand Opera House, Paris, is a fair type of the general French system of estimating. The various tradesmen in that case tendered at certain percentages below the Série Prix-de-la-Ville-de-Paris, the civic schedule of prices for b ders' work. When the work is completed it is measured by a surveyor ("métreur") employed by the tradesmen, who prepares a careful drawing of the work as executed, and the tradesman's account. These are both forwarded to the architect, who causes them to be examined by his surveyor ("vérificateur"). The scarcity of general contractors is the main reason for the estimating by each tradesman, but no doubt the system of working under a schedule of prices has found favour in the eyes of architects because of the facilities it offers for the modification of the design of a building during its erection; but the effect upon the architect must be most pernicious.

SYSTEMS OF ESTIMATING.

Mr. Francis Hooper, in an interesting paper read before the Architectural Association, "Architectural Education and Practice in France," says:—

"In public works the architect prepares the working-drawings and details, together with the *devis-descriptif*, or specification, which comprises a general description of the work to be executed, as of the materials and workmanship to be employed, and these are transferred to a *vérificateur*, or measuring surveyor, who is frequently one of the permanent staff of the department—requiring the work.

"In private enterprises, the architect not only prepares the plans and specifications, but also the bills of quantities, when such are required by the form of contract adopted.

"There are many methods of tendering, as in England, but two only are in general use, viz. (1) the *marché-à-forfait*, a lump sum, based on the plans and specifications without quantities; and (2) the *marché-au-rabais*, a tender based on bills of quantities prepared by the architect or at his expense, priced according to a recognised *serie des prix*, or schedule, by the architect himself or the *vérificateur*, and consisting of a uniform percentage either above or below the scheduled prices. In the former method viz. the *marché-à-forfait*, for the purpose of obviating any dispute in the pricing of variations, the architect provides in the "Conditions of Contract" that such prices shall accord with a published schedule cited by him, and upon which the contractor tenders his percentage either in excess or as a discount; whilst in the latter, viz. *marché-au-rabais*, the tender is based upon bills of quantities and not on the drawings, so that there is no difficulty in arriving at a final settlement, the work being measured as it proceeds, and vouchers of weight, &c., transmitted to the architect of all items the measurement of which would be impracticable on completion.

"These systems, which, as far as my knowledge goes, are rarely adopted in England except for the periodical contracts of certain public bodies, appear to merit some attention, as they afford the architect a basis for accurately estimating the cost of any alterations to his plans, and thus of constantly checking the expenditure, whilst the contractor is saved risk in respect of insufficient quantities, or clerical errors in the pricing of innumerable items, a careful examination of the plans and specification being sufficient to acquaint him with the general character of the work for which he tenders.

"In Paris and in all chief French provincial towns, schedules of prices are drawn up and revised every three or four years by the municipal architect as the basis of all the municipal contracts, and these schedules are available for private work, copies being sold for the use of municipal contractors and others.

"The final settlement of accounts is arrived at by the "*métreur*" or estimating clerk of the contractor who meets the "*vérificateur*," or measuring surveyor employed by, or conjointly with, the architect.

"The method of tendering by separate trades is generally adopted in large undertakings, although involving considerably increased responsibility on the architect, and is advocated on the ground that a contractor with a staff of men belonging to a single trade is more likely to be proficient in his work than one who undertakes all trades alike. In France, too, with universal suffrage, there is more tendency to encourage the small tradesmen than is the case in England whether, however, the position of the individual craftsman is in any degree bettered by such a course is with me an unsettled question."

The Glasgow system of obtaining tenders was described by Mr. Honeyman in his paper, "Bills of Quantities, their proper relation to Contracts," read before the Royal Institute of British Architects, May 19th, 1879. "In Glasgow the schedules have superseded the specification altogether, and it may almost be said that a specification is never referred to in the contract at all. The architect selects the measurer, who, with the aid of the drawings and specifications, or such substitute for that as he may get, prepares a detailed schedule of quantities. Copies of this are issued by the architect to selected tradesmen, who are invited to tender by a certain day. These schedules are returned to the architect with rates filled in at each quantity, the amount extended, and the total summed up at the end. A letter accompanies, or more generally is attached to, the schedule, in which the builder offers to execute the work in accordance with the drawings, and 'to the extent of the schedule,' for the sum brought out by the addition of the extended prices, it being further provided that the whole of the work shall be measured after it is finished, and whether it turns out to be more or less than estimated, the cost shall be determined by the rates contained in the schedule; or, where these do not exactly apply, by others strictly in proportion to them. Having considered the various offers, the architect writes on behalf of the proprietor accepting the one which is preferred, and that completes the transaction; in nine cases out of ten there is no more formal contract. When the work is in progress, and when it is finished, the measurer measures it, and prepares a final measurement applying the schedule rates to the various items, and so bringing out the total sum to which the contractor is entitled. This document is examined by the architect, and if he is satisfied that it is compiled in accordance with the estimate, he signs it as a final certificate. Half of the cost of the original schedule and subsequent measurement is deducted from the contractor's accounts, and the proprietor pays the full amount—that is ostensibly the half, but in reality the whole."

The practice in Edinburgh is thus stated:—

"The Edinburgh system resembles that of England in many respects, the principal differences being these: Contractors never have anything to do with the selection of the surveyor or measurer, as he is called, who is always employed by the architect; the measurer is paid by the proprietor, and the schedule is referred to in the contract. It is lodged with the architect with the various items cashed and summed up, and the rates contained in it regulate the cost of additions or omissions which may be ordered in the course of the work."

Both of these arrangements compare so unfavourably with the London practice that comment would be superfluous.

Some architects still entertain a strong prejudice against the supply of quantities by a surveyor, and uniformly endeavour to obtain tenders without them; there is very little doubt that as builders become more aware of their true interest this practice will disappear. It is surely unjust to impose upon a number of builders the trouble and expense of preparing estimates on the small chance of obtaining a contract. The builders whose office staff, and pecuniary position best qualify them to have quantities prepared in their own office, either refuse to tender on such conditions or employ a surveyor to prepare them, adding his charges to their estimate. The smaller men who are invited, prepare their own estimates, and are more frequently than not rewarded by loss; the builder who makes the greatest mistake in the way of omission being usually the successful competitor. The desire of the average client for a larger quantity of work for his money than he has a right to have, and a fallacious idea of saving expense, has led to this extensive trading upon the incapacity of the small builder. But though the architect proposes that no quantities shall be supplied, it is frequently disposed in another manner. If the work be advertised, quantities are in most cases prepared by a surveyor ostensibly acting for a builder who proposes to tender, and who, while preparing his quantities, takes the opportunity of offering them to the other builders who may call to inspect the drawings, and who are glad to avoid the trouble of preparing their own. Most frequently one of the men who tenders on his quantities is successful, and in the result the building owner pays much more than the charges of a practitioner employed in a regular way, and loses many of the advantages which a bill of quantities prepared under different circumstances would have afforded him. There is much need that the duties, obligations, and charges of the quantity surveyor should be authoritatively formulated, and this could be best done by the Institute of Surveyors. Although the progress towards this end had been slow, the steps which have been already taken are well calculated to effect this result.

In 1871, the general conference of architects discussed the existing condition of quantity surveyors and quantities, and appointed a committee to report upon various questions relating thereto. They circulated the following list of questions:—

Employment of Surveyors.

"1. Is it desirable that the practice generally adopted in the case of large works (whereby a surveyor is appointed to represent the employer and another, the builders—such surveyors being jointly responsible to the builder for the accuracy of the quantities) should be in any way modified?

"2. It was suggested, as a convenient course for general adoption, that the quantities should be prepared by a surveyor nominated by the architect, and who would be responsible to the employer for his accuracy; the builder being relieved from any responsibility in regard thereto.

"3. The adoption of this course would go far towards disconnecting the surveyor from the builder, and making him the agent and adviser of the employer in the matter of quantities, &c. Would this be a system advantageous and desirable for general adoption?

"4. A suggestion that the bills of quantities should form part of the contract was well received: architects in leading practice mentioned that they had for many years adopted such a course, and found it to work well, and to be equitable to both employer and builder. Are there any valid objections to the introduction of such a system; the dimensions on which the bills are founded being, in such a case, placed in the hands of both architect and builder?

"5. It appears to be not unusual for some architects, especially in the provinces, to furnish the bills of quantities for works to be carried out under their own superintendence. However convenient this practice may be in some instances, are not special precautions necessary? Should not the bills in such instances, *invariably* form part of the contract? And from whom should the architect receive payment for such quantities?

"6. It was mentioned as being an ordinary and reasonable course that the responsibility for the accuracy of the quantities should be thrown upon the builder, by fixing a time, say one month from the acceptance of the tender, during which he might prove the quantities, but after which no objection would be allowed. Might not great injustice be committed under such a system?

"7. It would be desirable to secure more general uniformity of practice with regard to the taking out of quantities. How can this be best attained?"

The result of the answers to these questions, and the deliberations of the committee, was the following valuable report:—

EMPLOYMENT OF SURVEYORS: REPORT OF SPECIAL COMMITTEE, 14TH JUNE, 1872.

"The wide range of the subject, and the varied opinions which had been expressed at the conference, rendered it difficult for the committee to determine how best to proceed; after having collected a certain amount of information, they prepared a series of inquiries which were circulated in the profession throughout the country, and to which sixty replies were received: these replies evince a wide diversity of opinion, arising from local circumstances and individual varieties of practice, and have satisfied your committee that for the present it will not be practicable to lay down any fixed rule for the guidance of the architect in dealing with 'quantities.' Each case must be governed by local or personal considerations, and the committee have therefore deemed it expedient to make their recommendations and expressions of opinion so general as to meet these varying conditions.

"The object to be attained by the employment of surveyors to take out the quantities of a building is to afford the builders who are to tender one uniform basis for competition; and to define more exactly and accurately than can frequently be done by the general drawings and specification, the exact amount and nature of the work to be executed. As no employer should desire to obtain from his builder more work or greater value, and should not obtain less, than was included in the estimate, while on the other hand he should have the greatest facility to secure due allowance in the case of omission, *it would appear reasonable* that the bills of quantities, which should express in an exact form the intentions of the architect as set out in his general drawings and specification, ought to form a part of the contract, and be dealt with as a recognised exposition of the responsibilities of both employer and builder.

"The established practice in London in the case of large and public works, by which one surveyor is nominated by the architect to represent the employer, and another by the builders to act for them, giving the construction of the work and the elucidation of the architect's ideas the advantage of the experience of two professional men, who consider each question from different points of view, has apparently for many years worked well; though in many cases the advantage derived from the engagement of two can only be considered nominal. There is an understanding that the surveyors so employed are responsible to the builder for the sufficiency of the quantities supplied, and the general drawings and specification prepared by the architect being sufficient, the employer is guaranteed against any excess of cost, and the builder is held safe against loss from errors of quantity. On this system your committee do not think it necessary to make further remark, although some objections have been made to it by leading members of the profession, who prefer to adopt the course next mentioned.

"Many eminent architects have adopted the course of nominating a surveyor who shall prepare the quantities on his own responsibility, and as far as the builder is concerned such quantities become practically a part of the contract. Your committee thoroughly appreciate the great advantages which the architect (and no less the employer) may derive from the employment of a surveyor acquainted with all the details of design peculiar to the architect and with his general manner of proceeding, and also how such an arrangement facilitates proceedings, when time is insufficient for the preparation of drawings and specification in such full detail as would be necessary if a stranger were employed as a surveyor. They are therefore of opinion that this system may in such cases be advantageously adopted, provided always that the builder be relieved from any responsibility as regards the quantities, and that the bills be considered as representing the work to be done. The successful working of such a system must depend entirely on the ability and position of the surveyor employed, his relation to the architect, and the extent of the confidence which the builders tendering may repose in him.

"It appears to be an ordinary custom in the provinces for the architect to supply the quantities for the carrying out of his own designs; where this is done, it should be with the knowledge and concurrence of the employer, and the quantities should form a part of the contract. The architect accepts the duty of providing the quantities, and should not attempt to evade his responsibility by throwing on the builder, as is sometimes done, the labour and risk of checking them, a course which your committee believe to be unsatisfactory, and in some cases likely to lead to great injustice.

"Your committee are of opinion that the practice of making the bills of quantities part of the contract has not been fully considered. Recognised or not, the quantities should be invariably referred to as the interpretation of the general drawings and specification, and in all cases where they are supplied to the builder by the architect, or by a surveyor in whose nomination the builder has had no part, and who is not responsible to the builder, they should form part of the contract. At the same time, when once it is admitted that these documents are to form part of the contract, the necessity for the employment of more than one surveyor in their preparation vanishes.

"The more general adoption of quantities, the extension to country work of the system of measurement usual in London, the gradual modification of local terms and usages, and, not least, the discussions consequent on such meetings as the conference, will all tend towards uniformity of practice; and as builders become more familiar with the London system of measurement, and the public better acquainted with the nature and bearings of the questions, as to the employment of surveyors, and both recognise when such acquaintance increases, the special advantages to be derived from the adoption of one or the other system of employing them, architects and surveyors will find it to the interests of their employers to adopt a uniform practice.

"H. CURREY,
"C. FOWLER,
"J. JENNINGS,
"J. T. KNOWLES,
"W. PAPWORTH,
"ARTHUR CATES, } *Acting Secs. to*
"T. M. RICKMAN, } *Committee.*"

Various discussions have since occurred at the Royal Institute of Architects, at the Architectural Association, and at the Institute of Surveyors, at which last the discussion on the interesting paper read by Mr. Saunders, "On Quantities and Quantity Practice," is of great interest both to architects and surveyors. The paper by Mr. Rickman, on "Building Risks and their incidences," and by Mr. F. Turner, on "The Law affecting Quantity Surveyors," may also be consulted with advantage. The list of propositions read by Mr. Arthur Cates embodies, in the writer's opinion, a very just view of the relations of the quantity surveyor so far as they are at present established.

One of the questions which has much exercised the professional mind may be here briefly referred to. Shall the architect take out his own quantities? A few considerations relating thereto may tend to settle the wavering mind.

The student should first decide whether he will be an expert architect or an expert surveyor, the average man cannot be both.

The standard of attainment in all professions is steadily rising, and all are tending in the direction of division into special branches.

The varied knowledge and accomplishments in which the capable architect should excel, leaving out those parts of the work of the profession which are gravitating in the quantity surveyor's direction, are very extensive, and calculated to heavily tax the powers of the finest minds.

Further, an expert quantity surveyor is the result of long training and diligent attention, such attention as the majority of architects cannot afford to devote to that branch of the work, and it is worth the consideration of the profession how much of the success of our leading architects is referable to the fact of their association with capable quantity surveyors in whose ability and rectitude they have trusted, and thus left themselves free to carry on the more essentially architectural parts of their work; it is a coincidence worth notice that the rising importance of quantity surveyors has been accompanied by a notable improvement in the artistic quality of our buildings.

Various considerations have been adduced as arguments in favour of the architect's taking out his own quantities. One, that it is such good practice for young architects, and probably it is; the object, however, of a bill of quantities is not to give young architects practice, but to save building owners and builders loss. "That it will teach him construction." The architect should learn construction thoroughly before he commences to take off quantities.

"That an architect knows what he requires in a building much better than any quantity surveyor employed by him can know." To this it may be answered that the cleverest architects frequently repeat themselves, and that a quantity surveyor who is used to a certain architect's methods of procedure can interpret his intentions with surprising accuracy.

The architect who professes to take out his own quantities, in the majority of cases does not do so, he leaves it to a clerk, not always with happy results; and the uncertainty as to these results in some cases leads him to introduce eccentric clauses into the bill which shall protect him from possible consequences.

If the architect, however, prefers to prepare his own quantities, he at least owes it to all the parties concerned that they shall be workmanlike and just, and that they shall really be what they profess to be, i.e. a complete schedule of the materials and labour of every kind, and of every element of the contract which affects the price, and he will avoid such clauses as the following, which have been extracted from quantities prepared by architects.

"One month will be allowed the builder to examine the quantities, and after the expiration of that time no question will be allowed as to their accuracy."

"The drawings and specifications are open to the inspection of the builder, who may check the accuracy of the quantities, but after the tender has been accepted no question in respect of the quantities will be allowed to be raised."

"The quantities are condensed as much as possible, and allowance must be made in price for all minor matters and appurtenances, and no claim shall afterwards be made on account thereof, or for any mistake or variation therein, as the builder must include sufficient for the works to be finished complete, the quantities being net."

"Laying on water from main to cistern, pipes, ball cock, etc., complete and all fees."

"The contractors are referred to the specification and drawings for more full explanation." (Part of the "more full explanation" was the fact that the whole of the joinery described in the bill as deal was described in the specification as pitch pine!)

"Reduced brickwork of hard approved stocks, including all cutting, splays, arches, external and internal, and every item of labour."

"Supply centering throughout" (a building estimated at over 8000*l*.).

Such instances might be multiplied, but they are sufficient to illustrate a kind of practice which is not uncommon, and which is preferred by some on the ground that it helps them "to dispense with unnecessary detail."

Quantity Surveyors' Charges.

The charges of quantity surveyors vary considerably. In the country, the architect commonly supplies quantities for his works, and rarely charges less than 2½ per cent. As they are in such cases prepared by a clerk in his office whose knowledge is very often superficial, their quality as a general rule is poor. The disputes about quantities have generally arisen under these circumstances. The cases of Stevenson v. Watson, Young v. Blake, Ford and Co. v. Bemrose and Sons, Toxteth Guardians v. Ellison and Son, go to prove this.

The question of incompetence is an important one, but besides this, there is the danger that the architect's position should pervert that impartiality which is so necessary to the quantity surveyor, whose most important function is the interpretation of the contract documents with rigid justice to both building-owner and builder.

The charge for quantities should depend upon the size and quality of the building.

Warehouses in which the work is usually simple, and large collections of buildings like hospitals, workhouses or lunatic asylums, in which parts are frequently repeated, are usually charged at the lowest rate.

Small new buildings, repairs, additions and alterations, works of decoration, are usually charged at the highest rate.

Sometimes a charge by percentage would be so small, that it would be ridiculous payment for the work of an expert, and then the charge should be by time.

In some contracts, a large proportion of the total consists of provisional sums; and some public bodies and some architects stipulate that the agreed percentage for the quantity surveyor's work should exclude these sums.

As the ordinary rates of charges are presumed to include these, the rate should be increased if they form a large proportion of the contract sum, and are omitted.

In many estimates, the settlement of the amount to be provided depends upon the calculations of the quantity surveyor, and such work as he does to arrive at a result should certainly be paid for.

In many cases of provisions, that part of the work which the surveyor measures as to be done by the general contractor, requires an amount of careful investigation quite disproportionate to that which would be required if he also measured the part excluded from the calculation for commission.

In agreeing to the rate of percentage to be charged, the lithography of copies of the quantities should be excluded, and this is reasonable, as the architect may wish to send out an extraordinary number of copies. Sometimes a public body will ask for thirty or more.

The lithographer of quantities allows a discount of sometimes as much as 20 per cent., but 15 per cent. is more generally the rate. This the surveyor keeps for his trouble, in examining and correcting the transfers and copies. This has been considered by some to be of the nature of an illicit commission, and some surveyors have refused to accept it, but the practice is now so well known and the trouble of the examination before referred to often so considerable that the surveyor should not scruple to accept it, or attempt to conceal it.

The quantity surveyor's charges are as a rule included in the amount of the tender, and it is usually stipulated that they shall be paid by the contractor out of the first instalment he receives. Sometimes, however, the building owner prefers to pay the surveyor's charges directly, and in such case no mention would be made of them in the summary of the bill of quantities.

As in law, the liability for the payment for quantities, if the charge is included in the bill, becomes an obligation of the builder as soon as a contract is signed, the building owner may not pay the surveyor's charges, and deduct them from the contract amount, unless there is a condition which empowers him to do so.

Such a condition has been adopted in those last issued by the Royal Institute of British Architects (July 1895) as follows:—

"14. *Bills of Quantities, Expenses of.*—The fees for the bills of quantities and the surveyor's expenses (if any) stated therein shall be paid by the contractor to the surveyor named therein, out of, and immediately after receiving the amount of the certificate or certificates in which they shall be included.

"The fees chargeable under Clause 13 shall be paid by the contractor, before the issue by the architect of the certificate for the final payment. If the contractor fails or neglects to pay as herein provided, then the employer shall be at liberty, and is hereby authorized, to do so on the certificate of the architect; and the amount so paid by the employer shall be deducted from the amount otherwise due to the contractor."

The time and trouble involved in the settlement of variations on a contract, is sometimes so great that no ordinary percentage will make the business a profitable one. The smallness of some of the variations which the builder insists upon measuring, the trouble and time of investigating the truth of opposing contentions or the eliciting evidence of the true facts, and the often long arguments about prices, take up so much time that the architect when he examines the completed account rarely realizes its extent. So far as profit is concerned, the surveyor would commonly be better off without such work, but in most cases the surveyor's duty to his employer is clear—it is not a question of profit, but obligation, as no surveyor can so well adjust an account as he who prepared the quantities.

The charges of the surveyor acting for the building owner, are added to the total of the account, and are paid by him to the builder, who afterwards pays such surveyor.

Some architects adhere to the old practice of asking the builder at the completion of a contract to deliver an account, which he entrusts to the surveyor for examination. Such work is best charged by time.

It is not usual to allow a builder anything for the cost he may have incurred in the preparation of an account, although he may employ an independent surveyor to prepare it for him.

There are cases, however, in which a large additional amount of new building is involved beyond the original intention and the builder employs a surveyor to measure and value it. Under these circumstances the charges of the builder's surveyor should be allowed, as he has been put to an expense which the contract did not contemplate, and which is not an ordinary establishment charge.

The ordinary rates are as follows:—

Quantities or measuring finished work:

	Calculated on the amount of estimate. Per cent.
Preparing quantities and supplying a draft bill for new works, or for repairs, or for alternations of an old building not comprising much new additional building not exceeding 1000*l.* in value	2½
Over 1,000*l.* and not exceeding 5,000	2
Over 5,000*l.* „ „ 20,000*l.*	1¾
Over 20,000*l* „ „ 50,000*l.*	1½
Above 50,000*l.*	1¼

For pricing bills of quantities, one-half per cent. is a reasonable charge, and if the surveyor knows his business, little enough; but much pricing of quantities for builders is done below this rate.

When the builder sends the amount of the surveyor's charges to the surveyor, who has generally been informed of the amount of the contract, he may calculate the amount which should have been included in the estimate in the following manner. If he has a copy of the priced bill, the calculation will be unnecessary.

The surveyor's claim is for the amount the builder should have included; sometimes his calculation is for an insufficient amount.

As 100*l*. + rate per cent. : 100*l*. :: amount of tender, minus lithography and expenses.

Example.—Surveyor's charge 2¼ per cent.; contract 1500*l*.; lithography and expenses 10*l*.

$$102\tfrac{1}{4} : 100l. :: 1490l.$$

```
            100      £  s.  d.
      _____
 102¼ ) 14,900 ( 1453 13  2
                  36  6 10    2¼ per cent.
                _____
                1490  0  0
```

```
                        £  s.  d.
Claim .. .. .. .. ..   36  6 10
Lithography and expenses 10  0  0
                       _____
                       46  6 10
```

or divide the amount of tender minus lithography and expenses.

$$1490l. \div 41 = 36l.\ 6s.\ 10d.$$

The charges ordinarily made on the adjustment of variations on works of moderate size, are as follows:—

Measuring and valuing additions 2 to 2¼ per cent. on the total
Ditto omissions 1½ to 2 per cent. on the total

Copies of account are charged in addition to the foregoing.
On large works, the charges accepted are often less.
The Metropolitan Asylums Board, paid on some of their large hospitals.—

1½ per cent. on additions.
¾ per cent. on omissions.

This rate is too low, but was probably accepted by the surveyors with a lively hope of future favours.

When work is abandoned after quantities have been prepared, the question of the reduction of the charges often arises, and the argument is used that the charge is partly for the responsibility of accuracy, and should consequently be reduced. As a rule this is a very small element of the charge, and if the original rate is low, cannot fairly be considered; $\frac{1}{8}$ per cent. off reasonable current rates is ample.

If the building goes on to an altered plan, the surveyor may commonly make a condition that if he reduces his charge, he shall be employed to prepare the quantities of the building under the new scheme.

INDEX.

ABBREVIATIONS, 293
Abstract, example of, 4
— example of bricklayer's, 307
Abstracting, 297
— description of, 7
— general rules for, 301
— omissions and additions, 422
— one trade at a time, 297
— precautions against errors of omission, 43
Abstracts, order of, 301
— reducing of, 7
Accounts, adjustment of building, 419
— disputed, 433
Accuracy, surveyor's responsibility for, 779
Air bricks or gratings, measurement of setting of, 88
"Allow," operation of word, 426
— use of word, 499
Alteration, works of, 45
Alterations, examples of items of bill of, 381
— preamble to bill of, 323
Alternative estimates, form of bill, 359
American walnut and wainscot, relative value, 672
Angle brackets, valuation of, 620
— plates to slate slabs, 143
— T or H steel or iron bearers, measurement of, 257
Angles and splays in Portland cement, measurement of, 242
— of battering facing, measurement of, 99
Apex stone, example of taking off an, 852
Apron lining, measurement of, 206
Aprons, lead, measurement of, 217
Arbitration, 756
— accounts prepared, each in a different way, in, 758
— and ordinary legal process, 757
— builder's evidence in, 759

Arbitration, legal assessor in, 758
— precautions witness should observe in, 759
— production of builder's books in, 759
Arbitrator, appointment of, 757
— nominated by Surveyors' Institution or R.I.B.A., 757
— person most suitable for, 757
Arbitrators, essentials in procedure before, 758
Arch, example of taking off a stone, 868
Arches gauged, valuation of, 547
— measurement of gauged, 93
— — of relieving, 81
— — of trimmer, 81
— rough segmental or semicircular, measurement of, 81
— value of relieving, 540
Architect and builder, surveyor's relations with, 435
Architraves, wooden, measurement of, 185
Archway, example of taking off cement finishings of, 889
Archways across passages, how to measure, 62
Area gratings, measurement of, 253
Artificial stone, measurement of, 136
— — preamble to bill of, 345
Artisans' dwellings, price per cubic foot, 464
"As last," use of words, 17
"As provision," use of words, 44
Ash, value of, 591
Ashes, cost of, 507
Ashlar, measurement of stone, 121
Asphalte damp-proof course, measurement of vertical, 79
— paving, measurement of, 79
— skirting, measurement of, 79
Attendance on bellhanger, 264
— on electric bell fitter, 721
— on electrician, 265

916   QUANTITY SURVEYING.

Attendance on engineer, 280
— on gasfitter, 263, 720
— on heating engineer, 290
— on joiner, 208
— on speaking-tube maker, 266
Attendances by joiner, 649
— how arranged, 44
— percentage of contract sum for, 498
Averaging, 14
— examples of, 300
— labour saved by, 300

BACK joint on stone, measurement of, 115
Balusters, wooden, measurement of, 206
— valuation of framed iron, 715
Balustrades and cornice in cement, example of taking off, 887
— and entablature, example of taking off a stone, 856
— iron, measurement of, 254
Bar iron, ordinary sizes of, 713
Bases, stone, measurement of, 116
Basketing, 49
— valuation of, 513
Bath fittings, wooden, measurement of, 200
— stone and labour, relative value to Portland, 564
Baths and wash-houses, price per cubic foot, 464
— measurement of, 226
Battening to walls, measurement of, 161
Battens for slating, measurement of, 157
Battering face, measurement of brick, 95
— facing, measurement of angles of, 99
Beads, sets of, measurement of, 187
Beam filling, measurement of, 85
Bed and point frames, measurement of, 86
— — — — valuation of, 544
Beds and joints, average of, 105
— — — on stones, measurement of, 110
— and joints to each cubic foot of stone, 557
Bell-boards, measurement of, 264
Bellhanger, 263
— attendance on, 264
— preamble to bill of, 353
— suggestions for abstract, 306
Bells and fittings, trade discount off, 721
— average price per pull, 721

Bells, electric, measurement of, 265
— example of collection of, 828
Bending glass, measurement of, 270
Bends in drain pipes, valuation of, 522
— in lead pipes, valuation of, 688
Bevelling glass, measurement of, 270
Bill, errors and discrepancies in, 366
— headings, 318
— names and addresses of special manufacturers should appear in, 22
— of Quantities, amount of detail in, regulated by quality of work, 3
— — — check by builder of items of, 3
— — — conditions as to checking priced, 795
— — — every item of any value should appear in, 3
— — — every item in, should be priced, 3
— — — example of endorsement of, 377
— — — form of a bill of, 368
— — — legitimate function of, 367
— — — use of priced, 444
— — — value of priced, 442
— of Variations, form of, 430
Billing, description of process of, 7
— directions for, 315
— work in feet, 21
Bills, examination of transfers of, 366
Binding of dimensions, 12
Birdsmouth, fair cut, value of, 548
— measurement of brick, 82
— or squint quoin, value of rough, 542
Boarding and firrings of flats, valuation of, 621
— to cupolas, example of dimensions of, 878
— to dormer cheeks, measurement of, 156
— to flats, measurement of, 157
— to roofs, measurement of, 156
— — — valuation of, 621
— to walls and ceilings, measurement of, 197
Boiler setting, 693
Boilers, steam, setting of, measurement of, 288
Bolts, espagnolette, measurement of, 213
— flush, measurement of, 218
— for doors, at per inch customary charge, 677
— iron, measurement of, 213
— screw, measurement of, 258
— valuation of iron screw, 716
— value of fixing, 716
Books, capacity of measuring, 12

INDEX.

Borrowed lights, measurement of, 181
Bosting for carver, measurement of, 122
Boundary walls, dimensions of, keep separate, 11
Box in ground for water pipes, valuation of, 691
Bracketing for cornices, measurement of, 165
— to cornices, valuation of, 619
Branch joints of lead pipes, 224
Breaking stone, valuation of, 516
Breweries, price per cubic foot, 465
— — — quarter, 466
Brick backing to stonework, 62
— filling to openings, 62
— on edge coping, measurement of, 98
— — — — valuation of, 548
— oversailing courses, measurement of, 85
— paving, measurement of, 77
— pavings, valuation of, 535
— pilasters, measurement of, 98
— sewers, measurement of, 70
— — preamble to bill of, 328
— vaulting, measurement of, 70
Brick walls, fair both sides to describe, 68
Bricklayer, constants of labour, 755
— modes of measurement, 61
— number of bricks per day laid by, 530
— preamble to bill of, 339
— suggestions for abstracts of, 303
Brick-nogged partitions, measurement of, 155
Brick-nogging, measurement of, 76
— value of, 583
Bricks, cost of, 527
— number laid per day by a bricklayer, 530
— of unusual size, 63
— proportion of mortar to, 527
— saving in number of, 528
— waste on, 527
Bricksetter, northern practice, 802
Brickwork, allowance for raking cuttings on, 64
— circular on plan, measurement of, 69
— cut and bonded to old, 62
— example of taking off, 836
— extra only in cement, 531
— for carving, measurement of, 95
— in groined vaults, measurement of, 70
— joints struck fair, value, 532
— measurement of projections of, 66
— proportions of value to finished building, 466

Brickwork, valuation of a rod of reduced work in cement, 530
— — in mortar, 530
— various directions for measurement of, 67
— with battering face, measurement of, 69
Bridgewater Trust, erection of building for, 897
Bridging pieces, valuation of, 626
Brilliant-cut plate glass, measurement of, 272
British polished plate glass, measurement of, 270
Broseley tiling, valuation of a square of, 581
Builder attempts to meet errors in estimate by doing bad work, 2
— ejectment of, 435
Builder's evidence in arbitration, 759
— account, charges of surveyor for preparing, 911
— price books, 443
— refuse to tender without quantities, 2
Building cases, special court for, 760
— completion of, after ejectment of first contractor, 435
— contract, clauses to make quantities part of, 762
— — quantities part of, 760
— enlarged, cubing not reasonable for, valuing, 428
— materials, supply of, to London, 454
— various methods of contract for, 756
Buildings, approximate cost of, 459
— in carcase, completion of, 26
— price per foot cube, 459
— various, cost per cubic foot, 464
Burnell and Ellis, 775
Burnt ballast, cost of, 509
— clay, cost of burning, 528

CALLIPERS, use of, 223
Cantilevers of slate, measurement of, 142
Canvas and battens for paper, measurement of, 273
Canvasing to window backs, 188
Carcase to finishings, proportion of, 466
Carcasing, inferior quality of deals and battens for, 595
Carpenter and joiner, northern practice, 808
— constants of labour, 738
— example of abstract of, 809
— measurement of, 148

Carpenter, preamble to bill of, 346
— suggestions for abstract, 304
Carpentry, preparation for valuation of, 585
— sources of information, 585
Carriage, allowance for, 428
— of materials, 479
Cartage, 485
— rates of, 485
Carver, measurement of bosting for, 122
Carving, measurement of, 123
— models for, 123
— on brickwork, measurement of, 95
Casement fastenings, measurement of, 214
Casements and frames, measurement of, 187
— generalisations, 641
— in Honduras mahogany, valuation of, 668
— iron or steel, measurement of, 259
— valuation of deal, 640
Casing metal-work with concrete, measurement of, 55
Cast iron, patterns for, 708
— — remarks on, 709
Casting money columns, precautions in, 424
Catalogue of timber sale, extract from, 587
Cates on quantities, 777
Cathedral rolled glass, measurement of, 272
— sheet glass, measurement of, 272
Ceiling joists, valuation of, 615, 625
— — measurement of, 154
— ribs, plaster moulded, measurement of, 238
— with moulded ribs, setting out of, 239
Ceilings and soffits of plaster, measurement of, 235
— flueing, valuation of lath, plaster, float and set, 698
— valuation of lath, plaster, float and set, 697
Cement, cost of Portland, 508
— filleting, measurement of, 83
— floated face, measurement of, 76
— mortar, per cubic yard, valuation of, 529
— paving, measurement of, 77
Centering to floors, measurement of, 158
— to openings, measurement of, 158
— to trimmers, measurement of, 159
— to vaults, measurement of, 157

Centering valuation of, 615
Cesspool, example of taking off wooden, 885
Cesspools, measurement of, 89
— — of digging to, 56
— — of wooden, 166
— valuation of deal, 627
— valuation of extra labour and solder to, 684
— — of small brick, 522
Chamfer on brick, value of, 541
— on stone, measurement of, 115
Channel pipes, measurement of, 59
Channels, measurement of slate, 142
Charge for lithography of quantities, 909
Charges on provisional sums, 909
— quantity surveyor's, 908
Chase cut and parge, measurement of, 84
— — and parget, value of, 542
— — pipe, measurement of, 83
Checking of dimensions, abstracts and bills, 7
Chimney and bearing bars, measurement of, 252
— bar, valuation of iron, 714
— — separate bill for, 64
— furnace, 64
— gutters, flashings, etc., example of collection of, 826
— pieces of stone and slate, measurement of, 120
— pieces, slate, measurement of, 148
— pots, measurement of, 89
— shafts, diagram useful in measuring, 61
— stack, facings of, example of taking off, 844
Church fittings, measurement of, 209
Churches, price per cubic foot, 465
— — — sitting, 466
Circles, deduction of, how written, 65
— how to write, dimensions of, 65
Circular and skewback cutting on facing, measurement of, 99
— circular work on stone, measurement of, 113
— — — sunk on stone, measurement of, 114
— beds and joints on stone, measurement of, 111
— rib to a roof truss, example of taking off, 876
— ribs to roof trusses, measurement of, 163
— — — — valuation of fir framed in, 614

INDEX.

Circular segments, how written, 65
— slating per square, valuation of, 573
— sunk joint on stone, measurement of, 112
— work on stone, measurement of, 113
— — sunk on stone, measurement of, 113
Cisterns, cost of slate, 577
— slate, measurement of, 143
— wooden, measurement of, 200
— wrought-iron, measurement of, 225
Clay, cost of, 507
— puddle, measurement of, 48, 54
Clean up and straighten groove in terra-cotta for lead lights, 102
Clearing away rubbish, etc., cost of, 498
Cleats, measurement of wooden, 169
— who supplies iron, 707
Clerk of works, cost of, 497
Close cut and mitred hip to slating, measurement of, 139
Coal plates, measurement of, 247
Coffer dams, measurement of, 47
Coffered ceilings in plaster, measurement of, 239
Collection of bells, 828
— of chimney gutters, flashings, etc., 826
— of damp proof course, 820
— of lead flashings, 825
— of linings, 823
— of papering, 824
— of plaster cornices, 825
— of plastering, 824
— of quarter partitions, 821
— of rain-water pipes, 827
— of skirtings, 822
— of stoves and chimney pieces, 820
— of timbers, 821
— of walls, 51
— one available for various kinds of work, 14
Collections, dangers of, 9
— examples of, 820
Column, example of taking off a stone, 864
— of cast iron and its base, example of taking off, 891
Columns, iron, measurement of, 245
— measurement of brick cores to, 85
— valuation of cast iron, 710
— wooden, measurement of, 203
Common-place book, 448
Completion of buildings in carcase, 26
Concrete and digging, example of measurement of, 52
— buildings, measurement of, 104
— — preamble to bill of, 342

Concrete casing to metal work, measurement of, 55
— floors, make good soffites of, 55
— in fireproof floors, measurement of 55
— in trenches, measurement of, 55
— to receive paving, levelling and making up, 55
— valuation of, 517
Conditions of contract, 436
— — by iron merchant, 707
— — — for payment of surveyor's charges by building owner, 910
Cones, how to write dimensions of solidity of, 15
— how to write dimensions of surfaces of, 15
Connections of iron or steel joists, measurement of, 258
Constants of labour, remarks on, 732
Contract for building, various methods of, 756
— French system of, 899
— quantities part of, 388
Contractions in quotations, 487
— used in timber trade, 586
Contractor, Sir William Chambers as a, 897
Contractor's risk, items at, 499
Contracts by separate tradesmen, 896
— in Edinburgh, 901
— in Glasgow, Honeyman on, 900
— on schedule of prices, 897
Conversion of timber for carcasing, analysis of, 602
Coping, brick on edge, measurement of, 98
— — — valuation of, 548
Copings, measurement of stone, 126
— of ornamental bricks, measurement of, 98
Copper and lead, relative value of, 694
— cramps to stone, measurement of, 118
— dowels and mortises in stone, measurement of, 118
— lids, measurement of, 203
— measurement of, 232
— price of, 694
— use of, 694
Coppers, measurement of setting of, 87
Coppersmith, 233
— preamble to bill of, 355
Corbels of brick, measurement of, 100
— stone, measurement of, 116
Core rail iron, measurement of, 254
Cores to columns, measurement of brick, 85

QUANTITY SURVEYING.

Cornices and string courses of brick measurement of, 95
— bracketing for, measurement of, 165
— example of collection of plaster, 825
— moulds for plaster, 702
— of deal, measurement of, 209
— plaster, generalisations, 701, 702
— — measurement of, 237
— — valuation of, 702
Cost of cartage of timber, 589
— of work in each trade, record of, 754
Cottages, price per cubic foot, 464
Counterclaims when exorbitant claims are made, 429
Counterlathing, measurement of, 234
Countersunk heads to rivets, measurement of, 251
Country allowance for workmen, 456
— houses, price per cubic foot, 464
Cover stone, measurement of, 125
Coves in plaster, measurement of, 237
Cradling, measurement of deal, 166
— valuation of deal, 619
Cramps, iron, in stone, measurement of, 118
Credits bill, 363
— — form of, 364
— suggestions on, 291
Crosses, finials, etc., measurement of stone, 122
Cube stone, measurement of, 106
Cubic content of building compared with quantity of certain material, 755
Cubing of buildings, 459
Cupboard fronts, measurement of, 198
Cupolas, example of dimensions of boarding to, 878
Curbs, measurement of stone, 126
"Customary square" for boarding and flooring, 597
Cut and parge chase, measurement of, 84
— — pin ends, value of, 539
Cutting and bonding new brickwork to old, 62
— — — valuation of, 541
— — — new walls to old, measurement of, 84
— — pinning landings, etc., valuation of, 538
— — — measurement of, 83
— — — miscellaneous items, 88
— — — shelf-edge, measurement of, 83
— to slate valleys, valuation of 574,
Cylinder, how to write dimensions of solidity, 16

Dado rails, measurement of, 209
Dados, wooden, measurement of, 178
Damp-proof course, example of collection of, 820
— — valuation of slate, 575
— courses, measurement of, 82
— — valuation of, 534
Day account, form of, 431
— — schedule for labour prices in, 360
— accounts, abstracting of, 430
— — charge for mortar in, 528
— — price of small quantities of material in, 754
— — pricing of, 754
— — time charged in, 458
— — treatment of, 425
— work, billing of, 429
Deal and hardwood, percentage of difference of value, 653
— and pitch-pine, labour and materials, relative value, 656
— for joinery, valuation of thicknesses, 630
— table of thicknesses at a given price per standard, 609
Deals and battens, quality of, 595
— and timber, respective merits of, 592
— table of relative values of standards of, 596
— valuation of a St. Petersburg standard of, 608
"Deduct and add," use of words, 21
Deductions from brickwork, miscellaneous, 65
Deficient quantities, 751
Detached buildings, dimensions of, keep separate, 11
Detail drawings, supply of, 9
Details, surveyor should make, 44
Diagram useful in measuring chimney shafts, 61
Diapers of brick, measurement of, 95
Dictation of dimensions, 20
Digging and concrete, example of measurement of, 52
— measurement of, 49
— per yard, valuation of, 512
— to cesspools, measurement of, 56
— to trenches, example of taking off, 836
— — — measurement of, 50
— value of, 510
— — of additional throws, 512
— various kinds of, 511
— — rules for measuring, 49
Dimension books for office work not desirable, 7
— paper, designations of columns of, 6

INDEX.

Dimensions, alterations in, 25
— booking of, 421
— collection of, to proceed in the same direction, 8
— description of position of, 8
— — of squaring of, 7
— dictation of, 20
— distinct sets of, for different blocks, 10
— emphasise changes in, 21
— examination of, 12
— heading for set of, 17
— headings in, 20
— index to, 20
— indexing of, 12
— length, breadth, depth, order of, 8
— numbering columns of, 12
— of boundary walls, outbuildings, etc., kept separate, 11
— of circles, how to write, 15
— of cubic content of a building, 17
— of plan to be preferred, 8
— of pyramids, how to write, 15
— of solidity of cones, how to write, 15
— — — of cylinder, how to write, 16
— — — of spheres, how to write, 15
— of surface of cone, how to write, 15
— — — of spheres, how to write, 16
— of triangles how to write, 15
— omissions from, 385
— precautions in alterations of, 11
— should be written distinctly, 8
— use of red ink for "timesing," 10
— with two descriptions, 16
— write title of work on each sheet of, 12
Directions for measurement in published schedules, 24
Discounts, fraudulent, 428
Disputed accounts, 433
— — form of statement of, 434
Distempering, measurement of, 243, 279
District surveyor, charges of, 504
— surveyor's fees, 44
Dock charges on timber, 588
Docks in London, 480
Documents, binding of, 12
— precautions as to, 12
Door frames, measurement of, 194
Doors, 191
— example of taking off, 878
— framed and braced, measurement of, 191
— in deal, generalisations, 646
— — — valuation of framed and braced, 645
— — — valuation of panelled, 643

Doors in mahogany, valuation of square framed, 667
— in Sequoia, valuation of square framed, 675
— in Spanish mahogany, valuation of square framed, 669
— in teak, valuation of square framed, 672
— in white wood, valuation of square framed, 674
— ledged, measurement of, 191
— panelled, measurement of, 192
Dormer cheeks, boarding of, measurement of, 157
— — valuation of, 623
"Dotting on," definition of, 13
Draft bill to be read before lithography, 366
Drain pipes, cost of laying, 520
— — cost of tested, 520
— — discounts off, 519
— — net prices of, 521
— — published lists of, 519
— — quantity carried by various vehicles, 519
Draining-boards, measurement of, 203
Drains and laying, valuation of, 521
— iron, measurement of, 59
— — preamble to bill of, 338
— measurement of, 57
— — of digging for, 53
— — preamble to bill of, 336
Drawings, examination and comparison of, 9
— examine before taking off quantities, 26
— figuring of, by surveyor, 10
— in pencil, surveyor's objections to, 22
— insurance of, 22
— marking openings when taken off, 11
— reproduction of, 18
— surveyor to make notes of alterations of, 10
Dredging and removing, measurement of, 46
Dressers, measurement of, 202
Dressing lead to mouldings, valuation of, 685
Drip, short lengths of, valuation of, 625
— valuation of, 624
Driving belts, measurement of, 285
Dwelling-houses, price per cubic foot, 464

EARTH, separately stating carting of, 50
Earthwork, proportion of fillers to getters, 515

Earthworks, example of taking off large, 829
— distinctions in valuation of large, 514
— proportion of pickmen to shovellers, 515
Eaves board, measurement of, 162
— gutter, valuation of iron, 712
— gutters, current prices of iron, 711
— — iron, items numbered, 246
— — — measurement of, 245
Edinburgh, contracts in, 901
Ejectment of builder, 435
Electric bell fitter, attendance on, 721
— bells, average price per push, 721
— — preamble to bill of, 353
— lighting, measurement of, 266
— — preamble to bill of, 354
Electrical engineer, attendance on, 268
Electrician, attendance on, 265
Elm and wainscot, relative value of labour on, 676
— in thickness, cost of, 676
— uses of, 676
— value of, 591
"Elsewhere taken," use of words, 21
Embossing glass, measurement of, 270
Enamelled slate, cost of, 579
Ends cut and pinned, valuation of, 589
— of centres fitted to jambs, measurement of, 160
Engine beds, measurement of, 287
Engineer, attendances on, 280
English oak, conversion of, 591
— — value of, 591
Enrichments, cost of plaster, 702
— plaster carton-pierre, etc., remarks on, 700
— — measurement of, 288
Errors in estimates, 387
— in quantities, treatment of, 389
Establishment charges, 452
— — of London County Council, 453
Estimate, checking the prices of, 387
— liability to error in pricing an, 2
Estimates based on rough quantities, 471
— errors in, 387
— for builders, pricing on abstract or dimensions, 754
— pricing by surveyor, 479
— unfair to builder to require him to prepare quantities for, 2
Estimating, ancient, 896
— by a rate per square, 461
— by number of rods of brickwork, 466
— by the price per cubic foot, 459

Estimating in the seventeenth century, 897
— present system of, 896
Estimator should take off with facility 754
Examples of taking off, 829
Excavation after pulling down, 48
— profit on, 514
— voids in, 48
Excavator, constants of labour, 735
— materials for, 505
— modes of measurement, 46
— northern practice, 801
— suggestions for abstract, 303
Excess in quantities, percentage of, 22
Excessive quantities, 500
Exorbitant claims, counterclaims, 429
Expansion boards, measurement of, 56
External walls, example of taking off, 837
"Extra for," use of words, 21
Extra hoisting of stone work, 105
— labour and solder to cesspools, valuation of, 684
"Extra only," use of words, 21
Extra work, written orders for, 419
Eyelets in walls, measurement of, 89

FACING, measurement of flint, 135
— of glazed brick, measurement of, 100
— of Kentish rag, measurement of, 135
— — — — valuation of, 551
— of rubble, valuation of, 553
Facings, distinctions of value of, 544
— in bands, measurement of, 94
— of stone, measurement of, 134
— rules for measurement of brick, 92
— valuation of picked stock, 547
— — of red brick, 544
Faience, measurement of, 103
Fanlights, measurement of, 181
Fascias, measurement of wooden, 164
Feather-edged springer, valuation of, 625
Felt, measurement of, 162
— pads, measurement of, 259
— valuation of roofing, 620
Fences, measurement of wooden, 160
Fibrous plaster, measurement of, 244
— — valuation of, 699
Figured rolled glass, measurement of, 272
Figures, precautions in writing, 424
Filleting and counter-lathing, measurement of, 166
— measurement of cement, 83
— soffites of trimmers, valuation of, 626

INDEX.

Fillets for joinery, valuation of, 632
— valuation of, 625
— — of deal, rough, 611
Filling in and ramming earth, value of, 513
Finger plates, measurement of, 215
Finished sizes of joinery, 171, 172
Finishings, refer to dimensions of deductions of openings, 11
Fir fitted to iron, measurement of, 152
— framed in floor, valuation of, 607
— — floors, measurement of, 151
— — — valuation of, 612
— — in roof trusses, valuation of, 613
— — in roofs, measurement of, 153
— girders, measurement of, 151
— in quarter partitions, measurement of, 154
— roof trusses, measurement of, 154
Fire for plumber's work, 681
— insurance, cost of, 492
— mains, measurement of, 260
— — preamble to bill of, 352
Fireproof floors, measurement of concrete in, 55
Firrings to cupolas, measurement of, 163
— valuation of, 621
Fitted ends in joinery, 174
Fittings, wooden, billed as a separate section, 210
Flashing-boards, measurement of, 162
Flashings, example of collection of lead, 825
— lead, measurement of, 217
— measurement of raking out and pointing, 88
Flats, boarding to, measurement of, 157
— lead, measurement of, 217
— valuation of boarding and firrings to, 621
Flatting, measurement of, 277
Flint facing, measurement of, 135
Flitch plates, measurement of, 251
Floor tiling, measurement of, 78
Flooring, measurement of boarded, 175
Floors and partitions measured superficially, 150
— fir framed in, measurement of, 151
— labours measured lineally on, 176
— parquet, measurement of, 177
— valuation of deal, 640
— wood block, measurement of, 176
Flue pipes, measurement of, 84
— plates, measurement of, 84
— valuation of parget and core, 587
Flues, deductions of brickwork for, 64

Flues, parget and core, measurement of, 87
Foot boards to loop-hole frames, measurement of, 195
Footings, averaging of brick, 68
Foreman, cost of, 492
Foreman's time, when to allow, 426
Foundation plan, when necessary, 51
Foundations stepped, measurement of, 52
Founder and smith, 244
— — — constants of labour, 748
— — — example of abstract, 310
— — — preamble to bill of, 349
— — — remarks, 705
— — — suggestions for abstract, 306
Frames, bed and point, valuation of, 544
— in deal, generalisations, 648
— — — valuation of, 648
— measurement of bed and point, 86
— of wainscot on deal core, valuation of, 660
Framings in deal, generalisations, 646
— wooden, measurement of, 197
French method of contracting, 899
— polishing, measurement of, 277
"From Banker," stone, 426
"From Bench," joinery, 426
Functions of quantity surveyor in an arbitration, 756
Furnace chimneys, separate bill for, 64

GALVANIZING, current rates for, 708
— how charged, 707
Gantries, value of, 491
Gas at per point, 717
— meter, shelf for, measurement of, 201
— meters, measurement of, 262
— pipes, valuation of iron, 720
— tubing, proportion of, value of fittings to pipe, 717
Gasfitter, 261
— attendance on, 720
— constants of labour, 749
— preamble to bill of, 353
— remarks on value, 717
— suggestions for abstract, 306
— wages of boy, 720
Gasfitting as a separate contract, 720
— at per point, measurement of, 263
— items numbered, 262
Gasfittings in small quantity, 720
— trade discount off, 720
Gate posts, measurement of, 194
Gates, iron, measurement of, 255

QUANTITY SURVEYING.

Gates, wooden, measurement of, 193
Gauged arches, measurement of, 93
General directions to surveyors, 1
— items, northern practice, 800
Girder, iron or steel boxed, example of taking off, 894
Girders, fir, measurement of, 151
Glass, cost of reinstatement of, 723
— discount off, 721
— slates, measurement of, 139
Glazed brick facing, measurement of, 100
— bricks, cost of, 546
Glazier, constants of labour, 750
— example of abstract, 314
— items numbered, 271
— measurement of, 269
— preamble to bill of, 355
— suggestions for abstract, 306
Glazing by general contractor, 722
— by glass merchant, 722
Good and bad work, difference in value of, 443
Gothic arches, deductions of, how written, 65
— window, of stone, example of taking off, 866
Graining and varnishing, valuation of, 729
— current rates for, 729
— measurement of, 277
Granite, measurement of, 129
Granolithic stone, measurement of, 186
Gratings, iron, measurement of, 248
— valuation of framed iron, 715
Groin point, valuation of, 533
Groined roofs, measurement of, stone filling in to cells of, 121
Groove for flashing and burning in, measurement of, 116
— in stone for lead lights, measurement of, 116
Ground joists and sleepers, measurement of, 151
— — valuation of, 612
Grounds, measurement of, 179
— skeleton, measurement of, 179
Guard bars, measurement of, 253
— — valuation of iron, 714
Gulleys, valuation of, 522
Gusset pieces, measurement of wooden, 166
Gutter boards and bearers, example of taking off, 883
— — — — measurement of, 166
— — — — valuation of, 622
Gutters, lead, measurement of, 217
Gwyther v. Gaze, 772

Hair felt, cost of, 690
— — covering to pipes, valuation of, 690
Half-brick partitions, extra value of, 532
Half-sawing on stone, measurement of, 107
Half-timbering, measurement of, 155
Handrail, valuation of deal, 670
— valuation of mahogany, 670
— — — iron, 715
Handrails, generalisations, 670
— iron, measurement of, 253
— relative value of deal and mahogany, 670
— wooden, measurement of, 206
Hard dry rubbish, measurement of, 56
Headings of dimensions, 20
Hearths, measurement of stone, 125
Heating apparatus, as provision, 692
— engineer, attendance on, 290
Herring-bone and other strutting, measurement of, 167
— strutting, valuation of, 625
Hinges, cross-garnet, measurement of, 211
— iron, ornamental, measurement of, 213
— spring, measurement of, 212
— strap, measurement of, 211
Hip and valley tiles, valuation of, 583
— ends of tile, measurement of, 146
— how to measure length of, 16
— knobs, measurement of wooden, 169
— — of tile, measurement of, 146
— or valley, measurement of tile, 145
— hooks, measurement of, 146
Hoarding, value of, 490
Hoisting and fixing roof trusses, measurement of, 169
— extra of stonework, 105
Holes for pipes in tiling, measurement of, 146
— — — value of, 540
— in slate slabs, measurement of, 142
— in slating, measurement of, 139
— in walls for pipes, measurement of, 89
— in wood for pipes, measurement of, 201
— through facing, measurement of, 100
Hollow walls, measurement of, 70
— — valuation of, 531
Honeyman on contracts in Glasgow, 900
Hooper on architectural education and practice in France, 899
Hooping, measurement of iron, 85
Hospitals, price per bed, 465

Hot-water pipes, percentage for fittings, 692
— supply, domestic, 691
— — preamble to bill of, 348
— system, measurement of, 230
— work, rates of labour, 692
Hudson on surveyor's responsibility, 779

IMPORTED scantlings, 597
Index of dimensions, 12
— rerum, 447
— to dimensions, 20
Inspection pit, example of taking off, 848
— pits, designation of, 59
— — measurement of, 58
Insufficient quantities, surveyor's liability for, 763, 780, 784
Internal walls, example of taking off, 838
Invert blocks, measurement of, 76
Iron articles, by what mechanic fixed, 707
— bar, ordinary sizes of, 713
— barrel, trade discounts, 691
— butts, measurement of, 211
— casements, measurement of, 271
— columns, measurement of, 245
— dowels in stone, measurement of, 118
— hooping, measurement of, 85
— merchant, conditions of contract, 706
— or steel, current prices of structural, 707
— — — work, items numbered, 257
Ironfounder, northern practice, 815
Ironmonger, preamble to bill of, 347
Ironmongery, 211
— quality of, 676
— relative value of fixing to deal and hardwood, 677
— trade discounts, 678
Ironwork, cost of fixing structural, 707
— provision for, ornamental wrought, 716
— to carpentry, measurement of, 153
— valuation of fixing only, 627
Items at contractor's risk, 499
— not included in schedule, claim for value of, 428

JAMB linings, measurement of, 196
Jobbing, profit on, 452
— works, definition of, 459
Joggles on stone, measurement of, 116
Joiner, 170

Joiner and ironmonger, suggestions for abstract, 305
— constants of labour, 741
— preamble to bill of, 346
Joiner's deals, valuation of, 629
Joinery, materials for, 627
— short lengths of moulding, measurement of, 173

KEENE's cement, measurement of, 243
— — moulding, valuation of, 704
— — plain face, valuation of, 704
Kentish rag facing, measurement of, 135
— — value of, 551
Key blocks of brick, measurement of, 100
King heads, iron, measurement of, 247
— — valuation of iron, 710
Knot, stop and rub down, valuation of, 727

LABOUR, 453
— economy of, 458
— in position to ironwork, measurement of, 245
— on stone, as a percentage, 558
— on woodwork, relative value on various kinds, 596
— proportion of materials to, 467
Labourer's time, proportion of, required by plasterer, 696
— — — to each mechanic, 457
Labours on carpentry measured lineally, 167
— on plumbing measured lineally, 219
Ladders, wooden, measurement of, 201
Landings, measurement of cutting and pinning, 83
— — of stone, 124
— wooden, measurement of, 205
Lath and half, valuation of, 697
— plaster, float and set ceilings, valuation of, 697
— — float and set flueing ceilings, valuation of, 698
— — float and set partitions, valuation of, 697
Lathing done by lath-renders, cost of, 697
— for tiles, size of, 580
Laundry machinery, measurement of, 285
Lavatories, measurement of, 225
Lavatory tops, slate, measurement of, 143

QUANTITY SURVEYING.

Law cases, defining quantity surveyor's position and obligations, list of, 795
Lead, charge for, exchange of old for new, 680
— cost of laying, 680
— delivery by merchant, 680
— lights, fixing by manufacturer, 722
— distinctions of measurement of, 681
— gutter, example of taking off, 883
— gutters, valuation of, 683
— lights, current prices, 722
— — manufacturer's charges, 722
— — measurement of, 270
— pipes and fixing, valuation of, 682
— — bends in, 223
— — table of weight of, 686
— plugs and mortises in stone, measurement of, 119
— — in slate, measurement of, 142
— price of, 679
— soil pipes, weight required by London County Council, 682
— stepped flashings, valuation of, 683
— supply pipes, weights required by water companies, 682
— weighing old, 681
Legal assessor in arbitration, 758
Letters a.b., use of, 17
— use of reference, 17
Level and prepare old walls, measurement of, 85
— — — old walls to receive new work, value of, 538
Levelling ground, measurement of, 56
Levels, making up with earth, 50
Lime, cost of Has, 508
— — of stone, 508
— mortar, per cubic yard, valuation of, 529
Lime-whiting, measurement of, 76
— value of, 532
Lineal labours on stone, measurement of, 124
— or superficial dimensions of joinery, 174
Lining, apron, measurement of, 206
Linings, example of collection of, 823
— jamb, measurement of, 196
— window, measurement of, 184
Lintels of fir, measurement of, 151
Liquidated damages, claim for, 429
List of items for taking off, 18
— price and P.O., 46
Lithography of quantities, charge for, 909
—: trade discount on, 910
Load for various vehicles, 487
Local authorities, charges of, 502

Locks, measurement of, 215
Lodging money for workmen, 456
Lodgings, payment for workmen's, 429
Lock furniture, measurement of, 215
London County Council, establishment charges of, 453
Louvres, measurement of slate, 142
Lunatic asylums, per inmate, 466
— — price per cubic foot, 465

MACHINERY and pipes, measurement of, 279
— — — order of taking off, 280
— — — preamble to bill of, 357
Mahogany and deal, relative value, 666
— at auction, trade custom, 664
— average prices, 664
— dock-measurement of, 596
— in thicknesses, price of, 666
— uses of, 663
— various kinds of, 665
Making up levels with earth, 50
Maltings, price per cubic foot, 465
Manchester Society of Architects' recommendations as to quantities, 800
Mansions, price per cubic foot, 465
Marble, value of, 566
— — of labour on, 567
Marbling, measurement of, 277
Margin on stone, measurement of, 116
Marked bars, current price of, 713
Mason, constants of labour, 737
— modes of measurement, 104
— northern practice, 805
— preamble to bill of, 943
— suggestions for abstract, 304
Masonry schedule of the War Department, preamble to, 411
— waste in conversion of, 107
Materials delivered to Thames wharves, 480
— delivery of, by merchants, 483
— description of, as a preamble to a trade, 16
— carriage of, 479
— to labour, relative proportion of, 467
— ports of entry for, 481
Mats, measurement of sinkings for, 176
Matted felt covering to pipes, valuation of, 691
Measurement, modes of, 44
— of timber, 596
— various systems of, 13
Measurements, periodical, 420
Measuring books, capacity of, 12
Memory, do not trust to, 17
Merchant bar, current price of, 713

INDEX. 927

Metal articles in masonry, measurement of, 116
— lathing, measurement of, 237
Mitre to splay on stone, measurement of, 115
Mitred border to hearths, measurement of, 176
Mitres to chamfer on stone, measurement of, 117
— to moulding on stone, measurement of, 116
Modelling for terra-cotta, 102
Modes of measurement, 44
— — — prescribed for Houses of Parliament, 406
Moon v. Guardians of Witney Union, 771
Mortar, charge in day accounts for, 528
— cost of mixing, 528
— valuation of cement, 529
— — of lime, 529
Mortises for frames in stone, measurement of, 118
— in stone, measurement of, 117
Moulded angles or splays on facing, measurement of, 97
— courses on facings cut and rubbed, measurement of, 97
— — valuation of brick, 545
— work on stone, measurement of, 114
Moulding in Keene's cement, valuation of, 704
Mouldings, enlarged drawings of, 11
— in deal, valuation of, 648
— in panels, plaster, measurement of, 239
— in plaster or cement, generalisations, 700
— in Portland cement, generalisations, 703
— — — — measurement of, 241
— — — — valuation of, 703
— measurement of deal, 167
— of deal, trade discount, 678
— on brick, value of cut and rubbed, 546
— plaster, measurement of, 239
Muffled glass, measurement of, 272
Muranese glass, measurement of, 272

Nails, lengths, weights, and cost of, 601
Narrow widths, measurement of plastering in, 234
— — use of words, 21
Newels and balusters, iron, measurement of, 249
— valuation of cast iron, 710

Newels, wooden, measurement of, 206
Northern practice and London practice, comparison of, 816
— — bill to illustrate, 818
— — of surveyors, 800
Nosings to landings, measurement of, 206
Notching centres for keystones, measurement of, 160
Notes of cubic content of building, compared with quantity of certain materials, 755
— of prices, surveyor should collect and keep, 441
Numbering rooms on drawings, 10

Oak and Honduras mahogany, relative value of, 658
— in thicknesses, valuation of, 661
— uses of, 657
— value of English, 591
— waste in conversion of, 658
Official referees, quality of, 757
Oiling and rubbing, measurement of, 276
Old buildings, what to observe) on visiting, 26
— lead, allowance for waste, or tare on, 679
— walls, thickening of, 80
— work, value of painting on, 731
Omissions from dimensions, 385
Omitted work, profit on, 428
"On waste," definition of, 9
Openings, deduction of, example of taking off, 840
— filling up old, 62
Order of taking off, 25
— — — — old method, 797
Ornamental aprons of brick, measurement of, 100
— panels of brick, measurement of, 100
— zinc work, 694
Outer strings, measurement of, 205
Oversail, value of fair, 548
Oversailing courses of facings, measurement of, 97
— — measurement of brick, 84

Packing, merchant's charges for, 484
Pads as wood-bricks, valuation of, 624
— or pallettes, measurement of, 168
Painted glass, measurement of, 271
— work, labour on, 725
Painter, constants of labour, 750
— measurement of, 274

Painter, preamble to bill of, 356
— suggestions for abstract, 306
Painter's materials, current prices of, 725
— — in day accounts, charge for, 725
— — remarks on, 726
Painting, example of taking off, 895
— generalisations, 731
— items measured lineally, 276
— — numbered, 276
— on iron, value of, 726
— on old work, valuation of, 730
— on stone or cement, value of, 729
— sash squares, valuation of, 728
— second coat, valuation of, 727
— tenders for, 725
— third and following coats, valuation of, 727
— value of items numbered, 726
— — of lineal items, 726
Pantiling, measurement of, 148
Paper for surveyor's work, 6
— sizing and varnishing, measurement of, 273
Paperhanger, constants of labour, 751
— example of abstract, 314
— measurement of, 273
— preamble to bill of, 356
Paperhanging, current price of materials for, 732
— table of number of pieces for room of given size, 478
Paperhangings, allowance for waste, 732
Papering, example of collection of, 824
Parget and core flue, valuation of, 537
Parian cement, measurement of, 243
Parquet floors, measurement of, 177
Partitions and floors, measured superficially, 150
— brick-nogged, measurement of, 155
— plastering to, measurement of, 234
— valuation of lath, plaster, float and set, 697
Pattern books of wall paper, 731
Patterns for cast iron, 244, 708
Pavement lights, measurement of, 248
Paving, measurement of asphalte, 79
— — of brick, 77
— — of cement, 77
— — of tar, 77
— Yorkshire stone, cost of, 565
Pavings of stone, measurement of, 123
Payment of surveyor by time, 909
P.C. and list price, 46
— definition of letters, 428
Percentage, relative value of various woods expressed by a, 595

Perforations through stone, measurement of, 117
Picked stock facings, valuation of, 547
Picking-out mouldings, measurement of, 277
Picture rails, measurement of, 209
Pier cap, example of taking off a stone, 850
Pilasters on brick facing, measurement of, 98
— wooden, measurement of, 203
Piles, cutting off heads of, 47
— measurement of the driving of, 47
Piling, measurement of, 46
— preamble to bill of, 327
Pine per standard, valuation of, 650
— valuation of thicknesses of, 651
Pipe casing, measurement of, 179
— cost of fixing lead, 680
— hooks, value of, 720
Pipes and fixing, valuation of lead, 682
— copper, measurement of, 223
— lead, measurement of, 223
— measurement of flue, 84
— — of holes in wall for, 89
— protection of water, 690
— trade lists of cast iron heating, 693
— valuation of hot-water pipes, 692
— wrapping, measurement of, 229
Pitch pine, generalisations, 656
— — valuation of thicknesses of, 653
Plain face in Keene's cement, valuation of, 704
— — in Portland cement, valuation of, 703
— — on stone, measurement of, 109
— work on stone, measurement of, 112
Planing on fir, measurement of. 162
— on iron, measurement of, 245
Plaster, fibrous, measurement of, 244
Plasterer, 233
— and painter, northern practice, 812
— constants of labour, 747
— example of abstract, 310
Plasterer's materials, cost of, 696
Plasterer, preamble to bill of, 348
— proportion of labourer's time required by, 696
— suggestions for abstract, 305
Plastering in ceilings and soffites, measurement of, 235
— circular, measurement of, 239
— example of collection of, 824
— in narrow widths, measurement of, 234
— items numbered, 240
— measurement of, 233
— incised, measurement of, 236

Plastering, remarks on, 695
— sub-letting of, 696
— to partitions, measurement of, 234
— to walls, measurement of, 234
— uncertainty of cost of, 695
Plate glass, tariff of, 724
— — valuation of, 724
— racks, measurement of, 202
Plates, bedding of, 63
— measurement of fir, 151
Plugging, value of, 624
Plumber, 216
Plumber and glazier, northern practice, 814
— and zincworker, suggestions for abstract, 305
— attendances on, 226
— constants of labour, 745
— example of abstract, 309
— preamble to bill of, 348
Plumber's brass work, 224
— — — trade discount, 678
— — — trade lists of, 686
— mate, 681
— material, cost of, 681
— work in position, 681
Plumbing as a provision, 679
— as a separate contract, 680
— internal, 221
— items numbered, 224
— labours measured lineally, 219
— — numbered, 220
— low contract rates for, 681
— sub-contract for labour only, 680
— sub-letting of, 679
Pointing with cement, valuation of, 534
Points of compass, use of, 10
Portland cement, angles and splays, measurement of, 242
— — cost of, 508
— — items numbered, 242
— — plain face, valuation of, 703
— — plastering, measurement of, 240
— stone and labour, valuation of, 560
— — value of sawn, 563
Ports of entry for materials, 481
Post holes, measurement of, 56
Preamble to a schedule of prices, 393
Preambles of bills lithographed, 318
Precautions as to documents, 12
— as to timesing, 14
— in altering dimensions, 11
Preliminary bill, usual items of, 319
Preparation of walls, for papering, 732
Preparatory faces on stone, measurement of, 109
Priced bill of quantities, who shall examine, 794

Prices, 441
— acquiring, arranging and preserving notes of, 445
— fluctuations of, 441
— schedules of, 391
Pricing by surveyor, 445
— demands ability and judgment, 441
— of estimates by surveyor, 478
Priestley v. Stone, 763
Prime cost book, production of, as evidence of cost, 424
Priming, valuation of, 727
Profit, 450
— Mr. Lucas on builder's, 450
— on jobbing, 452
— on omitted work, 428
Protection of stonework, value of, 560
Provisional sums, adjustment of, 426
— — exclusion from surveyor's charges of percentage on, 909
Provisions, definition of, 502
— how to deal with, 45
— preamble to bill of, 325
— work incident to, 327
Puddling, cost of clay for, 507
Pugging, measurement of, 234
Pulling down, form of schedule of prices for, 415
— — methods of treatment, 414
Pumping, considerations as to, 509
Pumps, measurement of, 224
Putty, valuation of plasterer's, 701
Pyramids, how to write dimensions of, 15

QUANTITIES a part of the contract, 388
— always worth more than the surveyor's fees, 1
— applications preliminary to preparing, 26
— architect disclaiming liability for accuracy of, 761
— architect's prejudice against, 901
— arguments for architect taking out his own, 907
— builders refuse to tender without, 2
— by inexperienced surveyors mostly deficient, 14
— Cates on, 777
— clauses to avoid in bill of, 908
— delivery with tender of priced bill of, 3
— description of paper for, 6
— essential to production of good work, 1
— estimate based on rough, 471
— examination of priced bill of, 3

3 o

Quantities, excess in, 22
— extent of detail in taking off, 23
— general procedure in preparation of, 6
— importance of deposit of a priced copy, 2
— increase cost of building, 1
— insufficient, 751
— literature of, 23
— Manchester Society's recommendations as to, 800
— object to be attained by a bill of, 3
— part of contract, 760
— preparation of as a speculation, 2
— — by unskilful persons, 2
— rough methods in preparation of, 3
— sealed copy of, 3
— shall architect take out, 906
— should form part of the contract, 2
— speedy taking off, 42
— supply of, assists settlement of final accounts, 1
— treatment of errors in, 389
— various methods of taking off, 13
— what to observe on the drawings before taking off, 26
— when not supplied builder charges for a surveyor's fees, 2
Quantity surveyor, efficient training of, important to builder, 2
— — his functions in an arbitration, 756
— — should be familiar with law and trade customs, 760
— — what he should know, 6
— surveyor's charge for adjusting variation, 793
— — obligations, list of cases defining, 795
— surveyors, the law as it affects, 760
Quarter partition, example of taking off, 874
— partitions, example of collection of timber in, 821
— — measurement of, 154
Queries, sheet of, 17
Quoins, example of taking off stone, 854
— to rubble facing, valuation of, 553
Quotations, contractions in merchant's, 487

RAFTERS, cut ends to, measurement of, 169
Railings, iron, measurement of, 255
Railway rates, 481
— waggons, capacity of, 505

Rain-water pipes, current prices of iron, 711
— — example of collection of, 827
— — iron, items numbered, 246
— — — measurement of, 246
— — valuation of iron, 709
Rake out and point flashings, valuation of, 542
Raking cuttings in brickwork, allowances for, 64
— out and pointing flashings, measurement of, 83
— — joints and pointing soffites, value of, 534
Rates, deducing from a total, 423
— of original estimate, application of, to variations, 424
Rebated drips, measurement of, 166
Rebates on stone, measurement of, 115
Red ink, use of, for "timesing" complete set of dimensions, 10
— — used in writing prices from original estimate, 423
Reference letters, use of, 17
References, suggestions in, 429
Relacquering ironmongery, cost, 678
Relieving arches, measurement of, 81
— — valuation of, 540
Render and set, valuation of, 697
— float and set, valuation of, 697
— valuation of, 696
Restorations, 378
— examples of items in a bill of, 379
— points to observe in preparing quantities of, 380
Reveals in plaster, measurement of, 236
Ribs in thicknesses, measurement of wooden, 164
Richardson and Waghorn v. Beales, 776
Ridge and hips of slate, measurement of, 139
— measurement of tile, 146
Ridges and hips, lead, measurement of, 217
— stone, measurement of, 144
— valuation of slate, 575
— — of tile, 584
Rings, trade, 487
Riveted girders, steel or iron, measurement of, 251
Rivets, measurement of, 259
Rod iron, ordinary sizes of, 713
Roll, valuation of short lengths of, 626
Rolled joists, steel or iron, measurement of, 250
— plate glass, measurement of, 269

INDEX. 931

Rolls for lead, valuation of deal, 624
— measurement of hip and ridge, 167
Roof boarding, measurement of, 156
— — valuation of, 621
— glazing, 272
— trusses, iron or steel, measurement of, 256
— — measurement of wooden, 154
— — — of hoisting and fixing, 169
— — valuation of fir framed in, 613
Roofs, fir framed in, measurement of, 153
Rooms, numbering of, on drawings, 10
Rough boarding, price at docks, 608
— cast, measurement of, 236
— — valuation of, 698
— corbel, measurement of, 123
— cutting on brick, measurement of, 80
— — on brickwork, value of, 532
— oversail, value of, 541
— plate glass, measurement of, 269
— sunk face on stone, measurement of, 109
— — work on stone, measurement of, 111
Rubbish and ramming, measurement of, 56
— cost of hard dry, 509
Rubble walling, measurement of, 138
— — valuation of, 552
— — value of, 550
"Running joints" in lead supply pipes, 687
— — in soil pipes, 687
Rustic groove in stone, measurement of, 116

SADDLE bars, measurement of, 252
Sand and ballast, cost of, 505
— cost of washing, 518
— value of screening, 518
Sanitary apparatus, fixing of, 686
— — trade discount, 679
Sash centres, measurement of, 216
— fastenings, measurement of, 215
— frames, valuation of painting of, 728
— handles, measurement of, 216
— lifts, measurement of, 216
Sashes and frames, examples of taking off, 881
— — — generalisations, 643
— — — measurement of, 181
— — — valuation of deal, 641
— iron or steel, measurement of, 259
Sawing and waste in wood, stipulation of War Office, schedule as to, 595

Sawing mill, charges for, 597
— on stone, 109
— various woods, relative value, 599
Scaffolding, valuation of, 488
— value per rod of brickwork for, 530
Scales, precautions as to use of measuring, 8
Scantling lengths, extra value of Portland stone in, 560
Scarfings of timbers, measurement of, 152, 169
Schedule, claim for value of items not included in, 428
— for labour prices in day account, 360
— of prices, form of, 393
— — — form of tender, 404
— — — preamble to, 393
— — — work costs more when done by, 898
— rates, Matheson on, 451
Schedules of prices, 391
— — — analysis of, 392
— — — directions for measurement in official, 24
— — — of public departments, 391
Schools, price per scholar, 466
Screeds, plaster, measurement of, 237
Screws, in day account, 638
— lengths used for various thicknesses of wood, 687
— trade list of wood, 636
— value of, 633
Scrivener v. Pask, 776
Seasoning wood, 605
Selected deal, 175
Selenitic plastering, remarks on, 698
Semicircles, deductions of, how written, 65
Separate estimates, 359
Sequoia in thicknesses, cost of, 674
Série Prix de la Ville de Paris, 898
Setting air bricks or gratings, measurement of, 88
— coppers, measurement of, 87
— stoves and ranges, measurement of, 87
Sewers, measurement of brick, 70
Shafting and pulleys, measurement of, 284
Sheet glass, current prices of, 723
— — measurement of, 269
— — valuation of, 724
— — value in relation to size of squares of, 723
Shelf edge, measurement of cutting and pinning, 83
— for gas meter, measurement of, 201
Shelves, lattice, measurement of, 196

932   QUANTITY SURVEYING.

Shelves, measurement of stone, 125
— wooden, measurement of, 195
Shingles, valuation of oak, 585
Shipping-marks on wood, 593
Shop sashes, measurement of, 184
Shoring, measurement of, 290
— value of, 491
" Short lengths," use of words, 21
Shutters, boxing, measurement of, 188
— lifting, measurement of, 189
— revolving, measurement of, 190
— sliding, measurement of, 189
Silicate cotton, cost of, 690
— — packing to chases, valuation of, 691
Sill bars, measurement of, 214
Sills bedded hollow, measurement of, 87
— brick, measurement of, 99
Silvering glass, measurement of, 270
Sink stones, measurement of, 116
Sinks, lead lining, measurement of, 223
— stone, measurement of, 120
— wooden, measurement of, 200
Sites of buildings visits to, 25
Skeleton grounds, measurement of, 179
Skewback cutting, valuation of fair, 548
— — value of rough, 542
Skirting, example of collection of, 822
— in Portland cement, measurement of, 241
— — — — valuation of, 703
— in two pieces of mahogany, valuation of, 667
— — — — of Sequoia, valuation of, 675
— — — — of Spanish mahogany, valuation of, 669
— — — — of teak, valuation of, 671
— — — — of wainscot, valuation of, 658
— — — — of white wood, valuation of, 673
— — — — valuation of deal, 639
— measurement of asphalte, 79
— — of slate, 142
— valuation of deal torus, 639
— wooden, measurement of, 177
Skylight curbs, measurement of, 180
Skylights, sets of fillets to bottom rail, measurement of, 181
— sinkings to bottom rails of, measurement of, 180
— wooden, measurement of, 180
Slag wool, packing chases with, measurement of, 229
Slate cantilevers, measurement of, 142
— channels, measurement of, 142

Slate chimney pieces, measurement of, 143
— cisterns, cost of, 578
— — measurement of, 143
— dowels and mortises in stone, measurement of, 118
— enamelled, cost of, 579
— lavatory tops, measurement of, 143
— louvres, measurement of, 142
— mason, 141
— masonry, cost of labour on, 577
— — trade customs, 576
— skirtings, measurement of, 142
— slabs, cost in London of, 577
— — measurement of angle plates to, 143
— — — of, 141
— — — of holes in, 142
— — — of lineal labours on, 141
Slater and tiler, northern practice, 816
— constants of labour, 738
— tiler, slate and marble mason, suggestions for abstract, 304
Slates, cost of delivery of, 571
— value per thousand, 570
— various kinds of, 569
— waste on, 569
Slating battens, measurement of, 157
— — valuation of, 620
— in bands, measurement of, 138
— — — value of, 574
— measurement of, 137
— merchant's prices for, 576
— new trades rules for measuring, 140
— stone, measurement of, 144
— table of sizes, weights, nails, etc., 568
— to steep roofs, extra cost of, 572
— valuation of a square of Countess, 571
— — of eaves of, 574
— value of final repair of, 576
— vertical on walls, valuation of, 574
— weight of nails for, 570
— Westmoreland, measurement of, 138
Slopes, allowance in excavations for, 516
Slotted screws, measurement of, 216
Small quantities of material in day accounts, pricing of, 754, 755
— — percentage of allowances for, 755
— — profit should be largest on, 451
Smith, rate of wages, 713
Smith's work, proportion of labour to material, 717
Snow boards, measurement of, 167
Soakers, lead, measurement of, 218
— valuation of lead, 684
Soffites, measurement of wooden, 165

INDEX. 933

Soil and ventilating pipes, iron, measurement of, 229
— — — — lead, measurement of, 228
— nature of, when to be stated, 48
— pipes, cost of iron, 689
— — valuation of iron, 690
— — — of lead, 689
Solder, 682
— proportion of tin to lead in, 687
Soldered angle, valuation of, 685
— dots, valuation of, 685
— joints, quantity of solder required for pipes of various sizes, 683
— — valuation of, 688
Sound boarding, measurement of, 161
— — valuation of, 622
Spaced slating, valuation of a square of, 572
Speaking-tube maker, attendance on, 266
Speaking-tubes, measurement of, 265
Special manufacturers' names and addresses should appear in bill, 22
Specification clauses which neutralise each other, 440
— list of items to observe in writing a, 439
— correcting from dimensions, 14
— convenient way of writing a, 436
— numbering rooms facilitates writing of, 10
— read through after taking off, 18
— should embody every particular absent from the other documents, 436
— supplied by architect, correction of, 440
— preliminary examination of, 10
— written by quantity surveyor, 436
Spheres, how to write dimensions of solidity of, 15
— how to write dimensions of surface of, 16
Splay on brick facing, measurement of, 99
Sprockets, measurement of, 168
— valuation of, 626
Square framed doors in wainscot, valuation of, 659
Squaring dimensions, 296
Squint-quoin, measurement of brick, 82
— on facing, measurement of, 99
Stable fittings, iron, measurement of, 249
Stables, price per cubic foot, 465
— price per stall, 465
Stain, current prices of, 730
Staining and varnishing, measurement of, 277

Staining and varnishing, valuation of, 730
Staircases, example of taking off wooden, 880
— wooden, measurement of, 204
Stairs in deal, valuation of, 647
Stanchions and columns, cast iron, 709
— iron, measurement of, 245
Staves, use of oak, 657
Steam boilers, 281
— coils, measurement of, 284
— engines, 281
— pipes, cast iron, measurement of, 282
Steel joists, cost of cambering, 706
— — distinctions of price, 706
— — ordinary and compound relative prices, 706
— relative price of Belgian and English, 706
Stepped flashings, lead, measurement of, 217
Steps, measurement of stone, 127
Stevenson v. Watson, 780
Stone, allowance for waste on, 107
— elements of cost of, 553
— principles of valuation of, 555
— labour as a percentage on, 558
— cost of carriage of, 557
— — per foot cube in London, 558
— extra value in scantling lengths of Portland, 560
— Portland, valuation of, 560
— processes of working, 130
— proportion of beds and joints to each cubic foot of, 557
— — of waste on, 560
— slating, measurement of, 143
— walling, preamble to bill of, 337
— working by machinery, 558
— York, value of templates of, 564
Stones, numbering, 106
Stonework, brick backing to, 62
— rules for deductions of, 64
— value of protection of, 560
Stoppage of work, claims incident to, 429
Stopped ends to mouldings on stone, measurement of, 117
— work on stone, measurement of, 108
Stops to chamfer on stone, measurement of, 117
Stoves and chimney pieces, example of collection, 820
— and ranges, measurement of setting, 87
— measurement of, 248
— setting, valuation of, 543
Straps and bolts, measurement of, 252

String course, example of taking off a stone, 853
" Strong " and " middling " lead pipes, 685
Strudwick on surveyor's responsibility, 780
Strutting and planking, valuation of, 523
— — — to basements, measurement of, 58
— — — to holes, measurement of, 54
— — — to trenches, measurement of, 54
— and ribbing to traceried windows, measurement of, 159
— to stone lintels, measurement of, 159
Stucco, measurement of, 234
Sub-contractor's accounts, checking of, 430
Summary, form of, 361
Sundries, northern practice, 816
Sunk beds and joints on stones, measurement of, 111
— work on stone, measurement of, 113
Supervision of taking off, 18
Sureties, 329
Surface digging, measurement of, 49
Surveyor, obligation of, to protect building owner from claims for extra work, 3
— questions relating to employment of, 902
— report of committee on employment of, 908
— responsibility for accuracy, 779
— — Hudson on, 779
— — Strudwick on, 780
— should be familiar with building construction, 1
— — know value of leading items of each trade, 442
— — rectify structural mistakes, 293
Surveyor's charge for preparing builder's account, 911
— charges, architect liable for, 776
— — building owner liable for, 771, 772, 773
— — — — not liable for, 774, 776, 784
— — calculating amount of, 912
— — for measuring variations, 911
— — ordinary rates of, 911
— — payment by building owner of, 910
— — when work abandoned, reduction of, 912
— relation to architect and builder, 435

TABLES and formulæ, where to find, 1
Tacks, soldering on, 685
Taker off, notes by, 20
" Taking off," estimator should be expert at, 754
— — examples of, 829
— — expedients for quickly, 42
— — extent of detail in, 23
— — general arrangements to facilitate, 19
— — marks on openings, 11
— — order of, 25
— — precautions against errors of omission, 43
— — quantities, description of, 6
— — supervision of, 18
— — system of, should be always alike, 7
— — use of list of items for, 18
— — use of tracings when several men, 18
— — uniform system of, a protection from error, 7
— — various ideas of the order of, 797
— — — methods, 13
Tanks, cast iron, measurement of, 282
Tar paving, measurement of, 77
— — value of, 526
Tarpaulins, hire of, 489
Taylor v. Hall, 774
Teak in thicknesses, cost of, 671
— uses of, 671
Templates, stone, measurement of, 116
— value of York stone, 564
Tender, form of, 362
— on schedule of prices, form of, 404
— selection of builders for, 367
Terra-cotta, clean up reveals of, 102
— measurement of, 100
— modelling of, 102
— preamble to bill of, 340
— stipulations for delivery of, 102
— valuation of, 549
Thames, materials delivered in the, 480
The law as it affects quantity surveyors, 760
Theatres, price per auditor, 466
Thickening old walls, 80
Thresholds, measurement of stone, 128
Throat on stone, measurement of, 115
Tile and a half to verges, valuation of, 582
— hip or valley, measurement of, 145
— lathing, cost of, 581
— — size of, 580
— pavings, valuation of, 537
— pegs and pins, 584

INDEX. 935

Tile ridges, valuation of, 584
Tiler, 145
— constants of labour, 736
Tiles, cost of roofing, 579
— roofing, number required per square, 580
— valuation of hip and valley, 583
— waste on roofing, 581
Tiling, measurement of floor, 78
— — of wall, 78
— plain to roofs, measurement of, 145
— roof, bedding in lime and hair, 584
— trade rules for measuring, 147
— vertical plain, measurement of, 145
Timber and deals, respective uses, 592
— average content and limits of section and length, 588
— by what measure sold, 590
— cost of barking, 589
— — of cartage of, 589
— — of loading of, 587
— — of removal of English, 591
— extra cost when average exceeded, 588
— definitions of labour on, 606
— dock charges on, 588
— Gwilt's analysis of price per foot cube of, 604
— modes of purchase of, 588
— Peter Nicholson's analysis of price per foot cube of, 604
— ports of entry for, 587
— purchase from timber merchants of, 590
— railway rates for carriage of, 587
— specification of, 592
— valuation of, 605
— waste by sawing into scantlings, 594
— — in conversion of, 592
— — in sawing, 593
— — in slabbing, 594
— trade, contractions used, 586
Timbers, example of collection of, 821
— of small scantling, 612
— treatment by the surveyor of wrought, 601
"Timesing," definition of, 13
Torching, valuation of, 575
Tracery heads, measurement of stone, 121
Tracing paper, use of, for defining courses of masonry or terra-cotta, 11
Tracings when several men taking off, 18
Trade discounts, 444
— rules for measuring slating, 140
— — — — tiling, 147

Treads and risers, measurement of stone, 125
— — — wooden, measurement of, 204
Trench for pipe, measurement of, 58
Trenches, allowance in width in digging, 49
Triangles, how to write dimensions of, 15
Trimmer arch, valuation of, 533
— arches, measurement of, 81
Trimmers, filleting soffites of, valuation of, 626
Tubing, price list of iron, 718
— trade discounts off wrought iron, 717
Turning pieces, valuation of, 618

UNDERGROUND conveniences, price per cubic foot, 464
Underpinning directions for measuring, 63
— measurement of, 90
Uniformity of practice desirable, 6
Unnecessary descriptions to be avoided, 11
Urinals, measurement of, 226
Use and waste, 290

VALLEY boards, measurement of, 162
Valleys, lead, measurement of, 217
Valves, steam, measurement of, 283
Variations, applications of original contract rates to, 424
— before acceptance of tender, 383
— billing of, 422
— measurement of, preliminary arrangements, 419
— on contract orders for, 423
— pricing of items of, 386
— surveyor's charge for adjusting, 793
— who should price the bill of, 424
Varnishing, measurement of, 277
— valuation of, 729
Vaulting, valuation of brick, 533
Vehicles, usual load for various, 487
Venetian rippled glass, measurement of, 272
Ventilation and warming, 290
Verge to slating, measurement of, 139
— to tiling, measurement of, 146
Vertical damp-proof course, measurement of asphalte, 79
Vestries and boards of works, charges of, 502
Voids in heaps of stone broken to various sizes, 517

WAGES, London County Council's list of, 467
Waghorn v. Wimbledon Local Board, 773
Wainscot, in thicknesses, price of, 658
Waling pieces, measurement of, 47
Wall hooks, weight and cost of, 687
Wall strings, measurement of, 205
— tiling, measurement of, 78
Wall-paper, pattern-books of, 731
— trade discounts, 731
Wall-papers, current price for hanging, 732
Waller, constants of labour, 736
Walls, example of collection of, 51
— faced with stone, how to measure, 63
— when to deal with extra thicknesses of, 61
"Wants," definition of, 8
Warburton v. Llandudno Urban District Council, 794
Warehouses, price per cubic foot of, 465
Waste on stone, proportion of, 560
Waste-preventers, measurement of, 226
Watching and lighting, charges for, 504
Water companies' regulations, 686
— for works, cost of, 493
— meters, measurement of, 226
— quantity required for various items, 497
W.C. apparatus, measurement of, 225
— fittings, wooden, measurement of, 199
Weather boarding, measurement of, 161
Weatherings in Portland cement, measurement of, 242
Wedging up bases of columns or stanchions, 245
Wells, measurement of, 90
Westmoreland slating, measurement of, 138
Wheeling, 49
— value of, 512
White wood, American (or Canary) in thicknesses, cost of, 673

Whitening and colouring, measurement of, 243
Window backs and elbows, measurement of, 187
— boards, measurement of, 185
— example of taking off traceried, 866
— sills, measurement of stone, 129
— nosings, measurement of, 185
— opening in a hollow wall, example of taking off, 840
— — with brick dressings, example of taking off, 842
Winders, wooden, measurement of, 205
Wood block floors, measurement of, 176
— bricks, measurement of, 151
— seasoning of, 605
Woods of various kinds, relative value referred to a percentage, 595
Woodwork, relative value of labour on, 596
Work in difficult positions, 324
— in position, plumbers, 681
Workhouses, price per cubic foot, 465
— price per inmate, 465
Working rules, London trades, 471
Words in specification should be used always in the same sense, 436
Worsley Hall, erection of, 897
Writing, measurement of, 277
Wrought face on fir, measurement of, 162
— timber, treatment by surveyor of, 601

YORKSHIRE stone, its uses in London, 564
— — uses of, 123
Young v. Blake, 784

ZINC, cost of, 693
— flat, valuation of, 693
— measurement of, 231
Zincworker, 231
— preamble to bill of, 355
— wages of, 693
— constants of labour, 747

CPSIA information can be obtained
at www.ICGtesting.com
Printed in the USA
BVHW03s0123210418
513858BV00002B/89/P